Functional Organic Materials

Edited by
Thomas J. J. Müller
and Uwe H. F. Bunz

1807–2007 Knowledge for Generations

Each generation has its unique needs and aspirations. When Charles Wiley first opened his small printing shop in lower Manhattan in 1807, it was a generation of boundless potential searching for an identity. And we were there, helping to define a new American literary tradition. Over half a century later, in the midst of the Second Industrial Revolution, it was a generation focused on building the future. Once again, we were there, supplying the critical scientific, technical, and engineering knowledge that helped frame the world. Throughout the 20th Century, and into the new millennium, nations began to reach out beyond their own borders and a new international community was born. Wiley was there, expanding its operations around the world to enable a global exchange of ideas, opinions, and know-how.

For 200 years, Wiley has been an integral part of each generation s journey, enabling the flow of information and understanding necessary to meet their needs and fulfill their aspirations. Today, bold new technologies are changing the way we live and learn. Wiley will be there, providing you the must-have knowledge you need to imagine new worlds, new possibilities, and new opportunities.

Generations come and go, but you can always count on Wiley to provide you the knowledge you need, when and where you need it!

William J. Pesce
President and Chief Executive Officer

Peter Booth Wiley
Chairman of the Board

Functional Organic Materials

Syntheses, Strategies and Applications

Edited by
Thomas J. J. Müller and Uwe H. F. Bunz

WILEY-VCH Verlag GmbH & Co. KGaA

The Editors

Prof. Dr. Thomas J. J. Müller
Heinrich-Heine-Universität Düsseldorf
Institut für Organische Chemie
und Makromolekulare Chemie
Universitätsstr. 1
40225 Düsseldorf
Germany

Prof. Dr. Uwe H. F. Bunz
Georgia Institute of Technology
School of Chemistry and Biochemistry
Atlanta, GA 30332-440
USA

■ All books published by Wiley-VCH are carefully produced. Nevertheless, authors, editors, and publisher do not warrant the information contained in these books, including this book, to be free of errors. Readers are advised to keep in mind that statements, data, illustrations, procedural details or other items may inadvertently be inaccurate.

Library of Congress Card No.: Applied for

British Library Cataloguing-in-Publication Data:
A catalogue record for this book is available from the British Library.

Bibliographic information published by the Deutsche Nationalbibliothek
The Deutsche Nationalbibliothek lists this publication in the Deutsche Nationalbibliografie; detailed bibliographic data are available in the Internet at http://dnb.d-nb.de.

© 2007 WILEY-VCH Verlag GmbH & Co. KGaA, Weinheim

All rights reserved (including those of translation into other languages). No part of this book may be reproduced in any form – by photoprinting, microfilm, or any other means – nor transmitted or translated into a machine language without written permission from the publishers. Registered names, trademarks, etc. used in this book, even when not specifically marked as such, are not to be considered unprotected by law.

Printed in the Federal Republic of Germany.
Printed on acid-free paper.

Composition   Hagedorn Kommunikation, Viernheim
Printing   Strauss GmbH, Mörlenbach
Bookbinding   Litges & Dopf Buchbinderei GmbH, Heppenheim

ISBN   978-3-527-31302-0

Contents

Preface *XIII*

List of Contributors *XVIII*

Part I 3-D Carbon-rich π-Systems – Nanotubes and Segments

1   Functionalization of Carbon Nanotubes *3*
         Andreas Hirsch and Otto Vostrowsky
1.1   Introduction to Carbon Nanotubes – A New Carbon Allotrope *3*
1.2   Functionalization of Carbon Nanotubes *4*
1.3   Covalent Functionalization *5*
1.3.1   Halogenation of Carbon Nanotubes *5*
1.3.1.1 Fluorination of Carbon Nanotubes *5*
1.3.1.2 Chlorination of Carbon Nanotubes *7*
1.3.1.3 Bromination of MWCNTs *7*
1.3.1.4 Chemical Derivatization of "Fluoronanotubes" *7*
1.3.2   Oxidation of CNTs – Oxidative Purification *8*
1.3.2.1 Carboxylation of CNTs *8*
1.3.2.2 Defect Functionalization – Transformation of Carboxylic Functions *10*
1.3.3   Hydrogenation of Carbon Nanotubes *19*
1.3.4   Addition of Radicals *19*
1.3.5   Addition of Nucleophilic Carbenes *20*
1.3.6   Sidewall Functionalization Through Electrophilic Addition *21*
1.3.7   Functionalization Through Cycloadditions *21*
1.3.7.1 Addition of Carbenes *21*
1.3.7.2 Addition of Nitrenes *22*
1.3.7.3 Nucleophilic Cyclopropanation – Bingel Reaction *24*
1.3.7.4 Azomethine Ylides *24*
1.3.7.5 [4+2]-Cycloaddition – Diels–Alder Reaction *26*
1.3.7.6 Sidewall Osmylation of Individual SWCNTs *27*
1.3.8   Aryl Diazonium Chemistry – Electrochemical Modification of Nanotubes *27*

Functional Organic Materials. Syntheses, Strategies, and Applications.
Edited by Thomas J.J. Müller and Uwe H.F. Bunz
Copyright © 2007 WILEY-VCH Verlag GmbH & Co. KGaA, Weinheim
ISBN: 978-3-527-31302-0

1.3.9	Reductive Alkylation and Arylation of Carbon Nanotubes	29
1.3.10	Addition of Carbanions – Reactions with Alkyllithium	30
1.3.11	Covalent Functionalization by Polymerization – "Grafting To" and "Grafting From" 31	
1.4	Noncovalent Exohedral Functionalization – Functionalization with Biomolecules 32	
1.5	Endohedral Functionalization 42	
1.6	Conclusions 44	
1.7	Experimental 44	
	References 49	
2	**Cyclophenacene Cut Out of Fullerene** 59	
	Yutaka Matsuo and Eiichi Nakamura	
2.1	Introduction 59	
2.2	Synthesis of [10]Cyclophenacene π-Conjugated Systems from [60]Fullerene 63	
2.2.1	Synthetic Strategy 63	
2.2.2	Synthesis and Characterization of [10]Cyclophenacenes 64	
2.2.3	Structural Studies and Aromaticity of [10]Cyclophenacene 67	
2.2.4	Synthesis of Dibenzo-fused Corannulenes 69	
2.2.5	Absorption and Emission of [10]Cyclophenacenes and Dibenzo Fused Corannulenes 70	
2.3	Conclusion 72	
2.4	Experimental 72	
	References 78	

Part II Strategic Advances in Chromophore and Materials Synthesis

3	**Cruciform π-Conjugated Oligomers** 83	
	Frank Galbrecht, Torsten W. Bünnagel,	
	Askin Bilge, Ullrich Scherf and Tony Farrell	
3.1	Introduction 83	
3.2	Oligomers with a Tetrahedral Core Unit 85	
3.3	Oligomers with a Tetrasubstituted Benzene Core 91	
3.4	Oligomers with a Tetrasubstituted Biaryl Core 95	
3.5	Conclusion 113	
3.6	Experimental 113	
	Acknowledgments 115	
	References 116	

4	**Design of π-Conjugated Systems Using Organophosphorus Building Blocks** *119*	
	Philip W. Dyer and Régis Réau	
4.1	Introduction *119*	
4.2	Phosphole-containing π-Conjugated Systems *121*	
4.2.1	α,α'-Oligo(phosphole)s *123*	
4.2.2	Derivatives Based on 1,1'-Biphosphole Units *126*	
4.2.3	Mixed Oligomers Based on Phospholes with Other (Hetero)aromatics *129*	
4.2.4	Mixed Oligomers Based on Biphospholes with other (Hetero)aromatics *137*	
4.2.5	Mixed Oligomers Based on Phospholes with Ethenyl or Ethynyl Units *138*	
4.2.6	Polymers Incorporating Phospholes *140*	
4.2.7	Mixed Oligomers and Polymers Based on Dibenzophosphole or Dithienophosphole *143*	
4.3	Phosphine-containing π-Conjugated Systems *147*	
4.3.1	Polymers Based on *p*-Phenylenephosphine Units *147*	
4.3.2	Oligomers Based on Phosphine–Ethynyl Units *151*	
4.3.3	Mixed Derivatives Based on Arylphosphino Units *155*	
4.4	Phosphaalkene- and Diphosphene-containing π-Conjugated Systems *161*	
4.5	Conclusion *168*	
4.6	Selected Experimental Procedures *169*	
	References *172*	
5	**Diversity-oriented Synthesis of Chromophores by Combinatorial Strategies and Multi-component Reactions** *179*	
	Thomas J. J. Müller	
5.1	Introduction *179*	
5.2	Combinatorial Syntheses of Chromophores *180*	
5.2.1	Combinatorial Azo Coupling *181*	
5.2.2	Combinatorial Condensation Reactions *182*	
5.2.3	Combinatorial Cross-coupling Reactions *187*	
5.2.4	Combinatorial Coordination Chemistry *197*	
5.3	Novel Multi-component Syntheses of Chromophores *199*	
5.3.1	Multi-component Condensation Reactions *199*	
5.3.2	Multi-component Cross-coupling Reactions *204*	
5.4	Conclusion and Outlook *215*	
5.5	Experimental Procedures *215*	
	References *218*	

6	**High-yield Synthesis of Shape-persistent Phenylene–Ethynylene Macrocycles** *225*
	Sigurd Höger
6.1	Introduction *225*
6.2	Synthesis *227*
6.2.1	General *227*
6.2.2	The Kinetic Approach *227*
6.2.2.1	Statistical Reactions *227*
6.2.2.2	Template-controlled Cyclizations *238*
6.2.3	The Thermodynamic Approach *251*
6.3	Conclusion *254*
6.4	Experimental Procedures *255*
	References *258*

7	**Functional Materials via Multiple Noncovalent Interactions** *261*
	Joseph R. Carlisle and Marcus Weck
7.1	Introduction *261*
7.2	Biologically Inspired Materials via Multi-step Self-assembly *262*
7.3	Small Molecule-based Multi-step Self-assembly *265*
7.4	Polymer-based Self-assembly *275*
7.4.1	Main-chain Self-assembly *276*
7.4.2	Side-chain Self-assembly *279*
7.4.3	Macroscopic Self-assembly *287*
7.5	Conclusion and Outlook *288*
	References *289*

Part III Molecular Muscles, Switches and Electronics

8	**Molecular Motors and Muscles** *295*
	Sourav Saha and J. Fraser Stoddart
8.1	Introduction *295*
8.2	Mechanically Interlocked Molecules as Artificial Molecular Machines *297*
8.3	Chemically Induced Switching of the Bistable Rotaxanes *299*
8.3.1	A Bistable [2]Rotaxane Driven by Acid–Base Chemistry *300*
8.3.2	A pH-driven Molecular Elevator *301*
8.3.3	A Molecular Muscle Powered by Metal Ion Exchange *303*
8.3.4	Redox and Chemically Controlled Molecular Switches and Muscles *304*
8.3.4.1	Solution-phase Switching *305*
8.3.4.2	Condensed-phase Switching *306*
8.3.4.3	A Solid-state Nanomechanical Device *308*
8.4	Electrochemically Controllable Bistable Rotaxanes *309*
8.4.1	A Benzidine/Biphenol-based Molecular Switch *310*

8.4.2	Electrochemically Controlled Switching of TTF/DNP-based [2]Rotaxanes *311*	
8.4.2.1	Solution-phase Switching *311*	
8.4.2.2	Metastability of a Redox-driven [2]Rotaxane SAM on Gold Surfaces *311*	
8.4.2.3	A TTF/DNP [2]Rotaxane-based Electrochromic Device *313*	
8.4.2.4	A Redox-driven [2]Rotaxane-based Molecular Switch Tunnel Junctions (MSTJs) Device *314*	
8.4.3	A Redox and Chemically Controllable Bistable Neutral [2]Rotaxane *315*	
8.4.3.1	Electrochemical Switching *315*	
8.4.3.2	Chemical Switching Induced by Lithium Ion (Li^+) *316*	
8.5	Photochemically Powered Molecular Switches *316*	
8.5.1	Molecular Switching Caused by Photoisomerization *317*	
8.5.2	PET-induced Switching of an H-bonded Molecular Motor *318*	
8.5.3	MLCT-induced Switching of a Metal Ion-based Molecular Motor *319*	
8.5.4	A Photo-driven Molecular Abacus *320*	
8.6	Conclusions *322*	
	Acknowledgments *322*	
	References *323*	
9	**Diarylethene as a Photoswitching Unit of Intramolecular Magnetic Interaction** *329*	
	Kenji Matsuda and Masahiro Irie	
9.1	Introduction *329*	
9.2	Photochromic Spin Coupler *331*	
9.3	Synthesis of Diarylethene Biradicals *333*	
9.4	Photoswitching Using Bis(3-thienyl)ethene *335*	
9.5	Reversed Photoswitching Using Bis(2-thienyl)ethene *340*	
9.6	Photoswitching Using an Array of Photochromic Molecules *341*	
9.7	Development of a New Switching Unit *344*	
9.8	Conclusions *348*	
9.9	Experimental Procedures *348*	
	Acknowledgments *349*	
	References *350*	
10	**Thiol End-capped Molecules for Molecular Electronics: Synthetic Methods, Molecular Junctions and Structure–Property Relationships** *353*	
	Kasper Nørgaard, Mogens Brøndsted Nielsen and Thomas Bjørnholm	
10.1	Introduction *353*	
10.2	Synthetic Procedures *354*	
10.2.1	Protecting Groups for Arylthiols *354*	
10.2.1.1	Synthesis of Arylthiol "Alligator Clips" *354*	
10.2.2	One-terminal Wires *358*	
10.2.3	Two-terminal Wires *359*	
10.2.4	Three-terminal Wires *364*	

10.2.5	Four-terminal Wires	365
10.2.6	Caltrops	367
10.3	Electron Transport in Two- and Three-terminal Molecular Devices	368
10.3.1	Molecular Junctions	375
10.3.1.1	Scanning Tunneling-based Molecular Junctions	375
10.3.1.2	Conducting-probe Atomic Force Microscopy	379
10.3.1.3	Solution-phase Molecular STM Junctions	380
10.3.1.4	Break Junctions	381
10.3.1.5	Crossed Wires	382
10.3.1.6	Nanopore Junctions	382
10.3.1.7	Square-tip Junctions	383
10.3.1.8	Mercury Drop Junctions	384
10.3.1.9	Particle Junctions	384
10.3.1.10	Nanowire Junctions	385
10.3.1.11	Three-terminal Single-molecule Transistors	386
10.4	Summary and Outlook	387
10.5	Experimental	388
	References	389

11   Nonlinear Optical Properties of Organic Materials   *393*
Stephen Barlow and Seth R. Marder

11.1	Introduction to Nonlinear Optics	393
11.1.1	Introduction	393
11.1.2	Linear and Nonlinear Polarization	394
11.1.3	Second-order Nonlinear Optical Effects	396
11.1.4	Measurement Techniques for Second-order Properties, β and $\chi^{(2)}$	397
11.1.5	Third-order Nonlinear Optical Effects	399
11.1.6	Measurement Techniques for 2PA Cross-section, δ	401
11.2	Second-order Chromophores for Electrooptic Applications	404
11.2.1	Design of Second-order Chromophores: the Two-level Model	404
11.2.2	Other Chromophore Designs	409
11.2.3	Other Considerations	411
11.2.4	High-performance Electooptic Poled-polymer Systems	413
11.3	Design and Application of Two-photon Absorbing Chromophores	418
11.3.1	Essential-state Models for Two-photon Cross-section	418
11.3.2	Chromophore Designs	420
11.3.3	Applications of Two-photon Absorption	427
11.4	Appendix: Units in NLO	430
	Acknowledgments	431
	References	431

Part IV Electronic Interaction and Structure

12 Photoinduced Electron Transfer Processes in Synthetically Modified DNA *441*
Hans-Achim Wagenknecht

12.1 DNA as a Bioorganic Material for Electron Transport *441*
12.2 Mechanism of Hole Transfer and Hole Hopping in DNA *444*
12.3 Reductive Electron Transfer and Excess Electron Transport in DNA *446*
12.3.1 Strategies for the Synthesis of DNA Donor–Acceptor Systems *446*
12.3.2 Chromophore Functionalization of DNA Bases via Synthesis of DNA Building Blocks *448*
12.3.3 DNA Base Modifications via a Solid-phase Synthetic Strategy *452*
12.3.4 Chromophores as Artificial DNA Base Substitutes *454*
12.4 Results from the Electron Transfer Studies *456*
12.5 Outlook: Towards Synthetic Nanostructures Based on DNA-like Architecture *460*
References *463*

13 Electron Transfer of π-Functional Systems and Applications *465*
Shunichi Fukuzumi

13.1 Introduction *465*
13.2 Efficient Electron-transfer Properties of Zinc Porphyrins *467*
13.3 Efficient Electron-transfer Properties of Fullerenes *474*
13.4 Photoinduced Electron Transfer in Electron Donor-Acceptor Linked Molecules Mimicking the Photosynthetic Reaction Center *477*
13.5 An Orthogonal π-Donor-Acceptor Dyad Affording an Infinite CS Lifetime *485*
13.6 A Long-lived ET State Acting as an Efficient ET Photocatalyst *490*
13.7 Organic Solar Cells Using Simple Donor-Acceptor Dyads *494*
13.8 Organic Solar Cells Composed of Multi-porphyrin/C_{60} Supramolecular Assemblies *499*
13.9 Conclusion *506*
Acknowledgments *506*
References *507*

14 Induced π-Stacking in Acenes *511*
John E. Anthony

14.1 Introduction *511*
14.2 Anthracene *512*
14.2.1 Monosubstituted Anthracene *514*
14.2.2 Disubstituted Anthracene *517*
14.2.3 Edge-substituted Anthracenes (Anthracene Functionalized at the 1,8- or 1,8,9-Positions) *522*

14.2.4	Liquid Crystalline Anthracenes	525
14.2.5	Anthracene Self-assembly: Hydrogen Bonding	526
14.3	Tetracene (Naphthacene)	528
14.3.1	Ethynyltetracenes	530
14.3.2	Tetrasubstituted Tetracenes	532
14.4	Pentacene	534
14.5	Higher Acenes	540
14.6	Conclusion	541
	Acknowledgments	542
	References	542
15	**Synthesis and Characterization of Novel Chiral Conjugated Materials**	**547**
	Andrzej Rajca and Makoto Miyasaka	
15.1	Introduction	547
15.2	Synthetic Approaches to Highly Annelated Chiral π-Conjugated Systems	548
15.2.1	Helicenes	548
15.2.1.1	Photochemical Syntheses	549
15.2.1.2	Non-photochemical Syntheses	551
15.2.2	Double Helicenes and Chiral Polycyclic Aromatic Hydrocarbons	560
15.2.3	Tetraphenylenes and π-Conjugated Double Helices	563
15.3	Barriers for Racemization of Chiral π-Conjugated Systems	567
15.4	Strong Chiroptical Properties in Absorption, Emission and Refraction	570
15.4.1	Absorption and Emission	570
15.4.2	Refraction	572
15.5	Conclusion	574
	Acknowledgments	574
	References	574

Index 583

Preface

Functional organic materials constitute a combination of diverse fields that range from carbon nanotubes to self-assembled polymers but encompass also nonlinear optical effects, the organic solid state and the construction of chiral materials. We, as the two Editors, have tried to put together an overview over the development of modern functional organic materials in this book. The content is divided into four different sections, which deal with 3-D carbon-rich systems (I), strategic advances in synthesis (II), molecular muscles, switches and electronics (III) and the interplay of electronic interaction and structure (IV). Renowned experts in each field give an introduction to/overview of each of their selected thematic fields, making an easy entry for readers into the respective field. The book is intended both for active scientists working in the field of organic materials, in academe and in industry, and also for graduate students who are interested in this area. Owing to the daunting scope of the topic, the book is not intended to be comprehensive, but rather presents selected highlights that are deemed of particular importance.

In the first part, 3-D carbon-rich systems, *Hirsch* and *Vostrovski (Chapter 1)* describe the strategies for attaching chemical functionalities to CNTs to make these fascinating materials processable. Solubilization allows tailoring of the interactions of nanotubes with semiconductors and (bio)polymer matrices. The authors describe the chemical and physical functionalization of single-walled and multiple-walled carbon nanotubes, of individual tubes and of bundles. Covalent functionalization of the nanotubes' carbon scaffold can be performed at the tubes' termini, at the sidewalls and at defect sites of enhanced reactivity. Physical functionalization is a noncovalent attachment of functions, based on complexation and adsorption forces, van der Waals and π-stacking effects. The authors discuss the functionalization of nanotubes comprehensively and show how functionalized nanotubes connect to biologically active environs and advanced devices.

In *Chapter 2, Matsuo and Nakamura* describe the synthetic developments of a subgroup of bowl-shaped π-conjugated corannulenes and expand the topic to the discussion of the synthesis of a novel class of organic fluorophores derived from C_{60}, hoop-shaped condensed aromatic compounds. They are made by selective removal of ten sp^2 carbon centers out of conjugation from the north and south pole regions of C_{60}. The resulting 40-π-electron [10]cyclophenacenes excised

Functional Organic Materials. Syntheses, Strategies, and Applications.
Edited by Thomas J.J. Müller and Uwe H.F. Bunz
Copyright © 2007 WILEY-VCH Verlag GmbH & Co. KGaA, Weinheim
ISBN: 978-3-527-31302-0

from the equatorial region of the fullerene molecule are chemically stable, yellow-luminescent, non-conventional π-systems.

The second part of the book deals with strategic advances in the synthesis of organic materials. This part is ushered in by the contribution of *Scherf* and colleagues, who describe in *Chapter 3* the synthesis and characterization of cruciform, π-conjugated oligomers. The so-called cruciform approach leads to compounds with increased solubility and allows solution processing of such materials to produce homogeneous thin films and layers. Cruciform oligothiophene dimers have been used as active, semiconducting layer in solution-processed OFETs with a high hole mobility. Other examples of such cruciforms involve oligophenylenevinylene-type, oligophenyleneethynylene-type and oligophenylene-type cruciforms. The approach allows the introduction of donor or acceptor substituents leading to spatially addressable electronic properties in small molecule materials.

The introduction of the heavier main group elements, particularly phosphorus, is reviewed in *Chapter 4* by *Dyer* and *Réau*. The authors describe the use of P-containing building blocks (phospholes, phosphino groups or trivalent phospha- and diphospha-alkenes) for the design of conjugated systems. They illustrate the specific advantages offered by organophosphorus synthons compared with their widely used sulfur or nitrogen analogues. The possibility of chemically modifying P-centers provides a unique way to create structural diversity and to tailor conjugated systems for optoelectronic applications. The coordination ability of P-centers towards transition metals offers additional opportunities to build supramolecular architectures in which the π-systems can be organized in a defined manner and with significant changes to its electronic properties.

In *Chapter 5*, *Müller* provides an overview over multi-component processes, domino reactions and sequential transformations in diversity-oriented syntheses. The author demonstrates that multi-component reactions (MCRs), having found broad application in pharmaceutical high-throughput screening and lead finding, can also be used in the design and construction of functional π-electron systems (chromophores, fluorophores, electrophores). The author shows the developments of the last decade that led to adaptations of diversity-oriented approaches. New strategies, developments and perspectives of diversity oriented syntheses of functional π-electron systems are summarized and highlighted.

Most of the systems described in Chapter 5 contain small- or medium-sized or multinuclear benzenoid and non-benzenoid arenes. In *Chapter 6*, *Höger* gives an overview over the mastery of the synthesis of macro- and megacycles. He shows different approaches towards shape-persistent macrocycles and carefully examines and discusses selected examples that display the advantages and disadvantages of macrocycle synthesis under kinetic and thermodynamic control. The template approach (both supramolecular and covalent) towards functionalized rings is also discussed and introduces a strong motif of supramolecular chemistry, which is much further developed but in a more polymer-oriented topic, in the next chapter.

In *Chapter 7 Carlise* and *Weck* discuss the design and synthesis of multifunctionalized, architecturally controlled polymers as a prerequisite for a variety of future applications of polymeric materials. They review recent progress in polymer science that is based on Nature's use of self-assembly in the creation of biomaterials and that utilizes noncovalent interactions such as hydrogen bonding, ionic interactions, electrostatic interactions and metal coordination. These concepts have been employed to synthesize both main-chain and functionalized side-chain polymeric materials. The examples outlined in this chapter demonstrate the great potential of noncovalently functionalized polymers. Weck's group has laid the groundwork for future endeavors in this area.

Although supramolecular assembly is a necessary prerequisite for the construction of molecular machines, it is not sufficient. In Part III, concepts of molecular muscles, switches and molecular electronics are illuminated. In *Chapter 8*, on molecular motors and muscles, *Saha* and *Stoddart* survey bistable rotaxane-based molecular switches and motors, and also muscles that display piston-like linear translational motions, when triggered by chemical, electrical or photochemical inputs. The problematic issues that arise by miniaturization of semiconductor devices by a "top-down" approach can often be circumvented by noncovalent syntheses leading to a "bottom-up" approach, based on molecular recognition and self-assembly. Bistable catenanes and rotaxanes are capable of nanoscale mechanical movements of one component with respect to another in response to external stimuli – a property which makes them one of the best molecular systems that can be made to function like a switch or a machine. These relative movements within interlocked molecules, arising from changes in the noncovalent bonding interactions, can be initiated and controlled by three types of stimuli – chemical, electrical and light.

While the transformation of chemical, electrical and radiation energy into mechanical movement is important, the control of intramolecular magnetic interactions using light is yet another facet of the fascinating world of organic materials. In *Chapter 9*, *Matsuda and Irie* discuss and outline attempts to control intramolecular magnetic interactions by photoswitchable materials. The most convenient way to introduce a photoswitching function to molecular systems is to use photochromic units as spin couplers. A particularly interesting group of these couplers are diarylethenes, which reversibly change the π-conjugation and structure upon irradiation. These materials are successfully used to control the intramolecular magnetic interaction and can change these by a factor of >150-fold.

Matsuda and Irie's chapter on photonic switching gives an excellent transition to the final fourth part of the book, where electronic interaction and structure are discussed. In *Chapter 10*, *Nørgaard, Brøndsted Nielsen and Bjørnholm* review the current status of conductivity measurements through single molecules with particular emphasis on thiol end-capped π-conjugated oligomers. This field of research has been progressing rapidly over the last few years and aims at developing the methodology and fundamental understanding of sending current through individual molecules – towards the ultimate goal of constructing "single molecule" electronics. The first part of the chapter describes different synthetic approaches towards the fabrication of candidate molecules for these aims, followed by a survey of current experimental methods for measuring the electronic transmission properties of organic molecules in various two- and three-terminal device configurations.

Going from molecular electronics to molecular photonics, in the elegant *Chapter 11, Barlow* and *Marder* introduce the general reader to the origin of nonlinear optical effects in conjugated organic systems and to structure–property relationships for π-chromophores. The chapter focuses on two exciting areas of current interest. In the case of second-order nonlinear optics, recent developments in overcoming various challenges in the commercial application of organic chromophores in electrooptic switching devices are described. In the case of third-order nonlinear optics, the chapter summarizes recent breakthroughs in understanding the design of molecules with a high two-photon cross-section and surveys some of the applications that may be facilitated by the availability of these chromophores.

Whereas in all of the preceding chapters non-biological building blocks were used to construct organic materials, *Wagenknecht* demonstrates in *Chapter 12* that DNA is a viable bioorganic material for molecular electronics, since the unique self-assembling and predictable structure make DNA a superb architectural scaffold. For the investigation of photoinduced electron transfer processes through DNA, it is necessary to modify oligonucleotides with suitable chromophores: The different methods to introduce these into the building-block strategy via phosphoramidite chemistry and also protocols for post-synthetic oligonucleotide modifications are discussed in terms of their ability to prepare structurally and electronically well-defined DNA systems. The author discusses issues of electron transport through DNA on different length scales and concludes that efficient electron transport over longer distances using DNA (>20 nm) requires the development of DNA-based materials which contain the DNA-typical structural features but exhibit improved electron transport capabilities by engineering electronic interactions between – now artificial – bases in single- and double-stranded DNA molecules.

The concept of electron transfer in non-biological systems is deepened in *Chapter 13* by *Fukuzumi*, who reviews recent developments of electron-transfer systems of functional π-compounds focusing on the specific aspect of their supramolecular organization to construct efficient photovoltaic devices. Specifically, porphyrins and fullerenes exhibit excellent electron-transfer properties. Their small reorganization energies of electron transfer facilitate the design of artificial photosynthetic systems composed of functional π-compounds with fast charge separation but very slow charge recombination. The author discusses the two factors for efficient photocurrent generation. One is charge separation and the other one is charge carrier transport in thin films. Three-dimensional control of π-complexes between porphyrins and fullerenes contributes to both the efficient charge separation and excellent charge carrier transport. Efficient self-exchange electron transfer between porphyrin/porphyrin$^+$ and between fullerene/fullerene$^-$ is well established and leads in supramolecular clusters with interpenetrating networks to efficient hopping of holes and electrons in each network.

Charge transfer and charge migration in organic solids are also a critically important issue for the successful fabrication of organic competitive solid-state devices. In *Chapter 14, Anthony* unfurls the relationship between molecular packing and its consequences for linear acenes. Although many high aspect ratio aromatic systems adopt an edge-to-face arrangement referred to as "herringbone" packing, the arrangement predicted to yield the best electronic properties involves the face-to-face interaction of aromatic species, direct π-stacking. The separation between the π-faces of the molecules and even the degree of lateral or longitudinal offset between adjacent aromatic units can lead to dramatic changes in electronic communication between molecules. This chapter discusses modes for enhancing or enforcing π-stacking arrangements in linearly fused aromatic systems and discusses the effect of π-stacking on the properties of devices fabricated from these materials.

From electronic back to optical properties, in the final *Chapter 15, Rajca* and *Miyasaka* provide an overview of the synthesis of molecules with highly annelated, chiral, π-conjugated systems, primarily of the helicene-type. Tremendous progress has been made in the synthesis of helicenes; however, the corresponding helical, ladder-type polymers remain a significant and attractive challenge. In addition to the synthesis, the configurational stability (barriers for racemization) and chirooptical properties are described and discussed in depth in this chapter. In this context, perspectives for isotropic materials with molecule-based chirooptical properties are outlined.

We hope that this book, *Functional Organic Materials*, gives an effective overview of novel and important aspects in the field. We have not included the area of semiconducting polymers as there is a very recent book [Georges Hadziioannou and George G. Malliaras (Editors), *Semiconducting Polymers, Chemistry, Physics and Engineering*, 2nd edn., Wiley-VCH, Weinheim, 2006] that treats this topic in depth. Neither has this book covered the area of organic materials for light-emitting diodes, as there is an excellent recent monograph (Klaus Müllen and Ullrich Scherf (Editors, *Organic Light Emitting Devices, Synthesis, Properties and Application*, Wiley-VCH, Weinheim, 2005). We hope that readers will enjoy the herein presented developments in organic materials as much as the Editors did in assembling these chapters.

Atlanta and Düsseldorf, November 2006   Uwe H. F. Bunz
                                                    Thomas J. J. Müller

List of Contributors

John E. Anthony
Department of Chemistry
University of Kentucky
Lexington
KY 40506
USA

Stephen Barlow
School of Chemistry and Biochemistry
Georgia Institute of Technology
Atlanta
GA 30332
USA

Askin Bilge
Institut für Makromolekulare Chemie
Fachbereich C
Bergische Universität Wuppertal
Gaußstr. 20
42097 Wuppertal
Germany

Thomas Bjørnholm
Nano-Science Center
University of Copenhagen
Universitetsparken 5
2100 Copenhagen
Denmark

Mogens Brøndsted Nielsen
Nano-Science Center
University of Copenhagen
Universitetsparken 5
2100 Copenhagen
Denmark

Torsten W. Bünnagel
Institut für Makromolekulare Chemie
Fachbereich C
Bergische Universität Wuppertal
Gaußstr. 20
42097 Wuppertal
Germany

Joseph R. Carlisle
School of Chemistry and Biochemistry
Georgia Institute of Technology
Atlanta
GA 30332
USA

Philip W. Dyer
Department of Chemistry
Durham University
South Road
Durham, DH1 3LE
UK

Functional Organic Materials. Syntheses, Strategies, and Applications.
Edited by Thomas J.J. Müller and Uwe H.F. Bunz
Copyright © 2007 WILEY-VCH Verlag GmbH & Co. KGaA, Weinheim
ISBN: 978-3-527-31302-0

Tony Farrell
Color Technology
ADC, Room E 221a
GE Plastics
Plasticslaan 1
P.O. Box 117
4600 AC Bergen op Zoom
The Netherlands

Shunichi Fukuzumi
Department of Material
and Life Science
Graduate School of Engineering
Osaka University
Suita
Osaka 565-0871
Japan

Frank Galbrecht
Institut für Makromolekulare Chemie
Fachbereich C
Bergische Universität Wuppertal
Gaußstr. 20
42097 Wuppertal
Germany

Andreas Hirsch
Institut für Organische Chemie
Universität Erlangen-Nürnberg
Henkestr. 42
91054 Erlangen
Germany

Sigurd Höger
Kekulé-Institut für Organische
Chemie und Biochemie
Universität Bonn
Gerhard-Domagk-Str. 1
53121 Bonn
Germany

Masahiro Irie
Department of Chemistry and
Biochemistry
Graduate School of Engineering
Kyusyu University
Hakozaki 6-10-1, Higashi-ku
Fukuoka 812-8581
Japan

Seth R. Marder
School of Chemistry and Biochemistry
Georgia Institute of Technology
Atlanta
GA 30332
USA

Kenji Matsuda
Department of Chemistry and
Biochemistry
Graduate School of Engineering
Kyusyu University
Hakozaki 6-10-1, Higashi-ku
Fukuoka 812-8581
Japan

Yutaka Matsuo
Japan Science and Technology Agency
Nakamura Functional Carbon
Cluster Project, ERATO
4th Build, School of Science
The University of Tokyo
7-3-1 Hongo, Bunkyo-ku
Tokyo 113-0033
Japan

Makoto Miyasaka
Department of Chemsitry
University of Nebraska – Lincoln
Hamilton Hall 818E
Lincoln, NE 68588-0304
USA

List of Contributors

Thomas J. J. Müller
Institut für Organische Chemie
und Makromolekulare Chemie
Heinrich-Heine-Universität Düsseldorf
Universitätsstr. 1
40225 Düsseldorf
Germany

Eiichi Nakamura
Department of Chemistry
The University of Tokyo
7-3-1 Hongo, Bunkyo-ku
Tokyo 113-0033
Japan

Kasper Nørgaard
Nano-Science Center
University of Copenhagen
Universitetsparken 5
2100 Copenhagen
Denmark

Andrzej Rajca
Department of Chemsitry
University of Nebraska – Lincoln
Hamilton Hall 818D
Lincoln, NE 68588-0304
USA

Régis Réau
UMR "Sciences Chimiques
de Rennes" 6226
Université de Rennes
Campus de Beaulieu
35042 Rennes Cedex
France

Sourav Saha
Department of Chemistry and
Biochemistry
University of California, Los Angeles
405 Hilgard Avenue
Los Angeles, CA 90095-1569
USA

Ullrich Scherf
Institut für Makromolekulare Chemie
Fachbereich C
Bergische Universität Wuppertal
Gaußstr. 20
42097 Wuppertal
Germany

J. Fraser Stoddart
Department of Chemistry and
Biochemistry
University of California, Los Angeles
405 Hilgard Avenue
Los Angeles, CA 90095-1569
USA

Otto Vostrowsky
Institut für Organische Chemie
Universität Erlangen-Nürnberg
Henkestr. 42
91054 Erlangen
Germany

Hans-Achim Wagenknecht
Institute of Organic Chemistry
University of Regensburg
Universitätsstr. 31
93053 Regensburg
Germany

Marcus Weck
School of Chemistry and Biochemistry
Georgia Institute of Technology
901 Atlantic Drive, NW
Atlanta, GA 30332-0400
USA

Part I
3-D Carbon-rich π-Systems – Nanotubes and Segments

1
Functionalization of Carbon Nanotubes*

Andreas Hirsch and Otto Vostrowsky

*A List of Abbreviations can be found at the end of this chapter.

1.1
Introduction to Carbon Nanotubes – A New Carbon Allotrope

Carbon nanotubes (CNTs) are cylindrically shaped macromolecules consisting of graphene-structured carbon atoms with diameters from 1.5 to 10 nm and lengths ranging from 1 to several hundred μm [1]. The small tubules appear as concentrically nested multi-walled carbon nanotubes (MWCNTs) or as a single layer of carbon atoms, the single-walled carbon nanotubes (SWCNTs). Both MWCNTs and SWCNTs can be grown in tangled structures or ordered close-packed struc-

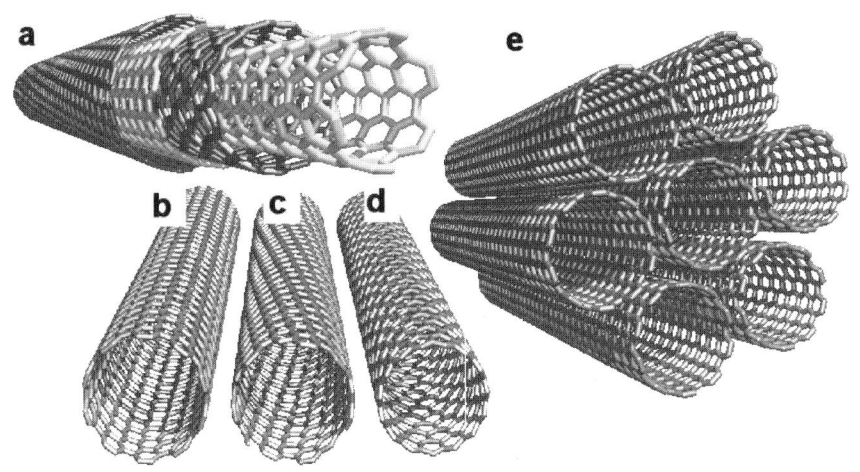

Fig. 1.1 Idealized representation of different structures of defect-free and opened carbon nanotubes: (a) concentric MWCNT; (b) "metallic" armchair [10,10] SWCNT; (c) helical chiral semiconducting SWCNT; (d) zigzag [15,0] SWCNT; (e) SWCNT bundle. The armchair (b) and zigzag tubes (d) are achiral.

Functional Organic Materials. Syntheses, Strategies, and Applications.
Edited by Thomas J.J. Müller and Uwe H.F. Bunz
Copyright © 2007 WILEY-VCH Verlag GmbH & Co. KGaA, Weinheim
ISBN: 978-3-527-31302-0

tures. As a consequence, within the SWCNTs achiral zigzag and armchair tubes and helical chiral nanotubes have to be distinguished because of different "rollings" of the graphene sheet into a cylinder (Fig. 1.1). SWCNTs typically exist as ropes or bundles of individual tubes.

Because of their diverse structure, one-third of the tubes are expected to possess metallic character and the remaining two-thirds to behave as semiconductors [2, 3]. CNTs represent potential candidates to be used in field emission [4–6] and nanoelectrical devices [6–10], components of electrochemical energy [11, 12] and hydrogen storage systems [13, 14] and as components in composite materials [15–17]. They represent the ultimate carbon fiber, exhibiting exceptional mechanical properties [18–21] by being up to 100 times stronger than steel [22].

Since their discovery by Iijima in 1991 [23], carbon nanotubes have become the subject of intense research activities [2, 3]. Today they are widely recognized as *the* essential contributors to nanotechnology. However, their lack of solubility and their multidisperse dimensions present a considerable barrier towards processing and usage of their promising property profile for technological applications.

1.2
Functionalization of Carbon Nanotubes

Attaching chemical functionalities to CNTs can improve their solubility and allow for their manipulation and processability [24]. The chemical functionalization can tailor the interactions of nanotubes with solvents, polymers and biopolymer matrices. Modified tubes may have physical or mechanical properties different from those of the original nanotubes and thus allow tuning of the chemistry and physics of carbon nanotubes. Chemical functionalization can be performed selectively, the metallic SWCNTs reacting faster than semiconducting tubes [25].

In dealing with functionalization, one has to distinguish between functionalization of SWCNTs and MWCNTs, between covalent and noncovalent functionalization, sidewall functionalization and defect functionalization, chemical functionalization, biofunctionalization and exohedral and endohedral functionalization (Fig. 1.2) [24]. Covalent functionalization is based on covalent linkage of functional entities onto the nanotube's carbon scaffold at the termini of the tubes or at their sidewalls. Covalent sidewall functionalization is associated with a change in hybridization from sp^2 to sp^3 and a loss of conjugation. Defect functionalization takes advantage of oxygenated sites and structural defects such as pentagon and heptagon irregularities in the hexagon graphene framework. Noncovalent functionalization is mainly based on supramolecular complexation. A special case of functionalization is the endohedral functionalization of CNTs, i.e. the filling of the tubes with atoms or small guest molecules.

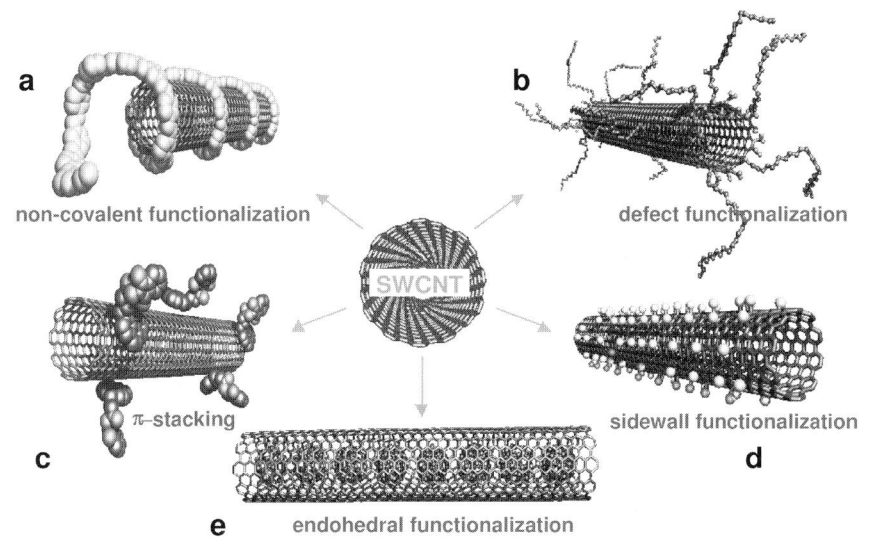

Fig. 1.2 Different possibilities for the functionalization of SWCNTs: (a) noncovalent exohedral functionalization with polymers; (b) defect-group functionalization; (c) noncovalent exohedral functionalization with molecules through π-stacking; (d) sidewall functionalization; (e) endohedral functionalization, in this case C_{60}@SWCNT.

1.3
Covalent Functionalization

Addition chemistry has developed into a promising tool for the modification and derivatization of the surface of nanotubes [24, 26]. However, it is difficult to achieve chemoselectivity and regioselectivity control of addition reactions, requiring hot addends such as arynes, carbenes, radicals, nitrenes or halogens under drastic reaction conditions.

1.3.1
Halogenation of Carbon Nanotubes

1.3.1.1 Fluorination of Carbon Nanotubes

Because of the low reactivity of the surface of CNTs, fluorination was taken into consideration as one of the first sidewall functionalization reactions [27]. Fluorine as the most electronegative element in its elemental form is a powerful oxidizer. Mickelson et al. reported in 1998 extensive controlled and nondestructive sidewall fluorination of SWCNTs (Fig. 1.3) [28]. The functionalized F-SWCNTs dissolved well in alcohol and gave long-living metastable solutions [29].

Most of the experimental work on fluorination has been performed using elemental fluorine diluted in an inert gas at elevated temperatures [28, 30, 31]. Comparative fluorination of open- and closed-end SWCNTs was performed with ele-

Fig. 1.3 Schematic depiction of a sidewall fluorinated SWCNT.

mental fluorine gas and the structural changes of the tubes were investigated [32]. Using iodine pentafluoride (IF$_5$), the highest degree of fluorination with a composition of about CF was achieved [33]. Low-temperature plasma processes do not involve the use of wet chemicals or exposure of the tubes to high temperature and generate very little chemical residue [34]. Recently, SWCNTs have been functionalized using CF$_4$ [35] or a CF$_4$ or SF$_6$ plasma produced by reactive ion etching (RIE) [36]. The surface of aligned MWCNT arrays was modified by a CF$_4$ RIE-plasma treatment and the effect on the field emission properties of the aligned MWCNT films was studied [34]. CF$_4$ microwave discharge was used to F-functionalize SWCNTs and the samples were characterized through spectroscopy, scanning electron microscopy (SEM) and X-ray methods [37]. A low-level fluorination of MWCNTs was achieved at ambient temperature under the influence of light using XeF$_2$ as the fluorine source [38]. An atmosphere of trifluoromethane or hexafluoropropene served as the functionalizing medium during ball-milling of SWCNTs at room temperature. Thus, F-SWCNTs with 0.3–3.5 wt. % of fluorine were obtained [39].

The major disadvantage of fluorine-functionalization is the high degree of addition of fluorine atoms, reflecting a great number of tube defects. Furthermore, the fluorinated material exhibits significant changes in spectroscopic properties, providing evidence for electronic perturbation. Fluorination results in a modification of the electronic structures of the tubes and changes the electrical resistance depending on the coverage and method of fluorine application. For fluorinated SWCNTs, evidence was shown for exfoliation of nanotube ropes and bundles into individual tubes. F-SWCNTs were found to be "cut" at the fluorinated sites during pyrolysis [40]. Especially for the production of composite materials, F-SWCNTs seem to be a rather promising material.

The fluorotubes are very soluble and open new routes to solution-phase chemistry. F-SWCNTs can serve as a starting material for a wide variety of chemical sidewall functionalizations (see Section 1.3.1.4). By treatment with hydrazine or LiBH$_4$–LiAlH$_4$, the majority of the covalently bound fluorine could be removed (Scheme 1.1) [29, 41], restoring most of the conductivity and spectroscopic properties of the pristine material.

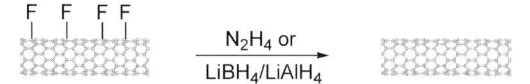

Scheme 1.1 Removal of covalently bound fluorine by treatment with hydrazine or lithium borohydride–lithium aluminum hydride.

1.3.1.2 Chlorination of Carbon Nanotubes

Electrolytic functionalization of MWCNTs succeeded with evolution of chlorine on an anode made from a foil of carbon nanotubes. Furthermore, oxygen-bearing functional groups such as hydroxyl and carboxyl are formed at the same time, enhancing the solvation of the tubes in water or alcohol [42]. Using another technique of functionalization, from ball-milling of purified MWCNTs in the presence of Cl_2 as reactant gas, cleavage of the tubes' C–C bonds and breaking of the tubes occurred and Cl-functionalized tubes were obtained [43]. Recently, Cl-SWCNTs were prepared by chemical vapor deposition of $CHCl_3$ or $Cl_2C=CCl_2$ and subsequent radical addition of chlorine atoms under ball-milling conditions at room temperature. This kind of functionalized tubes were estimated to contain 5.5–17.5 wt.% of chlorine [39]. A theoretical suggestion for controlled functionalization of SWCNTs with chlorine was presented by Fagan et al. in 2003 [44]. The idea was to substitute carbon network atoms of nanotubes by inserting Si atoms, thus gaining an sp^3-like stable defect center as trapping site for the chemisorption of other atoms or molecules. Because of the experimental evidence about the possibility of substitutional doping of fullerenes by Si atoms, it is conceivable to dope CNTs similarly with Si [44].

1.3.1.3 Bromination of MWCNTs

The low susceptibility of CNTs to bromination was utilized as a means of purification for MWCNTs contaminated by other carbon particles [45, 46].

1.3.1.4 Chemical Derivatization of "Fluoronanotubes"

By treating fluoronanotubes with strong nucleophiles such as Grignard reagents, alkyl- and aryllithium reagents, metal alkoxides, acyl peroxides, amines and diamines, the fluorine atoms can be replaced through substitution [29, 47–50].

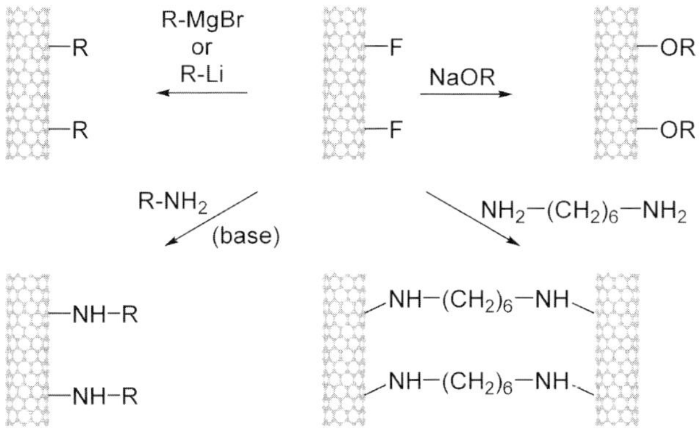

Scheme 1.2 Treatment of fluorinated tubes with strong nucleophiles and replacement of the fluorine substituents, leading to derivatized alkyl- and alkoxyl-nanotubes and amino-functionalized nanotubes as products.

Derivatized products such as alkyl-, aryl-, alkoxyl- and amino-nanotubes and cross-linked nanotubes are obtained (Scheme 1.2). F-SWCNTs prepared by the HiPCO process exhibited a higher degree of alkylation using alkyllithium reagents than fluorinated SWCNTs from the laser-oven method. Dealkylation occurred at 500 °C. 1-Butene and n-butane were formed during the thermolysis [51]. A number of OH group-terminated SWCNTs have been prepared by fluorine displacement reactions of F-SWCNTs with a series of diols and glycerol in the presence of alkali or with amino alcohols in the presence of pyridine as catalyst [52]. F-SWCNTs can be efficiently defluorinated with anhydrous hydrazine [29, 41, 48].

1.3.2
Oxidation of CNTs – Oxidative Purification

1.3.2.1 Carboxylation of CNTs
One of the milestones in nanotube chemistry was the "oxidative purification" of carbon nanotubes by liquid- or gas-phase oxidation. This leads to opening of the tubes' caps, formation of defects in the sidewalls and introduction of oxygen-bearing functionalities into the tubes [24]. Oxidation ("purification") was achieved applying boiling nitric acid [53, 54], sulfuric acid [54] or mixtures of both [55], "piranha" (sulfuric acid–hydrogen peroxide) [56] or gaseous oxygen [57, 58], ozone [59–62] or air as oxidant [63–65] at elevated temperatures or combinations of nitric acid and air oxidation [66]. Other oxidants used to functionalize CNTs are superacid $HF-BF_3$, OsO_4 and RuO_4, OsO_4-NaIO_4, H_2O_2, $K_2Cr_2O_7$, $KMnO_4$ plus phase-transfer catalyst (PTC), etc. [67–70]. The aim of this oxidative treatment, drawing its inspiration from well-known graphite chemistry [71, 72], is the oxidative removal of metal catalyst particles used in the synthesis of the tubes and of amorphous carbon as a synthesis byproduct [56].

Upon oxidative treatment, the introduction of carboxylic groups and other oxygen-bearing groups at the end of the tubes and at defect sites is promoted, decorating the tubes with a somewhat indeterminate number of oxygenated functionalities. However, because of the large aspect ratio of CNTs, considerable sidewall functionalization also takes place (Fig. 1.4). HNO_3 oxidation in combination with column chromatography and vacuum filtration led to a new purification method for SWCNTs, developed by Hirsch's group [53]. The oxidative treatment introduces defects on the nanotube surface [73], oxidizes the carbon nanotubes ("hole doping") and produces impurity states at the Fermi level of the tubes [74]. In addition to purifying the raw material by removal of impurities, oxidation can be used to cut ("etch"), shorten and open the CNTs [56, 75–78]. Cutting and shortening depend on the extent of the reaction and give rise to a new length distribution which can be determined by transmission electron microscopy (TEM). The cutting results in SWCNTs with open and oxygenated ends; the degree of oxygenated functionalities were determined spectroscopically [59, 60]. Since chemical reactivity is a function of curvature [79], the oxidative stability also depends on the tube diameter [49, 80]. Mawhinney et al. [59] showed that with 1.4 nm diameter

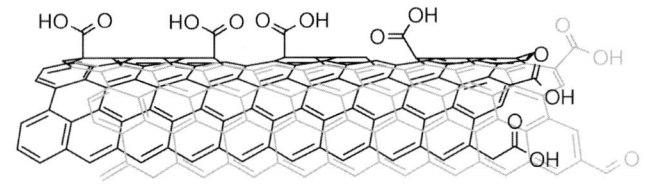

Fig. 1.4 Section of an oxidized SWCNT, reflecting terminal and sidewall oxidation.

SWNTs, room-temperature oxidation by ozone is confined to the end caps and to dangling bonds created by removal of the caps. A detailed study of diameter-dependent oxidative stability by Zhou et al. [80] recently confirmed a direct relationship between diameter and reactivity. Using the resonance-enhanced Raman radial breathing mode, the authors clearly showed that smaller diameter tubes are more rapidly air oxidized than larger diameter tubes [71]. In their review on the covalent chemistry of SWCNTs [24d], Bahr and Tour reported that the HiPCO process produced smaller diameter SWCNTs (ca. 0.7 nm) were more reactive towards ozone than larger diameter SWNTs formed by laser ablation. This allows for enrichment of large-diameter single-walled carbon nanotubes by using a mixed concentrated H_2SO_4–HNO_3 treatment [81].

Solution-phase mid-IR spectroscopy was used to assess the amount of functionality introduced into SWNTs by oxidation. For a [10, 10] SWNT containing 40 carbon atoms in the unit cell, 20 carboxylic acid groups at each end of the tubes were found. A perfect 100 nm long [10, 10] SWNT-COOH contains approximately 16 000 C atoms [40 × (1000/2.46)] and 40 carboxylic acid groups [82]. Zhang et al. [83] claim to have improved the efficiency of nanotube oxidation and functionalization by using $KMnO_4$ as oxidant and a PTC. A comparison between $KMnO_4$ oxidation with or without PTC resulted in a yield of about 35–40% of functionalized tubes/total weight for the reaction without PTC. For the PTC-catalyzed reaction, the yield of functionalized nanotubes was about 65–70%. However, PTC-functionalized CNTs displayed a higher concentration of –OH groups (~23%) and ~3.8% of –COOH groups only [83].

SWNTs oxidized by either acid or ozone treatment have been assembled on a number of surfaces, including silver [84], highly-oriented pyrolytic graphite (HOPG) [85] and silicon [86]. Heavily oxygenated ozonized SWCNTs have been used as ligands for the growth of CdSe and CdTe quantum dots, leading to the formation of nanotube–nanocrystal assemblies [87, 88]. To mitigate the problem of poor matrix-SWCNT connectivity and phase segregation in polymer–SWCNT hybrid materials, acid-treated oxidized and negatively charged SWCNTs were assembled layer-by-layer (LBL) with positively charged poly(ethylenimine) (PEI) polyelectrolyte [89]. After subsequent chemical cross-linking, a nanometer-scale composite with SWCNT loadings as high as 50 wt.% could be obtained with a tensile strength approaching that of ceramics [89].

SWCNT films (bucky paper) have been prepared using aqueous dispersions of SWCNTs containing 0, 3, 6 and 10 M nitric acid. With increasing acid concentration, the film tensile strength increased from 10 to 74 MPa and the tensile modulus from 0.8 to 5.0 GPa, whereas the d.c. electrical conductivity decreased [90]. The effect of oxidation time and ultrasonication on MWCNTs, the concentration of carboxylic groups being measured by potentiometric titration, was investigated [91]. A sonochemical treatment to promote the density of surface functional groups of MWCNTs was successfully employed during oxidation [92]. Distinctive direct sidewall functionalization of MWCNTs has been carried out using dilute nitric acid under supercritical water (SCW) conditions. The functionalization proceeded invasively from the outer to the inner graphitic layers of the MWCNTs, the resulting material being comprised of a functionalized amorphous carbon sheath covering the remaining inner nanotube [93]. Recently, the electrical conductivity of MWCNT–epoxy composites was investigated with respect to the oxidative treatment of the original MWCNTs [94]. Oxidized aligned MWCNT arrays, grown by chemical vapor deposition (CVD) on a platinum substrate and acid or air treated, were used to immobilize the enzyme glucose oxidase [95]. The enzyme immobilization allows for direct electron transfer from the enzyme to the transducer and can be used as an amperometric biosensor (cf. Section 1.3.7.4), to record electrically the conversion of glucose to gluconic acid [95].

1.3.2.2 Defect Functionalization – Transformation of Carboxylic Functions

Amidation – formation of CNT-carboxamides
The oxidative introduction of carboxylic functions to nanotubes provides a large number of CNT-functional exploitations and permits covalent functionalization by the formation of amide and ester linkages and other carboxyl derivatives [24]. Bifunctional molecules (diamines, diols, etc.) are often utilized as linkers. More illustrative examples are nanotubes decorated with amino-functionalized dendrimers, nucleic acids, enzymes, etc., and the formation of bioconjugates of CNTs [96].

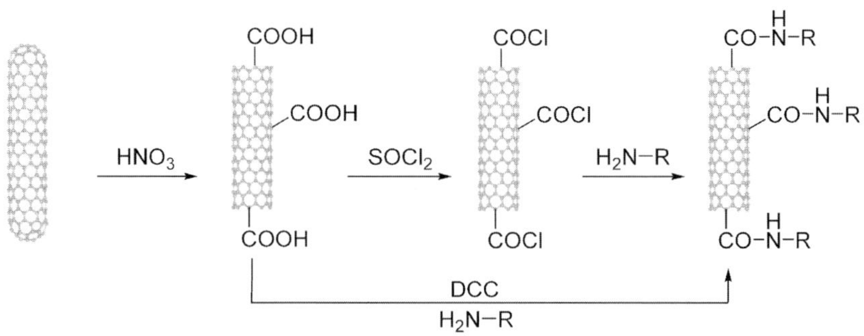

Scheme 1.3 (a) Schematic representation of oxidative etching ("cutting") of SWCNTs followed by treatment with thionyl chloride and subsequent amidation. The oxidations also occur at defect sites along the sidewalls; also other functionalities such as esters, quinones and anhydrides are formed. (b) DCC as coupling reagent.

CNT-carboxamides are conventionally obtained by amidation of CNT-acyl chlorides, derived from the reaction of carboxylated tubes with thionyl chloride [56, 97]. Similarly, carboxamide nanotubes have been prepared using dicyclohexylcarbodiimide (DCC) as condensing agent and allowing for the direct coupling of amines and carboxylic functions under neutral conditions (Scheme 1.3) [98, 99].

Haddon and coworkers were the first to report the functionalization of oxidatively treated SWCNTs with alkylamines and aniline derivatives [97, 100]. The conversion of the acid functionality to the N-octadecylamide led to the first shortened soluble SWCNTs [97]. The analysis of octadecylamido (ODA)-functionalized SWCNTs by solution-phase mid-IR spectroscopy gave about 50 wt. % of the acylamide functionality [101]. Tethering of a series of primary and secondary achiral and chiral amines, ranging from myrtanylamine to various nitroanilines, with SWCNTs and also covalent functionalization of SWCNTs with lipase enzymes was achieved by Wang et al. [102]. Functionalization of SWCNTs with N-(1-pyrenylmethyl)-1,5-diaminopentane resulted in the formation of pyrene-substituted SWCNT derivatives, the photochemical properties of which were studied by quenching experiments and laser flash photolysis [103]. Water solubilization of SWCNTs was achieved by amidation of SWCNT-COCl with glucosamine. Their solubility ranged from 0.1 to 0.3 mg mL^{-1} [104].

The reaction between toluene 2,4-diisocyanate and carboxylated MWCNTs afforded amido-functionalized nanotubes containing highly reactive isocyanate groups on their surface (Scheme 1.4). The amount of the isocyanate groups was determined by chemical titration and thermogravimetric analysis (TGA) [105]. The modified tubes may constitute promising components to prepare polymer–nanotube composites and coatings [106].

Exposure of CNT-acyl chlorides to $H_2N(CH_2)_{11}SH$ produced an amide linkage of the nanotubes to the alkanethiol [51]. Although the more nucleophilic thiols would be expected predominately to form thioesters, free thiols were shown to exist by atomic force microscopy (AFM) imaging of attached 10-nm gold nanoparticles [51]. Similarly, Liu et al. [107] achieved the thiolation of SWCNT pipes by reacting cysteamine [$HS(CH_2)_2NH_2$] with carboxyl-terminated nanotubes under carbodiimide conditions yielding $CNT–CONH(CH_2)_2SH$. The functionalized tubes could be assembled as monolayers on a gold surface via Au–S chemical bonding (Scheme 1.5) [107].

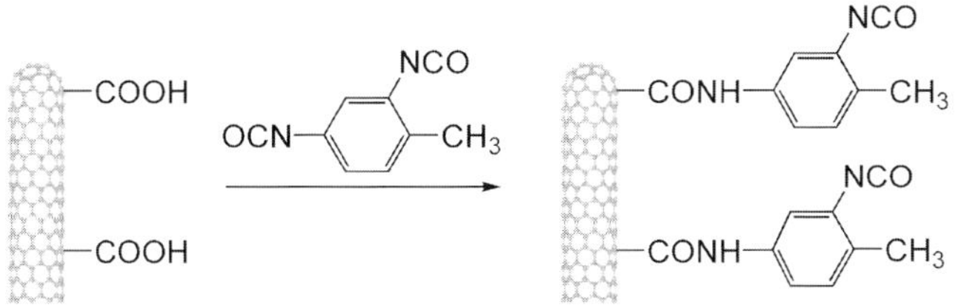

Scheme 1.4 Amido-functionalized CNTs containing highly reactive isocyanate groups at the tube surface.

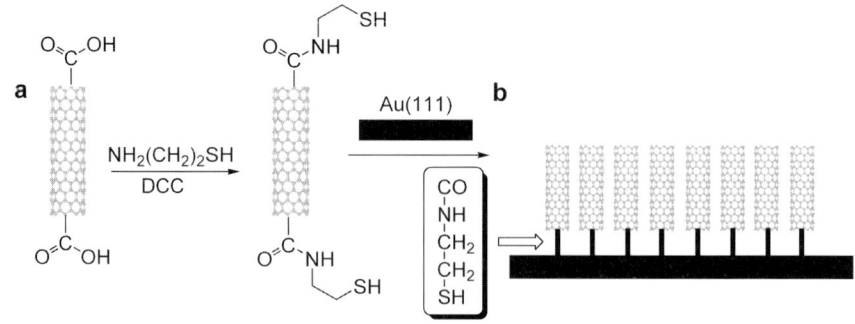

Scheme 1.5 Schematic diagrams of (a) the thiolization reaction of carboxyl-terminated CNTs with cysteamine ($NH_2CH_2CH_2SH$) and (b) the assembling structure of SWCNTs on gold.

Newkome-type dendrons were attached to the carbon scaffold of SWCNTs and MWCNTs by defect group functionalization [108]. First- and second-generation amine dendrons such as those depicted in Fig. 1.5 were condensed with the carboxyl groups of purified and opened SWCNTs and MWCNTs according to the carbodiimide technique [108]. These CNT derivatives can be expected to combine the characteristics of carbon nanotubes with those of dendrimers, potential building blocks for supramolecular, self-assembling and interphase systems.

Upon reacting SWCNT-acyl chlorides with α,ω-diamines such as tripropylenetetramine as molecular linker and subsequent diamide formation with another SWCNT, Roth and coworkers [109, 110] and Kiricsi et al. [111] succeeded in the interconnection of tubes and the formation of carbon nanotube junctions. End-to-end (Scheme 1.6a) and end-to-side nanotube interconnections (Scheme 1.6b) were formed and observed by AFM. Statistical analyses of the AFM images showed around 30% junctions in functionalized material [42].

A gas-phase derivatization procedure was employed for direct amidization of oxidized SWCNTs with simple aliphatic amines. In some cases a minor amount

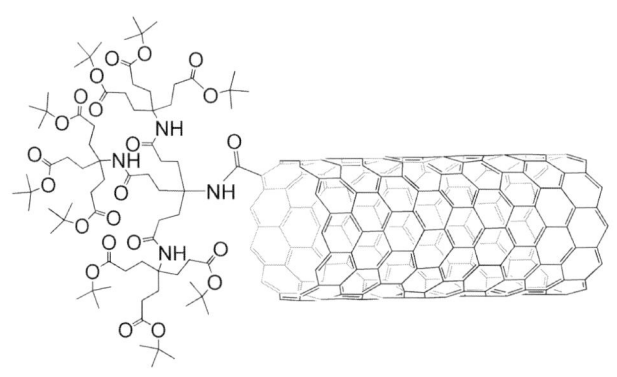

Fig. 1.5 Modification of oxidized CNTs with Newkome-type amino dendrimers.

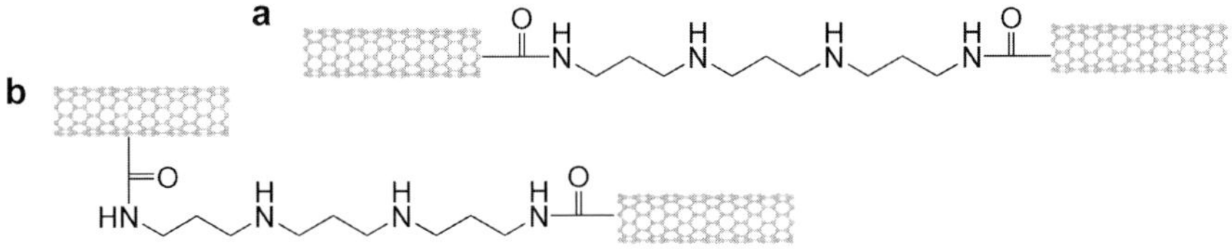

Scheme 1.6 Preparation of carboxamide junctions between SWCNTs: (a) end-to-end junctions and (b) sidewall-to-end junction between individual tubes.

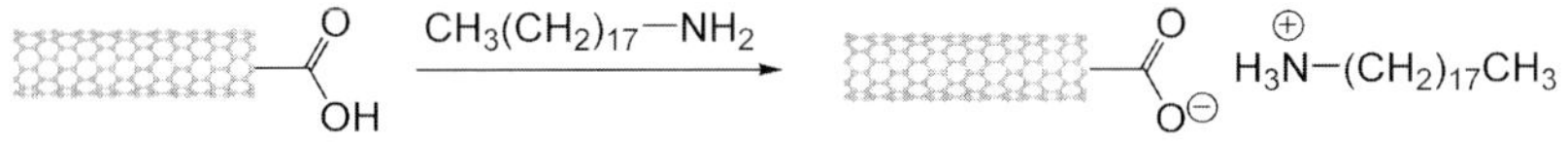

Scheme 1.7 Gas-phase amidation of oxidatively purified SWCNTs with octadecylamine (ODA) to obtain zwitterionic functionalization products.

of chemically formed amides in addition to a larger portion of physisorbed amines was observed [112]. Full-length oxidatively purified SWCNTs were rendered soluble in common organic solvents by noncovalent (zwitterionic) functionalization (Scheme 1.7) in high yield [100, 113]. About 4–8% of the SWCNT C-atoms can be functionalized by octadecyl amine in this way [100, 113, 114], offering a simple route to solubilize SWCNTs.

AFM micrographs showed that the majority of the thick SWCNT rope bundles were exfoliated into small ropes (diameter 2–5 nm) and individual nanotubes with lengths of several micrometers during the dissolution process. Multiwavelength laser Raman scattering spectroscopy and solution-phase IR spectroscopy were used to characterize the library of SWCNTs produced in current preparations. The average diameter of metallic tubes was found to be smaller than that of semiconducting nanotubes in the various types of preparations [112]. Such zwitterion-functionalized SWCNTs were length-separated and size-fractioned by gel permeation chromatography (GPC) by Chattopadhyay et al. [115], AFM being the method to determine the length distribution/fraction.

HiPCO-SWCNTs were oxidized in a UV–O_3 gas–solid interface reaction and subsequently assembled on a rigid oligo(phenylenethynylene) self-assembled monolayer (SAM). In a "chemical assembly", based on condensation between the carboxylic acid functionalities of the O_3-oxidized SWNTs and the amine functionalities of the SAMs, SWCNT-amides were formed in ordered arrays [116].

Carboxylic groups positioned at the open ends of SWCNTs were coupled to amines to form AFM probes with basic or hydrophobic functionalities by Wong et al. [117] (Scheme 1.8). Force titrations recorded between the ends of the SWCNT–AFM tips and hydroxy-terminated SAMs confirmed the chemical sensitivity and robustness of the AFM tips. Images recorded on patterned SAM allowed real molecular-resolution imaging [117].

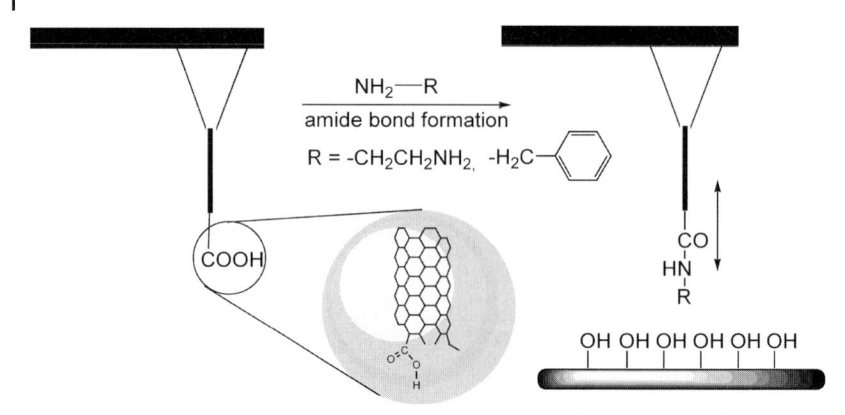

Scheme 1.8 Schematic illustration of an SWCNT force microscope probe and modification of an oxidized SWCNT tip by coupling an amine RNH$_2$ to a terminal –COOH. The probe is able to sense specific interactions between the functional group R and surface –OH groups.

With the amido functionalization, also new routes were opened to the covalent linkage of oligomers and polymers, dendrimers, peptides and biopolymers and to the formation of bioconjugates of carbon nanotubes. α,ω-Diaminopoly(ethylene glycol) and long-chain ethers of hydroxyaniline were attached to CNTs via amide bonds [118, 119]. Poly(ethylene glycol) (PEG) was grafted to shortened SWCNTs by SOCl$_2$ activation and amidation with PEG-monoamine [120]. Mono-amino-terminated poly(ethylene oxide) (PEO) was grafted onto SWCNTs by amide formation of SWCNT-COCl and the aggregation behavior of the PEO-SWCNTs was investigated in solution and as Langmuir–Blodgett (LB) films [121]. Amino-terminated polystyrene (PS) was grafted onto oxidatively cut nanotubes via amide formation [122], purified MWCNTs were covalently functionalized with the amino copolymer poly(propionylethyleneimine-co-ethylenimine) by amidation of CNT-carbonyl chlorides and by heating CNT-carboxylic acids in the presence of the amino polymer, respectively [123]. A multifunctionalized nanotube was used by Holzinger et al. as the core in the synthesis of first-, second- and third-generation Frechet- and Newkome-type dendrimers, using DCC and water-soluble EDC as condensing agents [124]. Nanotubes were functionalized by bovine serum albumin (BSA) via diimide-activated amidation; the BSA conjugates obtained were highly water soluble [125, 126]. Results from characterizations showed intimate association with the tubes and the bioactivity of the CNT–BSA conjugate was proven by a protein microdetection assay [125]. Research is under way to investigate the biocompatibility of chemically inert carbon nanotubes by immobilization of biopolymers and proteins at the tubes. The protein transferrin, tagged with a fluorescent label, was immobilized by covalent amide formation with CNT-COOHs in the presence of carbodiimide EDC and sulfo-N-hydroxysuccinimide (sulfo-NHS) [127]. Amino-terminated DNA strands were used to functionalize the open ends and defect sites of oxidized SWCNTs [128].

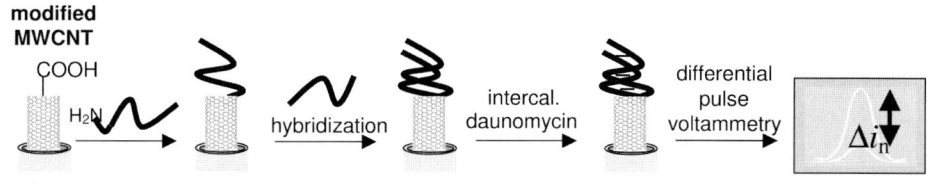

Scheme 1.9 Schematic representation of the enhanced electrochemical detection of DNA hybridization based on an MWCNT-COOH constructed DNA biosensor.

Haxani et al. reported carbodiimide-assisted amidation of SWCNT-COOHs with oligonucleotides and the preparation of a highly water-soluble adduct. Fluorescence imaging of individual nanotube bundles showed that the SWCNT–DNA adducts hybridized selectively with complementary strands [129]. A multistep route to the formation of covalently linked SWCNTs and DNA oligonucleotides was developed by Baker et al. and the covalent linkage proven by X-ray photoemission spectroscopy (XPS) [130]. The nanoscience group from Delft University had developed a technique to couple SWCNTs covalently to peptide nucleic acids (PNA), an uncharged DNA analog, and to hybridize the conjugate with complementary DNA. The recognition properties imparted to SWCNTs by oligonucleotide adducts could be used to program the attachment of CNTs to each other and to substrate features on which monolayers of complementary sequences are self-assembled [131]. From oxidized MWCNT-COOHs a glassy carbon electrode (GCE) was fabricated. In the presence of a water-soluble coupling reagent, oligonucleotide probes with an amino group at the 5′-phosphate end were covalently attached, to be used as a DNA biosensor (Scheme 1.9). Nucleotide-hybridization was performed and the specific nucleotide assembly was detected with the redox intercalator daunomycin as indicator [132].

The reduction of the amido functions of amide-solubilized MWCNTs by LiAlH$_4$ afforded the corresponding hydroxy-substituted CNTs, confirmed by FT-IR and XPS studies [133]. No morphology change of the nanotubes after reduction could be observed by Raman spectroscopy [133].

Esterification – formation of CNT-esters
Acyl chloride-functionalized SWCNTs are also susceptible to reactions with other nucleophiles, e.g. alcohols. Haddon's group reported the preparation of soluble ester-functionalized carbon nanotubes SWCNT-COO(CH$_2$)$_{17}$CH$_3$ (Fig. 1.6a) obtained by esterification with octadecanol [134]. The syntheses of soluble polymer-bound and dendritic ester-functionalized SWCNTs have been reported by Riggs et al. by attaching poly(vinyl acetate-co-vinyl alcohol) (Fig. 1.6b) [135] and hydrophilic and lipophilic dendron-type benzyl alcohols [119], respectively, to SWCNT-COCl (Fig. 1.6c). These functional groups could be removed under basic and acidic hydrolysis conditions and thus additional evidence for the nature of the attachment was provided [119, 136].

Fig. 1.6 (a) SWCNT-ester SWCNT-COO(CH$_2$)$_{17}$CH$_3$; (b) ester from SWCNT-COOH and poly(vinyl acetate-co-vinyl alcohol); (c) dendritic-type benzyl alcohol ether.

Ester-functionalized MWCNTs with terminal thiol groups were synthesized by converting the sodium salt of oxidatively purified MWCNTs with 2-bromoethanethiol in the presence of a phase-transfer agent. Gold-nanoparticles have been successfully self-assembled on the MWCNT-COO(CH$_2$)$_2$SH tubes to fabricate nanocomposites [137]. Labeling of SWCNTs with fluorescence probes was accomplished via esterification of the tubes by oligomerically tethered pyrene derivatives. The fluorescence and fluorescence excitation results showed that the tethered pyrenes form "intramolecular" excimers by π–π-interactions (Scheme 1.10). The pyrene monomer and excimer emissions were significantly quenched by the attached SWCNTs. The quenching was explained in terms of the nanotube serving as acceptor for excited-state energy transfer from the tethered pyrene moieties [138].

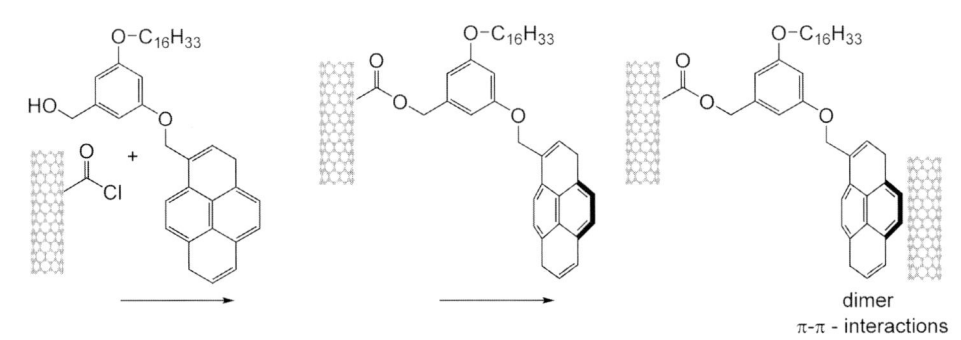

Scheme 1.10 Labeling of SWCNTs with fluorescence probes via esterification by derivatized pyrenes and formation of "intertubulary" dimers and excimers by π–π interactions.

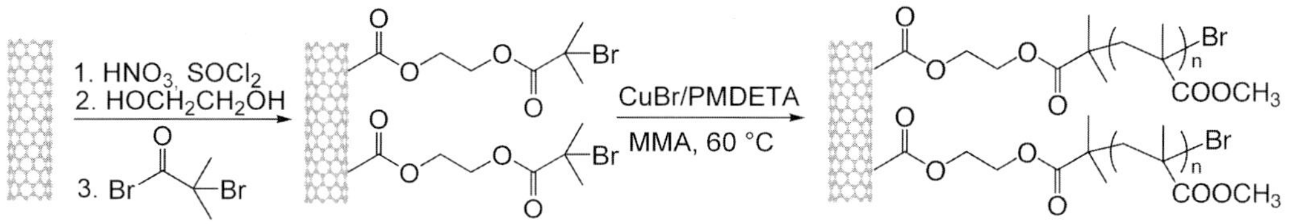

Scheme 1.11 Grafting of polymers via atom transfer radical polymerization (ATRP) using an ester linkage to oxidized MWCNTs. Formation of MWCNT-COCl, reaction with ethylene glycol and with 2-bromo-2-methyl-propionyl bromide and grafting polymerization with methyl methacrylate (MMA).

A general strategy for grafting of polymers from MWCNTs via atom transfer radical polymerization (ATRP) using an ester linkage to oxidized MWCNTs was described by researchers at Shanghai Jiao Tong University [139]. Four steps included the formation of MWCNT-COCl, reaction with ethylene glycol, subsequent reaction with 2-bromo-2-methylpropionyl bromide to initiate radical addition sites and finally grafting polymerization with methyl methacrylate (MMA) in the presence of CuBr–pentamethyldiethylenetriamine (PMDETA) (Scheme 1.11) [139]. Polymer brushes with SWCNTs as backbones were similarly synthesized by grafting n-butyl methacrylate with 2-hydroxyethyl 2′-bromopropionate-initiated SWCNTs and ATRP polymerization under CuCl–bipy catalysis [140].

The formation of covalently PEG-grafted SWCNT hybrid material was achieved by the reaction of SWCNT-COCl with hydroxyl-terminated PEG in various solvents. Two different self-assembling morphologies, depending on the quality of the solvent, were discovered by high-resolution transmission electron microscopy (HR-TEM) [141].

Thiolation – formation of SWCNT-$(CH_2)_n$SH

In 2003, Lim et al. [142] succeeded in a direct thiolation of the open ends of SWCNTs via successive carboxylation (H_2SO_4–HNO_3; H_2O_2–H_2SO_4; sonication), $NaBH_4$ reduction, chlorination with $SOCl_2$ and thiolation (Na_2S–NaOH) (Scheme 1.12). The intermediates and the final products were verified by FT-IR and NMR spectroscopy [142].

The thiolated CNTs were adsorbed on micron-sized silver and gold particles and gold surfaces to study the interactions between the thiol groups and the noble

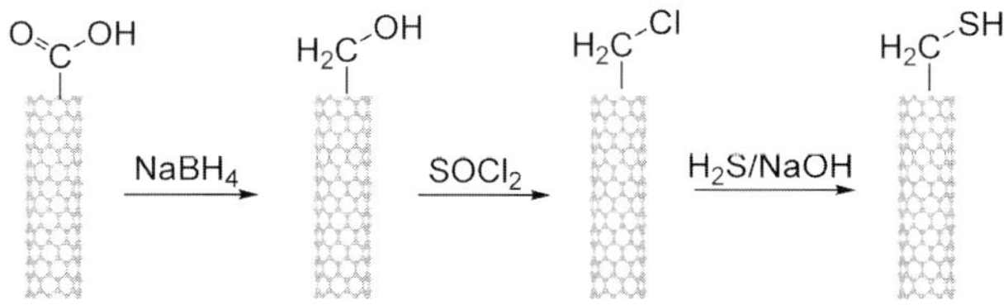

Scheme 1.12 Thiolation of CNTs by modification of carboxylated tubes.

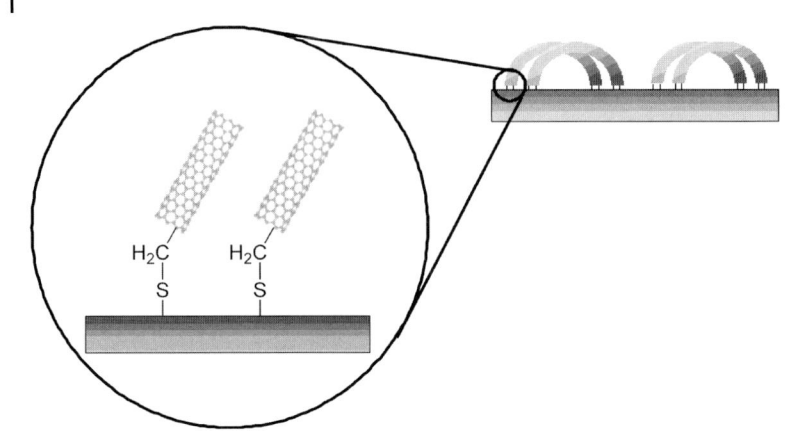

Fig. 1.7 Schematic representation of the vertical view image of the thiolated CNTs with thiol groups at both ends, binding to a gold surface. The flexible tube body of SWCNTs allows the CNT to conform its geometry for maximum binding energy at the expense of the bending energy. The result is a "bow-type" bundle of the thiolated CNTs.

metals. The thiol–metal adhesion was studied by SEM, AFM, wavelength-dispersive electron spectroscopy and Raman spectroscopy. A new type of bonding between the CNT and a noble metal surface was proposed (Fig. 1.7) that involved a bow-type SWCNT with its two ends attached to the metal surface [142].

Silylation of oxidized CNTs
A novel chemical functionalization method for MWCNTs through an oxidation and silylation process was reported in 2002. Purified and oxidatively functionalized MWCNTs were reacted with 3-mercaptopropyltrimethoxysilane, the CNT surface being joined to the organosilane moieties through OH groups [143]. Similarly, MWCNTs were functionalized by KMnO$_4$ oxidation under PTC catalysis and subsequent reaction with the hydrolysis product of 3-methacryloxypropyltrimethoxysilane (3-MPTS) (Scheme 1.13). The O-silyl-functionalized MWCNTs were characterized by FT-IR spectroscopy and energy-dispersive spectroscopy

Scheme 1.13 Reaction of oxidatively functionalized MWCNTs with the hydrolysis product of 3-methacryloxypropyltrimethoxysilane (3-MPTS).

(EDS), SEM and TEM analysis [70]. The method allowed different organo-functional groups to be attached to MWCNTs, improving their compatibility with specific polymers for producing CNT-based composites.

A fluorinated octyltrichlorosilane was reacted with the carboxylic moieties of oxidized MWCNTs and, after reduction with the corresponding alcohol groups. The modification was confirmed by XPES and TGA [144].

1.3.3
Hydrogenation of Carbon Nanotubes

Several methods for the hydrogenation of SWCNTs have been described. Dissolved metals acted as the reducing agents. Chen et al. first reported the Birch-like reduction of SWCNTs using lithium in diaminoethane [79a]. TEM micrographs showed corrugation and disorder of the nanotube walls due to hydrogenation and the formation of C–H bonds was suggested. The average hydrogen content of SWCNTs after lithium–ammonia hydrogenation was determined by Pekker et al. from TGA/MS and corresponded to a composition of $C_{11}H$ [145]. Hydrogenation occurred even on the inner tubes of MWCNTs, as shown by the chemical composition and the overall corrugation [145].

Owens and Iqbal [146] succeeded in an electrochemical hydrogenation of open-ended SWCNTs synthesized by CVD. Sheets of SWCNT bucky paper were used as the negative electrode in an electrochemical cell containing aqueous KOH solution as electrolyte. The authors claimed to have incorporated up to 6 wt. % of hydrogen into the tubes, determined by laser Raman IR spectroscopy and hydrogen release by thermolysis at 135 °C under TGA conditions [146]. However, the stability of exohydrogenated carbon nanotubes and the low temperature of hydrogen release at 135 °C [146] is contradictory with the 400–500 °C reported elsewhere [79a, 145].

Hydrogen bound to SWCNTs should not be released until *ca.* 500 °C, indicating robust attachment [145]. Theoretical first-principles total energy and electronic structure calculations of fully exohydrogenated zigzag and armchair SWCNTs (C_nH_n) point to crucial differences in the electronic and atomic structures with respect to hydrogen storage and device applications. C_nH_ns were estimated to be stable up to a radius of a [8,8] CNT, with binding energies proportional to $1/r$. By calculation, zigzag nanotubes were found to be more likely to be hydrogenated than armchair tubes with equal radius [147].

1.3.4
Addition of Radicals

Perfluorinated alkyl radicals, generated by photoinduction from heptadecafluorooctyl iodide, were added to SWCNTs and the perfluorooctyl-derivatized CNTs obtained (Scheme 1.14). No difference in the solubility of the fluoroalkyl-substituted nanotubes and the starting materials was observed [148]. A pathway to the radical functionalization of CNTs' sidewalls was predicted by classical molecular dy-

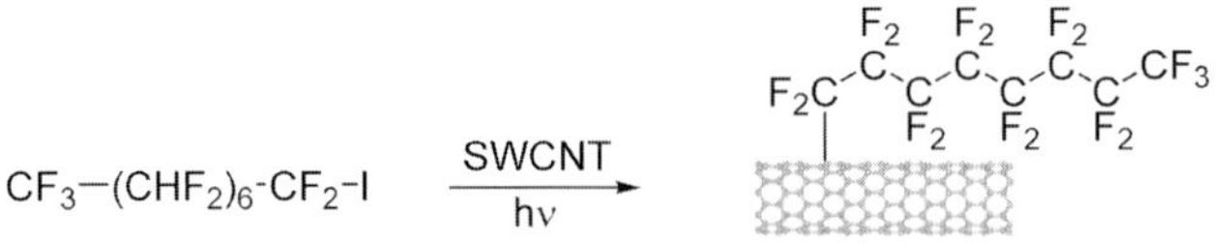

Scheme 1.14 Addition of heptadecafluorooctyl radicals obtained from irradiation of heptadecafluorooctyl iodide.

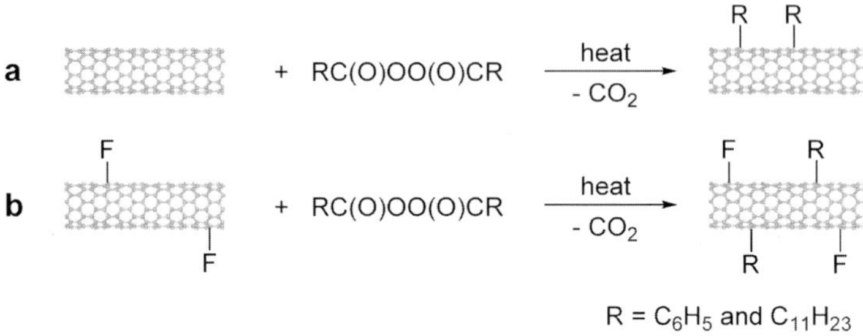

Scheme 1.15 Functionalization of SWCNTs and fluorinated F-SWNTs with benzoyl (R = C_6H_5) and lauroyl (R = $C_{11}H_{23}$) peroxides.

namics simulations of the bombardment of a bundle of SWCNTs by CH_3 radicals [149].

Pristine SWCNTs and their fluorinated derivatives, F-SWCNTs, were reacted with organic peroxides to functionalize their sidewalls covalently by attachment of free radicals (Scheme 1.15). The tubes' reactivity towards radical addition was compared with that of corresponding polyaromatic and conjugated polyene π-systems [150, 151]. The characterization of the functionalized SWCNTs and F-SWCNTs was performed by Raman, FT-IR and UV/Vis/NIR spectroscopy and also by TGA/MS, TGA/FT-IR and with TEM measurements. The solution-phase UV/Vis/NIR spectra showed complete loss of the van Hove absorption band structure, typical of functionalized SWCNTs [150].

1.3.5
Addition of Nucleophilic Carbenes

The reaction of a nucleophilic dipyridyl imidazolidene (DPI) with the electrophilic SWCNT π-system to give zwitterionic polyadducts was reported in 2001 by Holzinger et al. [148]. DPI was generated from the corresponding dipyridyl imidazolium system by deprotonation. Each covalently bound imidazolidene addend bears a positive charge, one negative charge/addend is transferred to the delocalized tube surface and a stable 14 π-system is obtained (Scheme 1.16) [148].

Sufficiently derivatized nanotubes were soluble in DMSO, allowing the separation of insoluble, unreacted and insufficiently functionalized SWCNTs [148]. The n-doping of the tubes surface offers a new way to modify the tube properties and control the electronic properties.

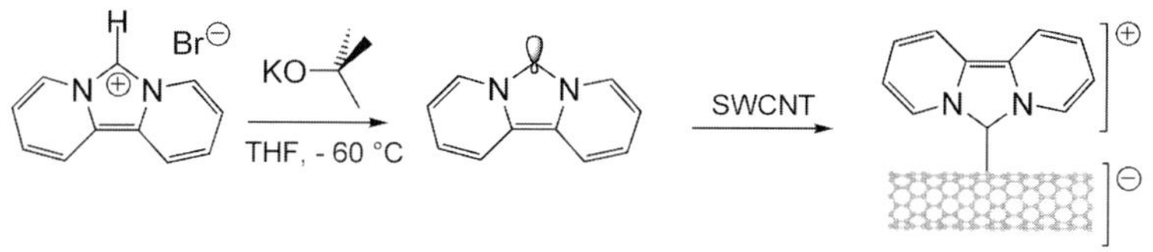

Scheme 1.16 Sidewall functionalization by addition of nucleophilic dipyridyl imidazolidene to the electrophilic SWCNT π-system.

1.3.6
Sidewall Functionalization Through Electrophilic Addition

In 2002, Tagmatarchis et al. [152] reported the modification of SWCNTs through electrophilic addition of $CHCl_3$ in the presence of $AlCl_3$. From hydrolysis of the so-produced labile chlorinated intermediate species, hydroxy-functionalized SWCNTs were obtained, the coupling of which with propionyl chloride led to the corresponding SWCNT propionate esters (Scheme 1.17) [152].

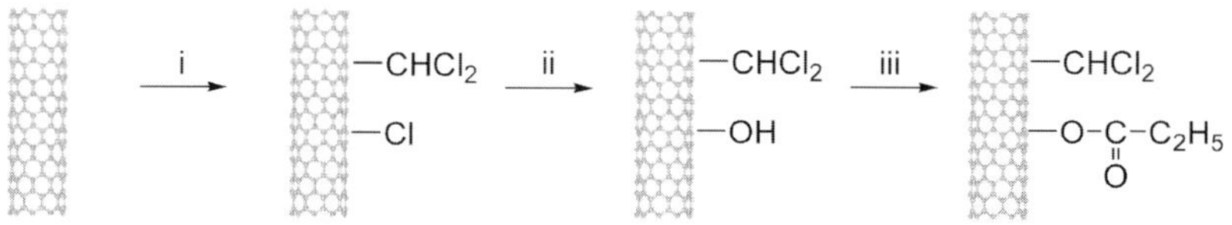

Scheme 1.17 Electrophilic addition of $CHCl_3$ to SWCNTs (i), followed by substitution of chlorine (ii) and esterification (iii): (i) $CHCl_3$, $AlCl_3$; (ii) OH^-, MeOH; (iii) C_2H_5COCl.

1.3.7
Functionalization Through Cycloadditions

1.3.7.1 Addition of Carbenes

In the course of a study on organic functionalization of CNTs, Haddon's group discovered in 1998 that dichlorocarbene was covalently bound to soluble SWCNTs (Scheme 1.18) [97]. Originally, the carbene was generated from chloroform with potassium hydroxide [79a] and later from phenyl(bromodichloromethyl)mercury [97]. However, the degree of functionalization was as low as 1.6 at.% of chlorine only, determined by XPS [153].

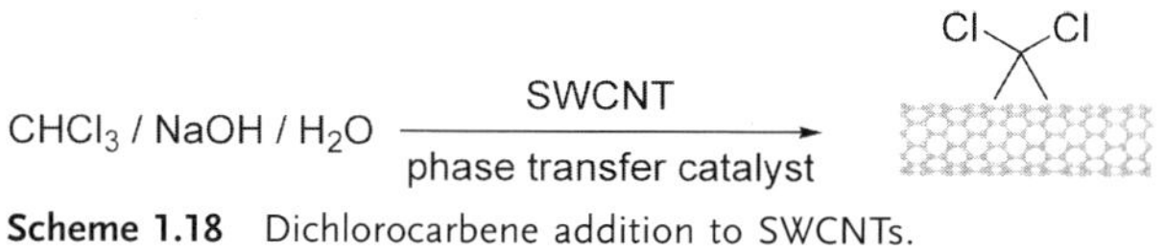

Scheme 1.18 Dichlorocarbene addition to SWCNTs.

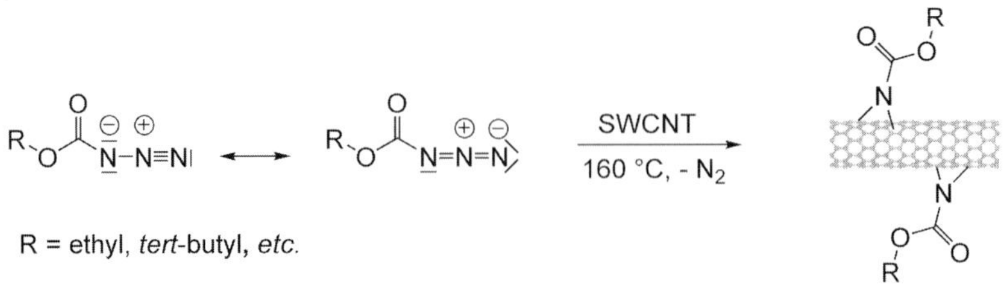

Scheme 1.19 Sidewall functionalization of SWCNTs by [2+1]-cycloaddition of alkoxycarbonyl nitrenes obtained from azides.

1.3.7.2 Addition of Nitrenes

Sidewall functionalization of SWCNTs was achieved via the addition of reactive alkyloxycarbonyl nitrenes obtained from alkoxycarbonyl azides. The driving force for this reaction is the thermally-induced N_2 extrusion. The nitrenes generated attack nanotube sidewalls in a [2+1]-cycloaddition forming an aziridine ring at the tubes sidewalls (Scheme 1.19).

With this technique, a broad range of aziridino-SWCNTs was obtained by our group by cycloadding addends such as alkyl chains, aromatic groups, crown ethers and oligoethylene glycol units (Fig. 1.8) [148, 154].

Nitrene additions led to a considerable increase in solubility in organic solvents; the highest solubility of 1.2 mg mL^{-1} in DMSO and TCE was found for SWCNT ethylene glycol–crown ether adducts. AFM and TEM revealed that the formation

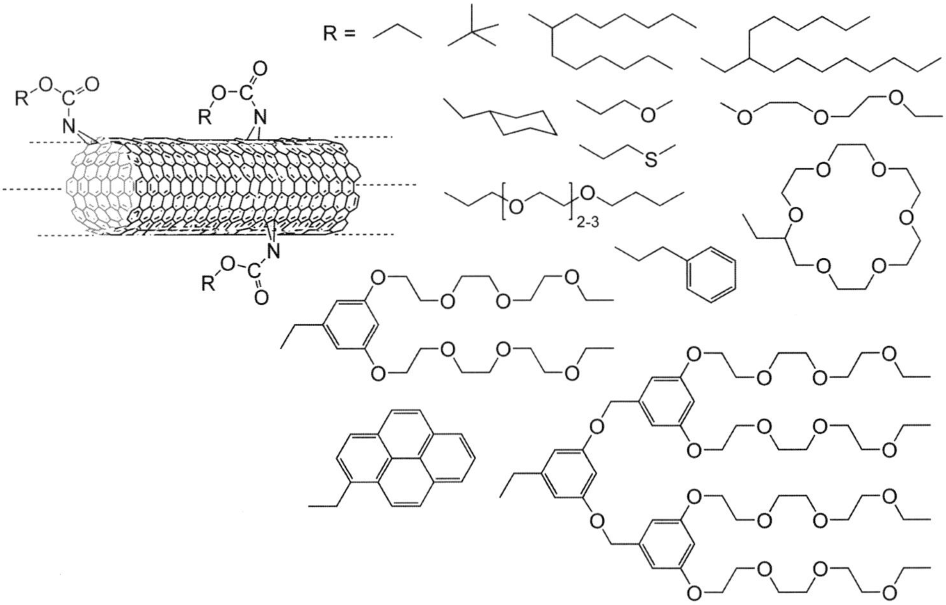

Fig. 1.8 Sidewall functionalization of SWCNTs via addition of (R)-oxycarbonyl nitrenes.

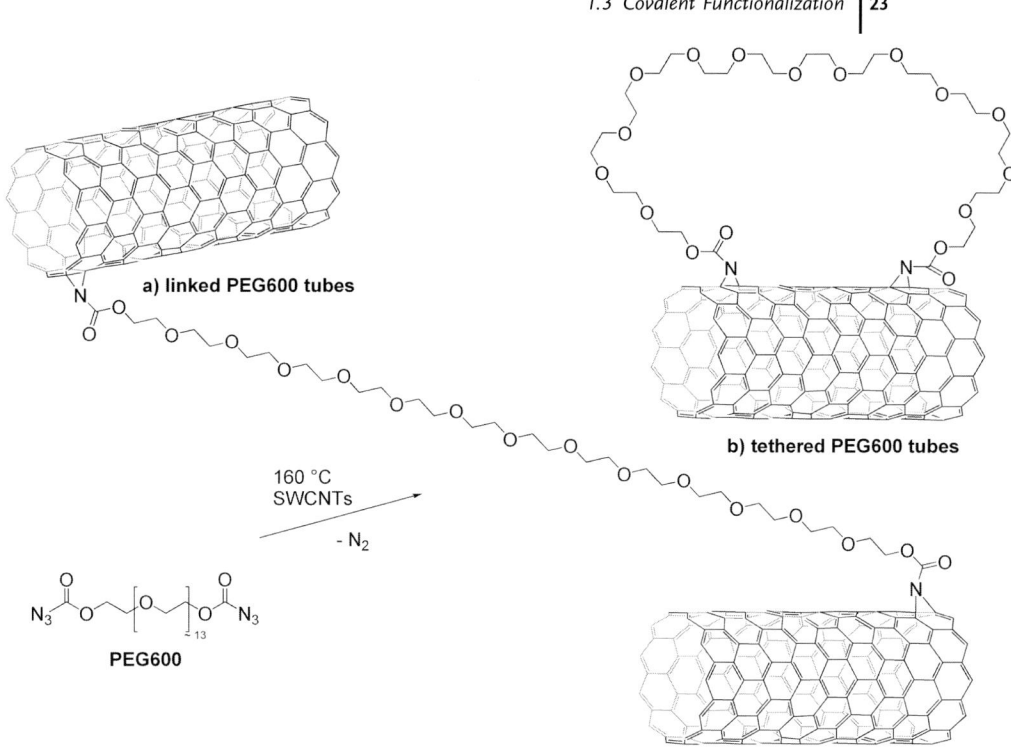

Scheme 1.20 PEG-tethered SWCNTs and interconnections between individual SWCNTs through cycloaddition of α,ω-polyethylene glycol dinitrenes as the molecular linker.

of thin bundles with typical diameters of 10 nm. The presence of the bundles in solution was supported by ^{1}H NMR spectroscopy; the elemental composition of the functionalized SWCNT was determined by XPS [155, 156]. The use of Raman and electron absorption spectroscopy (UV/Vis/NIR) showed that the electronic properties of the SWCNTs were mostly retained after functionalization, indicating less than 2 at.% addend per C-atom of the tube sidewalls [154]. Nitrene addition to nanotubes of different origins and production methods was compared with that to unfunctionalized pristine tubes by investigating their XPS, Raman and UV/Vis/NIR spectra and TEM images [155].

Tethered SWCNTs and interconnections between individual SWCNTs using α,ω-dinitrenes as the molecular linker (Scheme 1.20) were reported the first time in 2003 by Holzinger et al. [157]. The bisnitrenes were derived from thermolysis of poly(ethylene glycol) PEG 600 bisazidocarbonate; the covalent linkage was characterized by SEM studies, XPS, X-ray diffractograms and Raman spectra [157].

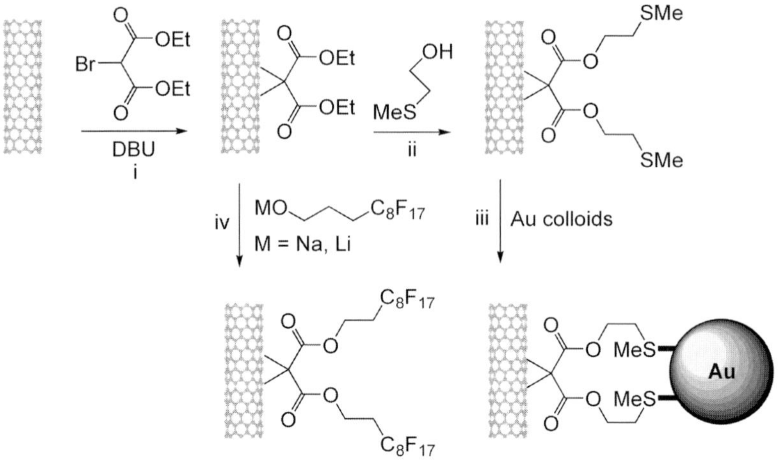

Scheme 1.21 Schematic representation of the cyclopropanation of SWCNTs and the introduction of chemical markers for AFM visualization and ^{19}F NMR spectroscopy and XPS: (i) diethyl bromomalonate, DBU, room temperature; (ii) 2-(methylthio)ethanol, diethyl ether; (iii) preformed 5-nm gold colloids; (iv) sodium or lithium salt of 1H,1H,2H,2H-perfluorodecan-1-ol.

1.3.7.3 Nucleophilic Cyclopropanation – Bingel Reaction

Fullerenes are known to react easily with bromomalonates in the Bingel reaction [158] to form cyclopropanated methanofullerenes [159]. The equivalent transformation was performed by Coleman et al. [160] with purified SWCNTs and diethyl bromomalonate as addend. They developed a chemical tagging technique which allows the functional groups to be visualized by AFM. The cyclopropanated methano-SWCNT derivatives were transesterified with 2-(methylthio)ethanol and the cyclopropane groups were "tagged" by exploiting gold–sulfur interactions using preformed 5-nm gold colloids (Scheme 1.21) [160].

Gold colloids were observed both on the sides and at the ends of the nanotubes, indicating sidewall and termini modification. To confirm further the derivatization of the SWNTs, a perfluorinated marker was introduced by transesterification to allow the nanotubes to be probed by ^{19}F NMR spectroscopy and XPS [160].

1.3.7.4 Azomethine Ylides

Prato and coworkers succeeded in the 1,3-dipolar addition of azomethine ylides to CNTs [161]. Treatment of pristine SWCNTs with an aldehyde and an N-substituted glycine derivative resulted in the formation of substituted pyrrolidine moieties on the SWCNT surface (Scheme 1.22). The approach works effectively with both SWCNTs, prepared by several different methods, and MWCNTs. The pyrrolidino-functionalized CNTs were sufficiently soluble in common organic solvents and were characterized by several spectroscopic techniques and TEM [161]. The

$R^2\text{-CHO} + R^1\text{-NHCH}_2\text{COOH}$ SWCNT →(DMF, 130 °C)   [pyrrolidine-functionalized SWCNT with R^1 on N and R^2]

Scheme 1.22 1,3-Dipolar addition of azomethine ylides to SWCNTs and MWCNTs.

azomethine ylide functionalization was also used to purify HiPCO SWCNTs from metal nanoparticles and carbon impurities [162].

The covalent functionalization of CNTs with azomethine ylides allows for a number of other functionalities to be immobilized onto the SWCNT surface [163]. Amidoferrocenyl-functionalized SWCNTs appeared to be efficient anion receptors for the redox recognition of $H_2PO_4^-$ [164]. A ferrocene-modified glycine precursor was used to functionalize the CNT sidewalls with ferrocene units (Fig. 1.9). A subsequent electron transfer was observed from ferrocene to the SWCNTs on photoexcitation, opening up ways to the use of ferrocene-SWCNT hybrids in solar energy applications [165]. Ferrocenyl-functionalized SWCNTs of that kind were coupled to glucose oxidase (GOx) within an amphiphilic polypyrrole matrix for the amperometric catalytic detection of glucose as a glucose biosensor [166].

Functionalized N-triethylene glycol pyrrolidino-CNTs (Fig. 1.10a) allowed electrochemistry and quantum chemical calculations to be carried out to investigate the bulk electronic properties [167]. Functionalization obviously modified the electronic state of pristine CNTs; however, some of the metallic character was retained and the overall electron density of states (DOS) was not strongly affected [167]. Pyrrolidino-SWCNTs and -MWCNTs bearing a free amino-terminal N-oligoethylene glycol moiety formed supramolecular associates with plasmid DNA through ionic interactions. The complexes were able to penetrate within cells. SWCNTs

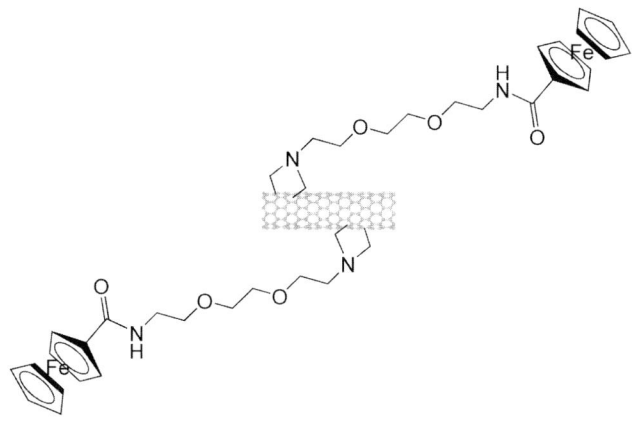

Fig. 1.9 1,3-Dipolar cycloaddition of ferrocene-modified azomethine ylide to SWCNTs in order to functionalize CNT sidewalls with electron donor units.

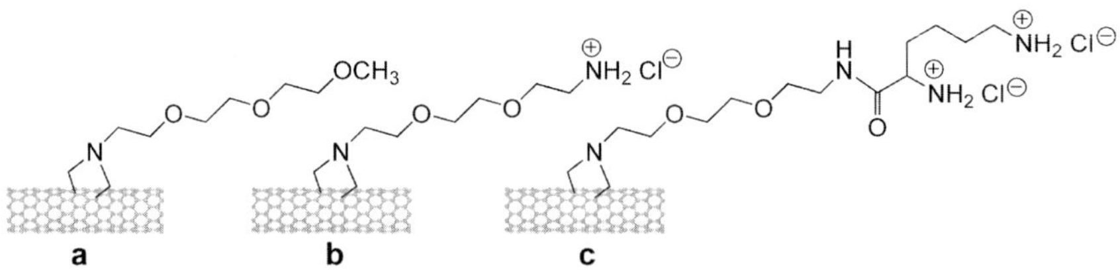

Fig. 1.10 (a) Functionalized N-(triethylene glycol)pyrrolidino-CNTs to carry out electrochemistry and quantum chemical calculations; (b) and (c) two types of cationically functionalized SWCNTs to bind with synthetic oligodeoxynucleotides.

complexed with plasmid DNA were able to allow for higher uptake of DNA and gene expression *in vitro* than could be achieved with individual DNA only [168]. The bindings of two types of cationically functionalized SWCNTs (Fig. 1.10b and c) with synthetic oligodeoxynucleotide (ODN) immunostimulatory CpG motif were compared by Bianco et al. [169]; the results demonstrated the potential of functionalized CNTs to enhance the immunostimulatory properties of ODN CpGs and the advantage for an effective delivery of ODN CpGs into target cells.

A series of amino acids, fluorescent probes and bioactive peptides have been covalently linked to functionalized CNTs through the terminal amino group of oligoethylene glycol functionalities (Fig. 1.10b). Using fragment condensation or selective chemical ligation of Cys-thiol groups, mono- and dipeptide-functionalized CNTs were obtained and their potential as drug delivery vehicles, for gene delivery and for delivery of antigens was evaluated [170].

1.3.7.5 [4+2]-Cycloaddition – Diels–Alder Reaction

The first Diels–Alder [4+2]-cycloaddition functionalization of CNTs has been reported recently [171]. Ester-functionalized SWCNTs were reacted with *o*-quinodimethane, generated *in situ* from benzo-1,2-oxathiin-2-oxide under microwave irradiation (Scheme 1.23). This technique opened up a new access to a novel family of modified carbon nanotubes.

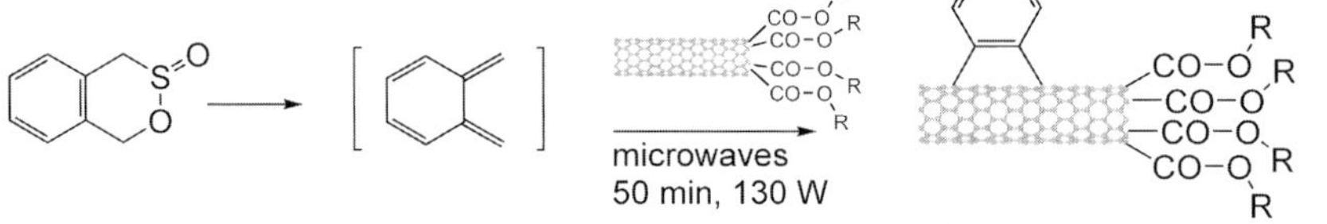

Scheme 1.23 Diels–Alder [4+2]-cycloaddition reaction of functionalized SWCNTs with *o*-quinodimethane under microwave irradiation.

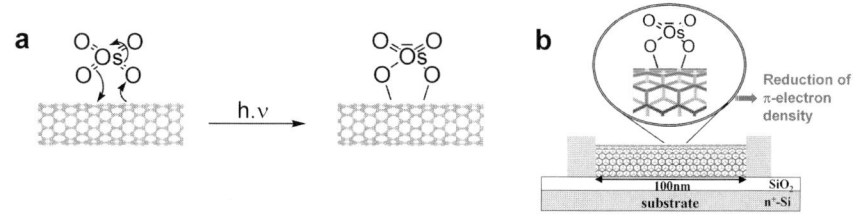

Scheme 1.24 (a) UV light-induced reversible sidewall osmylation of SWCNT and (b) electrical monitoring of the resistance at the single tube level.

1.3.7.6 Sidewall Osmylation of Individual SWCNTs

UV light-induced osmylation of carbon nanotubes was observed by exposure of SWCNTs to OsO_4 vapor [172] and also in organic solvents (Scheme 1.24) [173]. OsO_4 addition to CNTs resulted in a pronounced increase in their electrical resistance by up to several orders of magnitude, which was electrically monitored at the individual tube level. The addition was reversible and the original resistance was restored [172]. The major products of osmylation in organic media were SWCNTs decorated with OsO_2 particles, resulting in extended tube aggregation [173].

1.3.8
Aryl Diazonium Chemistry – Electrochemical Modification of Nanotubes

Electrochemistry has become an elegant tool for the functionalization of CNTs. Applying a constant potential (potentiostatic) or a constant current (galvanostatic) to a CNT electrode immersed in a suitable reagent solution, highly reactive radical species can be generated. These reactive species readily functionalize CNTs or self-polymerize, forming a polymer coating of the tubes. In 2001, Tour's group grafted a series of phenyl functionalities by electrochemically coupling of aryldiazonium salts onto an SWCNT bucky paper electrode [174]. The reaction mechanism follows a one-electron reduction of the diazonium salt and subsequent addition of reactive aryl radicals (Scheme 1.25).

Scheme 1.25 Grafting of phenyl functionalities on to a SWCNT bucky paper electrode by electrochemically coupling of aryldiazonium salts; one-electron reduction of the diazonium salt and addition of aryl radicals.

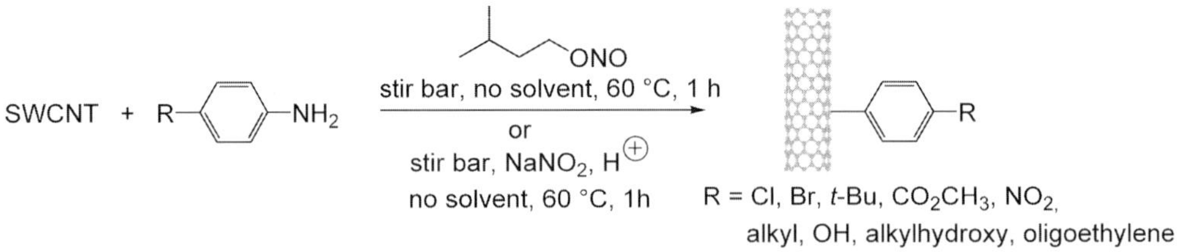

Scheme 1.26. Solvent-free functionalization of carbon nanotubes performed with various substituted anilines and isoamyl nitrite or $NaNO_2$–acid. R can be Cl, Br, NO_2, CO_2CH_3, alkyl, OH, alkylhydroxy, oligoethylene.

Based on TGA and elemental analysis, up to one in 20 nanotube carbons (5 %) has been found to possess an aryl addend [174]. For the preparation of PS-CNT nanocomposites, the same authors functionalized SWCNTs by in-situ generation of a diazonium compound from 4-(10-hydroxydecyl)aminobenzoate [175]. From rheology data it was suggested that the reinforcement and dispersibility of the thus functionalized SWCNT-composites were improved over PS composites with pristine SWCNTs.

Basic studies on diazonium-CNT chemistry led to two very efficient techniques for SWCNT derivatization: solvent-free functionalization [176] and functionalization of individual (unbundled) nanotubes [175]. With the solvent-free functionalization (Scheme 1.26), heavily functionalized and soluble material is obtained and the nanotubes disperse in polymer more efficiently than pristine SWCNTs [176]. With the second method, aryldiazonium salts react efficiently with the individual (unbundled) HiPCO produced and sodium dodecyl sulfate (SDS)-coated SWCNTs in water. The resulting functionalized tubes (one addend in nine tube carbons) remained unbundled throughout their entire lengths and were incapable of re-roping. [175].

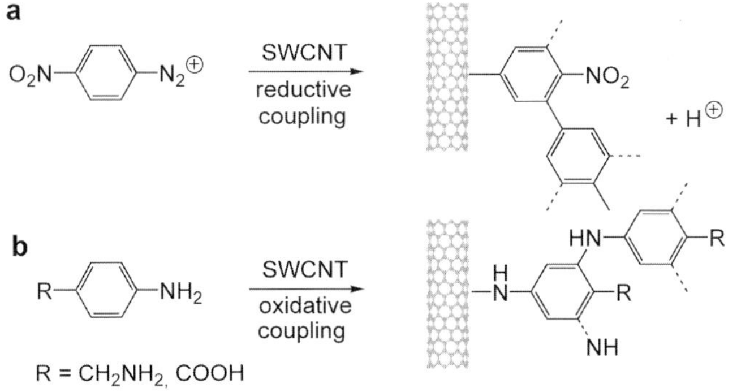

Scheme 1.27 Electrochemical modification of SWCNTs with aryldiazonium compounds: (a) reductive coupling of preformed p-nitrophenyldiazonium salts to SWCNTs and (b) anodic oxidation with in situ-generated diazonium compounds from anilines. The dotted lines mark positions of further linkages that could be formed during the growth of a polymerized layer of phenyl units on the SWCNTs.

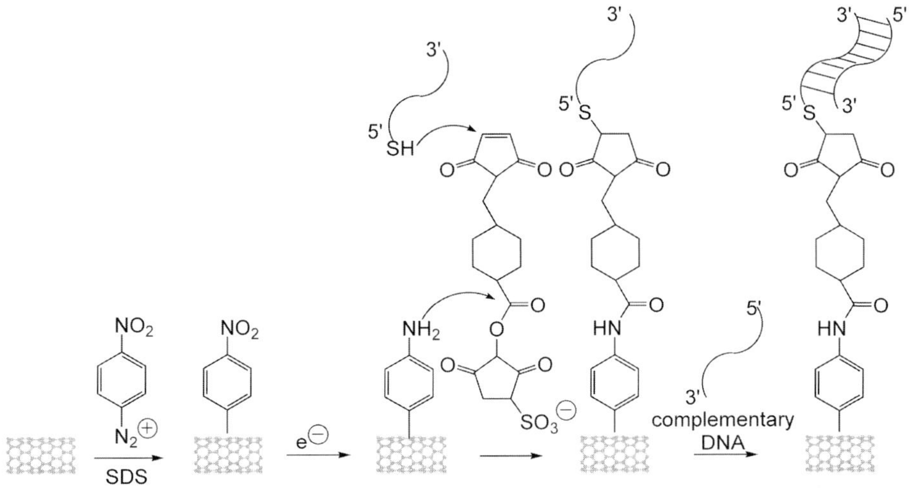

Scheme 1.28 Electrochemical addressing of nitrophenyl-substituted nanotubes, reduction and subsequent linking with DNA forming an array of DNA–CNT hybrid nanostructures.

In 2002, Burghard and coworkers described an elegant method for the electrochemical modification of individual SWCNTs [177]. To address electrically individual SWCNTs and small bundles, the purified tubes were deposited on surface-modified Si/SiO$_2$ substrates and subsequently contacted with electrodes, shaped by electron-beam lithography. The electrochemical functionalization was carried out in a miniaturized electrochemical cell. The electrochemical reduction was achieved by reduction of 4-NO$_2$C$_6$H$_4$N$_2^+$BF$_4^-$ in DMF with NBu$_4^+$BF$_4^-$ as the electrolyte (Scheme 1.27a), anodic oxidation was accomplished with aromatic amines in dry ethanol with LiClO$_4$ as the electrolyte salt (Scheme 1.27b) [177].

Four different aryldiazonium salts have been used to functionalize SWCNTs through electrochemical reduction. By XPS and Raman diffusion measurements, the growth of aryl chains on the sidewalls of the nanotubes was observed [178]. Electrically addressable biomolecular functionalization of SWCNT electrodes and vertically aligned carbon nanofiber electrodes with DNA was achieved by electrochemically addressing (reduction) of nitrophenyl substituted nanotubes and nanofibers. Subsequently, the resulting amino functions were covalently linked to DNA forming an array of DNA–CNT hybrid nanostructures (Scheme 1.28) [179].

1.3.9
Reductive Alkylation and Arylation of Carbon Nanotubes

Reductive alkylation of carbon nanotubes using lithium and alkyl halides yields sidewall-functionalized nanotubes soluble in common organic solvents. Billups' group prepared dodecylated SWCNTs from raw HiPCO tubes using lithium and dodecyl iodide in liquid ammonia and demonstrated the occurrence of exten-

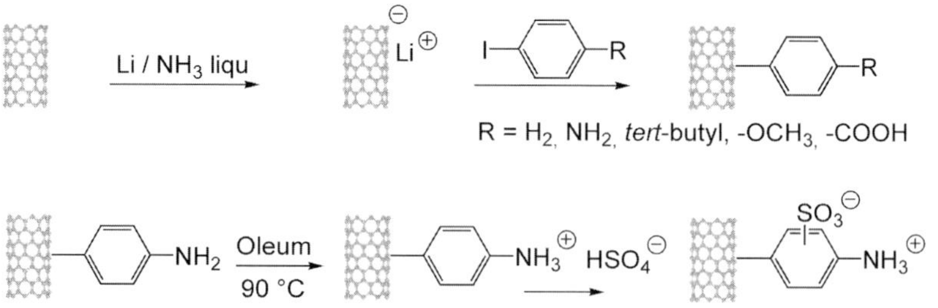

Scheme 1.29 Reductive arylation of SWCNTs with iodobenzene derivatives, conversion to zwitterions by sulfonation.

sive debundling of the tubes [180]. Alkyl radicals were identified to be the intermediates in the alkylation step by GC/MS analysis of hydrocarbons formed as by-products. Very recently, the same group reported the preparation of aryl-functionalized SWCNTs by treating SWCNTs with lithium in liquid NH_3 and subsequent reaction with aryl iodides (Scheme 1.29). Zwitterions were obtained from the aniline-derivatized SWCNTs by sulfonation, which were found to exhibit moderate solubility (50 mg L^{-1}) in methanol [181].

1.3.10
Addition of Carbanions – Reactions with Alkyllithium

Viswanathan et al. introduced carbanions onto SWCNT surfaces by treating the tubes with *sec*-butyllithium, providing initiating sites for the polymerization of styrene [182]. Recently, Chen et al. reported that on treating SWCNTs with *sec*-butyllithium and reacting the carbanions formed with CO_2 under oxygen-free and anhydrous conditions, the SWCNTs were alkylated and carboxylated [183] (Scheme 1.30). The so-derivatized SWCNTs could be dispersed in water to give a nearly transparent solution of 0.5 mg mL^{-1}; zeta potentiometric titrations indicated a feature similar to a zwitterionic polyelectrolyte [183].

MWCNTs have been functionalized with *n*-butyllithium by Blake et al. and subsequently the modified CNTs were covalently bonded to a chlorinated polypropylene [184]. The polypropylene-grafted MWCNTs were used to produce ultra-strong CNT polymer composites.

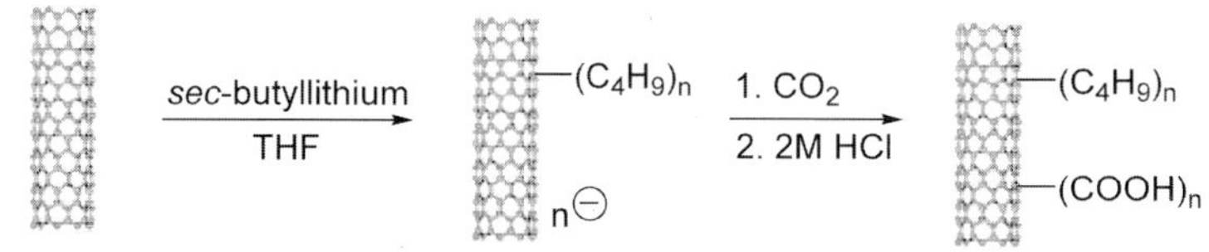

Scheme 1.30 Addition of carbanions, alkylation and carboxylation of SWCNTs: treatment with *sec*-butyllithium and subsequent reaction with CO_2.

1.3.11
Covalent Functionalization by Polymerization – "Grafting To" and "Grafting From"

Grafting macromolecules to the tips and onto the convex walls of CNTs has been explored for several years, performed using the "grafting to" approach via reactions such as esterification and amidization [136, 185]. The "grafting from" approach has been employed to graft polymer chains onto solids from polymer-functionalized CNTs. PS was grafted from MWCNTs [186] and PS and PVK polymer chains were successfully grafted from the surface of SWCNTs using anionic polymerization [182, 187]. By ATRP, Gao et al. grafted covalently poly(methyl methacrylate) (PMMA) and PS onto MWCNTs [188]. Ejaz et al. successfully grafted polymer chains onto a solid surface [189]. Hao et al. grafted PS from the convex walls of MWCNTs by *in situ* ATRP and prepared hybrid nanostructures [190]. SWCNTs were functionalized with PS by both grafting to and grafting from methods by Qin et al. [191]. The grafting addend PS–N_3 was synthesized by ATRP of styrene followed by end-group transformation and subsequent addition to SWCNT (Scheme 1.31a). The grafting from functionalization was achieved by ATRP of styrene using 2-bromopropionate groups immobilized as initiator (Scheme 1.31b) [191].

Polymer-linked MWCNT nanocomposites were prepared by reversible addition fragmentation chain transfer (RAFT). The RAFT reagent was successfully grafted on to the surface of MWCNTs and PS chains were grafted from MWCNTs via RAFT polymerization [192]. By covalently linking acyl chloride functions of functionalized MWCNTs with living polystyryllithium, Huang et al. succeeded in the preparation of polystyrene-functionalized MWCNTs (Scheme 1.32) [193].

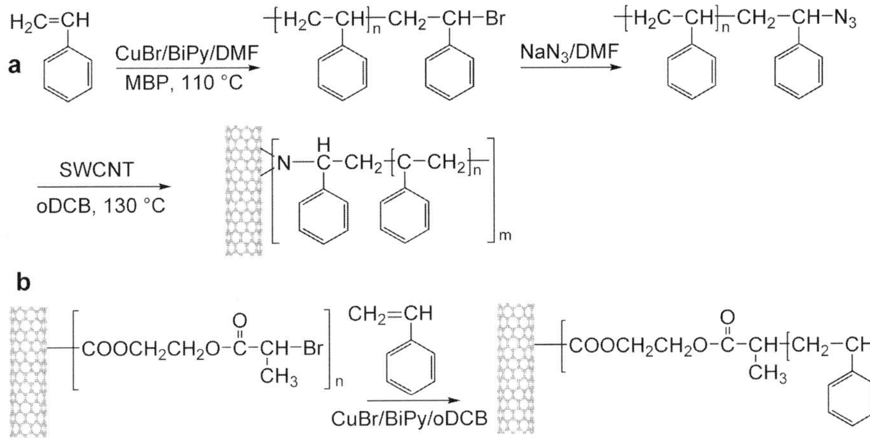

Scheme 1.31 "Grafting to" and "grafting from" of SWCNTs with PS. (a) The addend PS-N_3 was synthesized by ATRP of styrene, end-group transformation and addition to SWCNT. The grafting from functionalization was achieved by ATRP of styrene using 2-bromopropionate groups as initiator.

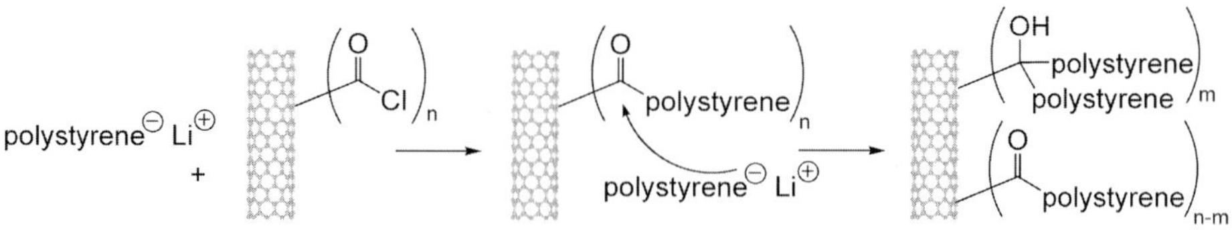

Scheme 1.32 Covalently linking of MWCNT-acyl chlorides with living polystyryllithium for the preparation of polystyrene-functionalized MWCNTs.

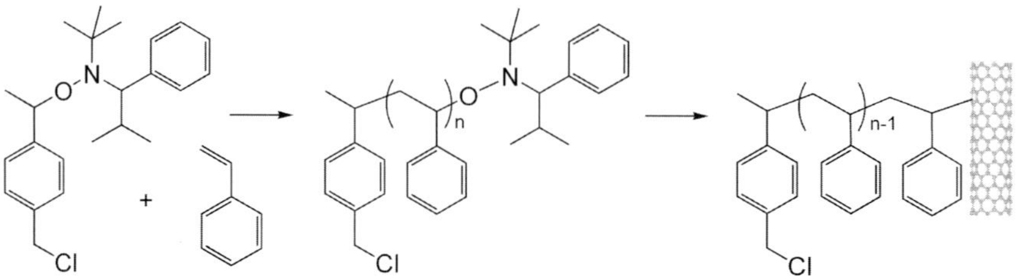

Scheme 1.33 Preparation of PS by nitroxide-mediated "living" free-radical polymerization and its utilization for the functionalization of shortened SWCNTs.

Homopolymer PS and block copolymer poly(*tert*-butyl acrylate)-*b*-styrene, prepared by nitroxide-mediated "living" free-radical polymerization, were utilized for the functionalization of shortened SWCNTs through a radical coupling reaction (Scheme 1.33) [194].

1.4
Noncovalent Exohedral Functionalization – Functionalization with Biomolecules

Noncovalent functional strategies to modify the outer surface of CNTs in order to preserve the sp^2 network of carbon nanotubes are attractive and represent an effective alternative for sidewall functionalization. Some molecules, including small gas molecules [195], anthracene derivatives [196–198] and polymer molecules [118, 199], have been found liable to absorb to or wrap around CNTs. Nanotubes can be transferred to the aqueous phase through noncovalent functionalization of surface-active molecules such as SDS or benzylalkonium chloride for purification [200–202]. With the surfactant Triton X-100 [203], the surfaces of the CNTs were changed from hydrophobic to hydrophilic, thus allowing the hydrophilic surface of the conjugate to interact with the hydrophilic surface of biliverdin reductase to create a water-soluble complex of the immobilized enzyme [203].

The planar purple dye thionine (Fig. 1.11) was found to show strong noncovalent interactions with the sidewalls of individual CNTs. The noncovalent modification with thionine enriched the surface of the tubes with NH_2 groups, opening

1.4 Noncovalent Exohedral Functionalization – Functionalization with Biomolecules

Fig. 1.11 Thionine and phenosafranine.

up possibilities for anchoring other molecular species such as proteins and semiconductive nanoparticles [204]. The cationic phenazine dye phenosafranine showed self-assembly to defect sites of MWCNTs. Charge-transfer complex formation was observed, associated with charge transfer from phenosafranine to electron-accepting sites of the MWCNTs [205].

Images of the assembly of surfactants on the surface of CNTs were obtained by a French group using TEM [206]. Above the critical micellar concentration, SDS formed supramolecular structures consisting of rolled-up half-cylinders on the nanotube surface (Fig. 1.12). Depending on the symmetry and the diameter of the CNTs, the formation of rings, helices and double helices was observed [206]. Similar self-assemblies have been obtained with synthetic single-chain lipids designed for the immobilization of histidine-tagged proteins. At the nanotube–water interface, permanent assemblies were produced from mixed micelles of SDS and different water-insoluble double-chain lipids after dialysis of the surfactant [206].

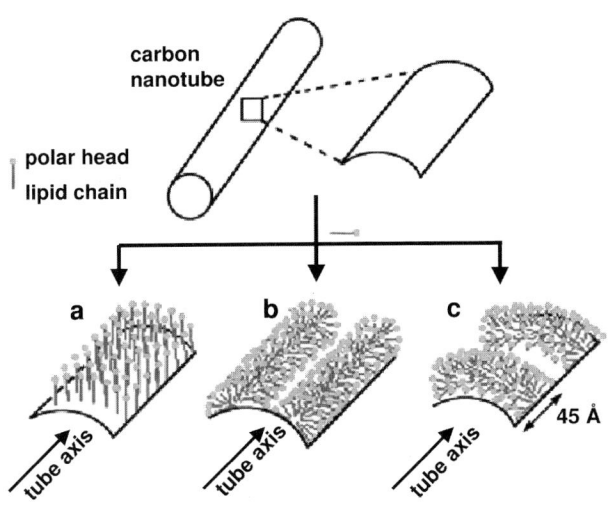

Fig. 1.12 Organization of the SDS molecules on the surface of a CNT. (a) Adsorbed perpendicular to the surface, forming a monolayer; (b) organized into half-cylinders oriented parallel to the tube axis; (c) half-cylinders oriented perpendicular to the tube axis. Copyright; reproduced with permission from *Science* International, UK [206].

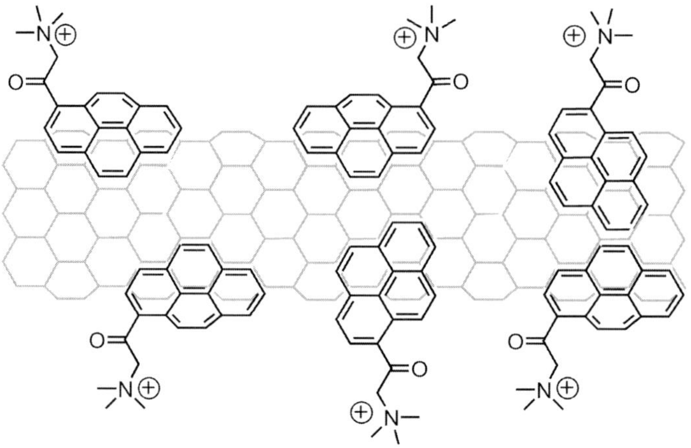

Fig. 1.13 Water-soluble SWCNTs via noncovalent sidewall functionalization with a pyrene-carrying ammonium ion.

Biological functionalization of nanomaterials has become to be of significant interest in recent years owing to the possibility of developing detector systems. Noncovalent immobilization of biomolecules on carbon nanotubes motivated the use of the tubes as potentially new types of biosensor materials [207–210] (a review on carbon nanotube based biosensors was recently published by Wang [211]). So far, only limited work has been carried out with MWCNTs [207–210]. Streptavidin was found to adsorb on MWCNTs, presumably via hydrophobic interactions between the nanotubes and hydrophobic domains of the proteins [210].

Sonication of SWCNTs in an aqueous solution of a pyrene-carrying ammonium ion (Fig. 1.13) gave a transparent dispersion/solution of the tubes. The dispersion was characterized by TEM, UV/Vis absorption, fluorescence and ^{1}H NMR spectroscopy and the results evidenced the interaction of the tube sidewalls with the pyrene moiety [212].

Simultaneous noncovalent and covalent functionalizations occur in the "intertubulary" dimers and excimers already depicted in Scheme 1.10. While the pyrene tether is covalently bound through an ester linkage with one individual SWCNT, an SWCNT dimer is formed by π–π interactions of the polyaromatic pyrene system with a neighboring CNT [138].

Proteins adsorb individually, strongly and noncovalently along the nanotube lengths [213, 214]. The resulting nanotube–protein conjugates are readily characterized by AFM. Several metalloproteins and enzymes have been bound on both the sidewalls and the termini of SWCNTs [215]. Although coupling can be controlled through variation of tube oxidative preactivation chemistry, careful control experiments and observations made by AFM suggested that immobilization is physical and does not require covalent bonding [213, 214]. Two enzymes, α-chymotrypsin (CT) and soybean peroxidase (SBP), were adsorbed on SWCNTs

1.4 Noncovalent Exohedral Functionalization – Functionalization with Biomolecules

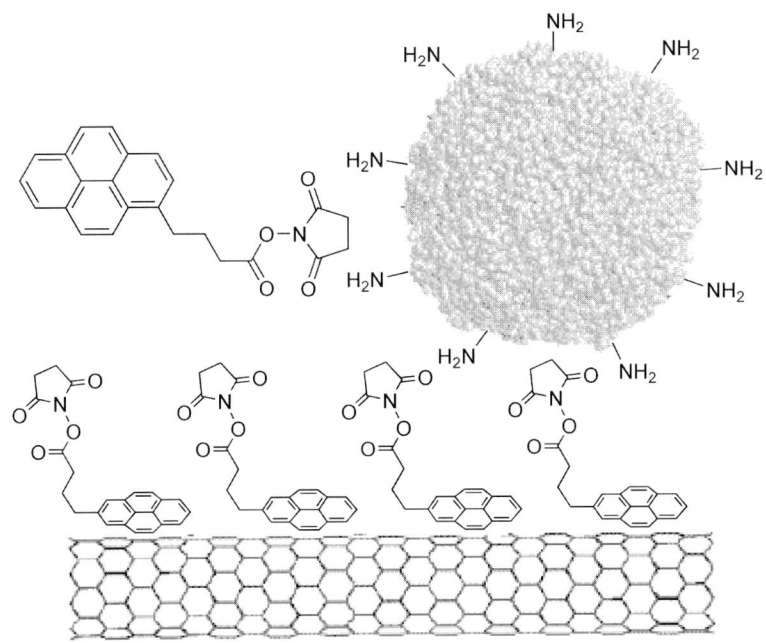

Fig. 1.14. 1-Pyrenebutanoic acid succinimidyl ester irreversibly adsorbed on to the sidewall of a SWCNT via π-stacking. Amino groups of a protein react with the anchored succinimidyl ester to form amide bonds for protein immobilization. Reprinted with permission from [196]. Copyright (2001) American Chemical Society.

[215]. Both enzymes underwent structural changes upon adsorption; the SBP retained up to 30% of its native activity upon adsorption, whereas the adsorbed chymotrypsin retained only 1% of its native activity. AFM images of the adsorbed enzymes indicated that the SBP retained its three-dimensional shape whereas CT appeared to unfold on the SWCNT surface [215]. Prolonged incubation of SWCNTs with glucose oxidase (GOx) led to the coating of the tubes with the enzyme through nonspecific adsorption on the CNT sidewalls. The surface of the CNTs was converted from hydrophobic to hydrophilic by electrochemical, thermal and plasma oxidation treatment, providing the enhanced adsorption of antibodies [216].

The very effective π-stacking interactions between aromatic molecules and the graphitic sidewalls of SWCNTs were demonstrated by the aggregation of the bifunctional N-succinimidyl-1-pyrenebutanoate [196], irreversibly adsorbed on the hydrophobic surfaces of the SWNT. With these conjugates, the succinimidyl group was nucleophilically substituted with amino groups from proteins such as ferritin or streptavidin (Fig. 1.14) and caused immobilization of the biopolymers at the tube surfaces [196].

Multilayer polymeric shells surrounding CNTs have been formed by a layer-by-layer deposition of oppositely charged polyelectrolytes [217]. The CNTs were first

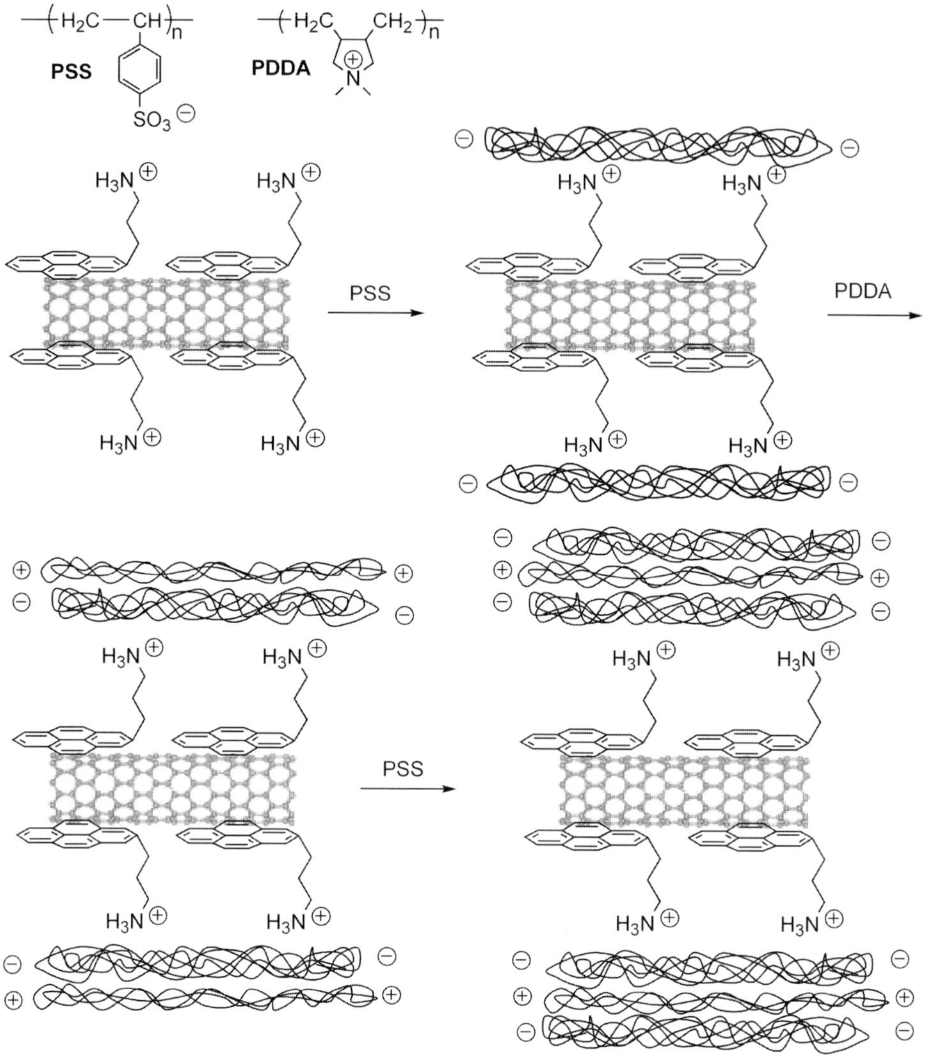

Fig. 1.15 Multilayer polymeric shells surrounding CNTs by a layer-by-layer (LBL) deposition of oppositely charged polyelectrolytes.

functionalized by cationic 1-pyrenepropylamine, followed by a stepwise deposition of negatively charged polystyrene sulfonate (PSS) and positively charged poly(diallyldimethylammonium chloride) (PDDA) (Fig. 1.15). The formation of nanometer-thick amorphous alternating PSS and PDDA layers was confirmed by TEM images and element mapping with energy-filtered TEM [217].

The π-stacking of SWCNTs with polyaromatics was used by Liu et al. [218] in the self-assembling of gold nanoparticles to soluble SWCNTs. 17-(1-Pyrenyl)-13-

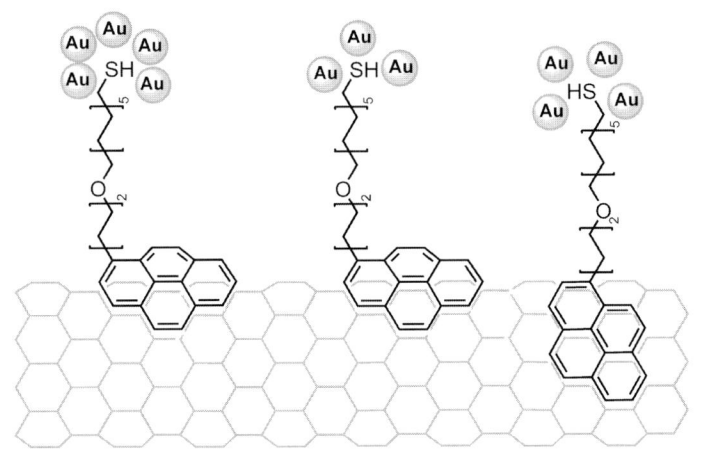

1.4 Noncovalent Exohedral Functionalization – Functionalization with Biomolecules | 37

Fig. 1.16 π-Stacking of 17-(1-pyrenyl)-13-oxaheptadecanethiol (PHT) with SWCNTs and self-assembling of gold nanoparticles from a colloidal gold solution.

oxaheptadecanethiol (PHT) was noncovalently attached on the nanotube surface by π–π interactions and subsequently treated with a colloidal gold solution (Fig. 1.16). Self-assembling of additional gold nanoparticles resulted in a dense coverage of the nanotube surface [218].

MWCNTs were functionalized with iron phthalocyanines (FePc) to improve the sensitivity towards hydrogen peroxide. A highly sensitive glucose sensor with an FePc-MWCNT electrode based on the immobilization of GOx on poly(o-aminophenol) (POAP)-electropolymerized electrode surface [219]. A hemin-modified MWCNT electrode to be used as a novel O_2 sensor was obtained by adsorption of hemin at MWCNTs and the electrochemical properties of the electrode were characterized by cyclic voltammetry [220].

Direct, nonsurfactant-mediated immobilization of metallothionein proteins [207–209] and streptavidin [210] at MWCNTs has also been carried out, the hydrophobic regions of the proteins probably being responsible for the adsorption. Specific affinity binding of proteins to unmodified SWCNT sidewalls was demonstrated by the adsorption of monoclonal antibodies, IgG, specific for C_{60} fullerenes, in aqueous solution. The affinity binding originated from the structural similarity of the tube sidewall graphite network and the C_{60} fullerene [221]. It was shown that the specific binding site of the IgG antibody is a domain of hydrophobic amino acids.

Recently, using electron-accepting CNTs and porphyrin derivatives as electron-donating components, a number of photoactive charge-transfer systems have been designed by Guldi and Prato, with great promise as biomimetic assemblies for photochemical energy conversion [222]. Van der Waals and electrostatic interactions served to integrate electron-accepting SWCNTs and a suitably functionalized light-harvesting polythiophene chromophore into nanostructured ITO elec-

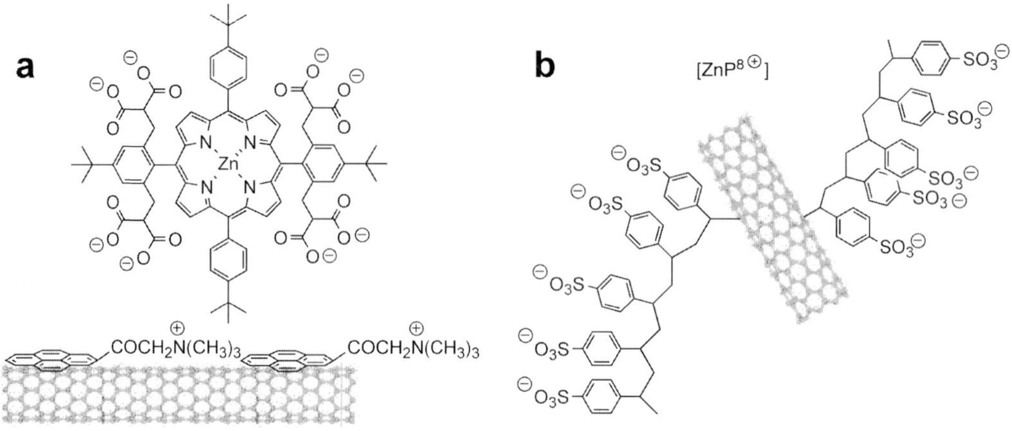

Fig. 1.17 Structures of SWCNT–pyrene$^+$–ZnP^{8-} and SWCNT–PSS$_n^-$–H$_2$P^{8+} nanohybrids.

trodes, exhibiting photoconversion efficiencies between 1.2 and 9.3% upon illumination [223]. Similar associative interactions represented the assembling forces in the integration of SWCNTs, negatively charged pyrene derivatives and metalloporphyrins MP^{8+} into functional nanohybrids. Upon photoexcitation, a rapid charge separation caused the reduction of the electron-accepting SWCNT and, simultaneously, the oxidation of the electron-donating MP^{8+}. The long-lived radical ion pairs exhibited lifetimes in the microsecond range, confirmed by transient absorption measurements [224]. Incorporation of SWCNT hybrids – noncovalently linked SWCNT–pyrene$^+$ and covalently linked SWCNT–PSS$_n^-$ – on semitransparent ITO electrodes led to suitable solar-energy conversion devices (Fig. 1.17) [225]. Novel donor–acceptor nanoassemblies have been prepared using pristine SWCNTs as electron acceptor components in "polymer wraps", with PMMA chains carrying tris(4-sulfonatophenyl)porphyrin (H$_2$P) subunits as excited-state electron donors [226].

Dieckmann et al. in 2003 described an amphiphilic α-helical peptide specifically designed to coat and solubilize CNTs and to control the assembly of the peptide-coated nanotubes into macromolecular structures through peptide–peptide interactions between adjacent peptide-wrapped nanotubes [227]. They claimed that the peptide folds into an amphiphilic α-helix in the presence of carbon nanotubes and disperses them in aqueous solution by noncovalent interactions with the nanotube surface. EM and polarized Raman studies revealed that the peptide-coated nanotubes assemble into fibers with the nanotubes aligned along the fiber axis. The size and morphology of the fibers could be controlled by manipulating the solution conditions that affect peptide–peptide interactions [227].

Pronounced noncovalent interactions were found between SWCNTs and anilines [228] and also between several types of alkylamines [229]. These interactions were detected by the change in electrical conductivity of SWCNTs upon adsorption of primary amines and by their high solubility (up to 8 mg mL^{-1}) in anilines.

Presumably, as in the case of C_{60} fullerenes [230], donor–acceptor complexes are formed, as the curvature present in both materials classes lends acceptor character to the corresponding molecular carbon networks [231]. The stable aniline solutions of SWNTs can be diluted with other organic solvents without causing precipitation of the tubes.

Polymers have also been used for the formation of supramolecular complexes of CNTs. Thus, the suspension of purified MWCNTs and SWCNTs in the presence of conjugated polymers such as poly(*m*-phenylene-*co*-2,5-dioctoxy-*p*-phenylenevinylene) (PmPV) in organic solvents led to hybrid systems, the polymer wrapping around the tubes [232–234]. The properties of these supramolecular compounds were markedly different from those of the individual components. For example, the SWCNT–PmPV complex exhibited a conductivity eight times higher than that of the pure polymer, without any restriction of its luminescence properties. AFM showed that the tubes were uniformly coated by the polymer. The small average diameter of the complexes (about 7.1 nm) suggested that most of the tube bundles were broken upon complex formation. The promising optoelectronic properties of the SWCNT–PmPV complexes have been used in the manufacture of photovoltaic devices [233]. Mono- to triple-layer devices of this CNT composite have been tested as electron transport layers in organic light-emitting diodes [234].

By coating with a conjugate polymer, Murphy et al. also succeeded in developing an experimental technique for a high-yield, nondestructive purification and quantification method for MWCNTs [235]. The polymer host selectively suspends nanotubes relative to impurities and, after removal by filtration, a 91 % pure CNT fraction was obtained [235]. The wrapping of SWNTs with polymers carrying polar side-chains such as polyvinylpyrrolidone (PVP) or PSS led to stable solutions of the corresponding SWCNT–polymer complexes in water [236], the SWCNT–PVP complex exhibiting liquid crystalline properties. The thermodynamic driving force for this complex formation is suggested to be the avoidance of unfavorable interactions between the apolar tube walls and the solvent water.

The wrapping of polymer ropes around MWCNTs occurs in a well-ordered periodic fashion [237]. The authors suggested that the polymer intercalated between the nanotubes, leading to unraveling of ropes and causing a decrease in interactions between the individual CNTs. Moreover, Raman and absorption studies suggested that the polymer interacts preferentially with CNTs of specific diameters or a specific range of diameters [238].

A molecular dynamics simulation in conjunction with experimental evidence was used to elucidate the nature of the interactions between polymer materials and CNTs [239]. Computational time was reduced by representing CNTs as a force field. The calculations indicated an extremely strong noncovalent binding energy. Furthermore, the correlation between the chirality of the nanotubes and mapping of the polymer on to the lattice was discussed [239].

Grafting shortened SWCNTs with PEG by microwave-assisted heating allowed for soluble derivatives of SWCNTs, and remarkably enhanced reaction rates were observed compared with conventional heating [240].

Sidewalls of CNTs were coated with PEG [241], preventing the nonspecific adsorption of proteins [242, 243]. The polymeric layer was subsequently used as an interface for the specific coupling or affinity binding of proteins [242]. By a diamine-terminated oligomeric PEG functionalization, the solubilization of as-prepared and purified SWCNTs was achieved [244]. The soluble tubes were characterized by spectroscopic, microscopic and gravimetric techniques.

A nonwrapping approach to noncovalent engineering of CNT surfaces by short and rigid conjugated polyarylenethynylenes was reported [199]. This technique allowed for the dissolution of various types of CNTs in organic solvents and the introduction of numerous neutral and ionic functionalities onto the CNT surfaces [199].

A family of poly[(m-phenylenevinylene)-co-(p-phenylenevinylene)]s (PamPV), functionalized in the synthetically accessible C-5 position of the meta-disubstituted phenylene rings, have been designed and synthesized [245]. They have been prepared both (1) by the polymerization of O-substituted 5-hydroxyisophthalaldehydes and (2) by chemical modifications carried out on polymers bearing reactive groups at the C-5 positions. PAmPV polymers solubilize SWNT bundles in organic solvents by wrapping themselves around the nanotube bundles. Specifically functionalized PAmPV derivatives wrapped around SWCNTs can form pseudorotaxanes along the walls of the CNTs in a periodic fashion. The formation of such polypseudorotaxanes has been investigated in solution by NMR and UV/Vis spectroscopy, and also on SiO_2 wafers in the presence of SWCNTs by AFM and surface potential microscopy [207].

The solubilization of oxidized carbon nanotubes has been achieved through derivatization using an amino-functionalized crown ether (Scheme 1.34). According to optical measurements, the resultant adduct allowed concentrations of dissolved CNTs of the order of 1 g L^{-1} in water and in methanol [246]. The CNT–crown ether adduct was readily redissolved in different organic solvents at substantially high concentrations. Characterization of the solubilized adducts was performed with ^1H NMR spectroscopy; ^7Li NMR spectroscopy was used to examine the ability of the crown ether macrocycle to bind Li^+ ions. Furthermore, the solutions were analyzed using UV/Vis spectroscopy, photoluminescence and FT-IR spectroscopy and were structurally characterized by AFM and TEM. The adduct formation likely resulted from noncovalent chemical interactions between the carboxylic groups of the oxidized tubes and amine moieties attached to the crown ethers [246].

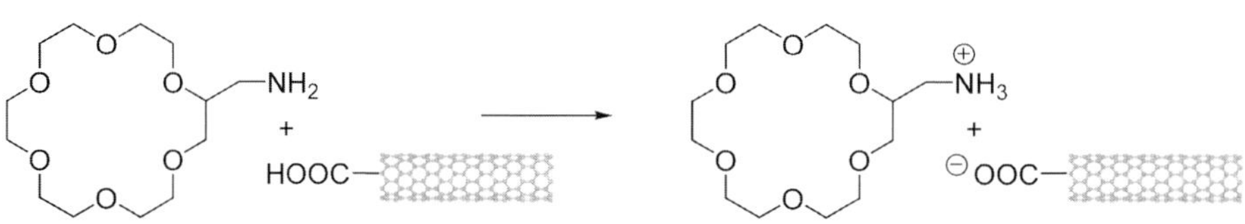

Scheme 1.34 Derivatization of SWCNTs with an amino-functionalized crown ether.

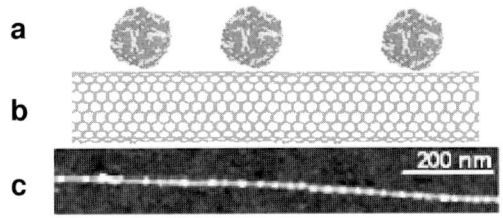

Fig. 1.18 Schematic illustration of (a) a globular protein, adsorbed on (b) a nanotube and (c) an AFM image, showing protein (bright dot-like structures decorating the line-like nanotube) nonspecifically adsorbed on a nanotube. Reprinted with the permission of the National Academy of Sciences of the USA [247].

SWCNTs were exploited as a platform to investigate surface–protein and protein–protein binding and to develop specific electronic biomolecular detectors [247]. The nonspecific binding on nanotubes (Fig. 1.18) was overcome by immobilization of poly(ethylene oxide) (PEO) chains. A general approach was then advanced to permit the selective recognition and binding of target proteins by conjugation of their specific receptors to PEO-functionalized nanotubes. This scheme, combined with the sensitivity of nanotube electronic devices, enables highly specific electronic sensors to be obtained for detecting clinically important biomolecules such as antibodies associated with human autoimmune diseases [247].

Prakash et al. [248] visualized individual CNTs by fluorescence microscopy through noncovalent labeling with conventional fluorophores (Fig. 1.19). Reversal of contrast in fluorescence imaging of the CNTs was observed when the labeling procedure was performed in a nonpolar solvent. The results were consistent with a CNT–fluorophore affinity mediated by hydrophobic interaction, the reverse-contrast images also provided a clear indication of the nanotube location [248].

Norbornene polymerization was initiated selectively on the surface of SWCNTs via a specifically adsorbed pyrene-linked ring-opening metathesis polymerization initiator (Fig. 1.20). The adsorption of the organic precursor was followed by cross-metathesis with a ruthenium alkylidene, resulting in a homogeneous noncovalent poly(norbornene) (PNBE) coating [249].

Fig. 1.19 Chemical structures of conventional fluorophores used for noncovalent labeling of individual CNTs for fluorescence microscopy.

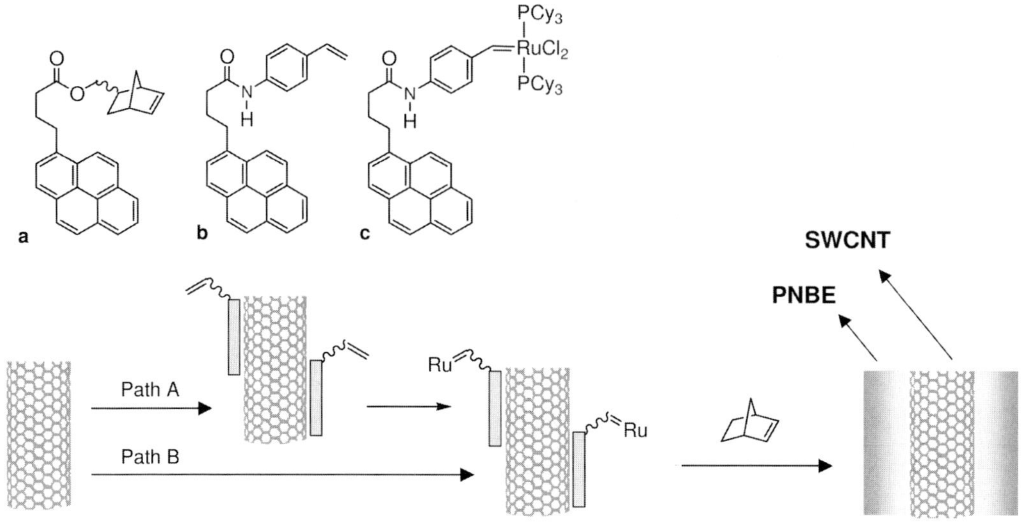

Fig. 1.20 Functionalization strategy for a polynorbornene (PNBE) coating of SWCNTs. Path A: adsorption of organic precursors (a) or (b) followed by cross-metathesis with a ruthenium alkylidene. Path B: adsorption of a pyrene-substituted ruthenium alkylidene (c).

Oligonucleotides and DNA molecules were nonspecifically bound to MWCNT sidewalls via nonspecific interactions and visualized by HR-TEM [208, 209]. Certain organizational properties of CNT–DNA systems have been reported [208, 250]. Nonspecific interactions of oligonucleotides with CNTs enhanced the polymerase chain reaction (PCR), due to the local increase in the reaction components on the surface of the CNTs [251]. HiPCO SWCNTs were dispersed into aqueous double-stranded DNA solutions forming stable solutions. With semiconducting DNA-wrapped HiPCO tubes, the first optical interband transitions of the wrapped tubes displayed a unique pH dependence [252]. When single-stranded DNA was used as dispersing agent and SWCNTs were dispersed, AFM investigations from this suspension showed isolation of the tubes from the pristine bundles and the removal of contaminating particles. Consequently, this biofunctionalization technique also represents a purification method for SWCNTs [253].

1.5
Endohedral Functionalization

Endohedral functionalization of CNTs is the filling of the tubes with various atoms or small molecules [254]. The internal cavity of CNTs (1–2 nm in diameter) provides space for the accommodation of guest molecules [254–260]. Even small proteins and other biomolecules [207, 208, 261] and oligonucleotides [262] have been trapped in the nanotube cavity. In 1998, Luzzi's group [263] for the first

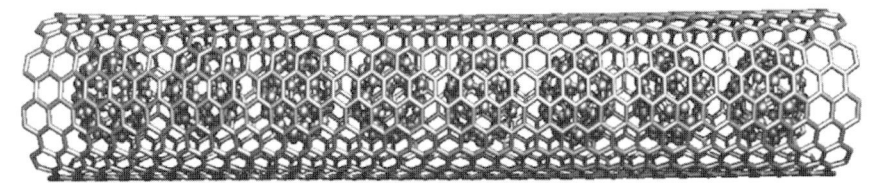

Fig. 1.21 C_{60}@SWCNT: computer simulation, rendered armchair SWCNT filled with C_{60} fullerenes.

time observed the formation of single-walled carbon nanotubes containing C_{60} fullerenes by means of HR-TEM and the term "fullerene peapods" was created for $(C_{60})_n$@SWCNTs (Fig. 1.21). The incorporation of fullerenes such as C_{60} [264, 265] or the metallofullerene Sm@C_{82} [266] is an impressive example of the endohedral chemistry of SWCNTs [254].

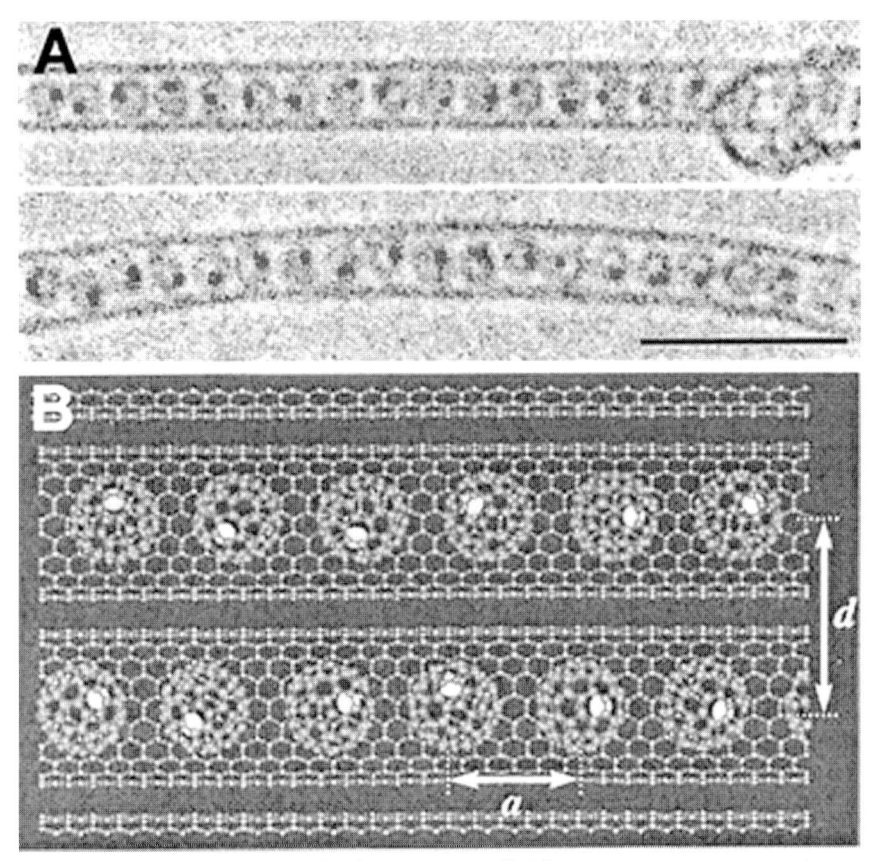

Fig. 1.22 (A) HR-TEM micrograph of an isolated SWCNT containing Gd@C_{82} (scale bar = 5 nm); (B) Schematic presentation of (Gd@C_{82})$_n$@SWCNT (distance between the metallofullerenes $a \approx 1.10$ nm; intertube distance in the bundles $b \approx 1.89$ nm). Reprinted with permission of the American Physical Society [267].

The incorporation of guest molecules can be achieved during their growth or is executed at defect sites and holes via wet chemistry, by surface diffusion and gas-phase transport. Encapsulated fullerenes tend to form chains that are coupled by van der Waals forces. Upon annealing, the encapsulated fullerenes coalesce in the interior of the SWCNTs, resulting in pill-shaped, concentric, endohedral capsules a few nanometers in length [265]. The progress of such reactions inside the tubes could be monitored in real time by use of HR-TEM [266].

In the class of metallofullerene peapods, in addition to Gd@C_{82}-containing peapods (Fig. 1.22) [267], also Dy@C_{82}-, La@C_{82}- and Sm@C_{82}-containing peapods and dimetallofullerene-containing peapods with peas of the kind of Ti_2@C_{80}, La_2@C_{80} and Gd_2@C_{92} have been synthesized (see the references in [254]).

In addition to doping with fullerenes, SWCNTs can also be filled with molecules such as ferrocene, chromocene, ruthenocene, vanadocene and tungstenocene dihydride. The filling with metallocenes occurred from the vapor phase with formation of collinear metallocene chains inside the nanotubes [268]. Also the filling with Zn diphenylporphyrin was successful and established from absorption spectra and Raman measurements [269].

1.6
Conclusions

These examples of functionalization of carbon nanotubes demonstrate that the chemistry of this new class of molecules represents a promising field within nanochemistry. Functionalization provides for the potential for the manipulation of their unique properties, which can be tuned and coupled with those of other classes of materials. The surface chemistry of SWCNTs allows for dispersibility, purification, solubilization, biocompatibility and separation of these nanostructures. Additionally, derivatization allows for site-selective nanochemistry applications such as self-assembly, shows potential as catalytic supports, biological transport vesicles, demonstrates novel charge-transfer properties and allows the construction of functional nanoarchitectures, nanocomposites and nanocircuits.

1.7
Experimental

In the following, the experimental details of different functionalization techniques are presented. Although the selection of the examples is rather arbitrary, an attempt has been made to give a general overview of the manifold methodologies of functionalization, each experimental description standing for a representative example of an individual functionalization method.

Fluorination of CNTs – synthesis of F-SWCNTs

Amounts of 1.5–10 mg of highly purified laser-ablated SWCNTs in the form of a bucky paper were placed in a temperature-controlled fluorination reactor. After purging in He at 250 °C, elemental fluorine, purified from HF by passing it over NaF pellets (HF trap), was introduced. The fluorine flow was gradually increased to a flow-rate of 2 sccm diluted in an He flow of 20 sccm. The fluorination was allowed to proceed for ~10 h, at which point the reactor was brought to room temperature and the fluorine flow was gradually lowered. When the fluorine flow had completely halted, the reactor was purged at room temperature for ~30 min before removing the fluorinated products. The F-SWCNTs consisted of approximately 30 at. % fluorine, determined by electron microprobe analysis (EMPA) [29].

Oxidative purification – carboxylation of SWCNTs

A 1.100-mg amount of SWCNT raw material was heated under reflux in 150 mL of 65 % HNO_3 for 3 h. Subsequently, the black solution was centrifuged and a black sediment remained at the bottom of the centrifuge jar. The clear, brownish yellow supernatant acid solution was decanted off. The sediment was treated by repeated resuspension in distilled water, followed by centrifugation and decantation of each of the supernatants. The remaining sediment was dispersed in distilled water and treated with two or three short (0.5 s) ultrasonic pulses. For permanent storage the solution was brought to pH 8–9 using potassium carbonate [53].

Amidation – formation of CNT-carboxamides

For an octadecylamide derivatization, 100 mg of dried, cut SWCNTs, 270 mg ODA and 210 mg of DCC were added to dry DMF and stirred under Ar for 24 h. After the reaction, the reaction mixture was filtered off over 0.2-µm pore size membrane filters, washed with ethanol and acetone and dried in vacuum for 12 h. The thus derivatized ODA-SWCNTs turned out to be fairly soluble in ODCB (~1 mg mL^{-1}), the suspensions were stable over at least several months under ambient conditions and an estimated 4 % of the SWCNT carbons were functionalized [99].

Addition of nucleophilic carbenes – imidazolidene-SWCNTs

A 200-fold excess of freshly prepared and filtered (Celite) solution of the nucleophilic carbene dipyridylimidazolidene was added to a dispersion of SWCNTs in THF at –60 °C. After stirring at –60 °C for 3 h, the reaction mixture was slowly warmed to room temperature and then diluted with ethanol. The precipitated functionalized nanotubes were isolated by centrifugation and washed with ethanol [148].

Addition of nitrenes – ethoxycarbonylaziridino-SWCNTs

Purified SWCNTs are dispersed in 1,1,2,2-tetrachloroethane (TCE) in an ultrasonic bath under a nitrogen atmosphere over several hours. The suspension is heated to 160 °C and a 200-fold excess of ethyl azidoformate as nitrene precursor is added dropwise. After thermally induced N_2 extrusion, nitrene addition results in the formation of ethoxycarbonylaziridino-SWCNTs which precipitate after a short time. Work-up proceeds by centrifugation and washing of the insoluble residue with diethyl ether [148, 154].

Functionalization via the Bingel reaction
A 1.10-mg amount of purified SWCNT material was annealed under vacuum (10^{-3} mbar) at 1000 °C for 3 h prior to use. To a suspension of this material in dry ODCB, 1.8 mmol of diethyl bromomalonate and 3.3 mmol of DBU were added. The mixture was allowed to react for 2 h under stirring. The modified SWCNTs were isolated from the reaction mixture by filtration and washed thoroughly with ODCB followed by ethanol [160].

Azomethine ylides – functionalization with amino acids
N-(N-Boc-8-amino-3,6-dioxaoctylglycine and paraformaldehyde were added to a suspension of CNTs in DMF and the mixture was heated at 130 °C for 96 h. After separation of unreacted material by filtration and subsequent evaporation of the solvent, the residue was diluted with chloroform and washed with water. The organic phase was dried and the solvent was evaporated. The N-Boc-aminotriethylene glycol-functionalized CNTs were isolated by precipitation with diethyl ether and filtration, and the filter cake was subsequently washed several times with diethyl ether. To a solution of such functionalized CNTs in DCM, gaseous HCl was bubbled through to remove the protecting group (Boc) at the chain end. The CNT ammonium chloride salt precipitated during the acid treatment. After removal of the solvent, the brown solid was dissolved in methanol and precipitated with diethyl ether. To derivatize these SWCNTs with N-protected glycine, Fmoc-Gly-OH was activated with N-hydroxybenzotriazole (HOBT) and diisopropylcarbodiimide (DIC) in DMF–DCM for 15 min and added to a suspension of the in DCM, previously neutralized with diisopropylethylamine (DIEA). After stirring at room temperature for 2 h, the coupling reaction was terminated and the solvent was completely evaporated. The raw material was dissolved in DCM and the derivatized SWCNTs were reprecipitated several times by addition of diethyl ether [170].

Functionalization by Diels–Alder cycloaddition
An amount of 1.20 mg (0.9 mmol) of pentyl carboxylate-functionalized SWCNTs (Scheme 1.23) was mixed with 150 mg (0.9 mmol) of 4,5-benzo-1,2-oxathiin-2-oxide in 40 mL of ODCB and the mixture was irradiated in a focused microwave reactor at 150 W for 45 min. The ODCB was removed by vacuum distillation and the residue was purified by washing several times with pentane and diethyl ether to obtain benzobutenylene-functionalized SWCNT as a dark-brown solid [171].

Functionalization with *in situ*-generated diazonium compounds
About 8 mg of HiPCO SWCNTs, purified by oxidation in wet air at 250 °C, were sonicated for 10 min in 10 mL of ODCB. To this suspension was added a solution of the aniline derivative (2.6 mmol, ~ 4 equiv. per mole of carbon) in 5 mL of acetonitrile. After transfer to a septum-capped reaction tube and bubbling with nitrogen for 10 min, 4.0 mmol of isoamyl nitrite were quickly added. The septum was removed and replaced with a Teflon screw-cap and the suspension was stirred at 60 °C for 15 h. (CAUTION! Considerable pressure was generated in the vessel due to the nitrogen evolved. This was alleviated by partially unscrewing the cap for venting every 30 min for the first 3 h). After cooling to 45 °C, the suspension was diluted with 30 mL of DMF, filtered over a PTFE (0.45 µm) membrane and washed extensively with DMF. Repeated sonication in and washing with DMF were used for purification of the material [176a].

Electrochemical functionalization – reductive coupling of diazonium salts
The electrochemical functionalization of SWCNTs, deposited on an Si/SiO_2 substrate, with 4-nitrophenyl groups succeeded by the reduction of 4-nitrobenzene-diazonium salt in a mini-electrochemical cell with platinum counter and (pseudo-)reference electrodes. A probe needle was used to make contact with one of the Au–Pd bonding pads on the substrate and allowed the application of an electric potential to the electrode (–1.3 V vs. Pt for >30 s) and to any SWCNTs underneath the electrode [177b].

Reductive alkylation – synthesis of 2-[2-(2-methoxyethoxy)ethyl]-MWCNTs
Into a dry and nitrogen-purged three-necked flask, equipped with a gas inlet and a condenser, 500 mL of anhydrous ammonia were condensed at –70 °C; 2 g (87 mmol) of sodium were added and a dark-blue solution was formed. To this mixture, 50 mg (4.2 mmol of carbon) of MWCNTs were added and a black solution of the tubes was obtained. Stirring was continued for 1 h. Subsequently, 1.63 mL (12 mmol) of 1-bromo-2-(2-methoxyethoxy)ethane was added dropwise and the reaction mixture was stirred overnight to remove the ammonia slowly. Water (100 mL) was carefully added to the black solid. After acidification with 20 mL of dilute (10 %) HCl, the nanotubes were extracted with 200 mL of hexane and washed with water. The organic phase was filtered through a 200-nm PTFE membrane filter and washed with ethanol and THF. The resulting black solid was dried in a vacuum oven at 50 °C overnight [270].

Addition of carbanions – syntheses of *tert*-butyl-H-SWCNTs and of *tert*-butyl-SWCNTs
In a nitrogen-purged flask, equipped with a gas inlet and a pressure compensator, 20 mg of HiPCO tubes (1.7 mmol of carbon) was dispersed in 50 mL of anhydrous benzene. To this dispersion 2.5 mL of a 1.7 M solution of *tert*-butyllithium (4.25 mmol) in hexane were added dropwise over a period of 10 min. Subsequently, the suspension was additionally stirred for 1 h at room temperature and the SWCNT dispersion turned into a black homogeneous solution. The solution was stirred for a further 1 h and subsequently quenched by the addition of

dilute HCl. The resulting dispersion was diluted with 100 mL of acetone and filtered through a 0.2-µm PTFE membrane filter and washed with ethanol and THF. The resulting black solid was dried in vacuum at 50 °C overnight [270]. Instead of quenching the negatively charged intermediates tert-butyl-SWCNT$\bar{\text{s}}$ to yield the reduced tert-butyl-H-SWCNTs, the intermediates could subsequently be reoxidized by bubbling oxygen through the homogeneous black dispersion. Thus, tert-butyl-SWCNTs were formed, and the workup was achieved by addition of cyclohexane, purging with water and diluted HCl, membrane filtration, washing with THF, MeOH and EtOH, and drying under vacuum at 50 °C [271].

Covalent functionalization by polymerization – polystyrene grafted to SWCNTs
Nitroxide-terminated PS samples were prepared by "living" free-radical polymerization utilizing an alkoxyamine unimolecular initiator. A 10-mg amount of purified and shortened sWCNTs was dispersed in anhydrous DMF and 500 mg of nitroxide-terminated PS along with a catalytic amount of acetic anhydride (50 µL) were added. After bubbling with N_2 for 30 min, the mixture was immersed in a 125 °C oil-bath and stirred under Ar for 3 days. Upon cooling, the reaction mixture was filtered through a 200-nm pore-diameter PTFE membrane and washed with THF and DCM (200 mL each) to remove unreacted PS. The residual black solid was peeled away from the membrane and dried at 50 °C under vacuum overnight [194].

π-Stacking of 1-pyrenebutanoic acid succinimidyl ester on SWCNT sidewalls
SWCNTs were suspended on meshed gold grids according to known methods and a grid sample was incubated in a 1-pyrenebutanoic acid succinimidyl ester solution (6 mM in DMF or 1 mM in methanol) for 1 h at room temperature, after which the sample was rinsed three times in pure DMF or methanol [196].

Noncovalent functionalization – incubation with surfactants and lipids
A 1-mL aqueous solution of SDS [1 % by weight, concentration greater than the critical micellar concentration (CMC)], was sonicated with 1 mg of SWCNTs or MWCNTs. Similarly, an aqueous solution (1 mg mL^{-1}) of an amphiphilic monochain lipid reagent, e.g. an octadecanoyl moiety linked with nitrilotriacetic acid, was sonicated with 1 mg of MWCNTs for 3 min [206].

Endohedral functionalization – encapsulating metallofullerenes –
(Gd@C$_{82}$)$_n$@SWCNTs
Gd@C$_{82}$ was generated by arc discharge using a Gd–graphite rod and isolated by a multistage HPLC technique; SWCNT bundles were prepared by pulsed-laser evaporation. The doping of Gd@C$_{82}$ into the inner hollow space of SWCNTs was carried out in a sealed glass ampoule at 500 °C for 24 h. Prior to the introduction of SWCNTs to the ampoule, the SWCNTs were heated in dry air at 420 °C for 20 min [267].

List of Abbreviations

AFM	atomic force microscopy	PMMA	poly(methyl methacrylate)
ATRP	atom transfer radical polymerization	PS	polystyrene
		PSS	polystyrene sulfonate
CNT	carbon nanotube	PTC	phase-transfer catalysis
CVD	chemical vapor deposition	PVP	polyvinylpyrrolidone
DBU	1,8-diazabicyclo[5.4.0] undecene	RBM	Raman breathing mode
		SCW	supercritical water
DCC	dicyclohexylcarbodiimide	SDS	sodium dodecyl sulfate
DCM	dichloromethane	SEM	scanning electron microscopy
DMF	dimethylformamide		
DOS	density of states	STM	scanning tunneling microscopy
DPI	dipyridylimidazolidene		
GCE	glassy carbon electrode	SWCNT	single-walled carbon nanotube
HiPCO	high pressure CO conversion		
		TEM	transmission electron microscopy
MMA	methyl methacrylate		
MWCNT	multiwalled carbon nanotube	TGA	thermogravimetric analysis
		THF	tetrahydrofuran
ODA	octadecylamine	XPS	X-ray photoemission spectroscopy

References

1. H. W. Zhu, C. L. Xu, D. H. Wu, B. Q. Wie, R. Vajtai, P. M. Ajayan, *Science* **2002**, *296*, 884.
2. M. S. Dresselhaus, G. Dresselhaus, P. C. Eklund (eds.), *Science of Fullerenes and Nanotubes*, Academic Press, San Diego, **1996**.
3. M. S. Dresselhaus, G. Dresselhaus, P. Avouris (eds.), *Carbon Nanotubes: Synthesis, Structure Properties and Applications*, Springer-Verlag, Berlin, **2001**.
4. W. A. de Heer, A. Chatelain, D. A. Ugarte, *Science* **1995**, *270*, 1179.
5. H. J. Dai, J. H. Hafner, A. G. Rinzler, D. T. Colbert, R.E. Smalley, *Nature* **1996**, *384*, 147.
6. S. J. Tans, A. R. M. Verschueren, C. Dekker, *Nature* **1998**, *393*, 49.
7. R. Martel, T. Schmidt, H. R. Shea, T. Hertel, P. Avouris, *Appl. Phys. Lett.* **1998**, *73*, 2447.
8. H. T. Soh, C. F. Quate, A. F. Morpurgo, C. M. Marcus, J. Kong, H. Dai, *Appl. Phys. Lett.* **1999**, *75*, 627.
9. J. Kong, N. R. Franklin, C. Zhou, M. G. Chapline, S. Peng, K. Cho, H. Dai, *Science* **2000**, *287*, 622.
10. H. W. Ch. Postma, T. Teepen, Z. Yao, M. Grifoni, C. Dekker, *Science* **2001**, *293*, 76.
11. G. Maurin, Ch. Bousquet, F. Henn, P. Bernier, R. Almeirac, B. Simon, *Chem. Phys. Lett.* **1999**, *312*, 14.
12. P. Chen, X. Wu, J. Lin, K. L. Tan, *Science* **1999**, *285*, 91.
13. H. Dai, *Surf. Sci.* **2002**, *500*, 218.
14. C. Liu, Y. Y. Fan, M. Liu, H. T. Cong, H. M. Cheng, M. S. Dresselhaus, *Science* **1999**, *286*, 1127.
15. C. Pirlot, I. Willems, A. Fonseca, J. B. Nagy, J. Delhalle, *Adv. Eng. Mater.* **2002**, *4*, 109.
16. T. Liu, T. V. Sreekumar, S. Kumar, R. H. Hauge, R. E. Smalley, *Carbon* **2003**, *41*, 2440.
17. A. R. Bhattacharayya, T. V. Sreekumar, T. Liu, S. Klimar, L. M. Ericson, R. H.

Hauge, R. E. Smalley, *Polymer* **2003**, *44*, 2373.
18. W. E. Wong, P. E. Sheehan, C. M. Lieber, *Science* **1997**, *277*, **1971**.
19. V. N. Popov, V. E. Van Doren, M. Balkanski, *Phys. Rev. B* **2000**, *61*, 3078.
20. D. Qian, E. C. Dickey, R. Andrews, T. Rantell, *Appl. Phys. Lett.* **2000**, *76*, 2868.
21. J. P. Salvetat, A. J. Kulik, J. M. Bonard, G. A. D. Briggs, T. Stoeckli, K. Metenier, S. Bonnamy, F. Beguin, N.A. Burnham, L. Forro, *Adv. Mater.* **1999**, 11, 161.
22. R. Saito, G. Dresselhaus, M. S. Dresselhaus (eds.), *Physical Properties of Carbon Nanotubes*, Imperial College Press, London, **1998**.
23. S. Iijima, *Nature*, **1991**, *354*, 56.
24. (a) S. B. Sinnott, *J. Nanosci. Nanotechnol.* **2002**, *2*, 113; (b) A. Hirsch, *Angew. Chem. Int. Ed.* **2002**, *41*, 1853; (c) S. Niyogi, M. A. Hamon, H. Hu, B. Zhao, P. Bhowmik, R. Sen, M. E. Itkis, R. C. Haddon, *Acc. Chem. Res.* **2002**, *35*, 1105; (d) J. L. Bahr, J. M. Tour, *J. Mater. Chem.* **2002**, *12*, 1952; (e) A. Hirsch, O. Vostrowsky, *Top. Curr. Chem.* **2005**, *245*, 193.
25. C. A. Dyke, J. M. Tour, *J. Phys. Chem. A* **2004**, *108*, 11151.
26. S. Banerjee, T. Hhemraj-Benny, S.S. Wong, *Adv. Mater.* **2005**, *17*, 17.
27. T. Nakajima, S. Kasamatsu, Y. Matsuo, *Eur. J. Solid State Inorg. Chem.* **1996**, *33*, 831.
28. E. T. Mickelson, C. B. Huffman, A. G. Rinzler, R. E. Smalley, R. H. Hauge, J. L. Margrave, *Chem. Phys. Lett.* **1998**, *296*, 188.
29. E. T. Mickelson, I. W. Chiang, J. L. Zimmerman, P. J. Boul, J. Lozano, J. Liu, R. E. Smalley, R. H. Hauge, J. L. Margrave, *J. Phys. Chem. B* **1999**, *103*, 4318.
30. P. E. Pehrsson, W. Zhao, J. W. Baldwin, C. Song, J. Liu, S. Kooi, B. Zheng, *J. Phys. Chem. B* **2003**, *107*, 5690.
31. K. F. Kelly, I. W. Chiang, E. T. Mickelson, R. H. Hauge, J. L. Margrave, X. Wang, G. E. Scuseira, C. Radloff, N. J. Halas, *Chem. Phys. Lett.* **1999**, *313*, 445.
32. S. Kawasaki, K. Komatsu, F. Okino, H. Touhara, H. Kataura, *Phys. Chem. Chem. Phys.* **2004**, *6*, 1769.
33. A. Hamwi, H. Alvergnat, S. Bonnamy, F. Beguin, *Carbon* **1997**, *35*, 723.
34. Y. Zhu, F. C. Cheong, T. Yu, X. J. Xu, C. T. Lim, J. T. L. Thong, Z. X. Shen, C. K. Ong, Y. J. Liu, A. T. S. Wee, C. H. Sow, *Carbon* **2005**, *43*, 395.
35. N. O. V. Plank, L. Jiang, R. Cheung, *Appl. Phys. Lett.* **2003**, *83*, 2426.
36. N. O. V. Plank, R. Cheung, *Microelectron. Eng.* **2004**, *73–74*, 578.
37. B. N. Khare, P. Wilhite, M. Meyyappan, *Nanotechnology* **2004**, *15*, 1650.
38. E. Unger, M. Liebau, G. S. Duesberg, A. P. Graham, F. Kreupl, R. Seidel, W. Hoenlein, *Chem. Phys. Lett.* **2004**, *399*, 280.
39. R. Barthos, D. Mehn, A. Demortier, N. Pierard, Y. Morciaux, G. Demortier, A. Fonseca, J. B. Nagy, *Carbon* **2005**, *43*, 321.
40. Z. Gu, H. Peng, J. L. Zimmerman, W. Ivana, V. N. Khabashesku, R. H. Hauge, J. L. Margrave, *Abstr. Pap. 223rd ACS National Meeting*, Orlando, FL, FLUO-012.
41. P. R. Marcoux, J. Schreiber, P. Batail, S. Lefrant, J. Renouard, G. Jacob, D. Albertini, J. Y. Mevellec, *Phys. Chem. Chem. Phys.* **2002**, *4*, 2278.
42. E. Unger, A. Graham, F. Kreupl, M. Liebau, M. Hoenlein, *Curr. Appl. Phys.* **2002**, *2*, 107.
43. Z. Konya, I. Vesselenyi, K. Niesz, A. Kukovecz, A. Demortir, A. Fonseca, J. Delhalle, Z. Mekhalif, J. B. Nagy, A. A. Koos, Z. Osvath, A. Kocsonya, L. P. Biro, I. Kiricsi, *Chem. Phys. Lett.* **2002**, *360*, 429.
44. S. B. Fagan, A. J. R. da Silva, R. Mota, R. J. Baierle, A. Fazzio, *Phys. Rev. B* **2003**, *67*, 033405.
45. Y. K. Chen, M. L. H. Green, J. L. Griffin, J. Hammer, R. M. Lago, S. C. Tsang, *Adv. Mater.* **1996**, *8*, 1012.
46. P. X. Hou, S. Bai, Q. H. Yand, C. Liu, H. M. Cheng, *Carbon* **2002**, *40*, 81.
47. P. J. Boul, J. Liu, E. T. Mickelson, C. B. Huffman, L. M. Ericson, I. W. Chiang, K. A. Smith, D. T. Colbert, R. H. Hauge, J. L. Margrave, R. E. Smalley, *Chem. Phys. Lett.* **1999**, *310*, 367.
48. V. N. Khabashesku, W. E. Billups, J. L. Margrave, *Acc. Chem. Res.* **2002**, *35*, 1087.

49. J. L. Stevens, A. Y. Huang, H. Peng, I. W. Chiang, V. N. Khabashesku, J. L. Margrave, *Nano Lett.* **2003**, *3*, 331.
50. L. Valentini, D. Puglia, I. Armentano, J. M. Kenny, *Chem. Phys. Lett.* **2005**, *403*, 385.
51. R.K. Saini, I. W. Chiang, H. Peng, R.E. Smalley, W.E. Billups, R.H. Hauge, J.L. Margrave, *J. Am. Chem. Soc.* **2003**, *125*, 3617.
52. L. Zhang, V. U. Kiny, H. Peng, J. Zhu, R. F. M. Lobo, J. L. Margrave, V. N. Khabashesku, *Chem. Mater.* **2004**, *16*, 2055.
53. M. Holzinger, A. Hirsch, P. Bernier, G. S. Duesberg, M. Burghard, *Appl. Phys. A* **2000**, *70*, 599.
54. G. U. Sumanasekera, J. L. Allen, S. L. Fang, A. L. Loper, A. M. Rao, P. C. Eklund, *J. Phys. Chem. B* **1999**, *103*, 4292.
55. A. G. Rinzler, J. Liu, H. Dai, P. Nikolaev, C. B. Huffman, F. J. Rodriguez-Macias, P- J. Boul, A. H. Lu, D. Heymann, D. T. Colbert, R. S. Lee, J. E. Fischer, A. M. Rao, P. C. Eklund, R. E. Smalley, *Appl. Phys. A* **1998**, *67*, 29.
56. J. Liu, A. G. Rinzler, H. Dai, J. H. Hafner, R. K. Bradley, P. J. Boul, A. Lu, T. Iverson, K. Shelimov, C. B. Huffman, F. Rodriguez-Macias, Y. S. Shon, T. R. Lee, D. T. Colbert, R. E. Smalley, *Science* **1998**, *280*, 1253.
57. K. Morishita, T. Takarada, *J. Mater. Sci.* **1994**, *34*, 1169.
58. K. Tohji, T. Goto, H. Takahashi, Y. Shinoda, N. Shimizu, B. Jeyadevan, I. Matsuoka, Y. Saito, A. Kasuya, T. Ohsuna, K. Hiraga, Y. Nishina, *Nature* **1996**, *383*, 679.
59. D. B. Mawhinney, V. Naumenko, A. Kuznetsova, J. T. Yates Jr., J. Liu, R. E. Smalley, *J. Am. Chem. Soc.* **2000**, *122*, 2383.
60. D. B. Mawhinney, V. Naumenko, A. Kuznetsova, J. T. Yates Jr., J. Liu, R. E. Smalley, *Chem. Phys. Lett.* **2000**, *324*, 213.
61. A. Kuznetsova, I. Popova, J. T. Yates, M. J. Bronikowski, C. B. Huffman, J. Liu, R. E. Smalley, H. Hwu, J. G. Chen Jr., *J. Am. Chem. Soc.* **2001**, *123*, 10699.
62. J. P. Deng, C. Y. Mou, C. C. Han, *Fullerene Sci. Technol.* **1997**, *5*, 1033.
63. P. M. Ajayan, S. Iijima, *Nature* **1993**, *361*, 333.
64. D. Ugarte, A. Chatelain, W. A. deHeer, *Science* **1996**, *274*, 1897.
65. J. F. Colomer, P. Piedigrosso, I. Willems, C. Journet, P. Bernier, G. VanTendeloo, A. Fonseca, J. B. Nagy, *J. Chem. Soc., Faraday Trans.* **1988**, *94*, 3753.
66. A. C. Dillon, K. M. Jones, T. A. Bekkedahl, C. H. Kong, D. S. Bethune, M. J. Heben, *Nature* **1997**, *386*, 377.
67. H. Hiura, T.W. Ebbesen, K. Tanigaki, *Adv. Mater.* **1995**, *7*, 275.
68. B. Satishkumar, A. Govindaraj, J. Mofokeng, G. N. Subbanna, C. N. R. Rao, *J. Phys. B: At. Mol. Opt. Phys.* **1996**, *29*, 4925.
69. K. C. Hwang, *J. Chem. Soc., Chem. Commun.* **1995**, 173.
70. D. S. Bag, R. Dubey, N. Zhang, J. Xie, V. K. Varadan, D. Lal, G. N. Mathur, *Smart Mater. Struct.* **2004**, *13*, 1263.
71. K. Kinoshita, *Carbon Electrochemical and Physicochemical Properties*, Wiley, New York, **1998**.
72. S. F. McKay, *J. Appl. Phys.* **1992**, 35, 1992.
73. M. Monthioux, B. W. Smith, B. Burteaux, A. Claye, J. E. Fischer, D. E. Luzzi, *Carbon* **2001**, *39*, 1251.
74. M. E. Itkis, S. Niyogi, M. Meng, M. Hamon, H. Hu, R. C. Haddon, *Nano Lett.* **2002**, *2*, 155.
75. E. Dujardin, T. W. Ebbesen, A. Krishnan, M. M. J. Treacy, *Adv. Mater.* **1998**, *10*, 611.
76. K. Tohji, H. Takahashi, Y. Shinoda, N. Shimizu, B. Jeyadevan, I. Matsuoka, Y. Saito, A. Kasuya, S. Ito, Y. Nishina, *J. Phys. Chem. B* **1997**, *101*, **1974**.
77. S. Bandow, S. Asaka, X. Zhao, Y. Ando, *Appl. Phys. A* **1998**, *67*, 23.
78. A. C. Dillon, T. Gennet, K. M. Jones, J. L. Alleman, P. A. Parilla, M. J. Heben, *Adv. Mater.* **1999**, *11*, 1354.
79. (a) Y. Chen, R. C. Haddon, S. Fang, A. M. Rao, P. C. Eklund, W. H. Lee, E. C. Dickey, E. C. Grulke, J. C. Pendergrass, A. Chavan, B. E. Haley, R. E. Smalley, *J. Mater. Res.* **1998**, *13*, 2423; (b) Z. Chen, W. Thiel, A. Hirsch, *ChemPhysChem* **2003**, *4*, 93.

80. W. Zhou, Y. H. Ooi, R. Russo, P. Papanek, D. E. Luzzi, J. E. Fischer, M. J. Bronikowski, P. A. Willis, R. E. Smalley, *Chem. Phys. Lett.* **2001**, *350*, 6.
81. Y. Yang, H. Zou, B. Wu, Q. Li, J. Zhang, Z. Liu, X. Guo, Z. Du, *J. Phys. Chem. B* **2002**, *106*, 7160.
82. M. A. Hamon, H. Hu, P. Bhowmik, S. Niyogi, B. Zhao, M. E. Itkis, R. C. Haddon, *Carbon'01, Int. Conf Carbon*, Lexington, USA, 14–19 July, **2001**, 872.
83. N. Zhang, J. Xie, V. Kvaradan, *Smart Mater. Struct.* **2002**, *11*, 962.
84. B. Wu, J. Zhang, Z. Wie, S. Cai, Z. Liu, *J. Phys. Chem. B* **2001**, *10*, 5075.
85. H. Yanagi, E. Sawada, A. Manivannan, L. A. Nagahara, *Appl. Phys. Lett.* **2001**, *78*, 1355.
86. D. Chattopadhyay, I. Galeska, F. Papadimitrakopoulos, *J. Am. Chem. Soc.* **2001**, *123*, 9451.
87. S. Banerjee, S.S. Wong, *Adv. Mater.* **2004**, *16*, 34.
88. S. Banerjee, S.S. Wong, *Chem. Commun.* **2004**, 1866.
89. A. A. Mamedov, N. A. Kotov, M. Prato, D. M. Guldi, J. P. Wicksted, A. Hirsch, *Nat. Mater.* **2002**, *1*, 190.
90. X. Zhang, T.V. Sreekumar, T. Liu, S. Kumar, *J. Phys. Chem. B* **2004**, *108*, 16435.
91. B. Kim, W. M. Sigmund, *Langmuir* **2004**, *20*, 8239.
92. Y. Xing, L. Li, C. C. Chusuei, R. V. Hull, *Langmuir* **2005**, *21*, 4185.
93. K. C. Park, T. Hayashi, H. Tomiyasu, M. Endo, M. S. Dresselhaus, *J. Mater. Chem.* **2005**, *15*, 407.
94. Y. J. Kim, T. S. Shin, H. D. Choi, J. H. Kwon, Y.-C. Chung, H. G. Yoon, *Carbon* **2005**, *43*, 23.
95. S. Sotiropoulou, N. A. Chaniotakis, *Anal. Bioanal. Chem.* **2003**, *375*, 103.
96. E. Katz, I. Willner, *ChemPhysChem* **2004**, *5*, 1084.
97. J. Chen, M. A. Hamon, H. Hu, Y. Chen, A. M. Rao, P. C. Eklund, R. C. Haddon, *Science* **1998**, *282*, 95.
98. B. Li, Y. F. Lian, Z. J. Shi, Z. N. Gu, Gaodeng *Xuexiao Huaxue Xuebao* **2000**, *21*, 1633.
99. F. Hennrich, S. Lebedkin, S. Malik, J. Tracy, M. Barczewski, A. Rösner, M. Kappes, *Phys. Chem. Chem. Phys.* **2002**, *4*, 2273.
100. M. A. Hamon, J. Chen, H. Hu, Y. Chen, M. E. Itkis, A. M. Rao, P. C. Eklund, R. C. Haddon, *Adv. Mater.* **1999**, *11*, 834.
101. M. A. Hamon, H. Hu, P. Bhowmik, S. Niyogi, B. Zhao, M. E. Itkis, R. C. Haddon, *Chem. Phys. Lett.* **2001**, *347*, 8.
102. Y. Wang, Z. Iqbal, S. V. Malhotra, *Chem. Phys. Lett.* **2005**, *402*, 96.
103. M. Alvaro, P. Atienzar, J. L. Bourdelande, H. Garcia, *Chem. Phys. Lett.* **2004**, *384*, 119.
104. F. Pompeo, D. E. Resasco, *Nano Lett.* **2002**, *2*, 369.
105. C. Zhao, L. Ji, H. Liu, G. Hu, S. Zhang, M. Yang, Z. Yand, *J. Solid State Chem.* **2004**, *177*, 4394.
106. C. Neumann, M. Fischer, L. Tran, J.G. Matisons, *J. Am. Chem. Soc.* **2002**, *124*, 13998.
107. Z. Liu, Z. Shen, T. Zhu, S. Hou, L. Ying, Z. Shi, Z. Gu, *Langmuir* **2000**, *16*, 3569.
108. A. Jung, Defektgruppen-Funktionalisierung von Kohlenstoffnanoröhren mit Newkome-Derimeren, Diploma Thesis, University of Erlangen-Nuremberg **2003**.
109. P. W. Chiu, G. S. Duesberg, U. Dettlaff-Weglikowska, S. Roth, *Appl. Phys. Lett.* **2002**, *80*, 2811.
110. U. Dettlaff-Weglikowska, J. M. Benoit, P. W. Chiu, R. Graupner, S. Lebedkin, S. Roth, *Curr. Appl. Phys.* **2002**, *2*, 497.
111. I. Kiricsi, Z. Konya, K. Niesz, A. A. Koos, L. P. Biro, *Proc. SPIE* **2003**, *5118*, 280.
112. V. O. Pokrovs'kyi, T. Y. Gromovyi, O. O. Chuiko, O. V. Basyuk, V. O. Basyuk, J. M. Saniger-Blesa, *Dop. Nat. Akad. Nauk. Ukr.* **2002**, 172.
113. J. Chen, A. M. Rao, S. Lyuksyutov, M. E. Itkis, M. A. Hamon, H. Hu, R. W. Cohn, P. C. Eklund, D. T. Colbert, R. E. Smalley, R. C. Haddon, *J. Phys. Chem. B* **2001**, *105*, 2525.
114. M. A. Hamon, J. Chen, H. Hu, M. E. Itkis, P. Bhowmik, S. M. Rozenzhak, A. M. Rao, R. C. Haddon, *Abstr. Pap. 220th National Meeting of the Am. Chem. Soc.*, Washington, D.C. **2000**, IEC-130.

115. D. Chattopadhyay, S. Lastella, S. Kim, F. Papadimitrakopoulos, *J. Am. Chem. Soc.* **2002**, *124*, 728.
116. L. Cai, J. L. Bahr, Y. Yao, J. M. Tour, *Chem. Mater.* **2002**, *14*, 4235.
117. S. S. Wong, A. T. Woolley, E. Joselevich, C. L. Cheung, C. M. Lieber, *J. Am. Chem. Soc.* **1998**, *120*, 8557.
118. Y. P. Sun, K. Fu, Y. Lin, W. Huang, *Acc. Chem. Res.* **2002**, *35*, 1096.
119. Y. P. Sun, W. Huang, Y. Lin, K. Fu, A. Kitaygorodskiy, L. A. Riddle, Y. J. Yu, D. L. Carroll, *Chem. Mater.* **2001**, *13*, 2864.
120. E. Menna, G. Scorrano, M. Mi, M. Cavallaro, F. Della Negra, M. Battagliarin, R. Bozio, F. Fantinel, M. Meneghetti, *ARKIVOC* **2003**, 64.
121. M. Sano, A. Kamino, J. Okamura, S. Shinkai, Langmuir **2001**,*17*, 5125.
122. H. T. Ham, C. M. Koo, S. O. Kim, Y. S. Choi, I. J. Chung, *Hwahak Konghak* **2002**, *40*, 618.
123. Y. Lin, A. M. Rao, B. Sadanadan, E. A. Kenik, Y. P. Sun, *J. Phys. Chem. B* **2002**, *106*, 1294.
124. (a) M. Holzinger, A. Hirsch, G. S. Duesberg, M. Burghard, Electronic Properties of Novel Materials – Molecular Nanostructures, *AIP Conf. Proc.* **2000**, *544*, 246; (b) M. Holzinger, Functionalization of single-walled carbon nanotubes, PhD Thesis, University of Erlangen-Nuremberg **2002**.
125. W. Huang, S. Taylor, K. Fu, Y. Lin, D. Zhang, T. W. Hanks, A. M. Rao, Y. P. Sun, *Nano Lett.* **2002**, *2*, 311.
126. K. Fu, W. Huang, Y. Lin, D. Zhang, T.W. Hanks, A.M. Rao, Y.-P. Suna, *J. Nanosci. Nanotechnol.* **2002**, *2*, 457.
127. S. Ravindran, S. Chaudhary, M. Ozkan, C. Ozkan, *Abstr. Pap. 225th ACS National Meeting*, New Orleans, 23–27 March, **2003**, IEC-161.
128. C. Dwyer, M. Guthold, M. Falvo, S. Washburn, R. Superfine, D. Erie, *Nanotechnology* **2002**, *13*, 601.
129. M. Hazani, R. Naaman, F. Hennrich, M. M. Kappes, *Nano Lett.* **2003**, *3*, 153.
130. S. E. Baker, W. Cai, T. L. Lasseter, K. P. Weidkamp, R. J. Hamers, *Nano Lett.* **2002**, *2*, 1413.
131. K. A. Williams, P. T. M. Veenhuizen, B. G. de la Torre, R. Eritjia, C. Dekker, *Nature* **2002**, *420*, 61.
132. H. Cai, X. Cao, Y. Jiang, P. He, Y. Fang, *Anal. Bioanal. Chem.* **2003**, *375*, 287.
133. L. Liu, Y. Qin, Z. X. Guo, D. Zhu, *Carbon* **2003**, *41*, 331.
134. M. A. Hamon, H. Hui, P. Bhowmik, H. M. E. Itkis, R. C. Haddon, *Appl. Phys. A* **2002**, *74*, 333.
135. J. E. Riggs, Z. Guo, D. L. Carroll, Y. P. Sun, *J. Am. Chem. Soc.* **2000**, *122*, 5879.
136. K. Fu, W. Huang, Y. Lin, L. A. Riddle, D. L. Carroll, Y. P. Sun, *Nano Lett.* **2001**, *1*, 439.
137. J. Hu, J. Shi, S. Li, Y. Quin, Z.-X. Guo, Y. Song, D. Zhu, *Chem. Phys. Lett.* **2005**, *401*, 352.
138. (a) L. Qu, R. B. Martin, W. Huang, K. Fu, D. Zweifel, Y. Lin, Y. P. Sun, C. E. Bunker, B. A. Harruff, J. R. Gord, L. F. Allard, *J. Chem. Phys.* **2002**, *117*, 8089; (b) L. Qu, R. B. Martin, W. Huang, K. Fu, D. Zweifel, Y. Lin, C. E. Bunker, B. A. Harruff, J. R. Gord, L. F. Allard, Y. P. Sun, *Proc. Electrochem. Soc.* **2002**, *12*, 563.
139. H. Kong, C. Gao, D. Yan, *J. Am. Chem. Soc.* **2004**, *126*, 412.
140. S. Qin, D. Qin, W.T. Ford, D.E. Resasco, J. E. Herrera, *J. Am. Chem. Soc.* **2004**, *126*, 170.
141. D.-H. Jung, Y.K. Ko, H.-T. Jung, *Mater. Sci. Eng. C* **2004**, *24*, 117.
142. J. K. Lim, W. S. Yun, M. Yoon, S. K. Lee, C. H. Kim, K. Kim, S. K. Kim, *Synth. Met.* **2003**, *139*, 521.
143. C. Velasco-Santos, A. L. Martinez-Hernandez, M. Lozada-Cassou, A. Alvarez-Castillo, V. M. Castao, *Nanotechnology* **2002**, *13*, 495.
144. L. Vast, G. Philippin, A. Destree, N. Moreau, A. Fonseca, J.B. Nagy, J. Delhalle, Z. Mekhalif, *Nanotechnology* **2004**, *15*, 781.
145. S. Pekker, J. P. Salvetat, E. Jakab, J. M. Bonard, L. Forro, *J. Phys. Chem. B* **2001**, *105*, 7938.
146. F. J. Owens, Z. Iqbal, *23rd Army Science Conference*, Session L/LP-11, 2–5 December **2002**, Orlando, FL; http://www.asc2002.com/summaries/l/LP-11.pdf.
147. T. Yildirim, O. Gülseren, S. Ciraci, *Phys. Rev. B* **2001**, *64*, 075404.
148. M. Holzinger, O. Vostrowsky, A. Hirsch, F. Hennrich, M. Kappes, R. Weiss, F.

Jellen, *Angew. Chem. Int. Ed.* **2001**, *40*, 4002.
149. B. Ni, S. B. Sinnott, *Phys. Rev. B* **2000**, *61*, 16343.
150. H. Peng, P. Reverdy, V. N. Khabashesku, J. L. Margrave, *Chem. Commun.* **2003**, 362.
151. H. Peng, L. B. Alemany, J.L. Margrave, V. N. Khabashesku, *J. Am. Chem. Soc.* **2003**, *125*, 15174.
152. N. Tagmatarchis, V. Georgakilas, M. Prato, H. Shinohara, *Chem. Commun.* **2002**, 2010.
153. W. H. Lee, S. J. Kim, J. G. Lee, R. C. Haddon, P. J. Reucroft, *Appl. Surf. Sci.* **2001**, *181*, 121.
154. M. Holzinger, J. Abraham, P. Whelan, R. Graupner, L. Ley, F. Hennrich, M. Kappes, A. Hirsch, *J. Am. Chem. Soc.* **2003**, *125*, 8566.
155. J. Abraham, P. Whelan, A. Hirsch, F. Hennrich, M. M. Kappes, A. Vencelova, R. Graupner, L. Ley, M. Holzinger, D. Samaille, P. Bernier, *Electronic Properties of Novel Materials: Molecular Nanostructures* (eds H. Kuzmany, J. Fink, M. Mehring, S. Roth), Proc XVIIth Int. Winterschool, Kirchberg, Austria, March 2003, World Scientific, Singapore, **2003**, 291.
156. R. Graupner, A. Vencelova, L. Ley, J. Abraham, M. Holzinger, A. Hirsch, F. Hennrich, M. Kappes, *Electronic Properties of Novel Materials: Molecular Nanostructures* (eds H. Kuzmany, J. Fink, M. Mehring, S. Roth), Proc XVIIth Int. Winterschool, Kirchberg, Austria, March 2003, World Scientific, Singapore, **2003**, 120.
157. M. Holzinger, J. Steinmetz, D. Samaille, P. Bernier, M. Glerup, L. Ley, R. Graupner, European Network Meeting, Functionalization of Single-walled Carbon Nanotubes, *AIP Conf. Proc.*, **2003**, 87.
158. C. Bingel, *Chem. Ber.* **1993**, *126*, 1957.
159. (a) X. Camps, A. Hirsch, *J. Chem. Soc., Perkin Trans. 1* **1997**, 1595; (b) A. Hirsch, O. Vostrowsky, *Eur. J. Org. Chem.* **2001**, 829.
160. K. S. Coleman, S. R. Bailey, S. Fogden, M. L. H. Green, *J. Am. Chem. Soc.* **2003**, *125*, 8722.
161. (a) V. Georgakilas, K. Kordatos, M. Prato, D. M. Guldi, M. Holzinger, A. Hirsch, *J. Am. Chem. Soc.* **2002**, *124*, 760; (b) D. Tasis, N. Tagmatarchis, V. Georgakilas, C. Gamboz, M. R. Soranzo, M. Prato, C. R. Chimie **2003**, *6*, 597; (c) V. Georgakilas, N. Tagmatarchis, D. Voulgaris, M. Prato, A. Kukovecz, H. Kuzmany, A. Hirsch, F. Zerbetto, M. Melle-Franco, Structural and Electronic Properties of Molecular Nanostructures, *AIP Conf. Proc.* **2002**, *633*, 73.
162. V. Georgakilas, D. Voulgaris, E. Vazquez, M. Prato, D.M. Guldi, A. Kukovecz, H. Kuzmany, *J. Am. Chem. Soc.* **2002**, *124*, 14318.
163. N. Tagmatarchis, M. Prato, *J. Mater. Chem.* **2004**, *14*, 437.
164. A. Callegari, M. Marcaccio, D. Paolucci, F. Paolucci, N. Tagmatarchis, D. Tasis, E. Vazquez, M. Prato, *Chem. Commun.* **2003**, 2576.
165. D. M. Guldi, M. Marcaccio, D. Paolucci, F. Paolucci, N. Tagmatarchis, D. Tasis, E. Vazquez, M. Prato, *Angew. Chem. Int. Ed.* **2003**, *42*, 4206.
166. A. Callegari, S. Cosnier, M. Marcaccio, D. Paolucci, V. Georgakilas, N. Tagmatarchis, E. Vazquez, M. Prato, *J. Mater. Chem.* **2004**, *14*, 807.
167. M. Melle-Franco, M. Marcaccio, D. Paolucci, F. Paolucci, V. Georgakilas, D. M. Guldi, M. Prato, F. Zerbetto, *J. Am. Chem. Soc.* **2004**, *126*, 1646.
168. D. Pantarotto, R. Singh, D. McCarthy, M. Erhardt, J.-P. Briand, M. Prato, K. Kostarelos, A. Bianco, *Angew. Chem. Int. Ed.* **2004**, *43*, 5242.
169. A. Bianco, J. Hobeke, S. Godefroy, O. Chaloin, D. Pantarotto, J.-P. Briand, S. Muller, M. Prato, C. D. Partidos, *J. Am. Chem. Soc.* **2005**, *127*, 58.
170. (a) V. Georgakilas, N. Tagmatarchis, D. Pantarotto, A. Bianco, J.-P. Briand, M. Prato, *Chem. Commun.* **2002**, 3050; (b) A. Bianco, K. Kostarelos, C. D. Partidos, M. Prato, *Chem. Commun.* **2005**, 571.
171. J. L. Delgado, P. De la Cruz, F. Langa, A. Urbina, J. Casado, J. T. Lopez Navarrete, *Chem. Commun.* **2004**, 1734.
172. J. B. Cui, M. Burghard, K. Kern, *Nano Lett.* **2003**, *3*, 613.
173. S. Banerjee, S. S. Wong, *J. Am. Chem. Soc.* **2004**, *126*, 2073.

174. J. L. Bahr, J. Yang, D. V. Kosynkin, M. J. Bronikowski, R. E. Smalley, J. M. Tour, *J. Am. Chem. Soc.* **2001**, *123*, 6536.
175. (a) C. A. Mitchell, J. L. Bahr, S. Arepalli, J. M. Tour, R. Krishnamoorti, *Macromolecules* **2002**, *35*, 8825; (b) C. A. Dyke, J. M. Tour, *Nano Lett.* **2003**, *3*, 1215.
176. (a) J. L. Bahr, J. M. Tour, *J. Mater. Chem.* **2001**, *12*, 3823; (b) C. A. Dyke, J. M. Tour, *J. Am. Chem. Soc.* **2003**, *125*, 1156; (c) J. M. Tour, C. A. S. Dyke, *Chem. Eur. J.* **2004**, *10*, 812.
177. (a) S. E. Kooi, U. Schlecht, M. Burghard, K. Kern, *Angew. Chem. Int. Ed.* **2002**, *41*, 1353; (b) M. Knez, M. Sumser, A. M. Bittner, C. Wege, H. Jeske, S. Kooi, M. Burghard, K. Kern, *J. Electroanal. Chem.* **2002**, *522*, 70.
178. P. R. Marcoux, P. Hapiot, P. Batail, J. Pinson, *New J. Chem.* **2004**, *28*, 302.
179. C.-S. Lee, S. E. Baker, M.S. Marcus, W. Yang, M.A. Eriksson, R.J. Hamers, *Nano Lett.* **2004**, *4*, 1713.
180. F. Liang, A. K. Sadana, A. Peera, J. Chattopadhyay, Z. Gu, R. H. Hauge, W. E. Billups, *Nano Lett.* **2004**, *4*, 1257.
181. J. Chattopadhyay, A. K. Sadana, F. Liang, J.M. Beach, Y. Xiao, R. H. Hauge, W. E. Billups, *Org. Lett.* **2005**, *7*, 4067.
182. G. Viswanathan, N. Chakrapani, H. Yang, B. Wei, H. Chung, K. Cho, C. Y. Ryu, P. M. Ajayan, *J. Am. Chem. Soc.* **2003**, *125*, 9258.
183. S. Chen, W. Shen, G. Wu, D. Chen, M. Jiang, *Chem. Phys. Lett.* **2005**, *402*, 312.
184. R. Blake, Y. K. Gun'ko, J. Coleman, M. Cadek, A. Fonseca, J.B. Nagy, W.J. Blau, *J. Am. Chem. Soc.* **2004**, *126*, 10226.
185. S. Ravindran, S. Chaudhary, B. Colburn, M. Ozkan, C. S. Ozkan, *Nano Lett.* **2003**, *3*, 447.
186. M. S. P. Shaffer, K. Koziol, *Chem. Commun.* **2002**, 2074.
187. W. Wu, S. Zhang, Y. Li, J. Li, L. Qin, Z.X. Guo, L. Dai, C. Ye, D. Zhu, *Macromolecules* **2003**, *36*, 6286.
188. C. Gao, H. Kong, D. Yan, Electronic Properties of Synthetic Nanostructures, *AIP Conf. Proc.* **2004**, *723*, 193.
189. M. Ejaz, Y. Tsujii, T. Fukuda, *Polymer* **2001**, *42*, 6811.
190. K. Hao, G. Chao, Y. Deyue, *Macromolecules* **2004**, *37*, 4022.
191. S. Quin, D. Quin, W.T. Ford, D.E. Resasco, J.E. Herrera, *Macromolecules* **2004**, *37*, 752.
192. J. Cui, W.P. Wang, Y.Z. You, C. Liu, P. Wang, *Polymer* **2004**, *45*, 8717.
193. H.M. Huang, I.C. Liu, C.Y. Chang, H.C. Tsai, C.H. Hsu, R. C.C. Tsiang, *J. Polym. Sci. Part A: Polym. Chem.* **2004**, *42*, 5802.
194. Y. Liu, Z. Yao, A. Adronov, *Macromolecules* **2005**, *38*, 1172.
195. H. J. Dai, *Acc. Chem. Res.* **2002**, *35*, 1035.
196. R. J. Chen, Y. Zhang, D. Wang, H. Dai, *J. Am. Chem. Soc.* **2001**, *123*, 3838.
197. J. Zhang, J. K. Lee, R. W. Murray, *Nano Lett.* **2003**, *3*, 403.
198. E. V. Basiuk, E. V. Rybak-Akimova, V. A. Basiuk, D. Acosta-Najarro, J. M. Saniger, *Nano Lett.* **2002**, *11*, 1249.
199. J. Chen, H. Liu, W. A. Weimer, M. D. Halls, D. H. Waldeck, G. C. Walker, *J. Am. Chem. Soc.* **2002**, *124*, 9034.
200. S. Bandow, A. M. Rao, K. A. Williams, A. Thess, R. E. Smalley, P. C. Eklund, *J. Phys. Chem. B* **1997**, *101*, 8839.
201. G. S. Duesberg, M. Burghard, J. Muster, G. Philipp, S. Roth, *Chem. Commun.* **1998**, 435.
202. V. Krstic, G. S. Duesberg, J. Muster, M. Burghard, S. Roth, *Chem. Mater.* **1998**, *10*, 2338.
203. M. Panhuis, C. Salvador-Morales, E. Franklin, G. Chambers, A. Fonseca, J. B. Nagy, W. J. Blau, A. I. Minett, *J. Nanosci. Nanotechnol.* **2003**, *3*, 209.
204. Q. Li, J. Zhang, H. Yan, M. He, Z. Liu, *Carbon* **2004**, *42*, 287.
205. S.A. Curran, A.V. Ellis, A. Vijayaraghavan, P.M. Ajayan, *J. Chem. Phys.* **2004**, *120*, 4886.
206. C. Richard, F. Balavoine, P. Schultz, T. W. Ebbesen, C. Mioskowski, *Science* **2003**, *300*, 775.
207. S. C. Tsang, J. J. Davis, M. L. H. Green, H. A. O. Hill, Y. C. Leung, P. J. Sadler, *Chem. Commun.* **1995**, 2579.
208. S. C. Tsang, Z. Guo, Y. K. Chen, M. H. L. Green, H. A. O. Hill, T. W. Hambley, P. J. Sadler, *Angew. Chem.* **1997**, *109*, 2291; *Angew. Chem. Int. Ed.* **1997**, *36*, 2198.
209. Z. Guo, P. J. Sadler, S. C. Tsang, *Adv. Mater.* **1998**, *10*, 701.

210. F. Balavoine, P. Schultz, C. Richard, V. Mallouh, T. W. Ebbesen, C. Mioskowski, *Angew. Chem.* **1999**, *111*, 2036; *Angew. Chem. Int. Ed.* **1999**, *38*, 1912.
211. J. Wang, *Electroanalysis* **2005**, *17*, 7.
212. N. Nakashima, Y. Tomonari, H. Murakami, *Jpn. Chem. Lett.* **2002**, *6*, 638.
213. J. J. Davis, K. S. Coleman, B. R. Azamian, C. B. Bagshaw, M. L. H. Green, *Chem. Eur. J.* **2003**, *9*, 3732.
214. B.R. Azamian, J. J. Davis, K.S. Coleman, C. Bagshaw, M. L. H. Green, *J. Am. Chem. Soc.* **2002**, *124*, 12664.
215. S. S. Karajanagi, A. A. Vertegel, R.S. Kane, J.S. Dordick, *Langmuir* **2004**, *20*, 11594.
216. Y. Tzeng, T.S. Huang, Y.C. Chen, C. Liu, Y.K. Liu, *New Diamond Front. Carbon Technol.* **2004**, *14*, 193.
217. A.B. Artyukhin, O. Bakajin, P. Stroeve, A. Noy, *Langmuir* **2004**, *20*, 1442.
218. L. Liu, T. Wang, J. Li, Z. X. Guo, L. Dai, D. Zhang, D. Zhu, *Chem. Phys. Lett.* **2003**, *367*, 747.
219. J.-S.- Ye, Y. Wen, W.D. Zhang, H.F. Cui, G.Q. Xu, F.-S. Sheu, *Electroanalysis* **2005**, *17*, 89.
220. J.-S. Ye, Y. Wen, W.D. Zhang, H.-F. Cui, L.M. Gan, G.Q. Xu, F.-S. Sheu, *J. Electroanal. Chem.* **2004**, *562*, 241.
221. B. F. Erlanger, B. X. Chen, M. Zhu, L. Brus, *Nano Lett.* **2001**, *1*, 465.
222. (a) D. M. Guldi, M. Prato, *Chem. Commun.* **2004**, 2517; (b) D. M. Guldi, *J. Phys. Chem. B* **2005**, *109*, 11432.
223. G. M. A. Rahman, D. M. Guldi, R. Cagnoli, A. Mucci, L. Schenetti, L. Vaccari, M. Prato, *J. Am. Chem. Soc.* **2005**, *127*, 10051.
224. D.M. Guldi, G. M. A. Rahman, N. Jux, D. Balbinot, U. Hartnagel, N. Tagmatarchis, P. Prato, *J. Am. Chem. Soc.* **2005**, *127*, 9830.
225. D. M. Guldi, G. M. A. Rahman, M. Prato, N. Jux, S. Qin, W. Ford, *Angew. Chem. Int. Ed.* **2005**, *44*, 2015.
226. D. M. Guldi, H. Taieb, G. M. A. Rahman, N. Tagmatarchis, M. Prato, *Adv. Mater.* **2005**, *17*, 871.
227. G. R. Dieckmann, A. B. Dalton, P. A. Johnson, J. Razal, J. Chen, G. M. Giordano, E. Munoz, I. H. Musselman, R. H. Baughman, R. K. Draper, *J. Am. Chem. Soc.* **2003**, *125*, 1770.
228. Y. Sun, S. R. Wilson, D. I. Schuster, *J. Am. Chem. Soc.* **2001**, *123*, 5348.
229. J. Kong, H. Dai, *J. Phys. Chem. B* **2001**, *105*, 2890.
230. A. Hirsch, M. Brettreich, *Fullerenes – Chemistry and Reactions*, Wiley-VCH, Weinheim, **2005**.
231. M. A. Hamon, M. E. Itkis, S. Niyogi, T. Alvaraez, C. Kuper, M. Menon, R. C. Haddon, *J. Am. Chem. Soc.* **2001**, *123*, 11292.
232. S. A. Curran, P. M. Ajayan, W. J. Blau, D. L. Carroll, J. N. Coleman, A. B. Dalton, A. P. Davey, A. Drury, B. McCarthy, S. Maier, A. Strevens, *Adv. Mater.* **1998**, *10*, 1091.
233. A. Star, J. F. Stoddart, D. Steuerman, M. Diehl, A. Boukai, E. W. Wong, X. Yang, S. W. Chung, H. Choi, J. R. Heath, *Angew. Chem.* **2001**, *113*, 1771; *Angew. Chem. Int. Ed.* **2001**, *40*, 1721.
234. (a) J. N. Coleman, A. B. Dalton, S. Curran, A. Rubio, A. P. Davey, A. Drury, B. McCarthy, B. Lahr, P. M. Ajayan, S. Roth, R. C. Barklie, W. J. Blau, *Adv. Mater.* **2000**, *12*, 213; (b) P. Fournet, J. N. Coleman, D. F. O'Brien, B. Lahr, A. Drury, H. H. Hoerhold, W. J. Blau, *Proc. SPIE* **2002**, *4464*, 239.
235. R. Murphy, J. N. Coleman, M. Cadek, B. McCarthy, M. Bent, A. Drury, R. C. Barklie, W. J. Blau, *J. Phys. Chem. B* **2002**, *106*, 3087.
236. M. J. O'Connell, P. Boul, L. M. Ericson, C. Huffman, Y. Wang, E. Haroz, C. Kuper, J. Tour, K. D. Ausman, R. E. Smalley, *Chem. Phys. Lett.* **2001**, *342*, 265.
237. B. McCarthy, J. N. Coleman, S. A. Curran, A. B. Dalton, A. P. Davey, Z. Konya, A. Fonseca, J. B. Nagy, W. J. Blau, *J. Mater. Sci. Lett.* **2000**, *19*, 2239.
238. A. B. Dalton, C. Stephan, J. N. Coleman, B. McCarthy, P. M. Ajayan, S. Lefrant, P. Bernier, W. J. Blau, H. J. Byrne, *J. Phys. Chem. B* **2000**, *104*, 10012.
239. M. Panhuis, A. Maiti, A. B. Dalton, A. van den Noort, J. N. Coleman, B. McCarthy, W. J. Blau, *J. Phys. Chem. B* **2003**, *107*, 478.
240. F. Della Negra, M. Meneghetti, E. Menna, Fullerenes, *Nanotubes Carbon Nanostruct.* **2003**, *11*, 25.

241. G. X. Shen, Y. Y. Zhuang, C. J. Lin, *Prog. Chem.* **2004**, *16*, 21.
242. M. Shim, N. W. S. Kam, R. J. Chen, Y. Li, H. Dai, *Nano Lett.* **2002**, *2*, 285.
243. O.-K. Kim, J. Je, J. W. Baldwin, S. Kooi, P. E. Pehrsson, L. J. Buckley, *J. Am. Chem. Soc.* **2003**, *125*, 4426.
244. W. Huang, S. Fernando, A. F. Allard, Y. P. Sun, *Nano Lett.* **2003**, *3*, 565.
245. A. Star, Y. Liu, K. Grant, L. Ridvan, J. F. Stoddart, D. W. Steuerman, M. R. Diehl, A. Boukai, J. R. Heath, *Macromolecules* **2003**, *36*, 553.
246. M. G. C. Kahn, S. Banerjee, S. S. Wong, *Nano Lett.* **2002**, *2*, 1215.
247. R. J. Chen, S. Bangsaruntip, K. A. Drouvalakis, N. W. S. Kam, M. Shim, Y. Li, W. Kim, P. J. Utz, H. Dai, *Proc. Natl. Acad. Sci. USA* **2003**, *100*, 4984.
248. R. Prakash, S. Washburn, R. Superfine, R. E. Cheney, M. R. Falvo, *Appl. Phys. Lett.* **2003**, *83*, 1219.
249. F. J. Gomez, R. J. Chen, D. Wang, R. M. Waymouth, H. Dai, *Chem. Commun.* **2003**, 190.
250. E. Buzaneva, A. Karlash, K. Yakovkin, Y. Shtogun, S. Putselyk, D. Zherebetskiy, A. Gorchinskiy, G. Popova, S. Prilutska, O. Matyshevska, Y. Prilutsky, P. Lytvyn, P. Scharff, P. Eklund, *Mater. Sci. Eng. C: Biomimetic Supramol. Syst.* **2002**, *19*, 41.
251. D. Cui, F. Tian, Y. Kong, I. Titushikin, H. Gao, *Nanotechnology* **2004**, *15*, 154.
252. W. Zhao, K. Kelley, *Abstr. 56th Southeast Regional Meeting of the American Chemical Society*, Research Triangle Park, NC, November **2004**.
253. S. Taeger, O. Jost, W. Pompe, M. Mertig, Electronic Properties of Synthetic Nanostructures, *AIP Conf. Proc.* **2004**, *723*, 185.
254. O. Vostrowsky, A. Hirsch, *Angew. Chem.* **2004**, *116*, 2380; *Angew. Chem. Int. Ed.* **2004**, *43*, 2326.
255. W. Han, S. Fan, Q. Li, Y. Hu, *Science* **1997**, *277*, 1287.
256. J. Sloan, J. Hammer, M. Zwiefka-Sibley, M. L. H. Green, *Chem. Commun.* **1998**, 347.
257. E. Dujardin, T. W. Ebbesen, A. Krishnan, M. M. J. Treacy, *Adv. Mater.* **1998**, *10*, 1472.
258. K. Matsui, B. K. Pradhan, T. Kyotani, A. Tomita, *J. Phys. Chem. B* **2001**, *105*, 5682.
259. A. Govindaraj, B. C. Satishkumar, M. Nath, C. N. R. Rao, *Chem. Mater.* **2000**, *12*, 202.
260. M. Wilson, P. A. Madden, *J. Am. Chem. Soc.* **2001**, *123*, 2101.
261. J. J. Davis, M. L. H. Green, O. A. H. Hill, Y. C. Leung, P. J. Sadler, J. Sloan, A. V. Xavier, S. C. Tsang, *Inorg. Chim. Acta* **1998**, *272*, 261.
262. T. Ito, L. Sun, R. M. Crooks, *Chem. Commun.* **2003**, 1482.
263. B. W. Smith, M. Monthioux, D. E. Luzzi, *Nature* **1998**, *396*, 323.
264. B. W. Smith, M. Monthioux, D. E. Luzzi, *Chem. Phys. Lett.* **1999**, *315*, 31.
265. (a) B. W. Smith, D. E. Luzzi, *Chem. Phys. Lett.* **2000**, *321*, 169; (b) D.E. Luzzi, B. W. Smith, *Carbon* **2000**, *38*, 1751.
266. T. Okazaki, K. Suenaga, K. Hirahara, S. Bandow, S. Iijima, H. Shinohara, *J. Am. Chem. Soc.* **2001**, *123*, 9673.
267. K. Hirahara, K. Suenaga, S. Bandow, H. Kato, T. Ohkazaki, H. Shinohara, S. Iijima, *Phys. Rev. Lett.* **2000**, *85*, 5384.
268. F. Stercel, N. M. Nemes, J. E. Fischer, D. E. Luzzi, Making Functional Materials with Nanotubes, *Mater. Res. Soc. Symp. Proc.* **2002**, *706*, 245.
269. H. Kataura, Y. Maniwa, M. Abe, A. Fujiwara, T. Kodama, K. Kikuchi, H. Imahori, Y. Misaki, S. Suzuki, Y. Achiba, *Appl. Phys. A: Mater. Sci. Process.* **2002**, *74*, 349.
270. (a) J. Abraham, Functionalization of carbon nanotubes, Doctoral Thesis, University of Erlangen-Nuremberg, **2005**; (b) J. Abraham, A. Hirsch, unpublished results **2005**.
271. R. Graupner, J. Abraham, D. Wunderlich, A. Vencelova, P. Lauffer, J. Röhrl, M. Hundhausen, L. Ley, A. Hirsch, *J. Am. Chem. Soc.* **2006**, *128*, 6683.

2
Cyclophenacene Cut Out of Fullerene

Yutaka Matsuo and Eiichi Nakamura

2.1
Introduction

Planar polycyclic aromatic molecules [1] are an important class of functional organic materials extensively studied for applications in electronic devices. Fullerenes [2] or carbon nanotubes [3] are emerging classes of such materials because of their rich functions. An example is their multi-electron-accepting character [4] with small reorganization energy [5]. Other new families of functional benzenoid compounds have also been the subject of interest (Fig. 2.1). One class of compounds includes bowl-shaped arenes [6] such as **3**, that links planar benzenoid compounds and spherical fullerenes. Another class involves the hoop-shaped cyclic benzenoid compounds **1** and **2** that can formally be generated by rolling a one-dimensional graphite network into a ring. While synthesis of the former class has been recorded in the literature, the latter class still remains elusive.

Hoop-shaped benzenoids, molecular belts, would be an interesting class of compound because of their resonance structure and chemical reactivities, in addition to their potential utility in materials science. They have become the subject of recent interest in the broader scientific community because they comprise a unit structure of carbon nanotubes. Hoop-shaped benzenoids can be categorized into two fundamental structures (Fig. 2.2). An [*n*]cyclacene (**1**) is a circular [*n*]acene formed of *n* linearly annulated benzene rings with one edge of a benzene ring folding onto the other one. An [*n*]cyclophenacene (**2**) is derived from the corresponding [*n*]phenacene. In 1954, Heilbronner discussed cyclacene [7] and thereafter extensive theoretical and synthetic investigations have been performed.

A previous theoretical study of a series of [*n*]cyclacenes by Houk and Kim [8] suggested that the cyclacenes comprise a double, parallel "trannulene" structure [9]. The HOMO–LUMO gap was found to be very small and to disappear as the ring becomes larger, hence indicating that the molecules would be unstable.

A theoretical investigation of [*n*]cyclophenacenes was reported by Aihara, who discussed stability of [*n*]cyclophenacenes with a topological resonance energy (TRE) method [10] and proposed that the cyclophenacenes are aromatic and stable

Functional Organic Materials. Syntheses, Strategies, and Applications.
Edited by Thomas J.J. Müller and Uwe H.F. Bunz
Copyright © 2007 WILEY-VCH Verlag GmbH & Co. KGaA, Weinheim
ISBN: 978-3-527-31302-0

1D-benzenoid ⇒ **1** hoop   **2**

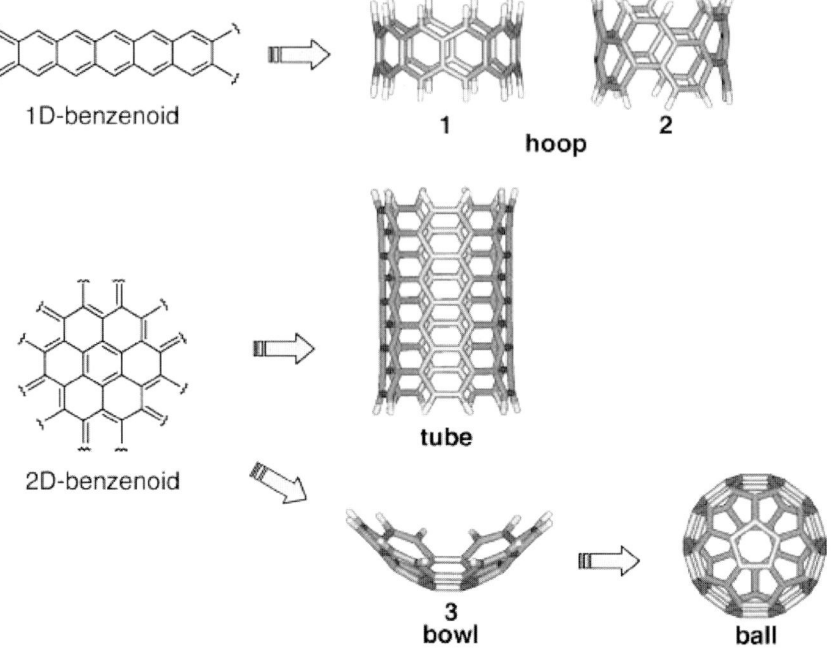

2D-benzenoid ⇒ tube

⇒ **3** bowl ⇒ ball

Fig. 2.1 Hoop- and bowl-shaped aromatics.

[n]acene (polyacene) —rolling→ [n]cyclacene (**1**)

[n]phenacene —rolling→ [n]cyclophenacene (**2**)

Fig. 2.2 Hoop-shaped aromatic molecules, [10]cyclacene and [10]cyclophenacene.

[11]. The aromaticity of [10]cyclophenacene was evaluated with the percentage resonance energy (%RE) [12] as 2.573, which indicates that [10]cyclophenacene is much more aromatic than [60]fullerene (%RE = 1.795) and less aromatic than benzene (%RE = 3.528). Türker also theoretically investigated the stability and electronic structure (UV/Vis band) of [n]cyclophenacenes with semi-empirical calculations (AM1 and ZINDO/S) and proposed that [n]cyclophenacenes with medium size (n = 8 and 10) can be synthesized [13]. Thus, based on the theoretical

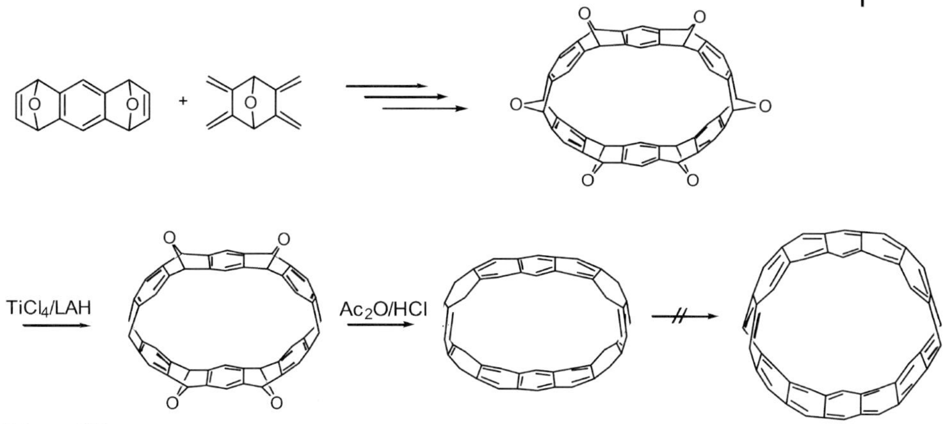

Scheme 2.1

studies, [*n*]cyclophenacenes are expected to be synthesized, rather than cyclacenes.

Attempts to synthesize the hoop-shaped molecules have been made several times during the last 50 years following Heilbronner's prediction. Because synthesis by simple rolling of a flat aromatic precursor into a hoop is expected to be difficult, owing to the strain of the hoop-shaped π-conjugated molecules, synthetic chemists have chosen other approaches that consist of the synthesis of partially conjugated cyclic molecules followed by aromatization. Stoddart's group attempted the synthesis of [*n*]cyclacenes (n = 12 and 14, where n represents the number of hexagons) via a multi-step Diels–Alder reaction between a bis-diene and a bis-dienophile (Scheme 2.1) [14]. Schlüter also reported the synthesis of a precursor to [8]cyclacene via a Diels–Alder reaction between a bis-diene and a diquinone (Scheme 2.2) [15]. In both examples, aromatization of the hoop-shaped precursors into the desired cyclacenes could not be achieved.

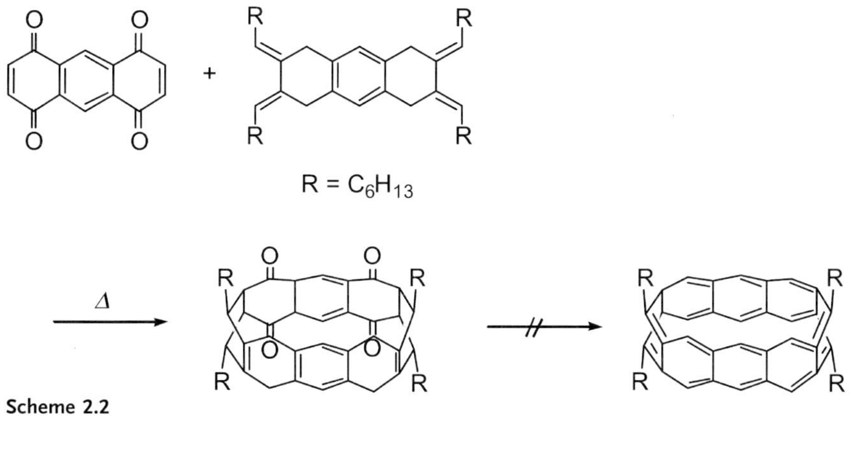

R = C_6H_{13}

Scheme 2.2

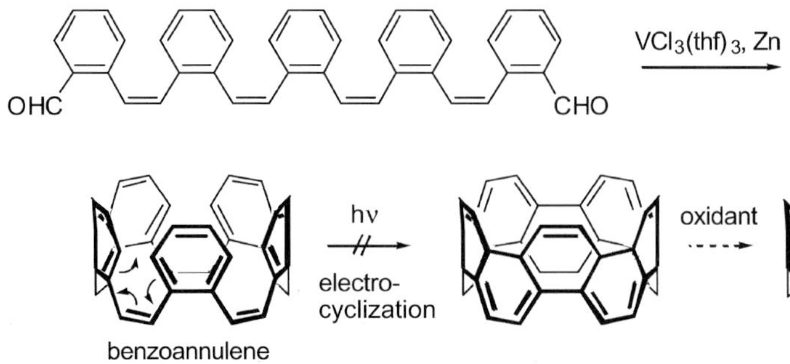

Scheme 2.3

The attempted synthesis of [n]cyclophenacenes has also been reported. Kuwatani and coworkers reported the synthesis of a pentabenzo[20]annulene, as a possible precursor to [10]cyclophenacene (Scheme 2.3) [16]. Light-induced electrocyclization of the precursor was unsuccessful because of isomerization of the (Z)-olefin into the (E)-olefin, which undergoes an undesirable transannular reaction [17]. St. Martin and Scott tried to synthesize [12]cyclophenacene, but attempts at the macrocyclic oligomerization of substituted naphthalene and phenanthrene building blocks by flash vacuum pyrolysis method were unsuccessful [18].

Selective destruction of the 60π-spherical conjugation of [60]fullerene has attracted considerable interest as a new approach to unconventional aromatic systems. For example, unique curved aromatic systems have been created by chemical modification of [60]fullerene (Fig. 2.3). Bowl-shaped 50π-electron conjugated systems have been obtained from [60]fullerene by regioselective addition of an organocopper reagent to five carbon atoms at the immediate neighbor of one pentagon [19] or by halogenation reactions (Fig. 2.3a) [20]. Open-cage fullerenes provide another new family of bowl-shaped π-electron conjugated systems (Fig. 2.3b–d) [21–23]. Trannulene systems have been obtained by the destruction of the π-electron conjugated system of [60]fullerene. Thus, treatment of the hexaanion of C_{60}^{6-} with 2-bromomalonate gives a hexa-adduct, $C_{60}[C(CH_3)(COO^tBu)_2]_6$ [24] (Fig. 2.3e), which possesses a trannulene structure and exhibits absorption in the near-infrared region (λ_{max} = 760 and 850 nm). Addition of the malonate to $C_{60}F_{15}$ affords an emerald green (λ_{max} = 658 nm) compound, $C_{60}F_{15}[CBr(COOEt)_2]_3$ [25] (Fig. 2.3f). Another trannulene π-conjugated system was synthesized by 30-fold chlorination of [60]fullerene with $SbCl_5$ [26] (Fig. 2.3g). The X-ray crystal structure of $C_{60}Cl_{30}$ shows a small bond alternation in the trannulene part (1.381 and 1.391 Å), as predicted by theoretical calculations [9].

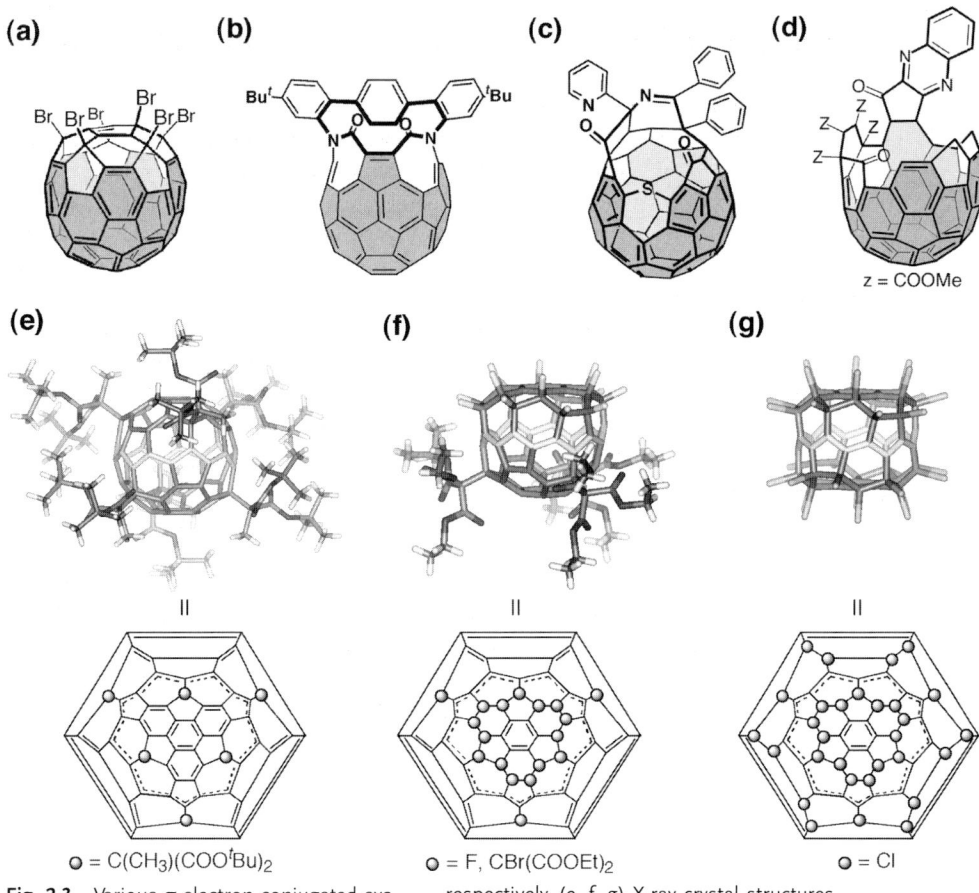

Fig. 2.3 Various π-electron conjugated systems constructed on the [60]fullerene skeleton. (a) $C_{60}Br_6$. (b, c, d) Open-cage fullerenes with a 14-, 13- and 18-membered orifice, respectively. (e, f, g) X-ray crystal structures and Schlegel diagrams of $C_{60}[C(CH_3)(COO^tBu)_2]_6$, $C_{60}F_{15}[CBr(COOEt)_2]_3$ and $C_{60}Cl_{30}$, respectively.

2.2 Synthesis of [10]Cyclophenacene π-Conjugated Systems from [60]Fullerene

2.2.1 Synthetic Strategy

The spherical π-electron conjugated system of [60]fullerene contains the [10]cyclophenacene system, which may be unraveled by removal of the top and bottom two [5]radialene structures (20π electrons) at both poles of [60]fullerene. The approach shown in Fig. 2.4 constitutes double application of the method that removes one radialene moiety out of conjugation to leave a 50π-electron bowl-shaped conju-

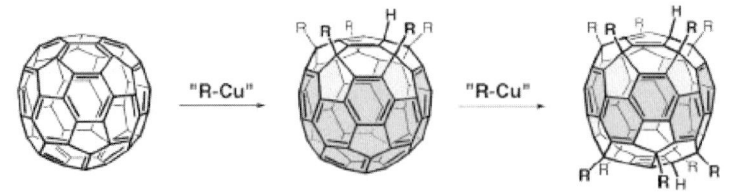

Fig. 2.4 A synthetic strategy to obtain [10]cyclophenacenes.

gated system (gray color in Fig. 2.4) [19]; that is, the penta-addition in the second round on the bottom 50π-electron part should produce the desired cyclic 40π-electron system [27, 28].

2.2.2
Synthesis and Characterization of [10]Cyclophenacenes

The synthesis started with $C_{60}Me_5H$ (**1**), which was treated first with KO^{t}Bu and then with *p*-toluenesulfonyl cyanide in benzonitrile to obtain $C_{60}Me_5CN$ (**2**) in 81% yield. Treatment of **2** with a phenylcopper reagent in the presence of N,N′-dimethylimidazolidinone (DMI) afforded [10]cyclophenacene, $C_{60}Me_5(CN)Ph_5H$ (**3**), in 14% isolated yield (Fig. 2.5) as a mixture of three regioisomers owing to the relative position of the cyano group and the bottom hydrogen atoms (Fig. 2.6). The chemical yield was slightly improved (18%) by the use of 1,4-dicyclohexyl-1,4-diaza-1,3-butadiene instead of DMI. The poor yield (14–18%) of **3** was due to formation of sideway adducts **4** and **5** (32% yield) (see below). The reaction of the methylcopper reagent to the nitrile **2** was much less clean was hoped for. The nitrile group in **3** was reductively removed by treatment with lithium naphthalenide in benzonitrile to obtain [10]cyclophenacene, $C_{60}Me_5Ph_5H_2$ (**6**), in 82% yield. This compound was found not to give any EPR signals (solid, at 4 K). In agreement with its closed-shell aromatic character, the 40π-electron system was found to be chemically stable.

To avoid possible perturbation of the electronic properties of cyclophenacene by the top and bottom cyclopentadiene π-system, we converted the sp^2 hybridized carbon atoms in the cyclopentadiene moieties into sp^3. Treatment of **6** with potassium hydride followed by exposure to molecular oxygen afforded the penta-oxygenated product $C_{60}Me_5Ph_5O_3(OH)_2$ (**7**). The molecular structure of **7** was unambiguously determined by X-ray crystallographic analysis (Fig. 2.7), which revealed that the top cyclopentadienyl anion was oxidized into a hydroxy cyclopentadiene bis(epoxide) [29], whereas the bottom cyclopentadienyl anion was oxidized into a hydroxy cyclopentadiene mono(epoxide) (Fig. 2.5). The hoop-shaped 40π-electron conjugated system remained intact during all these chemical transformations.

Deprotonation of the cyclopentadiene in **3** with KO^{t}Bu in THF afforded an anion, $[K(THF)_n][C_{60}Me_5(CN)Ph_5]$, which reacted with [PdCl(π-allyl)]$_2$ to give a pal-

2.2 Synthesis of [10]Cyclophenacene π-Conjugated Systems from [60]Fullerene

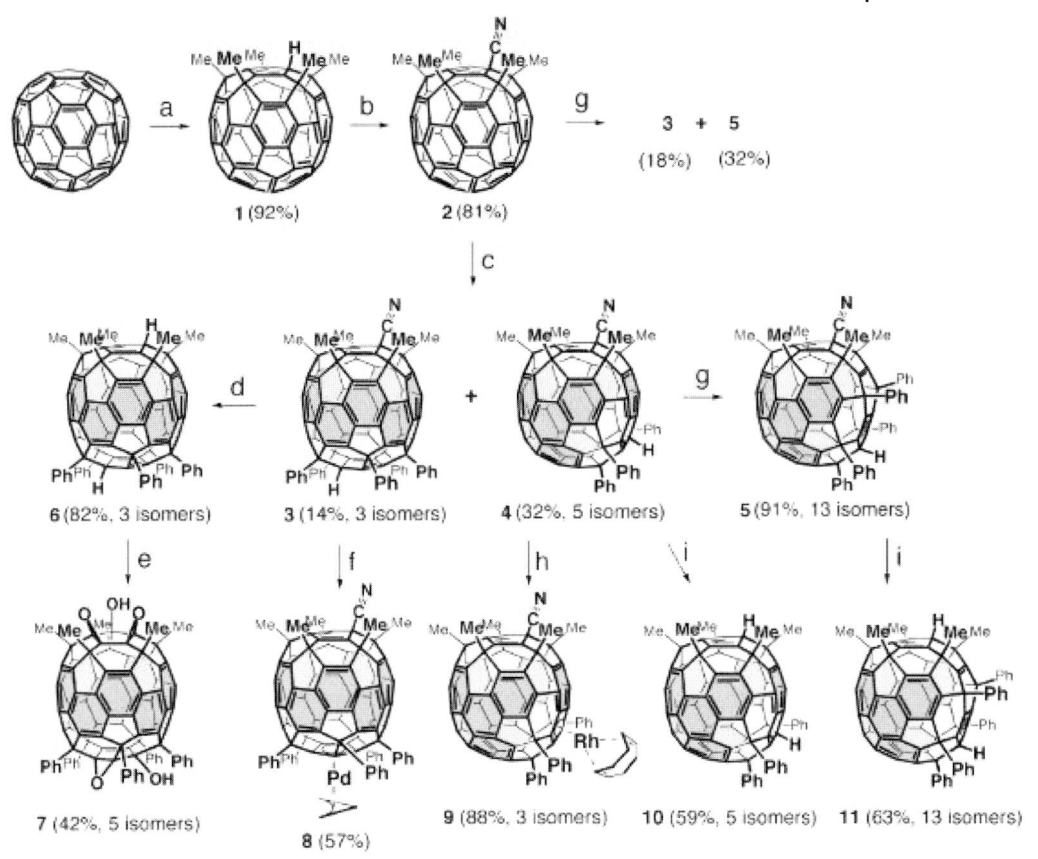

Fig. 2.5 Synthesis of [10]cyclophenacene derivatives. Reaction conditions: a, (1) MeMgBr, CuBr·SMe$_2$, DMI in THF–1,2-dichlorobenzene, (2) NH$_4$Cl–H$_2$O; b, (1) KO^{t}Bu in THF, (2) MeC$_6$H$_4$SO$_2$CN in PhCN; c, (1) PhMgBr, CuBr·SMe$_2$, DMI in THF, (2) NH$_4$Cl–H$_2$O; d, (1) Li[naphthalene] in PhCN, (2) NH$_4$Cl–H$_2$O; e, (1) KH in THF, (2) under air in THF; f, (1) KO^{t}Bu in THF, (2) [PdCl(π-allyl)]$_2$; g, (1) PhMgBr, CuBr·SMe$_2$, 1,4-dicyclohexyl-1,4-diaza-1,3-butadiene in THF–1,2-dichlorobenzene, (2) HCl–H$_2$O; h, (1) KO^{t}Bu in THF, (2) [RhCl(cyclooctadiene)]$_2$; i, (1) Li[naphthalene] in PhCN, (2) EtOH, (3) NH$_4$Cl–H$_2$O. For **3**, **4**, **5**, **6**, **7**, **9**, **10** and **11**, isomers were always formed according to the relative stereochemistry of the top and the bottom pentagons (only one isomer is shown in Fig. 2.5).

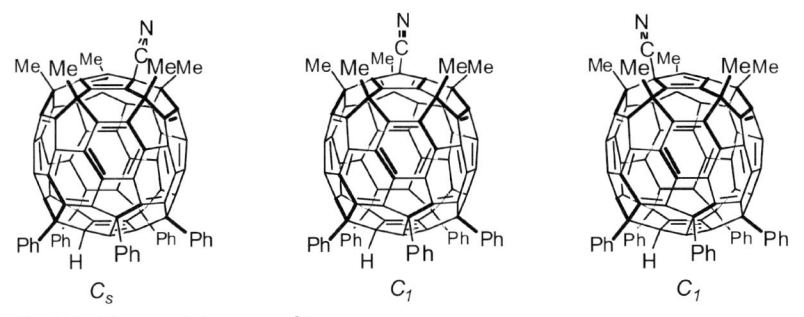

Fig. 2.6 Three regioisomers of **3**.

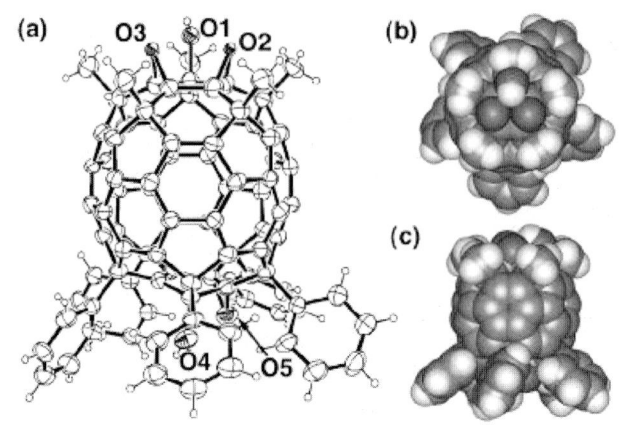

Fig. 2.7 Crystal structure of **7**·(C$_6$H$_5$Cl)$_{1.5}$. C$_6$H$_5$Cl molecules found in the unit cell and disordered oxygen atoms are omitted for clarity. (a) ORTEP drawing; (b) top view of CPK model; (c) side view of CPK model.

ladium complex, Pd[C$_{60}$Me$_5$(CN)Ph$_5$](π-allyl) (**8**) [30], as a single isomer in 57% yield. The ^{13}C NMR data showed the C_s symmetric structure of **8**, whose signals attributable to the cyclophenacene carbon atoms appeared at 144–158 ppm. The crystal structure of **8** (Fig. 2.8) gave the first experimental data for studies of the structure and properties of [10]cyclophenacene.

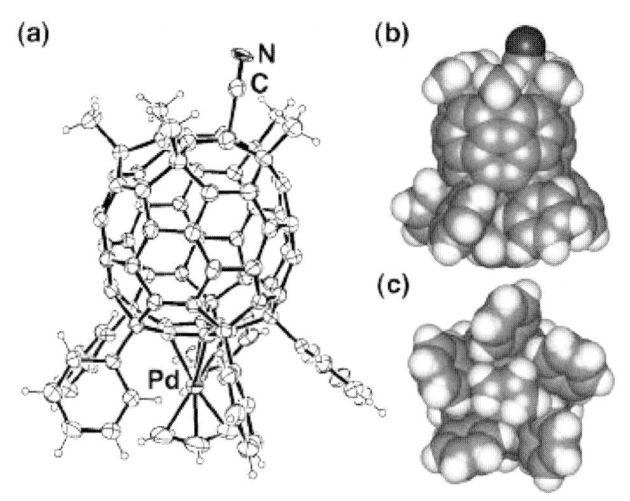

Fig. 2.8 Crystal structure of **8**·C$_6$H$_5$Cl. (a) ORTEP drawing of **8**·C$_6$H$_5$Cl molecules found in the unit cell; disordered cyano groups are omitted for clarity. (b) Side view of CPK model. (c) Bottom view of CPK model.

2.2.3
Structural Studies and Aromaticity of [10]Cyclophenacene

X-ray crystallographic structures of **7** and **8** provided information on the structure of the first hoop-shaped π-conjugated system (Table 2.1; see Chart 2.1 for the location of bonds a–g). Although bonds on the edge (a and g) are rather short (1.36–1.37 Å) compared with those on the equator (1.40–1.44 Å), the degree of bond alternation in the cyclophenacene part of **7** and **8** is smaller (single bond vs. double bond: 1.36–1.37 vs. 1.43–1.44 Å) than that in [60]fullerene (1.36 vs. 1.47 Å) [31], 1,3-butadiene (1.35 vs. 1.47 Å) [32] and the 20π-electron cyclic all-cis-polyacetylene $C_{20}H_{20}$ (1.36 vs. 1.46 Å, B3LYP/6–31G*-optimized). This experimental structure obviously does not conform to an ideal graphitic structure, which was assumed in most of the previous theoretical studies of carbon nanotube-like materials [33].

The structures of the model compounds $C_{40}H_{20}$ (**A**) and $C_{60}H_{12}$ (**B**) (Chart 2.1) were optimized with quantum mechanical calculations at various levels of theory (semiempirical, Hartree–Fock and hybrid density functional methods) and were found to reproduce the experimental data very well (within ca. 1.5%, Table 2.1). The optimized structures and experimental structures of **A** and **B** also match well with the structure obtained from an estimation of the Pauli bond order [34] assuming equal contributions of all 125 canonical resonance structures.

The nucleus-independent chemical shift (NICS) [35] is a useful measure of the magnetic shielding effect of aromatic ring currents. Analysis of the NICS values for six-membered rings of cyclophenacene in the model compounds **A** and **B** indicates that the hoop-like 40π-electron system is aromatic (Fig. 2.9, NICS = –8.6 and –11.5 to –12.0 respectively) and the other rings in **B** are nonaromatic (NICS = –1.3 to 0.3). The center of gravity (CG; NICS = –7.3 and –11.6) is predicted to be

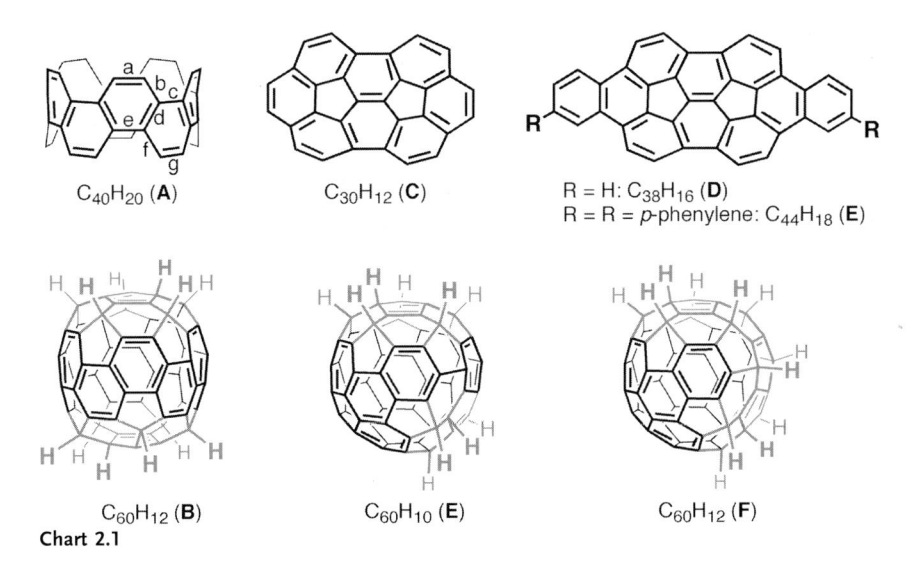

$C_{40}H_{20}$ (**A**)

$C_{30}H_{12}$ (**C**)

R = H: $C_{38}H_{16}$ (**D**)
R = R = p-phenylene: $C_{44}H_{18}$ (**E**)

$C_{60}H_{12}$ (**B**)

$C_{60}H_{10}$ (**E**)

$C_{60}H_{12}$ (**F**)

Chart 2.1

Table 2.1 Experimental data from X-ray crystallographic analysis and theoretical optimized structures of model compounds **A** and **B**

Compound	Bond length (Å)[a,b]						
	a	b	c	d	e	f	g
7 (X-ray)[c]	1.37(2)	1.44(2)	1.43(1)	1.40(2)	1.43(1)	1.45(1)	1.36(1)
8 (X-ray)[c]	1.36(1)	1.45(1)	1.43(1)	1.40(2)	1.44(2)	1.44(2)	1.37(2)
A (HF)	1.35	1.44	1.43	1.38	–	–	–
A (B3LYP)	1.37	1.44	1.45	1.42	–	–	–
A (PM3)	1.36	1.43	1.44	1.40	–	–	–
B (HF)	1.35	1.44	1.43	1.38	–	–	–
B (B3LYP)	1.37	1.45	1.44	1.39	–	–	–
B (PM3)	1.37	1.44	1.44	1.40	–	–	–

a Bond coding (a–g) is shown in Chart 2.1.
b Standard deviations in parentheses.
c a, b, c, d, average lengths of five equivalent bonds;
 e, f, g, average lengths of 10 equivalent bonds.

subject to an aromatic shielding effect (a value experimentally proved by ^{3}He NMR experiments) [36]. We ascribe the difference in the NICS values between **A** and **B** to the effect of the σ skeleton, i.e. the six-membered rings in the cyclophenacene part of **B** are much more planar than those of **A**, as judged by comparison of the dihedral angles between the two planes including the bonds [a and b] and [d and e] on the 10 hexagons (22.9° for **A** and 2.9° for **B**). The aromatic character of the cyclophenacene **A** and **B** is related to its low chemical reactivity. The 40π-electron cyclic benzenoid molecules are stabilized through conjugation, as has been predicted [11].

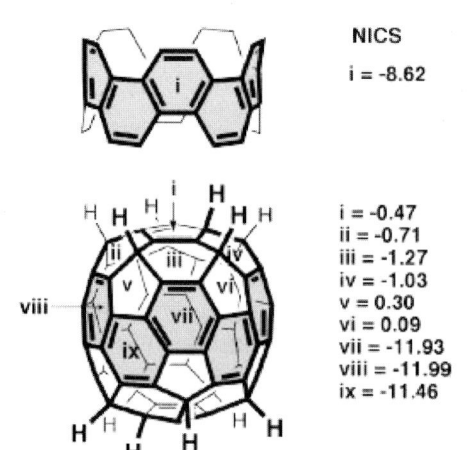

NICS

i = –8.62

i = –0.47
ii = –0.71
iii = –1.27
iv = –1.03
v = 0.30
vi = 0.09
vii = –11.93
viii = –11.99
ix = –11.46

Fig. 2.9 NICS values in **A** and **B**. Dark-colored hexagons indicate aromatic rings.

2.2.4
Synthesis of Dibenzo-fused Corannulenes

We removed the top and bottom radialene units of the fullerene molecule to generate [10]cyclophenacene as described above. An alternative view of [60]fullerene may generate hemispherical bowl-shaped aromatic systems [1c, 37]; for instance, fused corannulene **C** ($C_{30}H_{12}$, Chart 2.1), can be identified as half of fullerene [38].

Thorough analysis and separation of a product mixture of the reaction of **2** and phenylcopper reagents allowed us to identify two sideway adducts, $C_{60}Me_5(CN)Ph_3H$ (**4**) and $C_{60}Me_5(CN)Ph_5H$ (**5**) (Fig. 2.5), whose π-electron conjugated systems represent the dibenzo-fused corannulene (**D** in Chart 2.1) and its phenylene-bridged derivative (**E** in Chart 2.1). Optimization of the conditions by the use of DMI as an additive led to the production and isolation of **4** in 32% yield. Whereas **4** was found not to react further with the phenylcopper reagent under the same conditions, it reacted smoothly in the presence of 1,4-dicyclohexyl-1,4-diaza-1,3-butadiene to afford **5** in 91% isolated yield. Alternatively, treatment of **2** with the phenylcopper reagent in the presence of 1,4-dicyclohexyl-1,4-diaza-1,3-butadiene directly afforded **5** in 32% yield. The cyano group in **4** and **5** was removed by lithium naphthalenide reduction in benzonitrile to give diprotio products **10** and **11**, respectively. Rigorous structural identification of **4** and **5** was hampered by the presence of mixtures of isomers, but it was finally achieved for **4** through the synthesis and crystallographic analysis of a rhodium derivative (Fig. 2.10). Compound **4** was converted into the corresponding potassium salt, which was treated with $[RhCl(cod)]_2$ to afford a rhodium 1,5-cyclooctadiene complex, $Rh[C_{60}Me_5(CN)Ph_3](cod)$ (**9**), in 88% yield. The complex **9** is an η^5-indenyl metal complex similar to that reported for a [70]fullerene derivative [39].

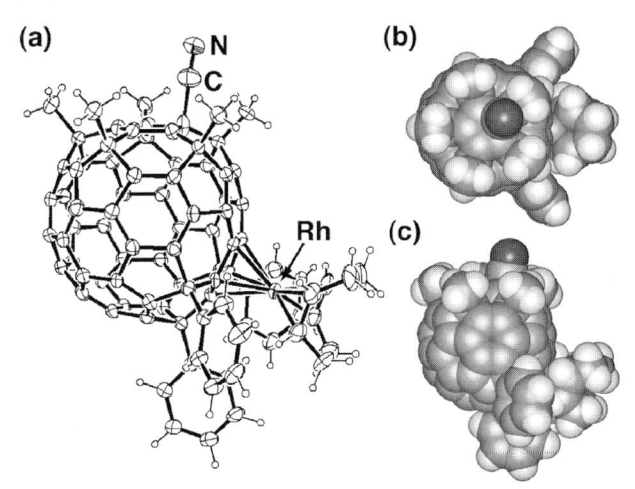

Fig. 2.10 Crystal structure of **9**·(THF)$_3$. (a) ORTEP drawing. THF molecules found in the unit cell and disordered cyano group due to a mixture of three regioisomers are omitted for clarity. (b) Top view of CPK model. (c) Side view of CPK model.

The X-ray crystallographic analysis of **9** provided experimental structural data of the new corannulene-type aromatic system. The bond alternation found in **9** closely follows the pattern found in an X-ray crystal structure of tetramethyl fused corannulene [38c]. Such agreement of the bond alternation patterns indicates that the fullerene framework and the dibenzo conjugation added to the parent corannulene structure have a small structural effect in spite of the rather large difference in hybridization of the carbon atoms between **9** {p-orbital axis vector (POAV) angles [40] = 8.0–12.8°} and **C** (POAV = 0.0–11.0°).

2.2.5
Absorption and Emission of [10]Cyclophenacenes and Dibenzo Fused Corannulenes

The absorption spectra of the [10]cyclophenacene derivatives **6** and **7** are similar to each other, showing a maximum at ca. 260 nm with a broad absorption that extends to ca. 500 nm (Fig. 2.11a). Both [10]cyclophenacene derivatives emit bright

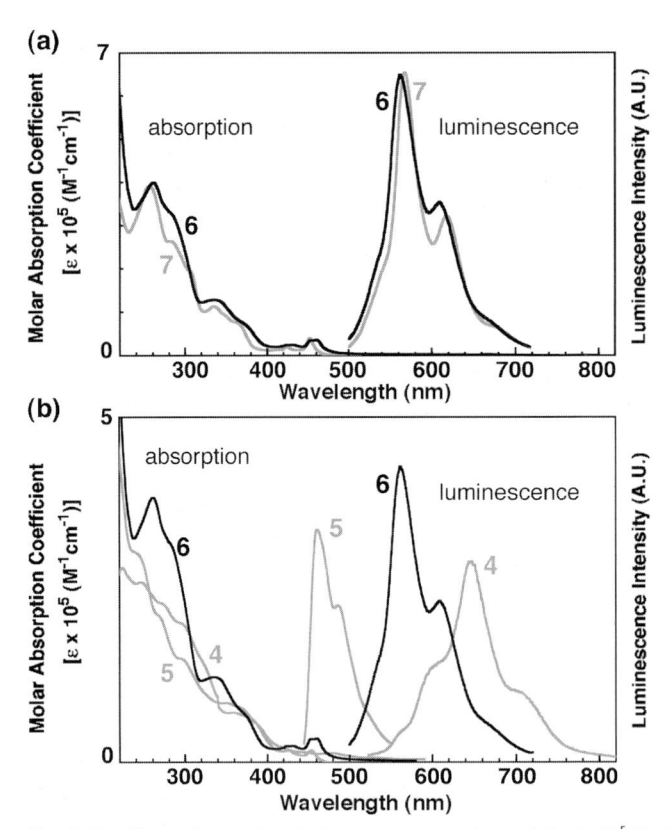

Fig. 2.11 Absorption and emission spectra of [10]cyclophenacene derivatives and sideway adducts in cyclohexane. (a) Spectra for **6** and **7**; (b) spectra for **4**, **5** and **6**. Concentration is 1.0×10^{-5} M. The intensities of emission spectra of **4** and **5** are magnified $\times 8$ and $\times 2$, respectively.

2.2 Synthesis of [10]Cyclophenacene π-Conjugated Systems from [60]Fullerene

yellow light (bimodal, λ_{max} = ca. 560 and ca. 620 nm) with quantum efficiency Φ = 0.10 (in cyclohexane, irradiation at 366 nm, rhodamine B as standard). This luminescent quantum efficiency is much higher than those of [60]fullerene (Φ = 0.00032) [41], several mono-adducts (Φ = 0.0006–0.0012) (Fig. 2.12) [42] and fullerene–hexapyrrolidine having a supercyclophane-type π-electron conjugated system (Φ = 0.024) (Fig. 2.12) [43]. We expect that the photophysical properties of the compounds can be modified by changing addends on the [60]fullerene skeleton [44, 45].

The sideway adduct **5** exhibits blue–green fluorescence with a maximum at 460 nm (Φ = 0.028, Fig. 2.11b) and the p-phenylene-bridged compound **4** emits red light with a maximum at 649 nm (Φ = 0.012). The maximum wavelength of **4** is ca. 100 nm blue-shifted and that of **5** is ca. 100 nm red-shifted, as compared with **5**. The large shift of the maximum wavelength between **4** and **5** indicates that cyclic conjugation through the p-phenylene bridge in **4** affects its photoluminescent properties. The wavelengths of the luminescence are in qualitative agreement with the HOMO–LUMO gaps calculated for simplified model compounds **B** (3.366 eV), **F** (3.417 eV) and **G** (2.965 eV) (structures are shown in Chart 2.1). Metal complex **8** is not luminescent, probably because of quenching of the excited state by the transition metal.

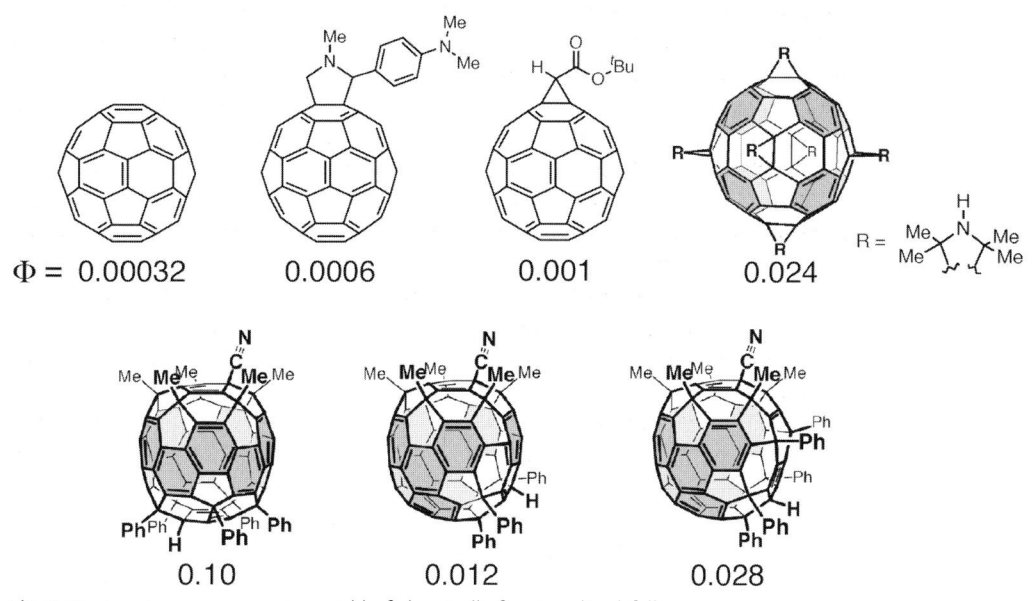

Fig. 2.12 Luminescence quantum yield of chemically functionalized fullerenes.

2.3
Conclusion

A series of hoop-shaped cyclic benzenoids, [10]cyclophenacenes, have been synthesized by selective interruption of π-conjugation out of the [60]fullerene π-electron array. The synthesis of the [10]cyclophenacene compounds provided the first information on the high-precision structure and the chemical and physical properties of such hoop-shaped cyclic benzenoids. The compounds were found to be stable, aromatic and luminescent. This last property is intriguing in view of recent reports on luminescent carbon nanotubes [46, 47]. Selective destructions of the 60π-spherical conjugation of [60]fullerene also gave two new varieties of bowl-shaped π-conjugated aromatic systems, dibenzo-fused corannulenes **5** and their phenylene-bridged derivatives **4**. Synthesis of transition metal complexes [48] and construction of supramolecular architectures [49, 50] using these materials are subjects of our current studies.

Since the [10]cyclophenacene comprises a unit structure of the [5, 5] armchair carbon nanotubes, the present data provided the basis analysis for the chemical reactivities of carbon nanotubes [27b]. The data on the various experimental and theoretical parameters of **7**, **8** and **A** thus gave us useful information on the structures and properties of finite length carbon nanotubes, $C_{50}H_{20}$, $C_{60}H_{20}$, $C_{70}H_{20}$, $C_{80}H_{20}$, etc., and also on their chemical reactivities [51].

2.4
Experimental

General

All manipulations involving air- and moisture-sensitive compounds were carried out using the standard Schlenk technique under nitrogen or argon. Hexane, toluene, THF and THF-d_8 were distilled from Na–K alloy and thoroughly degassed by trap-to-trap distillation. Chloroform-d was distilled from CaH_2 and thoroughly degassed by trap-to-trap distillation. o-Dichlorobenzene was distilled from CaH_2 under nitrogen and stored over molecular sieves. Benzonitrile was distilled from P_2O_5 under nitrogen and stored over molecular sieves. KO^{t}Bu in THF was purchased from Aldrich and used as received. 1,4-Dicyclohexyl-1,4-diaza-1,3-butadiene was prepared according to the literature [52].

Preparation of $C_{60}Me_5H$ (1)

A white suspension of CuBr·SMe$_2$ (4.28 g, 20.8 mmol) in THF (30 mL) was treated at 25 °C with a solution of MeMgBr (1.0 M, 20.8 mL, 20.8 mmol) in THF and with DMI (2.25 mL, 20.8 mmol). To the resulting yellow suspension, a solution of C_{60} (500 mg, 0.694 mmol) in o-dichlorobenzene (50 mL) was added. After stirring for 1 h at 23 °C, the reaction was quenched with NH$_4$Cl (aq.) (0.5 mL). The mixture was diluted with toluene (250 mL) and filtered through a pad of silica gel. The red eluate was concentrated to a small volume and then precipitated with

Et$_2$O (300 mL). The precipitate was washed with EtOH and diethyl ether and dried *in vacuo* to obtain **1** (507 mg, 92% yield, 95% purity determined by HPLC). IR (KBr), 2957, 2918, 2856, 1575, 1548, 1521, 1444, 1417, 1370, 1343, 1324, 1285, 1262, 1235, 1200, 1170, 1146, 1127, 1092, 1054, 1034, 1011, 953, 830, 807, 745, 683 cm^{-1}; ^{1}H NMR (400 MHz, CDCl$_3$), δ 2.30 (s, 6H, CH$_3$), 2.32 (s, 6H, CH$_3$), 2.42 (s, 3H, CH$_3$); ^{13}C NMR (100 MHz, CDCl$_3$), δ 26.91, 27.26, 32.86, 50.98, 51.12, 53.13, 59.22, 142.89, 143.52, 143.74, 144.04, 144.08, 144.28, 144.64, 144.82, 145.27, 145.37, 146.39, 146.55, 146.66, 147.49, 147.53, 147.69, 147.76, 147.93, 148.00, 148.26, 148.32, 148.38, 149.57, 153.70, 153.74, 154.02, 157.17; APCI-MS (positive), *m/z* = 796 [M$^+$].

Synthesis of C$_{60}$Me$_5$CN (2)

A solution of KO^{t}Bu (0.69 mL, 0.69 mmol, 1.0 M in THF) was added to a suspension of **1** (500 mg, 0.63 mmol, 94% purity) in PhCN (10 mL) at 23 °C. The color of the reaction mixture changed from red to black. A solution of MeC$_6$H$_4$SO$_2$CN (3.0 mL, 0.25 M in PhCN, 0.75 mmol) was added to the mixture. After stirring for 10 min, the reaction was quenched with HCl (aq.) (0.20 mL). The solvent was removed and resulting red solid was dissolved with CS$_2$. Purification on a pad of silica gel (CS$_2$ and toluene as eluent) afforded the fractions containing **2**. The fractions were evaporated to small volume. Precipitation with EtOH and then washing with Et$_2$O and hexane afforded **2** (416 mg, 81% yield, 99% HPLC purity). Slow diffusion of ethanol to a methylene chloride solution of **2** gave a single crystal suitable for X-ray analysis. IR (KBr), 2963, 2920, 2860, 2229 (CN), 1641, 1547, 1445, 1418, 1374, 1264, 1239, 1200, 1106, 684, 658, 552, 522 cm^{-1}; ^{1}H NMR (400 MHz, CDCl$_3$), δ 2.38 (s, 6H, CH$_3$), 2.39 (s, 6H, CH$_3$), 2.64 (s, 3H, CH$_3$); ^{13}C NMR (100 MHz, CDCl$_3$), δ 25.36, 26.80, 32.35, 51.12, 51.37, 52.62, 55.46, 118.08, 143.02, 143.07, 143.93, 144.03, 144.10, 144.31, 144.35, 144.50, 145.10, 145.27, 146.71, 146.83, 146.85, 147.77, 147.91, 147.98, 148.03, 148.15, 148.31, 148.33, 148.50, 148.56, 148.68, 151.75, 152.24, 152.98, 156.37; HRMS (APCI-TOF, negative), *m/z* calcd. for C$_{66}$H$_{15}$N$_1$ (M$^-$) 821.12045, found 821.12261.

Synthesis of [10]cyclophenacene, C$_{60}$(CN)Me$_5$Ph$_5$H (3), and *p*-phenylene-bridged dibenzo-fused corannulene, C$_{60}$(CN)Me$_5$Ph$_3$H (4)

To a white suspension of CuBr·SMe$_2$ (3.75 g, 18.3 mmol) in THF (31 mL) was added a solution of PhMgBr (0.93 M in THF, 20.0 mL, 18.6 mmol) at 23 °C and then DMI (2.05 mL, 25.4 mmol) was added immediately. To the resulting dark yellow solution, a solution of **2** (515 mg, 0.608 mmol) in *o*-dichlorobenzene (50 mL) was transferred though a cannula. After stirring for 24 h at 23 °C, the reaction was stopped with the addition of saturated NH$_4$Cl (aq.) (0.20 mL). The mixture was diluted with toluene (200 mL) and filtered though a pad of silica gel. The solvent of the orange eluate was evaporated and the orange residue was purified by silica gel column chromatography [toluene–CS$_2$ (3:97)]. The fractions containing **3** were collected and the solvent was evaporated to a small volume. Precipitation with MeOH afforded **3** (111 mg, 14% isolated yield) as a mixture of three stereoisomers. The fractions containing **4** were collected and the solvent was evaporated to a small volume. Precipitation with MeOH afforded **4** (205 mg, 32% isolated yield) as a mixture of five stereoisomers. NMR and IR spectra were measured for the mixture of the stereoisomers.

3: IR (ReactIR, diamond prove), 3056, 3026, 2963, 2922, 2860, 2229 (CN), 1600, 1493, 1445, 1316, 1210, 1116, 1032, 1003, 950, 913, 684, 658 cm^{-1}; ^{1}H NMR (400 MHz, CDCl$_3$), δ 2.44–2.49 (s, 12H, CH$_3$), 2.66-2.68 (s, 3H, CH$_3$), 5.421, 5.424 and 5.448 (s, 1H, three signals of C$_{60}$-H), 7.12–7.27, 7.32–7.38, 7.43-7.47, 7.61–7.70 and 7.82–7.89 (m, 25H, C$_6$H$_5$); HRMS (FAB-MS, positive), m/z calcd. for C$_{96}$H$_{41}$N$_1$ [M]$^+$ 1207.3230, found 1207.3236.

4: IR (ReactIR, diamond prove), 3059, 3027, 2962, 2923, 2863, 2229, 1593, 1583, 1493, 1445, 1375, 1297, 1243, 1156, 1031, 922, 759, 692, 676 cm^{-1}; ^{1}H NMR (400 MHz, CDCl$_3$), δ 2.03–2.19, 2.38–2.50, 2.68–2.76 and 3.02 (s, 15H, CH$_3$), 5.04, 5.07, 5.08, 5.09 and 5.12 (s, 1H, five signals of C$_{60}$-H), 7.16–7.38, 7.42–7.52, 7.57–7.82 and 8.12–8.16 (m, 15H, C$_6$H$_5$); ^{13}C NMR (125 MHz, CDCl$_3$), δ 24.99–27.32, 31.58–33.22, 50.62–53.22, 54.82–54.84, 55.53–55.83, 57.45–57.65, 59.97–60.88, 118.65–118.70, 127.16–129.07, 134.01–160.33; HRMS (APCI-TOF, negative), m/z calcd. for C$_{84}$H$_{30}$N$_1$ [M − H]$^-$ 1052.2378, found 1052.2396.

Synthesis of dibenzo-fused corannulene, C$_{60}$(CN)Me$_5$Ph$_5$H (**5**).
Method A: synthesis from **2**
To a brown suspension of CuBr·SMe$_2$ (150 mg, 0.73 mmol) and 1,4-dicyclohexyl-1,4-diaza-1,3-butadiene (161 mg, 0.73 mmol) in THF (2.0 mL) was added a solution of PhMgBr (0.97 M in THF, 0.75 mL, 0.73 mmol) at 23 °C. To the resulting yellow suspension, a solution of **2** (20 mg, 23 µmol) in o-dichlorobenzene (2.0 mL) was transferred though a cannula. After stirring for 40 h at 40 °C, the reaction was stopped by addition of an aqueous HCl. The mixture was diluted with toluene (20 mL) and filtered though a pad of silica gel. The orange eluate was concentrated and purified by HPLC [Nomura Chemical, RPFullerene, 250 mm, toluene–acetonitrile (2:3)]. The fractions containing **5** were collected and the solvent was evaporated to small volume. Precipitation with MeOH afforded **5** (9.5 mg, 32% isolated yield with 98% HPLC purity) as a mixture of stereoisomers. NMR and IR spectra were measured for the mixture of the stereoisomers. IR (ReactIR, diamond prove), 3056, 3024, 2962, 2921, 2863, 2227 (CN), 1598, 1582, 1492, 1445, 1374, 1262, 1096, 1075, 1030, 753, 694 cm^{-1}; ^{1}H NMR (500 MHz, CDCl$_3$), δ 1.05–1.30, 1.47–1.77, 2.02–2.25, 2.36–2.67 and 2.82 (s, 15H, CH$_3$), 4.303, 4.343, 4.349, 4.414, 4.559, 4.618, 4.670, 4.719, 4.713, 4.723, 4.741 and 4.778 (s, 1H, C$_{60}$-H), 6.85–8.00 (m, 25H, C$_6$H$_5$); HRMS (APCI-TOF, negative), m/z calcd. for C$_{96}$H$_{40}$N$_1$ [M − H]$^-$ 1206.3161, found 1206.3142.

Method B: synthesis from **4**
To a brown suspension of CuBr·SMe$_2$ (64 mg, 0.31 mmol) and 1,4-dicyclohexyl-1,4-diaza-1,3-butadiene (69 mg, 0.31 mmol) in THF (2.0 mL) was added PhMgBr (0.97 M in THF, 0.32 mL, 0.31 mmol) at 24 °C. A solution of **4** (11 mg, 10 µmol) in o-dichlorobenzene (2.0 mL) was transferred to the yellow phenylcopper reagent though a cannula. After stirring for 14 h at 40 °C, the reaction was quenched with

saturated HCl (aq.) (10 μL). The mixture was diluted with toluene (20 mL) and filtered though a pad of silica gel. The yellow eluate was evaporated to small volume. Precipitation with MeOH afforded **5** (12 mg, 91% isolated yield with 100% HPLC purity) as a mixture of stereoisomers.

Synthesis of [10]cyclophenacene, $C_{60}Me_5Ph_5H_2$ (**6**)
To a solution of **3** (80 mg, 66 mmol) in PhCN (20 mL) in a Schlenk tube was added a solution of lithium naphthalenide (2.6 mL, 0.76 M in THF, 2.0 mmol). The color of the solution changed immediately from yellow to dark red, indicating decyanation and formation of a bis-cyclopentadienyl dianion. After stirring for 1 h at 23 °C, the reaction mixture was quenched by addition of saturated NH_4Cl (aq.) (0.10 mL). The color of the resulting solution changed immediately from dark red to light red, indicating the protonation of the bis-cyclopentadienyl dianion took place. The mixture was diluted with toluene (100 mL) and filtered though a pad of silica gel. The red eluate was evaporated to a small volume and precipitation with MeOH afforded a crude product. This mixture was purified by HPLC [Nacalai Tesque, Buckyprep, 250 mm, toluene–2-propanol (7:3)]. The fraction containing **6** was collected and evaporated to a small volume. Precipitation with MeOH afforded **6** (65 mg, 82% yield) of 98% purity by HPLC as a mixture of three isomers. NMR and IR spectra were measured for the mixture of the stereoisomers. IR (ReactIR, diamond prove), 3064, 3024, 2959, 2918, 2857, 1599, 1521, 1492, 1445, 1370, 1345, 1318, 1262, 1210, 1185, 1158, 1076, 1031, 950, 922, 812, 789, 752, 727, 682, 660 cm^{-1}; ^{1}H NMR (400 MHz, THF-d_8), δ 2.37–2.50 (s, 15H, CH_3), 4.847, 4.853 and 4.868 (s, 1H, three signals of C_{60}-H), 5.527 and 5.545 (s, 1H, three signals of C_{60}-H), 7.00–7.20, 7.30–7.33, 7.45–7.50, 7.60–7.70 and 7.89–7.91 (m, 25H, C_6H_5); UV/Vis spectra (cyclohexane, ε), 261 (33 400), 278 (32 600), 334 (10 900), 380 (2500), 425 (2000), 454 (2900), 460 (3000) nm; HRMS (FAB, positive), m/z calcd. for $C_{95}H_{42}$ (M$^+$) 1182.3356, found 1182.3320.

Synthesis of the penta-oxygenated compound $C_{60}Me_5Ph_5O_3(OH)_2$ (**7**)
Compound **6** (10 mg, 8.5 mmol) and KH (3.2 mg, 80 mmol) were dissolved in THF (5.0 mL) in a Schlenk tube. The color of the solution changed immediately from yellow to blackish brown, indicating the formation of a bis-cyclopentadienyl dianion, $[K(thf)_n]_2[C_{60}Me_5Ph_5]$. The dianion was allowed to react with oxygen in air to afford oxidized products. This mixture was purified by column chromatography on silica gel (30–70% toluene in hexane) to obtain **7** (4.5 mg, 42% yield) as a mixture of isomers. Single crystals suitable for X-ray structure analysis were obtained by slow diffusion of MeOH to a saturated chlorobenzene solution of **7**. IR (ReactIR, diamond prove), 3501 (OH), 3484 (OH), 3084, 3058, 3031, 3025, 2963, 2923, 2860, 1598, 1583, 1494, 1446, 1373, 1365, 1345, 1323, 1224, 1070, 1019, 927, 807, 752, 723, 685 cm^{-1}; ^{1}H NMR (400 MHz, CDCl$_3$), δ 2.14–2.46 (s, 15H, CH_3), 3.86–3.91 (s, 1H, OH), 4.59–4.60 (s, OH), 7.13–7.17 7.26–7.35, 7.42–7.48, 7.55–7.57, 7.78–7.80, 7.86–7.90 and 8.09–8.12 (m, 25H, C_6H_5); UV/Vis spectra (cyclohexane, ε) 256 (32 600), 282 (22 000), 335 (9600), 414 (1000),

424 (1600), 452 (3300) nm; HRMS (APCI-TOF, positive), m/z calcd. for $C_{95}H_{41}O_5$ $(M-1)^+$ 1261.2954, found 1261.29370.

Synthesis of Pd[C_{60}(CN)Me$_5$Ph$_5$](π-allyl) (8)
A THF solution of KO^{t}Bu (0.50 M, 73 µL, 36 µmol) was added to a solution of 3 (40 mg, 33 µmol) in THF (4.0 mL). The color of the solution changed from yellow to black, indicating formation of [K(thf)$_n$][C_{60}(CN)Me$_5$Ph$_5$]. [Pd(π-allyl)Cl]$_2$ (7.3 mg, 20 µmol) was added to the solution. After stirring at 26 °C for 10 min, the reaction was quenched with saturated NH$_4$Cl (aq.) (0.10 mL). The crude mixture was diluted with toluene (10 mL) and the organic layer was collected. After drying over MgSO$_4$, the solution was concentrated in vacuo. The resulting orange solid was purified by HPLC [Nacalai Tesque, Buckyprep, 250 mm, toluene–hexane (3:7)]. The fraction containing 8 was collected and evaporated to a small volume. Precipitation with MeOH afforded 8 (26.0 mg 57 % yield). Recrystallization from a two-layered solution of chlorobenzene and ethanol gave a single crystals of 8. ^{1}H NMR (400 MHz, CDCl$_3$), δ 2.17 [d, J_{H-H} = 12.0 Hz, 2H, allyl-CH$_2$(anti)], 2.41 (s, 6H, CH$_3$), 2.44 (s, 6H, CH$_3$), 2.66 (s, 3H, CH$_3$), 3.14 [d, J_{H-H} = 6.4 Hz, 2H, allyl-CH$_2$(syn)], 4.78 (ddt, 1H, allyl-CH), 7.16–7.30 (m, 10H + 5H, o,m-C$_6$H$_5$), 7.84–7.86 (m, 10H, p-C$_6$H$_5$,); ^{13}C NMR (100 MHz, CDCl$_3$), δ 25.46, 26.70, 31.55, 50.99, 51.41, 52.92, 55.42, 57.12, 59.28, 59.36, 59.39, 99.87, 118.96, 121.28, 121.45, 122.32, 127.00, 127.63, 128.24, 143.90, 144.16, 144.27, 144.34, 144.38, 144.46, 144.93, 145.52, 146.02, 146.41, 146.63, 146.66, 146.88, 147.76, 151.09, 151.27, 153.37, 153.40, 154.05, 154.26, 154.59, 155.07, 156.41, 158.04; HRMS (APCI-TOF, positive), m/z calcd. for $C_{99}H_{45}N_1{}^{105}Pd_1$ (M$^+$) 1352.26028, found 1206.25577.

Synthesis of Rh[C_{60}(CN)Me$_5$Ph$_5$](cod) (9)
To a solution of 4 (15 mg, 14 µmol) in THF (2.0 mL) was added a solution of KO^{t}Bu (17 µL, 1.0 M in THF, 17 µmol). The color of the solution changed from orange to black, indicating formation of [K(thf)$_n$][C_{60}(CN)Me$_5$Ph$_3$]. To the reaction mixture was added [RhCl(cod)]$_2$ (21 mg, 43 µmol) and the mixture was stirred for 10 min at 26 °C. Precipitation with MeOH afforded 9 (16 mg, 88 % yield) as a mixture of three isomers. Single crystals suitable for X-ray structure analysis were obtained by slow diffusion of methanol to a THF solution of 9. NMR and IR spectra were measured for the mixture of the stereoisomers. IR (ReactIR, diamond prove), 3058, 3028, 2961, 2921, 2864, 2831, 2227, 1599, 1494, 1445, 1373, 1297, 1243, 1155, 1033, 1004, 963, 921, 861, 745, 733, 695, 678 cm^{-1}; ^{1}H NMR (400 MHz, CDCl$_3$), δ 1.74–1.77 (m, 8H, cod), 1.78–1.80 (m, 2H, cod), 2.22–2.32, 2.45–2.54, 2.67–2.77 and 2.92 (s, 15H, CH$_3$), 3.42 (br, 2H, cod), 7.12–7.32, 7.47–7.67, 7.75–7.79 and 9.28–9.35 (m, 15H, C$_6$H$_5$); HRMS (APCI-TOF, positive), m/z calcd. for $C_{92}H_{42}N_1{}^{103}Rh_1$ (M$^+$) 1263.23723, found 1263.23883.

Synthesis of the p-phenylene-bridged dibenzo-fused corannulene C_{60}Me$_5$Ph$_3$H$_2$ (10)
To a solution of 4 (40 mg, 38 µmol) in PhCN (10 mL) in a Schlenk tube was added a solution of lithium naphthalenide (1.5 mL, 0.76 M in THF, 1.1 mmol) under an argon atmosphere. The color of the solution changed immediately from yellow to

blackish red, indicating the decyanation and formation of bis-cyclopentadienide. After stirring for 1 h at 23 °C, the reaction was quenched by addition of EtOH (0.10 mL) and saturated NH_4Cl (aq.) (0.05 mL). The color of the solution changed immediately from dark red to light red. The mixture was diluted with toluene (50 mL) and filtered though a pad of silica gel. The red eluate was evaporated to a small volume and precipitation with MeOH afforded a crude product. This mixture was purified by HPLC [Nacalai Tesque, Buckyprep, 250 mm, toluene–2-propanol (5:5)] under an argon atmosphere. The fraction containing **10** was collected and concentrated to a small volume. Precipitation with MeOH afforded **10** (23 mg 59% yield) with 98% HPLC purity as a mixture of five isomers. IR (ReactIR 1000, diamond prove), 3059, 3030, 2960, 2918, 2859, 1598, 1493, 1445, 1241, 1155, 1072, 1031, 921, 911, 748, 731, 714, 685, 660 cm^{-1}; ^{1}H NMR (400 MHz CDCl$_3$), δ 1.80–2.04, 2.34–2.39, 2.46–2.49, 2.60–2.61 and 2.74 (s, 15H, CH$_3$), 4.445, 4.458, 4.591, 4.615 and 4.686 (s, 1H, C$_{60}$-H), 4.991, 5.108, 5.029, 5.034 and 5.063 (s, 1H, C$_{60}$-H), 7.00–7.43, 7.45–7.52, 7.62–7.73, 7.75–7.79 and 8.12–8.15 (m, 15H, C$_6$H$_5$); APCI-MS (negative), m/z = 1027 [(M − 1)$^-$].

Synthesis of the dibenzo-fused corannulene C$_{60}$Me$_5$Ph$_5$H$_2$ (**11**)
To a solution of **5** (15 mg, 12 μmol) in PhCN (4.0 mL) in a Schlenk tube was added a solution of lithium naphthalenide (0.74 mL, 0.50 M in THF, 0.37 mmol). The color of the solution changed immediately from yellow to dark red, indicating decyanation and formation of a bis-cyclopentadienide. After stirring for 1 h at 23 °C, the reaction mixture was quenched by addition of EtOH and then saturated NH$_4$Cl (aq.) (0.10 mL). The color of the resulting solution changed immediately from dark red to light red, indicating that protonation of the bis-cyclopentadienyl dianion had taken place. The mixture was diluted with toluene (20 mL) and filtered though a pad of silica gel. The red eluate was evaporated to a small volume and precipitation with MeOH afforded a crude product. This mixture was purified by HPLC [Nacalai Tesque, Buckyprep, 250 mm, toluene–2-propanol (7:3)]. The fraction containing **11** was collected and concentrated to a small volume. Precipitation with MeOH afforded **11** (9.1 mg, 63% yield) with 95% HPLC purity as a mixture of isomers; IR (ReactIR, diamond prove), 3085, 3058, 3027, 2958, 2919, 2861, 1600, 1584, 1493, 1445, 1372, 1181, 1156, 1136, 1078, 1032, 1003, 928, 754, 694, 675 cm^{-1}; ^{1}H NMR (500 MHz, CDCl$_3$), δ 1.14–1.27, 1.37–1.64 and 2.04–2.68 (s, 15H, CH$_3$), 3.946, 3.974, 3.986, 4.103, 4.126, 4.313, 4.320, 4.557, 4.593 and 4.636 (s, 2H, C$_{60}$-H), 6.92–7.24, 7.31–7.36, 7.39–7.41, 7.42–7.47, 7.51–7.54, 7.57–7.66 and 7.71–7.82 (m, 25H, C$_6$H$_5$); HRMS (APCI, negative), m/z calcd. for C$_{95}$H$_{41}$ [(M − H)$^-$] 1181.3208, found 1181.3256.

References

1. (a) M. Müller, C. Kübel, K. Müllen, Chem. Eur. J. **1998**, *4*, 2099; (b) A. J. Beresheim, M. Müller, K. Müllen, Chem. Rev. **1999**, *99*, 1747; (c) M. D. Watson, A. Fechtenkötter, K. Müllen, Chem. Rev. **2001**, *101*, 1267.
2. H. W. Kroto, J. R. Heath, S. C. O'Brien, R. F. Curl, R. E. Smalley, Nature **1985**, *318*, 162.
3. S. Iijima, Nature **1991**, *354*, 56.
4. C. A. Reed, R. D. Bolskar, Chem. Rev. **2000**, *100*, 1075.
5. D. M. Guldi, P. Neta, K.-D. Asmus, J. Phys. Chem. **1994**, *98*, 4617.
6. (a) A. Sygula, P. W. Rabideau, J. Am. Chem. Soc. **2000**, *122*, 6323; (b) M. M. Boorum, Y. V. Vasil'ev, T. Drewello, L. T. Scott, Science **2001**, *294*, 828; (c) H. Sakurai, T. Daiko, T. Hirao, Science **2003**, *301*, 1878.
7. E. Heilbronner, Helv. Chim. Acta **1954**, *37*, 921.
8. (a) H. S. Choi, K. S. Kim, Angew. Chem. Int. Ed. **1999**, *38*, 2256; (b) K. N. Houk, P. S. Lee, M. Nendel, J. Org. Chem. **2001**, *66*, 5517.
9. (a) A. B. McEwen, P. v. R. Schleyer, J. Org. Chem. **1986**, *51*, 4357; (b) A. A. Folkin, H. Jiao, P. v. R. Schleyer, J. Am. Chem. Soc. **1998**, *120*, 9364.
10. (a) J. Aihara, J. Am. Chem. Soc. **1974**, *98*, 2750; (b) J. Aihara, Bull. Chem. Soc. Jpn. **1975**, *48*, 3637.
11. J. Aihara, J. Chem. Soc., Perkin Trans. 2 **1994**, 971.
12. J. Aihara, J. Am. Chem. Soc. **1992**, *114*, 865.
13. L. Türker, J. Mol. Struct. (Theochem) **1999**, *491*, 275.
14. (a) P. R. Ashton, G. R. Brown, N. S. Isaacs, D. Giuffrida, F. H. Kohnke, J. R. Mathias, A. M. Z. Slawin, D. R. Smith, J. F. Stoddart, D. J. Williams, J. Am. Chem. Soc. **1992**, *114*, 6330; (b) U. Girreser, D. Giuffrida, F. H. Kohnke, J. R. Mathias, D. Philp, J. F. Stoddart, Pure Appl. Chem. **1993**, *65*, 119.
15. A. Godt, V. Enkelmann, A. D. Schlüter, Angew. Chem. Int. Ed. **1989**, *28*, 1680.
16. (a) Y. Kuwatani, T. Yoshida; A. Kusaka; M. Iyoda, Tetrahedron Lett. **2000**, *41*, 359; (b) Y. Kuwatani, J. Igarashi, M. Iyoda, Tetrahedron Lett. **2004**, *45*, 359; (c) Y. Kuwatani, New Trends Struct. Org. Chem. **2005**, 155.
17. Y. Kuwatani, Annual Meeting of the Chemical Society of Japan, Tokyo, Abstr. No. 4H1-16, March **2002**.
18. H. M. St. Martin, L. T. Scott, 10th International Symposium on Novel Aromatic Compounds, San Diego, CA, Abstr. No. P77, August **2001**.
19. (a) M. Sawamura, H. Iikura, E. Nakamura, J. Am. Chem. Soc. **1996**, *118*, 12850; (b) H. Iikura, S. Mori, M. Sawamura, E. Nakamura, J. Org. Chem. **1997**, *62*, 7912; (c) M. Sawamura, M. Toganoh, Y. Kuninobu, S. Kato, E. Nakamura, Chem. Lett. **2000**, *29*, 270; (d) M. Sawamura, H. Iikura, T. Ohama, U. E. Hackler, E. Nakamura, J. Organomet. Chem. **2000**, *599*, 32; (e) E. Nakamura, M. Sawamura, Pure Appl. Chem. **2001**, *73*, 355.
20. (a) P. J. Krusic, E. Wassermann, P. N. Keizer, J. M. Morton, K. F. Preston, Science **1991**, *254*, 1183; (b) P. R. Birkett, P. B. Hitchcock, H. W. Kroto, R. Taylor, D. R. M. Walton, Nature **1992**, *357*, 479; (c) P. R. Birkett, A. G. Avent, A. D. Darwisch, H. W. Kroto, R. Taylor, D. R. M. Walton J. Chem. Soc., Chem. Commun. **1993**, 1230; (d) A. G. Avent, P. R. Birkett, J. D. Crane, A. D. Darwish, G. J. Langley, H. W. Kroto, R. Taylor, D. R. M. Walton, J. Chem. Soc., Chem. Commun. **1994**, 1463.
21. (a) Y. Rubin, Chem. Eur. J. **1997**, *3*, 1009; (b) Y. Rubin, Top. Curr. Chem. **1999**, *199*, 67; (c) Y. Rubin, F. Diederich, in Stimulaing Concepts in Chemistry, F. Vögtle, J. F. Stoddart, M. Shibasaki (eds.), Wiley-VCH, Weinheim, **2000**, 163–186; (d) G. Schick, T. Jarrosson, Y. Rubin, Angew. Chem., Int. Ed. **1999**, *38*, 2360; (e) Y. Rubin, T. Jarrosson, G.-W. Wang, M. D. Bartberger, K. N. Houk, G. Schick, M. Saunders, R. J. Cross, Angew. Chem., Int. Ed. **2001**, *40*, 1543; (f) S. Irle, Y. Rubin, K. Morokuma, J. Phys. Chem. A **2002**, *106*, 680.
22. (a) Y. Murata, M. Murata, K. Komatsu, Chem. Eur. J. **2003**, *9*, 1600; (b) Y. Murata, M. Murata, K. Komatsu, J. Am.

Chem. Soc. **2003**, *125*, 7152; (c) K. Komatsu, M. Murata, Y. Murata, *Science* **2005**, *307*, 238.
23. (a) S.-i. Iwamatsu, T. Uozaki, K. Kobayashi, S. Re, S. Nagase, S. Murata, *J. Am. Chem. Soc.* **2004**, *126*, 2669; (b) S.-i. Iwamatsu, S. Murata, Y. Andoh, M. Minoura, K. Kobayashi, N. Mizorogi, S. Nagase, *J. Org. Chem.* **2005**, *70*, 4820; (c) S.-i. Iwamatsu, S. Murata, *Synlett* **2005**, 2117.
24. T. Cantenewala, P. A. Padmawar, L. Y. Chiang, *J. Am. Chem. Soc.* **2005**, *127*, 26.
25. (a) X.-W. Wei, A. D. Darwish, O. V. Boltalina, P. B. Hitchcock, J. M. Street, R. Taylor, *Angew. Chem., Int. Ed.*, **2001**, *40*, 2989; (b) G. A. Burley, P. W. Fowler, A. Soncini, J. P. B. Sandall, R. Taylor, *Chem. Commun.* **2003**, 3042.
26. P. A. Troshin, R. N. Lyubovskaya, I. N. Ioffe, N. B. Shustova, E. Kemnitz, S. I. Troyanov, *Angew. Chem. Int. Ed.* **2005**, *44*, 234.
27. (a) E. Nakamura, K. Tahara, Y. Matsuo, M. Sawamura, *J. Am. Chem. Soc.* **2003**, *125*, 2834; (b) Y. Matsuo, K. Tahara, E. Nakamura, *Org. Lett.* **2003**, *5*, 3181; (c) Y. Matsuo, K. Tahara, M. Sawamura, E. Nakamura, *J. Am. Chem. Soc.* **2004**, *126*, 8725.
28. (a) L. T. Scott, *Angew. Chem. Int. Ed.* **2003**, *42*, 4133; (b) M. Randic, *Chem. Rev.* **2003**, *103*, 3449; (c) R. Hoffmann, *Int. J. Phil. Chem.* **2003**, *9*, 7.
29. H. Al-Matar, P. B. Hitchcock, A. G. Avent, R. Taylor, *Chem. Commun.* **2000**, 1071.
30. Y. Kuninobu, Y. Matsuo, M. Toganoh, M. Sawamura, E. Nakamura, *Organometallics* **2004**, *23*, 3259.
31. S. Liu, Y. Lu, M. M. Kappes, J. A. Ibers, *Science* **1991**, *254*, 408.
32. K. Kveseth, R. Seip, D. A. Kohl, *Acta Chem. Scand.* **1980**, *A34*, 31.
33. (a) K. Tanaka, H. Ago, T. Yamabe, K. Okahara, M. Okada, *Int. J. Quantum Chem.* **1997**, *63*, 637; (b) A. Rochefort, D. R. Salahub, P. Avouris, *J. Phys. Chem. B* **1999**, *103*, 641; (c) T. Sato, M. Tanaka, T. Yamabe, *Synth. Met.* **1999**, *103*, 2525; (d) T. Yamabe, M. Imade, M. Tanaka, T. Sato, *Synth. Met.* **2001**, *117*, 61; (e) L. Liu, C. S. Jayanthi, H. Guo, S. Y. Wu, *Phys. Rev. B* **2001**, *64*, 033414; (f) K. Kanamitsu, S. Saito, *J. Phys. Soc. Jpn.* **2002**, *71*, 483; (g) J. Cioslowski, N. Rao, D. J. Moncrieff, *Am. Chem. Soc.* **2002**, *124*, 8489.
34. W. C. Herndon, *J. Am. Chem. Soc.* **1974**, *96*, 7605.
35. (a) P. v. R. Schleyer, C. Maerker, A. Dransfeld, H. Jiao, N. J. R. van E. Hommes, *J. Am. Chem. Soc.* **1996**, *118*, 6317; (b) M. Bühl, A. Hirsch, *Chem. Rev.* **2001**, *101*, 1153.
36. M. Saunders, H. A. Jimenez-Vazquez, R. J. Cross, S. Mroczkowski, D. I. Freedberg, F. A. L. Anet, *Nature* **1994**, *367*, 256.
37. P. W. Rabideau, A. Sygula, *Acc. Chem. Res.* **1996**, *29*, 235.
38. (a) P. W. Rabideau, A. H. Abdourazak, H. E. Folsom, Z. Marcinow, A. Sygula, R. Sygula, *J. Am. Chem. Soc.* **1994**, *116*, 7891; (b) A. Sygula, P. W. Rabideau, *J. Am. Chem. Soc.* **1999**, *121*, 7800; (c) S. Hagen, M. S. Bratcher, M. S. Erickson, G. Zimmermann, L. T. Scott, *Angew. Chem. Int. Ed. Engl.* **1997**, *36*, 406; (d) G. Mehta, G. Panda, *Chem. Commun.* **1997**, 2081; (e) A. Sygula, Z. Marcinow, F. R. Fronczek, I. Guzei, P. W. Rabideau, *Chem. Commun.* **2000**, 2439.
39. M. Toganoh, Y. Matsuo, E. Nakamura, *J. Organomet. Chem.* **2003**, *683*, 295.
40. (a) R. C. Haddon, L. T. Scott, *Pure Appl. Chem.* **1986**, *58*, 137; (b) R. C. Haddon, *Acc. Chem. Res.* **1988**, *21*, 243; (c) R. C. Haddon, *Science* **1993**, *261*, 1545.
41. (a) D. Kim, M. Lee, *J. Am. Chem. Soc.* **1992**, 114, 4429; (b) M. Lee, O.-K. Song, J.-C. Seo, D. Kim, Y. D. Suh, S. M. Jin, S. K. Kim, *Chem. Phys. Lett.* **1992**, *196*, 325.
42. (a) Y.-P. Sun, G. E. Lawson, J. E. Riggs, B. Ma, N. Wang, D. K. Moton, *J. Phys. Chem. A* **1998**, *102*, 5520; (b) C. Luo, M. Fujitsuka, A. Watanabe, O. Ito, L. Gan, Y. Huang, C. H. Huang, *J. Chem. Soc., Faraday Trans.* **1998**, *94*, 527; (c) R. M. Williams, J. M. Zwier, J. W. Verhoeven, *J. Am. Chem. Soc.* **1995**, *117*, 4093; (d) S. K. Lin, L. L. Shiu, K. M. Chien, T. Y. Luh, T. I. Lin, *J. Phys. Chem.* **1995**, *99*, 105; (e) Y. Matsubara, H. Muraoka, H. Tada, Z. Yoshida, *Chem. Lett.* **1996**, 373.

43. G. Schick, M. Levitus, L. Kvetko, B. A. Johnson, I. Lamparth, R. Lunkwitz, B. Ma, S. I. Khan, M. A. Garcia-Garibay, Y. Rubin, *J. Am. Chem. Soc.* **1999**, *121*, 3246.
44. H. Isobe, H. Mashima, H. Yorimitsu, E. Nakamura, *Org. Lett.* **2003**, *5*, 4461.
45. R. Hamasaki, Y. Matsuo, E. Nakamura, *Chem. Lett.* **2004**, *33*, 328.
46. M. J. O'Connell, S. M. Bachilo, C. B. Huffman, V. C. Moore, M. S. Strano, E. H. Haroz, K. L. Rialon, P. J. Boul, W. H. Noon, C. Kittrell, J. Ma, R. H. Hauge, R. B. Weisman, R. E. Smalley, *Science* **2002**, *297*, 593.
47. N. Nagasawa, H. Sugiyama, N. Naka, I. Kudryashov., M. Watanabe, T. Hayashi, I. Bozovic, N. Bozovic, G. Li, Z. Li, Z. K. Tang, *J. Lumin.* **2002**, *97*, 161.
48. (a) M. Sawamura, Y. Kuninobu, E. Nakamura, *J. Am. Chem. Soc.* **2000**, *122*, 12407; (b) M. Sawamura, Y. Kuninobu, M. Toganoh, Y. Matsuo, M. Yamanaka, E. Nakamura, *J. Am. Chem. Soc.* **2002**, *124*, 9354; (c) Y. Matsuo, E. Nakamura, *Organometallics* **2003**, *22*, 2554; (d) M. Toganoh, Y. Matsuo, E. Nakamura, *Angew. Chem. Int. Ed.* **2003**, *42*, 3530; (e) M. Toganoh, Y. Matsuo, E. Nakamura, *J. Am. Chem. Soc.* **2003**, *125*, 13974; (f) Y. Matsuo, Y. Kuninobu, S. Ito, E. Nakamura, *Chem. Lett.* **2004**, *33*, 68; (g) Y. Matsuo, A. Iwashita, E. Nakamura, *Organometallics* **2005**, *24*, 89; (h) Y. Matsuo, E. Nakamura, *J. Am. Chem. Soc.* **2005**, *127*, 8457; (i) R. H. Herber, I. Nowik, Y. Matsuo, M. Toganoh, Y. Kuninobu, E. Nakamura, *Inorg. Chem.* **2005**, *44*, 5629; (j) Y. Matsuo, H. Isobe, T. Tanaka, Y. Murata, M. Murata, K. Komatsu, E. Nakamura, *J. Am. Chem. Soc.* **2005**, *127*, 17148.
49. (a) M. Sawamura, K. Kawai, Y. Matsuo, K. Kanie, T. Kato, E. Nakamura, *Nature* **2002**, *419*, 702; (b) Y. Matsuo, A. Muramatsu, R. Hamasaki, N. Mizoshita, T. Kato, E. Nakamura, *J. Am. Chem. Soc.* **2004**, *126*, 432.
50. E. Nakamura, H. Isobe, *Acc. Chem. Res.* **2003**, *36*, 807.
51. (a) A. Hirsch, *Angew. Chem. Int. Ed.* **2002**, *41*, 1853; (b) J. L. Bahr, J. M. Tour, *J. Mater. Chem.* **2002**, *12*, 1952; (c) S. Niyogi, M. A. Hamon, H. Hu. B. Zhao, P. Bhowmik, R. Sen, M. E. Itkis, R. C. Haddon, *Acc. Chem. Res.* **2002**, *35*, 1105; (d) J. Liu, A. G. Rinzler, H. Dai, J. H. Hafner, R. K. Bradley, P. J. Boul, A. Lu, T. Iverson, K. Shelimov, C. B. Huffman, F. Rodriguez-Macias, Y.-S. Shon, T. R. Lee, D. T. Colbert, R. E. Smalley, *Science* **1998**, *280*, 1253.
52. (a) J. M. Kliegman, R. K. Barnes, *Tetrahedron* **1970**, *26*, 2555; (b) J. M. Kliegman, R. K. Barnes, *J. Org. Chem.* **1970**, *35*, 3140; (c) G. van Koten, K. Vrieze, *Adv. Organomet. Chem.* **1982**, *21*, 151.

Part II
Strategic Advances in Chromophore and Materials Synthesis

3
Cruciform π-Conjugated Oligomers

Frank Galbrecht, Torsten W. Bünnagel, Askin Bilge, Ullrich Scherf and Tony Farrell

3.1
Introduction

Research into the design, characterization and application of organic semiconductors (OSCs) has intensified in the last decade. So much so that it has attracted significant support from industry, which has resulted in their expeditious arrival onto the market. Especially as components of large, flat-panel electroluminescent displays. Attractive aspects of such OSCs are the possibilities of manipulating the optical, electronic and morphological properties via chemical modification and the straightforward use of liquid-phase processing techniques such as inkjet printing or spin coating. The last feature should in principle allow for the deposition of organic layers on large, flexible substrates in a cost-effective manner. A recent review by Malliaras and Friend [1] offers a concise overview of the scope and limitations in the application of OSCs and a report by Forrest [2] gives detailed insight into the processing techniques used to realize OSC-based devices [2–8].

In addition to the crucial issues concerning production costs and long-term stability, color purity and device efficiency are the other major parameters concerning the applicability of light-emitting organic compounds in commercial devices. While the color of an organic light-emitting diode (LED) is closely related to the electronic structure of the emitting electronic system, the device efficiency is determined by the complicated interplay between various physical processes [charge transport, recombination efficiency, electroluminescence (EL) quantum yield, competing photophysical processes, etc.].

It has recently become clear that the subject of nanoscopic and macroscopic order in π-conjugated systems is extremely important, as it is the solid-state morphology that often determines the efficiency of electronic or optoelectronic devices. Current findings show that it can be advantageous to apply amorphous materials, e.g. for organic light-emitting diode (OLED) applications. Aggregation or excimer formation represent a serious drawback in organic-based LEDs and solid-state laser applications as they tend to promote nonradiative recombination processes [9]. Conversely, efficient intermolecular interaction can greatly enhance

Functional Organic Materials. Syntheses, Strategies, and Applications.
Edited by Thomas J.J. Müller and Uwe H.F. Bunz
Copyright © 2007 WILEY-VCH Verlag GmbH & Co. KGaA, Weinheim
ISBN: 978-3-527-31302-0

the charge carrier mobility. Accordingly, polymers and oligomers that encourage substantial intermolecular π–π overlap are often excellent charge-transporting components for organic field effect transistors (OFETs) [10].

Moreover, of crucial importance to the fabrication of organic based bulk heterojunction solar cells is the requirement for bicontinuous partitioning of the donor and acceptor phases to allow effective charge separation and transport to the electrodes [4]. Solution processing of donor–acceptor blends can lead to a controlled demixing of the components to form an interpenetrating network of the two phases on a nanometer scale. Therefore, controlling the spatial assembly and packing of both oligomers and polymers in the solid state is a worthwhile, although non-trivial, endeavor. This is dependent on the proficiency of the chemist to design and prepare structurally defined π-architectures and correlate the solid-state morphology with chemical structure.

As mentioned previously, organic π-conjugated materials are expected to be viable components in a diverse range of electronic and optoelectronic devices, including thin-film transistors (OFETs) [5, 8], electroluminescent diodes (OLEDs) [3, 6], lasers [7] and organic photovoltaic cells [4]. Phenylene-based oligomers and polymers are an intensively studied class of conjugated materials owing to their excellent thermal and chemical stability [11]. Furthermore, their optical properties make them especially promising candidates as efficient blue emitters in electroluminescent devices and organic solid-state lasers [12, 13]. Although initial investigations concentrated on the design of soluble polymeric materials, of late research using the oligomer approach has also intensified [14, 15]. Unfortunately, with increasing chain length of linear, unsubstituted oligophenyls their solubility decreases dramatically. On the other hand, an advantage of oligomers is that standard purification techniques can be applied for the facile removal of impurities and byproducts. Furthermore, knowledge of the exact constitution and conformation of oligomers allows precise structure–property correlations to be deduced.

The subject matter chosen for this chapter concerns the synthesis and characterization of branched oligomers containing nonlinear building blocks, namely compounds that form a cruciform or cross-like assemblage [16]. According to Shirota, disrupting the capacity of a material to planarize (and crystallize) enhances the likelihood of glass formation and the solubility [9]. Moreover, a shape that discourages crystallization and prevents strong intermolecular packing should in general enhance the solubility. Representative topologies of such molecular glasses include starburst structures [17–21], spiro-type oligomers [22–25] and oligomers with a tetraphenylmethane core [26–28]. The chapter is divided into three sections classifying the cruciforms according to the constitution of the core. To limit the length of the review, we have purposely excluded dealing with spiro-type cruciforms [29, 30].

3.2
Oligomers with a Tetrahedral Core Unit

The first type of cruciform or cross-like material that we present is based on the so-called "tetrahedral approach". The motivation for constructing π-conjugated materials using such tetrahedral cores (e.g. tetraarylmethane or -silane building blocks) stems from the perception that a bulky, sterically hindered architecture should be less prone to self-aggregation in the solid state and hence should lead to more efficient and morphologically stable photo- and electroluminescence.

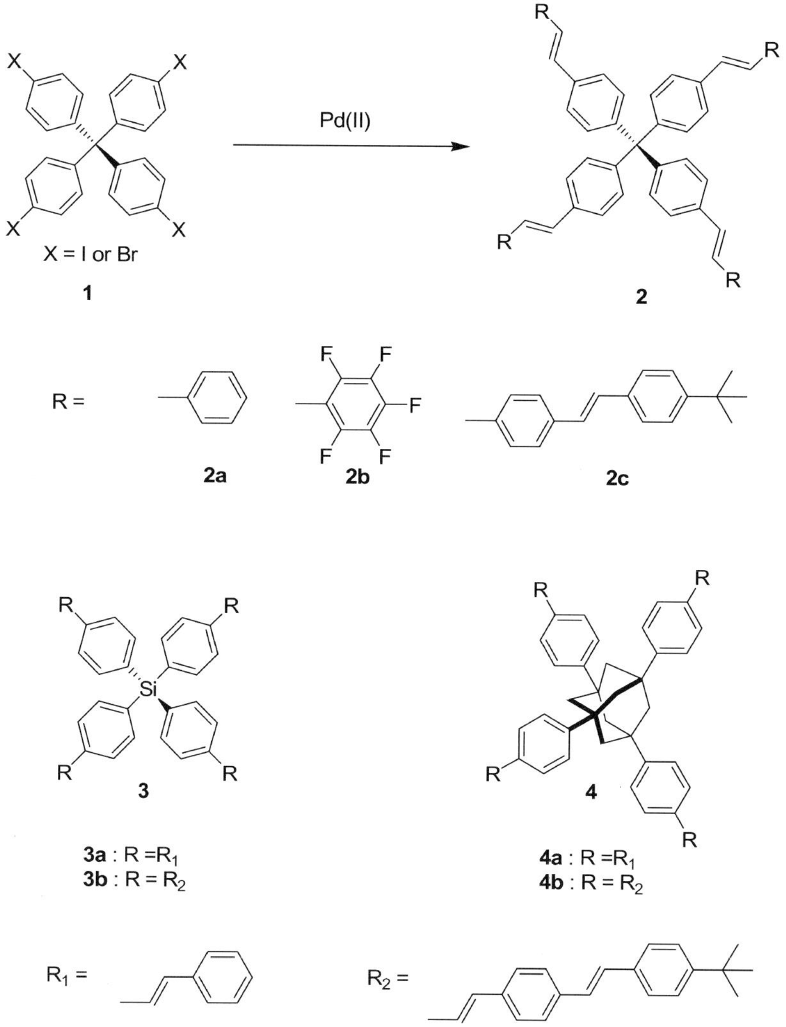

Scheme 3.1 Synthesis of cruciform tetrahedral oligomers according to Bazan and coworkers [31].

In 1998, Bazan and coworkers [31] synthesized a series of molecules containing four stilbenoid units arranged around a central sp^3-hybridized core. The molecules 2a–c were prepared via a palladium-catalyzed Heck coupling reaction as shown in Scheme 3.1. The optimal conditions for the coupling reaction to 2a was starting from the tetraiodoaryl educt 1 using $Pd(OAc)_2$ as the catalyst in the presence of an excess of styrene. However, only trace amounts of 2b and c were obtained when using this common $Pd(OAc)_2/P(o\text{-}tol)_3$ system. Furthermore, the reaction of the corresponding tetrabromoaryl derivative under analogous conditions also gave low yields.

Although it was possible to obtain the longer side-chain derivative 2c in moderate yield and good purity, Bazan's group decided to embark on a more detailed screening of the experimental conditions in order to prepare further molecules in the series [32]. By using the thermally stable palladacycle shown in Scheme 3.2 at elevated temperatures it was possible to improve the yield of 2c from 30 to 50%. These improved conditions were then used to prepare the tetraphenyladamantane and tetraphenylsilane oligomers 3a,b and 4a,b depicted in Scheme 3.1.

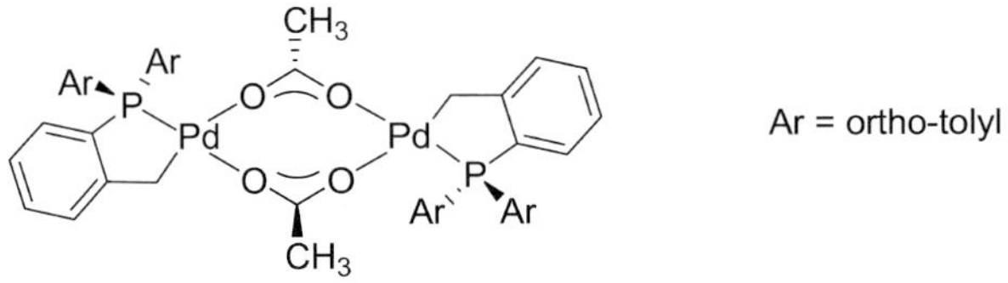

Scheme 3.2 Structure of a palladacycle used in the synthesis of the derivatives of 2, 3 and 4.

Suzuki-type coupling protocols were also evaluated as a method to prepare some related tetrahedral stilbenoid molecules and the optimal yields were achieved when using $Pd(dppf)Cl_2$ as the catalyst. The cruciform materials prepared using this strategy are shown in Scheme 3.3. Compounds 5a and b were obtained as a mixture of cis and trans isomers as identified by high-performance liquid chromatographic (HPLC) analysis. Nevertheless, the mixtures were readily converted into the all-trans isomers by irradiating the dissolved samples with a mercury lamp.

Bazan and co-workers intensively investigated structure–property relationships in the above series of tetrahedral oligophenylenevinylene materials. First, the optical properties of 2a and c were compared with those of stilbene and distyrylbenzene, which represent the individual π-conjugated "arms" of the tetrahedral array. The tetrastilbenoid oligomers 2a and c exhibit only a slight red shift in their optical spectra compared with the linear model compounds stilbene and distyrylbenzene, which was interpreted as a consequence of a weak π-delocalization through the sp^3-carbon centers. The thermal properties of the cruciform molecules were probed by differential scanning calorimetric (DSC) analysis. The smaller tetrahedral molecules 2a, 3a and 4a were all crystalline in nature with distinct melting

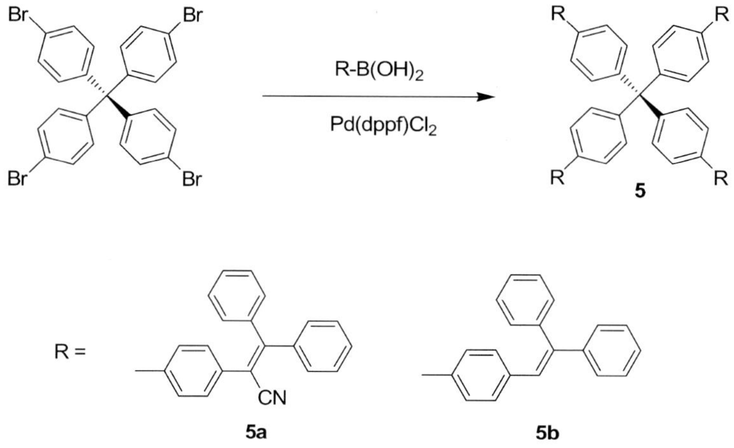

3.2 Oligomers with a Tetrahedral Core Unit

Scheme 3.3 Tetrahedral oligomers prepared in a Suzuki-type coupling, after Bazan and coworkers [32].

endotherms at 274, 179 and 306 °C, respectively. By contrast, in the case of the extended molecule **2c**, only a glass transition (T_g) at 175 °C was observed. Curiously, the DSC plot from a sample of all-*trans*-**2c** obtained after irradiation with a mercury lamp shows a distinctly higher T_g of 190 °C. This finding was attributed to the presence of the cis-linkages to give **2c** as a mixed isomeric sample. One can conclude, therefore, that also the isomeric purity is an important factor in determining the solid-state properties.

The tetrahedral arrangement generally reduces the interchain interactions, which leads to a reduced or suppressed crystallinity (Table 3.1). For example,

Table 3.1 Glass transition and melting temperatures of tetrahedral cruciform oligomers, according to Bazan and coworkers [32].

Compound	T_g (°C)	T_m (°C)
2a		274
3a	81	179
4a		306
2c	190	
3b	191	
4b	165	
5b	142	
5a	174	
6a	181	
7		320

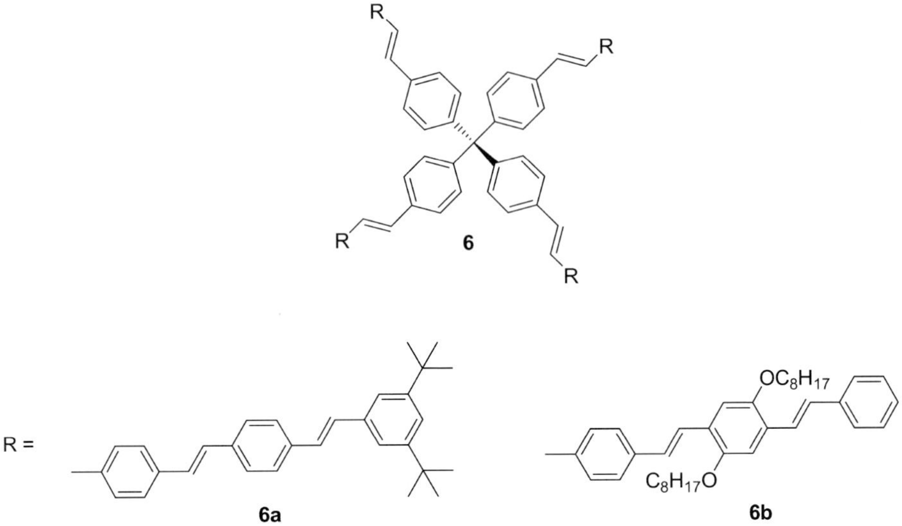

Scheme 3.4 Extended tetrahedral oligomers, after Bazan and coworkers [32].

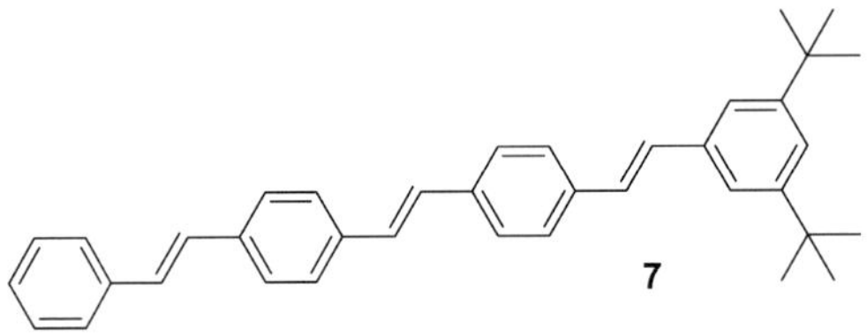

Scheme 3.5 Structure of the linear model compound **7**, after Bazan and coworkers [32].

the linear model compound **7** (Scheme 3.5) exhibits a melting transition at 320 °C whereas the corresponding oligomer **6a** (Scheme 3.4) only displays a glass transition at 181 °C with no observable melting transition, indicating an amorphous solid-state structure for **6a**.

In 2001, Chen and coworkers prepared another interesting series of tetraphenylmethane-based tetrahedral compounds. The initial step of the synthetic scheme involves the preparation of the tetrazole derivative **8** from the corresponding tetranitrile and its subsequent condensation with various benzoyl chlorides to give the cruciforms **9–11** depicted in Scheme 3.6. A modified procedure was developed to obtain the bipolar tetramers **16** and **17**, as shown in Scheme 3.7. The acyl chloride functional group was this time attached to the tetrahedral core unit and the tetrazole function was attached to the triphenylamine component. Once again DSC measurements illustrated the effectiveness of this architecture in inducing a stable glassy state in these cruciform materials.

The cruciforms **9**, **10** and **11** exhibit glass transitions at 175, 97 and 125 °C, respectively (Table 3.2). The three-dimensional arrangement of **9–11** also raised

Table 3.2 Thermal properties of **9**, **10** and **11** in comparison with PBD **12**, according to Chen and coworkers [33].

Compound	T_g (°C)	T_m (°C)
9	175	400
10	97	337
11	125	270
12	n.o.[a]	137

[a] Not observed.

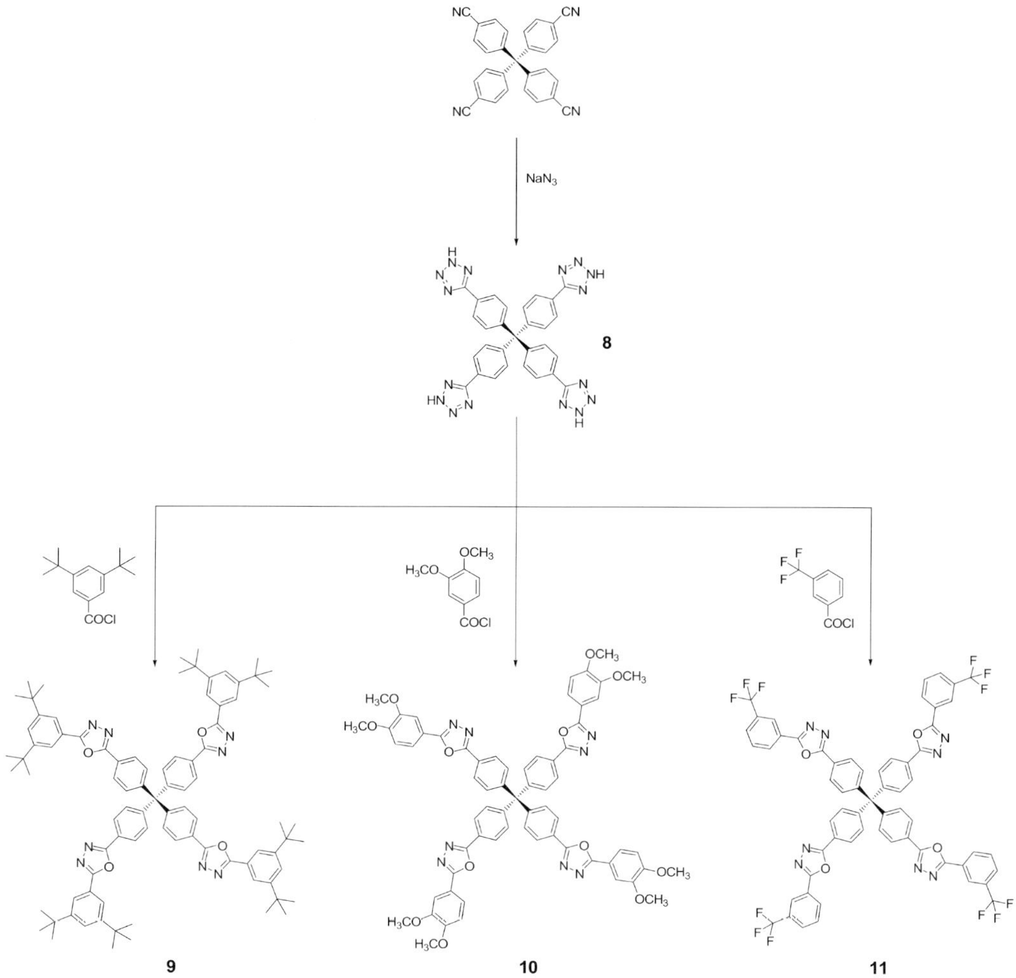

Scheme 3.6 Synthesis of tetrahedral diaryloxadiazole cruciforms according to Chen and coworkers [33].

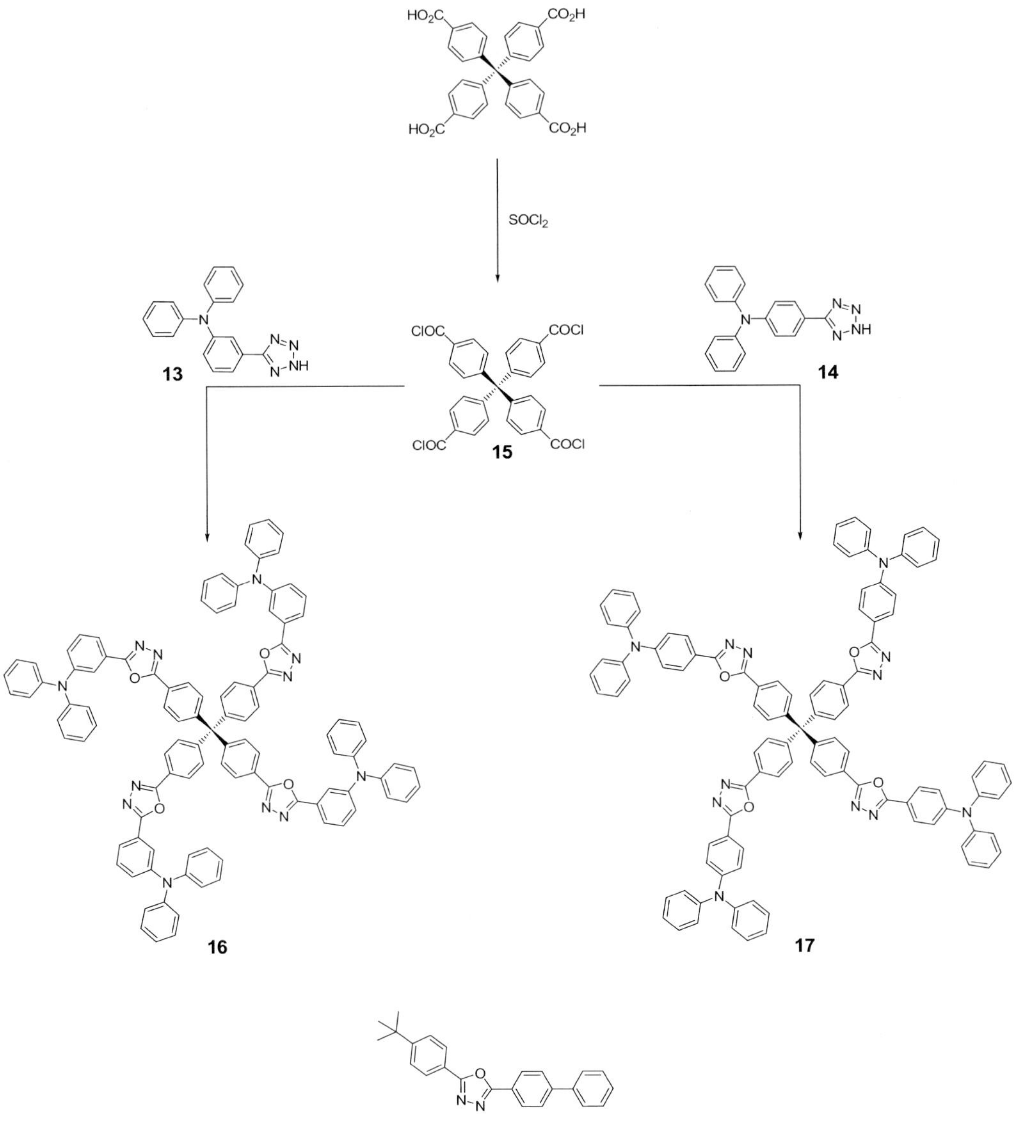

Scheme 3.7 Synthesis of the bipolar tetramers **16** and **17** according to Chen and coworkers [33]. For comparison, the structure of PBD **12** is also given.

their melting points relative to the linear counterpart 2-(4-biphenyl)-5-(4-*tert*-butylphenyl)-1,3,4-oxadiazole (PBD **12**). High T_g's (187–190 °C) were also observed for the bipolar molecules **16** and **17** with no indication of crystallinity [33].

3.3
Oligomers with a Tetrasubstituted Benzene Core

The next type of cruciform π-system to be discussed is constructed by replacing the central tetrahedral core described in Section 3.2 by a 1,2,4,5-tetrasubstituted benzene, leading to a rigid, two-dimensional arrangement of the four arms. Hence rotation is only possible within the arms. The synthetic scheme utilized by Bunz and coworkers for the preparation of cruciform phenylenevinylene/phenyleneethynylene oligomers is a two-step process as summarized in Scheme 3.8 [34]. The initial step is a Horner-type olefination of **18a–g** to give the diiododistyrylbenzene derivatives **19a–g**. A subsequent Sonogashira-type coupling between these diiodoaryls and a series of monosubstituted alkynes utilizing $(PPh_3)_2PdCl_2$ and CuI with piperidine or NEt_3 as base gave the desired cruciform phenyleneethynylene/phenylenevinylene oligomers **20a–g** in good yields of 50–80%.

The absorption and emission data for the cruciforms in solution were unremarkable. The absorption (λ_{max}) values generally follow the trends expected from the different substitution patterns. However, a more in-depth examination led to the finding that the solid-state optical properties vary substantially with the solid-state morphology (Table 3.3). The emissions of the samples **20d–g** were examined in the crystalline state and in thin films cast from $CHCl_3$ solutions, which were found to be amorphous. Samples **20d, f** and **g** exhibited a bathochromic shift of the emission maximum of ~20 nm of $\lambda_{max,\ em}$ on going from the amorphous (thin film) to the crystalline state whereas in **20e** a larger

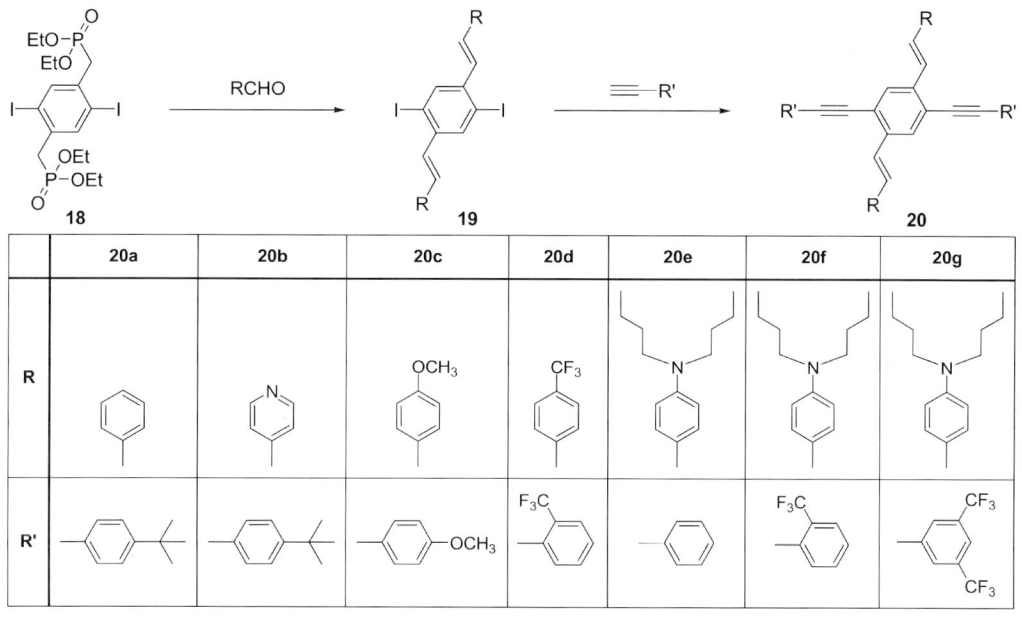

Scheme 3.8 Synthesis of cruciform phenyleneethynylene/phenylenevinylene oligomers, according to Bunz and coworkers [34].

Table 3.3 Optical properties of cruciforms **20d–g**, according to Bunz and coworkers [34] (all values in nm).

Compound	$\lambda_{max, abs}$ (film)	$\lambda_{max, em}$ (film)]	$\lambda_{max, em}$ (crystal)	$\lambda_{max, em}$ (solution)
20d	332	484	504	434 (CHCl$_3$)
20e	335, 440	560	601	498 (C$_6$H$_{14}$)
20f	340, 455	574	596	502 (C$_6$H$_{14}$)
20g	335, 425	586	607	579 (THF)

40-nm red shift was observed. The crystalline samples were thermally cycled (melting/cooling) and their optical spectra and X-ray diffraction (XRD) patterns recorded. Examination of the powder XRD patterns revealed that **20e** forms a glassy state after melting/cooling the sample. The loss of crystallinity was accom-

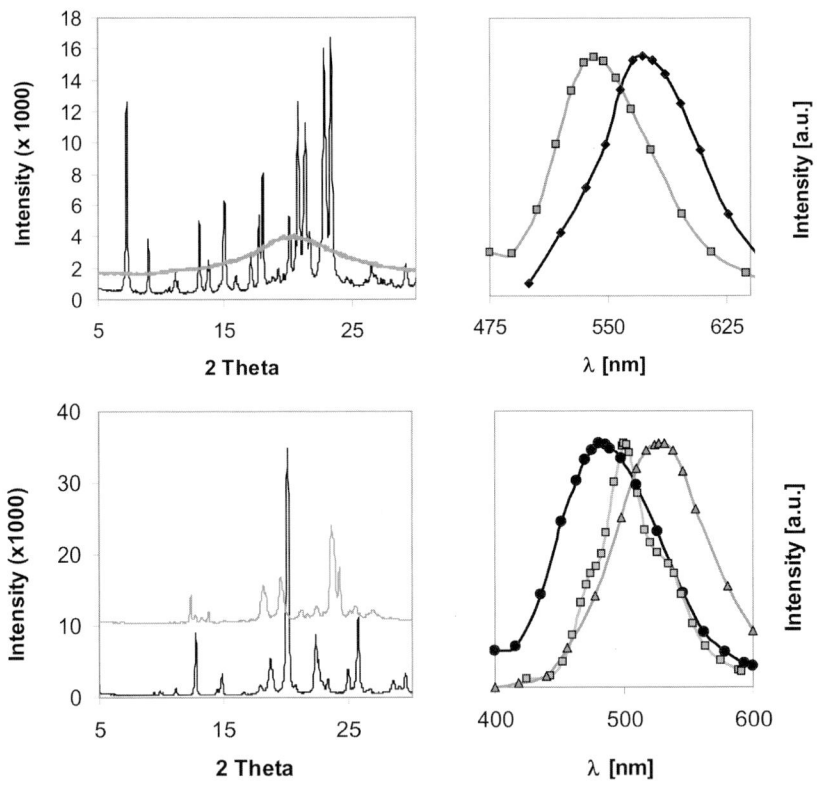

Fig. 3.1 XRD patterns and PL spectra of **20e** (top) with the crystalline state (top line/left picture: left line) and the amorphous state after melting/cooling; and of **20d** (bottom) with the crystalline state after melting/cooling (left picture: top line, right picture: right line) and the PL spectrum of an amorphous film, according to Bunz and coworkers [34].

panied by a 25-nm hypsochromic shift of $\lambda_{max, em}$. Interestingly, for **20d** a second crystalline state was observed after thermal cycling. The differences between the two crystalline states can be easily observed upon comparing the XRD patterns (Fig. 3.1). The three solid-state morphologies of **20d**, two crystalline states and one glassy phase, exhibit three different photoluminescence (PL) spectra as also shown in Fig. 3.1. The amorphous, glassy film of **20d** gave a $\lambda_{max, em}$ at 484 nm whereas the crystalline states emit with $\lambda_{max, em}$ values at 504 and 526 nm, respectively.

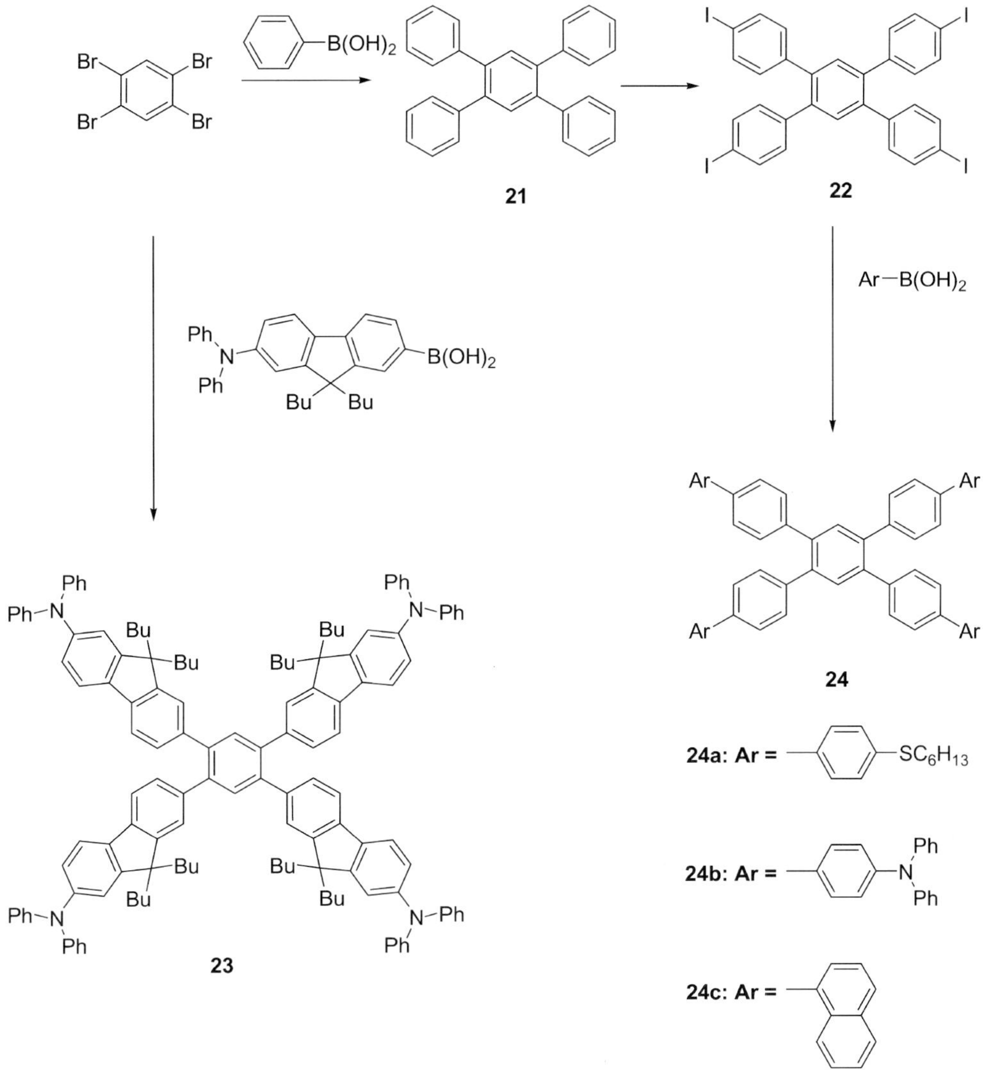

Scheme 3.9 Synthesis of cruciforms based on a 1,2,4,5-tetra-arylbenzene core, according to Li et al. [35].

3 Cruciform π-Conjugated Oligomers

Li et al. prepared a further series of cross-shaped oligoarylenes with a 1,2,4,5-tetraarylbenzene core and the thermal and optical properties of the cross-like architectures **23** and **24** were compared with those of their linear counterparts. The oligomers were synthesized via a Suzuki-type cross-coupling in the presence of a Pd(OAc)$_2$/P(o-tol)$_3$ catalyst system (Scheme 3.9). The tetraiodination of 1,2,4,5-tetraphenylbenzene **21** to **22** using HIO$_4$/I$_2$ was achieved with very good yields (80–90%). The extension of the cruciform molecule under Suzuki-type conditions gave the cruciforms **24a–c**. Interestingly, the direct reaction of 1,2,4,5-tetrabromobenzene with the relevant arylboronic acids led to **24a–c** in much lower yields. The exception here was the reaction of 1,2,4,5-tetrabromobenzene with 9,9-bis(n-butyl)-2-diphenylamino-7-fluorenylboronic acid in which **23** was isolated in a very good yield (77%).

The oligomers **23** and **24a–c** show enhanced solubility and increased glass transition temperatures when compared to their linear counterparts **25** and **26** (Scheme 3.10). No melting transition is observed for **23** whereas the linear molecule **25** (Scheme 3.10) melts at 254°C. The electronic absorption spectra of the cruciforms **23** and **24a–c** exhibit structureless, broad absorption peaks. Assuming a nonplanar ground state, the also featureless emission spectra indicate no significant increase of planarity upon photoexcitation. In addition, the cruciform molecules **23** and **24a–c** show lower photoluminescence quantum yields (41–70%)

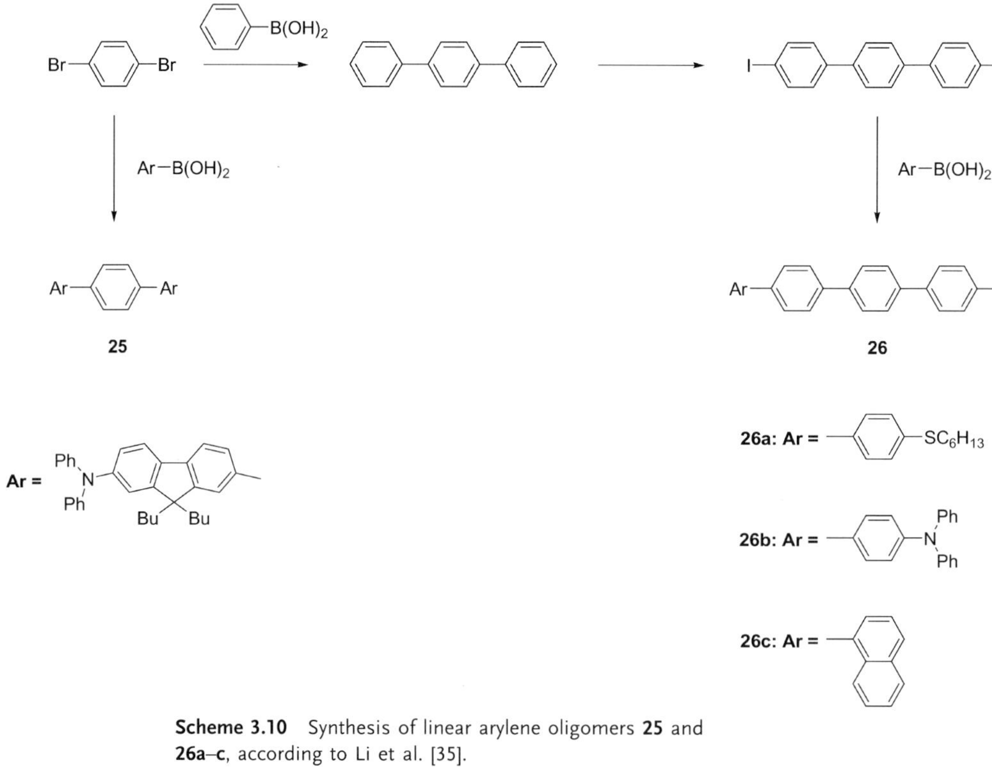

Scheme 3.10 Synthesis of linear arylene oligomers **25** and **26a–c**, according to Li et al. [35].

than their linear relatives **25** and **26a–c** (77–99 %). This finding is no doubt a consequence of a more distorted excited-state geometry.

3.4
Oligomers with a Tetrasubstituted Biaryl Core

In this section, we describe another attempt to manipulate the three-dimensional arrangement of conjugated oligomers by utilizing swivel-cruciform-type building blocks (based on binaphthyl, biphenyl or bithienyl cores) as depicted in Scheme 3.11. The main difference between these cross-like structures and the previous cruciform-type molecules is that in the swivel version there is, at least in principle, the possibility of rotation between the arms, leading to increased conformational freedom. Again, the primary motivation to deal with this configuration was to utilize these swivel-cruciform cores to generate nonplanar, noninteracting oligomers and enhance the likelihood of glass formation [9]. However, from our own work it quickly became clear that by manipulating the arm components and the pendant groups on these swivel-cruciforms it was possible in some cases to introduce a certain degree of intra- or intermolecular order biased by π–π interactions.

The swivel-cruciforms with binaphthyl cores will be described first. The attraction of the binaphthyl unit is manifold. Binaphthol has a versatile and well-developed substitution chemistry, which provides different functionalization patterns for the attachment of substituents [15]. This allows some tuning of the electronic and optical properties of the resulting conjugated oligomers. Moreover, the dihedral angle between the two adjacent naphthalene rings can range from 60 to 120° depending on the substitution pattern [36, 37], and this distorted nature of the binaphthyl unit should simultaneously facilitate the formation of a glassy amorphous state [38]. Additionally, enantiomerically pure binaphthol precursors are commercially available and conjugated oligomers with specific chiro-optical properties offer another intriguing facet to their potential. Applications being imagined for these chiral materials are, for example, light sources based on organics which directly emit circularly polarized light [39–41]. Such materials could be used in areas such as optical data storage and as background illumination for liquid crystal displays (LCDs) [42].

In the synthesis of the swivel-cruciform binaphthyl-based oligomers **27** and **28**, bromination of commercially available 1,1'-binaphthol followed by a Williamson reaction gives binaphthyl ethers with various alkoxy groups [43]. The dibromi-

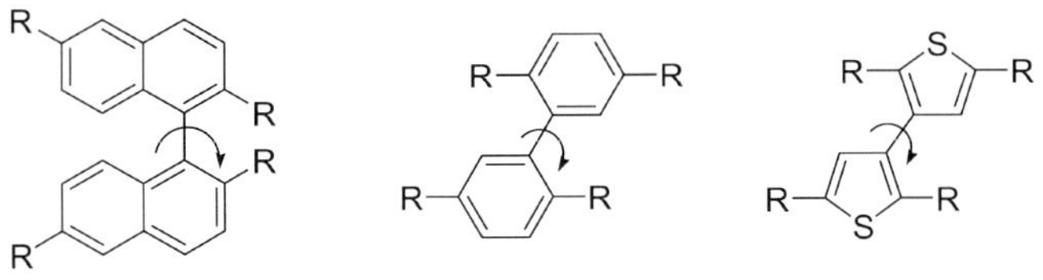

Scheme 3.11 Structure of swivel-cruciform core fragments

nated binaphthyl ethers were then reacted in Suzuki- or Heck-type procedures to give the sequence of oligomers **27** and **28** depicted in Scheme 3.12. The chiral oligomers were analyzed by chiral HPLC and found to have an enantiomeric purity in the range 80–90%. However, high reaction temperatures reduce the enantiomeric purity of the samples, especially during Heck-type coupling reactions. The model naphthalene compounds **29** and **30** were also synthesized in order to understand the effect of the binaphthyl core on the three-dimensional packing (Scheme 3.13).

The DSC traces from **27a–e** displayed T_gs in the range 160–400 °C with no evidence for crystallization. The choice of the alkoxy group had a profound effect on the T_g values. For example, on going from the methoxy compound **27e** (T_g = 137 °C) the glass transition temperature steadily decreases to T_g = 65 °C for the hexyloxy compound **27c**. In the short-chain ether derivatives there is no evidence that the enantiomeric purity plays a significant role in the thermal properties of the films. For example, for the *n*-butoxy series, the glass transition temperatures are found at 57, 52 and 54 °C for **28a**, (*R*)-**28a** and (*S*)-**28a**, respectively. However, this is not consistent throughout the series. In fact, in some cases, the stereochemistry was found to influence the thermal properties of the samples. For example, the T_g of the racemic mixture **28b** (T_g = 71 °C) is higher than that of either the *R*- or *S*-form (T_g = 52 and 55 °C, respectively). In the case of the hexyloxy derivative **28c**, (*R*)-**28c** shows T_g = 29 °C with no tendency to crystallize while the DSC of the racemic mixture of **28c** reveals T_g = 21 °C followed by some recrystallization (T_c = 75–100 °C) and T_m = 121 °C. It appears that the enantiomerically enriched sample is more effective in forming a stable glassy material. However, the thermal history of the sample is also important, and this was demonstrated by comparing the DSC traces of (*R*)-**28d** as it is repeatedly heated and cooled to 300 °C at a rate of 1 °C min^{-1}. A broad T_m at 260 °C is observed during the first heating cycle and upon cooling the T_c occurs at 188 °C. In the subsequent heating cycles a progressive degree of complexity is observed in the DSC traces, finally culminating after 20 cycles in a DSC trace which is identical with that of racemic **28d**.

Other swivel-cruciforms based on a 1,1'-binaphthyl core were prepared by Rajca et al. according to the procedure in Scheme 3.14 [44]. It is noteworthy that the solubility of the resulting materials in chloroform follows the order (*S*)-**31** (2 × 10^{-1} mol L^{-1}) > rac-**31** (8 × 10^{-4} mol L^{-1}) > **32** (2 × 10^{-5} mol L^{-1}). Although the energies of the π–π* transition for **31** and **32** were essentially identical, the extinction coefficient of **31** was approximately double that of the linear model compound **32**. Both findings demonstrate the well-documented function of the binaphthyl unit to act as a conjugation barrier with little electronic communication between naphthalene units.

Bazan and coworkers also prepared anthracene-substituted swivel cruciforms **33** with binaphthyl cores [44, 45]. They attached the anthracene units to the binaphthyl in the 6,6'-positions by a Suzuki-type cross-coupling reaction (Scheme 3.15). Based on the aforementioned findings that the alkoxy chain length influences the solid-state properties of these types of materials, they produced two de-

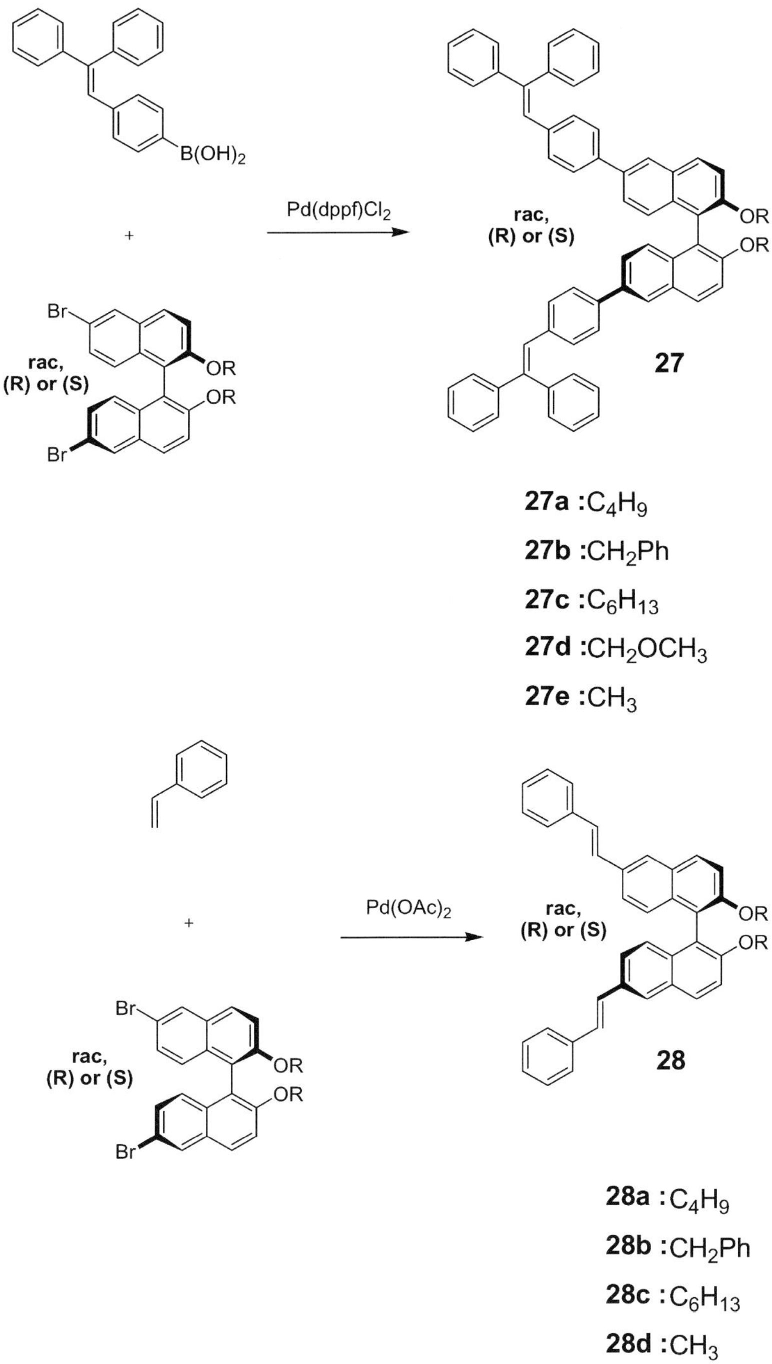

3.4 Oligomers with a Tetrasubstituted Biaryl Core | 97

27a : C_4H_9
27b : CH_2Ph
27c : C_6H_{13}
27d : CH_2OCH_3
27e : CH_3

28a : C_4H_9
28b : CH_2Ph
28c : C_6H_{13}
28d : CH_3

Scheme 3.12 Synthesis of swivel-cruciform binaphthyl-based oligomers **27** and **28**, according to Bazan and coworkers [43].

Scheme 3.13 Model compounds **29** and **30**, according to Bazan and coworkers [43].

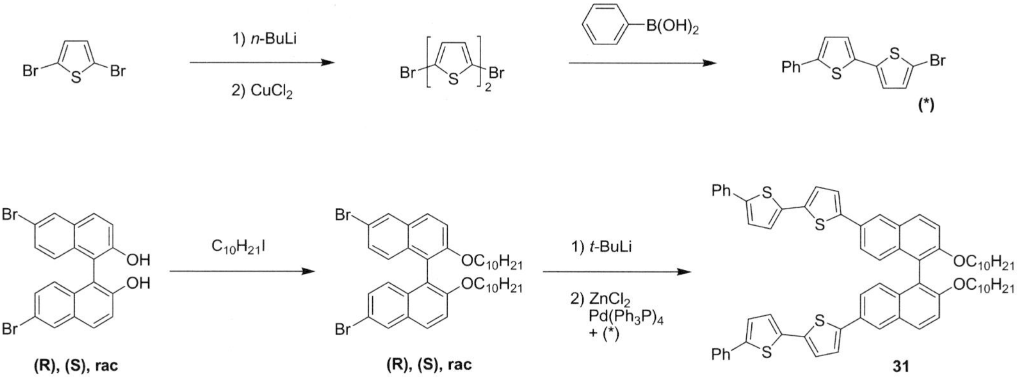

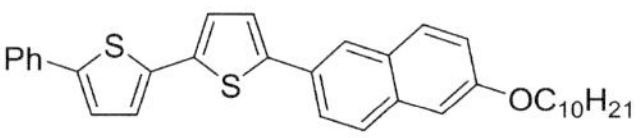

Scheme 3.14 Synthesis of binaphthyl-based swivel-cruciforms **31**, after Rajca et al. [44]. For comparison, the structure of a model compound **32** is also shown.

rivatives containing methoxy or hexyloxy substituents. As expected, the DSC measurements showed that the derivative with the smaller, less flexible methoxy unit leads to a higher T_g of 175 °C than the hexyloxy substituted molecule (T_g = 82 °C). OLED devices fabricated from these materials gave efficient blue–green emission with low turn-on voltage and good brightness [45].

Our first foray into the use of swivel-cruciform building blocks was also in the area of blue light-emitting polymers [46]. Polyfluorenes illustrate perfectly the sensitive interplay between molecular and supramolecular structure [13]. Particular interest has been devoted to the packing behavior of the polymer chains in solid-state films of poly(9,9-dioctylfluorene) (PFO). It has emerged that, in addition to a glassy phase (α-phase), an aggregated β-phase also occurs in PFO films, which is characterized by high intrachain order with a fixed angle of

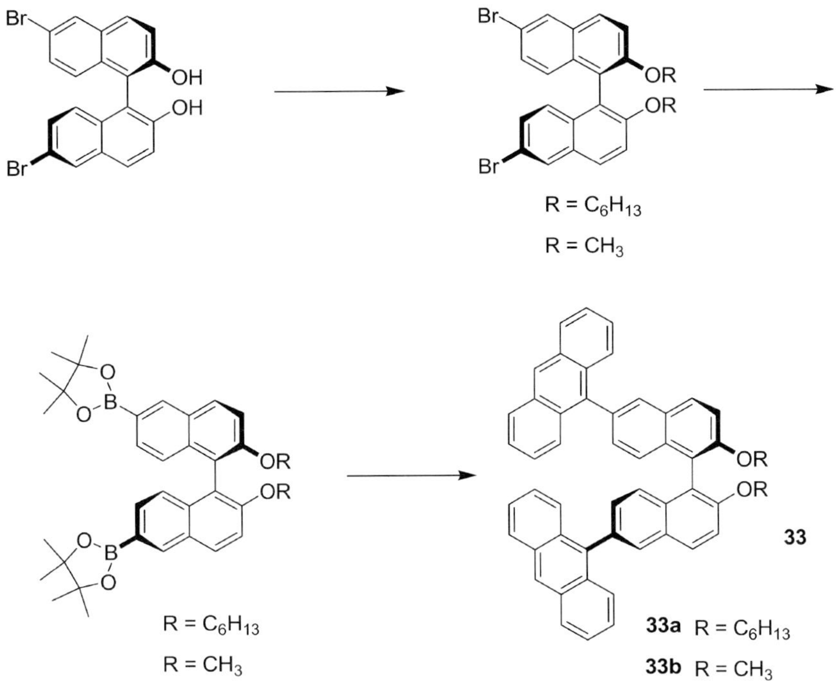

Scheme 3.15 Synthesis of anthracene-substituted swivel-cruciforms **33** with a binaphthyl core, according to Bazan and coworkers [45].

180° between adjacent monomer units [47]. Absorption and PL spectra of the β-phase are red shifted with respect to those of the glassy state or in dilute solution. For example, the 0–0 PL transition in the amorphous film is found at 429 nm whereas for thin films containing the more ordered β-phase this 0–0 transition is found at 438 nm [48]. Even minor fractions of the β-phase dominate the PL behavior of the polymer. A more drastic consequence of this agglomerated species in PFO films is a decrease in the PL quantum efficiency, which is attributed to the formation of stable quenching states (polarons) [49]. However, a high PL quantum efficiency in the solid state is a prerequisite when these materials are utilized as active media in applications such as OLEDs or organic-based solid-state lasers. Therefore, the elimination of the β-phase seemed a worthwhile endeavor for improving the performance of optical devices based on thin films of PFO.

Based on the perception that interchain interactions should be suppressed by introducing nonplanar units into the PFO chains, we decided to prepare random copolymers incorporating swivel-cruciform binaphthyl moieties into the polymer backbone [46]. Furthermore, as shown in the previous examples on low molecular weight chromophores by Bazan and coworkers, the distorted nature of the binaphthyl unit should simultaneously facilitate the formation of a glassy, amorphous state. PFO and random copolymers **34**, incorporating binaphthyl units,

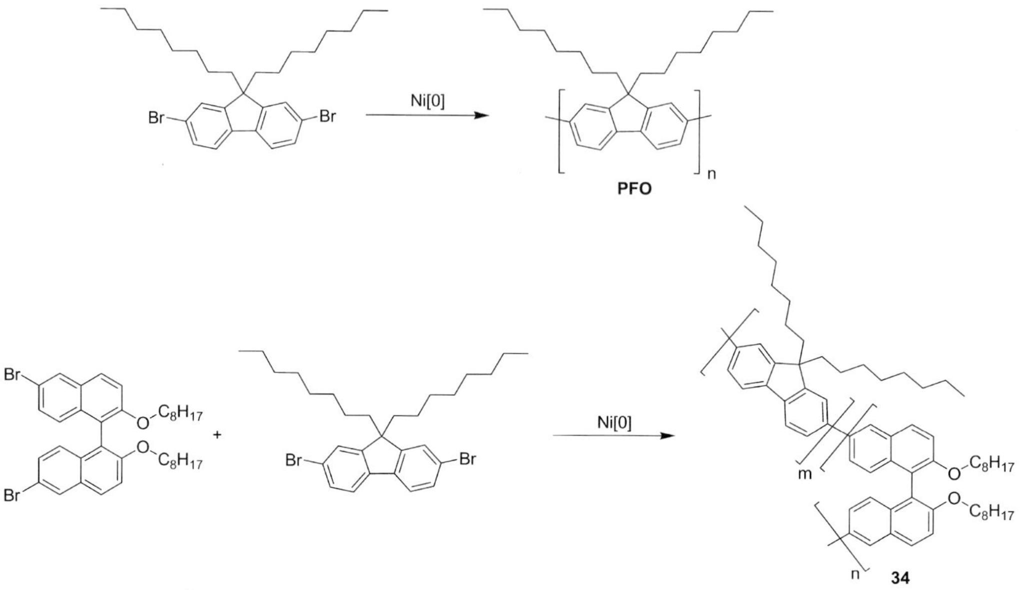

Scheme 3.16 Synthesis of PFO and random copolymers **34** incorporating binaphthyl moieties [46]

were prepared via nickel(0)-mediated Yamamoto-type coupling reactions (Scheme 3.16) by simply varying the relative ratios of monomers to comonomers. The actual amount of binaphthyl units incorporated was determined by comparing the relative intensities of the O–CH$_2$ protons at about $\delta = 3.9$ ppm from the alkoxy side-chains of the binaphthyl with those of the sum of the aryl protons.

The thermal properties of **PFO** and copolymer **34** were investigated by DSC in the range 0–300 °C with a heating and cooling rate of 10 °C min^{-1}. The incorporation of the binaphthyl unit has a dramatic effect on the thermal properties of the copolymers **34** even at low binaphthyl concentrations. The copolymer **34** with 4 % binaphthyl exhibits both a clear T_g at 78 °C and a melting transition into the nematic phase of $T_{LC} = 147$ °C. For comparison, PFO itself shows a melting transition into a nematic LC mesophase at $T_{LC} = 160$ °C in addition to a T_g at ~80 °C. For the two copolymers **34** with a higher binaphthyl content of 9.4 and 12.1 % only a T_g of 82 °C and 83 °C, respectively, was observed. The fact that the copolymers with the higher concentration of binaphthyl units form stable amorphous glasses with little discernible tendency to crystallize was not totally unexpected. Interestingly, the lasing threshold in second-order distributed feedback (DFB) lasers based on thin films of these random copolymers **34** steadily decreases with increasing binaphthyl concentration in the copolymer backbone to 3 µJ cm^{-2} for a binaphthyl concentration of ~12 % compared with a minimum lasing threshold of 11.7 µJ cm^{-2} for the homopolymer PFO. Therefore, the novel copolymers **34** provide a vast improvement for PFO-based optoelectronics.

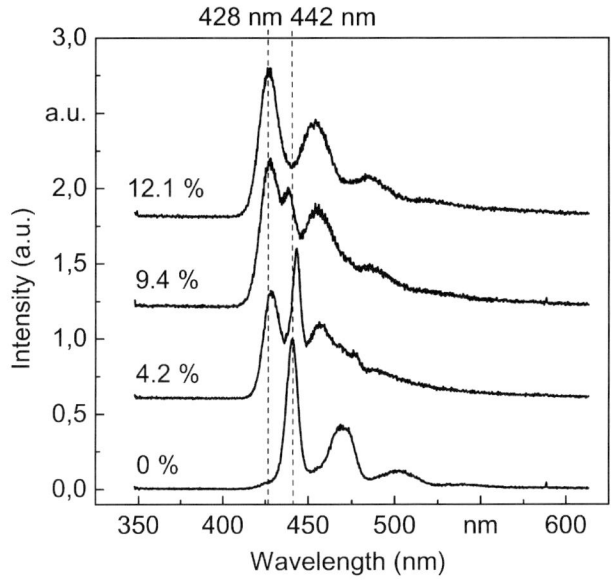

Fig. 3.2 Low-temperature (30 K) PL spectra of PFO and random copolymers **34** in thin films [46]

However, a clearer picture of the influence of the incorporated binaphthyl moieties on the optical properties of the copolymers **34** in thin films came from low-temperature (30 K) PL measurements. Figure 3.2 shows the PL features of a PFO film originating from the β-phase (bottom, 0%), with the main PL transition (0–0) at $\lambda_{max, em}$ = 442 nm. With increasing binaphthyl concentration, the PL contribution from the β-phase becomes weaker and is totally absent at a binaphthyl concentration of 12.1% with the PL emission now resembling that of the isolated polymer chain in solution.

While working on a related series of non-polymeric model compounds, we prepared the binaphthyl derivative **35** by the synthetic route depicted in Scheme 3.17 [34, 51] The yield was particularly poor (10%) but, nevertheless, a full characterization of **35** was achieved. The crystal structure of **35** has been elucidated from single crystals obtained from THF [51] and reveals that the molecule adopts a folded conformation in the solid state that is characterized by an almost parallel

Scheme 3.17 Synthesis of the binaphthyl-based swivel-cruciform **35** [50].

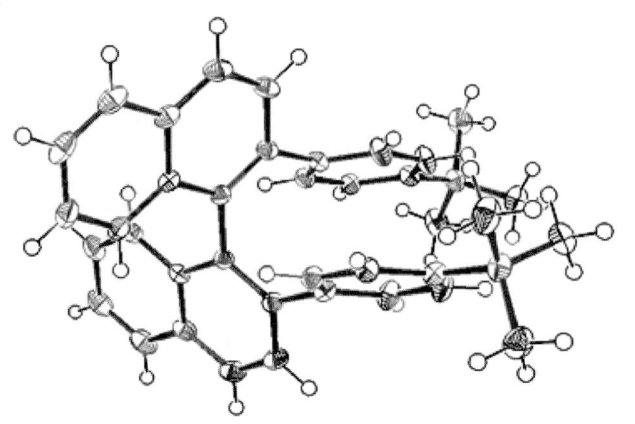

Fig. 3.3 Single-crystal structure of the binaphthyl model compound **35**.

orientation and a partial overlap of 4-*tert*-butylphenyl rings (Fig. 3.3). Offset parallel geometries are believed to be favorable for π–π interactions [52]. However, it should be noted that a purely geometric analysis of these distances gives only an indirect measure of electronic interactions [53].

In the case where a signal assignment is possible, a ^{1}H NMR study may be useful in identifying the driving forces, intramolecular π–π stacking or simply packing effects, which induce torsionally flexible molecules to adopt a folded structure. Figure 3.4 shows the aromatic region of the proton NMR spectrum of **35**. The most notable feature is the high-field position of two directly coupled doublets at δ = 6.18 and 6.77 ppm associated with the protons of the pendant 4-*tert*-butylphenyl units. This kind of shielding effect has been observed in conformationally rigid carbohelicenes and is due to the increased overlap of the aromatic rings [54–56]. In view of this, we propose that the pendant phenyl rings of **35** adopt a folded conformation also in dilute solution. This implies that the driving force for a folded conformation in the solid state results from intramolecular interactions.

However, as mentioned previously, the yield in the preparation of **35** was particularly poor and so an alternative path towards a swivel-cruciform arrangement was developed wherein the core of the cruciform was a tetrasubstituted biphenyl unit. For proof of concept, we synthesized the terphenyl dimer 2,5,2′,5′-tetra(4-*tert*-butylphenyl)-1,1′-biphenyl (**36**) via the strategy depicted in Scheme 3.18 [50]. Compound **36** was obtained in a nickel[0]-mediated (microwave-assisted) Yamamoto-type coupling of the chloroterphenyl derivative **37** in reasonable yields using THF as the solvent at temperatures of about 130 °C [57, 58].

The single-crystal structure of **36** revealed that it too adopts a folded solid-state structure (Fig. 3.5). Analysis of the ^{1}H–^{1}H COSY, ^{1}H–^{1}H COSYLR and ROESY spectra of **36** allow an unambiguous assignment of all the protons in the molecule. Figure 3.6 shows the aromatic region of the proton ^{1}H–^{1}H ROESY NMR spectrum of the terphenyl dimer 2,5,2′,5′-tetra(4-*tert*-butylphenyl)-1,1′-biphenyl (**36**). The most notable feature is again the high-field position of two directly

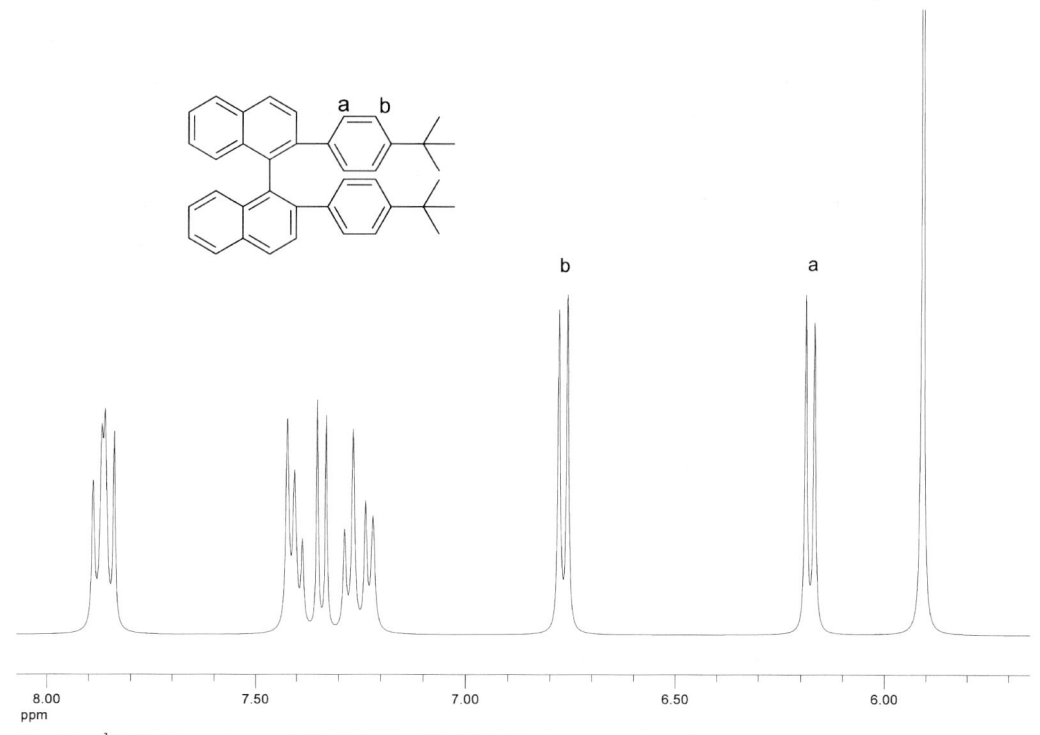

Fig. 3.4 ¹H NMR spectrum of **35** (C$_2$D$_2$Cl$_4$, 400 MHz) in the aromatic region.

Scheme 3.18 Synthetic approach towards the swivel-cruciform terphenyl dimer **36** [50].

coupled doublets. They are found at δ = 6.34 and 6.87 ppm and represent the four protons of the 4-*tert*-butylphenyl units. The spatial proximity between the upfield doublet at δ = 6.51 ppm and the proton signal at δ = 7.21 ppm from the central tetrasubstituted biphenyl unit is also manifested by the strong coupling between

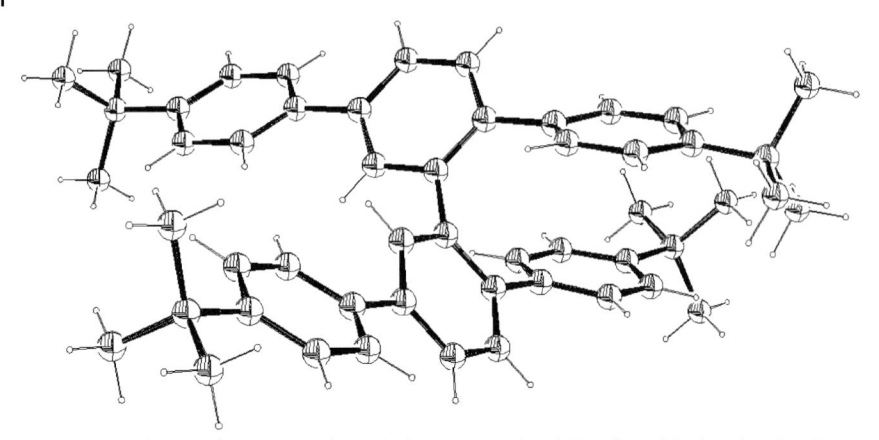

Fig. 3.5 Single-crystal structure of 2,5,2′,5′-tetra(4-*tert*-butylphenyl)-1,1′-biphenyl (**36**) [50].

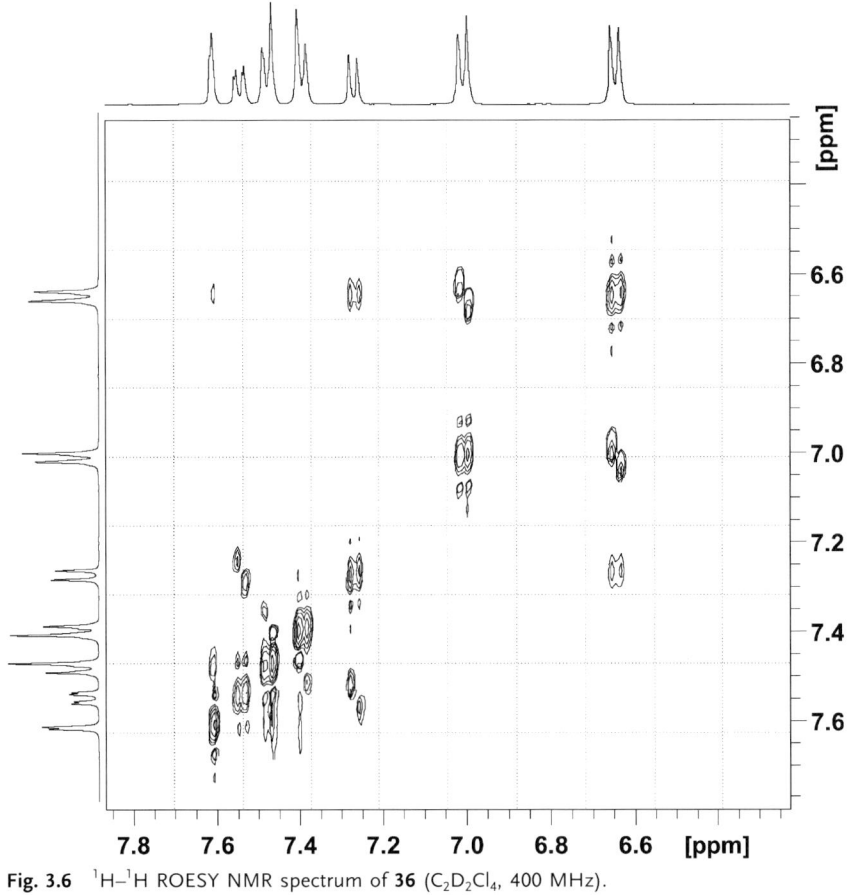

Fig. 3.6 ^{1}H–^{1}H ROESY NMR spectrum of **36** (C$_2$D$_2$Cl$_4$, 400 MHz).

3.4 Oligomers with a Tetrasubstituted Biaryl Core | 105

the signals in the 2D ROESY spectrum. Thus the interactions that induce this folding in the solid state are sufficiently strong to bias foldamer formation also in solution.

More extended cruciforms **40a** and **b** were prepared by reacting the diiodoterphenyl **38** with arylboronic acids to give the chloro-substituted precursors **39**. Dimerization of **39** leads to the cruciforms **40a** and **b** (Scheme 3.19). A complementary synthetic strategy was also developed wherein 2,5,2′,5′-tetra(4-iodophenyl)-1,1′-biphenyl (**41**) was prepared by treating the corresponding tetra(trimethylsilyl) derivative with ICl. The tetrasilyl precursor itself was prepared by the Yamamoto-type dimerization of **38** (Scheme 3.20). Since the tetrasilyl dimer was highly soluble in most organic solvents, a full spectroscopic analysis was possible, whereas the tetraiodo derivative **41** showed a drastic decrease in solubility. Nonetheless, it was possible to carry out several Suzuki-, Sonogashira- and Buchwald-type couplings with **41** to give a series of swivel-cruciforms **42–47** as depicted in

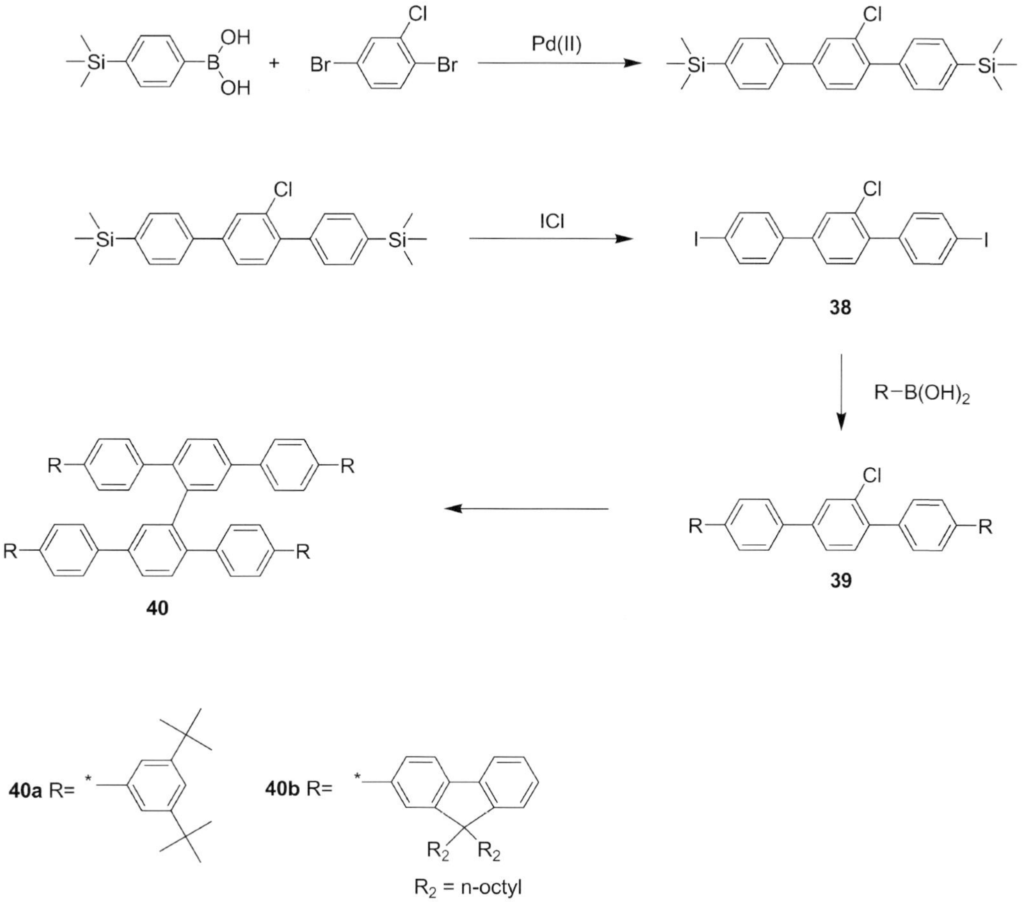

Scheme 3.19 Synthesis of the extended swivel-cruciforms **40a** and **b**.

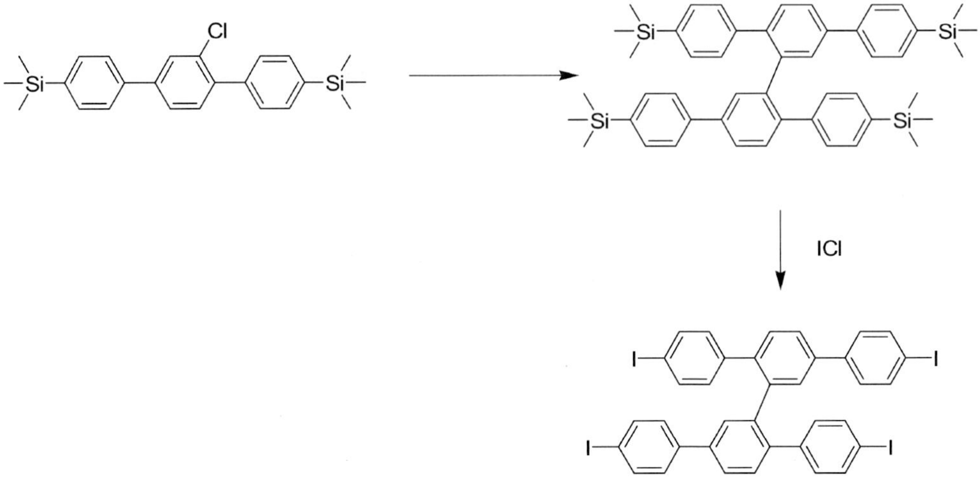

41

Scheme 3.20 Synthesis of the tetraiodo precursor **41**.

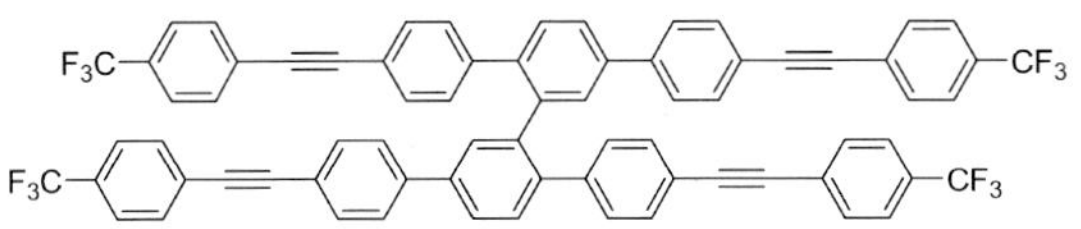

42

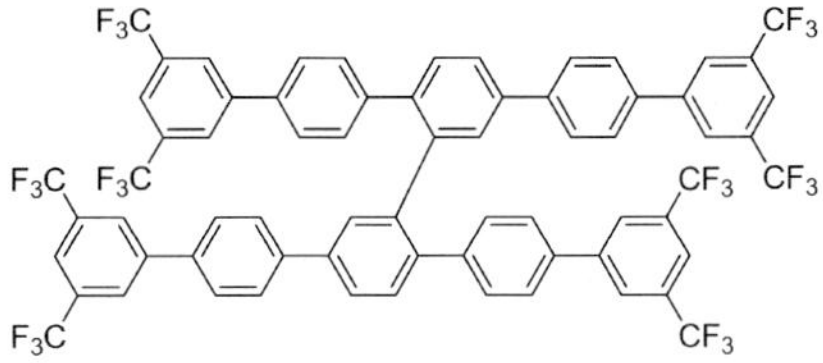

45

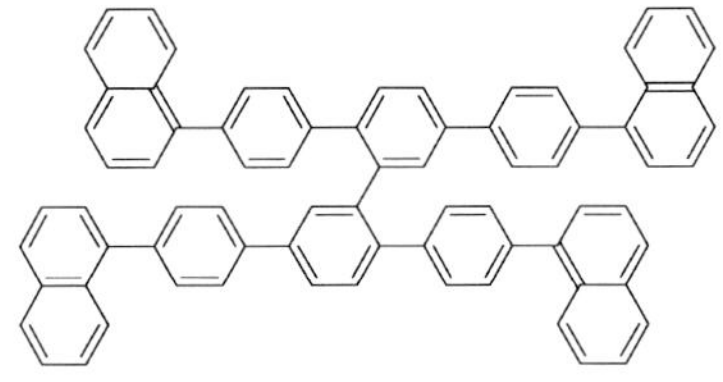

43

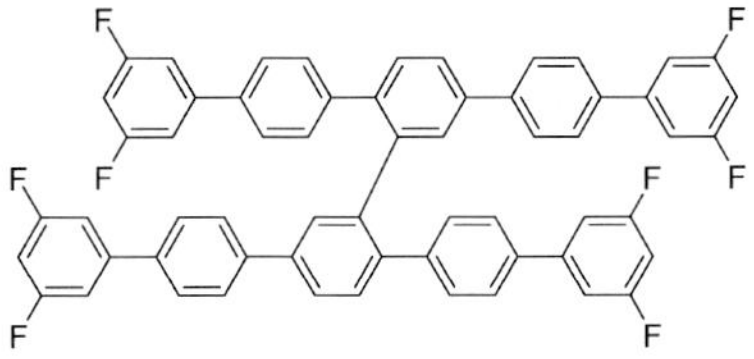

46

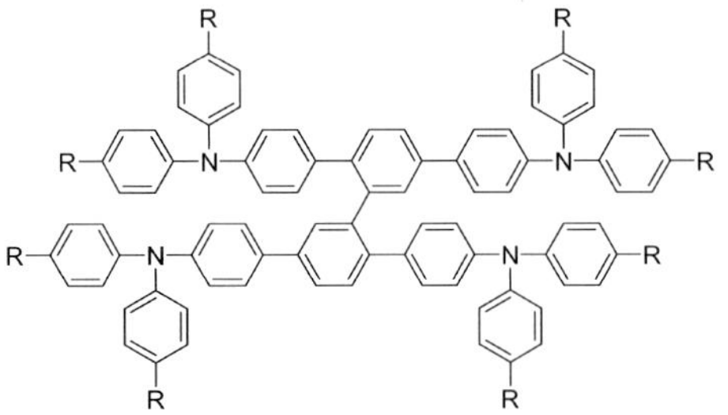

44

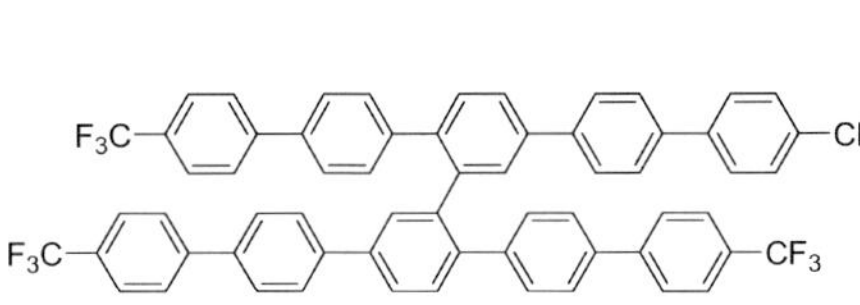

47

R = methyl, tert-butyl

Scheme 3.21 Structures of the other extended swivel cruciforms **42–47**.

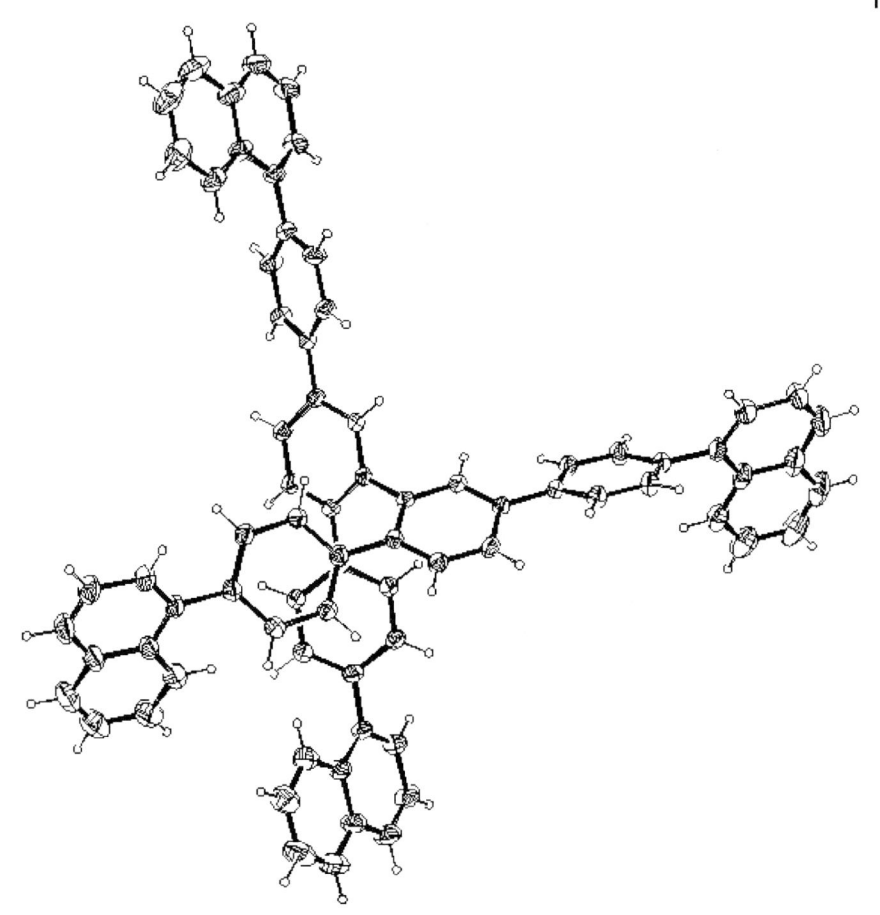

Fig. 3.7 Single-crystal structure of the tetranaphthyl derivative **43**.

Scheme 3.21. As example, the X-ray single-crystal structure of the tetranaphthyl derivative **43** is depicted in Fig. 3.7.

The thermal properties of the swivel cruciforms **40–47** were analyzed by DSC. Remarkably, the cruciforms **45**, **43**, **44** (R = *tert*-butyl), **46** and **42** show distinctly increased T_g values of 95, 130, 140, 130 and 225 °C, respectively. Only the fluorinated derivatives **45** and **46** display clear melting transitions in DSC with T_m = 228 and 257 °C, respectively.

The strong π–π interaction between the arms in some shorter swivel-cruciform molecules led us to investigate if such a molecular design might be advantageous for extended, soluble thienyl-based oligomers that form ordered thin films. For solution-processed organic field effect transistors (OFETs), wet processing seems crucial if the transistors are to be fabricated on large flexible substrates in a cost-effective manner. Thienyl-based oligomers such as α,α'-dihexyloligothiophenes are relatively facile to prepare and purify and exhibit appreciable mobili-

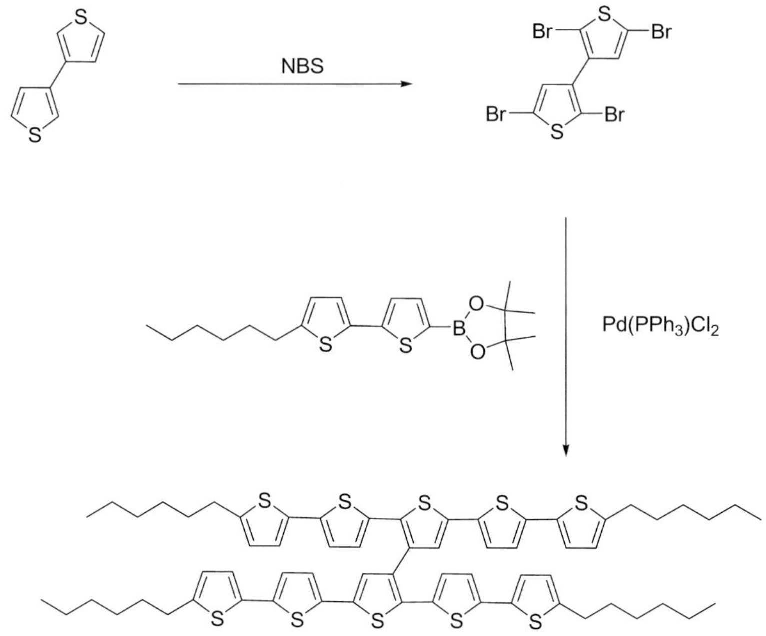

Scheme 3.22 Synthetic strategy for the preparation of a swivel-cruciform pentathiophene dimer **48**.

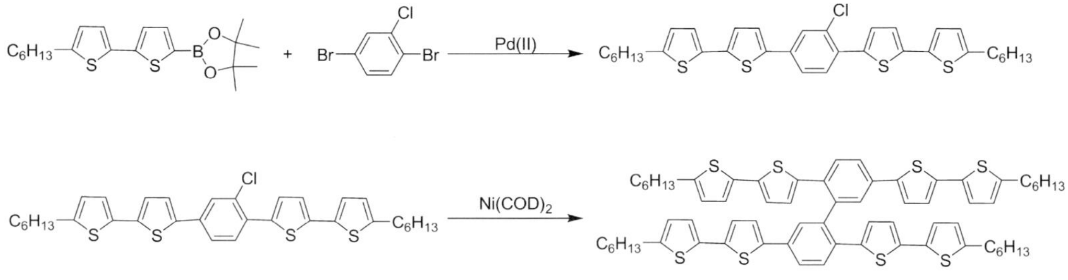

Scheme 3.23 Synthesis of the swivel-cruciform dimer **49** with dithienyl–phenylene–dithienyl arms.

ties when deposited by thermal evaporation in vacuum (<1 cm^2 V^{-1} s^{-1}) [59–64]. However, the solubility of such linear oligothiophenes, even with solubilizing terminal alkyl chains, is still problematic for liquid-based device fabrication and hence the number of wet-processed oligothiophene-based OFETs is relatively meager [65–69].

Hence we synthesized a related swivel-cruciform dimer **49** with dithienyl–phenylene–dithienyl arms by a synthetic strategy similar to the already described oligoaryl dimers (Scheme 3.23) while the all-thiophene (α,α'-dihexylpentathiophene) dimer **48** was prepared by attaching four dithienyl arms on to a 2,5,2',5'-tetrabromo-3,3'-bithiophene core in a Suzuki-type coupling (Scheme 3.22). Interestingly, the swivel-cruciform molecular shape enhances the solubility, which allowed for a detailed characterization of both cruciforms **48** and **49** by NMR spectroscopy [70]. DSC measurements of **48** and **49** showed first-order melting transitions at 196 and 174 °C, respectively. This thermal behavior indicates the propensity of the swivel-cruciforms to form a well-established crystalline phase.

Figure 3.8 shows the X-ray reflectivity scan from a thin film of **48** that was annealed at 120 °C for 5 min followed by slow cooling (1.3 K min^{-1}). Interestingly, the appearance of clearly resolved Bragg peaks (first- and second-order peaks of the interlayer distance) indicates that **48** forms a well-ordered layered structure in the solid state as typically observed for α,ω-dihexyloligothiophenes [71, 72]. Using Debye–Scherrer formalism, we calculated a domain size of about 25 nm. The d-spacing is ca. 3 nm, consistent with an almost upright orientation of the pentathiophene units, despite the covalent linkage between the arms. Kiessig thickness fringes are clearly visible up to the first diffraction peak (1st), indicating that **48** forms very smooth layers. This assumption is supported by atomic force microscopy (AFM) images of the films of the dimer **48** before and after thermal treatment (5 min, 120 °C), as depicted in Fig. 3.9. The average surface roughness was estimated to be ca. 1.8 nm for the as-deposited and ca. 4.7 nm for the annealed film of **48**.

OFETs with wet-deposited films of **48** as the active layer in bottom contact geometry were fabricated on highly doped n-type silicon wafers with the organic semiconductor layer (ca. 50 nm) deposited from chloroform solution [70]. OFETs made from **48** exhibited negative amplification, which is typical of

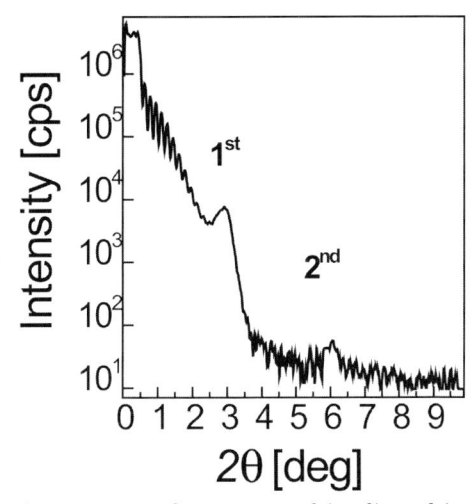

Fig. 3.8 X-ray reflectivity scan of thin films of the pentathiophene dimer **48** [70].

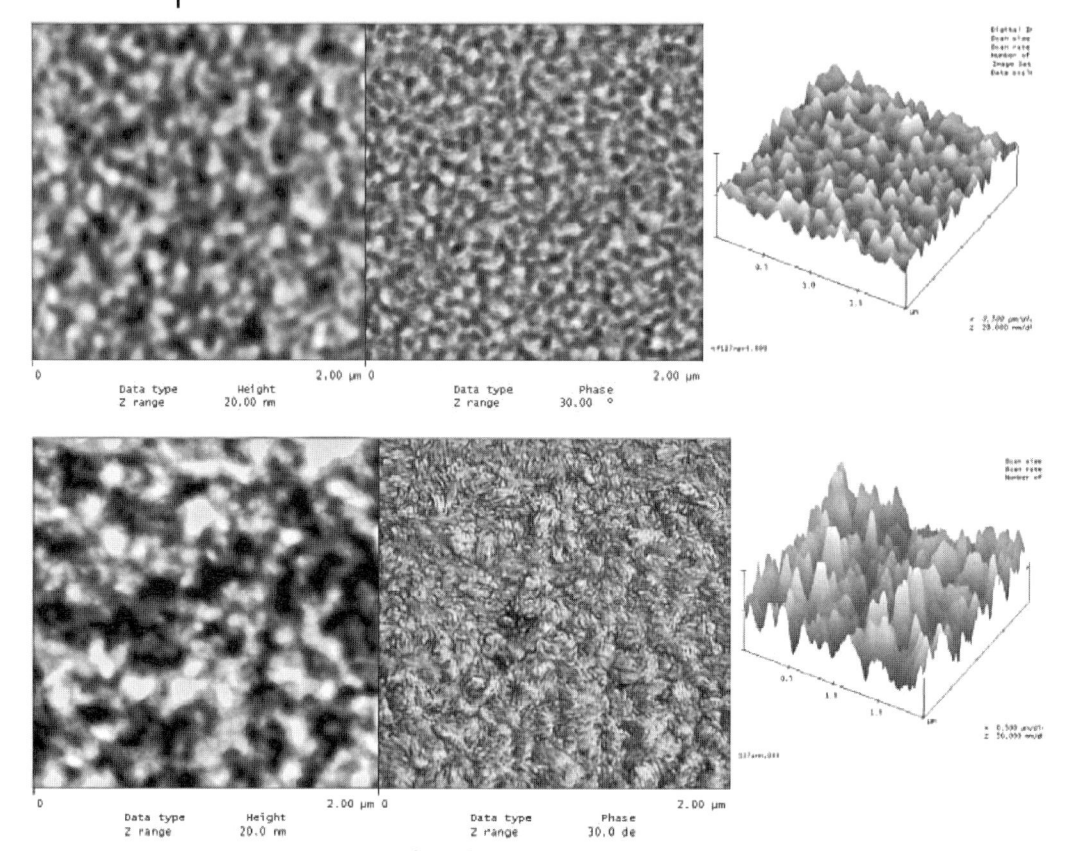

Fig. 3.9 AFM images of thin films of **48**. Top image, as-deposited films; bottom image, after annealing for 5 min at 120 °C (left, height image; right, phase image).

p-type semiconductors with well-defined linear and saturation regions. No hysteresis was observed in our measurements, which indicates very good current modulation and stability during operation. For non-annealed **48**, we obtained a field-effect mobility (μ_{FET}) of 9.6×10^{-3} cm^2 V^{-1} s^{-1} and a current on/off ratio of 6×10^4. The OFET device made from the annealed sample gave a slightly improved hole mobility ($\mu_{FET} = 0.012$ cm^2 V^{-1} s^{-1}) and an increased on/off ratio ($> 10^5$) [70]. These are among the highest values reported to date for wet-processed OFETs utilizing oligothiophenes [73]. Additionally, the turn-on voltages ($V_0 \approx 0$ V) are very close to zero for both devices, which is a highly desirable OFET property as such transistors have low power consumption.

A related series of oligophenylenevinylene dimers **50a–d** was prepared by Ma and coworkers and compared with the corresponding linear counterparts [74, 75]. The synthetic strategy towards **50a–d** is summarized in Scheme 3.24 and involves the dimerization of 1-bromo-2,5-dimethylbenzene in a Yamamoto-type coupling. The resulting tetramethylbiphenyl derivative **51** was subsequently reacted with NBS to give the tetra(bromomethyl) derivative **52**, which was then converted into the tetraphosphonium salt **53**. Compound **53** was finally reacted with the corresponding aromatic aldehydes in a Wittig-type carbonyl olefination to give the swivel-cruciforms **50a–d**.

The thermal properties of the dimers **50a–d** relative to the linear analogs were probed by DSC analysis. The "pure" phenylenevinylene cruciforms **50a** and **b** still form crystalline phases with T_m = 223 and ~258 °C, respectively. As the volume of the cruciforms is increased, the intermolecular interactions are reduced, which is documented by the occurrence of amorphous phases with glass transitions at 43 and 142 °C for **50c** and **d**, respectively. A corresponding cruciform trimer **54** was also synthesized by Ma and coworkers [76] following a related protocol, as illustrated in Scheme 3.25. The DSC of **54** confirms the previously discussed trend and **54** is fully amorphous and shows only a T_g at 118 °C [76].

The absorption and emission spectra of **50a**, **54** and the linear distyrylbenzene model compound **55** in dilute solution were rather similar. However, the solid-state emission maxima $\lambda_{max,\,em}$ of the swivel-cruciforms were red-shifted by 25 (**50a**) and 13 nm (**54**) relative to the model compound **55**. This may be a consequence of some electronic interaction between the arms within the partially pla-

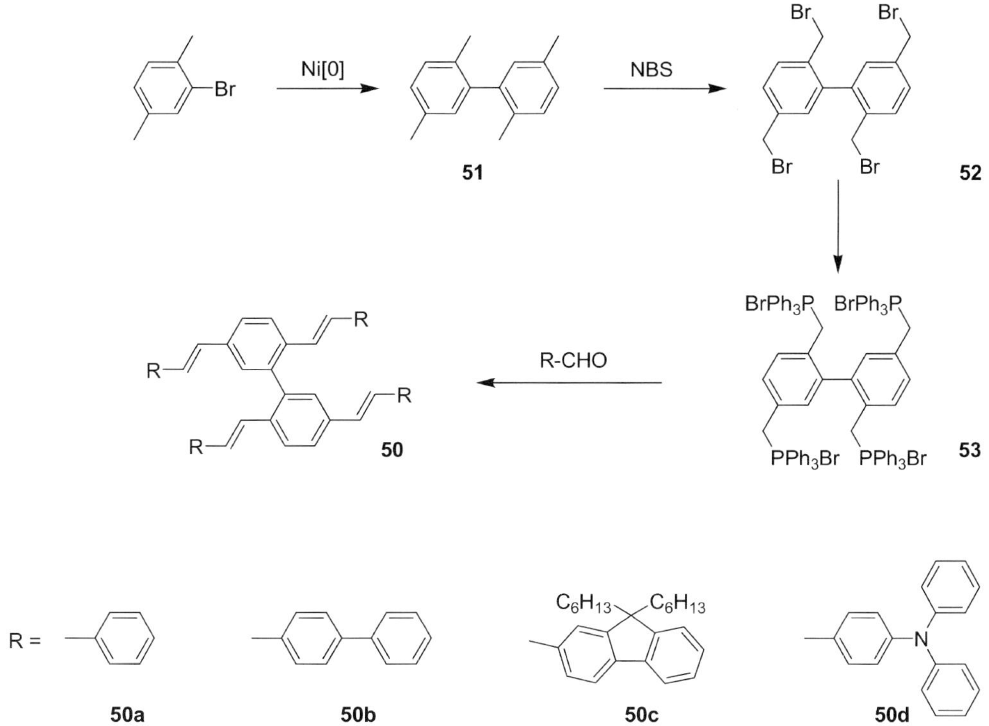

Scheme 3.24 Synthesis of swivel-cruciform oligophenylenevinylene dimers **50a–d**, according to Ma and coworkers [74, 75].

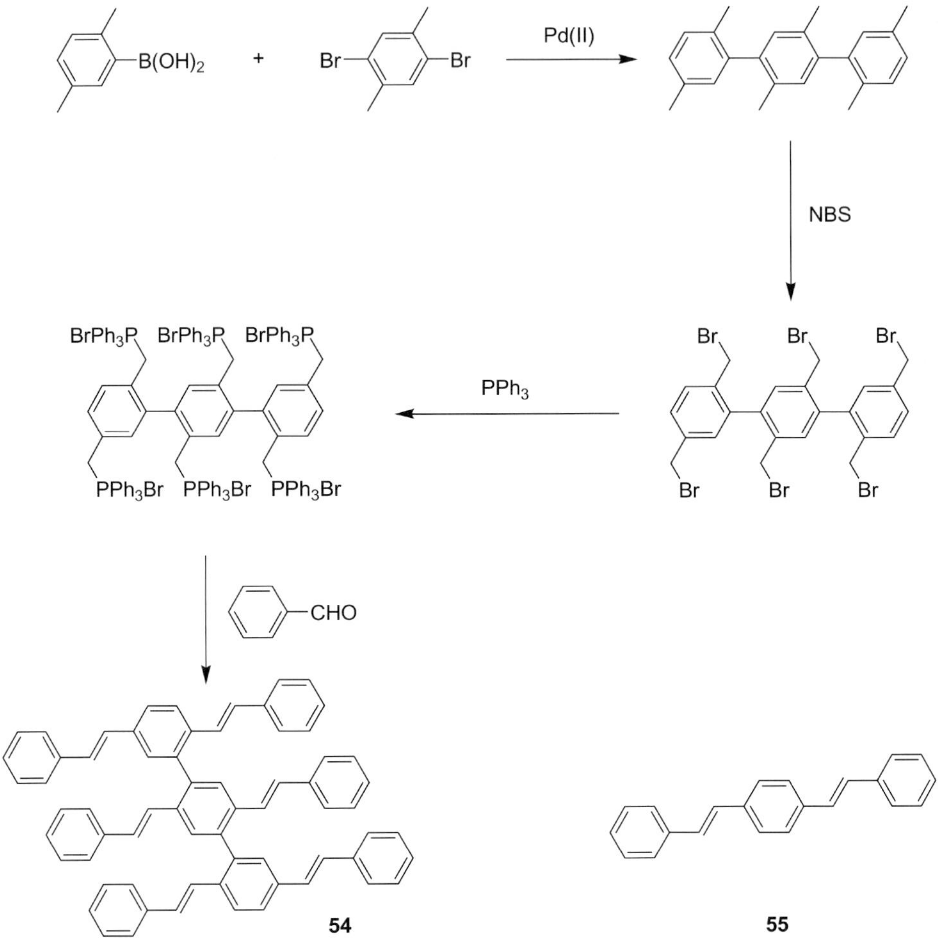

Scheme 3.25 Synthesis of a swivel-cruciform oligophenylenevinylene trimer **54** by Ma and coworkers [76]. For comparison, the structure of distyrylbenzene (**55**) is also given

narized solid-state structures of **50a** and **54** that is more pronounced for the dimer **50a**. The observed solid-state PL quantum efficiencies increase with increasing size and decreasing crystallinity of the cruciforms (**55**, 8%; dimer **50a**, 11%; trimer **54**, 13%). The amorphous triphenylamine-substituted dimer **50d** displays only a weak bathochromic shift in the PL emission maximum $\lambda_{max,em}$ on going from solution to the solid state. The solid-state PL quantum efficiency of **50d** is more than double that of the linear model compound **55**.

3.5
Conclusion

The previously outlined examples clearly demonstrate the application potential of cruciform π-conjugated oligomers. They combine the advantage of increased solubility with the possibility not only of tuning the electronic and optical properties of π-conjugated materials by design but also providing a means to manipulate the morphology of the materials when they are deposited in films. The increased solubility of the oligomers allows for wet processing of the materials by spin-coating or inkjet printing on flexible substrates. High T_g values of >150 °C can guarantee an optimum morphological stability of amorphous films, especially in organic light-emitting diode (OLED) applications, and is seen to be key for long-term stability and long operational lifetimes. Several oligomers based on a cross-like structure presented here exhibit amorphous states and many of the extended cruciforms have sufficiently high T_gs to be considered useful in such applications.

Nevertheless, other cruciform oligomers, especially those composed of oligothiophene segments, show a high tendency for crystallization, which is a crucial parameter for high charge carrier mobilities in organic field effect transistor (OFET) applications. High crystallinity, often realized after thermal post-treatment, in combination with an increased solubility, makes these cruciform oligomers promising candidates for solution-processed thin-film transistors. Such low-performance/low-cost electronic devices may find novel application areas in logistics (RFID tags) and security (smart labels).

Moreover, a variety of other applications in areas such as photovoltaic cells, sensors and solid-state lasers seem very promising and challenging. Future research will surely concentrate on tailor-made materials and hence the development of an understanding of the basic structure–property relation is crucial for a rational design of novel types of cruciform π-conjugated oligomers with improved material properties.

Finally, the examples discussed here cannot cover the whole field of cruciform oligomers; many further papers have been published, dealing with, e.g., cruciform tetra(ethynylbenzene) derivatives by Haley's group [77] and tetra(arylenevinylene)-based cruciform oligomers by Marks's group [78].

3.6
Experimental

All reactions were carried out under an argon atmosphere. The solvents used were of commercial p.a. quality. ^{1}H and ^{13}C NMR data were obtained on a Bruker ARX 400 spectrometer. Phase transitions were studied by differential scanning calorimetry (DSC) with a Bruker Reflex II thermosystem at a scanning rate of 10 °C min^{-1} for both heating and cooling cycles. UV–Vis and fluorescence spectra were recorded on a Jasco V-550 spectrophotometer and a Varian-Cary Eclipse spectrometer, respectively. Low-resolution mass spectra were obtained on a Varian

MAT 311A instrument operating at 70 eV [electron impact (EI) mode] and reported as m/z and relative intensity (%). Field desorption (FD) mass measurements were carried out on a ZAB 2-SE-FDP instrument. Microwave-assisted synthesis was performed using a CEM–Discovery monomode microwave system utilizing an IR temperature sensor and magnetic stirrer in sealed 10-mL glass vials with aluminum caps and a septum. All reactions were monitored and controlled using a personal computer.

Synthesis of 1-chloro-2,5-bis(4-*tert*-butylphenyl)benzene (**37**)
in a non-aqueous microwave-assisted protocol
A dried 10-mL microwave tube was charged with 1-chloro-2,5-dibromobenzene (0.1 g, 0.37 mmol), 4-*tert*-butylphenylboronic acid (0.14 g, 0.79 mmol), KOH (0.12 g, 2.14 mmol) and Pd(PPh$_3$)$_2$Cl$_2$ (0.013 g, 0.02 mmol) and sealed under argon with an aluminum cap with a septum. Dry THF (4 mL) was added via a syringe and the reaction mixture was irradiated with microwaves (300 W) for 10 min with air cooling to keep the temperature between 110 and 115 °C. The mixture was poured into water, extracted with dichloromethane and the extract was subsequently washed with water and brine, dried over MgSO$_4$ and the solvent removed by rotary evaporation. The residue was purified by column chromatography on silica gel with hexane–ethyl acetate (99:1) as eluent to give **37** in 95% yield. ^{1}H NMR (400 MHz, C$_2$D$_2$Cl$_4$, 80 °C): δ 7.64 (d, 1H, J = 1.9 Hz), 7.52 (m, 3H), 7.43 (m, 6H), 7.37 (d, 1H, J = 8.0 Hz), 1.34 (s, 9H), 1.33 (s, 9H) ppm. ^{13}C NMR (100 MHz, C$_2$D$_2$Cl$_4$, 80 °C): δ = 151.3, 150.7, 141.3, 138.9, 136.3, 136.0, 132.9, 132.1, 129.4, 128.4, 126.8, 126.2, 125.5, 125.3, 34.8, 34.8, 31.7, 31.6 ppm. FD-MS: m/z 377.0 (100.0). Elemental analysis: calculated for C$_{26}$H$_{29}$Cl: C, 82.84; H, 7.75. Found: C, 82.81; H, 7.46%.

Synthesis of 2,5,2′,5′-tetra(4-*tert*-butylphenyl)-1,1′-biphenyl (**36**)
A dried 10-mL microwave tube was charged with **37** (0.10 g, 0.27 mmol), Ni(COD)$_2$ (109 mg, 0.40 mmol), 2,2′-bipyridyl (62 mg, 0.40 mmol) and COD (43 mg, 0.40 mmol) and sealed under argon with an aluminum cap with a septum. Dry DMF (1 mL) and toluene (3 mL) were added via a syringe and the reaction mixture was irradiated with microwaves (300 W) for 12 min at ~220 °C. The mixture was poured into water and then extracted with dichloromethane. The organic phase was subsequently washed with 2 M HCl, water and brine and dried over Na$_2$SO$_4$. After the solvent had been removed by rotary evaporation, the residue was purified by column chromatography on silica gel with hexane–toluene (95:5) as eluent to give **36** in 82% yield. ^{1}H NMR (400 MHz, C$_2$D$_2$Cl$_4$, 80 °C): δ 7.63 (d, 2H, J = 1.7 Hz), 7.53 (dd, 2H, J = 8.2, 1.8 Hz), 7.49 (d, 4H, J = 8.4 Hz), 7.38 (d, 4H, J = 8.4 Hz), 7.21 (d, 2H, J = 8.1 Hz), 6.93 (d, 4H, J = 8.3 Hz), 6.51 (d, 4H, J = 8.5 Hz), 1.29 (s, 18H), 1.23 (s, 18H) ppm. ^{13}C NMR (100 MHz, C$_2$D$_2$Cl$_4$, 80 °C): δ 150.7, 148.9, 140.5, 139.8, 139.3, 137.8, 137.5, 130.6, 130.5, 128.9, 126.8, 126.1, 126.0, 124.8, 34.7, 34.5, 31.6 ppm. FD-MS: m/z 682.4 (100.0). Elemental analysis: calculated for C$_{52}$H$_{58}$: C, 91.44; H, 8.56. Found: C, 90.75; H, 8.12%.

Synthesis of α,α'-dihexylpentathiophene dimer **48**.

2,5,2',5'-Tetrabromo-3,3'-dithienyl: 3,3'-Dithienyl (1.03 g, 6.2 mmol) and NBS (5.52 g, 31 mmol) were added to a pre-dried 250-mL two-necked flask and placed under an argon atmosphere. Dry THF (100 mL) was added and the reaction mixture was stirred at room temperature for 2 days. The solvent was removed and the residue taken up with hexane and filtered. The solvent was removed from the filtrate and the residue was chromatographed on silica with hexane as eluent to give 2,5,2',5'-tetrabromo-3,3'-dithienyl in 31% yield. ^{1}H NMR (400 MHz, $C_2D_2Cl_4$): δ = 6.97 (s, 2H) ppm. ^{13}C NMR (100 MHz, $C_2D_2Cl_4$): δ = 135.2, 131.6, 111.5, 111.2 ppm. LR-MS (EI, 70 eV): m/z 482 (100). Elemental analysis: calculated for $C_8H_2Br_4S$: C, 19.94; H, 0.42; S, 13.31. Found: C, 19.90; H, 0.81; S, 13.23%.

α,α'-Dihexylpentathiophene dimer **48**: KOH (1.5 g, 26.7 mmol), 2,5,2',5'-tetrabromo-3,3'-dithienyl (0.2 g, 0.42 mmol) and $Pd(PPh_3)_2Cl_2$ (0.047 g, 0.067 mmol) were placed in a two-necked flask fitted with a reflux condenser and a pressure-equalized dropping funnel. The reaction flask was degassed and filled with argon. Subsequently, the flask was charged with dry THF (50 mL). The boronic ester 5-(4,4,5,5-tetramethyl-1,3,2-dioxaborolane-2-yl)-5'-*N*-hexyl-2,2'-bithiophene (1 g, 2.66 mmol) was placed in a Schlenk tube and dissolved in dry THF (10 mL). The solution was then added to the dropping funnel via a syringe. The mixture was heated to 60°C and the boronic ester solution added over a period of ca. 1 h. The reaction mixture was stirred under reflux for 24 h. After cooling to room temperature, the reaction solution was diluted with chloroform and the organic phase was washed successively with water, saturated EDTA solution and brine. The organic phase was dried over $MgSO_4$ and the solvent removed. The residue was chromatographed on silica gel with 10% toluene–hexane as eluent. After recrystallization from dichloromethane–heptane, the product **48** was obtained as a brown powder in 21% yield. ^{1}H NMR (400 MHz, $C_2D_2Cl_4$): δ = 7.04 (d, 2H, J = 3.7 Hz), 6.97 (s, 2H), 6.96 (d, 2H, J = 3.9 Hz), 6.92 (d, 2H, J = 3.5 Hz), 6.87 (d, 2H, J = 3.9 Hz), 6.81 (d, 2H, J = 3.8 Hz), 6.79 (d, 2H, J = 3.5 Hz), 6.63 (d, 2H, J = 3.5 Hz), 6.55 (d, 2H, J = 3.5 Hz), 2.69 (td, 8H, J = 7.6, 22.3Hz), 1.58 (m, 8H), 1.25 (m, 24H), 0.81 (td, 12H, J = 6.8, 9.3Hz) ppm. ^{13}C NMR (100 MHz, $C_2D_2Cl_4$): δ = 146.31, 146.13, 138.68, 137.76, 135.33, 135.19, 134.56, 134,53, 134,21, 134.0, 132.5, 126.9, 126.3, 125.15, 125.04, 124.87, 124.03, 124.0, 123.98, 123.55, 31.73, 31.71, 31.62, 31.6, 30.42, 30.38, 28.94, 28.92, 22.71, 22.7, 14.19 ppm. FD-MS: m/z 342.9 (100.0). Elemental analysis: calculated for $C_{64}H_{70}S_{10}$: C, 66.27; H, 6.08; S, 27.64. Found: C, 66.08; H, 6.89, S, 30.27S.

Acknowledgments

The authors would like to thank Achmad Zen and Dieter Neher, Potsdam University, Germany, for the OFET experiments and David J. Brauer, Wuppertal University, Germany, and Christian Lehmann, MPI for Coal Research, Mülheim a. d. Ruhr, Germany, for the single-crystal X-ray analyses.

References

1. Malliaras, G., Friend, R. *Phys. Today* **2005**, *58*, 53.
2. Forrest, S. R. *Nature* **2004**, *428*, 911.
3. Bernius, M. T., Inbasekaran, M., O'Brien, J., Wu, W. S. *Adv. Mater.* **2000**, *12*, 1737.
4. Hoppe, H., Sariciftci, N. S. *J. Mater. Res.* **2004**, *19*, 1924.
5. Horowitz, G. *J. Mater. Res.* **2004**, *19*, 1946.
6. Kraft, A., Grimsdale, A. C., Holmes, A. B. *Angew. Chem. Int. Ed.* **1998**, *37*, 402.
7. McGehee, M. D., Heeger, A. J. *Adv. Mater.* **2000**, *12*, 1655.
8. Sun, Y. M., Liu, Y. Q., Zhu, D. B. *J. Mater. Chem.* **2005**, *15*, 53.
9. Shirota, Y. *J. Mater. Chem.* **2005**, *15*, 75.
10. Katz, H. E. *Chem. Mater.* **2004**, *16*, 4748.
11. Schlüter, A. D. *Handbook of Conducting Polymers*, 2nd edn. Marcel Dekker, New York, **1998**, Chapter 8, p. 209.
12. Scherf, U. *J. Mater. Chem.* **1999**, *9*, 1853.
13. Scherf, U., List, E. J. W. *Adv. Mater.* **2002**, *14*, 477.
14. Martin, R. E., Diederich, F. *Angew. Chem. Int. Ed.* **1999**, *38*, 1350.
15. Müllen, K., Wegner, G. *Electronic Materials: the Oligomer Approach*. Wiley, New York, **1998**.
16. Klare, J. E., Tulevski, G. S., Sugo, K., de Picciotto, A., White, K. A., Nuckolls, C. *J. Am. Chem. Soc.* **2003**, *125*, 6030.
17. Li, J. Y., Ma, C. W., Tang, J. X., Lee, C. S., Lee, S. T. *Chem. Mater.* **2005**, *17*, 615.
18. Geng, Y. H., Fechtenkotter, A., Müllen, K. *J. Mater. Chem.* **2001**, *11*, 1634.
19. Pei, J., Wang, J. L., Cao, X. Y., Zhou, X. H., Zhang, W. B. *J. Am. Chem. Soc.* **2003**, *125*, 9944.
20. Shirota, Y. *J. Mater. Chem.* **2000**, *10*, 1.
21. Strohriegl, P., Grazulevicius, J. V. *Adv. Mater.* **2002**, *14*, 1439.
22. Shen, W. J., Dodda, R., Wu, C. C., Wu, F. I., Liu, T. H., Chen, H. H., Chen, C. H., Shu, C. F. *Chem. Mater.* **2004**, *16*, 930.
23. Spehr, T., Pudzich, R., Fuhrmann, T., Salbeck, J. *Org. Electr.* **2003**, *4*, 61.
24. Su, H. J., Wu, F. I., Shu, C. F. *Macromolecules* **2004**, *37*, 7197.
25. Wu, Y. G., Li, J., Fu, Y. Q., Bo, Z. S. *Org. Lett.* **2004**, *6*, 3485.
26. Deng, X. B., Mayeux, A., Cai, C. Z. *J. Org. Chem.* **2002**, *67*, 5279.
27. Liu, X. M., He, C. B., Huang, J. C., Xu, J. M. *Chem. Mater.* **2005**, *17*, 434.
28. Robinson, M. R., Wang, S. J., Bazan, G. C., Cao, Y. *Adv. Mater.* **2000**, *12*, 1701.
29. Salbeck, J. *Macromol. Symp.* **1997**, *125*, 121.
30. Pudzich, R., Fuhrmann-Lieker, T., Salbeck, J. *Spiro Compounds for Organic Electroluminescence and Related Applications*. Springer, Berlin, **2006**.
31. Oldham, W. J., Lachicotte, R. J., Bazan, G. C. *J. Am. Chem. Soc.* **1998**, *120*, 2987.
32. Wang, S. J., Oldham, W. J., Hudack, R. A., Bazan, G. C. *J. Am. Chem. Soc.* **2000**, *122*, 5695.
33. Yeh, H. C., Lee, R. H., Chan, L. H., Lin, T. Y. J., Chen, C. T., Balasubramaniam, E., Tao, Y. T. *Chem. Mater.* **2001**, *13*, 2788.
34. Wilson, J. N., Josowicz, M., Wang, Y. Q., Bunz, U. H. F. *Chem. Commun.* **2003**, 2962.
35. Li, Z. H., Wong, M. S., Tao, Y. *Tetrahedron* **2005**, *61*, 5277.
36. Jen, A. K. Y., Liu, Y. Q., Hu, Q. S., Pu, L. *Appl. Phys. Lett.* **1999**, *75*, 3745.
37. Pu, L. *Chem. Rev.* **1998**, *98*, 2405.
38. Benmansour, H., Shioya, T., Sato, G. C., Bazan, G. C. *Adv. Funct. Mater.* **2003**, *13*, 883.
39. Neher, D. *Macromol. Rapid Commun.* **2001**, *22*, 1365.
40. Grell, M., Knoll, W., Lupo, D., Meisel, A., Miteva, T., Neher, D., Nothofer, H. G., Scherf, U., Yasuda, A. *Adv. Mater.* **1999**, *11*, 671.
41. Jeukens, C. R. L. P., Jonkheijm, P., Wijnen, F. J. P., Gielen, J. C., Christianen, P. C. M., Schenning, A. P. H. J., Meijer, E. W., Maan, J. C. *J. Am. Chem. Soc.* **2005**, *127*, 8280.
42. Schadt, M. *Annu. Rev. Mater. Sci.* **1997**, *27*, 305.
43. Ostrowski, J. C., Hudack, R. A., Robinson, M. R., Wang, S. J., Bazan, G. C. *Chem. Eur. J.* **2001**, *7*, 4500.

44. Rajca, A., Wang, H., Pawitranon, V., Brett, T. J., Stezowski, J. J. *Chem. Commun.* **2001**, 1060.
45. Benmansour, H., Shioya, T., Sato, Y., Bazan, G. C. *Adv. Funct. Mater.* **2003**, *13*, 883.
46. Rabe, T., Hoping, M., Schneider, D., Becker, E., Johannes, H. H., Kowalsky, W., Weimann, T., Wang, J., Hinze, P., Nehls, B. S., Scherf, U., Farrell, T., Riedl, T. *Adv. Funct. Mater.* **2005**, *15*, 1188.
47. Grell, M., Bradley, D. D. C., Long, X., Chamberlain, T., Inbasekaran, M., Woo, E. P., Soliman, M. *Acta Polym.* **1998**, *49*, 439.
48. Chen, S. H., Su, A. C., Chen, S. A. *J. Phys. Chem. B* **2005**, *109*, 10067.
49. Ariu, M., Lidzey, D. G., Sims, M., Cadby, A. J., Lane, P. A., Bradley, D. D. C. *J. Phys.:Condens. Matter* **2002**, *14*, 9975.
50. Nehls, B. S., Galbrecht, F., Bilge, A., Brauer, D. J., Lehmann, C. W., Scherf, U., Farrell, T. *Org. Biomol. Chem.* **2005**, *3*, 3213.
51. Nehls, B. S., Galbrecht, F., Bilge, A., Scherf, U., Farrell, T. *Macromol. Symp.* **2006**, *239*, 21.
52. Hunter, C. A., Sanders, J. K. M. *J. Am. Chem. Soc.* **1990**, *112*, 5525.
53. Janiak, C. *J. Chem. Soc.,Dalton Trans.* **2000**, 3885.
54. El Abed, R., Ben Hassine, B., Genet, J. P., Gorsane, M., Marinetti, A. *Eur. J. Org. Chem.* **2004**, 1517.
55. Field, J. E., Hill, T. J., Venkataraman, D. *J. Org. Chem.* **2003**, *68*, 6071.
56. Paruch, K., Vyklicky, L., Katz, T. J., Incarvito, C. D., Rheingold, A. L. *J. Org. Chem.* **2000**, *65*, 8774.
57. Carter, K. R. *Macromolecules* **2002**, *35*, 6757.
58. Yamamoto, T., Fujiwara, Y., Fukumoto, H., Nakamura, Y., Koshihara, S. Y., Ishikawa, T. *Polymer* **2003**, *44*, 4487.
59. Crouch, D. J., Skabara, P. J., Heeney, M., McCulloch, I., Coles, S. J., Hursthouse, M. B. *Chem. Commun.* **2005**, 1465.
60. Facchetti, A., Mushrush, M., Katz, H. E., Marks, T. J. *Adv. Mater.* **2003**, *15*, 33.
61. Facchetti, A., Letizia, J., Yoon, M. H., Mushrush, M., Katz, H. E., Marks, T. J. *Chem. Mater.* **2004**, *16*, 4715.
62. Halik, M., Klauk, H., Zschieschang, U., Schmid, G., Ponomarenko, S., Kirchmeyer, S., Weber, W. *Adv. Mater.* **2003**, *15*, 917.
63. Tian, H. K., Wang, J., Shi, J. W., Yan, D. H., Wang, L. X., Geng, Y. H., Wang, F. *J. Mater. Chem.* **2005**, *15*, 3026.
64. Yoon, M. H., DiBenedetto, S. A., Facchetti, A., Marks, T. J. *J. Am. Chem. Soc.* **2005**, *127*, 1348.
65. Chang, P. C., Lee, J., Huang, D., Subramanian, V., Murphy, A. R., Frechet, J. M. J. *Chem. Mater.* **2004**, *16*, 4783.
66. Katz, H. E., Laquindanum, J. G., Lovinger, A. J. *Chem. Mater.* **1998**, *10*, 633.
67. Laquindanum, J. G., Katz, H. E., Lovinger, A. J. *J. Am. Chem. Soc.* **1998**, *120*, 664.
68. Mushrush, M., Facchetti, A., Lefenfeld, M., Katz, H. E., Marks, T. J. *J. Am. Chem. Soc.* **2003**, *125*, 9414.
69. Ponomarenko, S. A., Kirchmeyer, S., Elschner, A., Huisman, B. H., Karbach, A., Drechsler, D. *Adv. Funct. Mater.* **2003**, *13*, 591.
70. Zen, A., Bilge, A., Galbrecht, F., Alle, R., Meerholz, K., Grenzer, J., Neher, D., Scherf, U., Farrell, T. *J. Am. Chem. Soc.* **2006**, *128*, 3914.
71. Garnier, F., Yassar, A., Hajlaoui, R., Horowitz, G., Deloffre, F., Servet, B., Ries, S., Alnot, P. *J. Am. Chem. Soc.* **1993**, *115*, 8716.
72. Moret, M., Campione, M., Borghesi, A., Miozzo, L., Sassella, A., Trabattoni, S., Lotz, B., Thierry, A. *J. Mater. Chem.* **2005**, *15*, 2444.
73. Sirringhaus, H. *Adv. Mater.* **2005**, *17*, 2411.
74. He, F., Cheng, G., Zhang, H. Q., Zheng, Y., Xie, Z. Q., Yang, B., Ma, Y. G., Liu, S. Y., Shen, J. C. *Chem. Commun.* **2003**, 2206.
75. He, F., Xia, H., Tang, S., Duan, Y., Zeng, M., Liu, L. L., Li, M., Zhang, H. Q., Yang, B., Ma, Y. G., Liu, S. Y., Shen, J. C. *J. Mater. Chem.* **2004**, *14*, 2735.
76. He, F., Xu, H., Yang, B., Duan, Y., Tian, L. L., Huang, K. K., Ma, Y. G., Liu, S. Y.,

Feng, S. H., Shen, J. C. *Adv. Mater.* **2005**, *17*, 2710.
77. Marsden, J. A., Miller, J. J., Shirtcliff, L. D., Haley, M. M. *J. Am. Chem. Soc.* **2005**, *127*, 2464.
78. Kang, H., Zhu, P., Yang, Y., Facchetti, A., Marks, T. J. *J. Am. Chem. Soc.* **2004**, *126*, 15974.

4
Design of π-Conjugated Systems Using Organophosphorus Building Blocks

Philip W. Dyer and Régis Réau

4.1
Introduction

π-Conjugated oligomers and polymers (Fig. 4.1) have emerged as promising materials for application in flexible, lightweight and low-cost electronic devices such as organic light-emitting diodes (OLEDs) [1], field-effect transistors (FETs), plastic lasers and photovoltaic cells [2]. One of the most appealing attributes of these types of molecular material is their aptitude for exhibiting multifunctional properties, something that is nicely illustrated by their use in the fabrication of high-performance OLEDs, which are employed as devices for flat-panel displays in a host of commercial products (e.g. cell phones, digital cameras) [3]. To achieve an efficient single-layer OLED, in which a π-conjugated system is sandwiched between two electrodes, the organic material has to be able to transport charges and to be highly luminescent [1, 4]. In addition, the molecular material should also (i) be readily processable to give uniform and pinhole-free thin films, (ii) be thermally stable in order to engender durability and (iii) possess an amorphous morphology to prevent light scattering and crystallization-induced degradation. In order to achieve these diverse demands and capabilities from a single material, the only realistic approach is to tailor the properties of π-conjugated oligomers or polymers at the molecular level. Hence the development of advanced organic materials is directly linked to the ability of chemists to fine-tune the properties and the supramolecular organization of π-conjugated systems. This is possible by exploiting the enormous versatility and scope of organic and organometallic chemistry [1, 2, 5].

To achieve these goals, many different strategies have been used, which include modification of the nature of the conjugated segment itself (topology), introduction of electron-withdrawing or -donating groups and the grafting of bulky or functional side-chain substituents. However, perhaps the most powerful means of influencing and tailoring the physical properties of a material, by way of synthesis, is to vary the chemical composition of the conjugated backbone chain. For example, aromatic five- and six-membered rings have been widely used for the tailoring of π-conjugated systems since their electronic properties (aromatic char-

Functional Organic Materials. Syntheses, Strategies, and Applications.
Edited by Thomas J.J. Müller and Uwe H.F. Bunz
Copyright © 2007 WILEY-VCH Verlag GmbH & Co. KGaA, Weinheim
ISBN: 978-3-527-31302-0

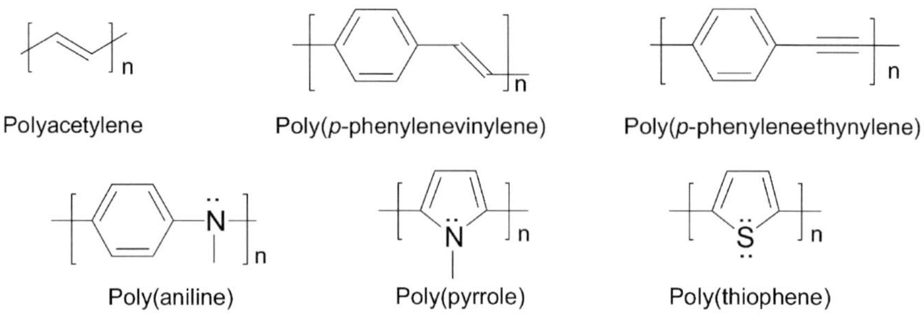

4 Design of π-Conjugated Systems Using Organophosphorus Building Blocks

Polyacetylene

Poly(p-phenylenevinylene)

Poly(p-phenyleneethynylene)

Poly(aniline)

Poly(pyrrole)

Poly(thiophene)

Fig. 4.1 Representative structures of π-conjugated systems.

acter, electron richness/deficiency, polarizability, etc.) can vary considerably according to their structure. Furthermore, these types of ring system provide not only excellent stability, but also offer significant synthetic flexibility: (i) their susceptibility towards electrophilic ring substitution allows for the introduction of pendent substituents with specific electronic and/or steric properties; and (ii) their halogen-substituted derivatives can be utilized in highly versatile catalytic methodologies for C–C bond formation (the so-called Kumada, Stille, Sonogashira, Suzuki–Miyaura and Heck couplings, for example).

Together, these synthetic methods allow for the gram-scale preparation of well-defined and monodisperse conjugated materials, in ways that allow for the control of both regio- and stereochemistry. This synthetic flexibility is crucial in allowing a reliable correlation of structure and properties to be established, via the synthesis of oligomers with increasing chain length and hence the optimization of the desired physical properties. However, each class of building block has its own advantages and disadvantages. One potential drawback of these types of five- and six-membered ring units is that their relatively high aromatic character results in a competition between π-electron confinement within the ring and delocalization along the conjugated chain [5d, 6]. This point reveals the possible conflicting properties of a given building block: although aromaticity provides stability and ease of functionalization, it can hinder access to low HOMO–LUMO gap materials.

The types of coupling methodology described above potentially allow for the preparation of a broad range of homo- and mixed oligomers and polymers with considerable structural variation. However, it is amazing to note that only a limited number of basic building blocks are commonly used, namely olefins, acetylenes, aromatic rings (benzene, naphthalene, etc.) and heterocyclopentadienes (thiophene, pyrrole, etc.) (Fig. 4.1) [1–5]. Hence the introduction of novel building blocks is clearly a basis for the further tailoring of π-conjugated systems. Of course, the challenge is not just simply to prepare new series of conjugated materials, but to introduce synthons that exhibit properties that the already well-established building blocks do not possess, in order to obtain innovative molecular architectures or unique electronic properties.

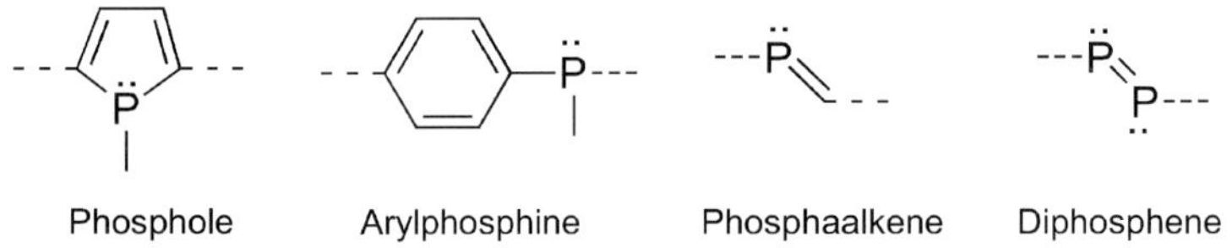

Phosphole   Arylphosphine   Phosphaalkene   Diphosphene

Fig. 4.2 P-based building blocks used for the construction of π-conjugated systems.

In contrast to sulfur- and nitrogen-based ring systems, which have been widely exploited for decades (Fig. 4.1), phosphorus-derived building blocks (Fig. 4.2) have only emerged in the late 1990s for the construction of π-conjugated materials [5j]. This situation is very surprising considering that the chemistry of these P-containing moieties is now well developed and that they exhibit properties that are markedly different from those of their N analogs [7]. This chapter intends to give an overview of the synthesis of various classes of π-conjugated oligomers and polymers incorporating P moieties (Fig. 4.2) and to illustrate the conceptual design and specific properties that result directly from the presence of the P atom. The presentation will follow a largely chronological order starting with phosphole (Fig. 4.2), which is still the most extensively used P unit for the synthesis of conjugated materials. Note that polyphosphazenes [8], which are the most familiar synthetic polymers incorporating phosphorus, will not be included in this review since they do not display the type of extended π-conjugation as sought in systems presented in Figs. 4.1 and 4.2.

4.2
Phosphole-containing π-Conjugated Systems

The first phospholes were described in the 1950s [9a–c], but synthetic methods combining high yields and diversity of substitution pattern were only discovered in the late 1960s [10a]. Hence the chemistry of phospholes is in its infancy compared with that of pyrrole and thiophene, compounds that were discovered during the 19th century. This is illustrated by the fact that the parent phosphole **1** (Scheme 4.1) was only characterized by NMR spectroscopy as recently as 1983 [10b]. Today, however, the chemistry (synthesis, reactivity, coordination behavior, etc.) of this P-based heterocycle has reached a much more mature state, allowing its use as a building block for the engineering of π-conjugated materials to be envisaged.

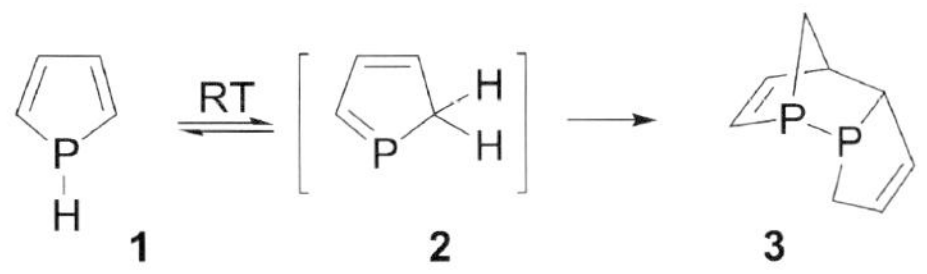

Scheme 4.1

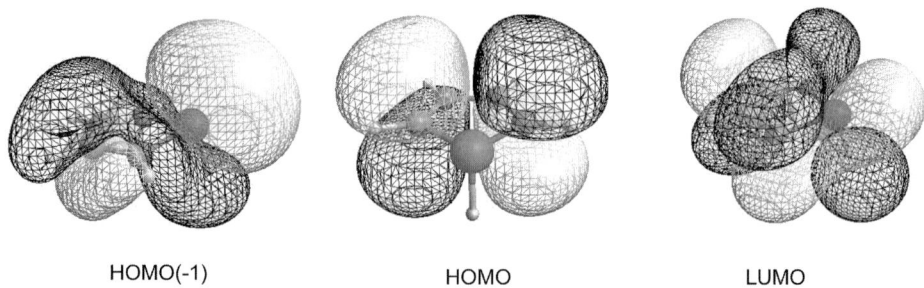

Fig. 4.3 B3LYP/6–31G⁺⁺ HOMO(–1), HOMO and LUMO of parent phosphole.

One of the major issues in understanding the properties of the phosphole ring was to determine its degree of aromaticity. This problem has long been debated and it is now accepted that, in marked contrast to pyrrole, phosphole exhibits a weak aromatic character [7, 11]. For example, the calculated aromatic stabilization energy (ASE) and nucleus-independent chemical shift (NICS) values are 7.0 kcal mol^{-1} and –5.3 for phosphole compared with 25.5 kcal mol^{-1} and –15.1 for pyrrole [11]. This lack of aromaticity is a consequence of two intrinsic properties of phospholes: the tricoordinate P atom adopts a pyramidal geometry and its lone pair exhibits a high degree of s character (Fig. 4.3). Together these two features prevent efficient interaction of the P lone pair with the endocyclic π-system. Although calculations have shown that planar phosphole would be more aromatic than pyrrole, this stabilization is not sufficient to overcome the high planarization barrier of the P atom (35 kcal mol^{-1}) [12], but is responsible for the reduced P-inversion barrier in phosphole (ca. 16 versus 36 kcal mol^{-1} for phospholanes [12a]). In fact, the weak aromatic character of the phosphole ring is due to hyperconjugation involving the exocyclic P–R σ-bond and the π-system of the dienic moiety [13]. As proposed for siloles (the heterole containing a tetrahedral Si center), which exhibit a similar σ–π interaction [14], this hyperconjugation is possible since the P atom adopts a tetrahedral geometry and the exocyclic P–R bonds are relatively weak.

These electronic properties (low aromatic character, σ–π hyperconjugation), which set phosphole apart from pyrrole and thiophene, have very important consequences. First, phosphole does not undergo electrophilic substitution, but does possess a reactive heteroatom. As a result, an electrophile will not attack the heteroatom C-α carbon atom, as observed with pyrrole, but the P atom instead (Fig. 4.4). Hence this P-containing ring has its own chemistry (synthetic routes, methods of functionalization, etc.) that cannot be predicted by simply extrapolating that of its aromatic S and N analogs. Second, phospholes make appealing building blocks for the tailoring of conjugated systems, since it is well established that conjugation will be enhanced for macromolecules built from monomer units that exhibit low resonance energies [5d, 6]. This phenomenon is nicely illustrated by theoretical work that showed that the energy gaps of oligo(phosphole)s are signif-

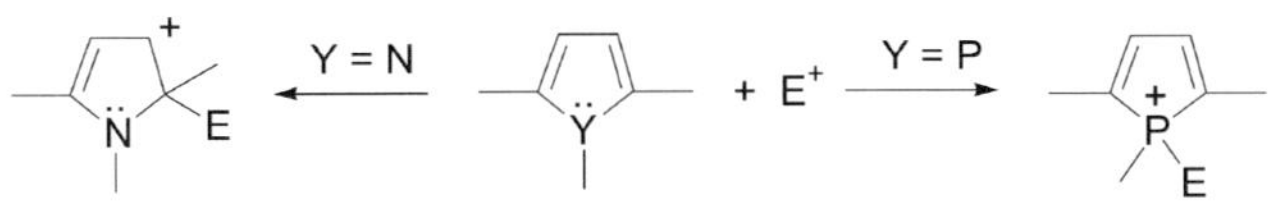

Fig. 4.4 Reactivity of pyrrole and phosphole towards electrophiles.

icantly lower than those of the corresponding oligo(pyrrole)s or oligo(thiophene)s [6]. Furthermore, the ease of functionalization of the P atom offers a unique way of creating structural diversity, including the preparation of transition metal complexes, from a single precursor. Lastly, if the chemistry of siloles is taken as a source of inspiration [14], the σ–π hyperconjugation associated with the phosphole ring should allow for the synthesis of multidimensional systems exhibiting σ–π conjugation (so-called "through-bond delocalization"). It will be demonstrated that all three of these features, which thiophene or pyrrole cannot offer, have been fully exploited in recent years for the preparation of phosphole-containing conjugated oligomers and polymers.

A last important issue that is crucial for the development of phospholes as building blocks for materials is their thermal stability and, to a lesser extent, their ease of handling and manipulation. The parent phosphole **1** can be observed at −100 °C; however, at room temperature, it isomerizes via a [1, 5]-sigmatropic shift to the 2H-phosphole **2**, which rapidly dimerizes to afford the endo derivative **3** (Scheme 4.1) [10]. The thermal stability of phospholes is considerably increased by the introduction of substituents on the ring. In particular, the nature of the P substituent has a dramatic influence since it is this group that migrates giving rise to the unstable 2H-phosphole **2** (Scheme 4.1). For example, phospholes bearing phenyl, cyclohexyl, cyano or alkoxy substituents at phosphorus are all stable at room temperature. Indeed, the decomposition temperatures of 2,5-diaryl-P-phenylphospholes, as estimated by thermogravimetric analyses, can reach 200 °C [15a]. Furthermore, these P-phenylphospholes are not moisture sensitive and can be purified by standard methods. In contrast, although P-amino- and P-halophospholes can be obtained at room temperature, they are readily hydrolyzed [15]. Just as for classical phosphines, phospholes have to be handled under an inert atmosphere in order to avoid oxidation of the P center. Notably, however, 2,5-diarylphospholes are generally stable enough to be purified and manipulated under air [15a].

4.2.1
α,α′-Oligo(phosphole)s

When considering the synthesis of phospholes, one has to forget most of the classical and powerful methods employed for the preparation of thiophenes and pyrroles. For example, Paal–Knorr condensation, direct *ortho*-lithiation, halogenation with NBS or I_2/Hg^{2+} and Vilsmeier–Haack formylation are not operative in phosphole chemistry. Likewise, no chemical or electrochemical oxidative polymerization

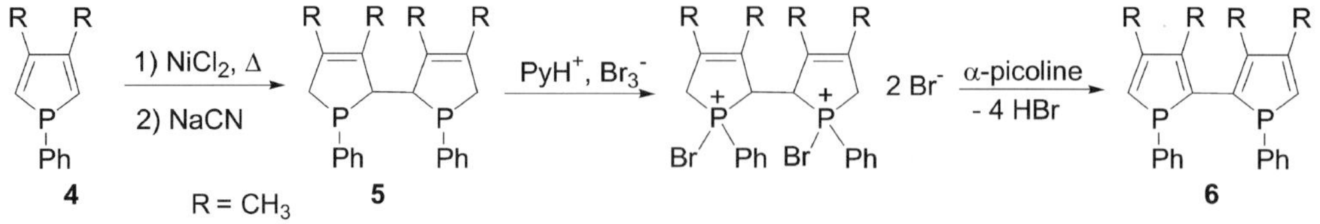

Scheme 4.2

of phospholes has yet been achieved. As a result, the preparation of α,α′-oligo(phosphole)s is a real synthetic challenge.

Mathey's group achieved a breakthrough in this field in the 1990s with the discovery of several synthetic routes to linear bi- and quaterphospholes. Initially, non-capped 2,2′-biphosphole **6** was prepared via a four-step sequence from 1-phenyl-3,4-dimethylphosphole **4** (Scheme 4.2), a starting material that is readily accessible in multi-gram quantities [16]. The Ni(II)-promoted reductive dimerization of phosphole **4**, followed by decomplexation, afforded the di-2-phospholene **5** (Scheme 4.2) [17]. Successive P-bromination and dehydrohalogenation gave rise to the target 2,2′-biphosphole **6**, which is stable enough to be purified by column chromatography [18a]. It is obtained as a mixture of diastereoisomers owing to the presence of two chirogenic P centers, which are in rapid equilibrium at room temperature since the barrier to inversion at phosphorus is low (ca. 16–17 kcal mol^{-1}) owing to the aromatic character of the planar transition state [18].

The second route to oligo(phosphole)s is based on the formation of 2-lithiophospholes. This reaction was a landmark in this area since it provided the first direct means of functionalization of the phosphole ring and opens the way to metal-catalyzed coupling reactions [19]. A quantitative bromine–lithium exchange transforms the 2-bromophospholes **7** into their highly reactive 2-lithio analogs **8** (Scheme 4.3). These 2-lithiophospholes undergo oxidative coupling leading to biphospholes **9** upon addition of copper(II) chloride (Scheme 4.3)

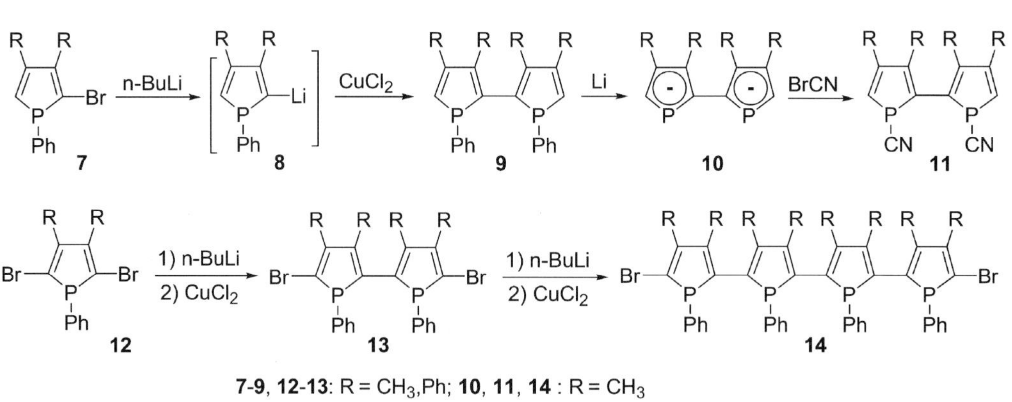

7-9, 12-13: R = CH$_3$,Ph; 10, 11, 14 : R = CH$_3$

Scheme 4.3

[19, 20]. The nature of the P substituents of **9** can be readily varied via a two-step sequence involving the generation of the phospholyl anion **10** by reductive cleavage of the exocyclic P–C bonds with lithium (Scheme 4.3). The dianion **10** can act as a nucleophilic reagent in the synthesis of new biphospholes, as illustrated with the preparation of **11** (Scheme 4.3) [18b]. The very efficient bromine–lithium exchange methodology has also been applied to the preparation of bromo-capped biphosphole **13** and quarterphosphole **14** using the dibromo precursor **12** (Scheme 4.3) [21]. These oligo(phosphole)s are obtained in good yields (e.g. **12** → **14**, 55 %) as mixtures of diastereoisomers.

This expedient and efficient synthetic methodology seems very attractive for the construction of longer, well-defined oligomers since **13** and **14** bear reactive termini. However, **14** is still the longest oligo(phosphole) known to date. No iterative coupling study has been undertaken, presumably because the preparation of the starting bromo precursors **7** and **12** is laborious and time consuming (Scheme 4.4) [19, 21], since it has been proved impossible to perform one-step bromination of the phosphole ring.

Unfortunately, the optical and electrochemical properties of these linear phosphole oligomers have not yet been elucidated, precluding an interesting comparison with related oligo(pyrrole)s. However, oligo(phosphole)s **9** (R = CH_3), **11** and **14** (Scheme 4.3) have been characterized by X-ray diffraction studies, which does give some insight into the properties of these systems.

In all cases, the phosphorus atoms of **9**, **11** and **14** adopt a pyramidal geometry, with the endocyclic P–C bond distances consistent with single bonds. Together, these geometric data show that, as observed for monophospholes, the P lone pair is not conjugated with the endocyclic diene framework. Furthermore, solid-state studies also revealed that the heterocyclic phosphole units are not coplanar. The dihedral angle between the two phosphole rings in **9** and **11** is about 46° [18b,c], whereas in **14** the twist angle between the two inner rings is 25.1° and that between the outer pair is 49.7° [21]. These rotational distortions should preclude these oligo(phosphole)s from possessing extended π-conjugated systems. However, these twists are probably due to packing effects in the solid state since the color of these compounds varies from pale yellow (**9**, **11**) to

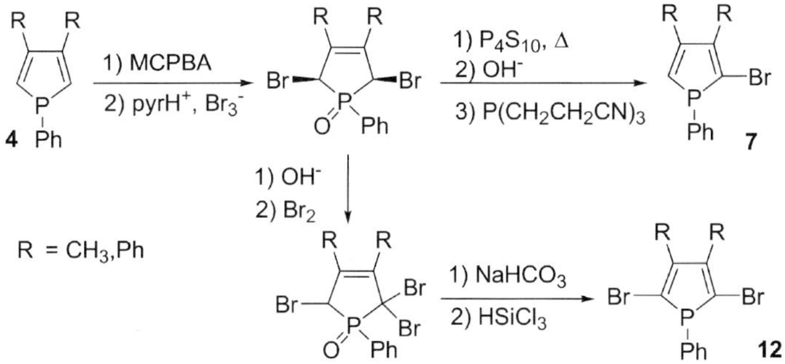

Scheme 4.4

R = CH_3, Ph

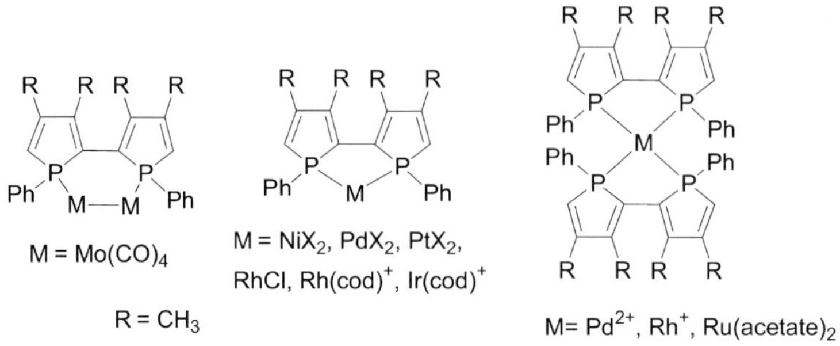

Scheme 4.5

M = Mo(CO)$_4$
R = CH$_3$

M = NiX$_2$, PdX$_2$, PtX$_2$, RhCl, Rh(cod)$^+$, Ir(cod)$^+$

M = Pd^{2+}, Rh$^+$, Ru(acetate)$_2$

deep orange (14), suggesting rather high λ_{max} values and, consequently, low optical HOMO–LUMO band gaps. The red shift observed on going from derivatives 9 and 11 to 14 hints that, in line with theoretical studies [6b,c], the energy gap of oligo(phosphole)s decreases with increasing chain length.

It is noteworthy that the P atoms of these phosphole oligomers retain the versatile reactivity observed with monophospholes [7]. In particular, biphosphole 9 exhibits a rich coordination chemistry, which allows for the preparation of a variety of neutral and cationic transition metal complexes (Scheme 4.5) [18a,b, 22]. These have been used as precursors for homogeneous catalysts [22]; however, their photophysical properties have not been studied.

It can be concluded that the size of the family of linear oligo(phosphole)s is somewhat limited to bi- and tetramers to date, although longer derivatives are potentially accessible via the efficient synthetic routes developed by Mathey's group. The synthesis of other oligo(phosphole)s and the elucidation of their photo-physical and electrochemical properties are still needed in order to establish reliable structure–property relationships.

4.2.2
Derivatives Based on 1,1'-Biphosphole Units

The α,α'-oligo(phosphole)s described in the previous section possess classical one-dimensional conjugated systems based on alternating double and single C–C bonds. The use of phospholes as building blocks allows for the variation in the topology of the conjugation pathway through the synthesis of 1,1'-biphospholes (Scheme 4.6). These derivatives date back to the early days of phosphole chemistry, the first member of the family having been prepared in 1979 (15, Scheme 4.6) [23a], with the three other examples 16–18 appearing in the early 1980s [23b,d]. The most common synthetic routes to 1,1'-diphospholes involve either the oxidative coupling of readily available phospholyl anions [23d] or the thermolysis of phospholes (Scheme 4.6). This last route implies the formation of transient P–H phospholes, via 1,5-shifts of R and H, followed by loss of H$_2$ via an as yet unexplained mechanism [10c, 23d]. The P atoms of 1,1'-biphospholes

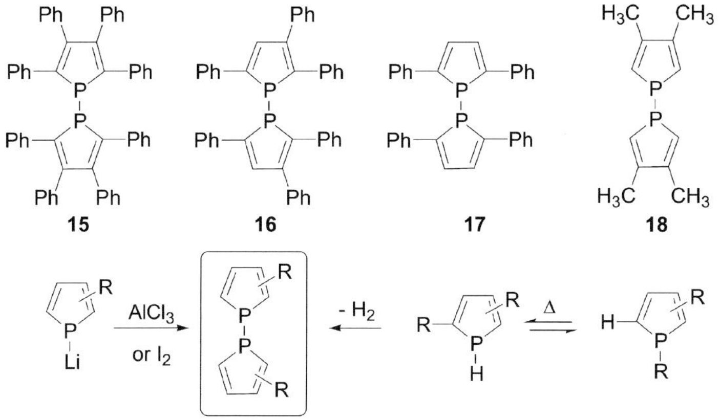

Scheme 4.6

retain a pyramidal shape, as observed for the corresponding monomeric phospholes [23e].

1,1′-Biphospholes have been mainly used as precursors for the synthesis of other P-containing species including mono- and diphosphaferrocenes [24]. Their electronic properties were recently investigated [25] with the aim of trying to observe any interaction between the π-systems via the P–P linkage resulting from the σ–π hyperconjugation occurring within the phosphole ring (Fig. 4.3). The fact that π-chromophores can effectively be conjugated via σ-skeletons [26], especially those exhibiting high polarizability and low σ–σ* energy gaps, was recognized by Hoffmann et al. in the late 1960s [26a,b]. This unusual conjugation pathway results in novel electronic properties that are starting to be exploited in materials science. For example, in oligo(1,1′-siloles) an effective interaction (σ*–π* conjugation) occurs between the Si–Si bridges and the butadienic moieties [27a,b], making these macromolecules very attractive chemical sensors for aromatic substrates [27c,d,g] and highly efficient electron-transporting materials for light-emitting diode materials [27e,f].

With a view to further exploitation of these unusual σ–π conjugated species, the mixed thiophene–phosphole derivative 19 (Fig. 4.5) has been prepared from the corresponding monomeric phospholyl anion as an air-stable powder [25]. As expected, theoretical calculations and UV–Vis data clearly establish a through-bond electronic interaction between the two 2,5-dithienylphospholyl moieties via the P–P bridge. One important consequence of this σ–π conjugation is that 19 possesses a smaller optical HOMO–LUMO gap than the corresponding 2,5-dithienylphosphole (see Section 4.2.3). It is interesting that the P atoms of 19 are reactive, allowing further chemical modification of the nature of the P–P bridge. Derivatives 20 and 21 are readily obtained using classical reactions, which exploit the nucleophilic behavior of the σ^3,λ^3-P centers of 19 (Fig. 4.5). These chemical modifications have an impact on the optical properties of the assemblies. For example, oxidation of one P center or complexation of the P atoms

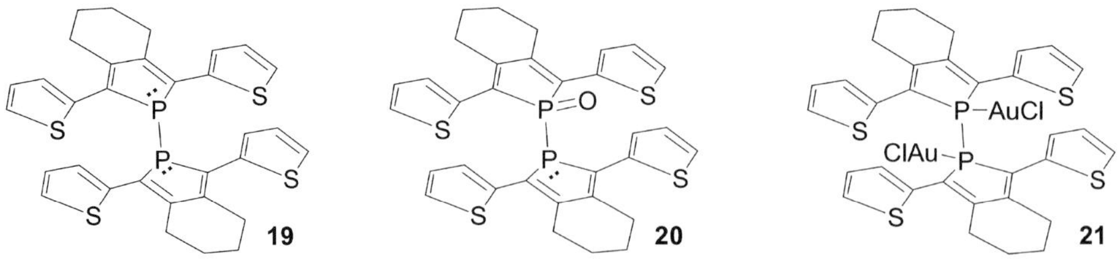

Fig. 4.5 Structures of thienyl-capped 1,1′-biphospholes and Au(I) derivatives.

with gold(I) results in bathochromic shifts of the band onset compared with 1,1′-biphosphole **19** [25].

The second route to 1,1′-diphospholes, involving [1, 5]-sigmatropic shifts and loss of H_2 (Scheme 4.6), conducted with 1-phenylphosphole **4**, afforded the macrocycle **22** (Scheme 4.7) [28a]. This compound has a remarkable structure with two α,α′-biphosphole cores connected by two P–P bonds. An X-ray diffraction study revealed that the P atoms are located trans with respect to the linking C–C bond [28b]. However, **22** is very probably flexible since, in the corresponding Mo(CO)$_5$ complex **23** (Scheme 4.7), they adopt mutually cis positions [28a]. As observed in the linear series, the two P rings of the α,α′-biphosphole moieties of **22** are not coplanar (twist angle, 44.2°) [28b]. The UV–Vis data for this unique type of macrocycle have not been reported, but its red color again suggests a low HOMO–LUMO separation due to σ–π conjugation.

Significantly, derivative **22** can be used as a precursor to other macrocycles featuring the 1,1′-diphosphole moiety. For example, the reductive cleavage of the two P–P bonds with sodium gave the dianion **24**, which, upon reaction with **22**, afforded the dianion **25** (Scheme 4.8) [28c]. Derivative **25** is a versatile nucleophile that reacts with tetrachloroethylene and dihaloalkanes to give the novel macro-

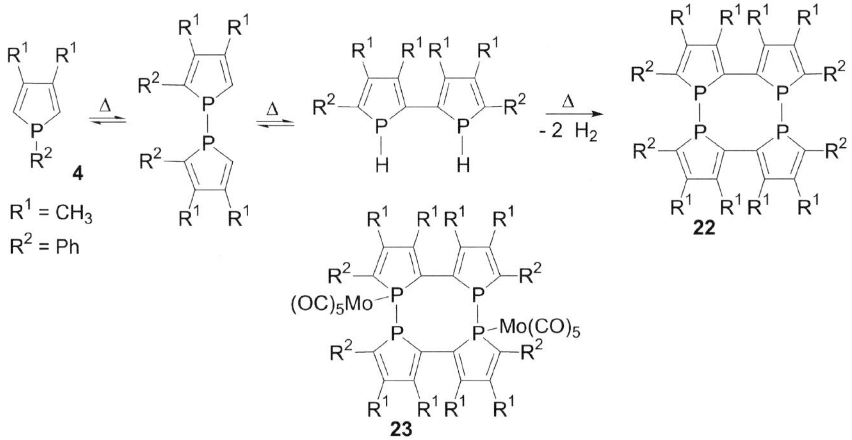

Scheme 4.7

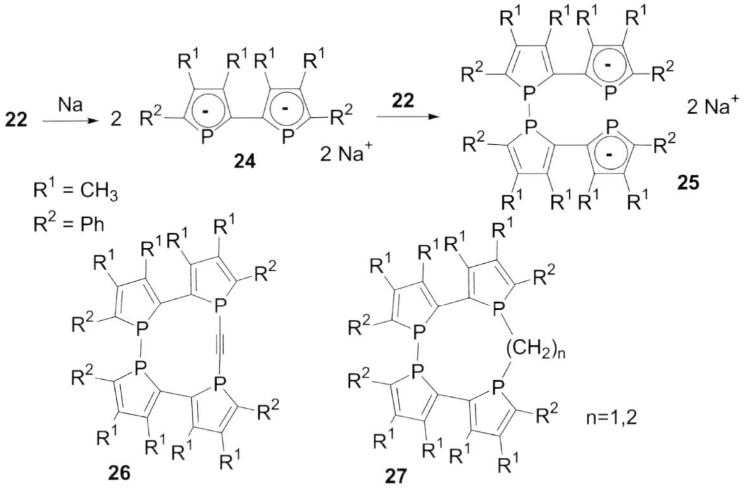

Scheme 4.8

cycles **26** and **27**, respectively [28c]. Compound **26** was characterized by an X-ray diffraction study [28c], which revealed that this fairly rigid 10-membered macrocyclic compound exhibits some strain, as shown by the nonlinearity of the P–C≡C–P unit [P–C–C, 170.2(8)° and 173.5(6)°].

The above-mentioned examples nicely illustrate the diversity of structures offered through the use of 1,1′-diphosphole building blocks. Importantly, it has been shown that these moieties exhibit σ–π conjugation, a type of electronic interaction that is still relatively rare. Together, these results show that this type of P-based moiety is a promising elementary unit for the engineering of conjugated systems.

4.2.3
Mixed Oligomers Based on Phospholes with Other (Hetero)aromatics

The first series of mixed macromolecules based on phospholes and other (hetero)-aromatics to be prepared were composed of 2,5-di(2-phosphole)thiophene or -furan derivatives. The key step in their synthesis is an electrophilic substitution in which the protected phosphole **28** acts as an electrophile towards electron-rich thiophene or furan rings, leading to the 3-phosphole adducts **29a,b** (Scheme 4.9) [29a]. These compounds were then transformed into the corresponding σ^3,λ^3-phosphole derivatives **30a,b** through a classical deprotection–bromination–dehydrohalogenation sequence (Scheme 4.9) [29]. Although these macromolecules evidently absorb in the visible region (**30a**, orange; **30b**, bright yellow), no UV–Vis data are available. An X-ray diffraction study performed on thioxo derivative **31b** showed that only one phosphole ring is coplanar with the central furan unit (twist angle, 3.3 ± 0.1 and 40.1 ± 0.1°). The two inter-ring C–C bond dis-

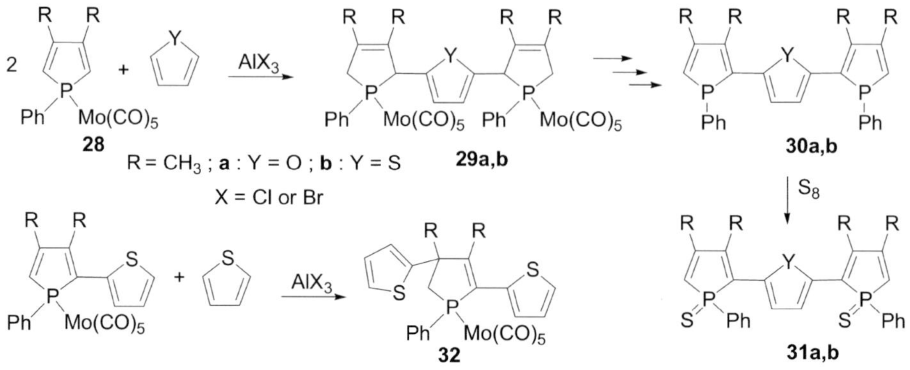

Scheme 4.9

tances [1.452(4) and 1.461(6) Å] lie between those observed for C–C single and double bonds, a feature that is in favor of a certain degree of delocalization [29a].

The first systematic investigation of phosphole-containing conjugated systems, which starts with model molecules and builds up to conductive polymers and materials suitable for OLED applications, was undertaken with 2,5-di(heteroaryl)-phospholes. In contrast, these derivatives are not accessible by the electrophilic substitution route described above for the preparation of **30**, since the second condensation step gives rise to 2,4-dithienylphospholene **32** (Scheme 4.9) [29a]. However, 2,5-di(heteroaryl)phospholes are readily accessible via a widely applicable organometallic route known as the Fagan–Nugent method [30]. The key to obtaining the desired 2,5-substitution pattern is to perform the metal-mediated oxidative coupling of diynes **33** possessing a $(CH_2)_3$ or a $(CH_2)_4$ spacer (Scheme 4.10) [15a, 31]. These diynes, including unsymmetrical variants, can be readily prepared using Sonogashira coupling [15a, 31]. The zirconacyclopentadiene intermediates **34** are extremely air and moisture sensitive derivatives that react with dihalophosphines to give the corresponding phospholes **35a–g** in medium to good yields (Scheme 4.10). This route is highly flexible since it not only allows electron-deficient and electron-rich rings to be introduced in the 2,5-positions, but also permits the nature of the P substituent to be varied. For example, the σ^3,λ^3-phospholes **35a** [31a] and **35d** [31b] bearing electron-deficient and electron-rich substituents, respectively, have been prepared using this approach. Together with the $W(CO)_5$-complex of **35g** [31b], they were subsequently characterized by X-ray diffraction studies. In spite of the different natures of the two 2,5-substituents (electronic disparity, shape, etc.), these compounds share some important structural features in the solid state. The twist angles between two adjacent rings are rather small [**35a**, 7.0 and 25.6°; **35d**, 12.5 and 16.7°] and the phosphorus atoms are strongly pyramidalized, as indicated by the sum of the CPC angles [**35a**, 299.3°; **35d**, 299.3°]. Note that these derivatives are the first phosphole-based conjugated systems that are planar in the solid state. The lengths of the C–C linkages between the rings [1.436(6)–1.467(8) Å] are in the range expected for Csp^2-Csp^2

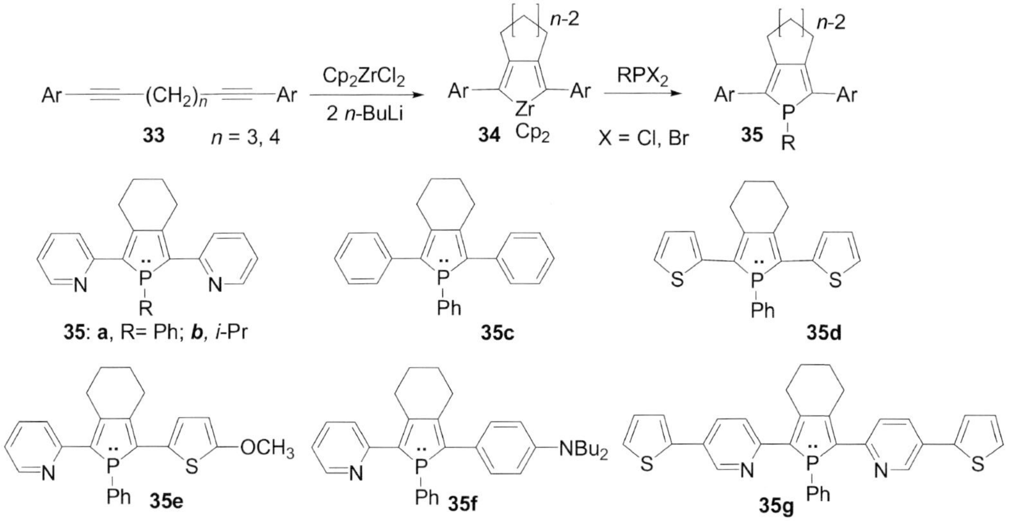

4.2 Phosphole-containing π-Conjugated Systems

Scheme 4.10

single bonds. Together these metric data suggest a delocalization of the π-system over the three heterocycles.

In solution, phospholes **35a–g** present broad absorptions in the visible region of the spectrum attributed to π–π* transitions [15a, 31]. The values of λ_{max} and the optical end absorption λ_{onset} depend on the nature of both the P and the 2,5-substituents. A red shift of the values of λ_{max} for P-arylphospholes is observed relative to those of the corresponding P-alkyl analogs (Table 4.1, **35a** vs. **35b**). An important bathochromic shift was recorded on replacing the phenyl groups either by 2-pyridyl (λ_{max} = 3 nm) or 2-thienyl rings (λ_{max} = 58 nm) (Table 4.1) [15a]. Collectively, these data suggest that the HOMO–LUMO gap gradually decreases in the series **35c** < **35a** < **35d**, a feature that was confirmed by high-level theoretical calculations [15a]. This variation was initially attributed to the fact that phospholes possess low-lying LUMO levels (high electron affinity), which favors intramolecular charge transfer from the electron-rich thienyl substituents [15a]. A more recent theoretical study proposed that the more pronounced π-conjugation in **35d** is due to a better interaction between the HOMO of the phosphole with the HOMO of the thiophene compared with that with pyridine [32]. It is interesting that the value of λ_{max} recorded for **35d** (412 nm) is considerably more red shifted than those of related 2,5-dithienyl-substituted pyrrole (322 nm), furan (366 nm) or thiophene (355 nm) [33]. These observations help to establish phospholes as valuable building blocks for the construction of co-oligomers exhibiting low HOMO–LUMO separations. This appealing property is probably due to the weak aromaticity of the phosphole ring, which favors an exocyclic delocalization of the dienic π-electrons.

Table 4.1 Physical properties of di(heteroaryl)phospholes **35a–d** and derivatives **36b** and **37–39**

Compound	$\lambda_{max}^{[a]}$ (nm)	$\lambda_{onset}^{[a]}$ (nm)	Log ε	$\lambda_{em}^{[a]}$ (nm)	Φ_f	$E_{pa}^{[b]}$ (V)	$E_{pc}^{[b]}$ (V)
35a	390	448	4.02	463	1.1×10^{-2}	+0.83	−2.45
35b	371	430	4.10	458	–	+0.79	−2.67
35c	354	430	4.20	466	14.3×10^{-2}	+0.69	−2.88
35d	412	468	3.93	501	5.0×10^{-2}	+0.40	–
36b	432	496	3.98	548	4.6×10^{-2}	+0.68	−1.95
37	442	528	3.92	593	0.8×10^{-2}	+0.92	−1.66
38	408	475	4.04	506	1.3×10^{-2}	+0.70	−2.20
39	428	500	4.18	544	14.0×10^{-2}	+0.82	−1.75

[a] In THF.
[b] In CH_2Cl_2, referenced to ferrocene/ferrocenium half-cell.

Varying the nature of the 2,5-substituents is also an effective way of tuning the emission and electrochemical behavior of phosphole-based extended π-electron systems. For example, di(2-pyridyl)phosphole **35a** emits blue light, whereas the emission of di(2-thienyl)phosphole **35d** is red shifted (Table 4.1). Note that the quantum yields in solution are relatively low, the highest (14.3×10^{-2}) being observed for 2,5-diphenylphosphole **35c** (Table 4.1). All the σ^3-phospholes **35** show redox processes that are irreversible at 200 mV s^{-1} [15a], with the redox potentials being related to the electronic properties of the phosphole substituents (Table 4.1). For example, derivative **35d** featuring electron-rich thienyl substituents is more easily oxidized than compound **35a**, which bears electron-deficient pyridyl groups.

One of the appealing properties of phosphole rings when creating structural diversity in these types of system is the versatile reactivity of the endocyclic heteroatom. Exploiting the nucleophilic behavior of the P center allows direct access to a range of new π-conjugated systems, including metal complexes, as illustrated with dithienylphosphole **35d** (Scheme 4.11). It is noteworthy that the bis(2-thienyl)phosphole moiety remains almost planar in the coordination sphere of W(CO)$_5$ (twist angles, 7.3 and 13.4°) and AuCl (twist angles, 3.8 and 13.5°) and that the bond lengths and angles are unchanged upon coordination [15a, 34]. However, these chemical modifications of the nucleophilic P center have a profound impact on the optical and electrochemical properties of the phosphole oligomers as a whole. For example, on going from the σ^3-phosphole **35d** to the neutral σ^4-derivative **36b** and the σ^4-phospholium salt **37**, a red shift is observed in their UV–Vis spectra, together with an increase and decrease in their oxidation and reduction potentials, respectively (Table 4.1). These trends can be rationalized on the basis that within this series the electronic deficiency of the P atom is gradually augmented [15a, 32]. Exploitation of this method of tailoring π-conjugated

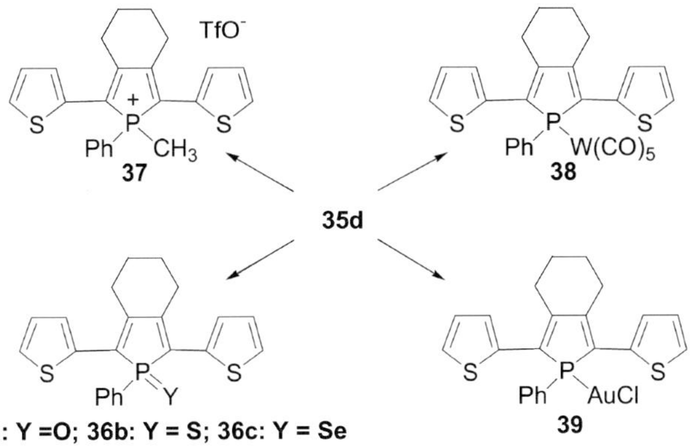

Scheme 4.11

36a: Y = O; 36b: Y = S; 36c: Y = Se

systems, which is simply not possible with pyrrole and thiophene units, has led to the optimization of the properties of thiophene–phosphole co-oligomers for use as materials suitable for OLED fabrication.

Single-layer OLEDs have been prepared through sublimation of the thermally stable thioxo derivative **36b** and gold complex **39** on to semi-transparent indium–tin oxide (ITO) anodes, with both substrates forming homogeneous thin films [34]. In contrast, the σ^3-phosphole **35d** decomposed under sublimation conditions. The device prepared using **36b** as the active organic layer exhibited yellow emission for a relatively low turn-on voltage of 2 V. The electroluminescence (EL) spectra of this device match those of the solid-state photoluminescence (PL) of **36b**, showing that the source of the EL emission band is from the phosphole derivative. In contrast, the emission exhibited by the single-layer device constructed using gold complex **39** covers the 480–800-nm region [34]. The low-energy emissions are probably due to aggregate formation in the solid state. It is clear that the gold centers play a key role in the formation of these aggregates, probably via aurophilic interactions [35], since no low-energy luminescence bands are observed with the thioxophosphole derivatives **36b**. The comparatively low maximum brightness (ca. 3613 cd m^{-2}) and EL quantum yields (ca. 0.16 %) of the single-layer OLEDs can be increased by nearly one order of magnitude by using a more advanced device, in which the organic layer consisting of **36b** or **39** was sandwiched between hole- and electron-transporting layers (α-NPD and Alq$_3$, respectively) [34].

Another effective way to further improve OLED performance and also to tune their color is to dope highly fluorescent dyes as guests into an emissive host matrix [36]. With this in mind, thioxophosphole **36b** was evaluated as a host material for DCJTB, the best red-emitting dopant for OLEDs reported to date [1d]. This device showed red emission from DCJTB, with an EL of 1.83 % and a maximum brightness of ca. 37 000 cd m^{-2} [34]. Notably, the external EL quantum efficiency of this device is unaffected by drive current density in the range 0–90 mA cm^{-2} [34]. This behavior is very promising since the efficiency of DCJTB-doped Alq$_3$

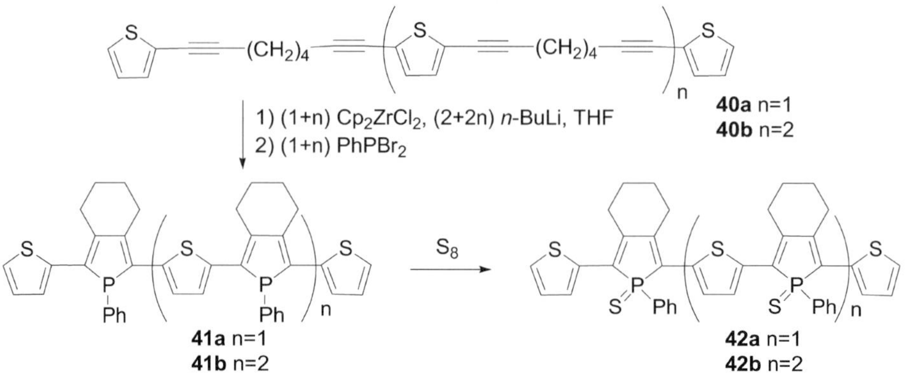

Scheme 4.12

devices usually decreases rapidly with the driving current, something that results from quenching effects of charged excited states of Alq$_3$ on the red dopants [36].

The evolution of optical and electrochemical properties with increasing chain length has been investigated for α,α'-thiophene–phosphole (Th-Phos) oligomers in order to gain a greater understanding of the characteristics of these novel π-conjugated systems. To this end, the well-defined oligomers **41a,b** were prepared using the Fagan–Nugent method from the corresponding bis- and tris-diynes **40a,b** (Scheme 4.12) [37]. Derivatives **41a,b**, and also their thiooxo-derivatives **42a,b**, exist as a mixture of diastereoisomers due to the presence of stereogenic P centers. They are air stable and soluble in common organic solvents (e.g. THF, CH$_2$Cl$_2$). Significantly, however, the yields of the various oligomers decrease dramatically with increasing chain length (Table 4.2), precluding the preparation of longer chain derivatives.

With respect to their electronic properties, the longest wavelength absorption, emission band and the oxidation potentials all gradually shift to lower energies with increasing chain length (Table 4.2). This reflects the expected decrease in the HOMO–LUMO gap upon increasing the chain length of the α,α'-(thio-

Table 4.2 Preparation and physical properties of oligo(α,α'-thiophene–thioxophosphole)s

Compound	Yield (%)	λ_{max}[a] (nm)	λ_{onset}[a] (nm)	Log ε	λ_{em}[a] (nm)	E_{pa}[b] (V)
36b ($n = 0$)	93	432	496	3.98	548	+0.68
42a ($n = 1$)	78	508	590	4.26	615	+0.45
42b ($n = 2$)	32	550	665	4.42	–	–

[a] In THF.
[b] In CH$_2$Cl$_2$, referenced to ferrocene/ferrocenium half-cell.

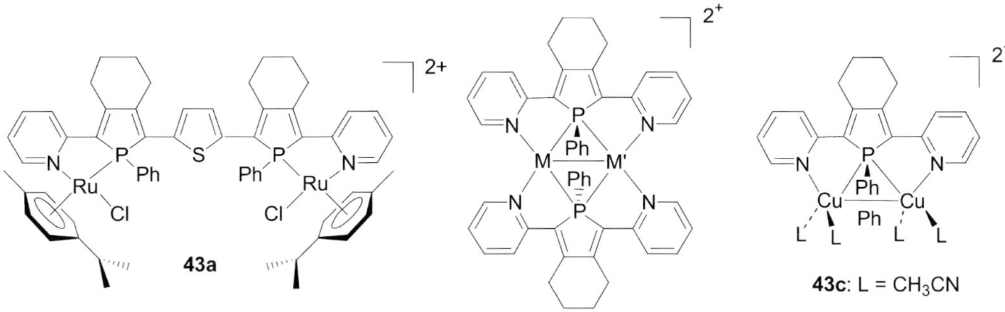

Scheme 4.13

43b: M = M' = Pd; M = Pd, M' = Pt

43c: L = CH$_3$CN

phene–phosphole) oligomers. It is therefore likely that the effective conjugation pathlength is much longer than 7 units for oligomers of these types. Notably, the value of λ_{max} for Th(Phos-Th)$_2$ **41a** (490 nm) is considerably red shifted compared with that of quinquethiophene (ca. 418 nm) [38], once again showing that replacing a thiophene subunit by a phosphole ring induces an important decrease in the optical HOMO–LUMO gap. It is also noteworthy that oligo(α,α'-thiophene–phosphole)s **42a,b** incorporating σ^4-thioxophospholes have smaller HOMO–LUMO gaps relative to their precursors **41a,b** based on σ^3-P rings. These results, along with the good stability and solubility of derivatives **41a,b** and **42a,b**, should encourage the search for new synthetic routes to longer oligo(α,α'-thiophene–phosphole)s.

The coordination of π-chromophores to transition metals is a powerful way to modify their characteristics and to engender novel properties. Two remarkable examples of this methodology are the use of cyclometalated (phenylpyridinato)Ir(II) complexes (highly efficient phosphorescent dopants for OLEDs due to the strong spin–orbit coupling caused by the heavy metal ion) [4d, 39] and homoleptic (aminostyrylbipyridine)Ru(II) complexes (powerful octupolar NLO-phores due to the octahedral geometry of the metal center) [40]. Since 2-pyridylphosphole moieties can act as 1,4-chelates towards transition metals, as illustrated with the synthesis of complexes **43a–c** (Scheme 4.13), it was of interest to examine the behavior of these conjugated P-containing oligomers in a similar way [41]. The UV–Vis spectrum of the free ligand and of the Ru(II) complex are almost identical [41a]. In contrast, for the Pd(I) dimer, low-energy UV–Vis absorptions assigned to charge transfer from the phosphorus–metal fragment to the pyridine ligands were observed [41c].

One key property of pyridylphosphole ligands is their heteroditopic nature. They possess two coordination centers with different stereo-electronic properties, which, in accordance with Pearson's antisymbiotic effect [42a], can control the orientation of a second chelating ligand in the coordination sphere of a square-planar d^8-metal center [42b]. This property has been exploited in order to control the in-plane parallel arrangement of 1D-dipolar chromophores [42d]. Phospholes

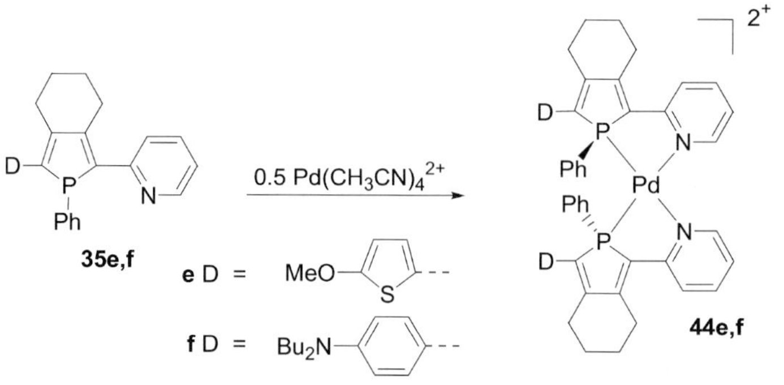

Scheme 4.14

35e,f (Scheme 4.14) have a typical "D–(π-bridge)–A" dipolar topology. They exhibit only moderate NLO activities ($\beta_{1.9\,\mu m}$, ca. 30×10^{-30} esu), something probably due to the weak acceptor character of the pyridine group [42d]. However, the potential of these dipoles in NLO is considerably increased by their P,N-chelate behavior. In accordance with the antisymbiotic effect, they undergo stereoselective coordination leading to a close parallel alignment of the dipoles on the square-planar d^8 Pd(II) template. Thus, the trans-effect can overcome the natural anti-parallel alignment tendency of 1D-dipolar chromophores at the molecular level. The noncentrosymmetric complexes **44e,f** exhibit fairly high NLO activities ($\beta_{1.9\,\mu m}$, ca. $170–180 \times 10^{-30}$ esu), something that is probably due to the onset of ligand-to-metal-to-ligand charge transfer, which contributes coherently to the second harmonic generation [42d].

Hence it is clear that synthetic strategies are available for the synthesis of low molecular weight, well-defined mixed oligomers incorporating phosphole moieties. Comparison with related derivatives containing thiophene or pyrrole rings has, without a doubt, shown that the specific properties introduced by the presence of the P-heterocycle (e.g. low aromaticity, reactive heteroatom, σ–π hyperconjugation) afford these systems unique properties and considerable scope for further fine-tuning. Significantly, the mixed phosphole-based oligomers described in this section are more than simple model molecules since some of these compounds have already been used as active layers in OLED devices. These results prove that there are no inherent problems associated with the development of phosphole-based conjugated systems that possess optoelectronic functions. Furthermore, they clearly demonstrate that exploiting the reactivity of the P center of phosphole-derived oligomers could provide an exciting and powerful means of both optimizing and developing opto-electronic materials for the future.

4.2.4
Mixed Oligomers Based on Biphospholes with other (Hetero)aromatics

Mathey and coworkers have prepared a series of dithienyl-capped biphospholes that are very attractive owing to the presence of electro-active termini and the potential diversity of structures that are available (e.g. linear, cyclic, P–C or P–P bonds). The synthetic strategies that afford these compounds are interesting, since they nicely illustrate some of the specific methods that can be used to prepare phosphole derivatives. For example, nucleophilic substitution at the P atom of phosphole **45** afforded the 1-bithienylphosphole **46** (Scheme 4.15) [43a]. Upon prolonged heating, this compound gave the cyclic tetramer **47** featuring biphosphole moieties, according to the mechanism depicted in Scheme 4.7. Reductive cleavage of the P–P bonds of **47** afforded the dianion **48**, which can subsequently be converted into the target dithienyl-capped biphospholes **49** and **50** (Scheme 4.15) [43a]. These derivatives are sensitive to oxidation and were isolated as adducts with sulfur or tungsten pentacarbonyl, derivatives **51** and **52**, respectively (Scheme 4.15).

An X-ray diffraction study of complex **52** showed that this mixed oligomer is not planar. The thiophene–thiophene, thiophene–phosphole and phosphole–phosphole twist angles are 6.35, 24.24 and 66.26°, respectively [43a]. Compound **51** un-

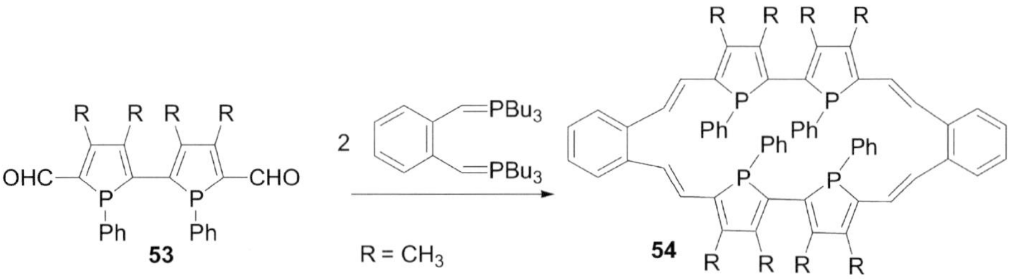

Scheme 4.16

dergoes two closely spaced successive one-electron reductions [43b]. Note that the reduced species **51**·⁻ and **51**⁻ are relatively stable on the cyclic voltammetry (200 mV s^{-1}) time-scale, suggesting the existence of extended delocalization in these compounds [43b].

Cyclic oligomers based on biphospholes are also available via Wittig reactions involving the 5,5′-bis(carboxaldehyde) **53** (Scheme 4.16) [44]. An X-ray diffraction study of **54** revealed that the four P-phenyl groups have an all-trans disposition and that this fully unsaturated macrocycle is distorted [44]. The cavity of the 24-membered heterocycle **54** is rather large, the diagonal distance between two P atoms reaching 6.1 Å.

4.2.5
Mixed Oligomers Based on Phospholes with Ethenyl or Ethynyl Units

Oligomers and polymers consisting of alternating aromatic building blocks and either ethenyl or ethynyl units {e.g. PPV, oligo(thienylenevinylene)s [45a,b], poly(p-phenyleneethynylene)s [5k, 45c]} have found numerous applications in the fields of OLEDs, nonlinear optics, sensors, polarizers for liquid crystal displays, etc. Furthermore, the controlled synthesis of aromatic-ethynyl chains is driven by the need for molecular wires for the construction of a variety of nanoarchitectures [5e,i, 46]. Surprisingly, very few derivatives incorporating phosphole rings linked by a double or a triple bond have been reported to date. Once again, this is mainly due to the fact that the synthetic routes, especially metal-catalyzed C–C coupling reactions involving aromatic building blocks, are not efficient with phospholes. For example, neither 2-bromo-5-iodophosphole **55** nor its dibromo analog **12** undergo Stille-type couplings with 1-stannylalkynes, and the Sonogashira coupling of **55** with phenylacetylene afforded derivative **56** in only 10% yield (Scheme 4.17) [47a].

The key to the synthesis of such target ethynyl- and ethenylphosphole derivatives is the mono- and dilithiation of 2,5-dibromophosphole **12**. The intermediate 2-lithio-5-bromophosphole reacts with arylsulfonylacetylene **57a** (Scheme 4.17) to gave rise to derivative **56**, which can be converted into α,α′-di(acetylenic)phosphole **58** according to the same general strategy as above. Although the yields of these reaction sequences are rather modest (typically around 30%), this synthetic approach allows the stepwise preparation of derivative **60** from **12** by employing trimethylsilyl-protected alkynes **57b** (Scheme 4.17) [47a]. Disappointingly, the modest yields of this overall methodology preclude its use for the preparation of longer oligomers or polymers starting from **12** or **59**.

X-ray diffraction analysis of the model compound **58** revealed that the C–C linkages between the P-heterocycle and the C≡C moieties are rather short [1.423(3)–1.416(3) Å] [47a]. These data suggest that the endocyclic dienic π-system of the phosphole unit is efficiently conjugated with the two acetylenic substituents. The presence of an extended π-conjugated system for **58** is also supported by its orange color. The P atom adopts a pyramidal geometry and the P–C bond distances are typical for P–C single bonds [1.815(2)–1.821(2) Å]. Hence, as observed for 2,5-diheteroarylphospholes, the lone pair of the P atom does not interact with the π-system. Note that derivatives **59** and **60** are orange, which is again suggestive of the presence of an extended π-conjugated system [47a,b].

To date, only the simplest member of the oligo(phospholylenevinylidene) family, namely derivative **62** (Scheme 4.17), is known [47b]. This compound can be obtained in high yield, as the *E*-isomer, by a McMurry coupling with aldehyde **61** (Scheme 4.17) [47b]. Note that the use of Ti salts is compatible with the presence of σ^3-P centers. No photophysical data for the orange mixed phosphole–ethenyl derivative **62** have been reported.

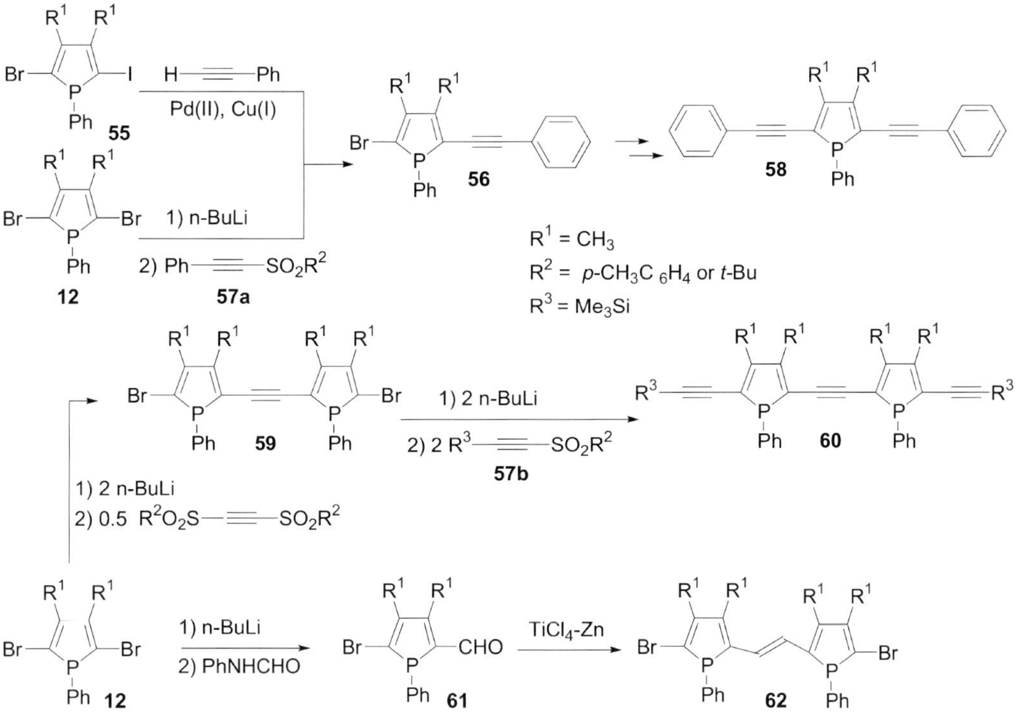

Scheme 4.17

In conclusion, very few mixed oligomers based on phospholes combined with ethenyl or ethynyl moieties are known. Initial, but so far incomplete, data suggest that the compounds that have been studied are fully conjugated. This feature should motivate the synthesis of longer oligomers that are potentially accessible using bromo-capped building blocks. However, further progress is currently hampered by the low efficiency of coupling reactions involving phosphole rings.

4.2.6
Polymers Incorporating Phospholes

To date, no homopolymers based on the phosphole motif are known. In contrast, a number of phosphole-containing copolymers have been reported. The first to be prepared was the biphenyl–phosphole derivative **65** obtained by Tilley and coworkers using the Fagan–Nugent methodology (Scheme 4.18) [48a]. Its synthesis involved the oxidative coupling of rigid diynes **63** with zirconocene, which proceeded in a non-regioselective way affording an 80:20 isomeric mixture of 2,4- and 2,5-connected metallacycles, respectively, in the polymer backbone of **64** (Scheme 4.18) [48a]. The reactive zirconacyclopentadiene moieties can subsequently be converted to the desired phospholes by adding dihalophosphine (Scheme 4.18) or to a range of other structures including dienes and benzene rings [48a–c].

Biphenyl–phosphole copolymer **65** is isolated as an air-stable and soluble powder exhibiting a rather high molecular weight (M_w = 16 000; M_n = 6200) according to gel permeation chromatographic (GPC) analysis. Although multinuclear magnetic resonance spectroscopy and elemental analysis support the proposed structure, the presence of a small number of diene defects is very likely. Polymer **65**

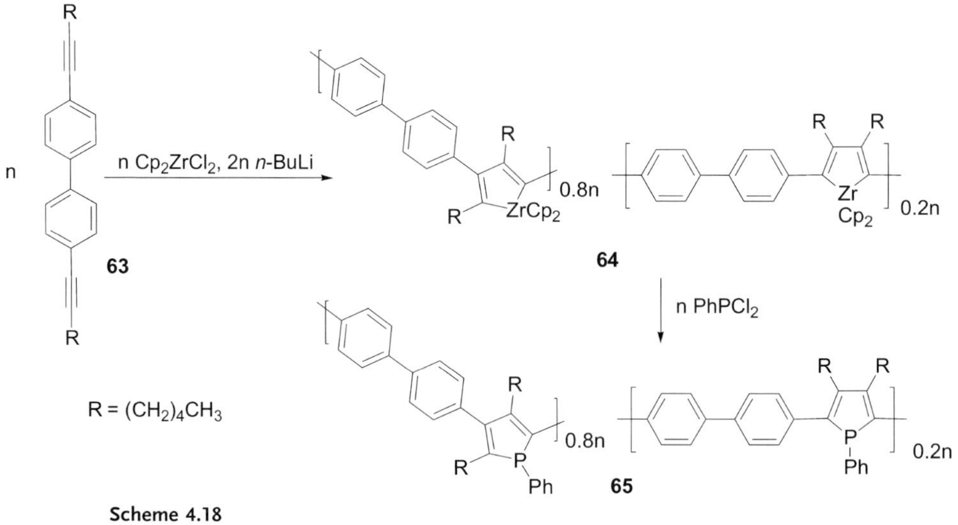

Scheme 4.18

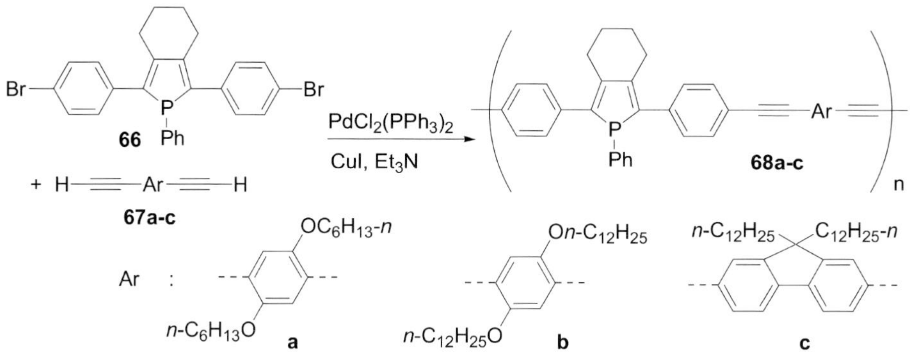

Scheme 4.19

exhibits a maximum absorption in its UV–Vis spectrum at 308 nm with a λ_{onset} value of 400 nm. These figures are consistent with a relatively high band gap, a feature probably due to a preponderance of cross-conjugated segments [48a]. Macromolecule **65** is photoluminescent; it emits in the bluish green region of the spectrum (470 nm) with a quantum yield reaching 9.2×10^{-2}.

A second type of π-conjugated polymer featuring a phosphole ring was obtained by Chujo et al. using the Heck–Sonogashira coupling of bromo-capped 2,5-(diphenyl)phosphole **66** [49a] with the diynes **67a–c** (Scheme 4.19) [49b]. Macromolecules **68a–c** are isolated in moderate to low yields as soluble powders, with low degrees of polymerization, ranging from 7 for **68c** to 15 for **68a**. The UV–Vis absorptions of **68a–c** are slightly red shifted in comparison with that of 2,5-diphenylphosphole, indicating the effective expansion of the π-conjugated system. The emission properties of these macromolecules depend on the nature of the comonomer. Green and blue emission are observed for **68a,b** and **68c**, respectively [49b]. However, the quantum yields in chloroform solution are modest (9–14%).

The most developed route to phosphole-containing polymers is the electropolymerization of thienyl-capped monomers, a process which involves the generation and coupling of radical cations [50]. Although this methodology is of great general interest for the preparation of electroactive materials based on thiophene or pyrrole monomers [5b,c, 50], it suffers from some significant drawbacks. The electrochemically-prepared polymers are often insoluble, preventing analysis by NMR or GPC. Furthermore, in addition to the desired α,α'-couplings that lead to fully conjugated polymers, α,β-couplings also take place [50]. The phosphole-containing monomers that have proved most successful in electropolymerization processes to date are depicted in Fig. 4.6. For example, the use of 2,5-(dithienyl)phosphole monomers **35–37** affords insoluble electroactive materials [15a, 31b]. The optimum polymerization potential E_{pol}, obtained by chronoamperometric investigations, depends on the nature of the P moiety (Table 4.3).

It is likely that the P atoms of the "as-synthesized" film obtained from the σ^3,λ^3-derivative **35** are protonated, at least partially, since the electropolymerization process generates protons [15a]. The polymers prepared with neutral phospholes

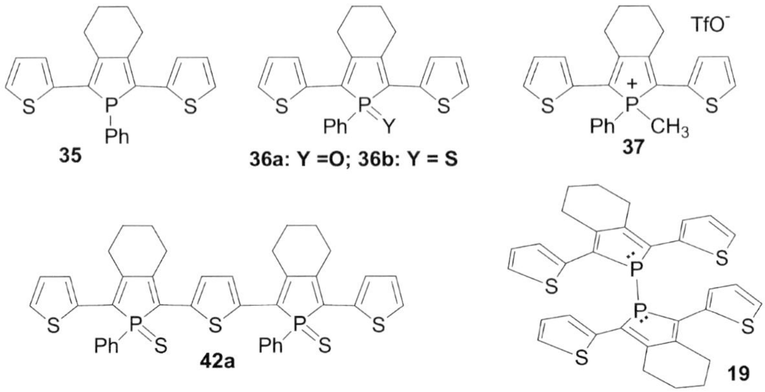

Fig. 4.6 Thienyl-capped monomers used in electropolymerization.

36a,b exhibited p- and n-doping characteristics with good reversibilities (>70%). Cationic poly-**37** also exhibits a reversible p-doping, but its electroactivity dramatically decreased upon reduction. It is noteworthy that these processes appeared at lower potentials than those of the corresponding monomers, suggesting that the electroactive materials possess much longer conjugation pathways and smaller band gaps than the monomers. This is confirmed by the fact that the values of λ_{onset} for the de-doped polymers were considerably red shifted compared with those observed for the corresponding monomers (Table 4.3) [15a,b]. A remarkable feature is that, as was observed for the P-containing monomers (Table 4.1), the electrochemical (doping range) and optical properties (λ_{max}, λ_{onset}) of these materials obtained by electropolymerization depend on the nature of the phosphorus moiety (Table 4.3). The electropolymerization of monomer **42a** (Fig. 4.6) also leads to an electro-active material presenting good reversible p-doping behavior [37]. The doping range (0.22–0.62 V) and λ_{max} of poly(**42a**) are comparable to those of poly(**36b**), as expected from their similar structures.

Table 4.3 Optimum electropolymerization potentials of monomers **35–37**, p-doping and n-doping potential ranges (V) and photophysical data for the corresponding dedoped polymers.

Monomer	$\lambda_{max}/\lambda_{onset}$ (nm)	E_{pol}[a] (V)	Polymer	p-Doping[a]	n-Doping[a]	λ_{max} (nm)	λ_{onset} (nm)
35	390/448	1.00	Poly-**35**	0.50 → 0.85	−0.60 → −1.00	463–567	724
36a	434/500	1.10	Poly-**36a**	0.25 → 0.60	−1.40 → −1.92	568	780
36b	432/496	1.15	Poly-**36b**	0.30 → 0.65	−1.80 → −2.42	529	750
37	442/528	1.20	Poly-**37**	0.40 → 0.70	−0.60	627	905

[a] Potentials referred to the ferrocene/ferrocenium half-cell.

Monomer **19**, exhibiting σ–π conjugation (see Section 4.2.2), can also be electropolymerized by scanning in the −0.5 to 0.8 V potential range [25]. The cyclic voltammograms show the appearance and the regular growth of a new reversible wave with a threshold potential of about −0.35 V. The shape of the new anodic wave and the regular growth of the two initial anodic peaks along the recurrent sweeps indicate the conductivity of the deposited material (no shift to more positive potentials and no decrease in intensity). Poly-**19** can also be obtained readily by potentiostatic oxidation at E_{pol} = 0.65 V. The threshold oxidation potential of poly-**19** (−0.35 V) is the least anodic encountered in the series of thiophene–phosphole copolymers generated by electropolymerization (Table 4.3). The UV–Vis spectrum of poly-**19** exhibits a large band with an unresolved maximum at about 594 nm and a high value of λ_{onset} (73 nm) indicating a narrow optical HOMO–LUMO gap. This result shows that low-gap electroactive materials based on 1,1′-biphosphole units can be readily obtained by electropolymerization [25]. The low band gap is probably due to the existence of σ–π conjugation within these polymeric materials.

In conclusion, for the most part, polymers incorporating phosphole units are still rare. However, the work described above demonstrates that they are accessible via a number of diverse synthetic routes including organometallic coupling or electropolymerization processes. The general property–structure relationships established with well-defined small oligomers have been shown to extend to the corresponding polymeric materials.

4.2.7
Mixed Oligomers and Polymers Based on Dibenzophosphole or Dithienophosphole

Dibenzophospholes **A** and dithienophospholes **B** and **C** (Fig. 4.7) do not display the typical electronic properties and reactivity patterns of phospholes, since the dienic system is engaged in the delocalized benzene or thiophene sextet [7, 10a, 51]. In fact, these building blocks have to be regarded as nonflexible diarylphosphines or as P-bridged diphenyl or dithienyl moieties.

Dibenzophosphole **A** was in fact the first type of phosphole to be prepared [9], but it has only very recently been used as a building block for the preparation of π-conjugated systems. In this regard, polymer **70** is obtained with a high polydispersity (Mn = 5 × 10^2; M_w = 6.2 × 10^3) by Ni-catalyzed homo-coupling of derivative **69** (Scheme 4.20) [52]. The presence of σ^3-P centers, which are potential donor sites for the Ni catalyst, does not prevent C–C bond formation. This macromolecule is photoluminescent in the solid state (λ_{em} = 516 nm), a property of potential interest for the development of OLEDs [52].

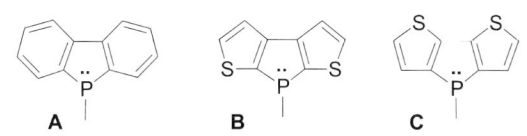

Fig. 4.7 Structures of dibenzophospholes **A** and dithienophospholes **B** and **C**.

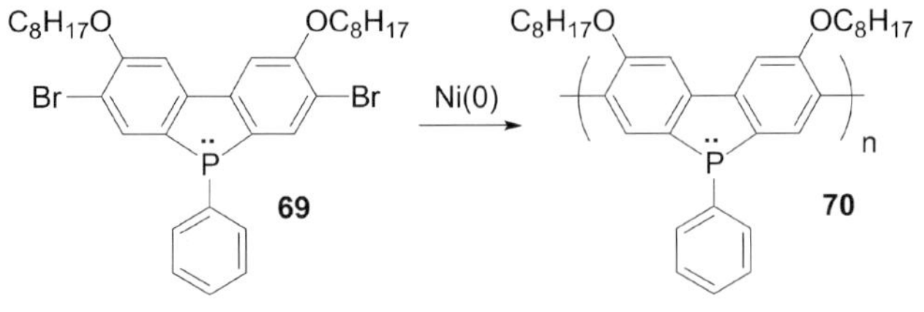

Scheme 4.20

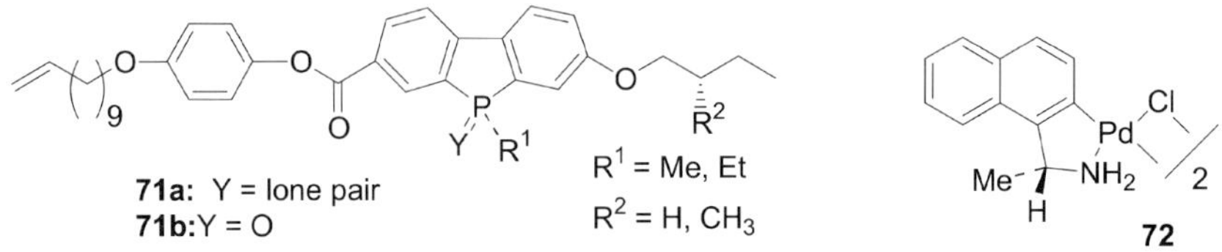

Scheme 4.21

- **71a**: Y = lone pair
- **71b**: Y = O
- R^1 = Me, Et
- R^2 = H, CH$_3$

Resolution of chiral σ³-benzophosphole **71a** (Scheme 4.21) was achieved by column chromatographic separation of the diastereoisomers obtained following its coordination to the chiral cyclometalated palladium(II) complex **72** [53]. This method nicely illustrates the advantages of having a P center present, which is able to coordinate to transition metals. The corresponding P-chiral dibenzophosphole oxide **71b** (Scheme 4.19) shows liquid crystalline behavior. Notably, the presence of a stereogenic P center is sufficient to generate a chiral cholesteric phase [53].

The dithienophosphole moiety **B** (Fig. 4.7) has been used to prepare cyclotriphosphazenes [54], but has not yet been investigated for the construction of π-conjugated systems. In contrast, dithieno[3,2-*b*:2′,3′-*d*]phosphole **C** [55] has recently been considered for such purposes by Baumgartner et al. This building block is structurally related to heteroatom-bridged bithienyl moieties (Fig. 4.8), which exhibit interesting properties such as reduced HOMO–LUMO band gaps and hole-transporting capability [56].

The dithienophospholes **74** are obtained in good yields by lithiation of dibromobithiophenes **73**, followed by addition of dihalo(aryl)phosphines (Scheme 4.22) [57]. The Lewis basicity of the P center of **74** enables a broad range of facile chemical modifications to be undertaken, as exemplified with the synthesis of derivatives **75–77** (Scheme 4.22) [57].

Analysis of the bond lengths of compounds **75a,b** and the square-planar complex **77** in the solid state reveals a high degree of π-conjugation of the fused sys-

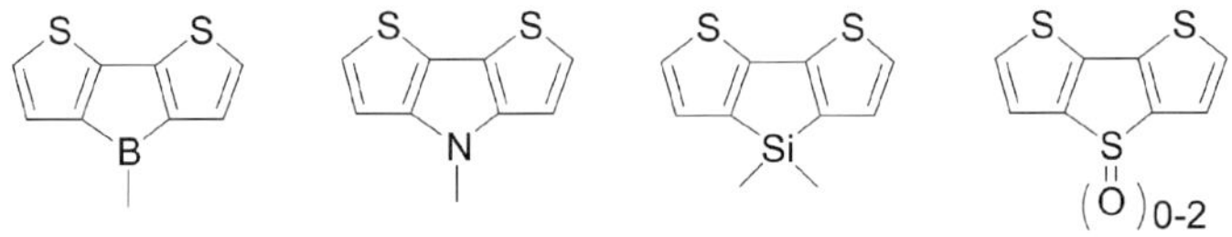

Fig. 4.8 Heteroatom-bridged bithienyl moieties.nAQ7n

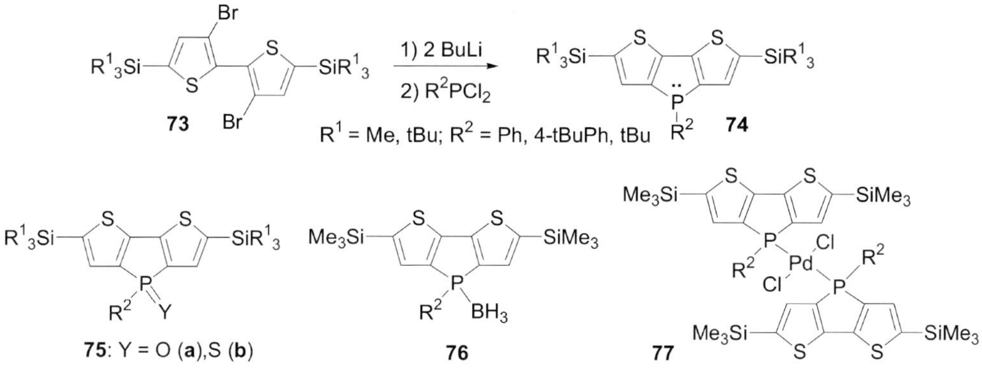

Scheme 4.22

tem and the involvement of the acceptor silicon substituents in this delocalization [57]. These conclusions were confirmed by theoretical calculations [57c]. Derivatives **74** show absorption maxima due to π–π* transitions in the UV region (Table 4.4). These data reveal, as expected for bridged dithienyl derivatives, rather high HOMO–LUMO separations. Chemical modification of the P center, leading to the formation of derivatives **75–77** (Scheme 4.22), induced a bathochromic shift in the value of λ_{max}, the most red-shifted value being that recorded for Pd complex **77** (384 nm) (Table 4.4) [57]. Note that modifying the substituents at the Si and P centers offers another tool for the further tuning of the values of λ_{max} for these systems. All these compounds exhibit strong blue photoluminescence with high quantum yields (Table 4.4). These PL quantum yields are far superior to those of 2,5-dithienylphospholes (Table 4.1), a feature that is probably due to the rigid structure of the dithienophospholes **74–77**.

Dithienophosphole units have been incorporated into polymers as either (i) side-chains or (ii) building block constituents of the extended polymer backbone itself. Polymers of the first type have been prepared by exploiting the ease with which a polymerizable styryl substituent may be introduced at the P atom of

Table 4.4 Physical properties of dithienophospholes **74** and derivatives **75–77**.

Compound	$\lambda_{max}^{[a]}$ (nm)	Logε	$\lambda_{em}^{[a]}$ (nm)	ϕ_f
74 (R^1 = CH$_3$; R^2 = Ph)	344	4.26	422	0.604
75a (R^1 = CH$_3$; R^2 = Ph)	374	4.09	460	0.556
75b (R^1 = CH$_3$; R^2 = Ph)	374	4.00	460	0.556
76	355	4.10	432	0.545
77	384	4.03	470	0.614

[a] In CH$_2$Cl$_2$.

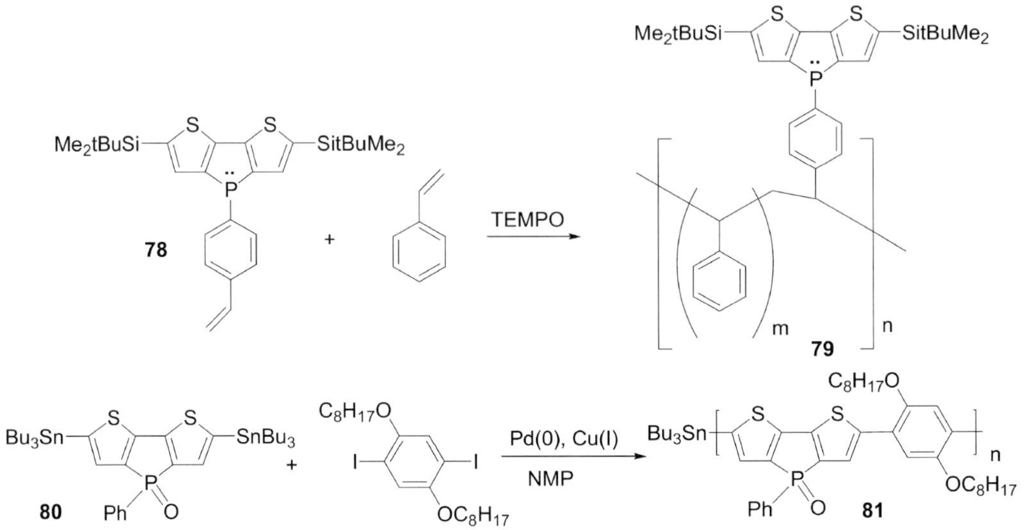

Scheme 4.23

dithienophospholes. Thus, the styryl derivative **78** was heated at 110 °C in the presence of styrene (ca. 1:30 ratio) and a catalytic amount of 2,2,6,6-tetramethylpiperidinyl-1-oxy (TEMPO) (Scheme 4.23) [57b]. This reaction gave polymer **79**, which was subsequently characterized and shown to have a high molecular weight (M_m = 147 650 g mol^{-1}) and a polydispersity of 2.46. The glass transition and decomposition temperatures of this material are 114.2 and 482.2 °C, respectively. Multinuclear magnetic resonance spectroscopy revealed that the polymer had the desired dithienophosphole:styrene ratio of 1:30. This material exhibits a blue fluorescence, its λ_{max} (352 nm) and λ_{em} (424 nm) values match those observed for the monomer **78** (Scheme 4.23). The P atoms can be oxidized with hydrogen peroxide, but surprisingly, the photophysical properties of the polymer are not affected by this chemical modification at phosphorus [57b].

The second strategy for the formation of polymers that incorporate the dithienophosphole fragment utilizes a Stille coupling involving tributyltin-capped oxo-dithienophosphole **80** in *N*-methylpyrrolidinone (Scheme 4.23) [57c]. The resulting polymer **81** has a low solubility in THF, preventing its analysis by GPC. The values of λ_{max} (502 nm) and λ_{em} (555 nm) for **81** are significantly red shifted compared with those of the corresponding monomer **80** (λ_{max} 379 nm; λ_{em} 463 nm).

It is clearly evident that the various series of conjugated polymeric systems based on dithienophospholes that have been prepared are promising candidates for optoelectronic applications since their optical properties can be easily varied in a variety of different ways (chemical modifications of the P center, variation of the Si and P substituents, etc.). Furthermore, their potential scope in a range of applications is broadened by the possibility of incorporating these conjugated derivatives into polymeric systems either as constituents of the main polymer backbone or as side-chains. Thus, dithienophospholes can potentially be utilized both in 'small molecule' and 'polymeric' approaches for the fabrication of OLEDs.

4.3
Phosphine-containing π-Conjugated Systems

Incorporating heteroatoms bearing lone pairs (Fig. 4.9) into π-conjugated backbones is a powerful means of tuning the electronic properties of polymers [58]. The heteroatom can participate in π-conjugation by virtue of its lone pair, in addition to being a site for chemical modification allowing for further structural and property diversification. This strategy has only recently been investigated with phosphorus. The derivatives presented in this section are subdivided according to the nature of the organic subunits (namely phenyl and ethynyl).

D   E   Y = O, S, Se

Fig. 4.9 Examples of polymers incorporating heteroatoms having a lone pair.

4.3.1
Polymers Based on p-Phenylenephosphine Units

Polyanilines **D** (Fig. 4.9) are amongst the oldest and best known photochromic materials. They are conveniently prepared by two different routes, namely polymerization of aqueous HCl solutions of anilines [58] or metal-mediated C–N bond formation from aryl halides or triflates with amines [58e,f]. Significantly, the first of these methods of polymer synthesis has not been reported with phosphorus-based monomers and it is likely that it will fail, owing to the profoundly different electronic properties of N and P centers. In contrast, the second route has been applied to the synthesis of poly(p-phenylenephosphine)s. Accordingly, polymers **82a–c** (Scheme 4.24) were obtained by palladium-catalyzed cross coupling of 1,4-diiodobenzene and primary aryl- and alkylphosphines [59a]. Owing to their good solubility, these materials were characterized by GPC and multinuclear magnetic resonance spectroscopy. They contain relatively short chains (M_n = 1000–4000) with narrow polydispersities (PDI = 1.3–1.5). The UV–Vis spectra of polymers **82a–c** show absorptions attributed to π–π transitions with values of λ_{max} ranging from 276 to 291 nm [59a]. The bathochromic shift observed on going from triphenylphosphine (λ_{max} = 263 nm) to 1,4-diphenylphosphinobenzene (λ_{max} = 275 nm) and then to **82b** (λ_{max} = 291 nm) suggests the presence of some extended π-delocalization involving the P lone pair in the latter material. However, the rather high band gap values are probably due to the fact that the

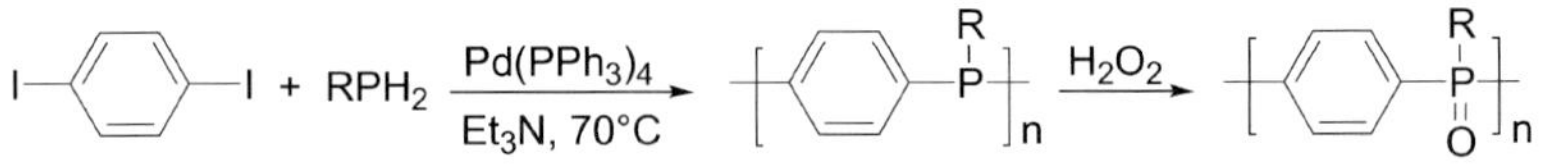

Scheme 4.24

	R	n	λ_{max} (nm)
82a	$CH_2CH(CH_3)_2$	10	278, 415
82b	Ph	7	291, 434
82c	$CH_2CH(CH_3)CH_2C(CH_3)_3$	14	276, 422

Scheme 4.25

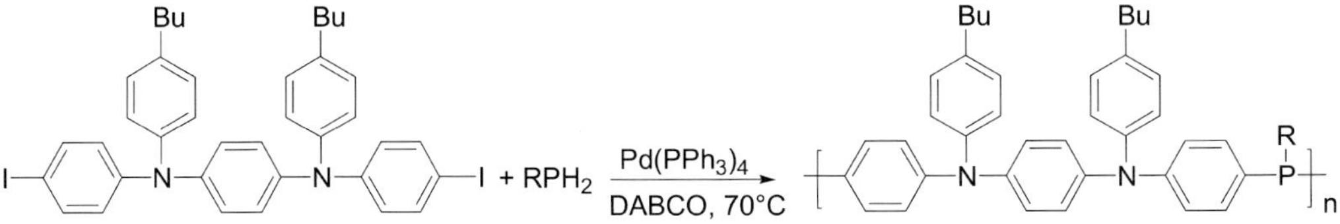

R = 2,4,4-trimethylpentyl

P atoms retain a tetrahedral geometry that prevents efficient conjugation of the phosphorus lone pair with the aryl groups. Interestingly, oxidation with $FeCl_3$ in the absence of oxygen afforded paramagnetic polymers characterized by UV–Vis absorptions that are considerably red shifted in comparison with those of **82a–c**. The high value of λ_{onset} (ca. 800 nm) supports the presence of an extension to the conjugation path through the P atoms. Note that the P atoms of **82a–c** are still reactive. The oxidized materials **83a–c** exhibit a number of new absorption bands that are red-shifted compared with those of their precursors **82a–c** [59a].

Utilizing the same general palladium-catalyzed C–P bond-forming reaction, several different types of alternating poly(p-phenylenephosphine)–polyaniline polymers have been prepared. Derivative **84** reacted with primary phosphine **85** (1:1 ratio) to produce copolymer **86a** (Scheme 4.25) [59b]. Approximately half of the material is soluble in THF, with GPC analysis of this soluble fraction revealing it to be of low molecular weight (M_n = 5000) with a narrow polydispersity (PDI = 1.6). The remaining insoluble material is assumed to have the same structure, but much higher molecular weight.

The two-step polycondensation of the iodobromo derivative **87** with phosphine **85** afforded polymer **88a** (Scheme 4.26) [59b]. The THF-soluble part (about one-third of the material) consists of oligomers of low molecular weight (M_w = 3000, n = 7, PDI = 1.5). Derivative **91a**, possessing a P:N ratio of 2 and 16 repeat units (M_w = 11 000, PDI = 1.9), was obtained via the same approach using bifunctional synthons **89** and **90** (Scheme 4.26). The higher molecular weight of **91a** vs. **88a** is simply due to its greater proportion of solubilizing 2,4,4-trimethylpentyl substituents. Note that it is very likely that the molecular weights, which are determined with reference to polystyrene standards, are underestimated [59b].

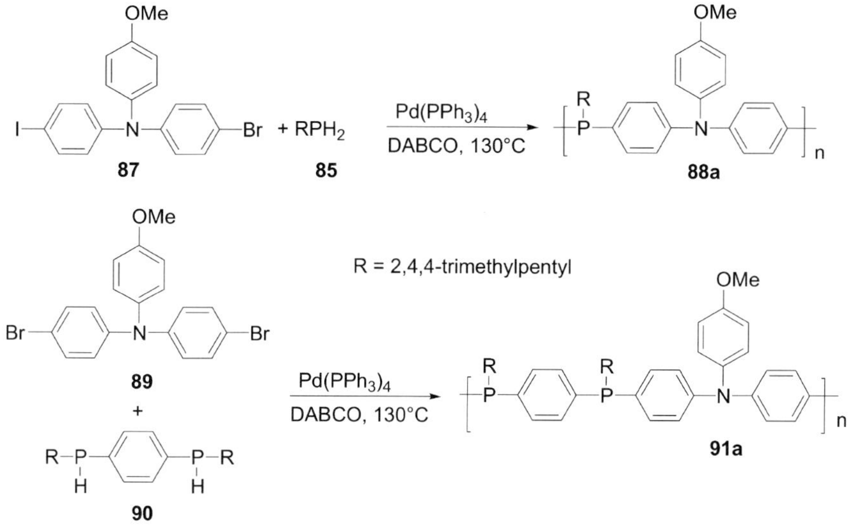

4.3 Phosphine-containing π-Conjugated Systems

Scheme 4.26

The corresponding phosphine oxide copolymers **86b**, **88b** and **91b** (Fig. 4.10) have also been prepared in quantitative yields using hydrogen peroxide. They have been fully characterized by IR and NMR spectroscopic methods [59b]. Their molecular weights, as estimated by GPC, are about half of those observed for the corresponding σ^3-derivatives, something that was assigned to a significant alteration in their conformation upon P oxidation.

Cyclic voltammetry revealed that the N-atoms of **86a**, **88a** and **91a** are oxidized at lower potentials than the trivalent P atoms. Comparison of these data with those observed with model compounds shows a very weak electronic delocalization via the P centers for copolymers **86a** and **91a**. In contrast, the low first oxidation potential observed for **88a** (Table 4.5) is assumed to result from an electronic communication between the N moieties through the connecting P centers [59b]. The equivalence of the oxidation potentials for the oxidized polymers (Table 4.5) suggests the presence of electronically isolated triarylamine fragments in these derivatives. Note that the involvement of the P lone pair in π-delocalization

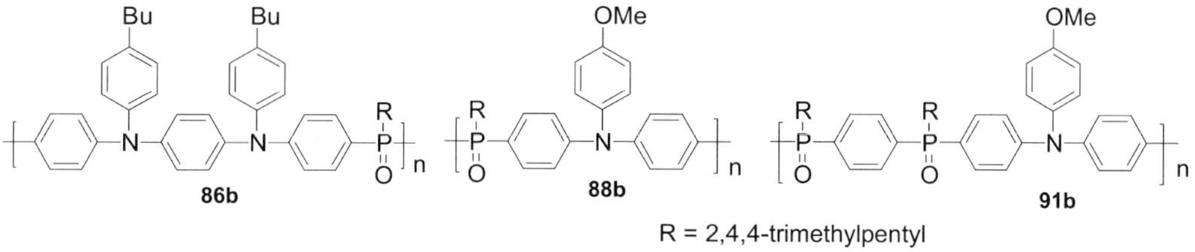

Fig. 4.10 Poly(p-phenylenephosphine oxide)-polyaniline polymers.

Table 4.5 Cyclic voltammetric data for copolymers 86a, 88a and 91a and for the corresponding oxides 86b, 88b and 91b[a].

	86a	88a	91a	86b	88b	91b
E_{ox1} (V)	0.15	0.09	0.5	0.36	0.67	0.63
E_{ox2} (V)	0.45	0.83	0.79	0.73	–	–
E_{ox3} (V)	0.83	–	–	–	–	–

[a] Ag/AgCl reference electrode in CH_2Cl_2.

along the backbone in **88a** is supported by the large shift in oxidation potentials observed upon its conversion to **88b** (Table 4.5). UV–Vis–NIR studies of these compounds are consistent with these general conclusions [59b].

In summary, linear phosphorus-containing poly(N-arylanilines) can be readily obtained by palladium-catalyzed C–P bond formation. Their physical properties strongly support electronic delocalization through σ^3-P centers along the polymer backbone, something that is switched off upon oxidation of the P atoms.

Biphenyl–phosphinidene polymers **93a,b** (Scheme 4.27) were prepared using Ni-catalyzed C–C bond-forming reactions with the bis(p-bromophenyl)phosphines **92a,b** [60a]. Disappointingly, this approach was found to be inappropriate for the formation of high molecular weight polymers. Thus, starting from **92a**, a dark-red insoluble solid was isolated, which was believed to be **93a**. Homo-coupling of phosphine **92b** bearing solubilizing groups afforded a low molecular weight (M_n = 1000), pale yellow material **93b** (Scheme 4.27). The color of **93b** is indicative of limited π-conjugation. This compound can be oxidized to afford **94b**. Note that the analogous P-Ph derivative **94c** can be obtained by Ni-catalyzed coupling of bis(p-chlorophenyl)phenylphosphine oxide **95** (Scheme 4.27) [60b]. The soluble material **94c** has a comparatively high molecular weight (M_n = 15 300) together

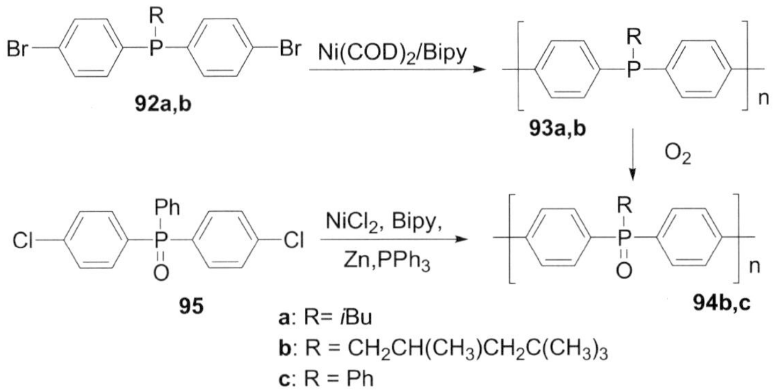

a: R = iBu
b: R = $CH_2CH(CH_3)CH_2C(CH_3)_3$
c: R = Ph

Scheme 4.27

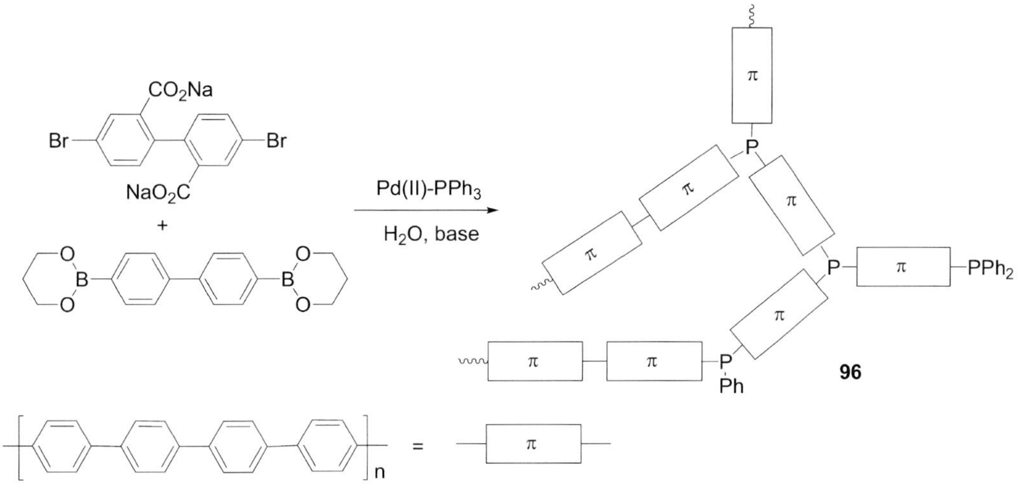

4.3 Phosphine-containing π-Conjugated Systems

Scheme 4.28

with a low molecular weight distribution (PDI = 1.6). It exhibits a high T_g (365 °C) with considerable thermal stability (<5% weight loss at 550 °C). As expected, its very low absorption maximum ($\lambda_{max} \approx 280$ nm) discounts the presence of an extended π-conjugated system involving the P-moieties.

An example of a branched (p-phenylene) incorporating phosphorus moieties, **96** (Scheme 4.28), was obtained adventitiously during attempts to synthesize soluble linear (p-phenylenes) using Pd-catalyzed Suzuki couplings [61]. In fact, the formation of the branched polymer **96** arises from aryl–aryl interchange taking place with the intermediate Pd(Ar)(I)(PPh$_3$)$_2$ complexes [61b]. It should be noted that although the concentration of the phosphine 'defects' is very low, they have a significant impact on the properties (e.g. molecular weight, viscosity) of the polymers.

4.3.2
Oligomers Based on Phosphine–Ethynyl Units

Phosphinoalkynes have been extensively used as rigid ligands in coordination chemistry [62]. In these cases, the role of the phosphino groups is simply to act as a two-electron donor towards transition metals, as exemplified by complexes **97** and **98** (Scheme 4.29) [62c]. When the ligands at Pt are chlorides, the phosphinobutadiyne rods undergo spontaneous [4 + 4] and [4 + 4 + 4] cycloadditions, leading to complexes **99** and **100** (Scheme 4.29) [62c]. These examples nicely illustrate how phosphinoalkynes can be used to build conjugated coordination oligomers or polymers. However, since the lone pair of the P atoms is clearly not involved in the conjugation, this type of derivative is not covered by this review.

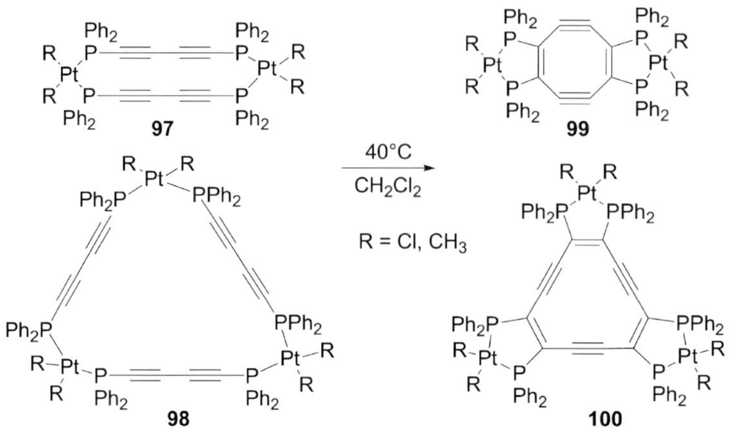

Scheme 4.29

Conjugated derivatives possessing phosphine–ethynyl units have mostly been prepared as building blocks for the synthesis of phosphapericyclines and polyphosphacyclopolyynes. The routes used to prepare these compounds belong to the classical synthetic arsenal of transformations associated with the alkyne and halophosphine functions [63]. Thus, treatment of dihalophosphine **101** with an excess of ethynylmagnesium bromide gave rise to a mixture of compounds **102** and **103** isolated in 53 and 3 % yield, respectively (Scheme 4.30) [64]. A double deprotonation of derivatives **102** and **103** followed by addition of 1.5 equiv. of P-electrophile **101** afforded triphospha[3]pericyclyne **104** and tetraphospha[4]pericyclyne **105** (Scheme 4.30). The UV–Vis spectra of the cyclic derivatives **104** and **105** showed strong absorption bands that extend out to nearly 300 nm, revealing that these P-heterocycles exhibit fairly strong cyclic electronic interactions [64].

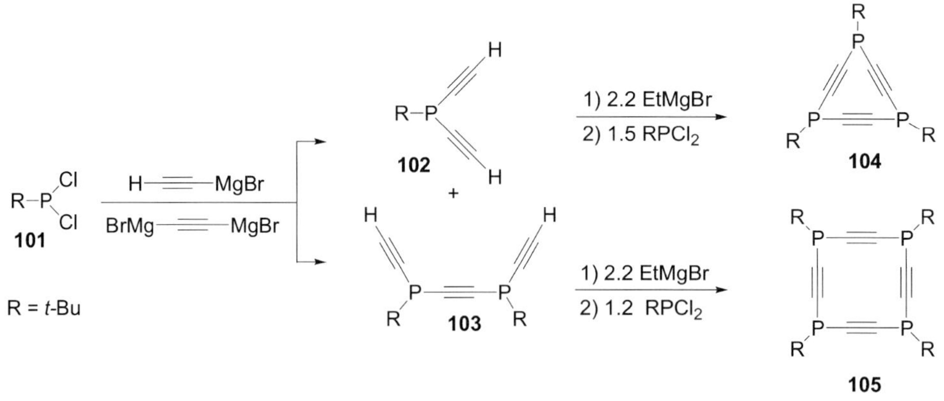

Scheme 4.30

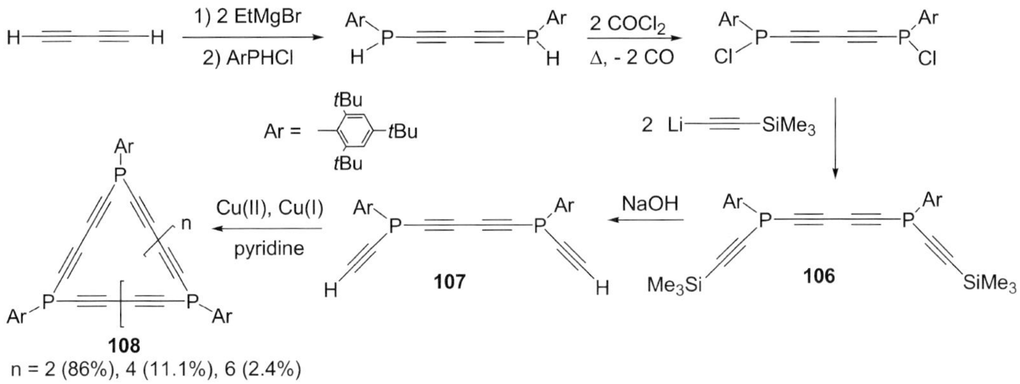

Scheme 4.31

108
n = 2 (86%), 4 (11.1%), 6 (2.4%)

A variety of linear ethynylphosphines, **106** and **107** (Scheme 4.31) and **109**–**112** (Scheme 4.32), have been prepared by Märkl et al. as synthons for the preparation of polyphosphacyclopolyynes **108** (Scheme 4.31) [65]. The ethynylphosphines **106** are obtained via a classical multi-step reaction sequence exploiting the electrophilic character of halophosphines (Scheme 4.31). The key steps involved in forming cyclic or linear oligomers are either Eglinton coupling of terminal alkyne moieties (Scheme 4.31) or a Cadiot–Chodkiewicz coupling involving bis-copper salts (Scheme 4.32). Derivatives **108** (Scheme 4.31) are obtained according to this random method as a mixture of di-, tri- and tetramers. Notably, these Cu-mediated C–C bond-coupling reactions are compatible with the presence of σ^3-P moieties. One drawback of the presence of several P atoms is that the resulting compounds are obtained as a mixture of isomers.

An interaction between the phosphorus lone pairs and the ethynyl units is supported by several facts. First, derivative **106** (Scheme 4.31) possesses an unusually low inversion barrier (15.5 vs. 35 kcal mol^{-1} for classical phosphines), indicating stabilization of the transient P-planar geometry. Second, the absorption maxima ($\lambda_{max} \approx 300$ nm) recorded for the heterocycles **108** (Scheme 4.31) and linear oligomers **109**–**111** (Scheme 4.32) ($\lambda_{max} = 210$–308 nm) are consistent with a degree of extended π-conjugation. However, the insolubility of the higher molecular weight, yellow oligomer **112** precluded UV–Vis analysis [65].

Thus, several routes to phosphine–ethynyl-based cyclic or linear oligomers are available and it is likely that these compounds possess extended conjugated systems involving the P lone pair. However, the development of this type of conjugated phosphorus system is hindered by the lack of efficient syntheses that allow for gram-scale preparations. This limitation is readily illustrated by the low-yielding preparation of dendron **113** (Scheme 4.33), something that precludes its use as a building block for higher generation dendrimers [66].

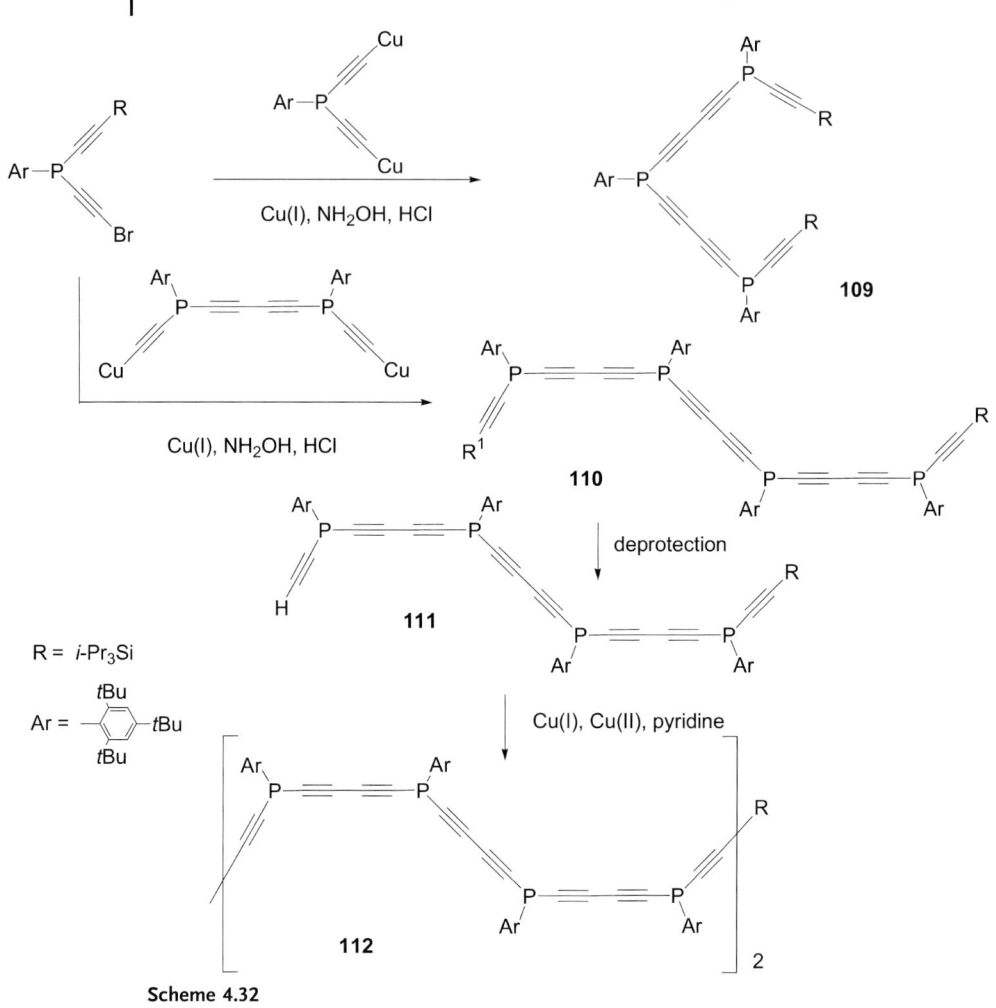

154 | *4 Design of π-Conjugated Systems Using Organophosphorus Building Blocks*

Scheme 4.32

R = *i*-Pr$_3$Si

Ar = 2,4,6-tri-*t*Bu-phenyl (tBu, tBu, tBu)

Scheme 4.33

4.3.3
Mixed Derivatives Based on Arylphosphino Units

Chromophores based on aryl moieties can be easily grafted on to halophosphines via simple nucleophilic substitutions, as illustrated by the synthesis of **114** (Scheme 4.34). As a result, numerous phosphines bearing extended π-conjugated substituents have been prepared. The aim of this section is not to give a comprehensive account of all derivatives that have been synthesized, but rather to illustrate how phosphorus fragments can be used either to influence or to organize the π-conjugated systems, with special emphasis upon fluorescence and NLO properties.

The readily available σ^3-phosphine **114** (Scheme 4.34) exhibits a broad absorption band at ~390 nm due to the π–π* transitions associated with the anthracene moieties, together with a band at 437 nm, probably due to extended π-conjugation through the P lone pair [67a]. This compound was prepared in order to investigate the changes in physical properties induced by increasing coordination number at the heteroatom, as observed previously with triarylboranes and triarylsilanes [67b–d]. The σ^4-derivatives **115** and **116** both possess tetrahedral geometry, whereas the σ^5-phosphorane exhibits a trigonal-pyramidal geometry with the three anthracenyl moieties adopting an equatorial disposition. The UV–Vis spectra of derivatives **115–117** are blue shifted relative to that of **114** ($\Delta\lambda_{onset}$, ca. 40–80 nm). This shift can be ascribed to the inductive effects of the phosphorus moieties and/or to a through-space interaction between the anthracene substituents [67a]. Moreover, it has been established that the fluorescence properties are highly dependent on the coordination number of the central P atom. The σ^3-phosphine **114** exhibits almost no fluorescence as a result of quenching by the P lone pair. The σ^4-derivatives **115** and **116** show weak fluorescence with relatively large Stokes shifts, again presumably as a result of through-space interactions between the anthracene substituents. In sharp contrast, the pentacoordinate compound **117** shows an intense fluorescence; the quantum yield is ~30–100 times greater than

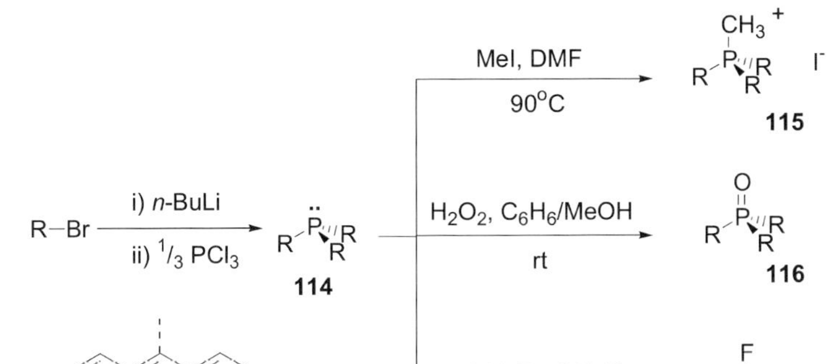

Scheme 4.34

156 *4 Design of π-Conjugated Systems Using Organophosphorus Building Blocks*

those of either **115** or **116**. Hence the fluorescence of the anthracene moiety is locked either by the presence of the P lone pair or by their pyramidal arrangement. These results give another clear example of property-control in conjugated systems through exploitation of phosphorus chemistry.

In order to investigate the optical excitations in multichromophore architectures, branched structures **120–122** (Scheme 4.35) bearing the same chromophore, but with different core units (C, N, P), have been investigated [68a]. The target P derivative **120** was prepared from a straightforward combination of a Heck coupling, to afford an intermediate functionalized stilbene phosphine oxide **118**, a Horner–Wittig reaction, yielding the phosphine oxide **119** and finally trichlorosilane reduction (Scheme 4.35).

It is assumed that the N-based system **121** is trigonal planar (in agreement with the planar structure of NPh$_3$ [68b]) and that the geometry of the central core P atom of **120** is more pyramidal than the C-derivative **122** [68a]. Examination of the UV–Vis spectral data reveals that the value of λ_{max} for the P-containing species **120** (376 nm) is red shifted relative to that for the C-cored compound **122** (λ_{max} = 325 nm), but blue shifted with respect to the N-based analog **121** (λ_{max} = 430 nm). This effect has been rationalized in terms of mesomeric effects. For the planar-cored compound **121**, an efficient overlap of the N lone pair with the adjacent carbon p-orbital gives rise to efficient conjugation with the chromophore substitu-

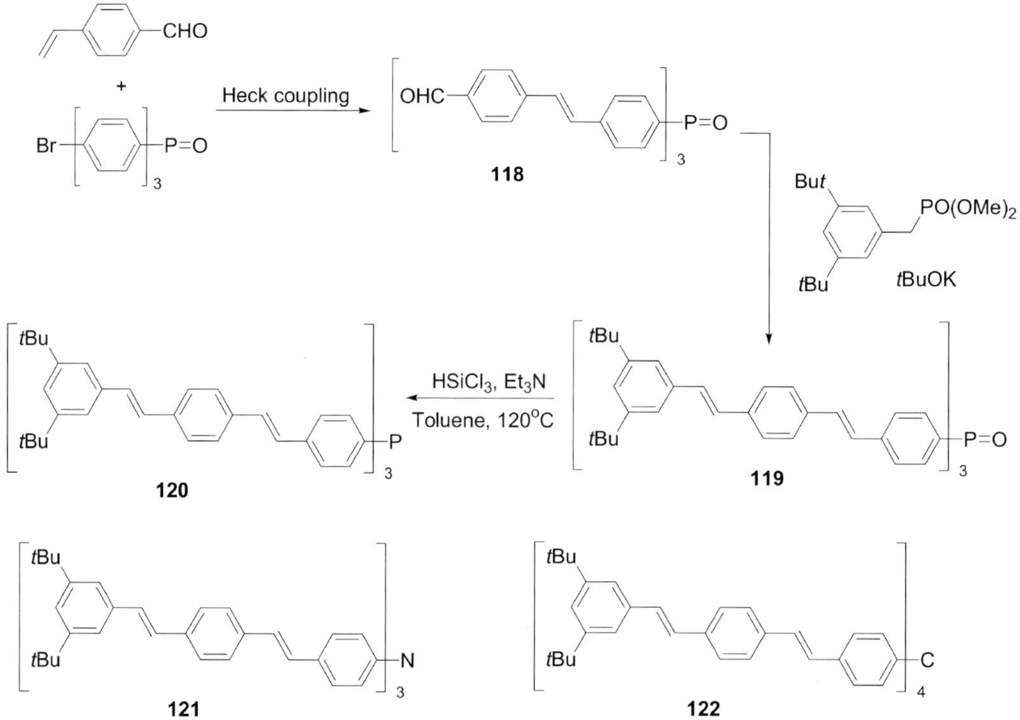

Scheme 4.35

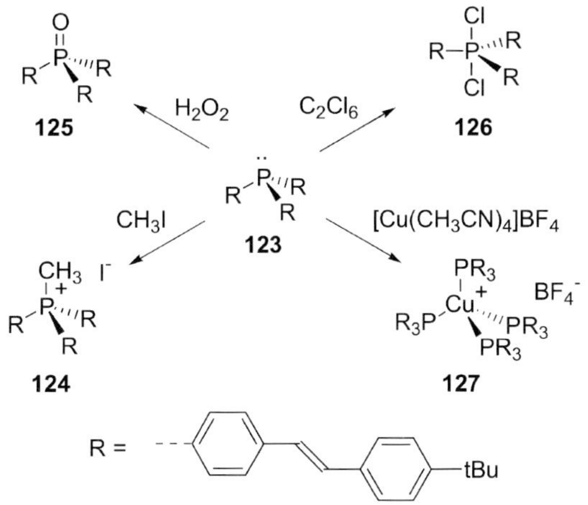

Scheme 4.36

ents, whereas for the larger pyramidal phosphorus of **120**, the overlap with the P lone pair will be significantly less efficient. Further studies to examine their fluorescence behavior revealed an incoherent hopping type of energy-transfer process, which dominates in the P derivative **120**. In contrast, a coherent mechanism is suggested for **121**.

Together, these data show that both the structural arrangement of dipoles about the central single atom core and the extent of electronic delocalization through the heteroatom have a direct impact on energy transfer in branched conjugated structures [68a]. This nicely illustrates the effect of replacing an N atom by a P atom, a modification that induces a dramatic alteration in the compound's optical properties. These changes arise from the differences in the preferred geometries and electronic characteristics of the two heteroatoms. Furthermore, the presence of a reactive P atom potentially gives access to a whole series of derivatives with a variety of geometries, oxidation states and coordination numbers, as illustrated by the related tris(4-styrylphenyl)phosphine **123** (Scheme 4.36) [68c,d]. As observed for anthracene-substituted derivatives **114–117** (Scheme 4.34), the fluorescence behavior of the styrylphenyl phosphorus compounds **123–127** varies greatly depending on the coordination number and oxidation state of the P center. Once again, the presence of a P lone pair in **123** is responsible for fluorescence quenching; the hypervalent species **126** is the most efficient fluorophore. Note that an investigation of the photophysical behavior of a structurally related phosphine bearing *p*-(*N*-7-azaindolyl)phenyl substituents has shown that, at 77 K, this P compound displays both a fluorescence band (λ_{max} = 372 nm) and a phosphorescence band [λ_{max} = 488 nm, lifetime 38(6) ms] [68e].

Diphenylphosphino moieties have been investigated as auxiliary donor groups for the tailoring of potential second- and third-order NLO-phores, although their π-donating ability is clearly much lower than that of diarylamino units. The P-based systems are prepared using classical synthetic transformations (e.g. Wittig

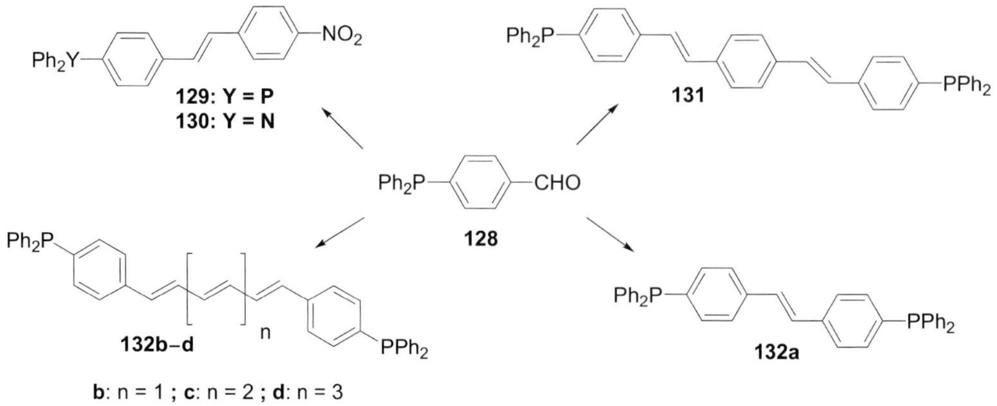

Scheme 4.37

b: n = 1 ; c: n = 2 ; d: n = 3

reactions, McMurry couplings) starting from the aldehyde function of p-(diphenylphosphanyl)benzaldehyde **128** (Scheme 4.37) [69]. As expected, the value of λ_{max} for the P-based dipole **129** is blue shifted relative to that of its N analog **130** (Scheme 4.37). This variation could be of interest in terms of a trade-off between transparency and NLO activity, both of which are important parameters for the engineering of valuable second-order NLO-phores [70]. The same blue shift was also observed between the centrosymmetric diphenylphosphino-capped chromophores **131** and **132a–d** (Scheme 4.37) upon replacing N by P [68]. Note that these compounds have been designed to exhibit third-order NLO properties [68]. As expected, systematically increasing the number of conjugated C–C double bonds in the series of polyenes **132a–d** (Scheme 4.35) led to a pronounced red shift in the values of λ_{max} (**132b**, 341 nm; **132d**, 418 nm).

Iminophosphoranes (phosphazenes), compounds with the general structure $R_3P=NR'$, are ylides possessing a highly polarized P=N bond [71]. With a view to optimizing push–pull NLO materials, they have been investigated as a new class of electron donors. Derivatives **133** and **134** (Scheme 4.38) are easily prepared by the so-called Staudinger reaction, which involves treating a tertiary phosphine with an organic azide (Scheme 4.38) [72]. The $\mu\beta$ product values of the resulting iminophosphoranes **133** and **134** are 310×10^{-48} and 1100×10^{-48} esu, respectively. These values are superior to that for p-nitroaniline (118×10^{-48}

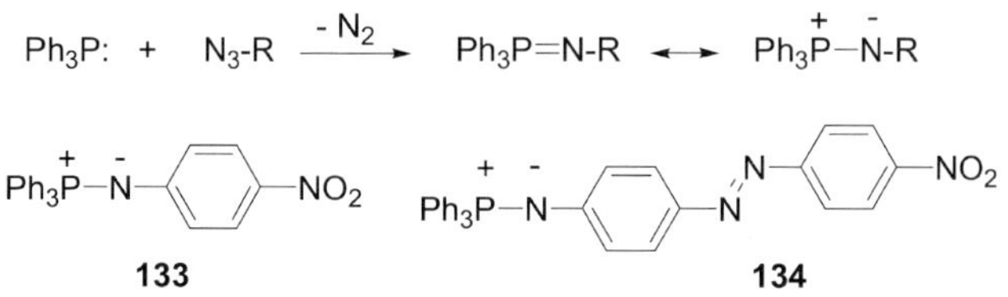

Scheme 4.38

esu), as expected from the significant donor ability of the iminophosphorane moiety, but are still modest compared with other efficient NLO-phores [70].

The search for efficient NLO-phores includes also the design of octupolar derivatives [40, 70a]. The advantages associated with this alternative class of NLO-phore include an improved nonlinearity–transparency trade-off and more facile noncentrosymmetric arrangement in the solid state due to the absence of dipolar moments. To this end, the 3-D chromophores **135** and **136** with C_3 and D_2 (approximate T) symmetry, respectively (Scheme 4.39), have been prepared according to classical synthetic routes [73a].

The UV–Vis data suggest that the subchromophores in **135** and **136** are near to being electronically independent. Derivative **135** has a small dipole moment and can be considered as an almost purely octupolar system, as is also the case for **136**. Compared with their tin analogs, the phosphorus derivatives **135** and **136** have higher β values owing to the more efficient acceptor nature of the phosphonium moiety [73a]. Furthermore, the NLO activity of the octupolar compound **135** is almost three times larger than that of the dipolar subchromophore **137**

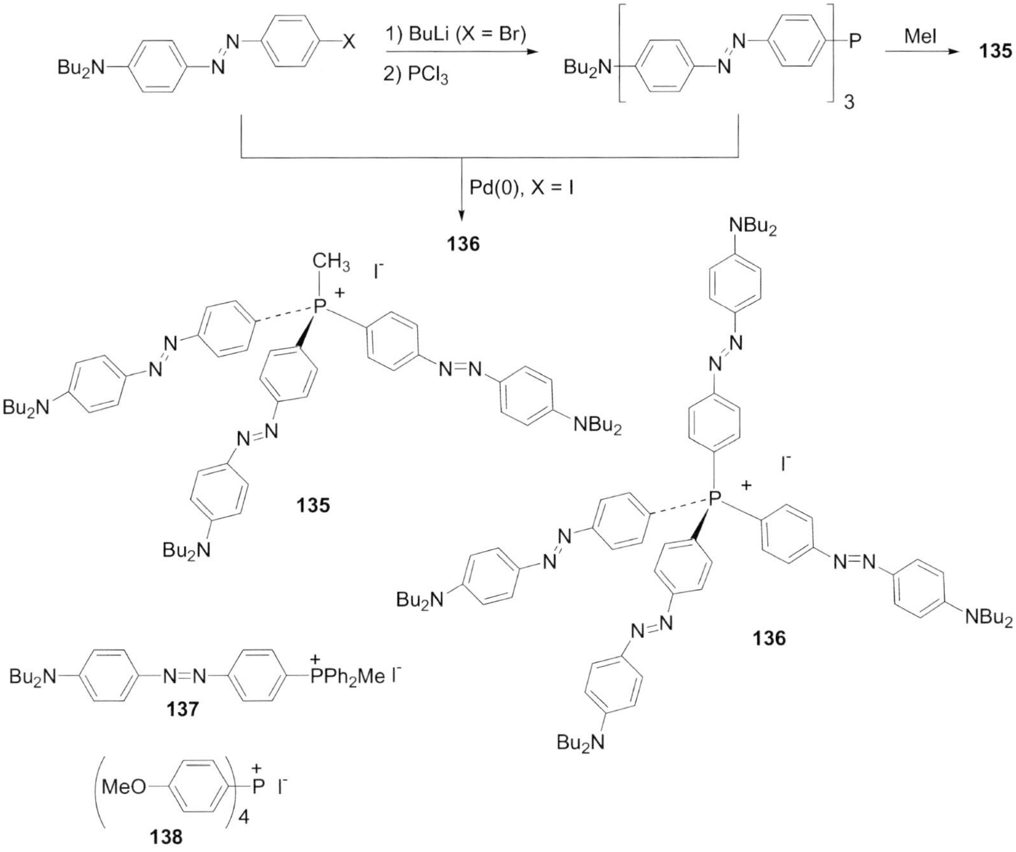

Scheme 4.39

(Scheme 4.39), with almost no cost in terms of transparency. The related octupolar phosphonium salt **138** (Scheme 4.39) was investigated with the aim of obtaining NLO-active crystals that remain transparent across all, or nearly all, of the visible region [73b]. Structural analysis of **138** revealed a weakly distorted ionic structure of the NaCl type. The tetrahedral phosphonium ion retains an almost pure octahedral symmetry in the solid state. The crystal is transparent throughout the visible region and exhibits moderate NLO behavior [73]. Together these examples very clearly illustrate the potential of phosphorus derivatives for the engineering of octupolar derivatives.

Structurally related P-branched multichromophores have also been designed for potential application in materials for OLED fabrication or as sensors. Notably, the branched tris(diphenylaminostilbenyl)phosphine **139** (Fig. 4.11) has been successfully used as a hole-transporting material in OLEDs [74a]. Not surprisingly, the values of λ_{max} for **139**, in solution (300 and 392 nm) and as a thin film (302 and 389 nm), are both blue shifted compared with those of its N analog **140** (309 and 410 nm). Note that in this series, substitution of P for N results in improved thermal stability. OLEDs having an ITO/**139**/Alq$_3$/Mg:Ag/Ag composition exhibit an EL efficiency of 0.13 % for a voltage of 11.8 V at a constant drive current of 13 mA cm^{-2} [74a]. This EL efficiency is higher than that obtained using the corresponding N derivative **140** (0.09 %), but rather low compared with devices utilizing NDP as the hole-transporting material.

Derivative **141** (Scheme 4.40) was designed as a fluorescent molecular sensor exploiting the presence of a triphenylphosphine oxide function, which is known to coordinate cations and the fluorescent phenylacetylene moieties [74b]. This compound is prepared in ca. 30 % yield according to the route depicted in Scheme 4.40. As observed with the structurally related styryl-substituted phosphine oxides **119** (Scheme 4.35), there is only an extremely weak (if any) interaction between the chromophores in the ground state for **141**. As expected, phosphine oxide **141** is strongly fluorescent with quantum yields varying from 0.71 to 0.89 % according to the nature of the solvent.

Hence it is clear that conjugated derivatives based on arylphosphino units are relatively easily prepared using either nucleophilic substitution with halophosphines or chemical modification of the phenyl moieties of arylphosphines. In most cases, the lone pair of the P atoms is only marginally involved in π-conjugation with the aryl substituents. Thus, the role of the P centers is mainly to organize the chromophores in a predictable way and to generate structural diversity via chemical modification at phosphorus. This general strategy has also been illustrated by the preparation of electrochemically generated polythiophenes

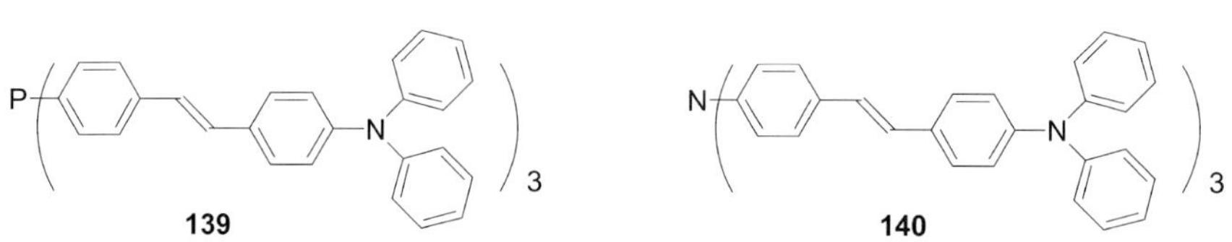

Fig. 4.11 Tris(diphenylaminostilbenyl)-phosphine and –amine.

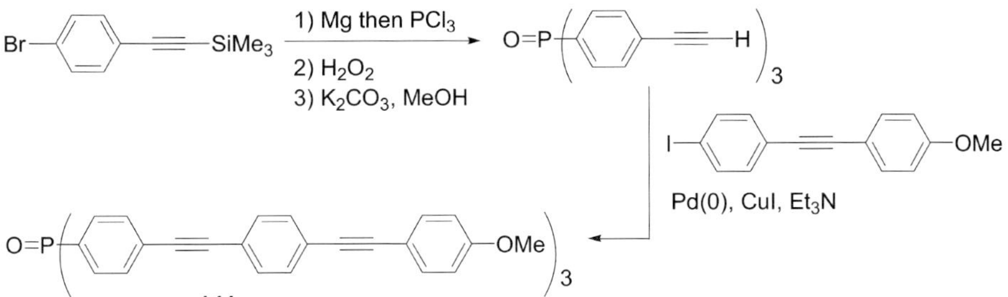

Scheme 4.40

that incorporate metal functionalities, with a view to modifying the electrochemical properties of the resulting macromolecules. A diphenylphosphino-functionalized terthiophene reacts with a source of palladium(II) to afford the cyclometalated complex **F** (Fig. 4.12). Subsequent electropolymerization gives palladium-containing polythiophene whose conductance and electronic properties are modified via inductive effects from the metal relative to those of the parent polythiophene [75].

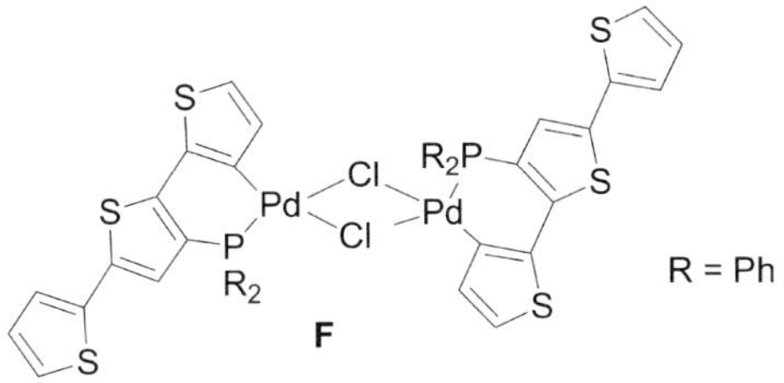

Fig. 4.12 Structure of cyclometallated complex **F**.

4.4
Phosphaalkene- and Diphosphene-containing π-Conjugated Systems

The properties and reactivity of low-coordinate carbon and phosphorus species are very similar in many regards [7]. Many of the parallels that exist between C=C and P=C bonds are due to the similar electronegativities of the two elements (C, 2.5; P, 2.2) and to the fact that, in marked contrast to imines, the HOMO of phosphaethylene is the π-bond and not the heteroatom lone pair [7a]. As a consequence, the P=C unit is almost apolar and its conjugative properties are comparable to those of the C=C bond [7, 76]. The concept that low-coordinate phosphorus behaves more like its diagonal relative carbon than its vertical congener nitrogen is now well accepted and has proved to be a fruitful tool for the design and synthesis of a raft of different compounds [7b]. Using this diagonal analogy, the simplest π-conjugated system incorporating phosphorus that can be envisaged would be poly(phosphaalkyne) **H** (Fig. 4.13), the P-containing analog of polyace-

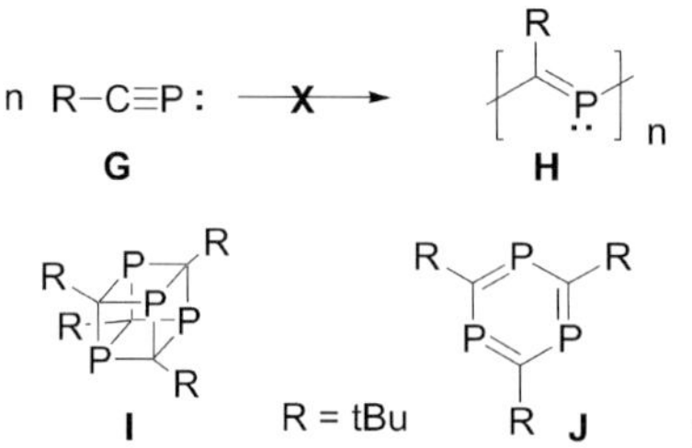

162 | *4 Design of π-Conjugated Systems Using Organophosphorus Building Blocks*

n R−C≡P: —X→ [structure H]ₙ

G   **H**

I R = tBu   **J**

Fig. 4.13 Phosphaalkyne oligomers.

tylene. Indeed, phosphaalkenes **G** that lack sterically demanding R substituents (e.g. those bearing merely R = H, Ph) can undergo thermally induced polymerization reactions [7a,b, 76b]. However, the resulting macromolecules feature mainly saturated trivalent P fragments with only some phosphaalkene moieties [76c]. In contrast, thermolysis of the more hindered tBuC≡P affords a mixture of a tetraphosphacubane **I** (Fig. 4.13) and other cage compounds [7a,b, 76b, 77a]. Alternatively, in the presence of metal complexes, several types of oligomer can be formed including 1,3,5-triphosphabenzene **J** (Fig. 4.13), tricyclic derivatives or cage compounds [7a,b, 76b, 77b,c].

Kinetically stabilized 1,3-diphosphabutadienes are known [76b, 77d], but the formation of oligomers or polymers **H** (Fig. 4.13) is probably hampered by the low thermodynamic stability of the P=C π-bond (43 vs. 65 kcal mol^{-1} for ethylene). For example, HP=CH$_2$ has a short half-life (1–2 min) and is unstable in the solid state at 77 K [77e,f]. Hence incorporation of an aromatic aryl group into the backbone of the polymers appeared to be an obvious strategy for increasing the thermodynamic stability of these π-conjugated systems. Furthermore, some steric protection would be provided by the aryl group's substituents, potentially overcoming the kinetic instability of the P=C moieties. Following this rationale, the first π-conjugated macromolecule containing phosphaalkene subunits was the PPV analog **144** (Scheme 4.41) [78]. This compound was prepared by thermolysis of the bifunctional derivatives **142** and **143**, a process involving thermodynamically favorable [1, 3]-silatropic rearrangements of intermediate acylphosphines to phosphaalkene moieties (Scheme 4.41) [76b]. Macromolecule **144** is soluble in polar organic solvents and, according to NMR measurements, is a mixture

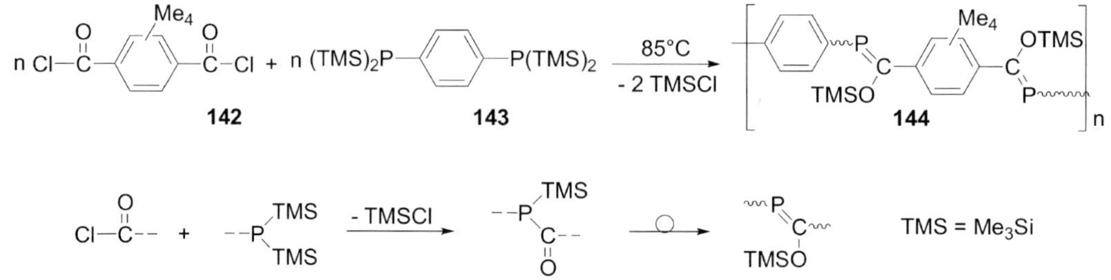

Scheme 4.41

TMS = Me$_3$Si

of Z- and E-isomers, with the degree of polymerization varying from 5 to 21. Remarkably, thermogravimetric analysis revealed that this polymer is stable up to 190 °C under an atmosphere of dry helium [78b].

The family of P=C-containing polymers was considerably broadened following the introduction of a highly efficient synthetic strategy based on intermediate "phospha-Wittig" reagents **L** [79], obtained by reacting the transient phosphinidene **K** with PMe$_3$ (Scheme 4.42). Polymers **147a–d** featuring different π-linkers were readily obtained by reacting dialdehydes **145a–d** with the bulky bis(dichlorophosphine) **146** (Scheme 4.42) [80a]. Derivatives **147a–c** are insoluble materials, whereas macromolecule **147d** is soluble and has been shown to contain an average of 12 phosphaalkene moieties per chain ($n = 6$) in an E-configuration. Remarkably, although **147d** decomposes slowly in solution, it is stable under air for 1 week in the solid state [80a].

Derivatives **144** (Scheme 4.41) and **147d** (Scheme 4.42) exhibit broad absorption bands, with values of λ_{max} of 328–338 and 445 nm, respectively, that presumably result from π–π* transitions. For comparison, model diphosphaalkenes **148** [78b] and **149** [80a] (Fig. 4.14) show absorption maxima at 314 and 445 nm, respectively. Hence the UV spectrum of polymers **144** is red shifted by only 20 nm compared with **148** whereas polymer **147d** and derivative **149** have identical values of λ_{max}. However, note that the bands arising from the polymeric derivatives extend into the visible region with values of the optical end absorption (λ_{onset}: **144**, ca. 400 nm; **147d**, ca. 540 nm), which are red shifted compared with those of the model compounds **148** and **149**. These data suggest that the P=C functions are effectively involved in the π-conjugation, but reveal rather limited effective conjugation path lengths for derivatives **144** and **147d**. This feature could be due to rotational disorder caused by the presence of the bulky aryl units, an idea supported by the X-ray diffraction analyses of **152** (Fig. 4.15), which reveals HCP–phenyl dihedral angles of 71 and 22° [80b]. The dramatic impact of such sterically induced noncoplanarity is also clearly reflected by the different values of λ_{max} obtained for **152** and

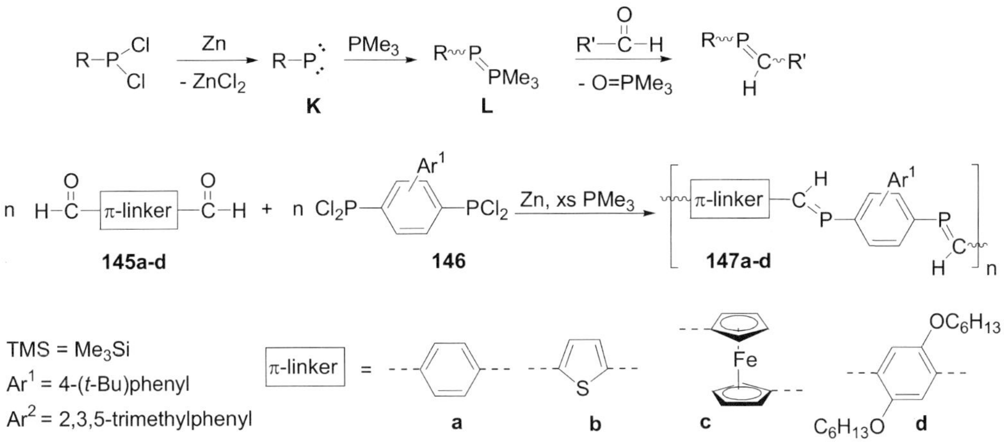

TMS = Me$_3$Si
Ar^1 = 4-(t-Bu)phenyl
Ar^2 = 2,3,5-trimethylphenyl

Scheme 4.42

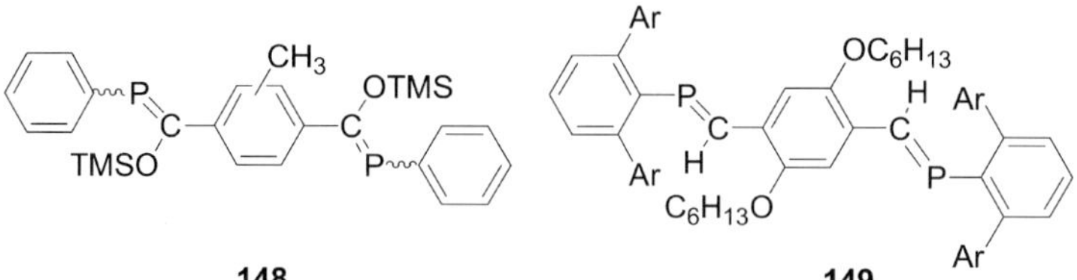

148

Ar = 2,3,5-trimethylphenyl
TMS = Me₃Si

149

Fig. 4.14 Strucutres of model diphospaalkenes.

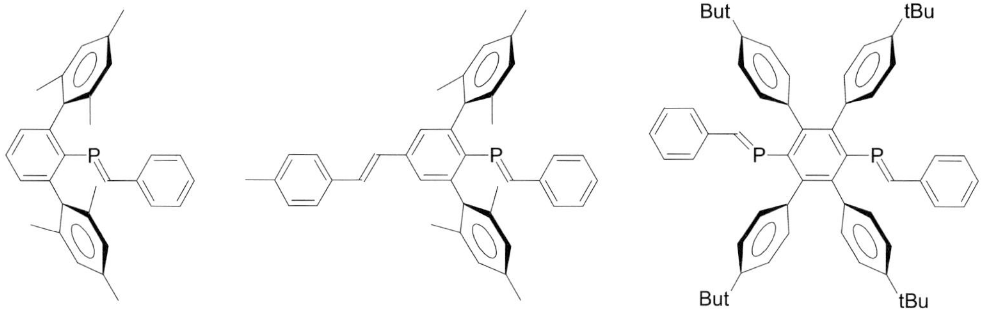

150: λ_{max} = 334 nm

151: λ_{max} = 341 nm

152: λ_{max} = 349 nm

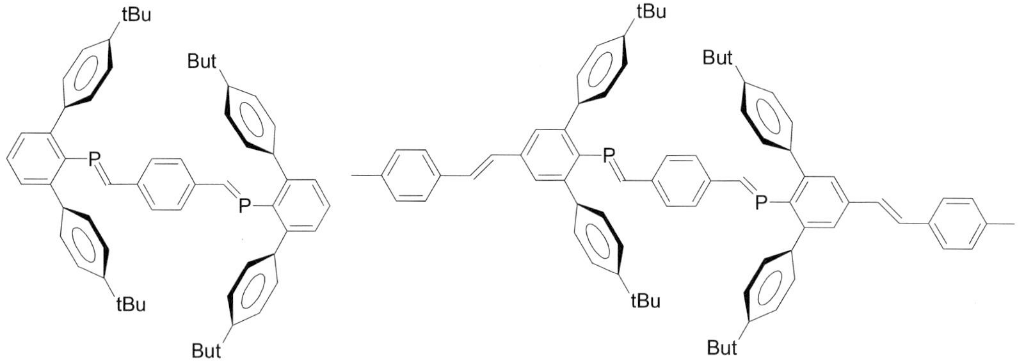

153: λ_{max} = 411 nm

154: λ_{max} = 417nm

Fig. 4.15 Conjugated systems incorporating phsophaalkene moieties.

153 ($\Delta\lambda_{max}$ = 57 nm, Fig. 4.15), which possess the same geometric conjugation path length [80c]. Note that the red shift observed upon comparing systems with a similar positioning of the sterically demanding units and increasing conjugation path length (**150/151/152; 153/154**) strongly supports the involvement of phosphaalkene moieties within the conjugated chain. Notably, polymer **147d** showed fluorescence with a broad emission centered around 530 nm [80a]. How-

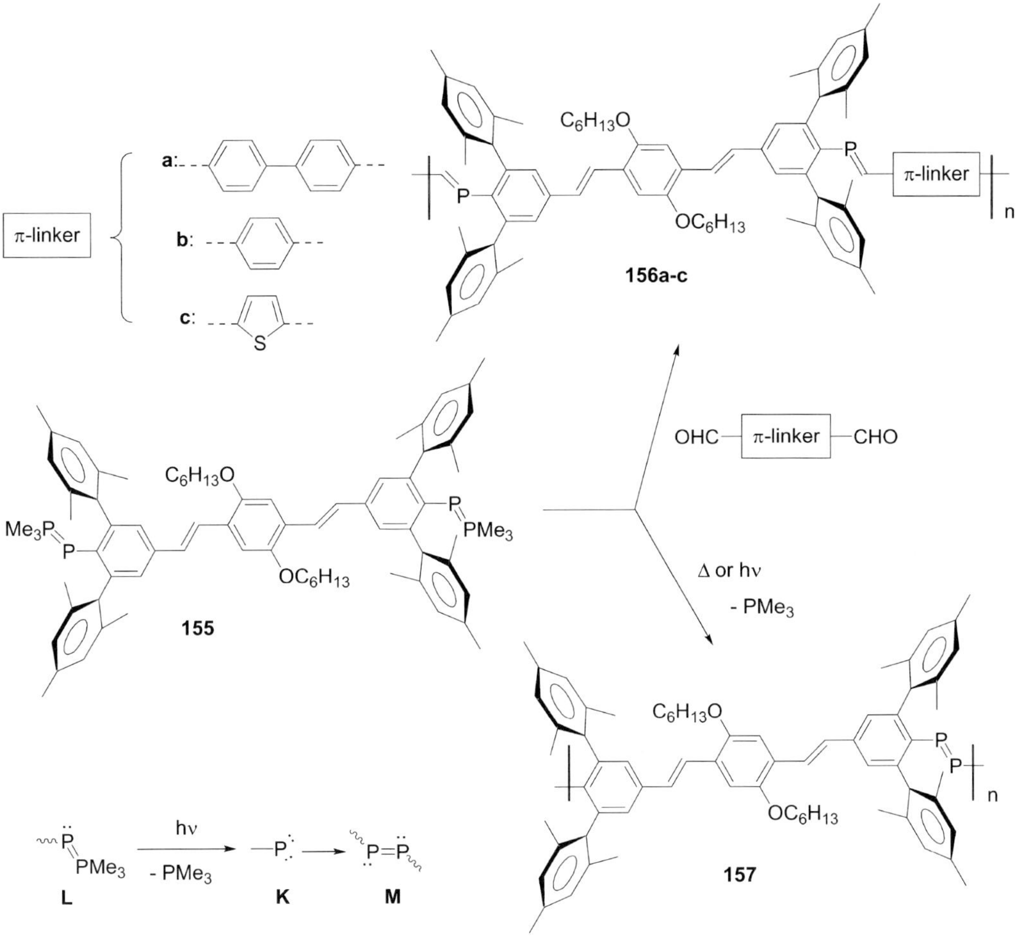

Scheme 4.43

ever, the fluorescence intensity is weak compared with those from its corresponding all-carbon analogs.

The formation of conjugated systems featuring P=P units is very challenging since this very reactive moiety needs significant steric protection in order to engender stability [76b, 81]. For such purposes, the sterically demanding 2,5-dimesitylphenyl substituent was employed by Smith and Protasiewicz [80d]. This bulky fragment was used initially to stabilize P=C units in polymers **156** that were prepared in 76–85 % yields via a phospha-Wittig reaction (Scheme 4.43). The phosphaalkenes adopt an *E*-configuration. The degree of polymerization is rather modest (M_n = 5000–7300) and the polydispersities vary from 1.9 to 2.3 [80d]. These derivatives are reasonably thermally stable in the absence of air or water. For example, polymers **156a** and **b** are unaffected by heating at 140 °C for 6 h under an inert atmosphere.

158: $\lambda_{\pi-\pi^*}$ = 372 nm

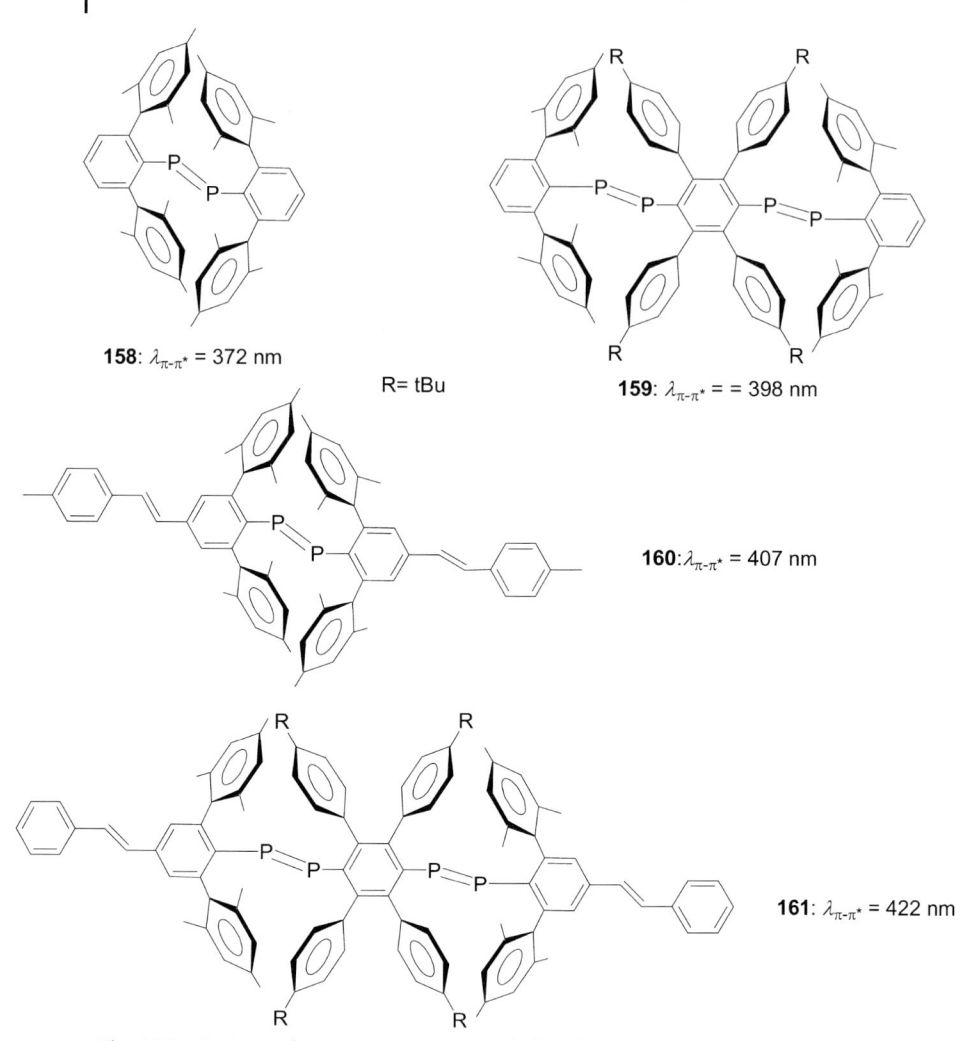

R= tBu

159: $\lambda_{\pi-\pi^*}$ = = 398 nm

160: $\lambda_{\pi-\pi^*}$ = 407 nm

161: $\lambda_{\pi-\pi^*}$ = 422 nm

Fig. 4.16 Conjugated systems incorporating diphosphene moieties.

Subsequently, following the successful isolation of stable phosphaalkene-containing polymers **156**, the use of the same substituent for the preparation of a related polymer featuring P=P moieties was attempted. It is known that diphosphenes **M** can be prepared by dimerization of transient phosphinidenes **J** generated by photolysis of phospha-Wittig reagents **L** (Scheme 4.43) [80e]. Photolysis at room temperature or thermolysis (neat, 250 °C, 2 min) of bifunctional compound **155** does indeed result in the formation of polymer **157** in near quantitative yield (Scheme 4.43) [80d]. This soluble material was characterized by NMR spectroscopy and GPC analysis, which revealed a rather low molecular weight (M_n = 5900). The UV–Vis spectrum of **157** shows a π–π* transition (435 nm) accompanied by an n–π* transition that is red shifted (481 nm). The $\lambda_{\pi-\pi^*}$ value for **157** is in the range observed for **156a–c** (416–435 nm) and only slightly blue shifted compared with poly(phenylenevinylene-*alt*-2,5-dihexyloxyphenylenevinylene) ($\lambda_{\pi-\pi^*}$ = 459 nm) [82]. These data indicate that diphosphene units support conjugation across extended systems. The validity of this assumption was enforced by the observed red shift within the series of compounds **158–161** of increasing chain length (Fig. 4.16) [80b]. Notably, in contrast to polymers **156a–c** featuring P=C units, the diphosphene-based derivative **157** is not fluorescent [80e].

Collectively, these studies show that oligomers incorporating phosphaalkene or diphosphene moieties are readily available by a variety of different routes and prove that the P=C and P=P units are capable of supporting conjugation across extended π-systems. Although structure–property relationships have still to be established, it is clear that macromolecules with higher degrees of polymerization and less sterically demanding substituents are exciting targets.

A last family of conjugated systems featuring P=C moieties that is worthy of note contains a fulvene structure. The reaction of dibromophosphaethene **162** with *tert*-butyllithium afforded the carbenoid **163**, which transformed into the deep-red 1,3,6-triphosphafulvene **164** (Scheme 4.44) [83a]. The latter compound can formally be regarded as a trimer of phosphanylidene carbene **166**, but the mechanism for its formation probably involves the generation of phosphaalkyne **167** (Scheme 4.44). The results of an X-ray diffraction study of the tungsten pentacarbonyl complex **165** showed that the triphosphafulvene framework is planar, as expected [83a]. Derivative **164** undergoes a reversible one-electron reduction at –0.68 V versus Ag/AgCl [83b]. Note that related 1,4-diphosphafulvenes (Fig. 4.17) having different substitution patterns were described recently [83c–e].

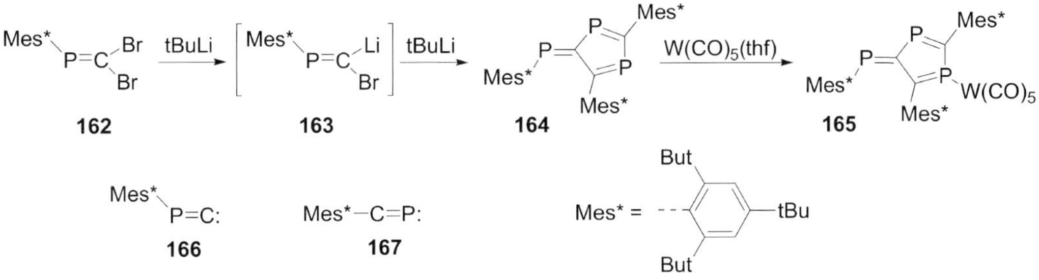

Scheme 4.44

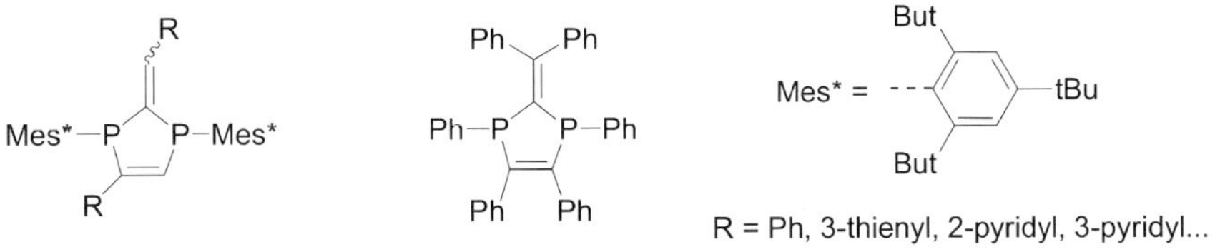

Fig. 4.17 Structures of 1,4-diphosphafulvenes.

4.5
Conclusion

The chemistry of π-conjugated systems incorporating P moieties has really only come to the fore following the pioneering work started in 1990 on phosphole-based oligomers and that on related phosphine–ethynyl co-oligomers. The last 5 years have seen a considerable expansion of the area with the synthesis of a plethora of novel P-based derivatives, together with the elucidation of their physical properties. Useful P-containing building blocks include heterocycles (e.g. phospholes), phosphino groups and low-valent phospha- and diphosphaalkenes. However, the chemistry of π-conjugated systems incorporating P units remains in its infancy, something that is nicely illustrated by the fact that the most recent review devoted to the synthesis and properties of conjugated molecular rods [5g] does not even mention organophosphorus building blocks!

Despite being a relatively young area of research, it has already been clearly established that organophosphorus derivatives offer specific advantages over their widely used sulfur and nitrogen analogs. Of particular interest is the possibility of chemically modifying P centers, something that provides a unique way to create structural diversity and to tune the physical properties of these phosphorus-based π-conjugated systems. This facet is of particular importance for the tailoring of organophosphorus materials for applications in optoelectronics. Moreover, the coordination ability of P centers towards transition metals offers manifold opportunities to build supramolecular architectures in which the π-systems can be organized in a defined manner. A great variety of synthetic routes have been used, but new strategies have to be developed in order to generate well-defined series of oligomers and polymers. Structure–property relationships have still to be established in order to exploit fully the potential of P moieties in the construction of conjugated frameworks.

A number of problems still remain that prevent the use of P-based π-conjugated systems from reaching real applications (e.g. thermal and chemical stability, solubility). However, the well-established and highly developed understanding of the chemistry of phosphorus-containing compounds, including both low-valent and heterocyclic species, will allow most of the remaining issues to be circumvented. Indeed, the possibility of using organophosphorus π-conjugated systems as materials for applications in the field of nonlinear optics, organic light-emitting diodes and conductive polymers has already been firmly demonstrated. Hence there is no inherent limitation associated with the use of conjugated phosphorus-containing materials for these important and appealing end uses. However, considering the richness and diversity of phosphorus chemistry, their true potential has not yet been fully exploited. It is very likely that new developments will occur in these areas in the near future, which will take full advantage of the specific properties of organophosphorus derivatives.

4.6
Selected Experimental Procedures

Synthesis of tetramer (22) (Scheme 4.7) [28a,c]
3,4-Dimethyl-l-phenylphosphole 4 (2.26 g, 12 mmol) and N,N'-dimethyl-4-bromoaniline (1.8 mmol) were heated in a sealed glass tube at 180 °C for 18 h. The mixture was cooled and treated with CH_2Cl_2. The insoluble red crystals of 22 were isolated by filtration and dried under vacuum. Yield, 40–45 %; m.p. >260 °C; ^{1}H NMR ($CDCl_3$), δ = 2.03 (br s, 12 H, Me), 2.19 (m, 12 H, Me), 7.40 (br s, 20 H, Ph); ^{31}P NMR, δ = −1 1.6; MS (70 eV), m/z 744 (M, l00 %), 558 (M − $C_{12}H_{11}P$, 41 %), 372 (M/2, 50 %).

Synthesis of 1-phenyl-2,5-di(2-thienyl)phosphole (35d) (Scheme 4.10) [15a]
To a THF solution (40 mL) of Cp_2ZrCl_2 (1.08 g, 3.70 mmol) was added dropwise, at −78 °C, a hexane solution of 2.5 M n-BuLi (3.10 mL, 7.70 mmol). The reaction mixture was stirred for 1 h at −78 °C and a THF solution (20 mL) of 1,8-di(2-thienyl)octadiyne (0.81 mL, 3.70 mmol) was added dropwise at this temperature. The solution was allowed to warm to room temperature and stirred for 12 h. To this solution was added, at −78 °C, freshly distilled $PhPBr_2$ (0.77 mL, 3.70 mmol). The solution was allowed to warm to room temperature and stirred for 4 h at 40 °C. The solution was filtered and the volatile materials were removed *in vacuo*. Compound 35d was isolated as an orange solid after purification on basic alumina (THF). Yield, 1.05 g, 75 %; R_F (alumina, THF) = 0.7; m.p. 134 °C; ^{1}H NMR (CD_2Cl_2): δ = 1.85 (m, 4H, C=CCH_2CH_2), 2.90 (m, 4H, C=CCH_2), 6.95 [ddd, $^3J(H,H)$ = 5.1 and 3.7 Hz, $^5J(P,H)$ = 1.0 Hz, 2H, H_4 thio], 7.08 [ddd, $^3J(H,H)$ = 3.7 and $^4J(H,H)$ = 1.1 Hz, $^4J(P,H)$ = 0.4 Hz, 2H, H_3 thio], 7.23 [ddd, $^3J(H,H)$ = 5.1 and $^4J(H,H)$ = 1.1 Hz, $^6J(P,H)$ = 1.2 Hz, 2H, H_5 thio], 7.25 (m, 3H, $H_{m,p}$ Ph), 7.46 [ddd, $^3J(H,H)$ = 7.8 and 1.6 Hz, $^3J(P,H)$ = 7.8 Hz, 2H, H_o Ph]; ^{13}C{^{1}H} ($CDCl_3$), δ = 23.1 (s, C=CCH_2CH_2), 29.3 (s, C=CCH_2), 124.8 [d, $^5J(P,C)$ = 2.0 Hz, C_5 thio], 125.6 [d, $^3J(P,C)$ = 10.0 Hz, C_3 thio], 126.8 (s, C_4 thio), 128.7 [d, $^3J(P,C)$ = 9.0 Hz, C_m Ph], 129.6 (s, C_p Ph), 133.5 [d, $^1J(P,C)$ = 13.0 Hz, C_i Ph], 133.9 [d, $^2J(P,C)$ = 19.0 Hz, C_o Ph], 135.7 (s, PC=C), 139.7 [d, $^2J(P,C)$ = 22.0 Hz, C_2 thio], 144.4 [d, $^2J(P,C)$ = 9.0 Hz, PC=C]; ^{31}P-{^{1}H} ($CDCl_3$), δ = +12.7; HRMS (EI), m/z found 378.0651 M^+; $C_{22}H_{19}PS_2$ calcd 378.0666; elemental analysis, calcd for $C_{22}H_{19}PS_2$ (378.49) C 69.82, H 5.06; found C 69.93, H 5.31 %.

Synthesis of polymer 81 (Scheme 4.23) [57c]
Catalytic amounts of [Pd(PPh$_3$)$_4$] (chiffres?) and CuI were added to a solution of dithienophosphole 80 (1.73 g, 1.5 mmol) and 1,4-diiodo-2,5-bis(octyloxy)benzene (0.88 g, 1.5 mmol) in N-methylpyrrolidinone (80 mL). The light-yellow reaction mixture was then stirred for 48 h at 200 °C, after which time the color changed to orange–red. The solution was then cooled to room temperature, the solvent evaporated under vacuum and the resulting amorphous solid taken up in a small amount of THF (ca. 5 mL). The suspension was filtered, precipitated into hexane and the residue dried under vacuum to yield 80 as a reddish brown pow-

der (1.1 g). ^{31}P{^{1}H} NMR (C$_2$D$_2$Cl$_4$), δ ≈ 14 ppm; ^{1}H NMR (C$_2$D$_2$Cl$_4$), δ ≈ 7.64–6.90 (br, Ar-H), 3.95 (br, OCH$_2$), 1.71–0.88 ppm (br m, alkyl-H); ^{13}C{^{1}H} NMR (C$_2$D$_2$Cl$_4$), δ ≈ 155.6, 144.0, 142.1, 130.5, 128.4 (br, C$_{Ar}$), 69.6 (br, OCH$_2$), 31.6, 29.1, 26.6, 26.1, 22.5, 14.1, 13.5 ppm (C$_{alkyl}$).

Synthesis of phosphonium salt **136** (Scheme 4.39) [73a]
Tris[4-(*N*,*N*-dibutylamino)azobenzen-4'-yl]phosphane (105 mg, 0.11 mmol), 4-(*N*,*N*-dibutylamino)-4'-iodoazobenzene (48 mg, 0.11 mmol) and Pd(OAc)$_2$ (0.4 mg, 0.0018 mmol) were stirred with 1.5 mL of oxygen-free, nitrogen-saturated *p*-xylene under an inert atmosphere for 20 h at 140 °C, resulting in the separation of a dark-red oil. The xylene solution was removed with a pipette and the residue purified by chromatography on silica gel with ethyl acetate–MeOH (10:0.5) as eluent. A concentrated solution of the product in CH$_2$Cl$_2$ was dropped into vigorously stirred petroleum ether. The dark-red precipitate was collected by filtration and dried *in vacuo*. Yield, 74 mg (48%); ^{1}H NMR (CDCl$_3$), δ = 8.09 (m, 8H, H3/5), 7.90 (m, 8H, H2'/6'), 7.75 (m, 8H, H2/6), 6.71 (m, 8H, H3'/5'), 3.41 (m, 16H, α-H), 1.65 (m, 16H, β-H), 1.40 (m, 16H, α-H), 0.99 (t, J = 7.2 Hz, 24H, δ-H); ^{13}C{^{1}H} NMR (CDCl$_3$), δ =157.9 [d, *J*(C,P) = 3.3 Hz, C4], 152.3 (C1'), 143.5 (C4'), 135.4 [d, *J*(C,P) = 11.1 Hz, C2/6], 126.8 (C2'/6'), 123.8 [d, *J*(C,P) = 13.7 Hz, C3/5], 116.2 [d, *J*(C,P) = 92.2 Hz, C1], 111.4 (C3'/5'), 51.1 (α-C), 29.6 (β-C), 20.3 (γ-C), 13.9 (δ-C); elemental analysis, calcd for C$_{80}$H$_{104}$IN$_{12}$P.H$_2$O (1409.69), C 68.16, H 7.58, N 11.92; found, C 68.17, H 7.58, N 11.66%; MS (PI-LSIMS), *m/z* 1263 (100%, K$^+$), 1571 (30%, [K.dibutylaminoazobenzenyl]); HRMS (FAB), calcd 1263.82140 (K), found 1263.82446.

Synthesis of polymer **144** (Scheme 4.41) [78]
All glassware was rinsed with Me$_3$SiCl and flame dried prior to use. Compounds **142** (0.601 g, 2.32 mmol) and **143** (1.00 g, 2.32 mmol) were mixed as finely ground powders and flame sealed *in vacuo* in a thick-walled Pyrex tube. The sample was placed in a preheated (85 °C) oven, whereupon the solids melted forming a colorless, free-flowing liquid. After 6–8 h, the mixture showed an increase in viscosity and was yellow. The reaction was monitored until the liquid was almost immobile (ca. 24 h) and the yellow–orange material was removed from the oven. The tube was broken, Me$_3$SiCl was removed *in vacuo* and the residue was dissolved in a minimum amount of THF (ca. 3 mL). The viscous solution was evenly distributed over the walls of the flask and cold hexane (ca. –30 °C) was added rapidly to precipitate the polymer as a yellow solid. The hexane-soluble fraction was removed, leaving polymer **144** (0.384 g, 35%) as a bright-yellow, glassy solid after drying *in vacuo*. ^{31}P NMR (CDCl$_3$), δ = 157–149 [br m, (*E*)–**144**], 138–124 [br m, (*Z*)–**144**], –137 ppm [br, P(SiMe$_3$)$_2$ end groups]; ^{29}Si NMR (CDCl$_3$), δ = 21.7–20.5 (br m), 18.4–17.0 (br m), 1.4 ppm [d; ^{1}*J*(Si, P) = 26 Hz, end groups]; ^{1}H NMR (CDCl$_3$), δ = 7.8–6.6 (br m; C$_6$H$_4$), 2.5–2.1 (br m, C$_6$(CH$_3$)$_4$], 0.5–0.5 [br m, Si(CH$_3$)$_3$]; ^{13}C{^{1}H} NMR (CDCl$_3$), δ = 211.9 [br, (*Z*)-C=P], 197.9 [br, (*E*)-C=P], 142.0 (br, *i*-C$_6$Me$_4$), 139.1 (br, *i*-C$_6$H$_4$), 132.4, 130.2 (br, *o*-C$_6$H$_4$, *o*-C$_6$Me$_4$), 18.6,

17.5 [br s, C$_6$(CH$_3$)$_4$], 0.7, 0.2 [br s, OSi(CH$_3$)$_3$]; IR (film), ν = 2955 (m), 2921 (m), 2849 (m), 1252 (vs), 1187 (s), 846 (vs).

Synthesis of polymer **156c** (Scheme 4.43) [80d]
To a solution of **155** (0.200 g; 0.171 mmol) in THF (15 mL) was added a solution of 2,5-thiophenedicarboxaldehyde (0.0241 g; 0.172 mmol) in THF (15 mL) over 1 h. Over the course of addition the reaction mixture changed progressively from violet to bright orange. After stirring overnight, additional 2,5-thiophenedicarboxaldehyde (0.0241 g) was added to quench any unreacted –P=PMe$_3$ end groups. Volatile components were removed *in vacuo* and the crude orange solid was rinsed with MeCN and *n*-pentane to yield **155** (0.178 g, 85.4%). ^{1}H NMR (C$_6$D$_6$), δ = 0.84 (t, 6H, J = 7 Hz), 1.20–1.40 (m, 8H), 1.40–1.55 (m, 4H), 1.75–1.90 (m, 4H), 2.04 (s, 24H), 2.09 (s, 3.7H, end group), 2.30 (s, 12H), 2.35 (s, 1.8H, end group), 4.00 (t, 4H, J = 7 Hz), 6.31 (s, 2H), 6.88 (s, 8H), 6.92 (s, 1.2H, end group), 7.05 (s, 2H), 7.08–7.21 (m, 2H), 7.23 (s, 4H), 7.40–7.55 (m, 2H), 8.28 (d, 2H, J = 20 Hz), 8.42 (d, 0.3H, J = 23 Hz, end group), 9.73 (s, 0.6H, terminal –CHO); ^{31}P NMR (CDCl$_3$), δ = 232.3 (s, internal P=C), 256.8 (s, terminal P=C).

Synthesis of diphosphene polymer **157** (Scheme 4.43) [80d]
A sample of **155** (0.10 g, 0.050 mmol) was heated under nitrogen at ~250° C for 2 min, during which time the material melted as it changed color from violet to red and a vapor was emitted. The material was cooled to room temperature, leaving a hard, transparent, red solid. The solid was stirred with acetonitrile (5 mL) to produce a fine suspension of the material, which was isolated by filtration and the solid dried *in vacuo* to give 0.091 g (99%) of **155** as a red powder. ^{1}H NMR (C$_6$D$_6$), δ = 0.82 (br, 6H), 1.14 (br, 8H), 1.32 (br, 4H), 1.53 (br, 4H), 1.96 (s, 24H), 2.21 (s, 12H), 3.73 (br, 4H), 6.81 (s, 8H), 7.09–7.23 (br m, 4H), 7.31 (s, 4H), 7.84 (d, 2H, J = 15 Hz); ^{31}P NMR (CDCl$_3$), δ = 493.5 (broad).

References

1. (a) A. Kraft, A. C. Grimsdale, A. B. Holmes, *Angew. Chem. Int. Ed.* **1998**, *37*, 402; (b) U. Mitschke, P. Bäuerle, *J. Mater. Chem.* **2000**, *10*, 1471; (c) A. P. Kulkarni, C. J. Tonzola, A. Babel, S. A. Jenekhe, *Chem. Mater.* **2004**, *16*, 4556; (d) C.-T. Chen, *Chem. Mater.* **2004**, *16*, 4389; (e) B. W. D'Andrade, S. R. Forrest, *Adv. Mater.* **2004**, *16*, 1585.

2. (a) K. Müllen, G. Wegner, *Electronic Materials: the Oligomer Approach*, Wiley-VCH, Weinheim, **1998**; (b) G. H. Gelinck, T. C. T. Geuns, D. M. de Leeuw, *Appl. Phys. Lett.* **2000**, *77*, 1487; (c) B. K. Crone, A. Dodabalapur, R. Sarpeshkar, A. Gelperin, H. E. Katz, Z. Bao, *J. Appl. Phys.* **2002**, *91*, 10140; (d) J. A. Rogers, Z. Bao, H. E. Katz, A. Dodabalapur, in *Thin-film Transistors*, C. R. Kagan, P. Andry (Eds.), Marcel Dekker, New York, **2003**, p. 377; (e) V. C. Sundar, J. Zaumseil, V. Podzorov, E. Menard, R. L. Willett, T. Someya, M. E. Gershenson, J. A. Rogers, *Science* **2004**, *303*, 1644; (f) Z. Bao, *Nat. Mater.* **2004**, *3*, 137; (g) N. Stutzmann, R. H. Friend, H. Sirringhaus, *Science* **2003**, *299*, 1881; (h) F. Garnier, R. Hajlaoui, A. Yassar, P. Srivastava, *Science* **1994**, *265*, 1684; (i) G. H. Gelinck, H. E. A. Huitema, E. van Veenendaal, E. Cantatore, L. Schrijnemakers, J. B. P. H. van der Putten, T. C. T. Geuns, M. Beenhakkers, J. B. Giesbers, B.-H. Huisman, E. J. Meijer, E. M. Benito, F. J. Touwslager, A. W. Marsman, B. J. E. van Rens, D. M. de Leeuw, *Nat. Mater.* **2004**, *3*, 106; (j) S. R. Forrest, *Nature* **2004**, *428*, 911; (k) E. J. Meijer, D. M. De Leeuw, S. Setayesh, E. van Veenendaal, B.-H. Huisman, P. W. M. Blom, J. C. Hummelen, U. Scherf, T. M. Klapwijk, *Nat. Mater.* **2003**, *2*, 678.

3. See, for example. (a) http://www.kodak.com/US/en/corp/display/index.jhtml; (b) http://www.research.philips.com/technologies/display/; (c) http://www.dupont.com/displays/oled/; (d) http://www.cdtltd.co.uk/; (e) http://www.universaldisplay.com/; (f) http://www.covion.com/index.html.

4. (a) C. W. Tang, S. A. Van Slyke, *J. Appl. Phys.* **1987**, *51*, 913; (b) C. W. Tang, S. A. Van Slyke, C. H. Chen, *J. Appl. Phys.* **1989**, *65*, 3610; (c) J. H. Burroughes, D. D. C. Bradley, A. R. Brown, R. N. Marks, K. Mackay, R. H. Friends, P. L. Burns, A. B. Holmes, *Nature* **1990**, *347*, 539; (d) M. A. Baldo, D. F. O'Brien, Y. You, A. Shoustikov, S. Sibley, M. E. Thompson, S. R. Forrest, *Nature* **1998**, *395*, 151.

5. (a) R. E. Martin, F. Diederich, *Angew. Chem. Int. Ed.* **1999**, *38*, 1350; (b) T. A. Skotheim, R. L.Elsenbaumer, J. R Reynolds, *Handbook of Conducting Polymers* (2nd edn.), Marcel Dekker, New York, **1998**; (c) H. S. Nalwa, *Handbook of Conductive Materials and Polymers*, Wiley, New York, **1997**; (d) J. Roncali, *Chem. Rev.* **1997**, *97*, 173; (e) F. J. M. Hoeben, P. Jonkheijm, E. W. Meijer, A. P. H. J. Schenning, *Chem. Rev.* **2005**, *105*, 1491; (f) F. Badubri, G. M. Farinola, F. Naso, *J. Mater. Chem.* **2004**, *14*, 11; (g) P. F. H. Schwab, J. R. Smith, J. Michl, *Chem. Rev.* **2005**, *105*, 1197; (h) M. B. Nielsen, F. Diederich, *Chem. Rev.* **2005**, *105*, 1837; (i) S. Szafert, J. A. Gladysz, *Chem. Rev.* **2003**, *103*, 4175; (j) M. Hissler, P. W. Dyer, R. Réau, *Coord. Chem. Rev.* **2003**, *244*, 1; (k) U. H. F. Bunz, *Chem. Rev.* **2000**, *100*, 1605; (l) B. J. Holliday, T. M. Swager, *Chem. Commun.* **2005**, 23.

6. (a) D. L. Albert, T. J. Marks, M. A. Ratner, *J. Am. Chem. Soc.* **1997**, *119*, 6575; (b) U. Salzner, J. B. Lagowski, P. G. Pickup, R. A. Poirier, *Synth. Met.* **1998**, *96*, 177; (c) J. Ma, S. Li, Y. Jiang, *Macromolecules* **2002**, *35*, 1109.

7. (a) F. Mathey, *Phosphorus–Carbon Heterocyclic Chemistry: the Rise of a New Domain*, Elsevier Science, Oxford, **2001**; (b) K. Dillon, F. Mathey, J.F. Nixon, *Phosphorus: the Carbon Copy*, Wiley, Chichester, **1998**; (c) L. Nyulaszi, *Chem. Rev.* **2001**, *101*, 1229; (d) F. Mathey, *Angew. Chem. Int. Ed.* **2003**, *42*, 1578.

8. (a) H. R. Allcock, *Chemistry and Applications of Polyphosphazenes*, Wiley-Interscience, New York, **2003**; (b) M. Gleria, R. De Jaeger, *Top. Curr. Chem.* **2005**, *250*, 165–251; (c) M. Escobar, Z. Jin, B. L. Lucht, *Org. Lett.* **2002**, *13*, 2213.

9. First dibenzophosphole (a) G. Wittig, G. Geissler, *Liebigs Ann. Chem.* **1953**, *580*, 44; first phospholes (b) E. H. Bray, W. Hübel, *Chem. Ind. (London)* **1959**, 1250; (c) E. H. Bray, W. Hübel, I. Caplier *J. Am. Chem. Soc.* **1961**, *83*, 4406.

10. (a) F. Mathey, *Chem. Rev.* **1988**, *88*, 429; (b) L. D. Quin, in *Phosphorus–Carbon Heterocyclic Chemistry: the Rise of a New Domain*, F. Mathey (Ed.), Elsevier Science, Oxford, **2001**, pp. 217, 307; (c) C. Charrier, H. Bonnard, G. de Lauzon, F. Mathey, *J. Am. Chem. Soc.* **1983**, *105*, 6871; (d) F. Mathey, *Acc. Chem. Res.* **2004**, *37*, 954.

11. (a) P. v. R. Schleyer, C. Maerker, A. Dransfeld, H. Jiao, N. J. R. van Eikema Hommes, *J. Am. Chem. Soc.* **1996**, *118*, 6317; (b) P. v. R. Schleyer, P. K. Freeman, H. Jiao, B. Goldfuss, *Angew. Chem. Int. Ed.* **1995**, *34*, 337.

12. (a) W. Egan, R. Tang, G. Zon, K. Mislow, *J. Am. Chem. Soc.* **1971**, *93*, 6205; (b) D. Delaere, A. Dransfeld, M. T. Nguyen, L. G. Vanquickenborne, *J. Org. Chem.* **2000**, *65*, 2631.

13. (a) W. Schäfer, A. Schweig, F. Mathey, *J. Am. Chem. Soc.* **1976**, *98*, 407; (b) E. Mattmann, F. Mathey, A. Sevin, G. Frison, *J. Org. Chem.* **2002**, *67*, 1208

14. (a) S. Yamaguchi, K. Tamao, *J. Chem. Soc., Dalton Trans.* **1998**, 3693; (b) F. Wang, J. Luo, K. Yang; J. Chen, F. Huang, Y. Cao, *Macromolecules* **2005**, *38*, 2253; (c) S. Yamaguchi, K. Tamao, *Chem. Lett.* **2005**, *34*, 2; (d) A. J. Boydston, B. L. Pagenkopf, *Angew. Chem. Int. Ed.* **2004**, *43*, 6336; (e) A. J. Boydston, J. Andrew, Y. Yin, B. L. Pagenkopf, *J. Am. Chem. Soc.* **2004**, *126*, 10350.

15. (a) C. Hay, M. Hissler, C. Fischmeister, J. Rault-Berthelot, L. Toupet, L. Nyulaszi, R. Réau, *Chem. Eur. J.* **2001**, *7*, 4222; (b) J. Hydrio, M. Gouygou, F. Dallemer, G. G. A. Balavoine, J-C. Daran, *Eur. J. Org. Chem.* **2002**, 675; (c) X. Sava, N. Mézaille, N. Maigrot, F. Nief, L. Ricard, F. Mathey, P. Le Floch, *Organometallics* **1999**, *18*, 4205.

16. A. Brèque, G. Muller, H. Bonnard, F. Mathey, P. Savignac, *Eur. Pat.* 41447, **1981**.

17. (a) F. Mercier, F. Mathey, J. Fischer, J. H. Nelson, *Inorg. Chem.* **1985**, *24*, 4141; (b) F. Mercier, F. Mathey, J. Fischer, J. H. Nelson, *J. Am. Chem. Soc.* **1984**, *106*, 425.

18. (a) F. Mercier, S. Holand, F. Mathey, *J. Organomet. Chem.* **1986**, *316*, 271; (b) O. Tissot, J. Hydrio, M. Gouygou, F. Dallemer, J.-C. Daran, G. G. A. Balavoine, *Tetrahedron* **2000**, *56*, 85; (c) O. Tissot, M. Gouygou, J.-C. Daran, G. G. A. Balavoine, *Chem. Commun.* **1996**, 2287.

19. E. Deschamps, F. Mathey, *Bull. Soc. Chim. Fr.* **1992**, *129*, 486.

20 T.-A. Niemi, P. L. Coe, S. J. Till, *J. Chem. Soc., Perkin Trans.* **2000**, 1519.

21. E. Deschamps, L. Ricard, F. Mathey, *Angew. Chem. Int. Ed. Engl.* **1994**, *33*, 1158.

22. (a) M. Gouygou, O. Tissot, J.-C. Daran, G. G. A. Balavoine, *Organometallics* **1997**, *16*, 1008; (b) J. Hydrio, M. Gouygou, F. Dallemer, G. G. A. Balavoine, J.-C. Daran, *Tetrahedron: Asymmetry* **2002**, *13*, 1097; (c) J. Hydrio, M. Gouygou, F. Dallemer, J.-C. Daran, G. G. A. Balavoine, *Eur. J. Org. Chem.*, **2002**, 675; (d) O. Tissot, M. Gouygou, J.-C. Daran, G. G. A. Balavoine, *Organometallics*, **1998**, *17*, 5927; (e) O. Tissot, M. Gouygou, F. Dallemer, J.-C. Daran, G. G. A. Balavoine, *Eur. J. Inorg. Chem.* **2001**, 2385; (f) T. Kojima, K. Saeki, K. Ono, Y. Matsuda, *Bull. Chem. Soc. Jpn.* **1998**, *71*, 2885.(g) O. Tissot, M. Gouygou, F. Dallemer, J.-C. Daran, G. G. A. Balavoine, *Angew. Chem. Int. Ed.* **2001**, *40*, 1076.

23. (a) E. W. Abel, C. Towers, *J. Chem. Soc., Dalton Trans.* **1979**, 814; (b) C. Charrier, H. Bonnard, F. Mathey, *J. Org. Chem.* **1982**, *47*, 2376; (c) C. Charrier, H. Bonnard, F. Mathey, D. Neibecker, *J. Organomet. Chem.* **1982**, *231*, 361; (d) S. Holand, F. Mathey, J. Fischer, A. Mitschler, *Organometallics* **1983**, *2*, 1234; (e) J. K. Vohs, P. Wie, J. Su, B. C. Beck, S. D. Goodwin, G. H. Robinson, *Chem. Commun.* **2000**, 1037.

24. (a) D. Carmichael, F. Mathey. *Top. Curr. Chem.* **2002**, *220*, 27–51; (b) F. Mathey, *J. Organomet. Chem.* **2002**, *646*, 15; (c) F. Nief, *Eur. J. Inorg. Chem.* **2001**, *4*, 891; (d) C. Ganter, *J. Chem. Soc., Dalton*

Trans. **2001**, 3541; (e) N. M. Kostic, R. F. Fenske *Organometallics* **1983**, *2*, 1008; (f) G. Frison, F. Mathey, A. Sevin, *J. Phys. Chem. A* **2002**, *106*, 5653; (g) D. Turcitu, F. Nief, L. Ricard, *Chem. Eur. J.* **2003**, *9*, 4916; (h) R. Shintani, G. C. Fu, *J. Am. Chem. Soc.* **2003**, *125*, 10778; (i) L. Weber, *Angew. Chem. Int. Ed.* **2002**, *41*, 563; (j) M. Ogasawara, T. Nagano, T. Hatashi, *Organometallics* **2003**, *22*, 1174; (k) L.-S. Wang, T. K. Hollis, *Org. Lett.* **2003**, *5*, 2543; (l) X. Sava, M. Melaimi, L. Ricard, F. Mathey, P. Le Floch, *New J. Chem.* **2003**, *27*, 1233; (m) G. de Lauzon, B. Deschamps, J. Fischer, F. Mathey, A. Mitschler, *J. Am. Chem. Soc.* **1980**, *102*, 994; (n) X. Sava, L. Ricard, F. Mathey, P. Le Floch, *Organometallics* **2000**, *19*, 4899.

25. C. Fave, M. Hissler, T. Karpati, J. Rault-Berthelot, V. Deborde, L. Toupet, L. Nyulaszi, R. Réau, *J. Am. Chem. Soc.* **2004**, *126*, 6058.

26. (a) R. Hoffmann, *Acc. Chem. Res.* **1971**, *4*, 1; (b) R. Hoffmann, A. Imamura, W. J. Hehre, *J. Am. Chem. Soc.* **1968**, *90*, 1499; (c) M. N. Paddon-Row, *Acc. Chem. Res.* **1994**, *27*, 18; (d) R. Gleiter, W. Schäfer, *Acc. Chem. Res.* **1990**, *23*, 369; (e) B. P. Paulson, L. A. Curtiss, B. Bal, G. L. Closs, J. R. Miller, *J. Am. Chem. Soc.* **1996**, *118*, 378.

27. (a) S. Yamaguchi, K. Tamao, in *The Chemistry of Organic Silicon Compounds*, Z. Rappoport, T. Appeloig (Eds.), Wiley, New York, **2001**, Vol. 3, p. 647; (b) S. Yamaguchi, R.-Z. Jin, K. Tamao, *J. Am. Chem. Soc.* **1999**, *121*, 2937; (c) H. Sohn, M. J. Sailor, D. Madge, W. C. Trogler, *J. Am. Chem. Soc.* **2003**, *125*, 3821; (d) H. Sohn, R. M. Calhoun, M. J. Sailor, W. C. Trogler, *Angew. Chem. Int. Ed.* **2001**, *40*, 2104; (e) T. Sanji, T. Sakai, C. Kabuto, H. Sakurai, *J. Am. Chem. Soc.* **1998**, *120*, 4552; (f) H. Sohn, R. R. Huddleston, D. R. Power, R. West, *J. Am. Chem. Soc.* **1999**, *121*, 2935; (g) A. Rose, Z. Zhu, C. F. Madigan, T. M. Swager, V. Bulovic, *Nature* **2005**, *434*, 7035.

28. (a) F. Mathey, F. Mercier, F. Nief, J. Fischer, A. Mitschler, *J. Am. Chem. Soc.* **1982**, *104*, 2077; (b) F. J. Fischer, A. Mitschler, F. Mathey, F. Mercier, *J. Chem. Soc., Dalton Trans.* **1983**, 841; (c) F. Laporte, F. Mercier, L. Ricard, F. Mathey, *J. Am. Chem. Soc.* **1994**, *116*, 3306.

29. (a) E. Deschamps, L. Ricard, F. Mathey, *Heteroat. Chem.* **1991**, *2*, 377; (b) E. Deschamps, F. Mathey, *J. Org. Chem.* **1990**, *55*, 2494.

30. (a) P. J. Fagan, W. A. Nugent, *J. Am. Chem. Soc.* **1988**, *110*, 2310; (b) P. J. Fagan, W. A. Nugent, J. C. Calabrese, *J. Am. Chem. Soc.* **1994**, *116*, 1880.

31. (a) D. Le Vilain, C. Hay, V. Deborde, L. Toupet, R. Réau, *Chem. Commun.* **1999**, 345; (b) C. Hay, C. Fischmeister, M. Hissler, L. Toupet, R. Réau, *Angew. Chem. Int. Ed.* **2000**, *39*, 1812; (c) M. Sauthier, F. Leca, L. Toupet, R. Réau, *Organometallics* **2002**, *21*, 1591; (d) C. Hay, M. Sauthier, V. Deborde, M. Hissler, L. Toupet, R. Réau, *J. Organomet. Chem.* **2002**, *643–644*, 494; (f) C. Fave, M. Hissler, K. Sénéchal, I. Ledoux, J. Zyss, R. Réau, *Chem. Commun.* **2002**, 1674.

32. D. Delaere, M. T. Nguyen, L. G. Vanquickenborne, *J. Phys. Chem. A* **2003**, *107*, 838.

33. (a) A. Hucke, M. P. Cava, *J. Org. Chem.* **1998**, *63*, 7413; (b) P. E. Niziurski-Mann, C. C. Scordilis-Kelley, T. L. Liu, M. P. Cava, R. T. Carlin, *J. Am. Chem. Soc.* **1993**, *115*, 887; (c) S. Yamaguchi, Y. Itami, K. Tamao, *Organometallics* **1998**, *17*, 4910.

34. C. Fave, T. Y. Cho, M. Hissler, C. W. Chen, T. Y. Luh, C. C. Wu, R. Réau, *J. Am. Chem. Soc.* **2003**, *125*, 9254.

35. (a) H. Schmidbaur, *Gold Bull.* **1990**, *23*, 11; (b) P. Pyykkö, *Chem. Rev.* **1997**, *97*, 597; (c) P. Pyykkö, *Angew. Chem. Int. Ed.*, **2004**, *43*, 4412; (d) W. W. W. Yam, K. K. W. Lo, W. K. M. Fung, C. R. Wang, *Coord. Chem. Rev.* **1998**, *171*, 17.

36. (a) T-H. Liu, C-Y. Iou, C. H. Chen, *Appl. Phys. Lett.* **2003**, *83*, 5241; (b) R. H. Young, C. W. Tang, A. P. Marchetti, *Appl. Phys. Lett.* **2002**, *80*, 874; (c) J. Feng, F. Li, W. Gao, G. Cheng, W.Xie, S. Liu, *Appl. Phys. Lett.* **2002**, *81*, 2935.

37. C. Hay, C. Fave, M. Hissler, J. Rault-Berthelot, R. Réau, *Org. Lett.* **2003**, *19*, 3467.

38. J. W. Sease, L. Zechmeister, *J. Am. Chem. Soc.* **1947**, *69*, 270.

39. (a) C. Adachi, M. A. Baldo, S. R. Forrest, S. Lamansky, M. E. Thompson, R. C. Kwong, *Appl. Phys. Lett.* **2001**, *78*, 1622; (b) R. J. Holmes, S. R. Forrest, Y. J. Tung, R. C. Kwong, J. J. Brown, S. Garon, M. E. Thompson, *Appl. Phys. Lett.* **2003**, *82*, 2422.
40. (a) C. Denault, I. Ledoux, I. D. W. Samuel, J. Zyss, M. Bourgault, H. Le Bozec, *Nature* **1995**, *374*, 339; (b) K. Sénéchal, O. Maury, H. Le Bozec, I. Ledoux, J. Zyss, *J. Am. Chem. Soc.* **2002**, *124*, 4560; (c) L. Viau, S. Bidault, O. Maury, S. Brasselet, I. Ledoux, J. Zyss, E. Ishow, K. Nakatani, H. Le Bozec, *J. Am. Chem. Soc.* **2004**, *126*, 8386; (d) O. Maury, H. Le Bozec, *Acc. Chem. Res.* **2005**, *38*, 691.
41. (a) C. Hay, M. Sauthier, V. Deborde, M. Hissler, L. Toupet, R. Réau, *J. Organomet. Chem.* **2002**, *643–644*, 494; (b) M. Sauthier, B. Le Guennic, V. Deborde, L. Toupet, J.-F. Halet, R. Réau, *Angew. Chem. Int. Ed.* **2001**, *40*, 228; (c) F. Leca, M. Sauthier, V. Deborde, L. Toupet, R. Réau, *Chem. Eur. J.* **2003**, *16*, 3785; (d) F. Leca, C. Lescop, E. Rodriguez, K. Costuas, J.-F. Halet, R. Réau, *Angew. Chem. Int. Ed.* **2005**, *44*, 2190.
42. (a) R. G. Pearson, *Inorg. Chem.* **1973**, *12*, 712; (b) M. Sauthier, F. Leca, L. Toupet, R. Réau, *Organometallics* **2002**, *21*, 1591; (c) F. Leca, C. Lescop, R. Réau, *Organometallics* **2004**, *23*, 6191; (d) C. Fave, M. Hissler, K. Sénéchal, I. Ledoux, J. Zyss, R. Réau, *Chem. Commun.* **2002**, 1674.
43. (a) O. M. Bevierre, F. Mercier, L. Ricard, F. Mathey, *Angew. Chem. Int. Ed.* **1990**, *29*, 655; (b) M. O. Bevierre, F. Mercier, F. Mathey, A. Jutand, C. Amatore, *New J. Chem.* **1991**, *15*, 545.
44. E. Deschamps, L. Ricard, F. Mathey, *J. Chem. Soc., Chem. Commun.* **1995**, 1561.
45. (a) J. J. Apperloo, J. M. Raimundo, P. Frère, P. Roncali, R. A. Janssen, *Chem. Eur. J.* **2000**, *9*, 1698; (b) I. Jestin, P. Frère, P. Blanchard, J. Roncali, *Angew. Chem. Int. Ed.* **1998**, *37*, 942; (c) S. Zahn, T. M. Swager, *Angew. Chem. Int. Ed.* **2002**, *41*, 4225.
46. For examples: (a) H. Jiao, K. Costuas, J. A. Gladysz, J.-F. Halet, M. Guillemot, L. Toupet, F. Paul, C. Lapinte, *J. Am. Chem. Soc.* **2003**, *125*, 9511; (b) Z. Zhu, T. M. Swager, *J. Am. Chem. Soc.* **2002**, *124*, 9670; (c) C. A. Breen, T. Deng, T. Breiner, E. L. Thomas, T. M. Swager, *J. Am. Chem. Soc.* **2003**, *125*, 9942.
47. (a) S. Holand, F. Gandolfo, L. Ricard, F. Mathey, *Bull. Soc. Chim. Fr.* **1996**, *33*, 133; (b) T.-A. Niemi, P. L. Coe, S. J. Till, *J. Chem. Soc., Perkin Trans.* **2000**, 519.
48. (a) S. S. H. Mao, T. D. Tilley, *Macromolecules* **1997**, *30*, 5566; (b) B. L. Lucht, S. S. H. Mao, T. D. Tilley, *J. Am. Chem. Soc.* **1998**, *120*, 4354; (c) B. L. Lucht, T. D. Tilley, *Chem. Commun.* **1998**, 1645.
49. (a) T. Takahashi, *Jpn. Pat. 2003261585*, **2003**; (b) Y. Morisaki, Y. Aiki, Y. Chujo, *Macromolecules* **2003**, *36*, 2594.
50. (a) P. Audebert, J.-M. Catel, G. Le Coustumer, V. Duchenet, P. Hapiot, *J. Phys. Chem. B* **1998**, *102*, 8661; (b) P. Audebert, J.-M. Catel, V. Duchenet, L. Guyard, P. Hapiot, G. Le Coustumer, *Synth. Met.* **1999**, *101*, 642.
51. (a) A. N. Hughes, D. Kleemola, *J. Heterocycl. Chem.* **1976**, *13*, 1; (b) W. Egan, R. Tang, G. Zon, K. Mislow, *J. Am. Chem. Soc.* **1971**, *93*, 6205.
52. S. Kobayashi, M. Noguchi, Y. Tsubata, M. Kitano, H. Doi, T. Kamioka, A. Nakazono, *Jpn. Pat. 2003231741*, **2003**.
53. (a) E. Duran, E. Gordo, J. Granell, D. Velasco, F. Lopez-Calahorra, *Tetrahedron Lett.* **2001**, *42*, 7791; (b) E. Duran, D. Velasco, F. Lopez-Calahorra, H. Finkelmann, *Mol. Cryst. Liq. Cryst.* **2002**, *381*, 43.
54. M. Taillefer, F. Plenat, C. Combes-Chamalet, V. Vicente, H. J. Cristau, *Phosphorus Res. Bull.* **1999**, *10*, 696.
55. J-P. Lampin, F. Mathey, *J. Organomet. Chem.* **1974**, *71*, 239.
56. (a) S. Kim, K.-H. Song, S. O. Kang, J. Ko, *Chem. Commun.* **2004**, 68; (b) K. Ogawa, S. C. Rasmussen, *J. Org. Chem.* **2003**, *68*, 2921; (c) J. Ohshita, M. Nodono, T. Watanabe, Y. Ueno, A. Kunai, Y. Harima, K. Yamashita, M. Ishikawa, *J. Organomet. Chem.* **1998**, *553*, 487; (d) G. Barbarella, L. Favaretto, G. Sotgiu, L. Antolini, G. Gigli, R. Cingolati, A. Bongini, *Chem. Mater.* **2001**, *13*, 4112; (e) G. Barbarella, M. Melucci, G. Sotgiu, *Adv. Mater.* **2005**, *17*, 1581.

57. (a) T. Baumgartner, *Macromol. Symp.* **2003**, *196*, 279; (b) T. Baumgartner, T. Neumann, B. Wirges, *Angew. Chem. Int. Ed.* **2004**, *43*, 6197; (c) T. Baumgartner, W. Bergmans, T. Karpati, T. Neumann, M. Nieger, L. Nuyalaszi, *Eur. J. Chem.* **2005**, *11*, 4684.
58. (a) A. G. MacDiarmid, J. C. Chiang, A. J. Epstein, *Synth. Met.* **1987**, *18*, 285; (b) F. Wudl, R. O. Angus, Jr., F. L. Lu, P. M. Allemand, D. F. Vachon, M. Novak, Z. X. Liu, A. J. Heeger, *J. Am. Chem. Soc.* **1987**, *109*, 3677; (c) S. A. Hay, *Macromolecules* **1969**, *2*, 107; (d) R. H. Baughman, J. L. Bredas, R. R. Chance, R. L. Elsenbaumer, L. W. Shacklette, *Chem. Rev.* **1982**, *82*, 209; (e) F. E. Goodson, S. J. Hauck, J. F. Hartwig, *J. Am. Chem. Soc.* **1999**, *121*, 7527; (f) R. A. Singer, J. P. Sadighi, S. L. Buchwald, *J. Am. Chem. Soc.* **1998**, *120*, 213.
59. (a) B. L. Lucht, N. O. St. Onge, *Chem. Commun.* **2000**, 2097; (b) Z. Jin, B. L. Lucht, *J. Am. Chem. Soc.* **2005**, *127*, 5586.
60. (a) Z. Jin, B. L. Lucht, *J. Organomet. Chem.* **2002**, *653*, 167; (b) H. Ghassemi, E. McGrath, *Polymer* **1997**, *38*, 3139.
61. (a) F. E. Goodson, T. I. Wallow, B. M. Novak, *Macromolecules* **1998**, *31*, 2047; (b) F. E. Goodson, T. I. Wallow, B. M. Novak, *J. Am. Chem. Soc.* **1997**, *119*, 12441.
62. (a) D. K. Johnson, T. Rakuchaisirikul, Y. Sun, N. J. Taylor, A. J. Canty, A. J. Carty, *Inorg. Chem.* **1993**, *32*, 5544; (b) A. J. Carty, N. J. Taylor, D. K. Johnson, *J. Am. Chem. Soc.* **1979**, *101*, 5422; (c) M. P. Martin-Redondo, L. Scoles, B. T. Sterenberg, K. A. Udachin, A. J. Carthy, *J. Am. Chem. Soc.* **2005**, *127*, 5038; (d) D. Xu, B. Hong, *Angew. Chem. Int. Ed.* **2000**, *39*, 1826; (e) J. Fornies, A. Garcia, J. Gomez, E. Lalinde, M. T. Moreno, *Organometallics* **2002**, *21*, 3733; (f) I. Ara, J. Fornies, A. Garcia, J. Gomez, E. Lalinde, M. T. Moreno, *Chem. Eur. J.* **2002**, *16*, 3698; (g) T. Baumgartner, K. Huynh, S. Schleidt, A. J. Lough, I. Manners, *Chem. Eur. J.* **2002**, *8*, 4622.
63. (a) P. J. Stang, F. Diederich, *Modern Acetylene Chemistry*, VCH, Weinheim, **1995**; (b) F. Diederich, P. J. Stang, R. R. Tykwinsky, *Acetylene Chemistry: Chemistry, Biology and Material Science*, VCH, Weinheim, **2005**.
64. L. T. Scott, M. Unno, *J. Am. Chem. Soc.* **1990**, *112*, 7823.
65. G. Märkl, T. Zollitsch, P. Kreitmeier, M. Prinzhorn, S. Reithinger, E. Eibler, *Chem. Eur. J.* **2000**, *6*, 3806.
66. V. Huc, A. Balueva, R.M. Sebastien, A.-M. Caminade, J.-P. Majoral, *Synthesis* **2000**, 726.
67. (a) S. Yamaguchi, S. Akiyama, K. Tamao, *J. Organomet. Chem.* **2002**, *646*, 277; (b) S. Yamaguchi, S. Akiyama, K. Tamao, *J. Am. Chem. Soc.* **2000**, *122*, 6793; (c) S. Yamaguchi, S. Akiyama, K. Tamao, *J. Am. Chem. Soc.* **2001**, *123*, 11372; (d) S. Yamaguchi, S. Akiyama, K. Tamao, *J. Organomet. Chem.* **2002**, *652*, 3.
68. (a) Y. Wang, M. I. Ranasinghe, T. Goodson, III, *J. Am. Chem. Soc.* **2003**, *125*, 9562; (b) M. Malagoli, J. Bredas, *Chem. Phys. Lett.* **2000**, *327*, 13; (c) R. C. Smith, J. D. Protasiewicz, *J. Chem. Soc., Dalton Trans.* **2000**, 4738; (d) R. C. Smith, M. J. Earl, J. D. Protasiewicz, *Inorg. Chim. Acta* **2004**, *357*, 4139; (e) Y. Kang, D. Song, H. Schmider, S. Wang, *Organometallics* **2002**, *21*, 2413.
69. L. G. Madrigal, C. W. Spangler, *Mater. Res. Soc. Symp. Proc.* **1999**, *561*, 75.
70. (a) T. Renouard, H. Le Bozec, *Eur. J. Inorg. Chem.* **2000**, *1*, 229; (b) S. Di Bella, *Chem. Soc. Rev.* **2001**, *30*, 355; (c) P. G. Lacroix, *Eur. J. Chem.* **2001**, 339.
71. (a) A. W. Johnson, W. C. Kaska, K. A. Astoja Stanzewski, D. A. Dixon, *Ylides and Imines of Phosphorus*, Wiley, New York, **1993**; (b) Y. G. Gololobov, L. F. Kasukhin, *Tetrahedron* **1992**, *48*, 1353.
72. K. V. Katti, K. Raghuraman, N. Pillarsetty, S. R. Kara, R. J. Gaulotty, M. A. Chartier, C. A. Langhoff, *Chem. Mater.* **2002**, *14*, 2436.
73. (a) C. Lambert, E. Schmälzlin, K. Meerholz, C. Bräuchle, *Chem. Eur. J.* **1998**, *4*, 512; (b) C. Bourgogne, Y. Le Fur, P. Juen, P. Masson, J. F. Nicoud, R. Masse, *Chem. Mater.* **2000**, *12*, 1025.
74. (a) A. B. Padmaperuma, G. Schmett, D. Fogarty, N. Washton, S. Nanayakkara, L. Sapochak, K. Ashworth, L. Madrigal, B. Reeves, C. W. Spangler, *Mater. Res. Soc. Symp. Proc.* **2001**, 621; (b) R. Métivier, R.

Amengual, I. Leray, V. Michelet, J.-P. Genêt, *Org. Lett.*, **2004**, *6*, 739.

75. (a) O. Clot, M. O. Wolf, B. O. Patrick, *J. Am. Chem. Soc.* **2000**, *122*, 10456; (b) O. Clot, M. O. Wolf, B. O. Patrick, *J. Am. Chem. Soc.* **2001**, *123*, 9963; (c) O. Clot, Y. Akahori, C. Moorlag, D. B. Leznoff, M. O. Wolf, R. J. Batchelor, B. O. Patrick, M. Ishii, *Inorg. Chem.* **2003**, *42*, 2704.

76. (a) L. Nyulaszi, T. Veszpremi, J. Reffy, *J. Phys. Chem.* **1993**, *97*, 4011; (b) M. Regitz, O. J. Scherer (Eds), *Mutiple Bonds and Low Coordination in Phosphorus Chemistry*, Georg Thieme, Stuttgart, **1990**; (c) D. A. Loy, G. M. Jamison, M. D. McClain, T. M. Alam, *J. Polym. Sci. Part A* **1999**, *37*, 129.

77. (a) T. Wettling, J. Schneider, O. Wagner, C. G. Kreiter, M. Regitz M, *Angew. Chem. Int. Ed.* **1989**, *28*, 1013; (b) F. Tabellion, A. Nachbauer, S. Leininger, C. Peters, F. Preuss, M. Regitz, *Angew. Chem. Int. Ed.* **1998**, *37*, 1233; (c) T. Wettling, B. Geissler, R. Schneider, S. Barth, P. Binger, M. Regitz, *Angew. Chem. Int. Ed. Engl.* **1992**, *31*, 758; (d) R. Appel, J. Hünerbein, N. Siabalis, *Angew. Chem. Int. Ed. Engl.* **1987**, *26*, 779; (e) M. J. Hopkinson, H. W. Kroto, J. F. Nixon, N. P. C. Simmons, *J. Chem. Soc., Chem. Commun.* **1994**, 513.

78. (a) D. P. Gates, *Top. Curr. Chem.* **2005**, *250*, 107–126; (b) V. A. Wright, D. P. Gates, *Angew. Chem. Int. Ed.* **2002**, *41*, 2389.

79. P. Le Floch, A. Marinetti, L. Ricard, F. Mathey, *J. Am. Chem. Soc.* **1990**, *112*, 2407.

80. (a) R. C. Smith, X. Chen, J. D. Protasiewicz, *Inorg. Chem.* **2003**, *42*, 5468; (b) S. Shah, T. Concolino, A. L. Rheingold, J. D. Protasiewicz, *Inorg. Chem.* **2000**, *39*, 3860; (c) R. C. Smith, J. D. Protasiewcz, *Eur. J. Inorg. Chem.* **2004**, 998; (d) R. C. Smith, J. D. Protasiewicz, *J. Am. Chem. Soc.* **2004**, *126*, 2268; (e) S. Shah, M. Cather Simpson, R. C. Smith, J. D. Protasiewicz, *J. Am. Chem. Soc.* **2001**, *123*, 6925.

81. (a) M. Yoshifuji, I. Shima, N. Inamoto, T. Hirotsu, J. Higushi, *J. Am. Chem. Soc.* **1981**, *103*, 4587; (b) M. Yoshifuji, *Top. Curr. Chem.* **2002**, *223*, 67–89.

82. Z. Bao, Y. Chen, R. Cai, L. Yu, *Macromolecules* **1993**, *26*, 5281.

83. (a) S. Ito, H. Sugiyama, M. Yoshifuji, *Angew. Chem. Int. Ed.* **2000**, *39*, 2781; (b) S. Ito, H. Miyake, H. Sugiyama, M. Yoshifuji, *Tetrahedron. Lett.* **2004**, *45*, 7019; (c) A. Nakamura, K. Toyota, M. Yoshifuji, *Tetrahedron* **2005**, *61*, 5223; (d) S. Ito, S. Sekigushi, M. Yoshifuji, *J. Org. Chem.* **2004**, *69*, 4181; (e) T. Cantat, N. Mézailles, N. Maigrot, L. Richard, P. LeFloch, *Chem. Commun.* **2004**, 1274.

5
Diversity-oriented Synthesis of Chromophores by Combinatorial Strategies and Multi-component Reactions*

Thomas J. J. Müller

*A List of Abbreviations can be found at the end of this chapter.

5.1
Introduction

The ideal synthesis [1] of functional molecules with all its boundary parameters is almost like squaring the circle. Consequently, synthetic chemists have sought and devised fruitful strategies that inevitably address the very fundamental principles of efficiency and efficacy. These encompass, besides the criteria of selectivity, i.e. chemo-, regio- and stereoselectivity, also, with increasing importance, economic and ecological aspects. Additionally, the intellectual challenge to create concise, elegant and conceptually novel syntheses has become a steadily accelerating driving force in both academia and industry. Therefore, in the past decade the productive concepts of multi-component processes, domino reactions and sequential transformations have considerably stimulated the synthetic scientific community [2, 3] In particular, these diversity-oriented syntheses [4] are challenges for synthetic efficiency and reaction design. Mastering unusual combinations of elementary organic reactions under similar conditions is the major conceptual challenge in engineering novel types of sequences. From a practical point of view, combinatorial chemistry [5] also offers manifold opportunities for directing diversity-oriented syntheses. Thus, the prospect of extending one-pot reactions into combinatorial and solid-phase syntheses [3d, 6] promises manifold opportunities for developing novel lead structures of pharmaceuticals, catalysts and even novel molecule-based materials.

By definition, multi-component reactions (MCRs) belong to the class of domino reactions [7]. Generally, domino reactions are regarded as sequences of uni- or bimolecular elementary reactions that proceed without intermediate isolation or workup as a consequence of the reactive functionality that has been formed in the previous step. In addition to uni- and bimolecular domino reactions that are generally referred to as "domino reactions", a third class is called multimolecular domino reactions or MCRs. Whereas uni- and bimolecular domino reac-

Functional Organic Materials. Syntheses, Strategies, and Applications.
Edited by Thomas J.J. Müller and Uwe H.F. Bunz
Copyright © 2007 WILEY-VCH Verlag GmbH & Co. KGaA, Weinheim
ISBN: 978-3-527-31302-0

tions inevitably cause a significant increase in the degree of molecular complexity, MCRs inherently lead to an increase in molecular diversity. Therefore, MCRs bear some significant advantages over uni- and bimolecular domino reactions. Besides the facile accessibility and high diversity of starting materials, multi-component syntheses promise high convergence and enormous exploratory potential. Besides a purist standpoint where all ingredients of MCRs have to be present from the very beginning of the process (MCR in a domino fashion), nowadays sequential (subsequent addition of reagents in a well-defined order without changing the conditions) and consecutive (subsequent addition of reagents with changing the conditions) one-pot reactions are as well counted to the class of MCRs [2, 3b,g]. Historically, there have been representatives known as name reactions for all subclasses of MCRs, but it was not until Ugi's ground-breaking work on the four-component coupling of an isonitrile, an aldehyde, an amine and an acid to give an α-acyloxycarboxamide, i.e. the Ugi–Passerini reaction or Ugi-4CR [2, 3b], that an eager search was initiated for novel MCRs.

Owing to the peptide-like structures obtained by Ugi- and Ugi-type reactions, isonitrile-based MCRs have a considerable impact on the development of biologically interesting target structures and, in particular, on the syntheses of heterocycles [8]. Although MCR technology has considerably fertilized pharmaceutically relevant lead finding, surprisingly, it is still in its infancy with respect to the application to functional π-electron systems, such as chromophores, fluorophores and redox–active molecules. On the other hand, the adaptation of combinatorial strategies has clearly experienced a significant increase within the last decade [9]. Therefore, this chapter reports on strategies, developments and perspectives of diversity-oriented syntheses of chromophores and functional π-electron systems.

5.2
Combinatorial Syntheses of Chromophores

Combinatorial chemistry plays the game of large numbers and has opened structural space of enormous extent [5]. This concept, applying high-speed synthesis in combination with high-throughput screening, has successfully been implemented in medicinal chemistry. Here, solid-phase and parallel syntheses and screening of novel biologically active compounds have set the stage for solving multi-parameter problems (reaction parameters, structural and functional variables) of molecule-based materials in a combinatorial fashion [10]. Although most materials properties are determined to a major extent by solid-state phenomena which are not readily predictable by theory, combinatorial chemistry offers rapid access to a greater number of structures, thus now allowing more detailed quantitative interpretation of structure–property relationships on a molecular basis. In particular, this holds true for chromophores and electrophores.

5.2.1
Combinatorial Azo Coupling

Azo dyes represent the largest class of colorants with widespread technical applications and they are easily accessible by azo coupling of diazonium salts with electron-rich (hetero)arenes. In a systematic study on the preparation and properties of resin-supported diazonium salts, Bradley and coworkers recently reported a convenient solution-phase combinatorial synthesis of azo dyes [11]. As an alternative to polymer-supported solid-phase synthesis, ion-exchange resins are well suited to allow selective removal of either excess reagents or the desired product or elimination of byproducts in a parallelization of classical ion-exchange purification methodology. Amberlyst A-26, which is functionalized with tetraalkylammonium groups, and Amberlyst A-15, a sulfonate acid-based resin, were sequentially applied to diazotization of the anilines 1 to generate the free diazonium salt 2 and the resin-supported diazonium salt 3 (Scheme 5.1). The process was completed by mixing an excess of the Amberlyst A-15-supported diazonium species 3 with phenols 4 at pH 9 to give azo dyes 5 of good purity. The validity of this approach was additionally demonstrated by the synthesis of a 6 × 6 library of azo dyes: six ion-exchange resin-immobilized diazonium salts were prepared from six anilines 1 and were treated with six phenols 4 in a parallel synthesis format (Fig. 5.1).

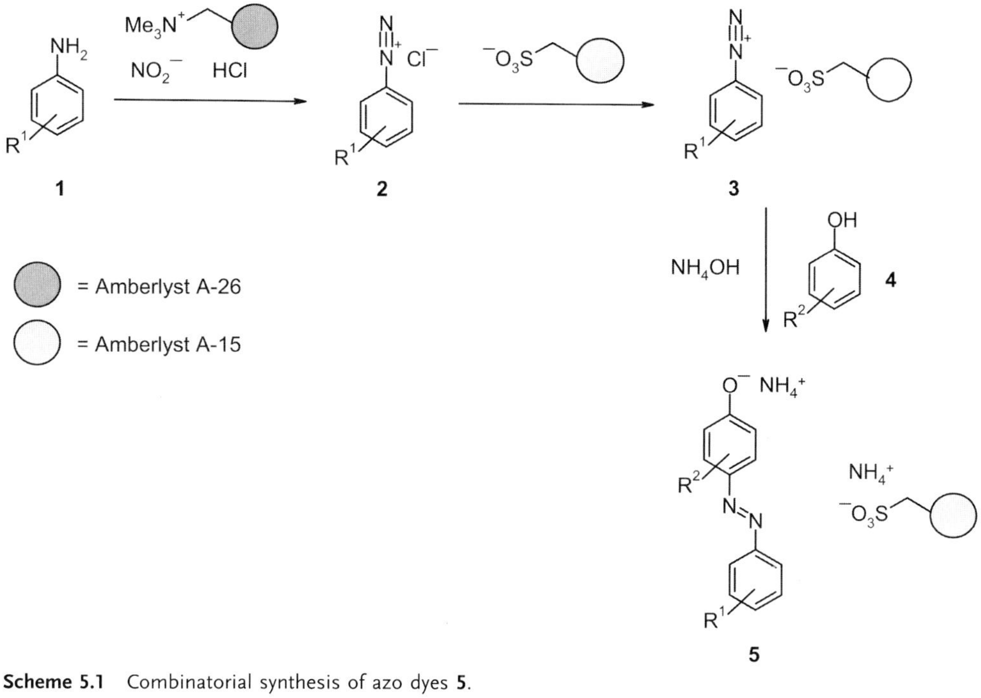

Scheme 5.1 Combinatorial synthesis of azo dyes 5.

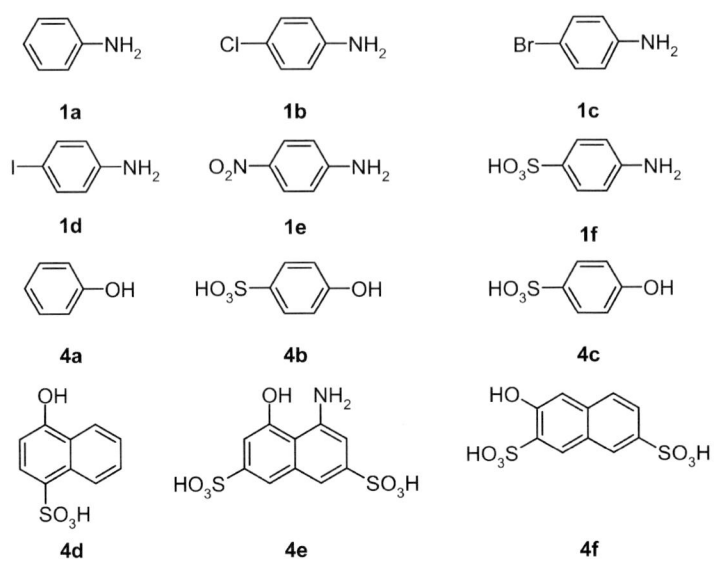

Fig. 5.1 Anilines **1** and phenols **4** for the combinatorial synthesis of the azo dyes **5**.

5.2.2
Combinatorial Condensation Reactions

Cyanines have been known for some time and have found use in a variety of applications [12], such as photosensitizers for color photography [13], as markers for flow cytometry [14] in studies and detection of nucleic acids [15] and as phototherapeutic agents [16]. Asymmetric cyanine dyes (Fig. 5.2) consist of two different heteroaromatic fragments conjugated by a mono- or polymethine chain. By varying the length of this chain, the photophysical properties of these dyes can be altered. Synthetically, cyanine dyes are easily accessible by condensation of methylene active α- and γ-methylpyridinium salts and amidinium or vinylogous amidinium salts as electrophiles.

An efficient combinatorial solid-phase synthesis of asymmetric cyanine dyes was developed by Isacsson and Westman [17] using a Rink amide polystyrene resin. The picolinium and lepidinium salts **6** were linked to the solid-phase resin by amide coupling, then the benzothiazole derivatives **7** were subsequently condensed with the coupled picoline and lepidine moieties to give the yellow to blue ($\lambda_{max, abs.}$ = 420–590 nm) asymmetric, fluorescent ($\lambda_{max, em.}$ = 480–650 nm) cyanine dyes **8** (Scheme 5.2, Fig. 5.3). As a consequence of restricted rotation upon intercalation, the fluorescence quantum yields increase significantly when these dyes are bound to DNA.

Closely related to cyanine dyes are stilbazolium salts that are based on the styryl scaffold. Stilbazolium dyes are a class of fluorescent, lipophilic cations that have been used as mitochondrial labeling agents and membrane voltage-sensitive

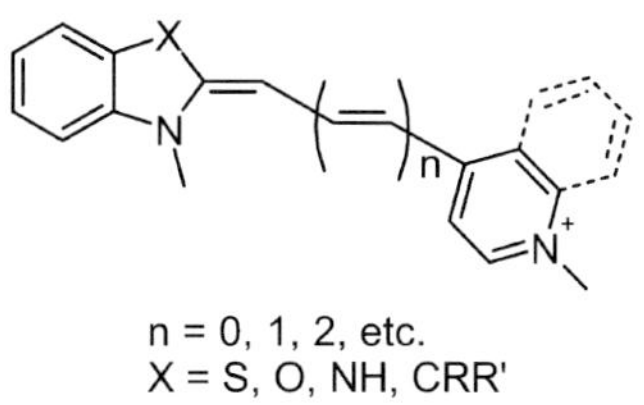
X
N
n
N+
n = 0, 1, 2, etc.
X = S, O, NH, CRR'

Fig. 5.2 General structure of asymmetric cyanine dyes.

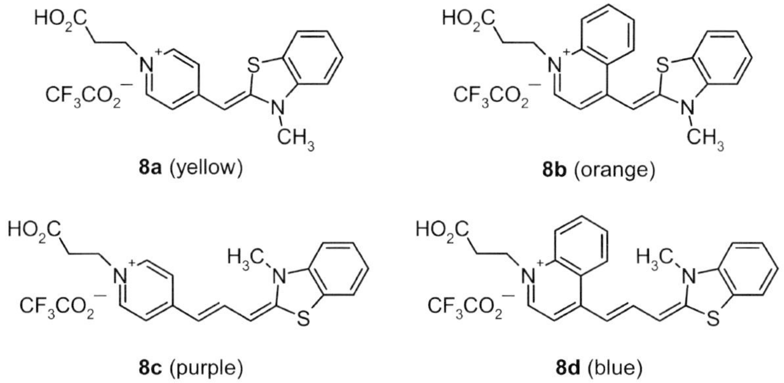
HO2C
CF3CO2−
N+
S
N
CH3
H3C
8a (yellow)
8b (orange)
8c (purple)
8d (blue)

Fig. 5.3 Asymmetric cyanine dyes **8** synthesized by solid-phase synthesis (colors in parentheses).

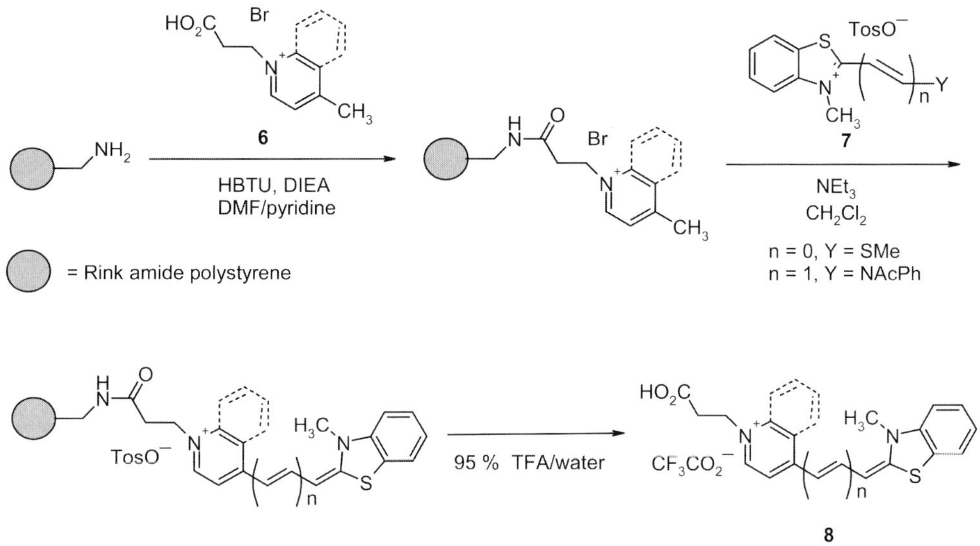
NH2
HO2C
Br
N+
CH3
6
HBTU, DIEA
DMF/pyridine
= Rink amide polystyrene
H
N
O
TosO−
S
N
CH3
n
Y
7
NEt3
CH2Cl2
n = 0, Y = SMe
n = 1, Y = NAcPh
H3C
95 % TFA/water
CF3CO2−
8

Scheme 5.2 General structure of asymmetric cyanine dyes.

Scheme 5.3 Diversity-oriented synthesis of stilbazolium dyes **11**.

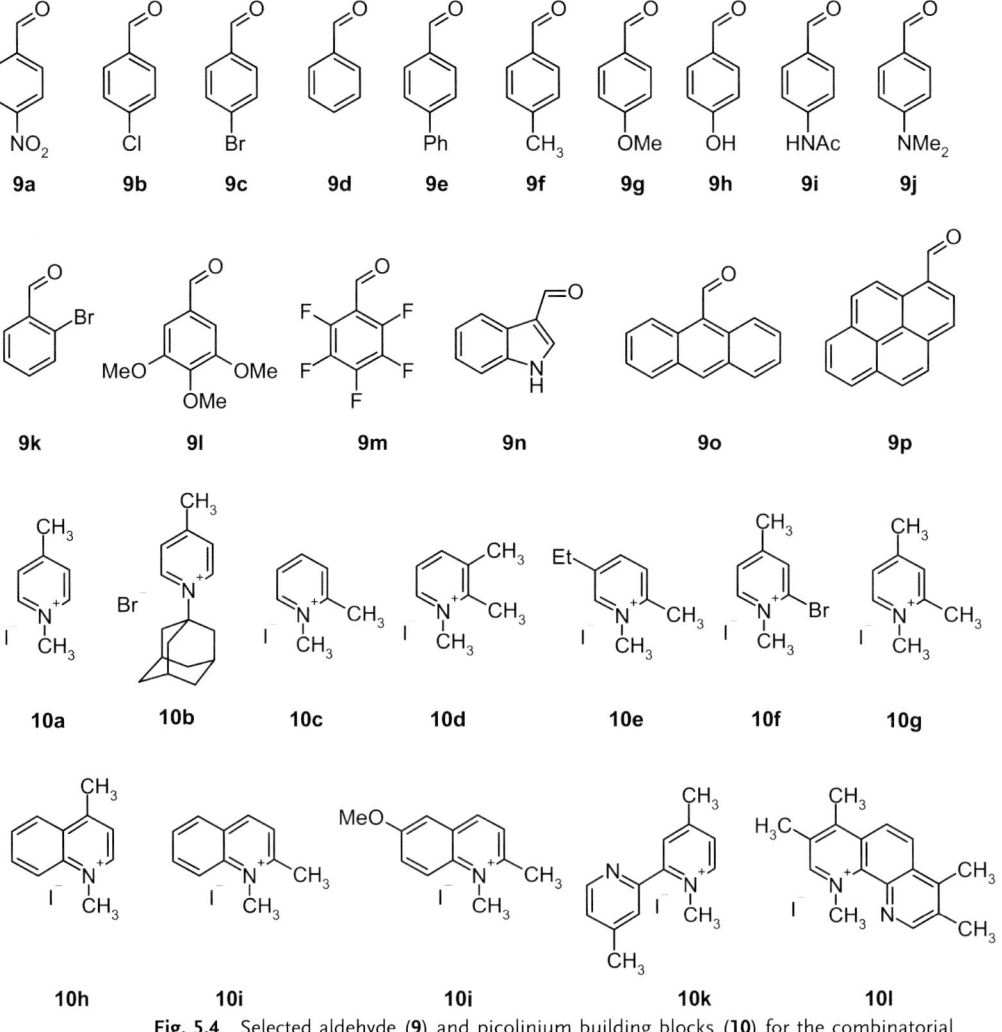

Fig. 5.4 Selected aldehyde (**9**) and picolinium building blocks (**10**) for the combinatorial synthesis of stilbazolium salts.

probes of cellular structure and function [18]. Condensation of (hetero)aromatic aldehydes **9** with CH-acidic α- or γ-picolinium salts **10** furnishes diversely substituted stilbazolium salts **11** (Scheme 5.3) [19].

Rosania et al. [19] have shown this straightforward preparation to be advantageous for the transposition to a diversity-oriented, combinatorial approach to an organelle-targeted fluorescent library. Therefore, the condensation of **9** and **10** (Fig. 5.4) with pyrrolidine as a catalyst was performed in 96-well plates and the dehydration reaction was accelerated by microwave irradiation for 5 min to give 10–90% conversion. The resulting library compounds were analyzed using an LC–MS system with diode-array and fluorescence detectors and a fluorescence plate-reader to determine the absorption and emission maxima and the emission colors.

Other members of the cyanine dye family that are readily accessible by condensation strategies are rhodacyanines **12**. In general, the rhodacyanine dyes **12** possess two different conjugated systems, a neutral merocyanine and a cationic cyanine moiety, consisting of three heterocyclic components, in which two end heteroaromatic rings, α and γ, flank a rhodanine moiety β (Fig. 5.5).

Since it is already known that the dyes **12** can be used as anticancer agents [20, 21], fluorescent dyes [22], and photosensitizers [23] and, in particular, as antimalarial agents [24], a combinatorial approach to a library of rhodacyanine compounds was suggested and developed by Ihara and coworkers [25]. Based on reported strategies [24, 26], the rhodacyanine **12** can be obtained with ease using a three-step synthetic sequence: (a) merocyanine formation by condensation of methylthioiminium salt **13** with rhodanine **14** to afford **15**, (b) activation by S-alkylation of **15** to give the corresponding thioimidate cation **16** and (c) cyanine formation by condensation of **16** with methyliminium salt **17** (Scheme 5.4, Fig. 5.6).

Each step was optimized first before transposing the sequence into a one-pot procedure where excess reagents were removed by filtration in a syringe-type filter before adding the next reactant. Starting from three different methylthioiminium salts **13**, rhodanine derivatives **14** and methyliminium salts **17** (Fig. 5.6) a 3 × 3 × 3 parallel synthesis was conducted to furnish 27 different kinds of rhodacyanines **12** (Table 5.1).

It is interesting to note that the rhodacyanines **12** display a range of colors both in the solid state and in the solution phase, depending strongly on their partial substructure.

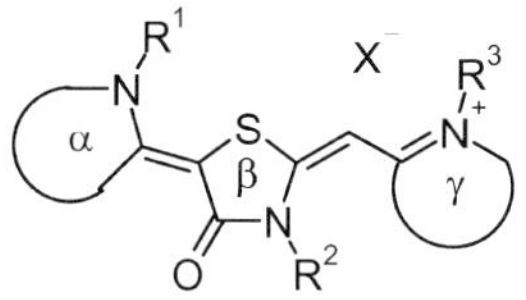

12

Fig. 5.5 General structure of the rhodacyanine dyes **12**.

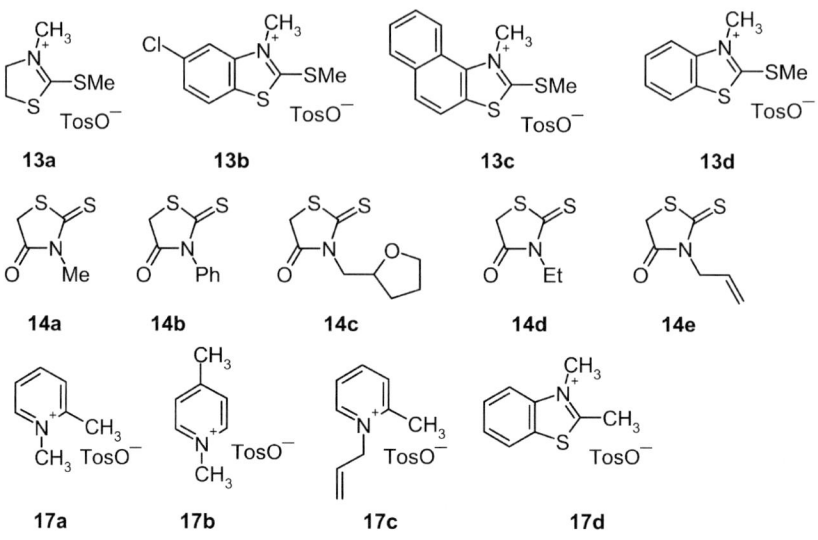

5 Diversity-oriented Synthesis of Chromophores

Fig. 5.6 Methylthioiminium salts **13**, rhodanine derivatives **14** and methyliminium salts **17** for the combinatorial synthesis of rhodacyanines **12**.

Scheme 5.4 General synthetic sequence for the rhodacyanine dyes **12**.

Table 5.1 One-pot, three-component combinatorial synthesis of rhodacyanines **12**: chemical yields and purities (parentheses)[a].

	14c	14d	14e	17
13b	12bca 22% (>95%)	12bda 28% (>95%)	12bea 27% (>95%)	17a
	12bcc 36% (>95%)	12bdc 18% (>95%)	12bec 26% (>95%)	17c
	12bcd 47% (>95%)	12bdd 44% (>95%)	12bed 49% (85%)	17d
13c	12cca 33% (>95%)	12cda 54% (80%)	12cea 33% (>95%)	17a
	12ccc 26% (>95%)	12cdc 53% (80%)	12cec 25% (90%)	17c
	12ccd 56% (>95%)	12cdd 64% (>95%)	12ced 53% (>95%)	17d
13d	12dca 3% (90%)	12dda 34% (>95%)	12dea 26% (>95%)	17a
	12dcc 3% (80%)	12ddc 30% (>95%)	12dec 26% (>95%)	17c
	12dcd 63% (>95%)	12ddd 58% (>95%)	12ded 60% (90%)	17d

[a] Purities were estimated by ^{1}H NMR spectroscopy.

5.2.3
Combinatorial Cross-coupling Reactions

The advent of metal-catalyzed cross-coupling reactions [27] has considerably revolutionized synthetic concepts for carbon framework constructions and has also had a significant impact on the synthesis of π-electron systems with extended conjugation [28]. Not only functional π-electron systems such as NLO materials [29] based on various scaffolds and conjugation motifs were made accessible in a broad range, but also conjugated oligomers and polymers [30] for detailed studies of charge transport, structure–property relationships and applications in photonic materials have become readily available with fine-tunable electronic properties. As a consequence of the mild reaction conditions, a major characteristic of Pd- and Ni-catalyzed cross-coupling reactions is their excellent compatibility with numerous polar functional groups. This aspect not only is important for the synthesis of chromophores by sp^2–sp^2 or sp–sp^2 bond-forming processes but also implies a vast potential for combinatorial and multi-component approaches (see Section 5.3.2). Furthermore, in materials sciences, where solid-state phenomena actually dominate the properties of molecule-based organic materials, struc-

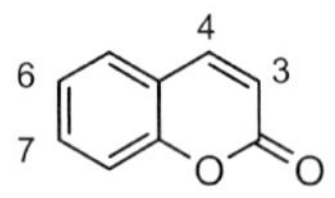

Fig. 5.7 Positions on the coumarin scaffold with major effects on the electronic properties.

ture–property relationships are often unpredictable and empirical approaches are better suited for the finding and development of novel lead structures. In 2001 Bäuerle's group implemented Pd-catalyzed coupling reactions into a combinatorial approach to novel organic materials, in particular coumarin dyes and oligothiophenes [31].

Coumarin dyes not only embody a pharmacologically intriguing class of substances [32] but also, owing to their intense fluorescence, they have attracted considerable interest as laser dyes [33], as fluorescent tags in biolabeling studies [34], as emitter layers in organic light-emitting diodes (OLEDs) [35], and as optical brighteners [36]. For establishing empirical structure–fluorescence efficiency correlations, it is reasonable to exploit the strong electronic influence of substituent variations at positions 3, 4, 6 and 7 on the coumarin framework (Fig. 5.7) [37].

Bäuerle and coworkers [38] suggested and applied Pd-catalyzed cross-coupling reactions of 3-bromocoumarin derivatives for the generation of coumarin libraries with a highly diverse substitution pattern. In the case of 3-bromocoumarin (**18**), the reaction conditions for combinatorial Heck vinylations with alkenes **19**, Suzu-

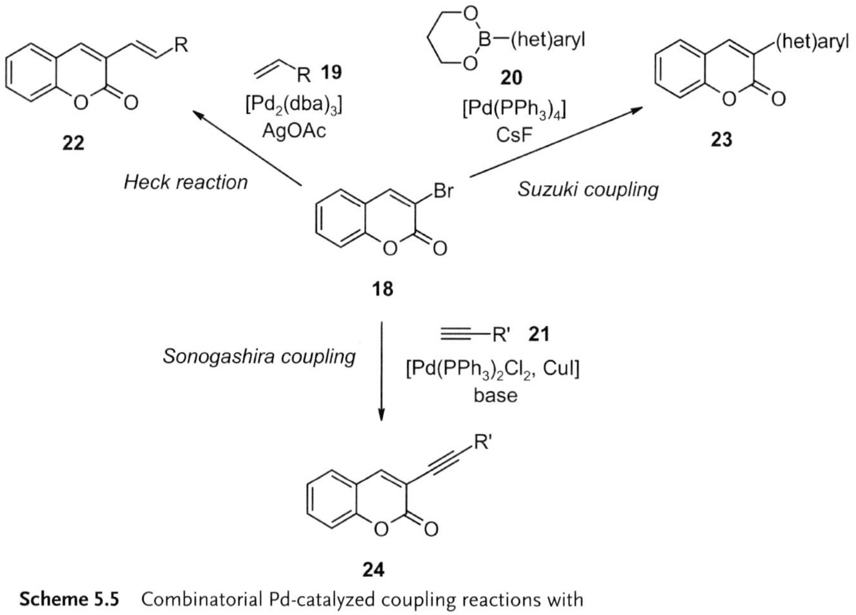

Scheme 5.5 Combinatorial Pd-catalyzed coupling reactions with 3-bromocoumarin (**18**) (for the substituents, see Fig. 5.8).

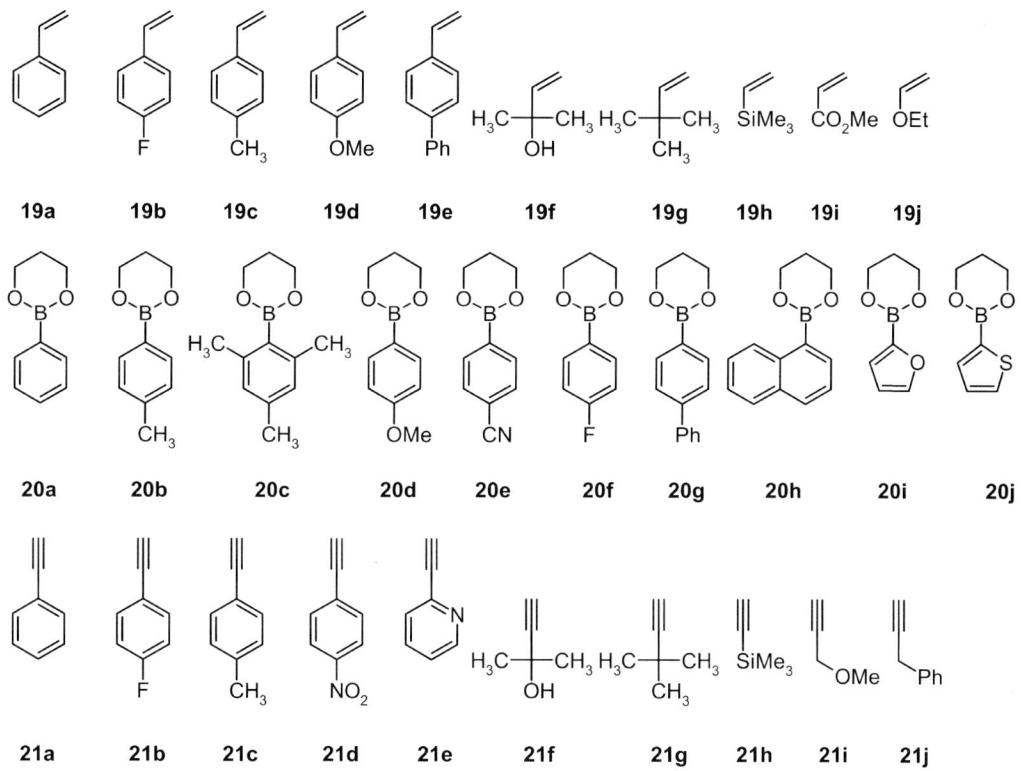

Fig. 5.8 Alkenes **19**, (hetero)aryl boronates **20** and terminal alkynes **21** applied in the combinatorial synthesis of 3-substituted coumarins.

ki arylations with (hetero)aryl boronates **20**, and Sonogashira alkynylations with terminal alkynes **21** were optimized, giving rise to 3-ethenyl- (**22**), 3-(hetero)aryl- (**23**) and 3-alkynyl-substituted coumarins **24** (Scheme 5.5, Fig. 5.8).

With optimized synthetic protocols, this combinatorial strategy was then transposed to 3-bromocoumarin derivatives **25** (Fig. 5.9) as coupling partners.

In a parallel synthesizer, electronically diverse libraries of more than 150 substituted coumarins **26** could be readily constructed in solution. Screening of the optical properties of the fluorophore ensemble gave rise to the identification of several library members with high fluorescence quantum yields (Fig. 5.10).

Likewise, combinatorial strategies can also be applied for the parallel synthesis and screening of oligothiophenes, one of the most investigated class of π-conjugated oligomers. The main efforts of Bäuerle and coworkers [39] were focused on regioregular head-tail-connected quater(3-arylthiophenes) **27** as targets (Fig. 5.11). As a consequence of their defined structure, these aryl-substituted oligomers are excellent model compounds for polydisperse poly(3-arylthiophenes), like the already intensively studied oligo(3-alkylthiophenes) [40]. In particular, the latter

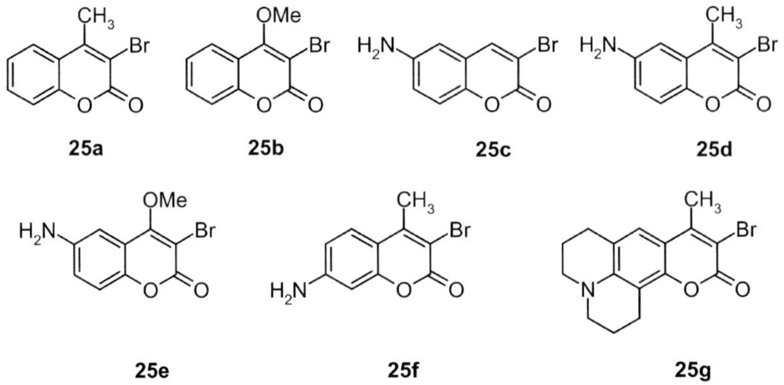

Fig. 5.9 3-Bromocoumarin derivatives **25** applied in the combinatorial synthesis of coumarins.

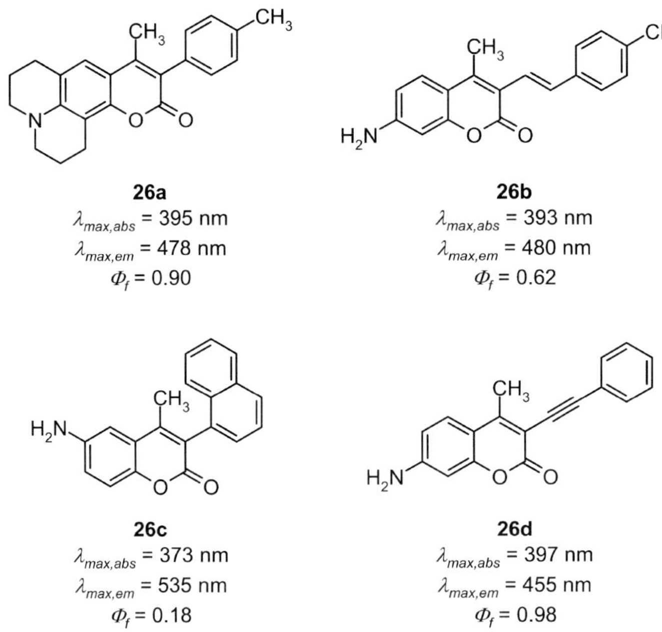

26a
$\lambda_{max,abs}$ = 395 nm
$\lambda_{max,em}$ = 478 nm
Φ_f = 0.90

26b
$\lambda_{max,abs}$ = 393 nm
$\lambda_{max,em}$ = 480 nm
Φ_f = 0.62

26c
$\lambda_{max,abs}$ = 373 nm
$\lambda_{max,em}$ = 535 nm
Φ_f = 0.18

26d
$\lambda_{max,abs}$ = 397 nm
$\lambda_{max,em}$ = 455 nm
Φ_f = 0.98

Fig. 5.10 3-Substituted coumarin derivatives **26** with high fluorescence quantum yields identified by screening of substance libraries.

have already been used successfully as active components in electronic devices such as organic field effect transistors [41].

The oligothiophenes **27** bear phenyl groups in the 3-position of the thiophene cores that are *para*-substituted with electronically diverse functional groups (R = CF$_3$, H, CH$_3$, OCH$_3$). Here, the phenyl spacers enhance the solubility of the oligomers and warrant the electronic communication between the elements of diver-

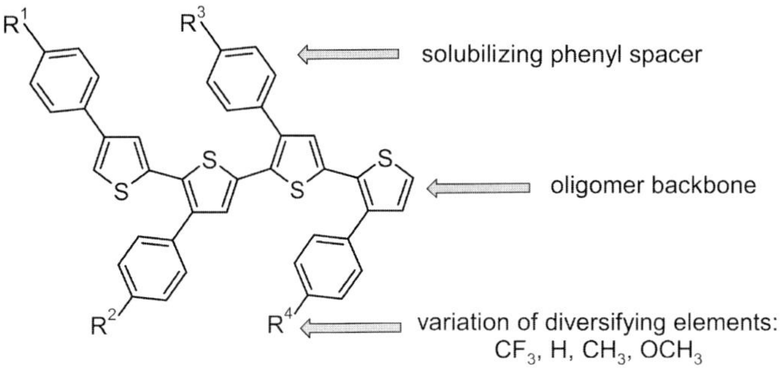

Fig. 5.11 Target structure of quater(3-arylthiophenes) **27**.

sity and the oligomer backbone. Furthermore, the electronic structure of the quaterthiophene should be susceptible by the diversifying substituents without altering the overall geometry of the *p*-aryl-substituted molecule. Therefore, a systematic investigation of substituent effects on the energy levels of the molecular orbitals and the establishment of structure–property relationships could be easily carried out with the oligomers **27**.

The solid-phase synthesis of the oligomers **27** is based on a repetitive Suzuki coupling–iodination strategy using both the parallel and the 'mix-and-split' techniques (Scheme 5.6). First, the chlorosilylthiophene **28** is reacted with hydroxymethyl-substituted polystyrene, binding the first of four arylthiophene fragments via a traceless linker to the polymer matrix [42]. Then, the coupled thiophene **29** is subjected to an iterative sequence of iodination (to give **30**) and Suzuki coupling with a thienylboronic ester **31** to furnish the dimer **32**. By mercuration-mediated iodination, the dimer **32** is transformed into the iodo derivative **33**, which is now extended to the trimer **34** by Suzuki coupling with the boronate **31**. Likewise, after iodination of **34**, the iodo trimer **35** is coupled with boronate **31** to give the resin-bound quater(3-arylthiophene) **36** that is liberated from the solid phase by reaction with trifluoroacetate in dichloromethane to furnish the target quater(3-arylthiophene) **27**.

After optimization of the synthetic protocol, a 256-membered library of quater(3-arylthiophene)s containing all permutations of the four elements of diversity was generated in microreactors where the beads containing the compounds were labeled with a radiofrequency code. Automated preparative HPLC finally furnished the pure quater(3-arylthiophenes) **27**, which were subjected to screening by absorption and emission spectroscopy and cyclic voltammetry. The latter was considerably facilitated by using an automated screening device. Whereas the optical properties (absorption and fluorescence) are influenced by the substitution pattern to only a minor extent, the first oxidation potentials stretch over a relatively large scope ranging from E_1^0 = 0.42–0.68 V and correlate reasonably well with the substituent descriptor $\Sigma\sigma_p^+$, defined as the sum of the Hammett substituent constants σ_p^+ of the corresponding substituents.

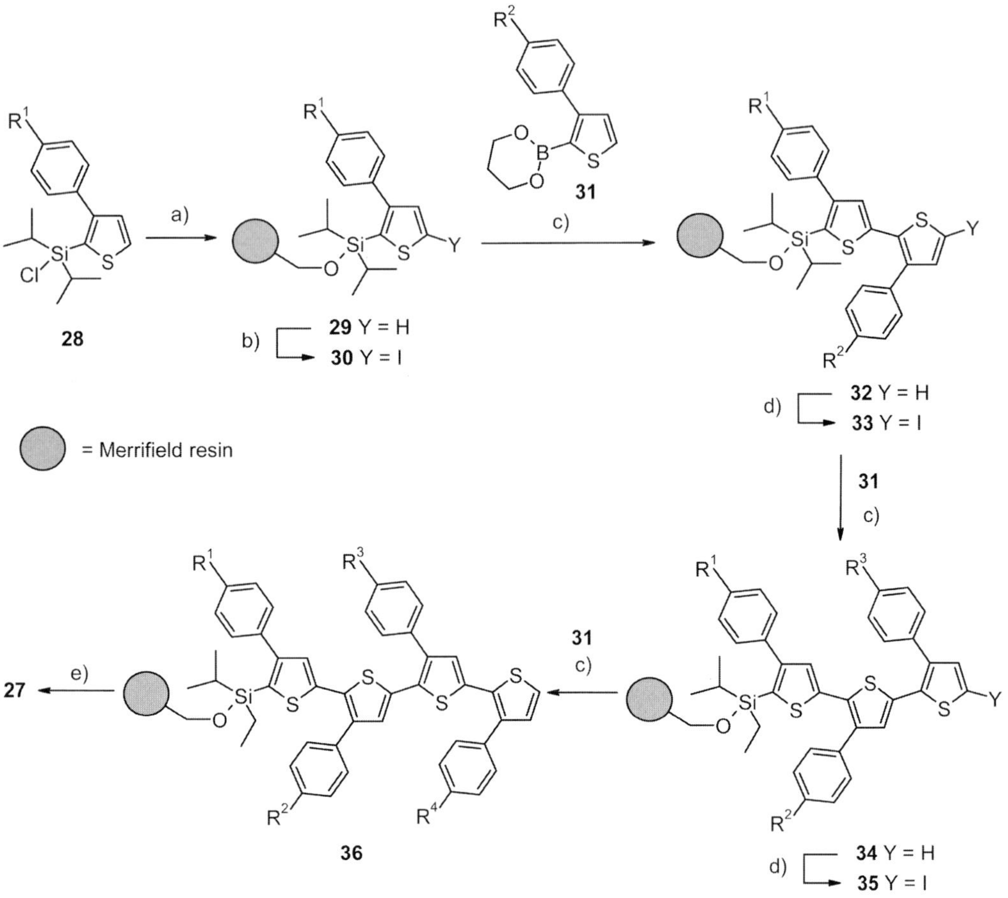

Scheme 5.6 Solid-phase synthesis of quater(3-arylthiophenes) **27**: (a) hydroxymethyl–substituted polystyrene, imidazole, dimethylformamide (DMF), 20 °C; (b) 1, LDA, THF, −60 °C; 2, I$_2$; (c) [Pd(PPh$_3$)$_4$], THF–H$_2$O, NaHCO$_3$ or Na$_2$HPO$_4$–NaH$_2$PO$_4$; (d) 1, Hg(OCOC$_5$H$_{11}$)$_2$, CH$_2$Cl$_2$, 20 °C; 2, I$_2$, CH$_2$Cl$_2$; (e) 10 % TFA, CH$_2$Cl$_2$, 20 °C.

In addition to well-defined oligomers, polymeric OLEDs have great potential for future display devices [43]. Polymeric OLEDs ideally combine color tuning through the polymer inherent change of molecular structure via variable length, substitution or degree of conjugation and outstanding mechanical and processability properties of polymers. Flexible thin films can be easily prepared from casting. For a full color display, red, green and blue colors are required and, therefore, fluorescent polymers with corresponding narrow emission wavelengths must be prepared. Several fluorescent conjugated polymers were investigated as suitable OLED precursors [44]. Depending on the nature of substituents of conjugated polymers, both color and light emission efficiency can be dramatically influenced and controlled [44a, 45]. Consequently, in the field of optoelectronics, innovative

5.2 Combinatorial Syntheses of Chromophores

Scheme 5.7 Synthesis of poly(arylene–ethynylenes) **39** by Sonogashira polymerization (for substituents, see Fig. 5.12).

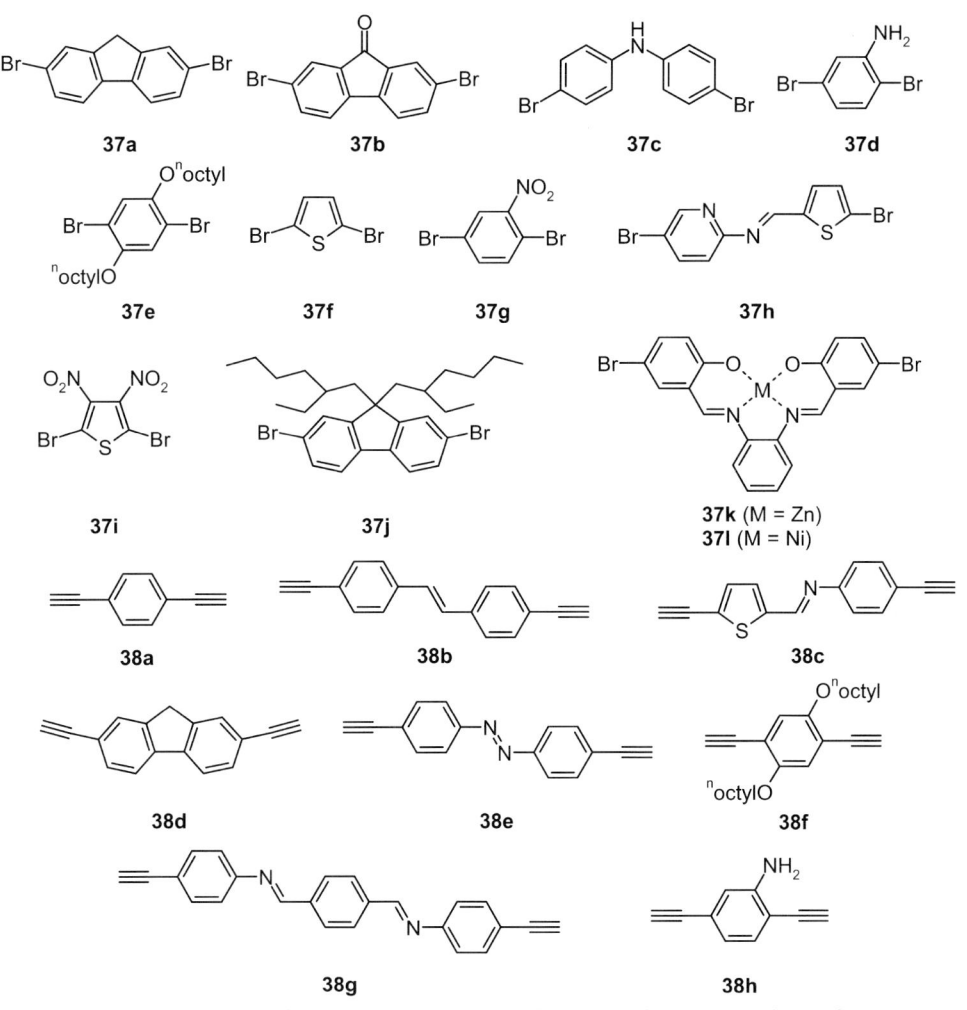

Fig. 5.12 Dihalides **37** and diethynyl compounds **38** applied in the combinatorial synthesis of poly(arylene–ethynylenes) **39**.

synthetic concepts and fast screening of a large number of polymer library compositions for the development of new generations of materials has become a crucial challenge.

Lavastre et al. [46] were the first to apply the concept of high-throughput screening (HTS) to the Sonogashira coupling reaction for (a) the rapid generation of sets of conjugated polymers and (b) the fast qualitative detection of new fluorescent polymers (Scheme 5.7). Owing to their interesting photoluminescence and high quantum yield of fluorescence [47], polymers derived from the poly(arylene–ethynylene) [45] family were prepared in a parallel manner by coupling a diversity of dihalogenated compounds **37** and diethynyl monomers **38** to furnish the poly(arylene–ethynylenes) **39** (Fig. 5.12). The average molecular weights of **39** lie in a range between 3000 and 6000 g mol^{-1}, which are in accordance with values reported in the literature.

The 8 × 12 library was simply irradiated with a hand-held UV lamp (365 nm) to discriminate easily fluorescent and nonfluorescent polymers and to visualize the corresponding emission color in solution. Then, with a spectrofluorimeter able to read 96-well plates for several excitation and emission wavelength combinations, the different excitation wavelengths were evaluated. By this procedure, new polymers showing green (**39a,b**, excitation at 460 nm, emission detection at 530 nm) or blue (**39c–e**, excitation at 360 nm, emission detection at 460 nm) emitting fluorescence were rapidly discovered (Fig. 5.13).

Combinatorial polymer synthesis in combination with HTS is suitable for the fast qualitative detection of new compounds showing specifications of light-emitting diodes as the fluorescence brightness and film-forming properties can be checked simultaneously.

Extended π-conjugated systems containing the 1,2-arylethene or stilbene scaffold and also the structurally related poly(arylene–vinylenes) have already established their utility as functional materials [44a, 48]. As a consequence, higher substituted derivatives such as 1,1,2-triarylethenes could be interesting targets with unexpected new properties.

Therefore, on the basis of efficient Pd-catalyzed triarylations to a vinylsilane platform, Yoshida and coworkers [49] rapidly prepared four types of structurally well-defined triarylethene-based extended π-systems (Fig. 5.14).

Appending a catalyst-directing 2-pyridyl group on silicon made it possible to overcome the otherwise difficult Heck reaction [50] of vinylsilanes [51]. Hence, using vinyl(2-pyridyl)silane (**40**) as a platform, aryl iodides can be regioselectively coupled to give the 2-arylethenyl(2-pyridyl)silane **41** that can be subsequently transformed by another Heck reaction into the 2,2-diarylethenyl(2-pyridyl)silane **42** (Scheme 5.8). Furthermore, both ethenylsilanes **41** and **42** can be reacted with aryl iodides under the conditions of a Hiyama coupling [52] to give diarylethenes **43** and triarylethenes **44** in a regio- and stereoselective manner.

This methodological approach to di- and triaryl-substituted ethenylene motifs is interesting for three reasons. First, all aryl groups assembled stem from readily available aryl iodides and diiodides (Fig. 5.15). Second, by applying the Heck–Hiyama sequel they can be installed at the desired position by the addition of

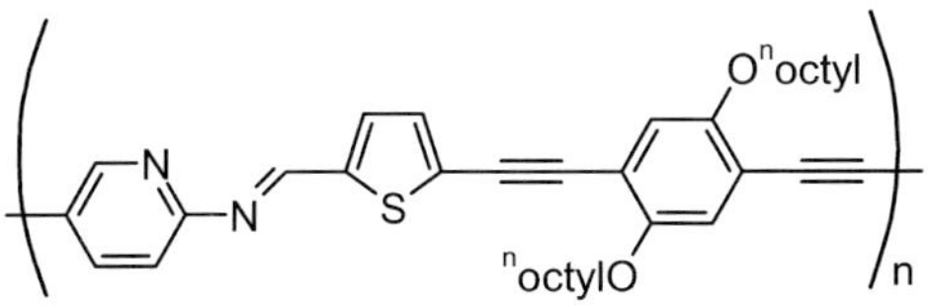

39a

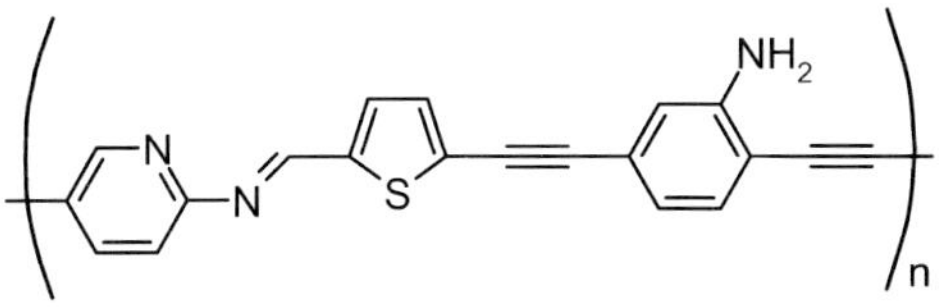

39b

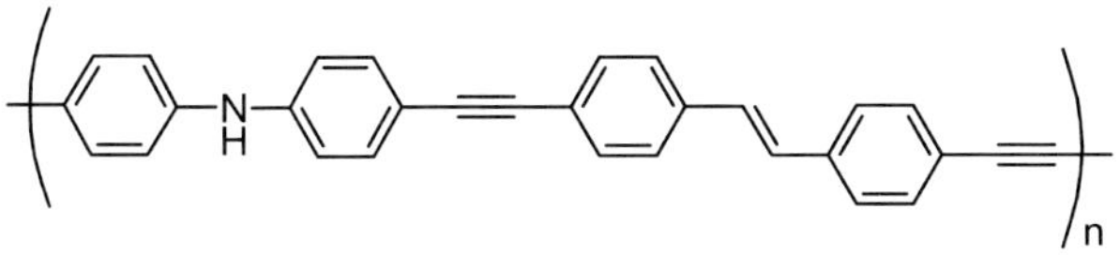

39c

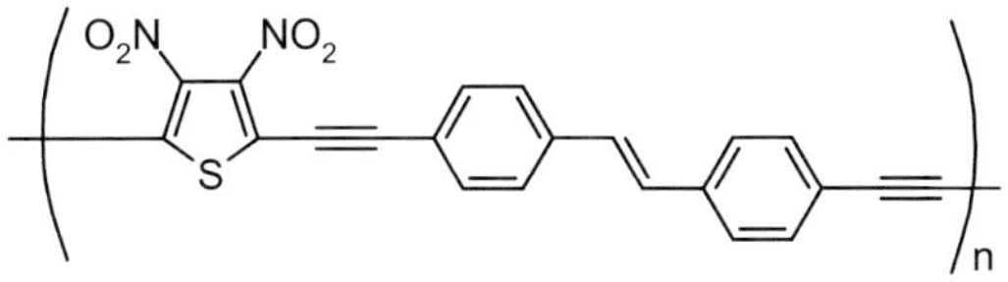

39d

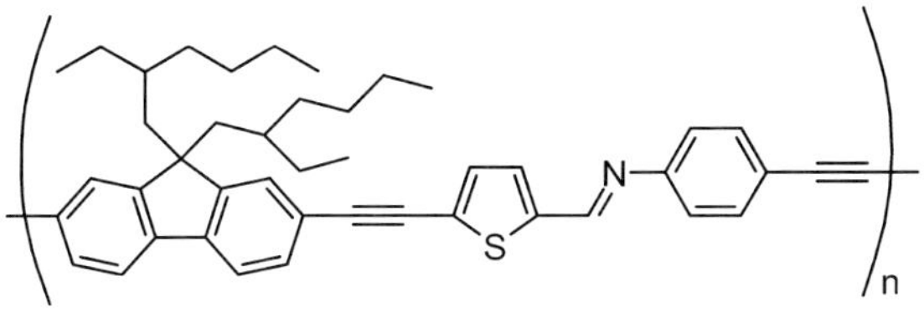

39e

Fig. 5.13 Green (**39a**,**b**) and blue (**39c–e**) fluorescent polymers by combinatorial synthesis and HTS.

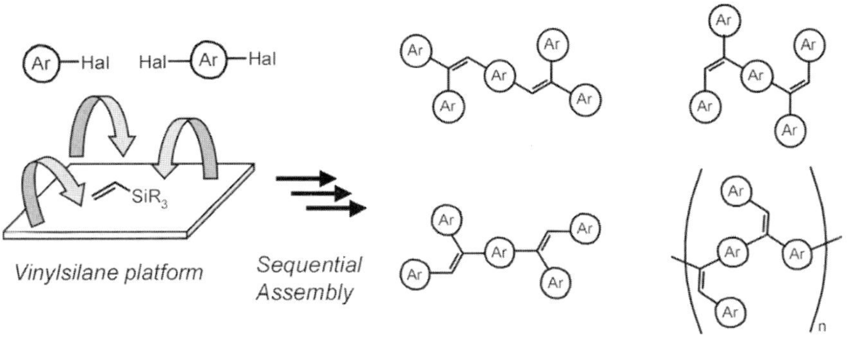

Fig. 5.14 Vinylsilane as a platform for Pd-catalyzed arylation to 1,1,2-triarylethene derivatives.

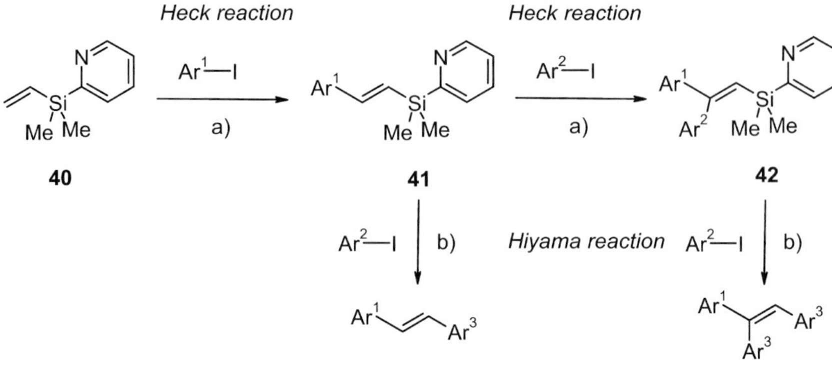

5 Diversity-oriented Synthesis of Chromophores

Scheme 5.8 Programmable and diversity oriented synthesis of diarylethenes **43** and triarylethenes **44**: (a) [Pd$_2$(dba$_3$)], P(2-furyl)$_3$, NEt$_3$, THF, 60 °C; (b) PdCl$_2$(PhCN)$_2$, Bu$_4$NF, THF, 60 °C.

aryl iodides in the appropriate order. Finally, a simple alteration of the order of addition in the sequence results in the production of all possible regio- and stereo-isomers of multisubstituted olefins. Therefore, this new synthetic strategy permits the assembly of π-systems, such as aryl groups, on to a carbon–carbon double bond in a programmable and diversity-oriented format. This approach elegantly overcomes the stereochemical ambiguities of carbon–carbon double bond-forming reactions such as Wittig and related reactions.

From a compound library containing 30 representatives of all four types of 1,1,2-triarylethenes (Fig. 5.14), it was possible to identify a number of interesting

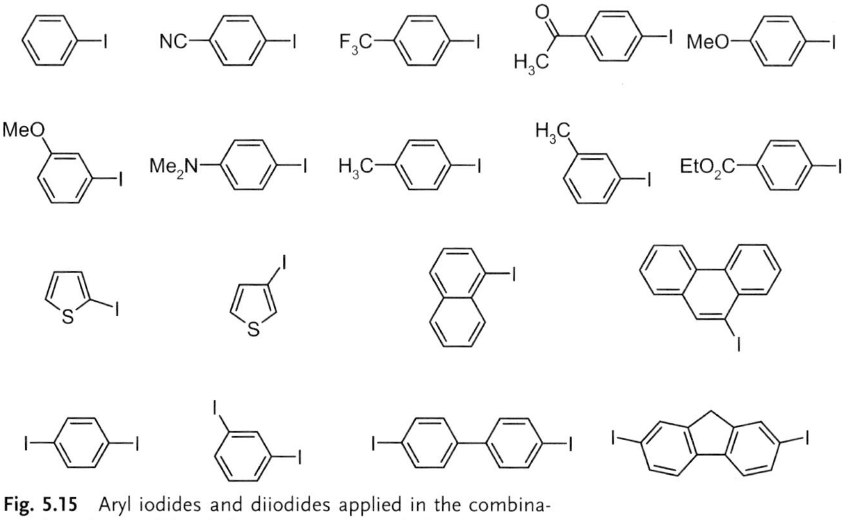

Fig. 5.15 Aryl iodides and diiodides applied in the combinatorial synthesis of diarylethenes **43** and triarylethenes **44**.

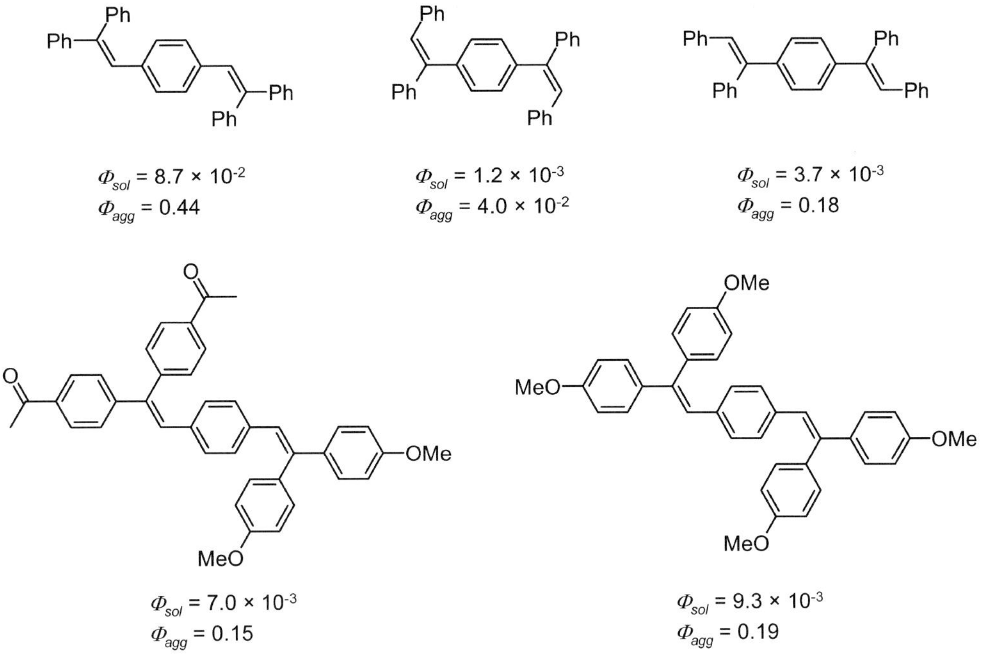

Fig. 5.16 Selected fluorophores **44** with aggregation-induced enhanced emission. Φ_{sol}, fluorescence quantum yield in dioxane; Φ_{agg}, fluorescence quantum yield in dioxane–water (20:80).

fluorescent materials, and also interesting fluorescence properties such as aggregation-induced enhanced emission (Fig. 5.16).

5.2.4
Combinatorial Coordination Chemistry

Transition metal complexes have emerged as promising candidates for applications in solid-state electroluminescent devices. Interestingly, these materials serve as multifunctional chromophores, into which electrons and holes are injected, then migrate and recombine to exhibit light emission [53]. In particular, the properties and net charge of transition metal can be tuned through ligation, thereby enabling facile, synthetic control over device performance as luminescent materials. Therefore, the design of novel complexes for OLED applications has become increasingly important in materials research [54]. Tailor-made transition metal complexes should emit across the visible spectrum and light emission should predominate over nonradiative decay. Chromophores that have been studied for single-layer devices predominantly involve osmium(II) [55], ruthenium(II) [53, 56] and rhenium(I) [57]. These materials emit primarily in the orange–red region of the spectrum (600–650 nm), but can be hardly tuned in their color. How-

198 | *5 Diversity-oriented Synthesis of Chromophores*

Scheme 5.9 Synthetic pathway for the preparation of ionic iridium(III) chromophores. The step in the box is explored through parallel synthesis.

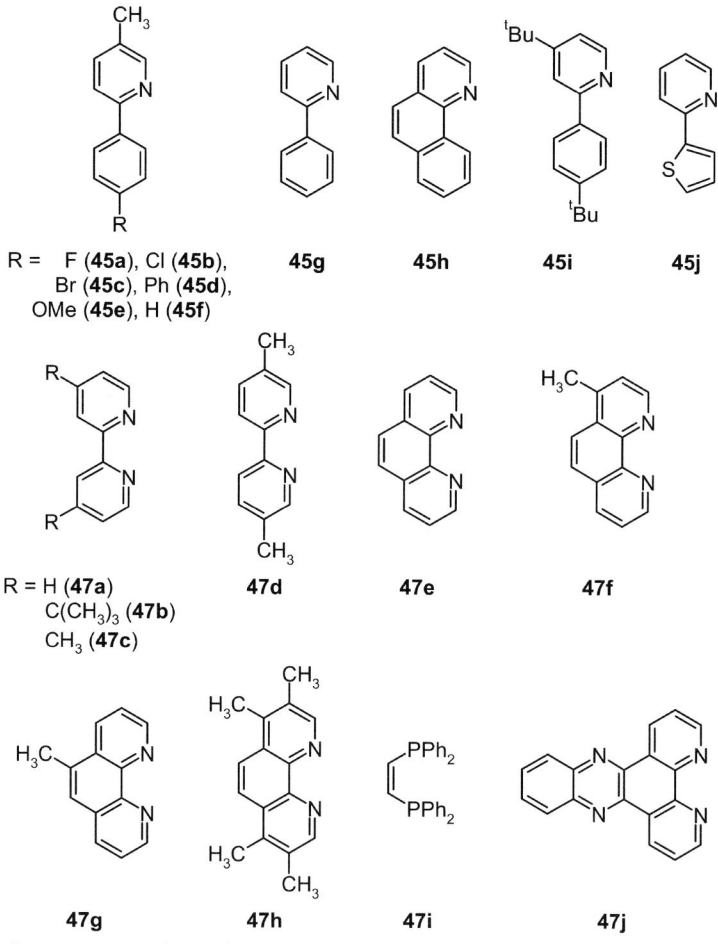

R = F (**45a**), Cl (**45b**), Br (**45c**), Ph (**45d**), OMe (**45e**), H (**45f**)

45g   **45h**   **45i**   **45j**

R = H (**47a**), C(CH$_3$)$_3$ (**47b**), CH$_3$ (**47c**)

47d   **47e**   **47f**

47g   **47h**   **47i**   **47j**

Fig. 5.17 Ten cyclometalating (**45**) and 10 neutral ligands (**47**) for the combinatorial synthesis of cationic iridium(III) (**49**) and ruthenium(II) complexes.

ever, a single layer of a mixed-ligand iridium(III) complex, [Ir(ppy)$_2$(dtbbpy)](PF$_6$) (where ppy = 2-phenylpyridine and dtbbpy = 4,4'-di-*tert*-butyl-2,2'-dipyridyl), displays an efficient yellow electroluminescence marking the highest emission energy from a single-layer device to date [57c].

Bernhard and coworkers [58] have addressed the discovery of ionic iridium(III) and ruthenium(II) complexes by combinatorial luminophore synthesis and screening (Scheme 5.9). Starting from iridium trichloride, cyclometalation with (hetero)arylpyridyl ligands **45** (Fig. 5.17) gives rise to the formation of binuclear iridium complexes **46**. Upon complexation with bidentate *N,N*- or *P,P*-ligands **47** (Fig. 5.17), the cationic complex **48** is formed, which upon anion metathesis with hexafluorophosphate is transformed into the target complex **49**. In this sequence, the step from **46** to **48** was performed in a traditional and a parallel manner, the latter leading to a library of 100 iridium and 10 ruthenium complexes.

Most impressive from the photophysical evaluation of the combinatorial libraries, a remarkable luminescent color versatility, spanning from blue to orange–red emission, can be observed. Furthermore, this set of data was well suited for correlations with static DFT calculations to establish whether it is feasible to predict the luminescent behavior of novel materials.

5.3
Novel Multi-component Syntheses of Chromophores

The covalent assembly of functional π-systems is a general synthetic principle and in some cases they can even be achieved in a multi-component fashion. One of the most impressive examples is the very elegant access to covalently linked donor–fullerene arrangements by 1,3-dipolar cycloadditions with *in situ*-generated azomethine ylids [59]. However, here only the multi-component *de novo* synthesis of the chromophore structures will be considered. The major developments have been achieved in condensation-based and cross-coupling strategies.

5.3.1
Multi-component Condensation Reactions

Merocyanine dyes **50–55** (Fig. 5.18) possess an electronic structure at the mesomeric center between neutral and zwitterionic electron distribution and have high polarizabilities and dipole moments and exhibit absorption spectra with sharp bands that give rise to exceptionally brilliant magenta hues [60].

A chromogenic system with these properties is highly favorable for several high-technology applications [61]. The cyanine-like narrow absorption band leads to very brilliant magenta hues suited for applications in digital photography and color copying. On the other hand, high dipole moments and polarizabilities along the conjugated chain permit unprecedented refractive index modulations in photorefractive materials [62]. Furthermore, the brilliancy and solid-state lumines-

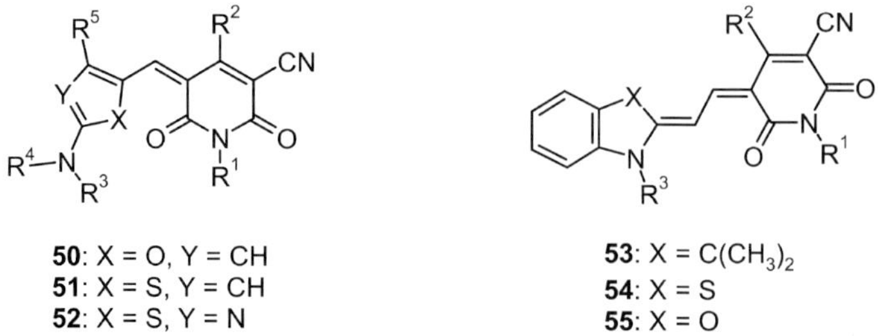

5 Diversity-oriented Synthesis of Chromophores

50: X = O, Y = CH
51: X = S, Y = CH
52: X = S, Y = N

53: X = C(CH$_3$)$_2$
54: X = S
55: X = O

Fig. 5.18 Merocyanine dyes with very brilliant magenta hues.

cence lead to especially bright colors on polyester and can therefore be suitable for textile coloration.

From the synthetic point of view, either traditional, i.e. stepwise, syntheses or by means of automated parallel synthesis as in the case of combinatorial chemistry pose the challenge of rapidly identifying most optimal representatives of the merocyanines **52–55**. However, based on retrosynthetic analysis (Scheme 5.10), Würthner and coworkers [60, 63] suggested an elegant concept based on a three-component synthesis of alkenylidene-2,6-dioxo-1,2,5,6-tetrahydropyridine-3-carbonitrile merocyanines **52–55**.

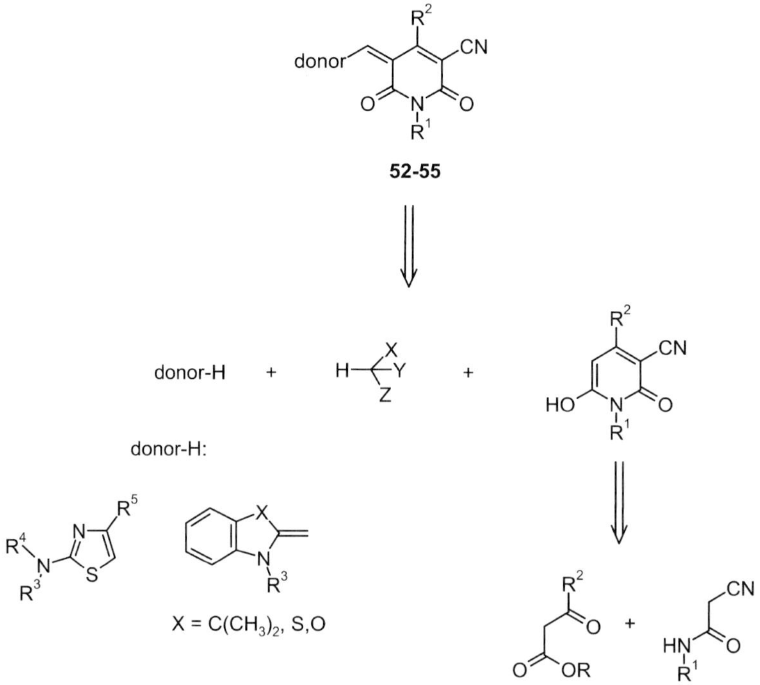

X = C(CH$_3$)$_2$, S, O

Scheme 5.10 Retrosynthetic analysis of merocyanine dyes **52–55**.

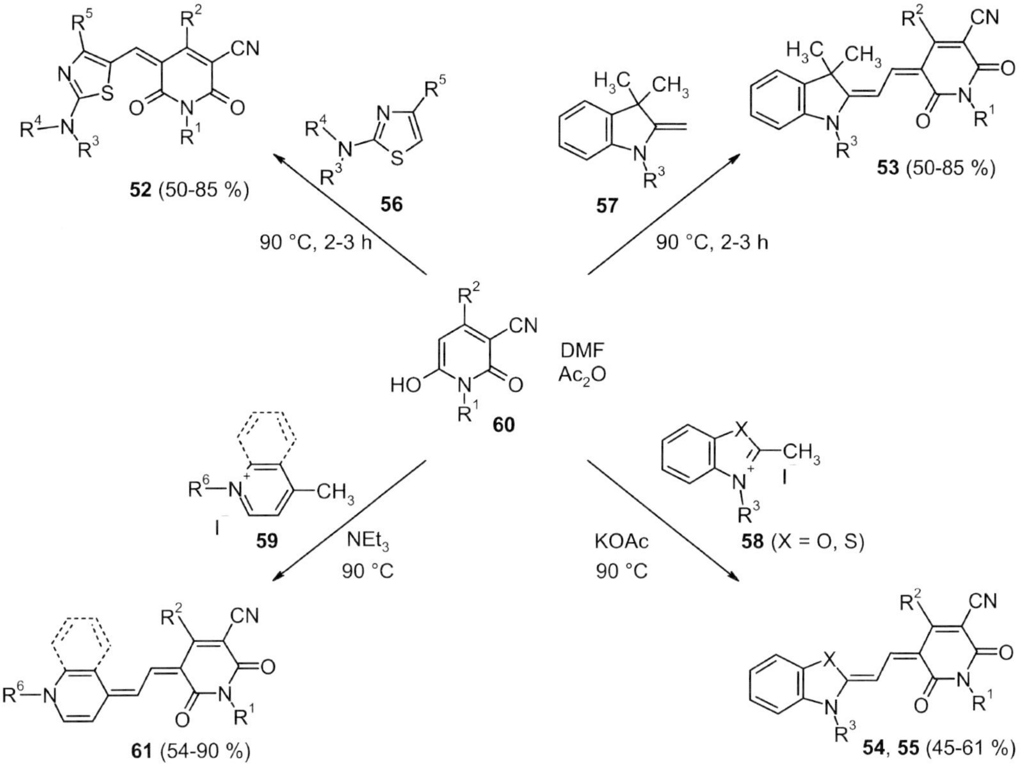

Scheme 5.11 One-pot, three-component synthesis of merocyanine dyes **52–55** and **61**.

R¹ = alkyl
R² = Me, Ph
R³, R⁴ = alkyl, sec. alkyl
R⁵ = Ph, tBu, neopentyl
R⁶ = Me, Et

This straightforward and highly efficient one-pot synthesis takes advantage of the fact that CH-acidic heterocycles, such as 1,2,5,6-tetrahydro-2,6-dioxo-3-pyridinecarbonitriles (that can be easily obtained by condensation of β-keto esters and cyanoacetamides) and electron-rich dialkylaminothiazoles or methylene bases and formic orthoester derivatives as formyl building blocks can be condensed to furnish merocyanine dyes. A formylating reagent very favorable for an MCR protocol is generated *in situ* from DMF and acetic anhydride. Therefore, on reacting electron-rich dialkylaminothiazoles **56** or methylene bases **57** (also heterocyclic methylene active salts **58** and **59** as precursors) with highly acidic 1,2,5,6-tetrahydro-2,6-dioxo-3-pyridinecarbonitriles **59** in the presence of the formylation–condensation system of DMF in acetic anhydride, the pure dyes **52–55** and **61** were obtained in 45–90 % yield directly from the reaction mixtures (Scheme 5.11).

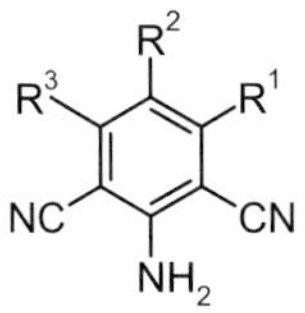

Fig. 5.19 2,6-Dicyanoanilines **62** are simple representatives of Acc–Do–Acc systems.

Acceptor–donor–acceptor (Acc–Do–Acc) systems are suitable for obtaining a long-lived charge separation upon photoinduced intramolecular charge transfer [64] and, in principle, these molecular systems have recently attracted attention for the development of single molecule based electronic devices [65].

2,6-Dicyanoanilines **62** (Fig. 5.19) are highly substituted benzene derivatives and can be considered as simple representatives of typical Acc–Do–Acc systems. However, synthetically persubstituted arenes are better synthesized by *de novo* benzene ring formation. Interestingly, the optical properties of the aniline derivatives **62** have only rarely been documented [66].

Therefore, based on the known cyclocondensation of arylidenemalonodinitriles and 1-arylethylidenemalonodinitriles in the presence of piperidine to give 1,6-dicyanoanilines [66a], Wang and coworkers [67] have developed a straightforward microwave irradiation-assisted pseudo-four-component synthesis of the anilines **62**. Thus, the reaction of (hetero)aromatic aldehydes **63**, acyclic and cyclic ketones **64** and 2 equiv. of malononitrile in the presence of triethylamine or piperidine furnishes, after 2 min of 300 W-irradiation power, 51–63 % of 2,6-dicyanoanilines **62** (Scheme 5.12).

From the set of products, two members, **62a** and **62b** (Fig. 5.20), with the most intense fluorescence were selected for photophysical evaluation. Inspection of the maximum emission wavelengths and fluorescence quantum yields in solvents with different polarities (CH_2Cl_2, MeOH and THF) reveals a significant solvochromicity of the emission maxima that influence the corresponding fluorescent quantum yields to only a minimal extent.

Additionally, this methodology was transposed to liquid-phase synthesis of 2,6-dicyanoanilines using poly(ethylene glycol) (PEG) as support, starting with a

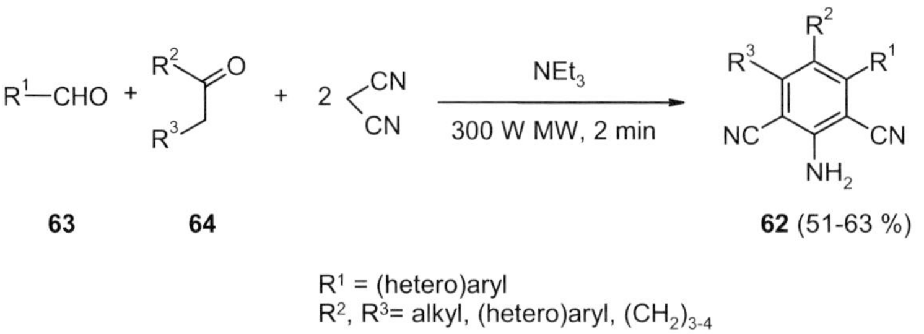

R^1 = (hetero)aryl
R^2, R^3 = alkyl, (hetero)aryl, $(CH_2)_{3-4}$

Scheme 5.12 One-pot, pseudo-four-component synthesis of 2,6-dicyanoanilines **62**.

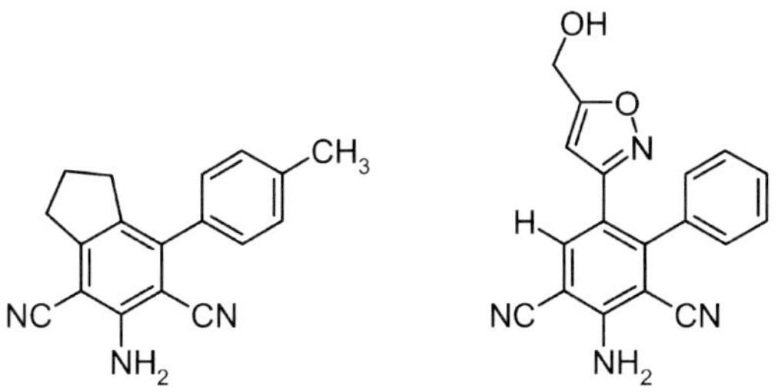

Fig. 5.20 Strongly fluorescent 2,6-dicyanoanilines **62a** and **62b**.

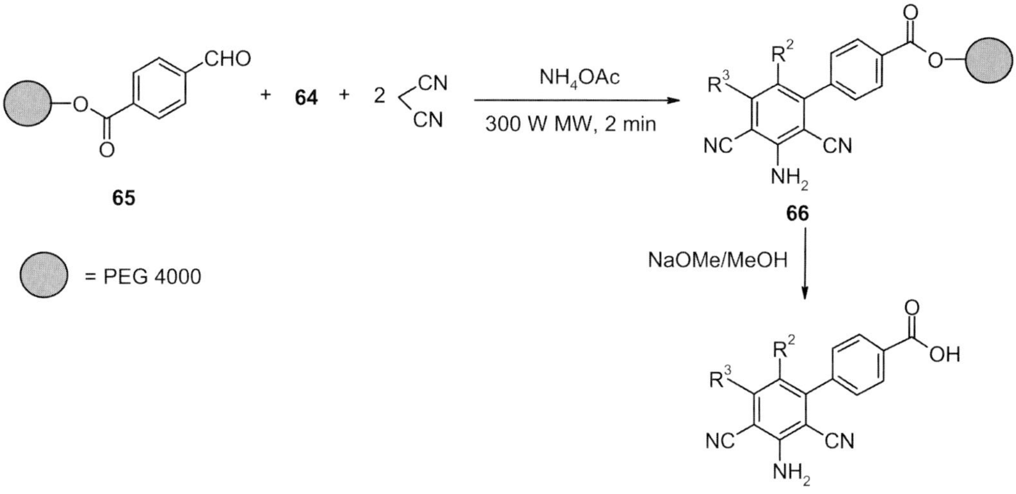

Scheme 5.13 Microwave-assisted, one-pot, pseudo-four-component liquid-phase synthesis of polysubstituted 2,6-dicyanoanilines **62**.

PEG-bound terephthalate monoaldehyde **65** (Scheme 5.13). The resulting PEG-bound 2,6-dicyanoanilines **66** were cleaved by NaOMe–MeOH to afford free polysubstituted 2,6-dicyanoanilines **62**. The products were isolated in good yields (65–82 %) and high purities (89–98 %).

Among chromophores with favorable fluorescence properties squaraines [68] (Fig. 5.21) display narrow and intense absorption bands and fluorescence emission with high quantum yields (Φ_f up to 0.9) both at long wavelengths ($\lambda_{max,abs}$, $\lambda_{max, em} > 600$ nm) [69]. These spectral characteristics render squaraines attractive for intracellular probing [70] and as photoreceptors [71] relying on intrinsic brightness [72]. Furthermore, squaraine chromophores exhibit strong and characteristic exciton interaction [73] due to H-dimer formation [74]. Hence squaraine dyes have been utilized for some chemosensor designs, for instance by covalently linking

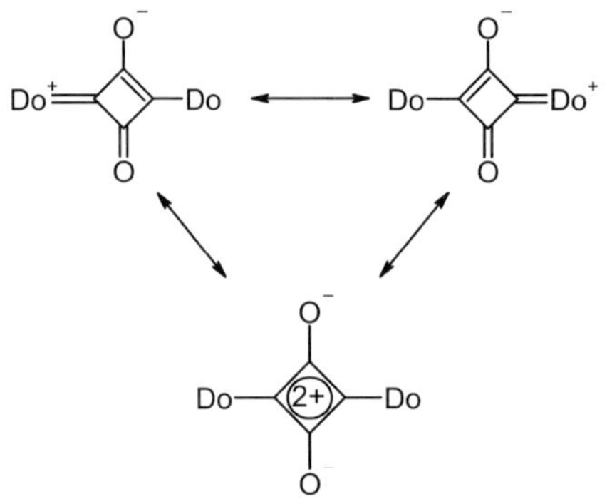

Fig. 5.21 General structure of squaraines.

them to crown ethers and also open-chain aminoethylenes and ethylene glycols [75].

In particular, nonconjugated polymers with alternating squaraine and receptor units represent a new class of sensory polymers suitable for signal transduction via conformationally induced and controlled exciton interaction. For this class of polymers, Block and Hecht [76] devised an elegant modular approach based on a one-pot, two-step procedure. Upon heating of acyclic and cyclic α,ω-diamines with phloroglucinol (67) with constant removal of generated water, the bifunctional bis(3,5-dihydroxyphenyl)-terminated tertiary α,ω-diamine building blocks 68–70 obtained are subsequently reacted with squaric acid (71) in the same vessel to furnish the polysquaraines 72–74 in 15–53% overall yield (Scheme 5.14).

Physical and chemical stimuli such as temperature, solvent polarity and addition of various cations induce conformational changes in the nonconjugated polysquaraines, leading to either preferential folding to, or unfolding from, chromophore H-dimers. The binding event is translated into a shift in the monomer to H-dimer equilibrium of the squaraine chromophores and can be conveniently visualized by UV–Vis and fluorescence spectroscopy and, therefore, can successfully be exploited for cation sensing.

5.3.2
Multi-component Cross-coupling Reactions

In Section 5.2.2, it was highlighted and discussed that Pd- and Ni-catalyzed cross-coupling reactions display excellent compatibility with numerous polar functional groups as a consequence of the mild reaction conditions. This is the key feature qualifying cross–cross-coupling methodologies for the development of novel multi-component and domino sequences. As a logical consequence, sequential catalysis [77], where one catalyst type is able to catalyze sequentially two or more equal or different coupling reactions, arises as a possible entry into multi-component cross-coupling reactions. Furthermore, transition metal-assisted sequential processes [78] can also be an initial or intermediate step in a sophisti-

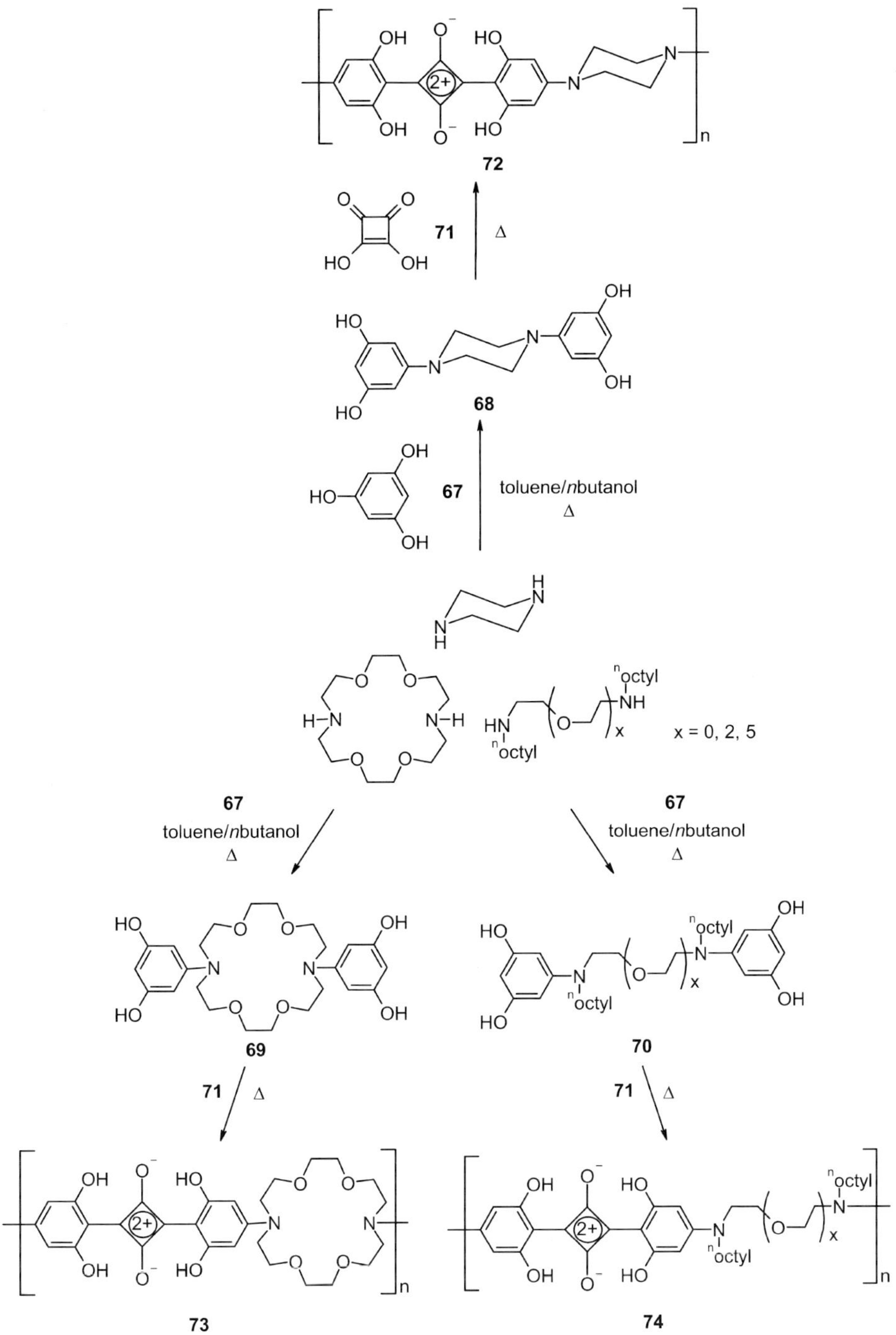

Scheme 5.14 Modular one-pot, three-component synthesis of poly(squaraines) from terminal α,ω-diamines.

cated sequence performed in a one-pot fashion. In this overview, these two major strategies will be highlighted and discussed.

Multiple Heck reactions are well suited for the development of consecutive multi-component reactions. In particular, vinylsilanes with a catalyst-directing 2-pyridyl group on silicon serve as an excellent platform for the combinatorial synthesis of 1,1,2-triarylethenes (see Section 5.2.3) [49]. In combination with Hiyama coupling, addressing the carbon–silicon functionality as organometallic portion this template was also involved in combinatorial syntheses of the target chromophores in a consecutive fashion. On this basis, Yoshida and coworkers [49] suggested and developed consecutive reactions in which palladium is sequentially applied for two subsequent Heck reactions in a one-pot fashion. For the efficient three-component reaction, the vinylsilane was appended with a 2-pyrimidyl substituent and the electron-rich bis[tris(*tert*-butyl)]phosphanepalladium(0) was deliberately chosen as the catalyst precursor. Therefore, the vinylsilane **75**

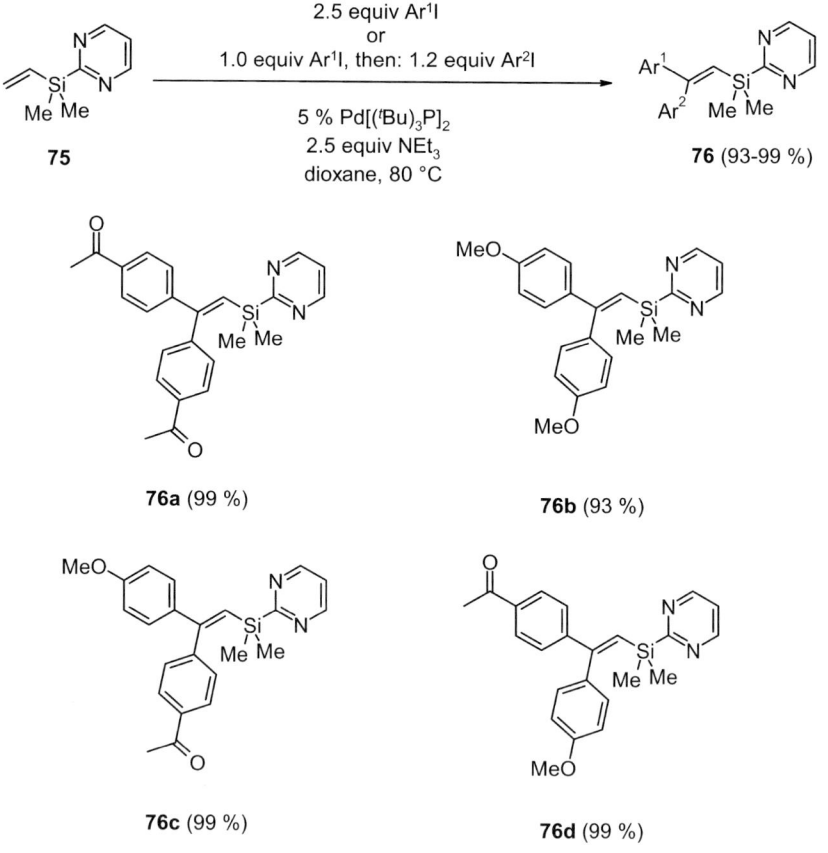

Scheme 5.15 One-pot, three-component synthesis of 2,2-di(hetero)aryl-substituted vinylsilanes **76**.

was sequentially reacted with 2.5 equiv. of the same (hetero)aryl iodide or with 1.0 and 1.2 equiv. of various aryl iodides in dioxane at 80 °C in the presence of bis[tris(*tert*-butyl)]phosphanepalladium(0) and triethylamine to furnish the 2,2-di(hetero)aryl-substituted vinylsilanes **76** in excellent yields and with virtually complete stereoselectivity (Scheme 5.15).

Interestingly, the concept of combining two Heck reactions and a cross-coupling reaction sequentially within the same vessel failed with vinylsilanes, but were successful if another vinyl organometallic was applied as a template. Starting with vinylpinacolyl boronate (**77**), Yoshida and coworkers [79] reacted 2 equiv. of (hetero)aryl halide in toluene at 80 °C in the presence of bis[tris(*tert*-butyl)]phosphanepalladium(0) and diisopropylamine to give the double Heck arylation product **78**, a boronate, which was not isolated (Scheme 5.16). Simply adding 1.1 equiv. of a second (hetero)aryl halide, sodium hydroxide and water concluded the sequence by a Suzuki cross-coupling and gave rise to the formation of 1,1,2-tri(hetero)arylethenes **79**.

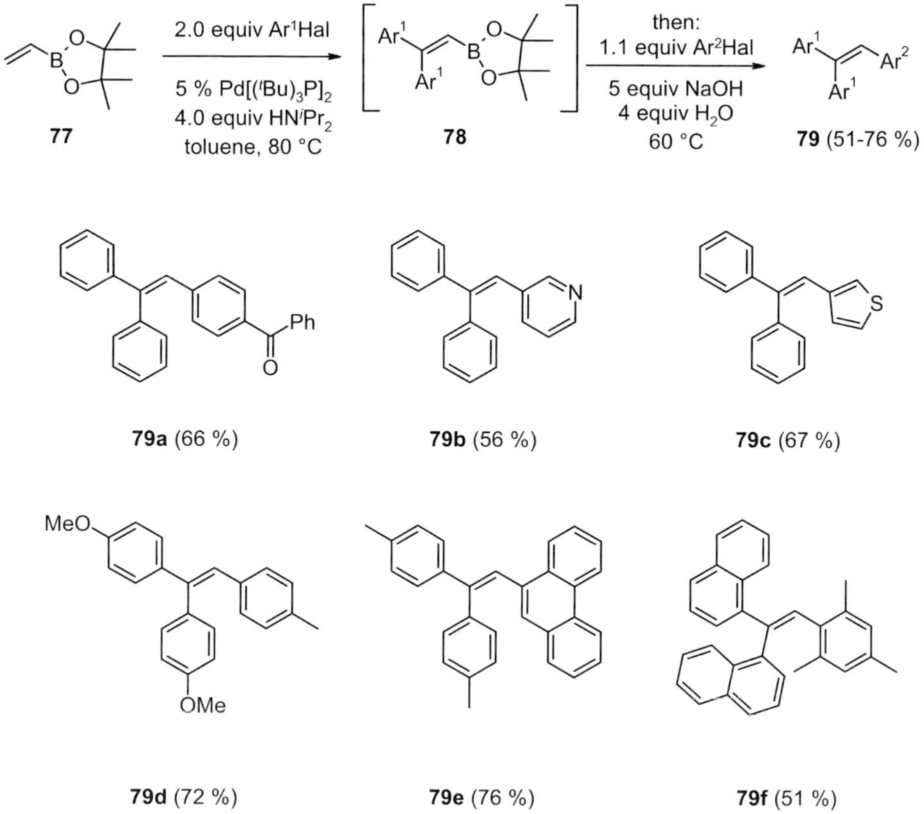

Scheme 5.16 One-pot, three-component synthesis of 1,1,2-tri(hetero)arylethenes **79**.

80a (60 %)   **80b** (61 %)

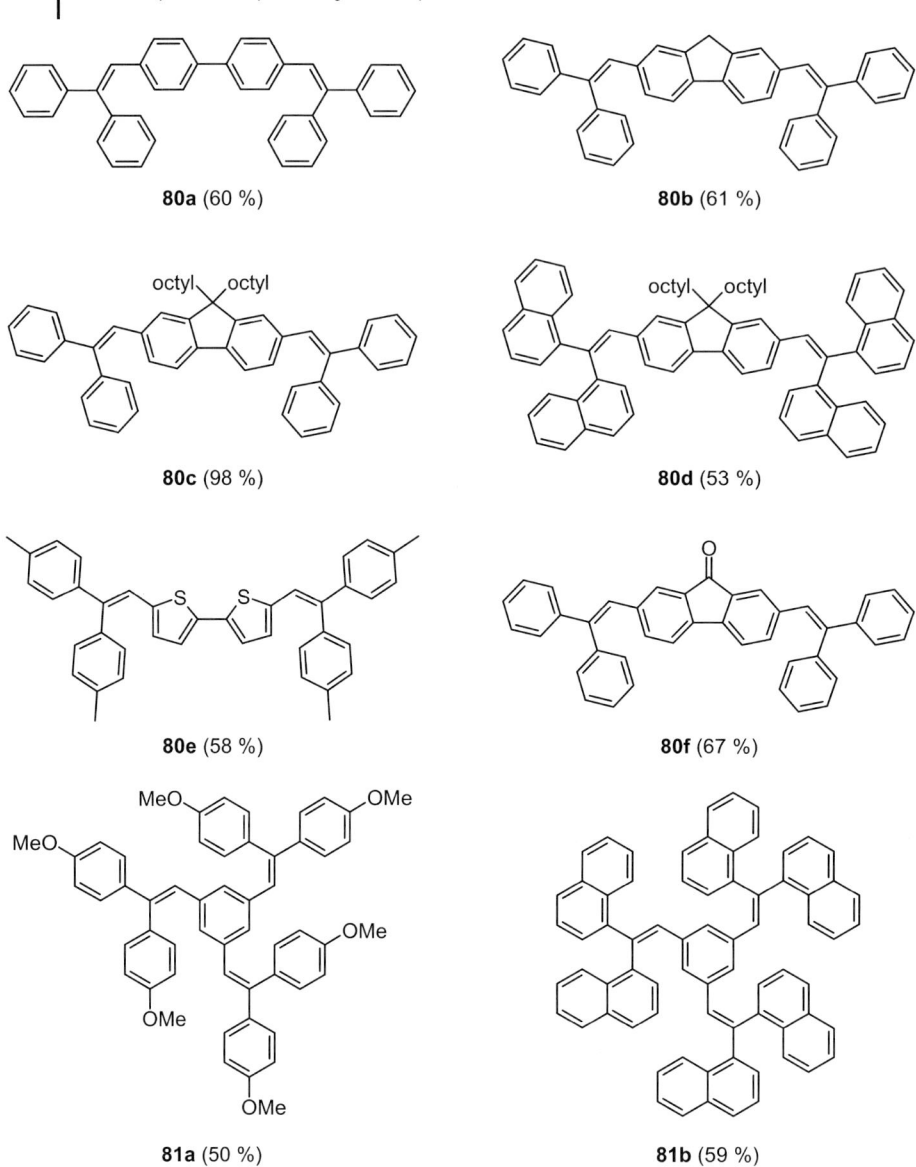

80c (98 %)   **80d** (53 %)

80e (58 %)   **80f** (67 %)

81a (50 %)   **81b** (59 %)

Fig. 5.22 One-pot, three-component synthesis of dumbbell-shaped and star-shaped 1,1,2-tri(hetero)arylethenes **80** and **81**.

Likewise, even more extended π-conjugation can be realized with the Heck–Suzuki one-pot reaction if dibromo- or tribromo(hetero)arenes are applied, now giving rise to dumbbell-shaped (80) and star-shaped extended π-systems 81 (Fig. 5.22). By using this method, several highly fluorescent materials were discovered, among them the three fundamental colors blue (80d, $\lambda_{max,\ abs}$ = 376 nm, $\lambda_{max,\ em}$ = 454 nm, Φ_f = 0.9), green (80e, $\lambda_{max,\ abs}$ = 429 nm, $\lambda_{max,\ em}$ = 523 nm, Φ_f = 0.11) and red (80f, $\lambda_{max,\ abs}$ = 465 nm, $\lambda_{max,\ em}$ = 599 nm, Φ_f = 0.01).

Since cross-coupling reactions are well suited for the development of sequential transformations in a multi-component fashion, new methodologies for diversity-oriented syntheses of chromophores could be based on sequential combinations of cross-coupling, i.e. catalytic organometallic reactions and organic basic reactions. Taking into account the excellent compatibility of polar functional groups that often dispenses with tedious protection–deprotection steps, Sonogashira coupling [80], a straightforward alkyne-to-alkyne transformation, is a highly favorable tool for devising novel synthetic strategies to obtain functional π-electron systems. Conceptually, the installation of a reactive functional group such as an alkyne with an electron-withdrawing substituent inevitably could result in an *in situ* activation of alkynes towards Michael-type addition, i.e. an entry to a coupling-addition sequence (Scheme 5.17).

Independently, the groups of Lin [81] and Müller [82] found that 2-alkynyl-5-nitrothiophenes 82 react very smoothly with secondary amines 83 to furnish intensely colored, highly solvochromic β-aminovinylnitrothiophenes 84 (Scheme 5.18).

β-Aminovinylnitrothiophenes 84, a novel type of push–pull chromophores, were closely investigated with respect to their NLO and thermal properties [82]. Hyper-Rayleigh scattering (HRS) measurements at a fundamental of 1500 nm re-

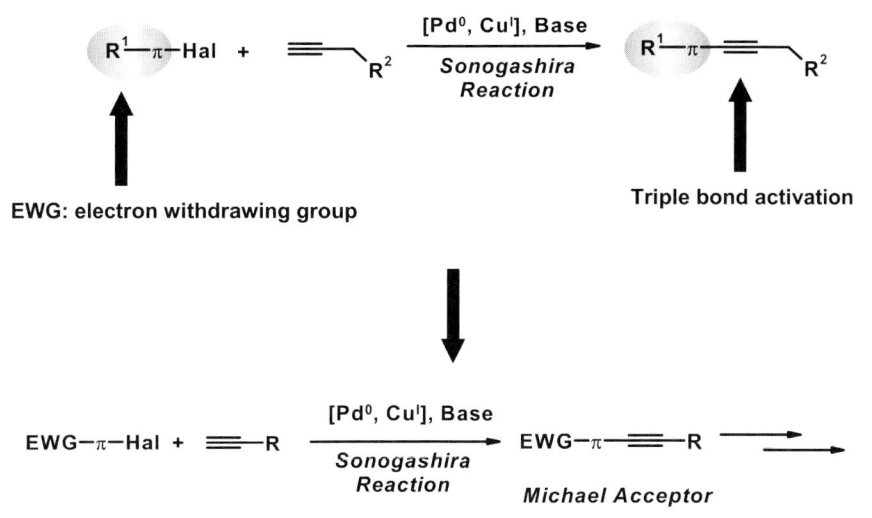

Scheme 5.17 Sonogashira coupling with electron-withdrawing halide compounds as a peculiar mode of alkyne activation and entry to coupling–Michael addition sequences.

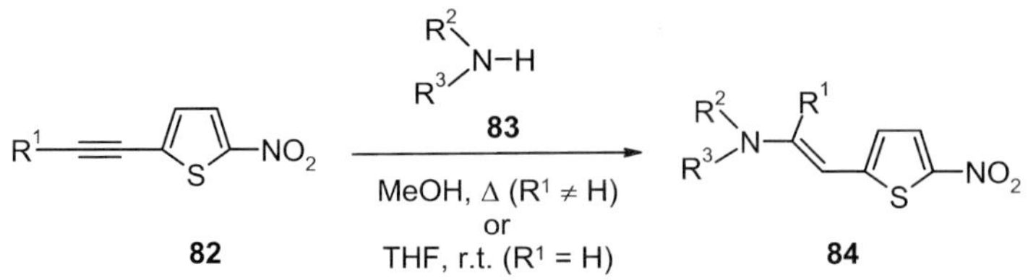

Scheme 5.18 Michael-type addition of secondary amines to nitrothienyl-substituted alkynes.

R^1 = H, aryl, ferrocenyl

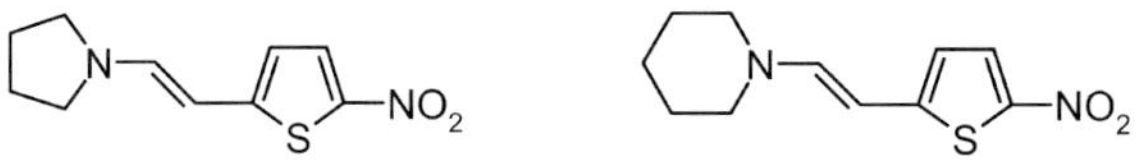

84a
λ_{max} (pentane) = 450 nm
λ_{max} (CHCl$_3$) = 520 nm
β^0 = 31 × 10^{-30} esu
$\beta^0 \mu / M_w$ = 1.34

84b
λ_{max} (pentane) = 443 nm
λ_{max} (CHCl$_3$) = 513 nm
β^0 = 29 × 10^{-30} esu
$\beta^0 \mu / M_w$ = 1.23

Fig. 5.23 Solvochromicity and NLO properties of selected β-aminovinylnitrothiophenes **84**.

vealed that the β-values are surprisingly large for such short dipoles (**84a**, β^0_{333} = 31 × 10^{-30} esu; **84b**, β^0_{333} = 29 × 10^{-30} esu; Fig. 5.23). With respect to the relatively low molecular mass, these chromophores display a rather favorable molecular figure of merit, $\beta^0 \mu / M_w$, where M_w is the molar mass.

Furthermore, selected push–pull chromophores **84** were investigated by differential scanning calorimetry (DSC), revealing relatively low glass transition temperatures, T_g, a favorable property for composites in photorefractive materials [83].

Based on the peculiar reactivity of nitrothienyl-substituted alkynes, Müller and coworkers [84] developed a one-pot, three-component coupling–aminovinylation sequence to give push–pull chromophores. Terminal alkynes **85** and sufficiently electron-deficient heteroaryl halides **86** were transformed under Sonogashira conditions into the expected coupling products, which were subsequently reacted with secondary amines **83** to furnish the push–pull systems **87** in good yields (Scheme 5.19). The critical step in this consecutive reaction is the addition of the amine to the intermediate internal acceptor substituted alkyne. According to semi-empirical and DFT calculations, the crucial parameters for the success of the amine addition are the relative LUMO energies and the charge distribution at the β-alkynyl carbon atom.

The concept of alkyne activation by Sonogashira coupling was successfully extended to the *in situ* generation of alkynones that are highly reactive and versatile synthetic equivalents of 1,3-dicarbonyl compounds. Interestingly, ynones can be readily synthesized in a catalytic fashion by Sonogashira coupling of an acyl chlor-

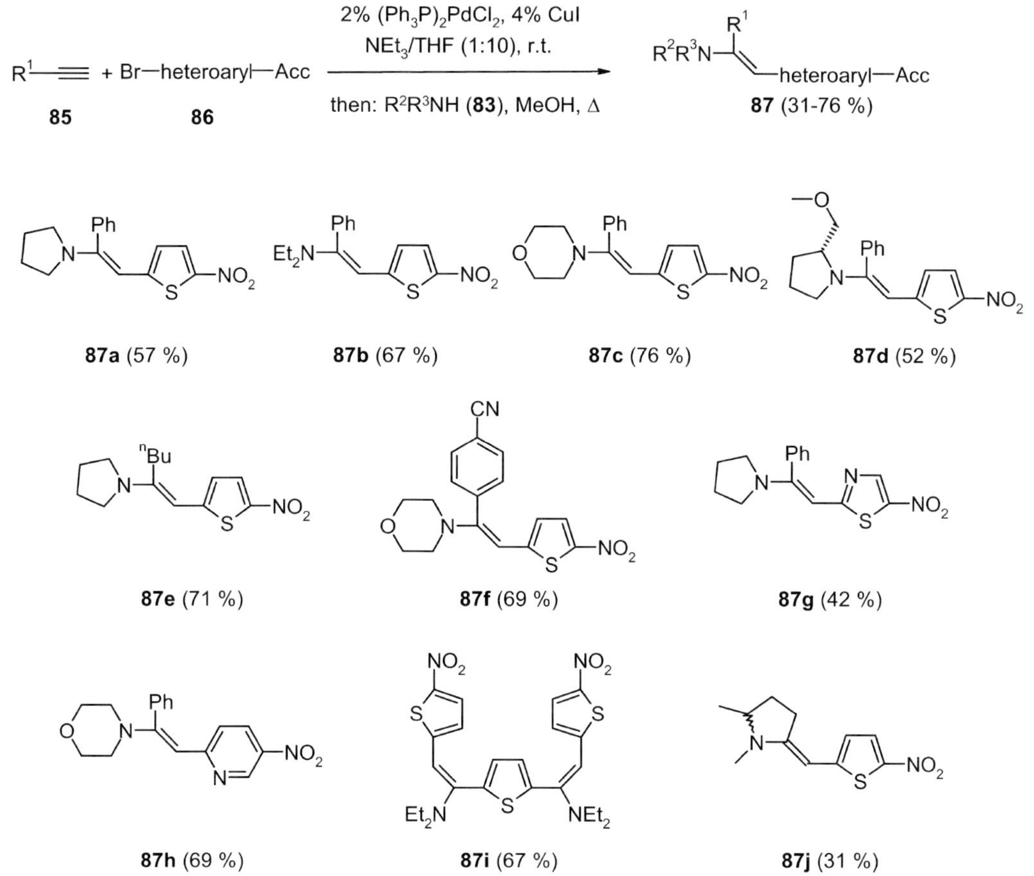

5.3 Novel Multi-component Syntheses of Chromophores | 211

Scheme 5.19 A coupling-aminovinylation sequence to β-aminovinyl heteroarenes.

ide with a terminal alkyne [85]. Therefore, the consecutive one-pot reaction principle of the coupling-addition sequence led to synthetic and methodological extension in enaminone and heterocycle synthesis [86]. Likewise, alkynones can also be reaction partners in pericyclic reactions such as cycloadditions. In particular, 1,3-dipolar cycloaddition [87] promises rapid access to many classes of conjugated and annelated five-membered heterocycles. Considering the well-established fluorescence properties of indolizines [88] and biindolizines [88b] and the steadily increasing importance of fluorophores in biolabeling and environmental trace analysis, new, efficient and diversity-oriented syntheses of fluorescent indolizines have become a methodological challenge.

In the context of a coupling–1,3-dipolar cycloaddition sequence, Müller and coworkers [89] developed a consecutive one-pot, three-component process to indolizines. Starting from (hetero)arenecarbonyl chlorides **88** and terminal alkynes **89** under Sonogashira conditions, the expected alkynones were formed (Scheme

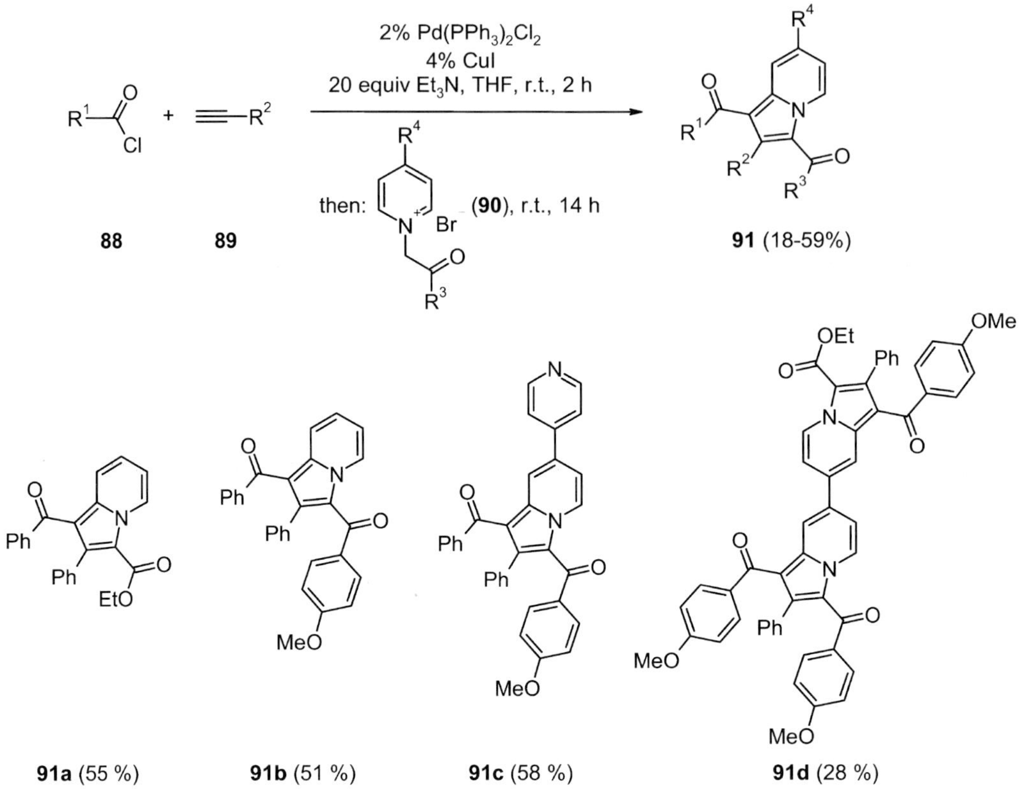

Scheme 5.20 One-pot, three-component coupling–1,3-dipolar cycloaddition synthesis of indolizines.

5.20). Under the amine basic conditions, the subsequently added 1-(2-oxoethyl)-pyridinium bromide derivatives **90** are transformed *in situ* into pyridinium ylides that undergo a [2 + 3] cycloaddition with the alkynone dipolarophiles present in the reaction mixture. The initial cycloadducts instantaneously aromatize to give rise to highly fluorescent indolizine derivatives **91**.

Fluorescence studies with pyridyl-substituted representatives revealed not only that indolizines and biindolizines are highly interesting fluorescence dyes but also that their fluorescence color can also be reversibly switched upon altering the pH of the medium.

Besides activating the triple bond towards Michael addition, the electron-withdrawing group introduced by Sonogashira coupling can also exert an activation of the remote propargyl position. This propargyl activation could, for instance, trigger an alkyne–allene isomerization; over the complete sequence, a coupling–isomerization reaction (CIR) would be the consequence (Scheme 5.21). Driven by a concluding tautomerization, the CIR of electron-deficient (hetero)aryl halides and 1-(hetero)arylpropargyl alcohols gives rise to the formation of 1,3-di(hetero)aryl

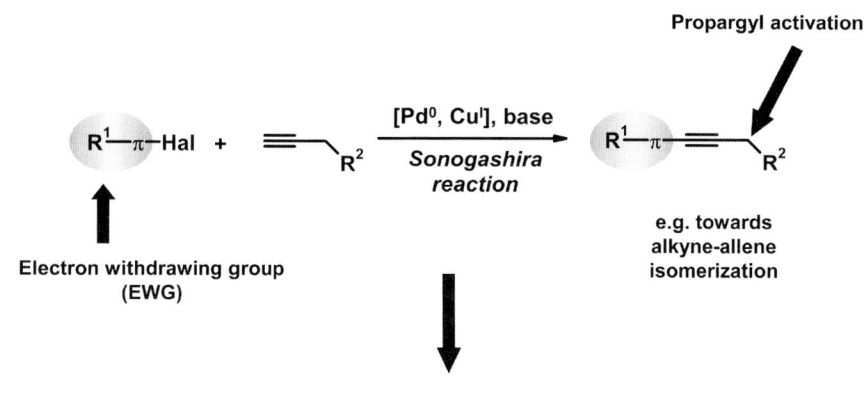

5.3 Novel Multi-component Syntheses of Chromophores

Scheme 5.21 Propargyl activation and coupling–isomerization reaction as an entry to novel MCRs.

propenones [90], i.e. chalcones that are as Michael acceptors suitable starting points for consecutive multi-component syntheses of heterocycles in a one-pot fashion [90, 91].

The CIR is exceptionally well suited as an entry to multi-component syntheses of aromatic heterocycles and on this basis Müller and coworkers [91f, 92] designed a coupling–isomerization–Stetter–Paal–Knorr sequence as a diversity-oriented approach to highly substituted furans and pyrroles (Scheme 5.22).

Scheme 5.23 CI–Stetter–Paal–Knorr synthesis of furans **95** and pyrroles **98**.

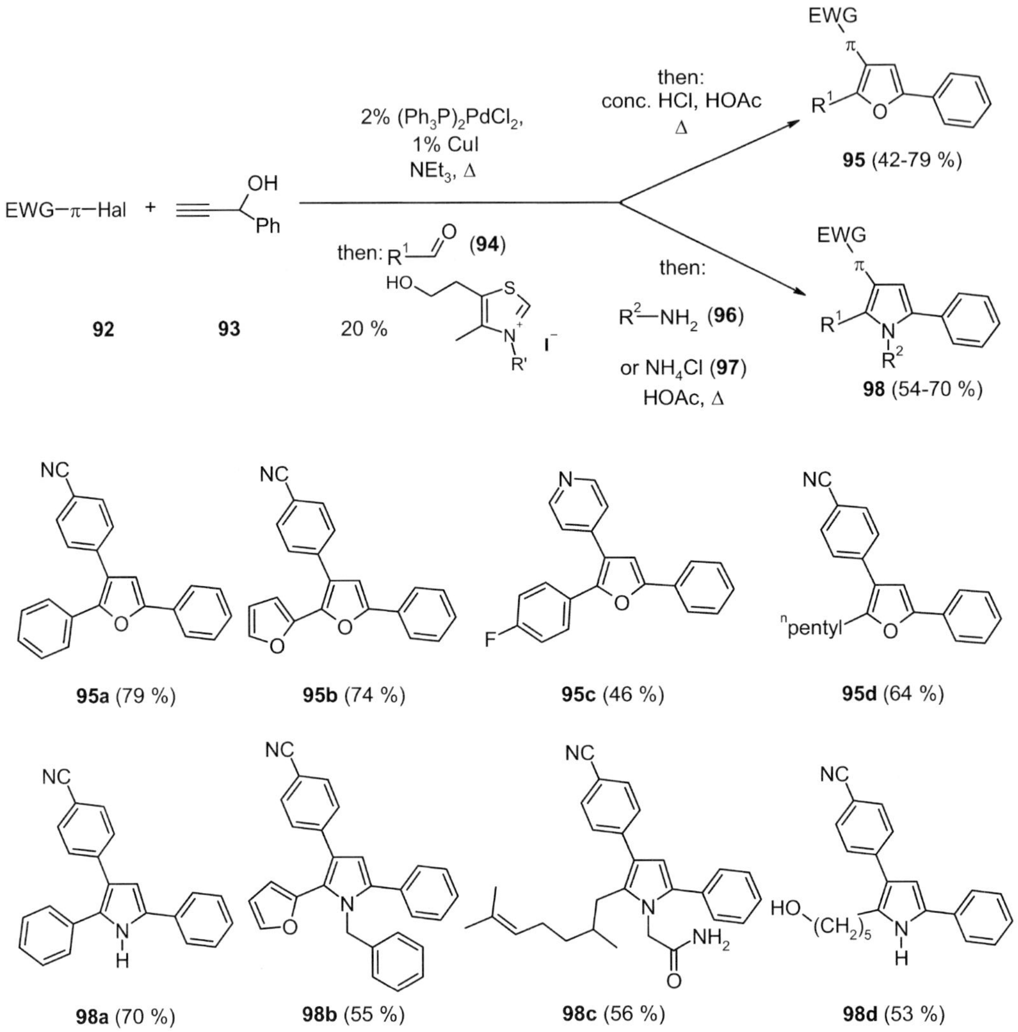

Scheme 5.22 Retrosynthetic concept for a three-component furan and four-component pyrrole synthesis.

This approach is particularly intriguing since it represents a consecutive combination of modern cross-coupling methodology and classical Michael addition–cyclocondensation, the latter still being of significant importance in the industrial processes of numerous heterocyclic pharmaceuticals. Therefore, the coupling-isomerization–Stetter–Paal–Knorr synthesis of furans and pyrroles begins with the CIR starting from electron-deficient (hetero)aryl halides **92** and 1-phenylpropyn-1-ol (**93**) (Scheme 5.23). Upon addition of aromatic or aliphatic aldehydes **94** and a thiazolium salt catalyst, the intermediate chalcone is subsequently transformed by a Stetter reaction into 1,4-diketones. Without isolation the 1,4-dicarbonyl compounds are reacted with concentrated hydrochloric acid and acetic acid in the same vessel to give 2,3,5-trisubstituted furans **95** or, upon subsequent addition of primary amines **96** or ammonium chloride (**97**) and acetic acid, 2,3,5-trisubstituted and 1,2,3,5-tetrasubstituted pyrroles **98** were obtained in good overall yields.

Since all novel furans **95** and pyrroles **98** exhibit a strong blue fluorescence with considerable Stokes shifts, where the absorption maxima are found in the range $\lambda_{max,\ abs}$ = 312–327 nm and the emission occurs at $\lambda_{max,\ abs}$ = 401–451 nm, this multi-component approach to fluorophores can be exploited for combinatorial optimization of emission properties.

5.4
Conclusion and Outlook

Diversity-oriented strategies have recently found application in the conception and development of functional π-electron systems, predominantly for chromophore and lead optimization for emitters in OLED. In combination with combinatorial and HTS methods, parallel syntheses have gained considerable importance and have become an essential tool in the hands of organic materials scientists. As a consequence, future developments will increasingly rely on novel combinatorial, parallel synthetic and HTS strategies, in particular if new molecular structures are created and evaluated *in vitro* where theoretical predictions are either impossible or incomplete. This also poses challenges for reaction design and new synthetic methodology, which no longer will remain a privilege of classical organic chemistry for natural product synthesis and medicinal and pharmaceutical applications. As a molecule-based molecular electronics emerged, functional π-electron systems with increasing structural complexity were synthesized and the quest for optimization of favorable electronic features and conformational properties necessitates the invention of rapid synthetic access methods. Hence multi-component synthetic strategies and complexity-enhancing domino reactions will receive increasing consideration for new lead structure finding in the field of chromophores and electroactive molecular materials. Diversity-oriented strategies for developing functional π-electron systems remain an intellectual and practical challenge and are a conceptional, synthetic, mechanistic and methodological "el dorado" for organic chemists.

5.5
Experimental Procedures

Synthesis of rhodacyanines **12** [25]
A mixture of the methylthioiminium salt **13** (0.04 mmol) and the rhodanine **14** (0.40 mmol) in a filter tube, capped with a septum at the bottom, was suspended in acetonitrile (1.6 mL). To the mixture was added NEt$_3$ at room temperature and

the resulting mixture was stirred for 3 h at the same temperature. By opening the bottom cap, the acetonitrile was removed through the filter. The precipitate was washed with acetonitrile twice and dried *in vacuo* to afford merocyanine **15** as a yellowish solid, which was used in following step without further purification. This merocyanine (in the same filter tube, capped with a septum at the bottom) was suspended in distilled DMF (0.40 mL). To the suspension was added methyl *p*-toluenesulfonate (1.2 mmol) and the mixture was heated for 4 h at 120 °C. After the mixture had cooled to room temperature, acetone was added. The precipitate was collected and washed with acetone to give **16** as orange crystals, which were used in the following step without further purification. To a mixture of **16** and the iminium salt **17** (0.40 mmol) in acetonitrile (4.0 mL) was added dropwise triethylamine (1.2 mmol) at 50 °C and the mixture was stirred for 3 h at the same temperature. After the mixture had cooled to room temperature, ethyl acetate was poured on to the resulting mixture. The precipitate was collected and washed with ethyl acetate to give the rhodacyanine **12** as crystals.

Parallel synthesis of 3-(hetero)arylchromenones **26**
(representative example: 7-amino-4-methyl-3-*p*-tolylchromen-2-one) [38]
An amount of 25.4 mg (0.1 mmol) of coumarin **25f**, 35.2 mg (0.2 mmol) of boronate **20b**, 121 mg (0.8 mmol) of CsF and 5.78 mg (5 mol%) of Pd(PPh$_3$)$_4$ were heated at 90 °C for 16 h under argon in dry dioxane. The solvent was removed *in vacuo*, the residue was dissolved in acetonitrile–water and the solution was filtered through an SPE syringe. Finally, purification was achieved by automated HPLC–MS (21.2 mg, 80%); m.p. 260–262 °C. ^{1}H NMR (400 MHz, DMSO-d_6, 25 °C), δ = 2.13 (s, 3 H), 2.33 (s, 3 H), 6.07 (br, 2 H), 6.45 (d, J = 1.9 Hz, 1 H), 6.60 (dd, J = 8.7, 1.9 Hz, 1 H), 7.12 (d, J = 7.9 Hz, 2 H), 7.20 (d, J = 7.9 Hz, 2 H), 7.44 (d, J = 8.7 Hz, 1 H); ^{13}C NMR (100 MHz, DMSO-d_6, 25 °C), δ = 16.3, 21.0, 98.5, 109.4, 111.6, 119.8, 126.8, 128.7 (2 C), 130.5 (2 C), 132.5, 136.6, 148.6, 152.8, 154.6, 160.8; UV–Vis (ethanol), λ_{max} (ε, L mol^{-1} cm^{-1}) = 360 nm (22 500); EI-MS (70 eV), m/z (%), 265 (88) M$^+$, 237 (100) [M$^+$ – CO].

Synthesis of an unsymmetrically diaryl-substituted vinylsilane **42** [49]
A mixture of dimethyl(2-pyrimidyl)vinylsilane (**40**) (171.4 mg, 1.04 mmol), 4-iodoacetophenone (246.4 mg, 1.00 mmol), triethylamine (253.0 mg, 2.50 mmol) and Pd(P^{t}Bu$_3$)$_2$ (25.6 mg, 50.1 µmol) in dry dioxane (2.0 mL) was stirred at 80 °C for 5 h under argon. To this mixture was added 4-iodoanisole (286.1 mg, 1.22 mmol) and the resulting mixture was further stirred at 80 °C for 19 h. After cooling the reaction mixture to room temperature, the catalyst and salts were removed by filtration through a short silica gel pad (EtOAc). The yield of **42** was estimated to be >99% by ^{1}H NMR of the crude mixture. Gel permeation chromatography (CHCl$_3$) of the crude mixture afforded 398.8 mg (99%) of the vinylsilane as a pale-yellow oil.

Hiyama coupling of a vinylsilane 42 to an extended π-system 44 [49]

To a mixture of 2,2-diphenyldimethyl-(2-pyrimidyl)vinylsilane (42) (95.6 mg, 0.30 mmol), 4,4′-diiodobiphenyl (41.1 mg, 0.10 mmol) and $PdCl_2(PhCN)_2$ (3.9 mg, 10.2 μmol) in dry THF (0.8 mL) was added a solution of Bu_4NF (0.30 mmol, 1.0 M) in THF at room temperature. The mixture was stirred at 60 °C for 4 h under argon. After cooling the reaction mixture to room temperature, the catalyst and salts were removed by filtration through a short silica gel pad (EtOAc). gel permeation chromatography ($CHCl_3$) of the crude mixture afforded 47.6 mg (92 %) of the extended π-system 44 as pale-yellow solid.

Three-component synthesis of the merocyanine dyes 52 and 53 [60]

A thiazole 56 (0.05 mol) or a methylene base 57 (0.05 mol), a hydroxypyridone (60) (0.05 mol) and dimethylformamide (0.075 mol, 5.5 g) were heated at 90 °C in acetic anhydride (20–30 mL) for about 3 h. The solid that precipitated upon cooling to room temperature was filtered off, washed thoroughly with 2-propanol and/or aqueous ethanol until the color of the filtrate changed from violet to red and subsequently dried in a vacuum-drying cabinet at 50 °C. For physical characterization, the dyes were recrystallized from acetic anhydride, toluene or toluene–hexane mixtures.

One-pot, three-component synthesis of the push–pull chromophore 87e by a Sonogashira coupling–aminovinylation sequence [84]

To a stirred mixture of 208 mg (1.00 mmol) of 86, 14 mg (0.02 mmol) of $Pd(PPh_3)_2Cl_2$ and 7 mg (0.04 mmol) of CuI in a mixture of 5 mL of THF and 1 mL of triethylamine under nitrogen was added dropwise over 10 min a solution of 0.13 mL (1.10 mmol) of 85 in 5 mL of THF. The reaction mixture was stirred at room temperature for 6 h until the complete consumption of 86 (monitored by TLC or GC–MS). Then, a solution of 0.17 mL (2.00 mmol) of pyrrolidine in 5 mL of methanol was added and the mixture was heated at reflux for 6 h until the complete conversion of the intermediate alkyne (monitored by TLC or GC–MS). The solvents were evaporated *in vacuo* and the residue was chromatographed over a short pad of aluminum oxide, eluting with dichloromethane, to furnish after recrystallization from hexane–chloroform 200 mg (71 %) of the analytically pure enamine 87e as crystals with a blue metallic luster, m.p. 100–101 °C.

One-pot, four-component synthesis of the pyrrole 98b by a coupling–isomerization–Stetter–Paal–Knorr sequence [92]

A stirred mixture of 364 mg (2.00 mmol) of *p*-bromobenzonitrile, 278 mg (2.10 mmol) of 1-phenylpropargyl alcohol and 4 mL of Et_3N was degassed for 5 min. Then, 28 mg (0.04 mmol) of $Pd(PPh_3)_2Cl_2$ and 14 mg (0.08 mmol) of CuI were added and the mixture was heated at reflux for 14 h. After cooling to room temperature, 230 mg (2.40 mmol) of furfural, 57 mg (0.20 mmol) of 3,4-dimethyl-5-(2-hydroxyethyl)thiazolium iodide and Et_3N (0.5 mL) were added and the mixture was heated at reflux for 10 h. After cooling to room temperature, glacial acetic acid (2.5 mL) and 857 mg (8.00 mmol) of benzylamine were added and the mixture was heated at reflux for 120 h. After cooling to room temperature, a saturated aqueous solution of NH$_4$Cl was added. The aqueous phase was extracted several times with Et$_2$O or EtOAc and the combined organic layers were dried with anhydrous Na$_2$SO$_4$ and filtered. The remaining solution was concentrated *in vacuo* and the residue was purified by chromatography on silica gel and recrystallized from ethanol to give 444 mg (55%) of the pure pyrrole **98b** as colorless crystals, m.p. 104–105 °C.

List of Abbreviations

Ar	aryl or heteroaryl as substituents	LUMO	lowest unoccupied molecular orbital
dba	dibenzylideneacetone	MeOH	methanol
DFT	density functional theory	m.p.	melting-point
EWG	electron-withdrawing group	NLO	nonlinear optics
HPLC	high-performance liquid chromatography	TBAF	tetrabutylammonium fluoride
LC–MS	liquid chromatography– mass spectrometry	TFA	trifluoroacetic acid
		THF	tetrahydrofuran
LDA	lithium diisopropylamide	UV	ultraviolet

References

1. P. A. Wender, S. T. Handy, D. L. Wright, *Chem. Ind.* **1997**, 765, 767–769.
2. For a recent monograph, see e.g. J. Zhu, H. Bienaymé (Eds.), *Multi-component Reactions*, Wiley-VCH, Weinheim, **2005**.
3. For reviews, see e.g; (a) H. Bienaymé, C. Hulme, G. Oddon, P. Schmitt, *Chem. Eur. J.* **2000**, *6*, 3321–3329; (b) A. Dömling, I. Ugi, *Angew. Chem.* **2000**, *112*, 3300–3344; *Angew. Chem. Int. Ed.* **2000**, *39*, 3168–3210; (c) I. Ugi, A. Dömling, B. Werner, B. *J. Heterocycl. Chem.* **2000**, *37*, 647–658; (d) L. Weber, K. Illgen, M. Almstetter, *Synlett* **1999**, 366–374; (e) R. W. Armstrong, A. P. Combs, P. A. Tempest, S. D. Brown, T. A. Keating, *Acc. Chem. Res.* **1996**, *29*, 123–131; (f) I. Ugi, A. Dömling, W. Hörl, *Endeavour* **1994**, *18*, 115–122; (g) Posner, G. H. *Chem. Rev.* **1986**, *86*, 831–844.
4. For reviews on diversity-oriented syntheses, see e.g. (a) S. L. Schreiber and M. D. Burke, *Angew. Chem.* **2004**, *116*, 48–60; *Angew. Chem. Int. Ed.* **2004**, *43*, 46–58; (b) M. D. Burke, E. M. Berger, S. L. Schreiber, *Science* **2003**, *302*, 613–618; (c) P. Arya, D. T. H. Chou, M. G. Baek, *Angew. Chem.* **2001**, *113*, 351–358; *Angew. Chem. Int. Ed.* **2001**, *40*, 339–346; (d) B. Cox, J. C. Denyer, A. Binnie, M. C. Donnelly, B. Evans, D. V. S. Green, J. A. Lewis, T. H. Mander, A. T. Merritt, M. J. Valler, S. P. Watson, *Prog. Med. Chem.* **2000**, *37*, 83–133; (e) S. L. Schreiber, *Science* **2000**, *287*, 1964–1969.
5. For leading reviews on combinatorial chemistry, see e.g. (a) G. Jung (Ed.), *Combinatorial Chemistry – Synthesis, Analysis, Screening*, Wiley-VCH, Weinheim, **1999**; (b) F. Balkenhohl, C. von dem Bussche-Hünnefeld, A. Lansky, C. Zechel, *Angew. Chem.* **1996**, *108*, 2437–2488; *Angew. Chem. Int. Ed. Engl.* **1996**, *35*, 2288–2337.
6. S. Kobayashi, *Chem. Soc. Rev.* **1999**, *28*, 1–15.

7. For reviews and classifications of domino reactions, see e.g. (a) L. F. Tietze, *J. Heterocycl. Chem.* **1990**, *27*, 47–69; (b) L. F. Tietze, U. Beifuss, *Angew. Chem.* **1993**, *105*, 137–170; *Angew. Chem. Int. Ed. Engl.* **1993**, *32*, 131–163; (c) L. F. Tietze, *Chem. Rev.* **1996**, *96*, 115–136.
8. For a recent review on isonitrile-based MCR, see e.g. J. Zhu, *Eur. J. Org. Chem.* **2003**, 1133–1144.
9. For reviews on combinatorial search for inorganic materials, polymers and catalysts, see e.g. (a) S. Senkan, *Angew. Chem.* **2001**, *113*, 332–341; *Angew. Chem. Int. Ed.* **2001**, *40*, 312–329; (b) B. Jandeleit, D. J. Schaefer, T. S. Powers, H. W. Turner, W. H. Weinberg, *Angew. Chem.* **1999**, *111*, 2648–2689; *Angew. Chem. Int. Ed.* **1999**, *38*, 2494–2532.
10. For actual reviews on combinatorial syntheses of functional π-electron systems, see e.g. (a) P. Bäuerle, *Nachr. Chem.* **2004**, *52*, 19–24; (b) C. A. Briehn, P. Bäuerle, *Chem. Commun.* **2002**, 1015–1023.
11. J. M. Merrington, M. James, M. Bradley, *Chem. Commun.* **2002**, 140–141.
12. J. Fabian, H. Nakazumi, M. Matsuoka, *Chem. Rev.* **1992**, *92*, 1197–1226.
13. W. West, P. B. Gilman, in *Theory of the PhotographicProcess*, T. H. James (Ed.), Macmillan, New York, **1977**, p. 277.
14. L. G. Lee, C.-H. Chen, L. A. Chiu, *Cytometry* **1986**, *7*, 508–517.
15. P. Selvin, *Science* **1992**, *257*, 885–886.
16. Z. Diwu, J. W. Lown, *Pharmacol. Ther.* **1994**, *63*, 1–35.
17. J. Isacsson, G. Westman, *Tetrahedron Lett.* **2001**, *42*, 3207–3210.
18. (a) R. Haugland, *Handbook of Fluorescent Probes and Research Chemicals*, 8th edn., Molecular Probes, Eugene, OR, **2001**; (b) J. Bereiter-Hahn, K. H. Seipel, M. Voth, J. S. Ploem, *Cell Biochem. Funct.* **1983**, *1*, 147–155; (c) J. Bereiter-Hahn, *Biochim. Biophys. Acta* **1976**, *423*, 1–14; (d) H. W. Mewes, J. Rafael, *FEBS Lett.* **1981**, *131*, 7–10.
19. G. R. Rosania, J. W. Lee, L. Ding, H.-S. Yoon, Y.-T. Chang, *J. Am. Chem. Soc.* **2003**, *125*, 1130–1131.
20. (a) M. Kawakami, K. Koya, T. Ukai, N. Tatsuta, A. Ikegawa, K. Ogawa, T. Shishido, L. B. Chen, *J. Med. Chem.* **1997**, *40*, 3151–3160; (b) M. Kawakami, K. Koya, T. Ukai, N. Tatsuta, A. Ikegawa, K. Ogawa, T. Shishido, L. B. Chen, *J. Med. Chem.* **1998**, *41*, 130–142.
21. MKT-077, one of the rhodacyanine dyes, has been subjected to phase I clinical investigation for the treatment of solid tumors: (a) D. J. Propper, J. P. Braybrooke, D. J. Taylor, R. Lodi, P. Styles, J. A. Cramer, W. C. J. Collins, N. C. Levitt, D. C. Talbot, T. S. Ganesan, A. L. Harris, *Ann. Oncol.* **1999**, *10*, 923–927; (b) C. D. Britten, E. K. Rowinsky, S. D. Baker, G. R. Weiss, L. Smith, J. Stephenson, M. Rothenberg, L. Smetzer, J. Cramer, W. Collins, D. D. Von Hoff, S. A. Eckhardt, *Clin. Cancer Res.* **2000**, *6*, 42–49.
22. (a) R. A. Jeffreys, E. B. Knott, *J. Chem. Soc.* **1952**, 4632–4637; (b) E. B. Knott, R. A. Jeffreys, *J. Chem. Soc.* **1952**, 4762–4775.
23. A. E. van Dormael, *Ind. Chim. Belge* **1953**, *18*, 1297–1302.
24. K. Takasu, H. Inoue, H.-S. Kim, M. Suzuki, T. Shishido, Y. Wataya, M. Ihara, *J. Med. Chem.* **2002**, *45*, 995–998.
25. K. Takasu, H. Terauchi, H. Inoue, H.-S. Kim, Y. Wataya, M. Ihara, *J. Comb. Chem.* **2003**, *5*, 211–214.
26. M. Kawakami, K. Koya, T. Ukai, N. Tatsuta, A. Ikegawa, K. Ogawa, T. Shishido, L. B. Chen, *J. Med. Chem.* **1997**, *40*, 3151–3160; (b) M. Kawakami, K. Koya, T. Ukai, N. Tatsuta, A. Ikegawa, K. Ogawa, T. Shishido, L. B. Chen. *J. Med. Chem.* **1998**, *41*, 130–142.
27. F. Diederich, P. J. Stang (Eds.), *Metal-catalyzed Cross-coupling Reactions*, Wiley-VCH, Weinheim, **1998**.
28. K. Müllen, G. Wegner (Eds.), *Electronic Materials: the Oligomer Approach*, Wiley-VCH, Weinheim, **1998**.
29. For reviews on NLO materials, see e.g. (a) N. J. Long, *Angew. Chem.* **1995**, *107*, 6–20; *Angew. Chem. Int. Ed. Engl.* **1995**, *34*, 21–38; (b) S. R. Marder, J. W. Perry, *Adv. Mater.* **1993**, *5*, 804–815; (c) W. Nie, *Adv. Mater.* **1993**, *5*, 520–545; (d) D. R. Kanis, M. A. Ratner, T. J. Marks, *Chem. Rev.* **1994**, *94*, 195–242; (e) T. J. Marks, M. A. Ratner, *Angew. Chem.* **1995**, *107*, 167–187; *Angew. Chem. Int. Ed. Engl.* **1995**, *34*, 155–173; (f) J. J. Wolff, R. Wortmann, *Adv. Phys. Org. Chem.* **1999**, *32*, 121–217.

30. U. Scherf, K. Müllen, *Synthesis* **1992**, 23–38.
31. M.-S. Schiedel, C. A. Briehn, P. Bäuerle, *J. Organomet. Chem.* **2002**, *653*, 200–208.
32. (a) K. R. Romines, J. K. Morris, W. J. Howe, P. K. Tomich, M.-M. Horng, K.-T. Chong, R. R. Hinshaw, D. J. Anderson, J. W. Strohbach, S.-R. Turner, S. A. Miszak, *J. Med. Chem.* **1996**, *39*, 4125–4130; (b) R. O'Kennedy, R. D. Thornes (Eds.), *Coumarins – Biology, Applications and Mode of Action*, Wiley, Chichester, **1997**.
33. (a) R. Raue, in *Ullmannns Encyclopedia of Industrial Chemistry*, 5th edn., Vol. A15, B. Elvers, S. Hawkins, G. Schulz (Eds.), VCH, Weinheim, **1990**, pp. 155–157; (b) R. S. Koefod, K. R. Mann, *Inorg. Chem.* **1989**, *28*, 2285–2290
34. (a) P. D. Edwards, R. C. Mauger, K. M. Cottrell, F. X. Morris, K. K. Pine, M. A. Sylvester, C. W. Scott, S. T. Furlong, *Bioorg. Med. Chem. Lett.* **2000**, *10*, 2291–2294; (b) M. Adamczyk, M. Cornwell, J. Huff, S. Rege, T. V. S. Rao, *Bioorg. Med. Chem. Lett.* **1997**, *7*, 1985–1988; (c) C. A. M. Seidel, A. Schulz, M. H. M. Sauer, *J. Phys. Chem.* **1996**, *100*, 5541–5553; (d) A. Adronov, S. L. Gilat, J. M. Fréchet, K. Ohta, F. V. R. Neuwahl, G. R. Fleming, *J. Am. Chem. Soc.* **2000**, *122*, 1175–1185; (e) K. H. Shaughnessy, P. Kim, J. F. Hartwig, *J. Am. Chem. Soc.* **1999**, *121*, 2123–2132.
35. (a) J. Kido, Y. Lizumi, *Appl. Phys. Lett.* **1998**, *73*, 2721–2723; (b) A. Niko, S. Tasch, F. Meghdadi, C. Brandstätter, G. Leising, *J. Appl. Phys.* **1997**, *82*, 4177–4182; (c) S. Tasch, C. Brandstätter, F. Meghdadi, G. Leising, G. Froyer, L. Athouel, *Adv. Mater.* **1997**, *9*, 33–36.
36. A. E. Siegrist, H. Hefti, H. R. Meyer, E. Schmidt, *Rev. Prog. Coloration* **1987**, *17*, 39–55.
37. (a) B. M. Krasovitskii, B. M. Bolotin (Eds.), *Organic Luminescent Materials*, Wiley-VCH,Weinheim, **1988**; (b) R. M. Christie, *Rev. Prog. Coloration* **1993**, *23*, 1–18; (c) N. A. Kuznetsova, O. L. Kaliya, *Russ. Chem. Rev.* **1992**, *61*, 683–696; (d) O. A. Ponmarev, E. R. Vasina, V. G. Mitina, A. A. Sukhorukov, *Russ. J. Phys. Chem.* **1990**, *64*, 518–521.
38. M.-S. Schiedel, C. A. Briehn, P. Bäuerle, *Angew. Chem.* **2001**, *113*, 4813–4816; *Angew. Chem. Int. Ed.* **2001**, *40*, 4677–4680.
39. C. A. Briehn, M.-S. Schiedel, E. M. Bonsen, W. Schuhmann, P. Bäuerle, *Angew. Chem.* **2001**, *113*, 4817–4820; *Angew. Chem. Int. Ed.* **2001**, *40*, 4680–4683.
40. (a) J. P. Ferraris, M. M. Eissa, I. D. Brotherston, D. C. Loveday, *Chem. Mater.* **1998**, *10*, 3528–3535; (b) R. D. McCullough, *Adv. Mater.* **1998**, *10*, 93–116.
41. H. Sirringhaus, J. P. Brown, R. H. Friend, M. M. Nielsen, K. Beechgard, B. M. W. Langeveld-Voss, A. J. H. Spiering, R. A. J. Janssen, E. W. Meijer, P. Herwig, D. M. de Leeuw, *Nature* **1999**, *401*, 685–688.
42. C. A. Briehn, T. Kirschbaum, P. Bäuerle, *J. Org. Chem.* **2000**, *65*, 352–359.
43. (a) J. S. Lewis, M. S. Weaver, *IEEE J. Sel. Top.s Quant. Electron.* **2004**, *10*, 45–57; (b) D. Metzdorf, E. Becker, T. Dobbertin, S. Hartmann, D. Heithecker, H.-H. Johannes, A. Kammoun, H. Krautwald, T. Riedl, C. Schildknecht, D. Schneider, W. Kowalsky, *Mater. Res. Soc. Symp. Proc.* **2003**, *769*, 113–124; (c) P. Gomez-Romero, *Adv. Mater.* **2001**, *13*, 163–174; (d) J. R. Sheats, *Science* **1997**, *277*, 191–192.
44. (a) A. Kraft, A. C. Grimsdale, A. B. Holmes, *Angew. Chem.* **1998**, *110*, 416–443; *Angew. Chem. Int. Ed.* **1998**, *37*, 403–428; (b) J. H. Burroughes, D. C. C. Bradley, A. R. Brown, R. N. Marks, K. Mackay, R. H. Friend, P. L. Burns, A. B. Holmes, *Nature* **1990**, *347*, 539–541; (c) G. Gustafsson, Y. Cao, G. M. Treacy, F. Klavetter, N. Colaneri, A. J. Heeger, *Nature* **1992**, *357*, 477–479.
45. (a) U. H. F. Bunz, *Chem. Rev.* **2000**, *100*, 1605–1644; (b) R. Giesa, *J. Macromol. Sci., Rev. Macromol. Chem. Phys.* **1996**, *C36*, 631–670.
46. O. Lavastre, I. Illitchev, G. Jegou, P. H. Dixneuf, *J. Am. Chem. Soc.* **2002**, *124*, 5278–5279.
47. (a) T. Mangel, A. Eberhardt, U. Scherf, U. H. F. Bunz, K. Müllen, *Macromol. Rapid Commun.* **1995**, *16*, 571–580; (b) H. Li, D. R. Powell, R. K. Hayashi, R. West, *Macromolecules* **1998**, *31*, 52–58;

(c) A. P. Davey, S. Elliott, O. O'Connor, W. Blau, *J. Chem. Soc., Chem. Commun.* **1995**, 1433–1434.

48. (a) J. R. Sheats, P. F. Barbara, *Acc. Chem. Res.* **1999**, *32*, 191–192; (b) R. H. Friend, R. W. Gymer, A. B. Holmes, J. H. Burroughes, R. N. Marks, C. Taliani, D. C. C. Bradley, D. A. Dos Santos, J. L. Brédas, M. Lögdlund, W. R. Salaneck, *Nature* **1999**, *397*, 121–128; (c) R. E. Martin, F. Diederich, *Angew. Chem.* **1999**, *111*, 1440–1469; *Angew. Chem. Int. Ed.* **1999**, *38*, 1351–1377; (d) J. L. Segura, N. Martín, *J. Mater. Chem.* **2000**, *10*, 2403–2435; (e) H. Meier, *Angew. Chem.* **1992**, *104*, 1425–1576; *Angew. Chem. Int. Ed. Engl.* **1992**, *31*, 1437–1456.

49. K. Itami, Y. Ohashi, J.-I. Yoshida, *J. Org. Chem.* **2005**, *70*, 2778–2792.

50. For reviews, see (a) I. P. Beletskaya, A. V. Cheprakov, *Chem. Rev.* **2000**, *100*, 3009–3066; (b) W. Cabri, I. Candiani, *Acc. Chem. Res.* **1995**, *28*, 2–7; (c) A. de Meijere, F. E. Meyer, *Angew. Chem.* **1994**, *106*, 2473–2506; *Angew. Chem. Int. Ed. Engl.* **1994**, *33*, 2379–2411; (d) R. F. Heck, in *Comprehensive Organic Synthesis*, B. M.Trost (Ed.), Pergamon Press, New York, **1991**, Vol. 4, Chapter 4.3.

51. (a) T. Kamei, K. Itami, J. Yoshida, *Adv. Synth. Catal.* **2004**, *346*, 1824–1835; (b) K. Itami, Y. Ushiogi, T. Nokami, Y. Ohashi, J. Yoshida, *Org. Lett.* **2004**, *6*, 3695–3698; (c) K. Itami, M. Mineno, N. Muraoka, J. Yoshida, *J. Am. Chem. Soc.* **2004**, *126*, 11778–11179; (d) K. Itami, T. Kamei, J. Yoshida, *J. Am. Chem. Soc.* **2003**, *125*, 14670–14671; (e) K. Itami, T. Nokami, Y. Ishimura, K. Mitsudo, T. Kamei, J. Yoshida, *J. Am. Chem. Soc.* **2001**, *123*, 11577–11585; (f) K. Itami, T. Nokami, J. Yoshida, *J. Am. Chem. Soc.* **2001**, *123*, 5600–5601; (g) K. Itami, K. Mitsudo, T. Kamei, T. Koike, T. Nokami, J. Yoshida, *J. Am. Chem. Soc.* **2000**, *122*, 12013–12014.

52. For reviews, see e.g. (a) Y. Hatanaka, T. Hiyama, *Synlett* **1991**, 845–853; (b) T. Hiyama, in *Metal-catalyzed Cross-coupling Reactions*, F. Diederich, P. J. Stang (Eds.), Wiley-VCH, Weinheim, **1998**, Chapter 10; (c) T. Hiyama, E. Shirakawa, *Top. Curr. Chem.* **2002**, *219*, 61–85; (d) S. E. Denmark, R. F. Sweis, *Chem. Pharm. Bull.* **2002**, *50*, 1531–1541.

53. (a) J. Slinker, D. Bernards, P. L. Houston, H. D. Abruña, S. Bernhard, G. G. Malliaras, *Chem. Commun.* **2003**, 2392–2399; (b) F. G. Gao, A. J. Bard, *J. Am. Chem. Soc.* **2000**, *122*, 7426–7427; (c) H. Rudmann, S. Shimada, M. F. Rubner, *J. Am. Chem. Soc.* **2002**, *124*, 4918–4921; (d) M. Buda, G. Kalyuzhny, A. J. Bard, *J. Am. Chem. Soc.* **2002**, *124*, 6090–6098; (e) S. Bernhard, J. A. Barron, P. L. Houston, H. D. Abruña, J. L. Ruglovsky, X. Gao, G. G. Malliaras, *J. Am. Chem. Soc.* **2002**, *124*, 13624–13628; (f) S. Lamansky, P. Djurovich, D. Murphy, F. Abdel-Razzaq, R. Kwong, I. Tsyba, M. Bortz, B. Mui, R. Bau, M. E. Thompson, *Inorg. Chem.* **2001**, *40*, 1704–1711.

54. (a) S. Lamansky, P. Djurovich, D. Murphy, F. Abdel-Razzaq, H. E. Lee, C. Adachi, P. Burrows, S. R. Forrest, M. E. Thompson, *J. Am. Chem. Soc.* **2001**, *123*, 4304–4312; (b) R. Pohl, V. A. Montes, J. Shinar, J. Pavel Anzenbacher, *J. Org. Chem.* **2004**, *69*, 1723–1725; (c) A. Tsuboyama, H. Iwawaki, M. Furugori, T. Mukaide, J. Kamatani, S. Igawa, T. Moriyama, S. Miura, T. Takiguchi, S. Okada, M. Hoshino, K. Ueno, *J. Am. Chem. Soc.* **2003**, *125*, 12971–12979.

55. S. Bernhard, X. Gao, G. G. Malliaras, H. D. Abruña, *Adv. Mater.* **2002**, *14*, 433–436.

56. P. Reveco, R. H. Schmehl, W. R. Cherry, F. R. Fronczek, J. Selbin, *Inorg. Chem.* **1985**, *24*, 4078–4082.

57. (a) X. Gong, P. K. Ng, W. K. Chan, *Adv. Mater.* **1998**, *10*, 1337–1340; (b) P. Spellane, R. J. Watts, A. Vogler, *Inorg. Chem.* **1993**, *32*, 5633–5636; (c) J. D. Slinker, A. A. Gorodetsky, M. S. Lowry, J. Wang, S. Parker, R. Rohl, S. Bernhard, G. G. Malliaras, *J. Am. Chem. Soc.* **2004**, *126*, 2763–2767.

58. M. S. Lowry, W. R. Hudson, R. A. Jr. Pascal, S. Bernhard, *J. Am. Chem. Soc.* **2004**, *126*, 14129–14135.

59. (a) M. Maggini, G. Scorrano, M. Prato, *J. Am. Chem. Soc.* **1993**, *115*, 9798–9788; (b) T. Da Rosa, M. Prato, F. Novello, M. Maggini, E. Banfi, *J. Org. Chem.* **1996**, *61*, 9070–9072; (c) M. Prato, M. Maggini, *Acc. Chem. Res.* **1998**, *31*, 519–526.

60. F. Würthner, R. Sens, K.-H. Etzbach, G. Seybold, *Angew. Chem.* **1999**, *111*, 1753–1757; *Angew. Chem. Int. Ed.* **1999**, *38*, 1649–1652.
61. (a) F. Würthner, R. Wortmann, R. Matschiner, K. Lukaszuk, K. Meerholz, Y. DeNardin, R. Bittner, C. Bräuchle, R. Sens, *Angew. Chem.* **1997**, *109*, 2933–2936; *Angew. Chem. Int. Ed. Engl.* **1997**, *36*, 2765–2768; (b) R. Sens, K. H. Etzbach, V. Bach, *Ger. Pat. Appl.* 4344116 (1993); *Chem. Abstr.* **1995**, *123*, 289602.
62. K. Meerholz, *Angew. Chem.* **1997**, *109*, 981–984; *Angew. Chem. Int. Ed. Engl.* **1997**, *36*, 945–948.
63. F. Würthner, *Synthesis* **1999**, 2103–2113.
64. (a) F. Dumur, N. Gautier, N. Gallego-Planas, Y. Sahin, E. Levillain, N. Mercier, P. Hudhomme, *J. Org. Chem.* **2004**, *69*, 2164–2177; (b) S. Depaemelaere, F. C. De Schryver, J. W. Verhoeven, *J. Phys. Chem. A* **1998**, *102*, 2109–2116; (c) Y. Xiao, X. H. Qian, *Tetrahedron Lett.* **2003**, *44*, 2087–2091.
65. (a) R. M. Metzger, C. Panetta, *New J. Chem.* **1991**, *15*, 209–221; (b) J. P. Lannay, *NATO ASI Ser., Ser. C*, **1991**, *343*, 321–328; (c) M. C. Petty, M. R. Bryce, D. Bloor (Eds.), *Introduction to Molecular Electronics*, Oxford University Press, New York, **1995**.
66. (a) J. Sepiol, P. Milart, *Tetrahedron* **1985**, *41*, 5261–5265; (b) J. Griffiths, M. Lockwood, B. Roozpeikar, *J. Chem. Soc., Perkin Trans. 2* **1977**, 1608–1610.
67. S.-L. Cui, X.-F. Lin, Y.-G. Wang, *J. Org. Chem.* **2005**, *70*, 2866–2869.
68. (a) A. Treibs, K. Jacob, *Angew. Chem.* **1965**, *77*, 680–681; *Angew. Chem. Int. Ed. Engl.* **1965**, *4*, 694–695; (b) for a review, see A. H. Schmidt, *Synthesis* **1980**, 961–994.
69. For reviews, see e.g. (a) S. Das, K. G. Thomas, M. V. George, *Mol. Supramol. Photochem.* **1997**, *1*, 467–517; (b) K.-Y. Law, *Mol. Supramol. Photochem.* **1997**, *1*, 519–584; (c) K.-Y. Law, *Chem. Rev.* **1993**, *93*, 449–486.
70. E. Terpetschnig, H. Szmacinski, A. Ozinskas, J. R. Lakowicz, *Anal. Biochem.* **1994**, *217*, 197–204.
71. K.-Y. Law, *J. Phys. Chem.* **1988**, *92*, 4226–4231.
72. Brightness is defined as product of molar absorptivity coefficient ε and fluorescense quantum yield Φ_f; see A. Minta, J. P. Y. Kao, R. Y. Tsien, *J. Biol. Chem.* **1989**, *264*, 8171–8178.
73. "Exciton interaction", commonly used in squaraine literature, describes the effect of electronic interactions between squaraine pairs, although it is strictly speaking a "pseudo-excimeric" interaction.
74. K. Liang, M. S. Farahat, J. Perlstein, K.-Y. Law, D. G. Whitten, *J. Am. Chem. Soc.* **1997**, *119*, 830–831.
75. (a) S. Das, K. G. Thomas, K. J. Thomas, P. V. Kamat, M. V. George, *J. Phys. Chem.* **1994**, *98*, 9291–9296; (b) U. Oguz, E. U. Akkaya, *Tetrahedron Lett.* **1997**, *38*, 4509–4512; (c) G. Dilek, E. U. Akkaya, *Tetrahedron Lett.* **2000**, *41*, 3721–3724; (d) A. Ajayaghosh, E. Arunkumar, J. Daub, *Angew. Chem.* **2002**, *114*, 1844–1847; *Angew. Chem. Int. Ed.* **2002**, *41*, 1766–1769; (e) C. R. Chenthamarakshan, A. Ajayaghosh, *Tetrahedron Lett.* **1998**, *39*, 1795–1798.
76. M. A. B. Block, S. Hecht, *Macromolecules* **2004**, *37*, 4761–4769.
77. For recent reviews, see e.g. (a) A. Ajamian, J. L. Gleason, *Angew. Chem.* **2004**, *116*, 3842–3848; *Angew. Chem. Int. Ed.* **2004**, *43*, 1766–1769; (b) J. M. Lee, Y. Na, H. Han, S. Chang, *Chem. Soc. Rev.* **2004**, *33*, 302–312.
78. For recent reviews on transition metal assisted sequential transformations, see e.g. (a) G. Balme, E. Bossharth, N. Monteiro, *Eur. J. Org. Chem.* **2003**, 4101–4111; (b) G. Battistuzzi, S. Cacchi, G. Fabrizi *Eur. J. Org. Chem.* **2002**, 2671–2681.
79. K. Itami, K. Tonogaki, Y. Ohashi, J. Yoshida, *Org. Lett.* **2004**, *6*, 4093–4096.
80. For lead reviews on Sonogashira couplings, see e.g. (a) S. Takahashi, Y. Kuroyama, K. Sonogashira, N. Hagihara, *Synthesis* **1980**, 627–630; (b) K. Sonogashira, in *Metal-catalyzed Cross-coupling Reactions*, F. Diederich, P. J. Stang (Eds.), Wiley-VCH, Weinheim, **1998**, 203–229; (c) K. Sonogashira, *J. Organomet. Chem.* **2002**, *653*, 46–49; (d) E.-I. Negishi, L. Anastasia, *Chem. Rev.* **2003**, *103*, 1979–2018.
81. I.-Y. Wu, J. T. Lin, C.-S. Li, W.-C. Wang, T. H. Huang, Y. S. Wen, T. Chow, C.

Tsai, *Tetrahedron* **1999**, *55*, 13973–13982.

82. T. J. J. Müller, J. P. Robert, E. Schmälzlin, C. Bräuchle, K. Meerholz, *Org. Lett.* **2000**, *2*, 2419–2422.

83. K. Meerholz, B. Kippelen, N. Peyghambarian, in *Electrical and Optical Polymer Systems*, D. L. Wise, G. E. Wnek, D. J. Trantolo, J. D. Gresser, T. M. Cooper (Eds.), World Scientific, Singapore, **1998**, pp. 571–631.

84. A. S. Karpov, F. Rominger, T. J. J. Müller, *J. Org. Chem.* **2003**, *68*, 1503–1511.

85. Y. Tohda, K. Sonogashira, N. Hagihara, *Synthesis* **1977**, 777–778.

86. (a) A. S. Karpov, T. J. J. Müller, *Org. Lett.* **2003**, *5*, 3451–3454; (b) A. S. Karpov, T. J. J. Müller, *Synthesis* **2003**, 2815–2826; (c) A. S. Karpov, T. Oeser, T. J. J. Müller, *Chem. Commun.* **2004**, 1502–1503; (d) A. S. Karpov, E. Merkul, T. Oeser, T. J. J. Müller, *Chem. Commun.* **2005**, 2581–2583.

87. (a) R. Huisgen, *Angew. Chem.* **1963**, *75*, 604–681; *Angew. Chem. Int. Ed.* **1963**, *2*, 565–632; (b) A. Padwa, *1,3-Dipolar Cycloaddition Chemistry*, Wiley, New York, **1984**.

88. (a) A. Vlahovici, M. Andrei, I. Druta, *J. Lumin.* **2002**, *96*, 279–285; (b) A. Vlahovici, I. Druta, M. Andrei, M. Cotlet, R. Dinica, *J. Lumin.* **1999**, *82*, 155–162; (c) H. Sonnenschein, G. Hennrich, U. Resch-Genger, B. Schulz, *Dyes Pigments* **2000**, *46*, 23–27.

89. A. V. Rotaru, I. D. Druta, T. Oeser, T. J. J. Müller, *Helv. Chim. Acta* **2005**, *88*, 1813–1825.

90. T. J. J. Müller, M. Ansorge, D. Aktah, *Angew. Chem.* **2000**, *112*, 1323–1326; *Angew. Chem. Int. Ed.* **2000**, *39*, 1253–1256.

91. Pyrimidines: (a) T. J. J. Müller, R. Braun, M. Ansorge, *Org. Lett.* **2000**, *2*, 1967–1970. Dihydrobenzo[*b*][1, 4]thiazepines: (b) R. U. Braun, T. J. J. Müller, *Tetrahedron* **2004**, *60*, 9463–9469; (c) R. U. Braun, K. Zeitler, T. J. J. Müller, *Org. Lett.* **2000**, *2*, 4181–4184. Substituted and annelated pyridines: (d) N. A. M. Yehia, K. Polborn, T. J. J. Müller, *Tetrahedron Lett.* **2002**, *43*, 6907–6910; (e) O. G. Dediu, N. A. M. Yehia, T. Oeser, K. Polborn, T. J. J. Müller, *Eur. J. Org. Chem.* **2005**, 1834–1858. Pyrroles: (f) R. U. Braun, K. Zeitler, T. J. J. Müller, *Org. Lett.* **2001**, *3*, 3297–3300.

92. R. U. Braun, T. J. J. Müller, *Synthesis* **2004**, 2391–2406.

6
High-yield Synthesis of Shape-persistent Phenylene–Ethynylene Macrocycles*

Sigurd Höger

*A List of Abbreviations can be found at the end of this chapter.

6.1
Introduction

The synthesis and investigation of shape-persistent macrocycles have attracted continuously increasing interest during the past several years [1]. The term shape-persistence in this context means that the average diameter of the compounds is equal to the contour length of the molecular backbone divided by π [1b]. This definition applies to a time or ensemble average and does not mean that the macrocycles are absolutely stiff or flat at all times. Indeed, gas-phase calculations show that the planar conformation of shape-persistent rings is less stable than slightly boat- or chair-like conformations [2]. Although a direct correlation between the conformation of the molecules in the gas phase and their structure in the solid state is questionable, single-crystal X-ray structures seem to confirm these data [3]. However, in contrast to flexible macrocycles, such as crown ethers, shape-persistent macrocycles do not collapse, regardless of the presence of a guest molecule in their interior. Scattering experiments in solution support this hypothesis [4], although the actual ring diameter is also in this case slightly influenced by the presence or absence of a guest molecule [5]. To prevent a collapse of structures with large lumens in the nanometer regime, the backbone of the structures must be made of rigid components, usually segments of macromolecules with a high persistence length [6]. Among these structures, rings containing phenylene–ethynylene, phenylene–butadiynylene or both moieties (phenylacetylene macrocycles, PAMs) are valuable candidates for the investigation of the supramolecular chemistry of the compounds for several reasons. First, the synthesis of the macrocycles is fairly easy and straightforward because palladium-catalyzed aryl–acetylene bond-formation protocols (Sonogashira–Hagihara coupling) show a high functional group tolerance and acetylene protective groups of different stability and polarity are well described [7, 8]. Second, the spacers between the aromatic rings prevent steric constraints between groups that are neces-

Functional Organic Materials. Syntheses, Strategies, and Applications.
Edited by Thomas J.J. Müller and Uwe H.F. Bunz
Copyright © 2007 WILEY-VCH Verlag GmbH & Co. KGaA, Weinheim
ISBN: 978-3-527-31302-0

Fig. 6.1 Orientations of side-groups relative to the molecular backbone of the macrocycle. From left to right: orthogonal, extra-annular, intra-annular or adaptable (i.e. the side-group orientation can vary depending on an external parameter).

sarily attached to the stiff backbone in order to ensure the tractability of the macrocycles. The orientation of the side-groups relative to the chain-stiff backbone either can be fixed (orthogonal, extra-annular or intra-annular [9]) or can vary according to an external parameter (adaptable [10]) (Fig. 6.1).

The actual shape and size of the molecules depend on the substitution pattern of the arylene units. Although pure para-substituted cyclic phenylene–arylenes have been investigated, e.g. **1**, most PAMs contain meta- or ortho-substituted aromatics at their corners and para-substituted aromatics for size expansion, respectively (**2**, Fig. 6.2) [11]. The position where the side-groups are attached to the backbone determines if they point to the inside (intra-annular substituents, I, I′), to the outside (extra-annular substituents, E, E′) or if they can change their orientation according to an external parameter (adaptable substituents, A, A′).

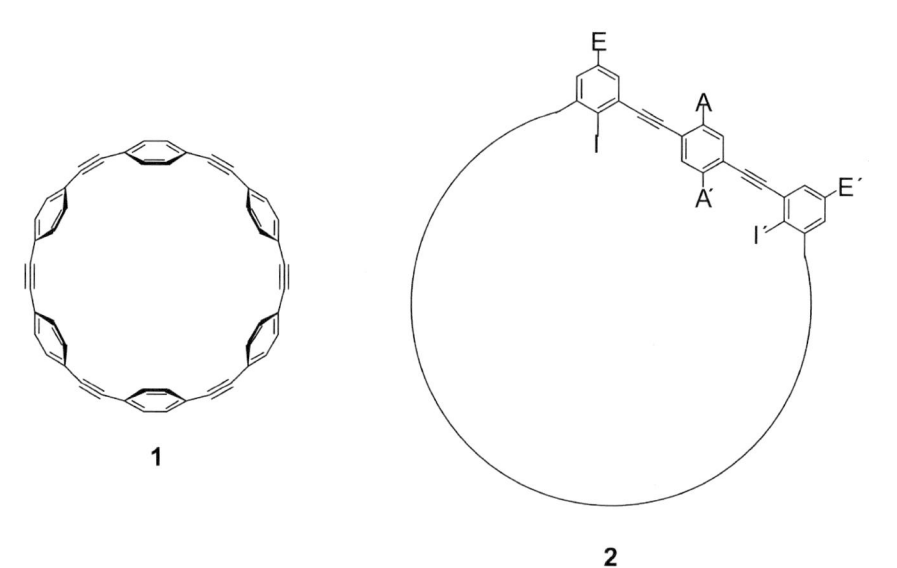

Fig. 6.2 Phenylacetylene macrocycle **1** (PAM) with pure para-substituted phenylenes and macrocycle **2** containing both para- and meta-substituted phenylenes. Additionally, the backbone may contain biphenyl (triphenyl) and/or butadiynylene (hexatriynylene) units.

6.2
Synthesis

6.2.1
General

Shape-persistent macrocycles are composed of an exact number of building blocks (e.g. phenylacetylene units), which may be differently functionalized. In principle, two approaches can be distinguished for the synthesis of the macrocycles: the kinetic and the thermodynamic approaches. In the kinetic approach, the bond formation is irreversible and often rings of different sizes are formed in the cyclization step, depending on the kinetics of the cyclization and competing oligomerization reaction. Since the bond-forming reactions are irreversible, the system does not have the ability to correct undesired bond formations. In the thermodynamic approach, however, the bond formation is reversible so that the system can undergo a self-healing process. As a result, a specific reaction product might be formed exclusively, if it is the most stable compound under the given conditions.

The kinetic approach towards PAMs is carried out mostly and has been used for the synthesis of numerous shape-persistent macrocycles. This approach will be discussed first. The thermodynamic approach has only recently been used for the construction of PAMs and will be discussed afterwards. The intention of this chapter is not a full coverage of all reported synthetic procedures towards shape-persistent macrocycles with nanometer-scale interiors. Rather, typical examples will point up the different existing procedures, including their advantages and disadvantages.

6.2.2
The Kinetic Approach

6.2.2.1 Statistical Reactions

In the kinetic approach, the building blocks of the macrocycle are connected to a linear oligomer that subsequently has to undergo an intramolecular bond formation in order to produce the cyclic compound. There are in principle two ways to perform this: either the oligomer is formed independently and then cyclized in a separate reaction vessel or oligomer formation and cyclization are performed in a one-pot reaction. Both approaches are described for a variety of shape-persistent macrocycles with different backbone structures as outlined in the examples below.

Staab and Neunhoeffer reported 1974 the synthesis of the *m*-cyclophane **4** by the intermolecular, sixfold Stephens–Castro coupling of the copper salt of *m*-iodophenylacetylene [12]. The advantage of this approach is the availability of the starting material **3** in large amounts, thus allowing the synthesis of **4** on an 800-mg scale, despite the fact that the yield in this reaction was only 4.6% (Scheme 6.1). Although nearly gram amounts of the macrocycle are available by this route, only the fully symmetrical (non-substituted) macrocycle was prepared. In 2000, Bunz

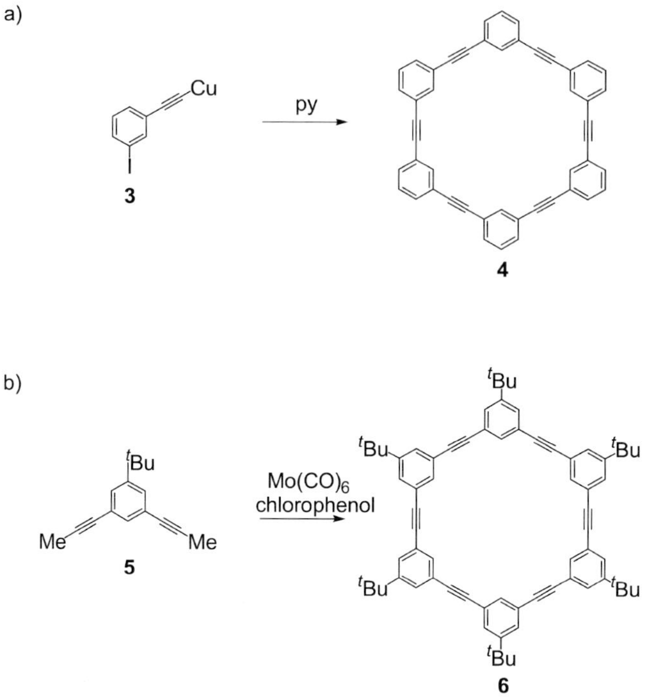

Scheme 6.1 (a) Staab's and (b) Bunz's syntheses of the cyclohexameric phenylacetylenes **4** and **6**.

and coworkers showed that alkyne metathesis is an alternative method for the preparation of a hexameric phenylacetylene macrocycle: Cyclooligomerization of **5** with an instant catalyst prepared from $Mo(CO)_6$ and 4-chlorophenol afforded **6** in 6% yield [13]. Also in this case, only the fully symmetrical product could be obtained.

A much higher yield in the synthesis of this compound was reported by Moore and Zhang in 1992 [14]. Instead of performing the formation of the ring precursor (and other oligomers) and the cyclization in a one-pot reaction, the α-iodo-ω-ethynyl precursor **7** was prepared separately and then in the final step cyclized under high-dilution conditions to give **8** in excellent 75% yield (Scheme 6.2). Similar results were obtained when larger macrocycles were prepared by this route. From precursor **9** the PAM **10** is accessible in 70% isolated yield.

Apart from the high yield in this step (along with a fairly easy product purification), this is the method of choice when macrocycles with different functional groups at the building blocks need to be prepared since the precursor is built up stepwise. This possibility of arranging different substituents at defined positions within the macrocycle is a fundamental requirement for detailed investiga-

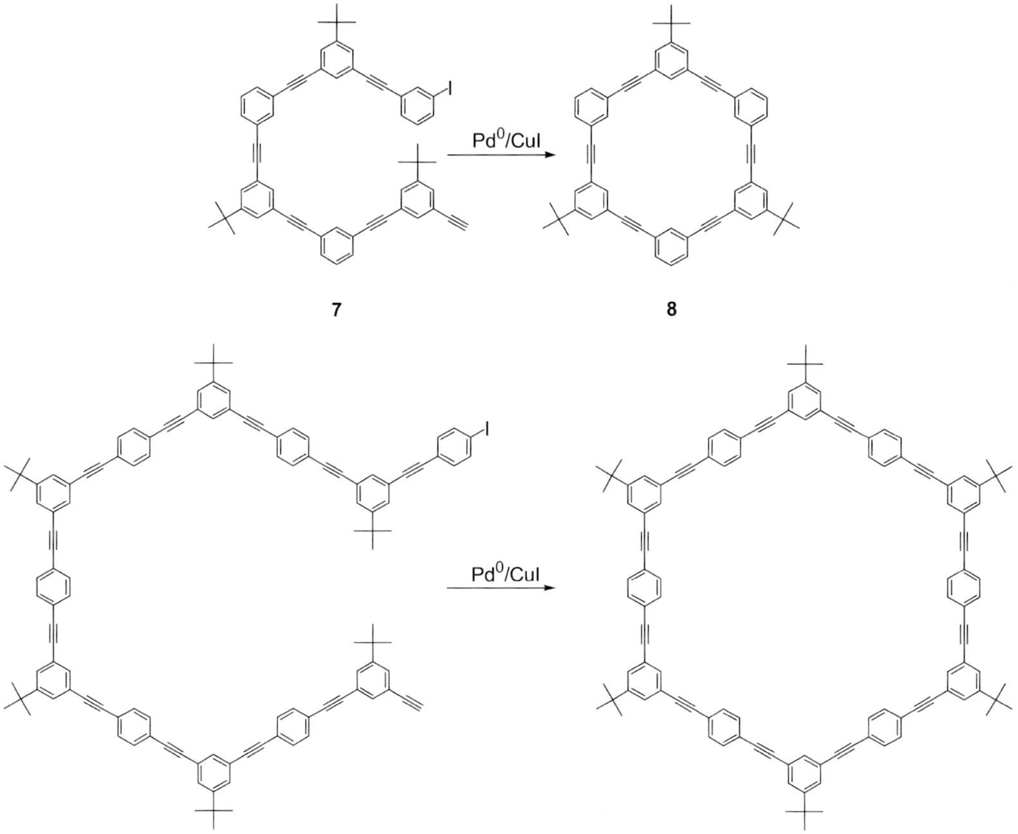

Scheme 6.2 Moore and Zhang's synthesis of the PAMs **8** and **10**.

tions of structure–property relationships in these materials. For example, rings with an alternating sequence of electron-donor and electron-acceptor units (e.g. **11**) were synthesized and compared with the isomeric macrocycle **12** in which the electronically different substituents are concentrated at opposite sides of the ring (Fig. 6.3a) [15]. This approach allows also the synthesis of cyclohexamers with a different orientation of the side-groups (**13**). Furthermore, cyclic pentamers (**14**), heptamers (**15**) and other defined oligomers are available in high yields, compounds that Staab and Neunhoeffer did not describe (Fig. 6.3b) [16]

However, the repctitive precursor synthesis is also responsible for the major disadvantage of this route: it requires numerous deprotecting/coupling reactions, as exemplified in Scheme 6.3. The key finding for the successful synthesis of **9** was that trimethylsilyl(TMS)acetylenes and 1-aryl-3,3-diethyltriazenes can act as

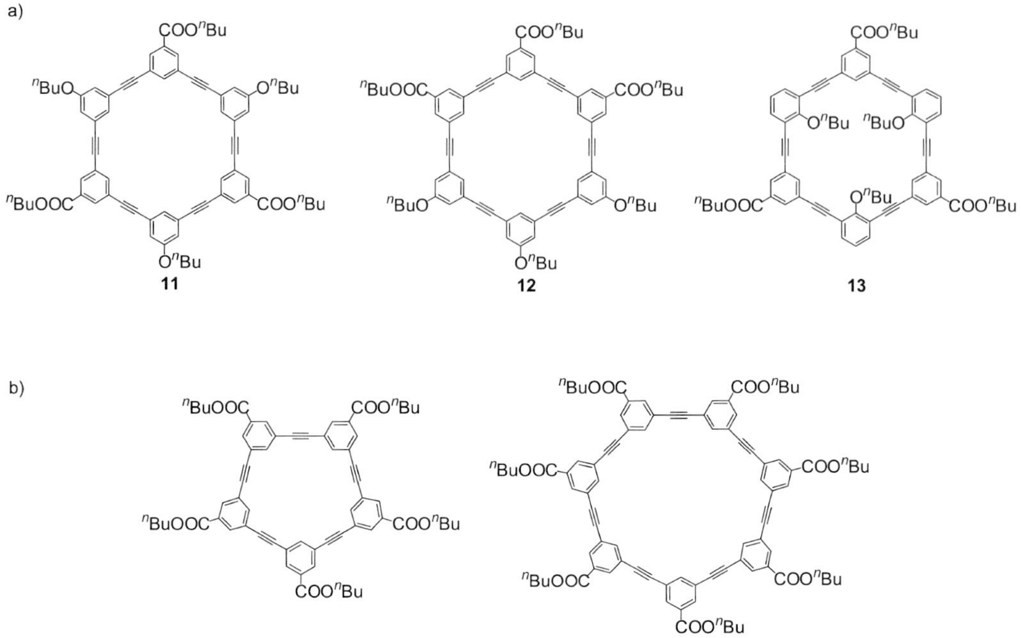

Fig. 6.3 (a) Isomeric PAMs with different arrangements of the functional groups; (b) cyclic pentamers and cyclic heptamers are also available in good yields (68 and 71 %) from the corresponding linear precursors.

complementary protective groups for terminal arylacetylenes and aryl iodides, respectively [17]. Moreover, both protecting groups are stable under the conditions of the Sonogashira–Hagihara coupling, the C–C bond-forming reaction for the precursor (and macrocycle) synthesis. Reaction of the bromide **16** with the acetylene **17** afforded the coupling product **18** in 86 % yield. Treatment of **18** with K_2CO_3 gave the terminal acetylene in nearly quantitative yield (98 %) and treatment of **18** with MeI formed the iodide **20** also in high yield of over 90 % (94 %). All subsequent coupling and deprotection reactions gave similar yields so that a variety of different macrocycles could be prepared, although each precursor synthesis required a series of steps.

Similar different approaches have also been described for phenylene, phenylene–butadiynylene- and other shape-persistent rings. Staab and Binning in 1967 prepared cyclic hexaphenylene **30** by an oxidative Grignard–coupling reaction using $CuCl_2$ and made the same observations as described before: The simple *m*-phenylene precursor **27** gave the cyclic product **30** in only 1.1 % yield whereas the same macrocycle was formed in 45 % yield when the reaction was performed with the hexaphenylene precursor **28**. The intermediate-sized bipheny-

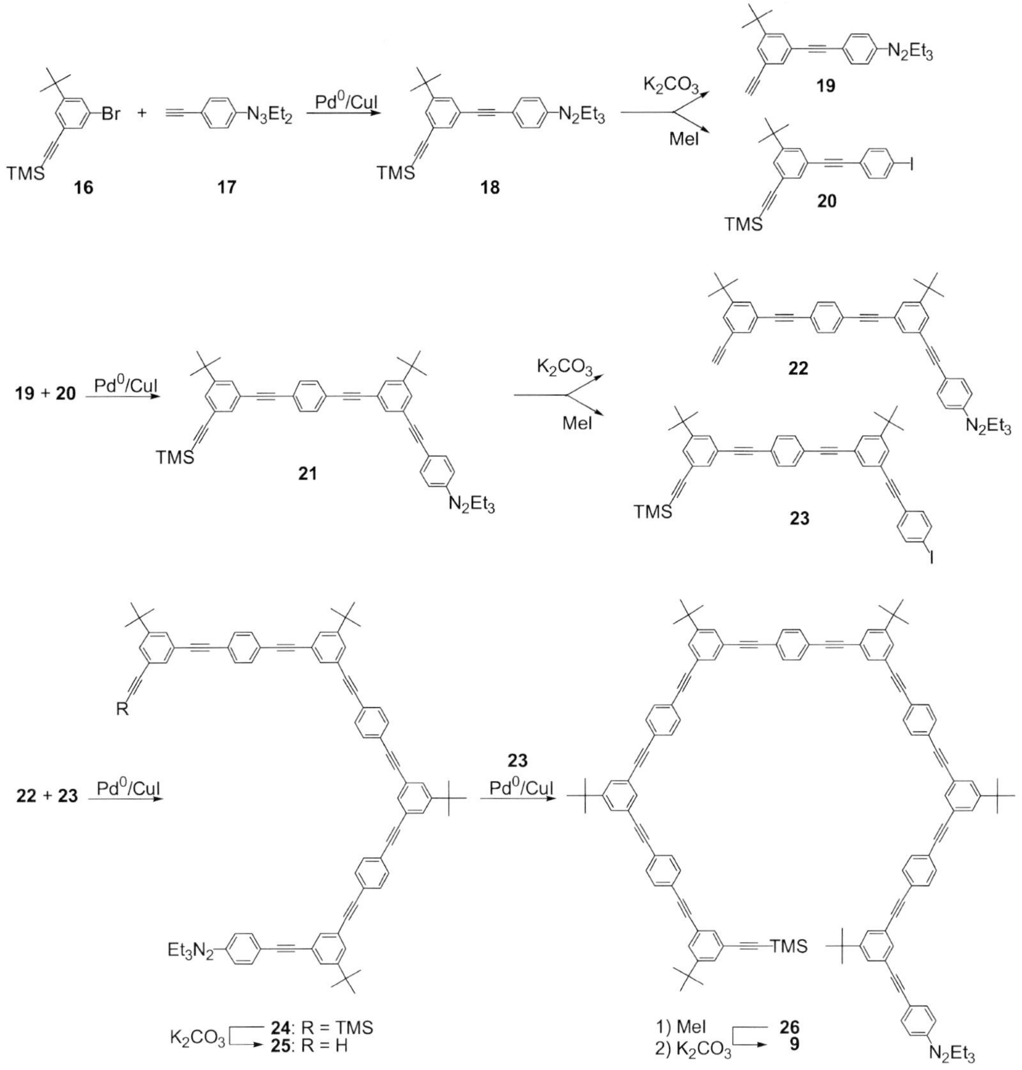

Scheme 6.3 Moore and coworkers' repetitive synthesis of linear phenylacetylene sequences capable of undergoing high-yield macrocyclizations.

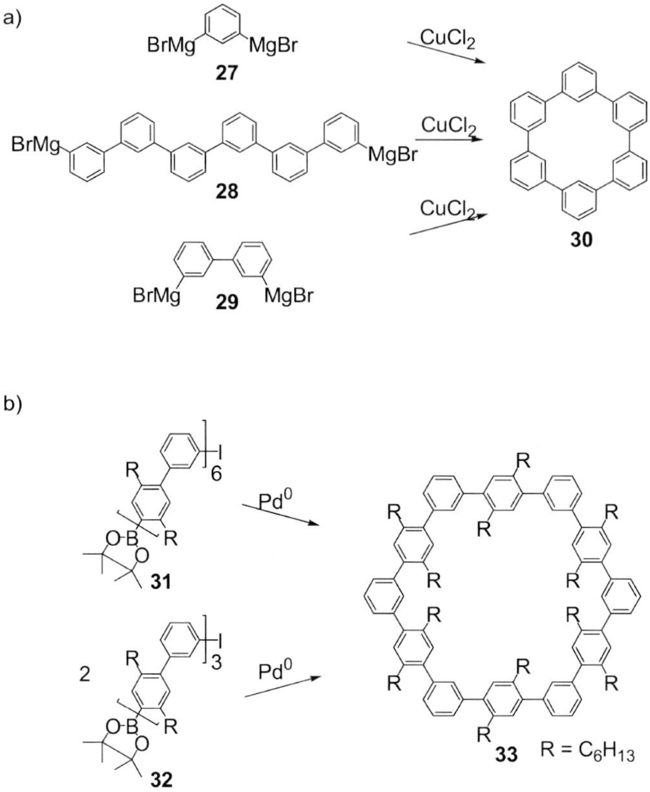

Scheme 6.4 (a) Staab and Binning's and (b) Schlüter and coworkers' syntheses of cyclic phenylene oligomers.

lene precursor **29** gave the macrocycle in 11% yield (Scheme 6.4a) [18]. Thirty years after that report, Schlüter and coworkers prepared larger phenylene macrocycles (**33**) [19]. Not surprisingly, similar to Staab and Binning, they obtained a high yield (85%) when an intramolecular reaction was performed starting from **31**, whereas the cyclodimerization of **32** gave the same ring in only 35% yield (Scheme 6.4b).

Again, the higher yield in these cyclization reactions had to be paid for with a more time-consuming precursor synthesis (Scheme 6.5). Boronic acid **34** was coupled with the bromoiodo compound **35** exclusively at the iodo position (96%). After iododesilylation of **36** (98%), **37** was again coupled with **34** to give **38** (87%). Compound **38** is the starting material for **39** (80%) and **40** (97%), which were coupled with Pd catalysis to give **41** (94%). Br–I exchange (93%) and transformation of the TMS group into a boronic ester (75%) gave **31**. Al-

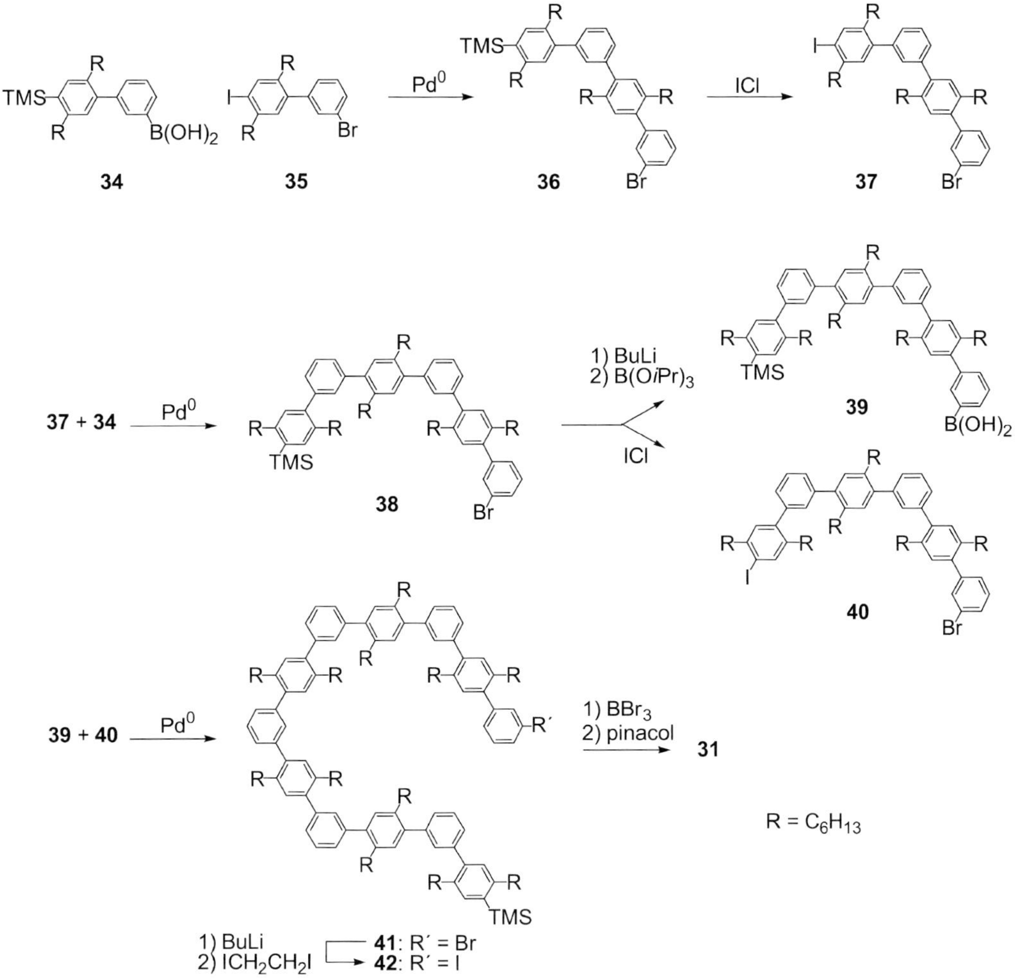

Scheme 6.5 Schlüter and coworkers' synthesis of phenylene oligomers capable of undergoing high-yield macrocyclizations.

though a multi-step synthesis, the high yields of Schlüter and coworkers' orthogonal reaction conditions allowed the synthesis of nearly 700 mg of **33** in one batch. This strategy has also been used for the preparation of larger phenylene macrocycles, e.g. **43** (Fig. 6.4).

Macrocycle synthesis based on oxidative Glaser coupling suffers from the same handicap [20, 21] For example, Tobe's group investigated the synthesis and properties of diethynylbenzene macrocycles in great detail [22]. The dimer unit **45** is ac-

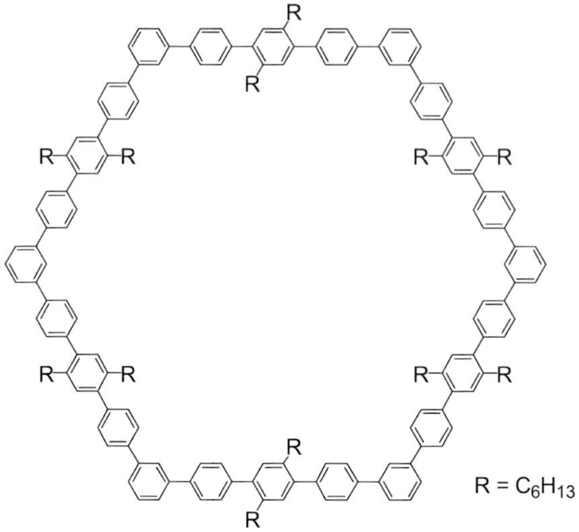

Fig. 6.4 Schlüter and coworkers' cyclotetraicosaphenylene.

R = C₆H₁₃

43

cessible by the oxidative coupling of the monoprotected bisacetylene **44** in good yield (96%). Carefully controlled deprotection of **45** allowed the isolation of the monoprotected bisacetylene **46** in 40% yield along with the diethynyl compound **47** (22%) and starting material. Oxidative intermolecular coupling of **47** under pseudo-high-dilution conditions gave the corresponding tetrakis(diethynylbenzene macrocycle) **48** (tetrakis-DBM), the cyclic dimer of **47** and the octakis-DBM **49** (tetramer of **47**) in 10% and 2% yields, respectively. However, the hexakis-DBM **53** (trimer of **47**) could not be isolated, which was not expected [23] Therefore, the stepwise synthesis of the macrocycles was employed in which the number of possible products is minimized. Conversion of **46** into the bromoacetylene **50** (90%) and intermolecular coupling with **47** gave the hexamer **51** in 44% yield. Deprotection (97%) and subsequent intramolecular cyclization of **52** afforded the cyclic hexamer **53** in 15% yield.

Höger and coworkers prepared a series of shape-persistent macrocycles with intra-annular, extra-annular or adaptable substituents by the intermolecular cyclodimerization of the corresponding bisacetylenic "half-rings" under pseudo-high-dilution conditions [10]. For example, cyclization of **54** gives a crude reaction product that contains about 60–65% of the cyclic dimer **55** as determined by gel permeation chromatography (GPC). The other 35–40% of the crude product belong to the cyclic trimer and tetramer and cyclic and/or non-cyclic oligomers and polymers. Compound **55** could be purified by recrystallization and was isolated in 40–50% yield (Scheme 6.7). Subsequent acid-catalyzed deprotection of the THP groups gives the macrocyclic amphiphile **56**.

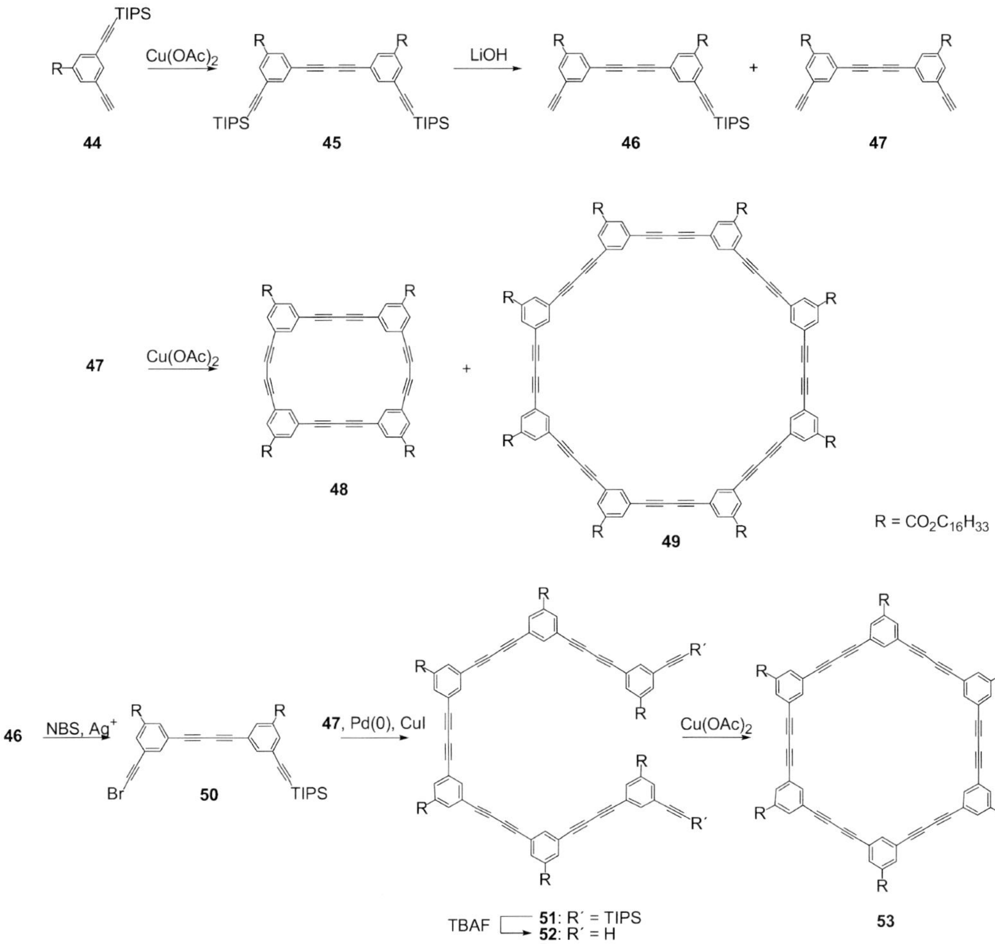

Scheme 6.6 Tobe and coworkers' statistical and stepwise synthesis of phenylene–butadiynylene macrocycles.

According to the observations described earlier, it is not surprising that the yield of the shape-persistent macrocycle **59**, which also contains non-polar and (toluate-protected) polar side-groups in an adaptable arrangement, is much lower. Although the reaction is performed under the same conditions as for the cyclodimerization of **54**, the crude reaction mixture in the cyclization of **58** contains only about 20–25 % of **59** (according to GPC analysis). Moreover, isolation of the shape-persistent ring was not possible, either by recrystallization or by column chromatography (Scheme 6.8) [24].

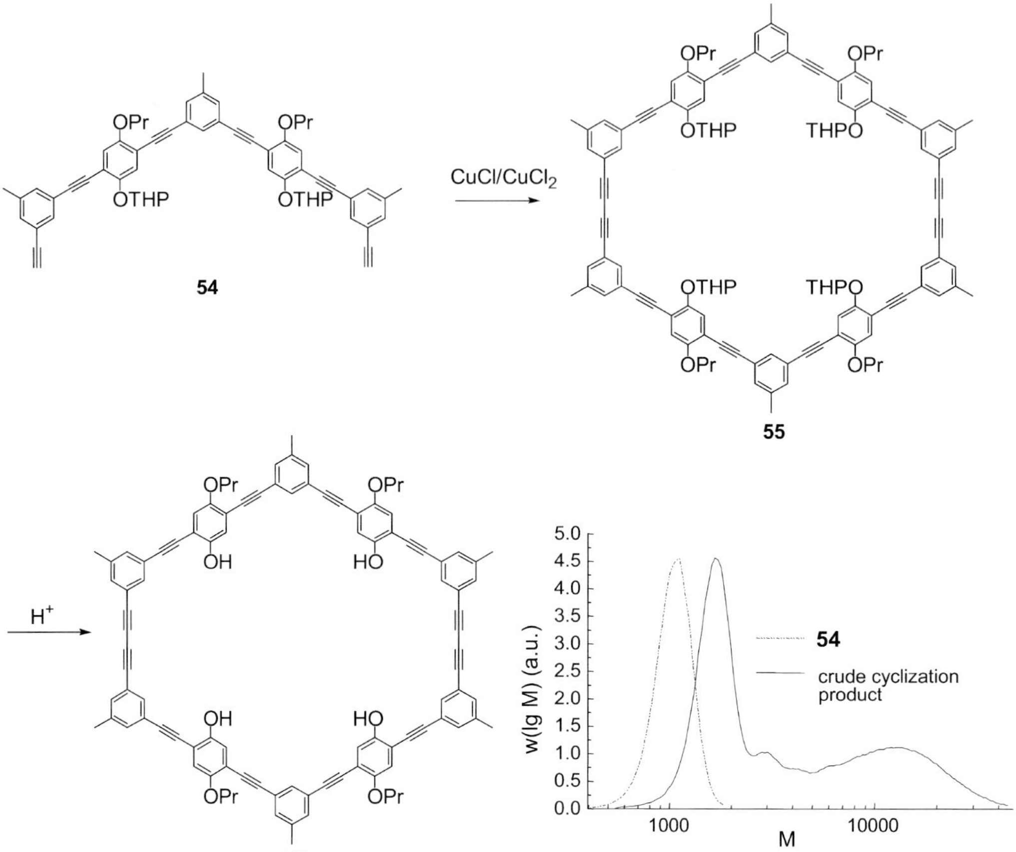

Scheme 6.7 Shape-persistent macrocycle **56** with an adaptable arrangement of the polar and non-polar side-groups by the intermolecular acetylene coupling of the corresponding bisacetylenic "half-ring" **54**. GPC of **54** and its crude cyclization product [molecular weights relative to polystyrene (PS) standards].

An interesting aspect of acetylene dimerization was recently added by Haley's group [25]. The outcome of the cyclization of the precursor **60** depends strongly on the catalyst used for the coupling reaction. Cu-mediated coupling conditions [Cu(OAc)$_2$, py] formed the macrocycle **61** as a sole product in 70% yield whereas acetylene homocoupling with [PdCl$_2$(dppe)] furnished **61a** exclusively (84%) (Scheme 6.9).

All examples described here and many more in the literature show that in general all protocols for the preparation of nanometer-scale macrocycles suffer from

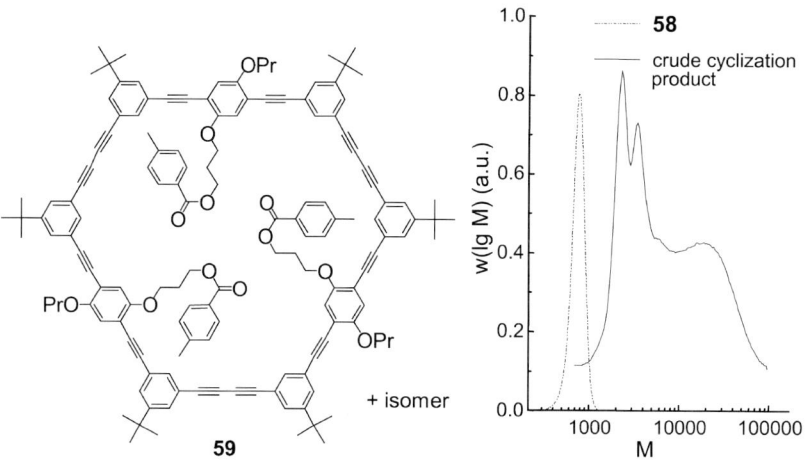

Scheme 6.8 Cyclization of **57** under high-dilution conditions (molecular weights relative to PS standards).

the same handicap, regardless of which specific CC coupling reaction is used for the cyclization: simple precursors are easily accessible but give low yields of the desired shape-persistent ring whereas high yields in the cyclization step require a time- and material-intensive multi-step precursor preparation. The method of choice always depends on the specific compound, the question of interest and how much effort can be put into the precursor synthesis.

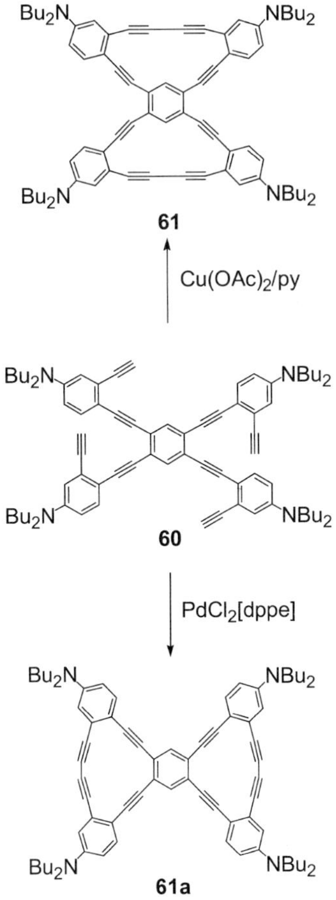

Scheme 6.9 Haley and coworkers' selective synthesis of the macrocycles **61** and **62**.

6.2.2.2 Template-controlled Cyclizations

An attractive way to overcome the problem described above is the use of templates [26]. According to Bush, "a template organizes an assembly of atoms, with respect to their loci, in order to achieve a particular linking of atoms" [27].

Sanders and coworkers have shown that the interaction between pyridine and Zn-porphyrins can be used for controlling the outcome of the Glaser coupling of **62**. They were able to show that ethynylphenyl-substituted Zn-porphyrins can be cyclodimerized or cyclotrimerized, depending on the specific pyridyl template present in the reaction medium [28]. With 4,4'-bipyridene (**63**), the cyclodimer **64** was obtained in 70% yield, whereas the presence of 1,3,5-tris(4-pyridyl)-triazine (**65**) supports the formation of the cyclotrimer **66** (50%) (Scheme 6.10).

Lindsey's group used the same methodology for the synthesis of a hexameric wheel of porphyrins that shows in the presence of appropriate guest molecules interesting light-harvesting properties [29]. Its synthesis starting from **67** and

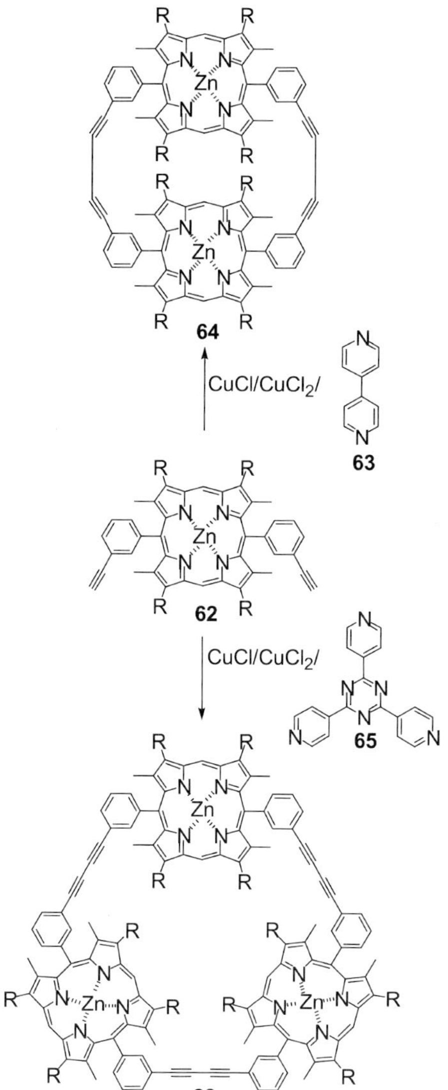

Scheme 6.10 Sanders and coworkers' template synthesis towards cyclic porphyrin dimers and trimers.

68 under Pd catalysis requires the presence of the template **69** in order to obtain **70** in 5.5 % yield (Scheme 6.11). The GPC and mass spectrometric analysis of the crude reaction products of the cyclization in the absence and presence of **69** showed the dramatic effect of the template on the reaction. Without a template a wide variety of different compounds were obtained and only a weak peak in the product regime could be detected in the mass spectrum. Attempts to isolate **70** from that mixture failed. However, the crude product of the template-supported reaction showed in GPC an intense product peak with an integrated inten-

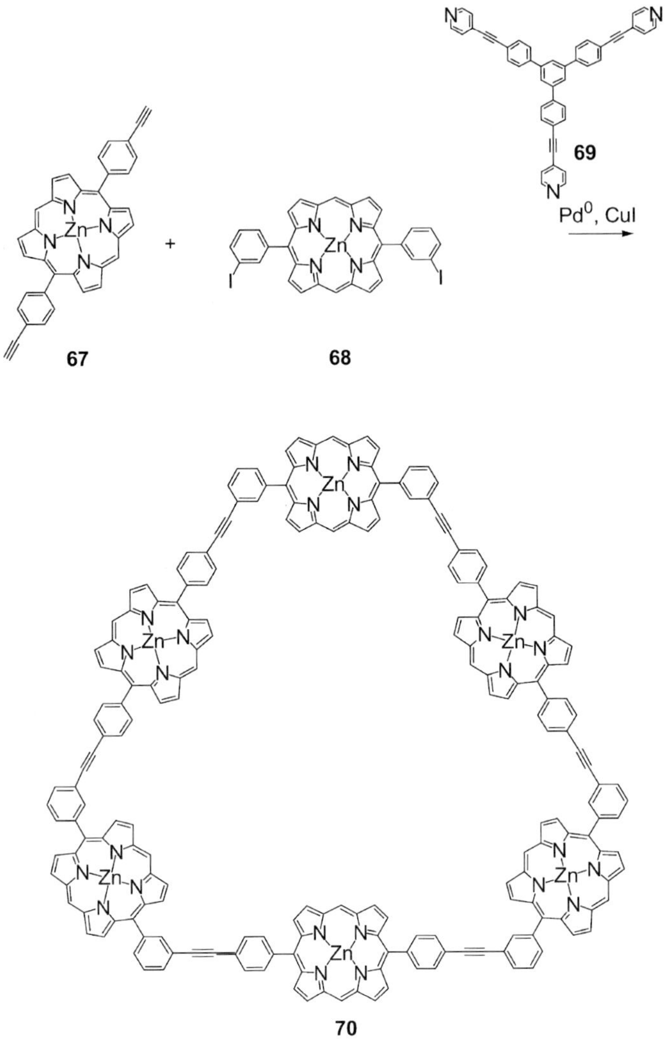

Scheme 6.11 Lindsey and coworkers' synthesis of a cyclic porphyrin hexamer.

sity of nearly 40% that is assigned to the product and a chromatographically similar side-product. Repeated column chromatography and additional recrystallizations are responsible for the product loss during the purification procedure. Nevertheless, the use of templates was the only way to obtain sufficient amounts of **70** in pure form for further investigations.

An interesting precoordination of bisacetylenes was reported by Bäuerle and coworkers. Reaction of the terthiophene diyne **71** with cis-Pt(dppp)Cl$_2$ (**72**) yielded the metallacycle **73** in 91% yield after chromatography [30]. Treatment of **73** with iodine led, with reductive 1,1-elimination, to the formation of the macrocycle

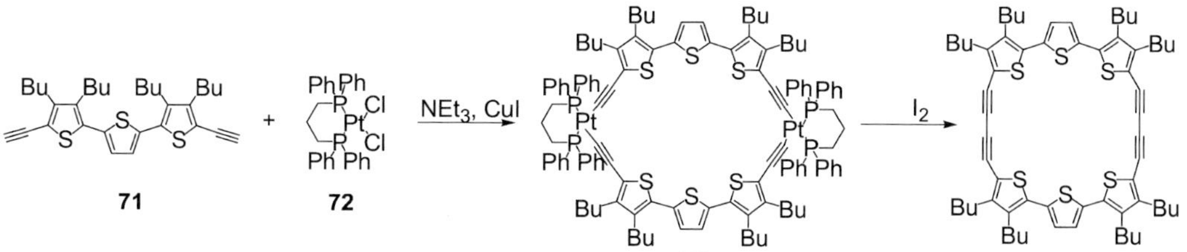

Scheme 6.12 Formation of butadiyne–terthiophene macrocycles via reductive elimination from metallamacrocycles.

74 in 54% isolated yield. Interestingly, this cyclic dimer was not observed when **71** was cyclized in a statistical reaction under high-dilution conditions.

Whereas "classical" templates bind noncovalently to the substrates, it is also possible to use covalently bound templates in order to obtain higher yields in cyclization reactions. Numerous examples in different areas of organic chemistry show that low yields of intermolecular reactions can be overcome by temporary intramolecularization of the reaction [31]. Intramolecularization means that an intermolecular process is made intramolecular. Achieving intramolecularization by temporary connection means that the reactants are covalently attached to one another through an atom or series of atoms that function as disposable tethers. Höger et al. showed that the covalent template approach can also be used for the synthesis of shape-persistent phenylene–ethynylene oligomers: Instead of protecting the hydroxy group of **57** as toluate, an esterification with trimesinic acid (**75**) (under Mitsunobu conditions) was performed (79%). When the template-bound hexaacetylene **76** is cyclized under the same conditions as used for the intermolecular reaction of **58**, i.e. pseudo-high-dilution conditions, GPC analysis of the crude cyclization product shows that it contains mostly the (template-bound) cyclotrimer **77** that could be isolated in 90% yield (Scheme 6.13) [24]. Base-promoted hydrolyses of the ester groups gave the shape-persistent macrocycle **78** with an adaptable arrangement of the polar and nonpolar side-groups in nearly quantitative yield (95%). It should be mentioned that in Scheme 6.13 only one of the two possible isomers of **77** and **78** are shown. Actually, according to the NMR spectra, the cyclization product is a 1:1 mixture of the compound displayed in Scheme 6.13 and the isomer in which one of the polar (and nonpolar) side-groups point in the opposite direction within the ring plane.

The outcome of this reaction was surprising and immediately the question arose of the origin of this high selectivity. Models indicate that the template and its connectors to the rigid parts of the macrocycle fit nearly perfectly into the cavity of the ring. Therefore, it might be possible that the terminal acetylenes in **76** are already in close proximity favoring cyclization over oligomerization. To prove this assumption, the length of the spacer between the rigid parts of the bisacetylenes and the template is dramatically enlarged from C_3 to C_{11} [24]. Nevertheless, the outcome of the cyclization reaction of **79** is nearly identical and **80** and

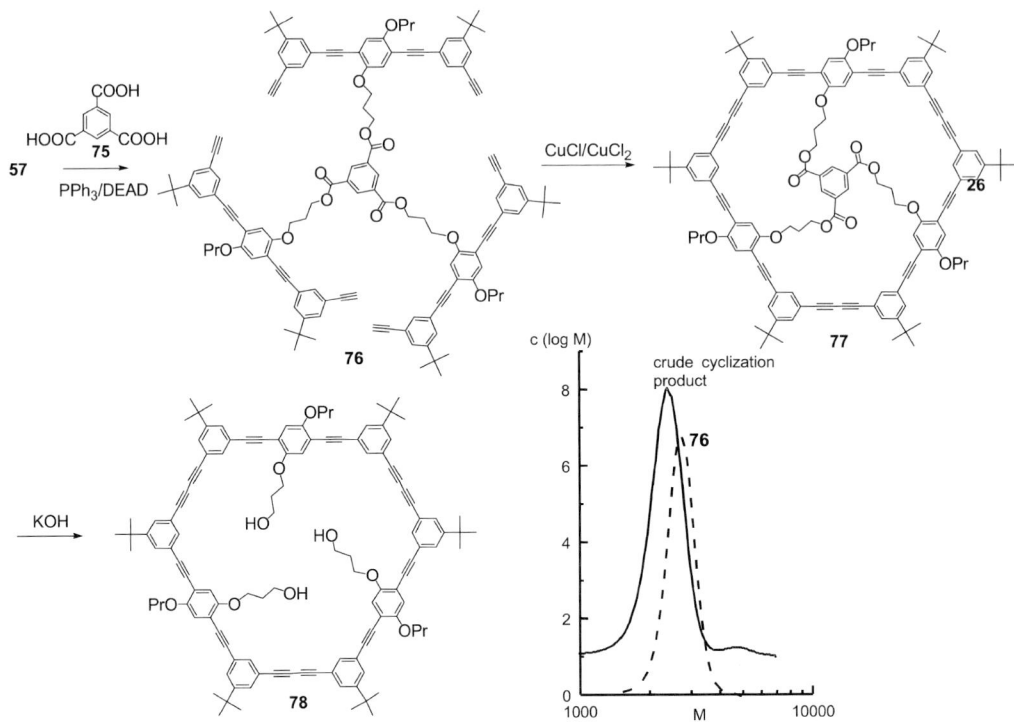

Scheme 6.13 Covalent template-supported cyclization of phenylacetylene oligomers; GPC of **76** and the crude cyclization product.

81 could be isolated in 84 and 95 % yield, respectively (Scheme 6.14). This result is important because it demonstrates that for a successful template-directed acetylene cyclization, preorganization of the acetylenes is not important. Rather, the reaction relies on the high concentration of the terminal acetylenes around the covalently bound template together with a low precursor concentration due to the high-dilution conditions.

Since a close proximity of the terminal acetylenes in the precursor is not necessary to obtain high yields of the template-bound shape-persistent ring, macrocycles with an extra-annular connection to the template are also accessible by this route. Macrocycles with extra-annular hydroxy groups are valuable building blocks for the preparation of coil–(rigid) ring–coil block copolymers that show an interesting aggregation behavior and yield in appropriate solvents supramolecular hollow polymer brushes [4, 32]. Therefore, it was desirable to investigate the scope and limits of template-supported methods in order to obtain higher yields in the cyclization reaction of half-rings.

As expected, the amount of shape-persistent macrocycle in the crude reaction product depends strongly on the length of the covalent tether (Scheme 6.15)

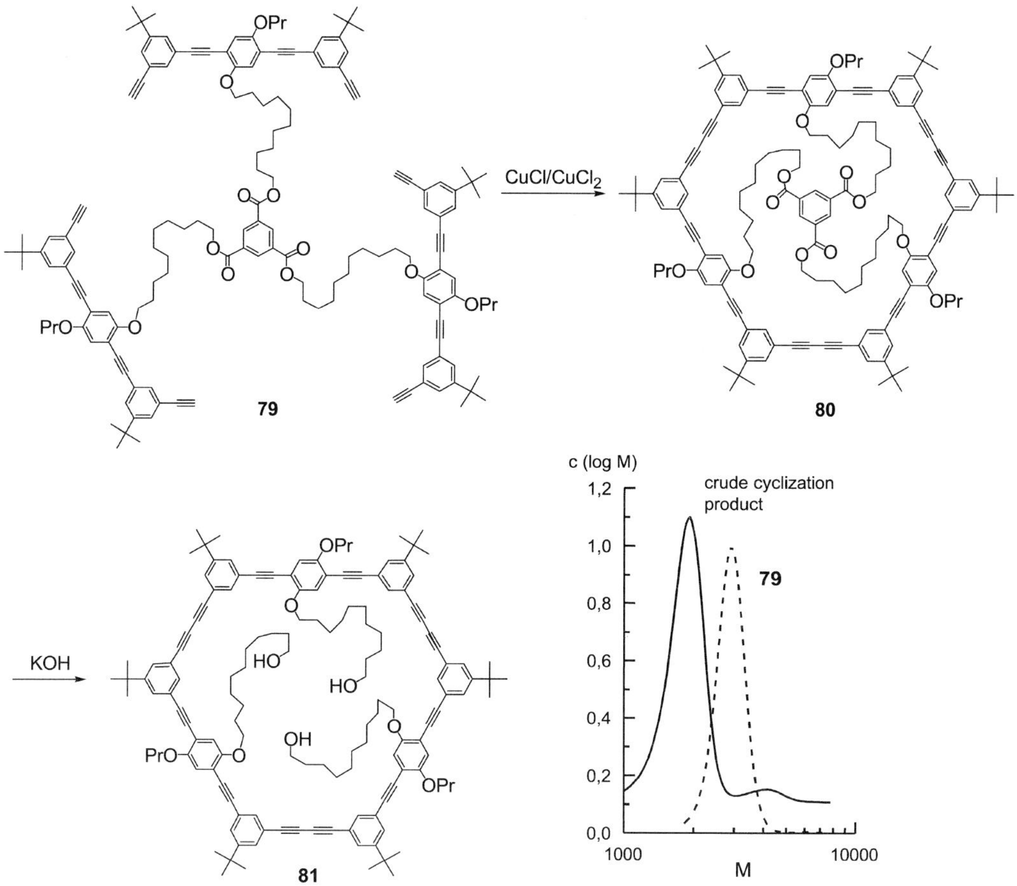

Scheme 6.14 Covalent template-supported cyclization of phenylacetylene oligomers with longer tethers; GPC of **79** and the crude cyclization product (molecular weights relative to PS standards).

[33]. With an appropriate spacer length between the rigid macrocycle and the template (**82a**), the crude reaction mixture contains about 90% of the (template-bound) macrocycle **83a**. The outcome of the reaction is independent of the size of the extra-annular alkyl groups, as a comparison between the reactions of **82a** and **82b** shows (Me vs. *t*-Bu). However, if the spacer length is reduced (**82c**), the yield of cyclic dimer **83c** drops to about 65%, which is comparable to the statistical reaction. The increasing amount of higher oligomers with shorter spacer length can be easily rationalized if the oxidative acetylene coupling is viewed as a stepwise process, as shown schematically in Scheme 6.16.

The macrocycle formation proceeds via the intermediate compound **A** that can undergo two different reactions: either intramolecular cyclization towards the desired macrocycle **M** or intermolecular reaction with additional **A** to produce the

82a,b,c   **a:** R = Me, n = 6; **b:** R = t-Bu, n = 6; **c:** R = t-Bu, n = 1

CuCl/CuCl$_2$ ↓

83a,b,c   **a:** R = Me, n = 6; **b:** R = t-Bu, n = 6; **c:** R = t-Bu, n = 1

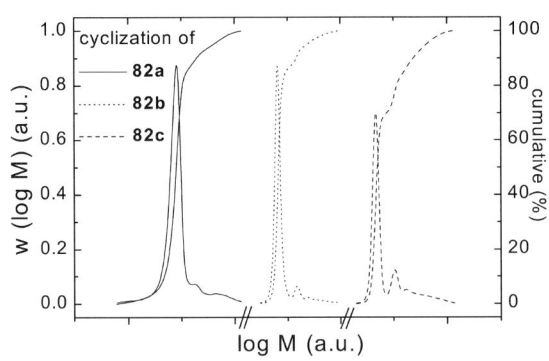

Scheme 6.15 Tether length dependence of the intramolecular dimerization of templated half-rings (molecular weights relative to PS standards).

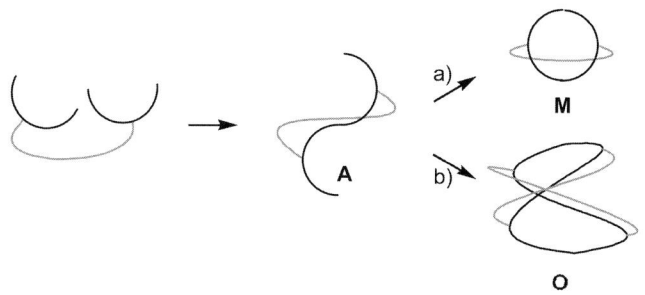

Scheme 6.16 The yield of the template-supported half-ring dimerization is limited when the template (light gray) in the intermediate **A** prevents fast intramolecular ring closure (through path a) and the intermolecular acetylene coupling (through path b) gains in importance.

tetrameric compound **O** and subsequently higher oligomers [34]. A spacer that is to short or rigid may build up excessive strain in the transition state of the formation of **M** and therefore facilitate the oligomerization (Scheme 6.16).

A similar result is obtained when a macrocycle with extra-annular carboxylic acid groups is prepared by template-supported cyclization (Scheme 6.17) [35]. The crude product of the cyclization of **84** contains only 70 % of the cyclic dimer **85**. However, after column chromatographic purification, the template-bound macrocycle could be crystallized from 1,2-dichloroethane (DCE) as solvate. The structure shows that the template length is sufficient to cross the whole macrocycle and no unusual bond lengths or bond angles build up extra strain. Nevertheless, already the first CC coupling towards intermediate **A** (Scheme 6.16) restricts the conformational freedom of the spacer. Therefore, alternatively (or additionally) to the arguments given above, it must also be taken into account that a to short or rigid tether can prevent the fast formation of **A** and therefore facilitate an intermolecular oligomerization that decreases the yield of the shape-persistent macrocycle. These results are in accordance with the design criteria generally accepted for intramolecularization due to temporary connection: The tether should be of sufficient length and flexibility so that the approach of the reacting centers is not hindered and side-reactions become more dominant.

An additional disadvantage of the use of covalently bound templates, that is not found in statistical reactions or by the use of supramolecular templates, is the requirement to attach the bisacetylenic building blocks of the macrocycle to the template. So far this has been an additional step in a linear reaction sequence, thus reducing the overall product yield. Whereas this extra step in the case of the cyclotrimerization is overcompensated by a considerable yield increase of the desired shape-persistent ring, this argument does not hold for the cyclodimerization where statistical reactions already give yields of the order of 50–60 %.

Another approach towards templated acetylenic ring precursors overcomes this problem. Instead of preparing the bisacetylenes first and attaching them to the template in an additional step, the complete synthesis of the bisacetylenic ring

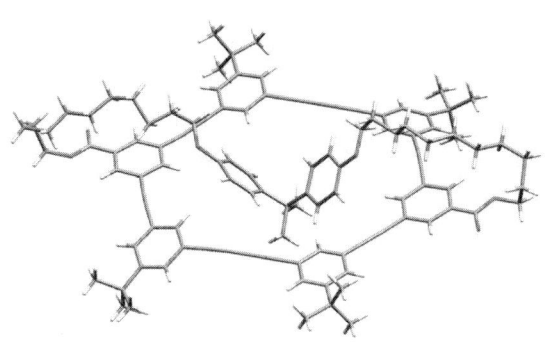

Scheme 6.17 Cyclization of the bisacetylene **84**. X-ray analysis of the templated macrocycle **85** (solvent molecules not displayed for clarity).

precursors is performed at the template (Schemes 6.18–6.20) [36]. Two corner pieces **86** of the ring which contain the functional groups are connected to the template **87** to yield the corresponding tetraiodide **88** (93%) (Scheme 6.18). Quadruple Sonogashira–Hagihara coupling with the monoprotected bisacetylene **89** gives the template-bound TIPS-protected tetraacetylene **90** (90%). The deprotection of the acetylenes is again performed at the template and the resulting tetraacetylene **91** (96%) is cyclized under pseudo-high-dilution conditions, as used in the previous reactions. As expected, the crude product of the reaction contains over 90% of the macrocycle **92**. Purification by column chromatography gave the template bound macrocycle in 92% isolated yield. Compound **92** could be deprotected to give the macrocyclic diol **93** with an intraannular arrangement of the functional groups (91%). This approach has several advantages: first, only a few steps are necessary to obtain the shape-persistent macrocycle; second, in all steps of the sequence the molecular weights of the compounds and along with that their physical properties change dramatically, so that the product purification is fairly simple and gives high isolated product yields; third, but of perhaps the greatest significance, is the fact that with the same monoprotected bisacetylenic building block **89**, non-symmetrical macrocycles are also available in comparably high yield if the reaction is performed with the non-symmetrical tetraiodide **98** (Scheme 6.19). It is of high practical significance that the chemical information that is necessary to obtain the asymmetric macrocycles **101** and **102** is introduced

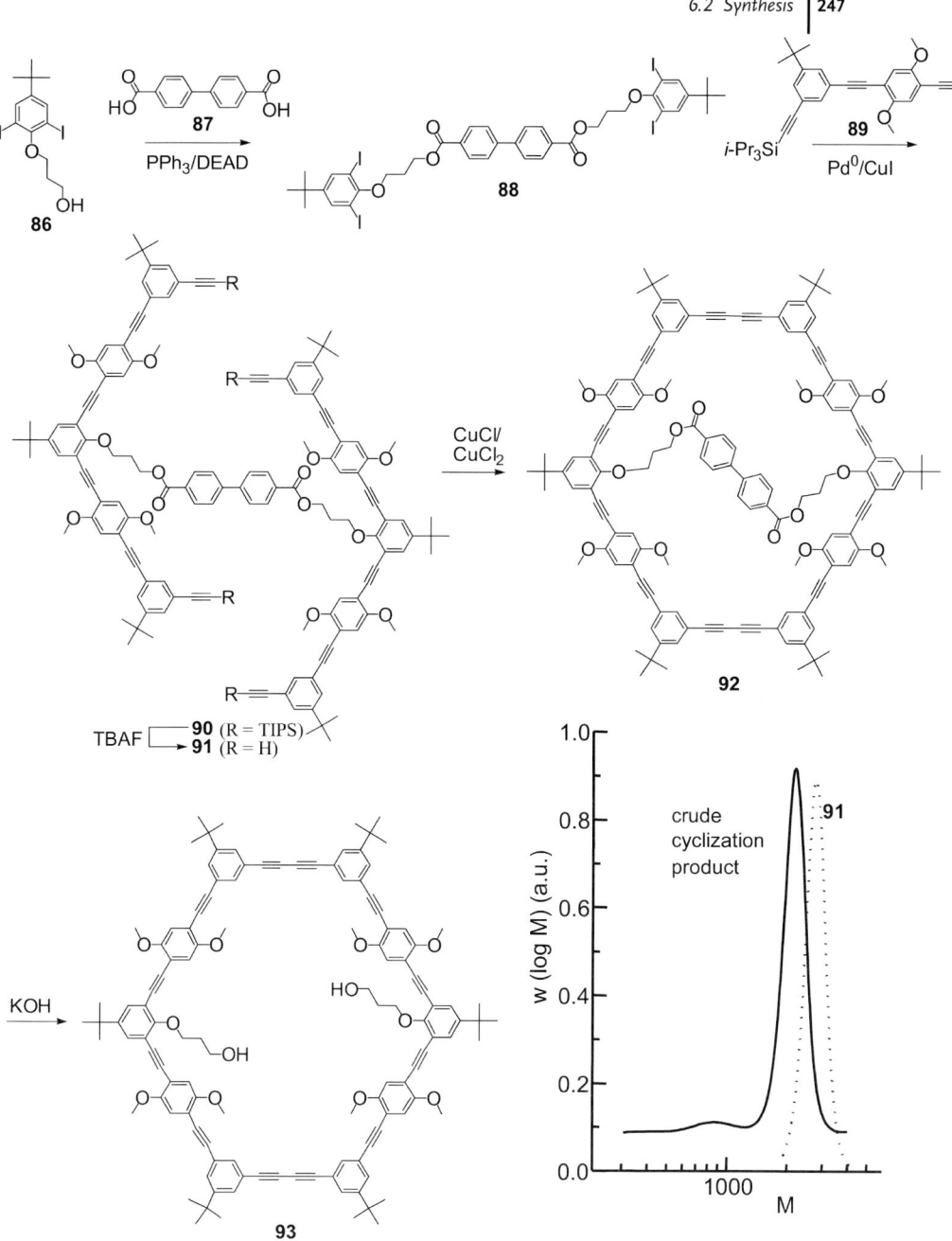

Scheme 6.18 Intramolecular cyclization leads to the macrocyclic diol **93** with intra-annular functional hydroxy groups. The GPC trace shows that **92** is nearly exclusively formed (molecular weights relative to PS standards).

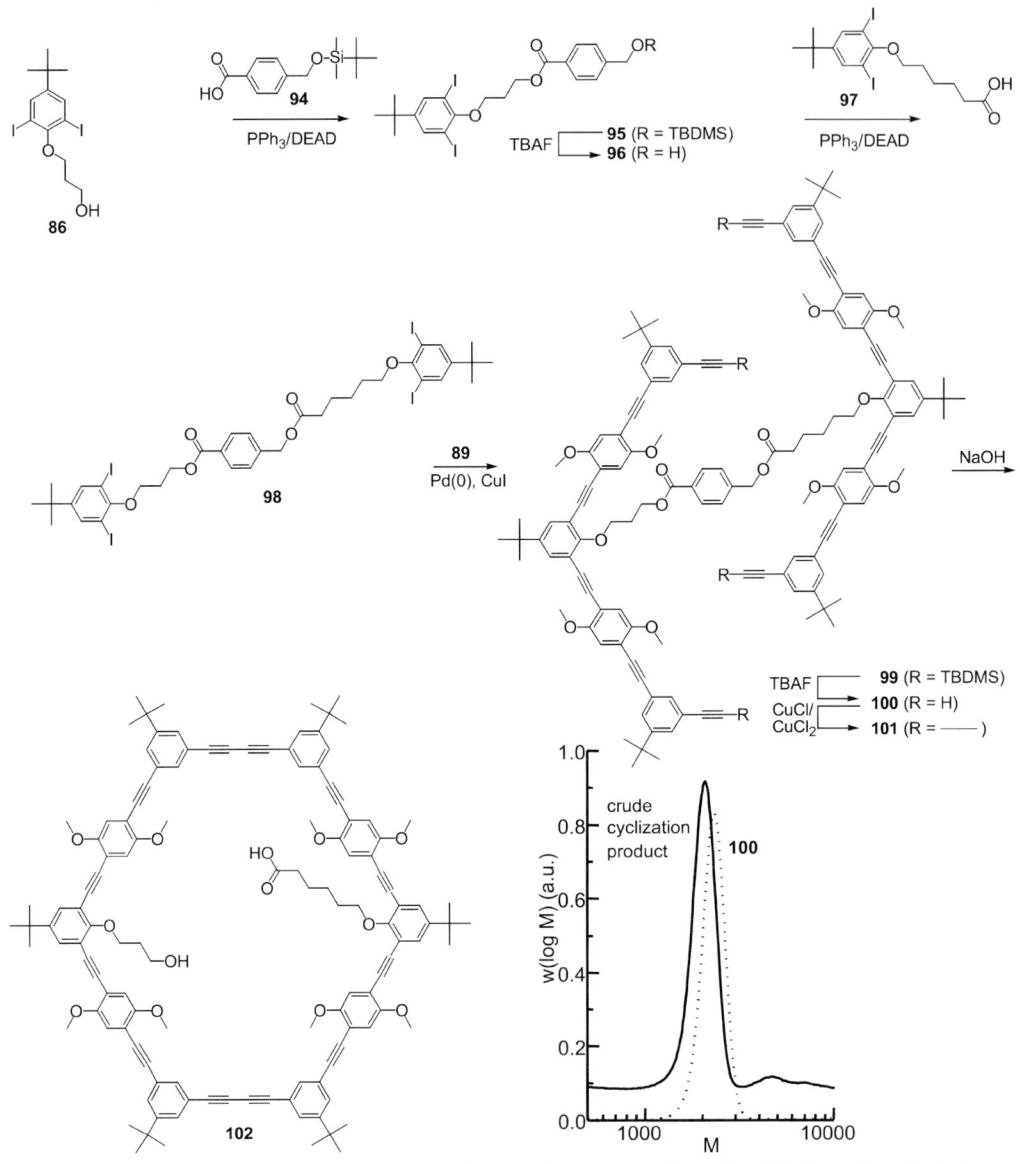

Scheme 6.19 Intramolecular cyclization that leads to the macrocyclic **102** with two different intraannular functional groups. The GPC trace shows that **101** is nearly exclusively formed (molecular weights relative to PS standards).

at the level of the low molecular weight tetraiodides. These are much easier to purify than high molecular weight ring precursors. Furthermore, all information for the synthesis of the non-symmetrical tetraiodides is summarized in common protective group chemistry textbooks.

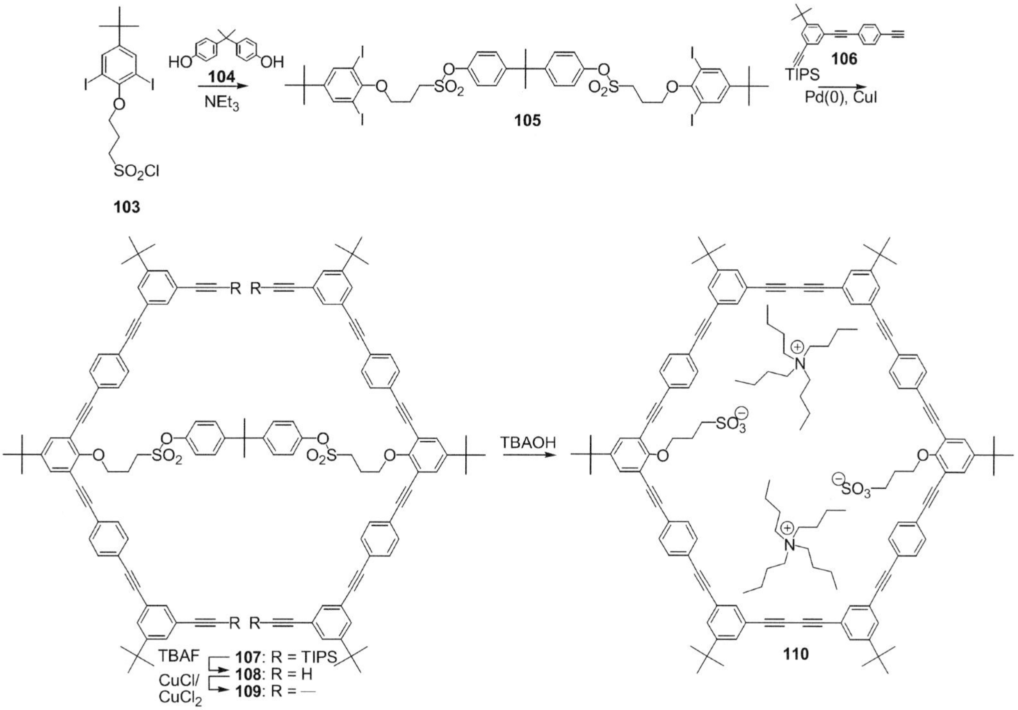

Scheme 6.20 Synthesis of a macrocyclic bissulfonate by template-supported Glaser coupling.

Similarly, the synthesis of a shape-persistent macrocycle with intra-annular sulfonate groups was performed (Scheme 6.20) [37]. First, the diiodide containing the sulfonate group was attached via the intermediate sulfonyl chloride **103** to the template (**104**) to give the templated tetraiodide **105** (81%). Quadruple Sonogashira–Hagihara coupling of the monoprotected bisacetylene **106** with the template (92%), deprotection of the TIPS groups (90%) and subsequent oxidative acetylene coupling yielded the template-bound macrocycle **109** (90%). Again, all reactions in this sequence gave high isolated yields owing to the fairly simple purification process. As in the synthesis described before, this holds also for the separation of **107** from the diacetylenic byproduct that is always present after Hagihara–Sonogashira coupling and sometimes troublesome to remove. Subsequent cleavage of the arylsulfonates using tetrabutylammonium hydroxide (TBAOH) gave the macrocyclic bis(tetrabutylammonium) disulfonate **110** in 85% yield.

High yields in the cyclization reaction are also of great importance for the product purification process. For example, the macrocyclic diester **112** obtained by a statistical cyclization of **111** could be isolated in 54% yield after column chromatography (Scheme 6.21) [38]. The similar macrocyclic diester **114** with additional extra-annular THP-protected phenol groups was also produced in about 60–65%

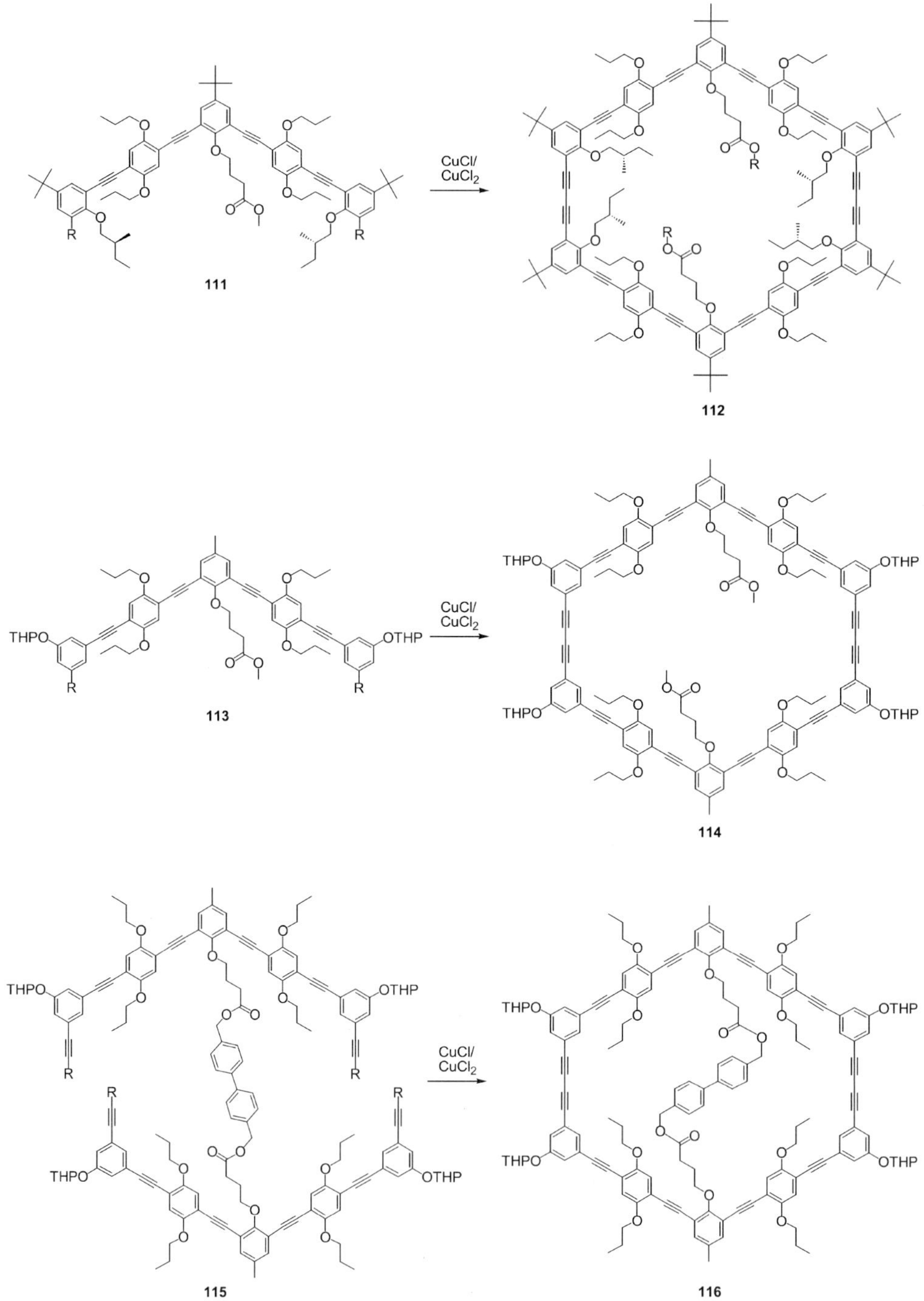

Scheme 6.21 Synthesis of macrocycles with intraannular ester groups by the statistical and by the template-supported cyclization reaction.

yield when **113** was cyclized under the same conditions (according to GPC analysis of the crude cyclization product). However, the purification by tedious column chromatography allowed the isolation of **114** in only 29% yield [39a]. Nevertheless, the additional extra-annular (protected) phenol groups in **114** allow a further alkylation (after THP deprotection) of the macrocycles that finally leads to interesting liquid crystalline compounds and structures that can be used for surface functionalization in the nanometer regime [2, 39]. Therefore, the alternative synthesis based on a covalent template approach was explored, and indeed, the cyclization of **115** gave the template-bound macrocycle **116** in 81% isolated yield [40].

These results show unambiguously that templates are an attractive alternative to the tedious precursor synthesis commonly used for the preparation of shape-persistent macrocycles. They allow the preparation of symmetrically and even of nonsymmetrically functionalized PAMs in high yields. Moreover, the increased amount of the shape-persistent macrocycle in the crude product simplifies the purification procedure in most cases. However, both the supramolecular and the covalent template approach require the presence of appropriate functional groups that bind (are bound) to the template.

6.2.3
The Thermodynamic Approach

A different approach towards shape-persistent macrocycles is macrocycle preparation under thermodynamic control, i.e. if the desired structure is the most stable under the given conditions and the cyclization reaction is reversible, one might be able to obtain a single macrocyclic product in quantitative yield although the reaction starts with small building blocks [41, 42].

Imine formation is a convenient way to form macrocycles in good to high yields [43, 44]. In recent years the synthesis of arylene–ethynylene macrocycles containing imine bonds has been described in which the cyclization step is Schiff base formation. For example, Zhao and Moore prepared the imine-containing macrocycle **119** in nearly quantitative yield (>95%) by the condensation of the diamine **117** and the dialdehyde **118** (Scheme 6.22) [45]. Although also in this case both starting materials (half-rings) needed to be prepared separately by a multi-step sequence, the yield in the cyclization step is incomparably higher than in statistical reactions.

Recently, shape-persistent imine-containing macrocycles have also been obtained from smaller building blocks. Based on the results of Nabeshima and coworkers [46], Gallant and MacLachlan prepared the soluble 30-membered ring **122** in 70% yield by the thermodynamically favored [3 + 3] cyclocondensation of the corresponding dialdehyde **120** and diamine **121** (Scheme 6.23) [47]. The extension of this reaction sequence towards the preparation of larger (acetylene-containing) macrocycles is also possible. Condensation of the corresponding dialdehydes and diamines gave the macrocyclic products **123** and **124** in a one-pot, template-free synthesis in 68 and 40% yields, respectively [48]. Therefore, large,

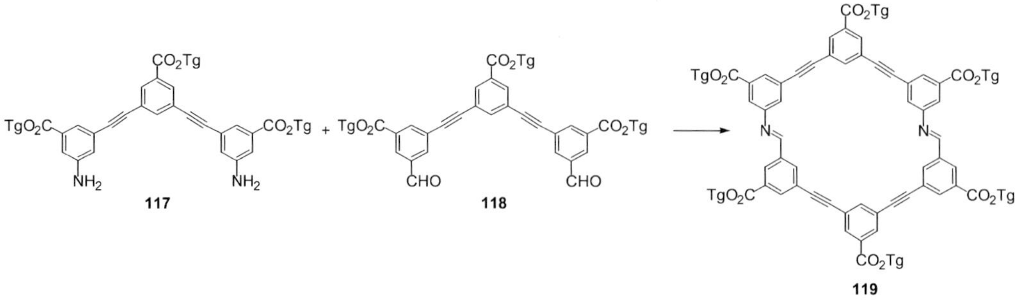

252 6 High-yield Synthesis of Shape-persistent Phenylene–Ethynylene Macrocycles

Scheme 6.22 High-yield synthesis of Schiff base-containing phenylene–ethynylene macrocycles.

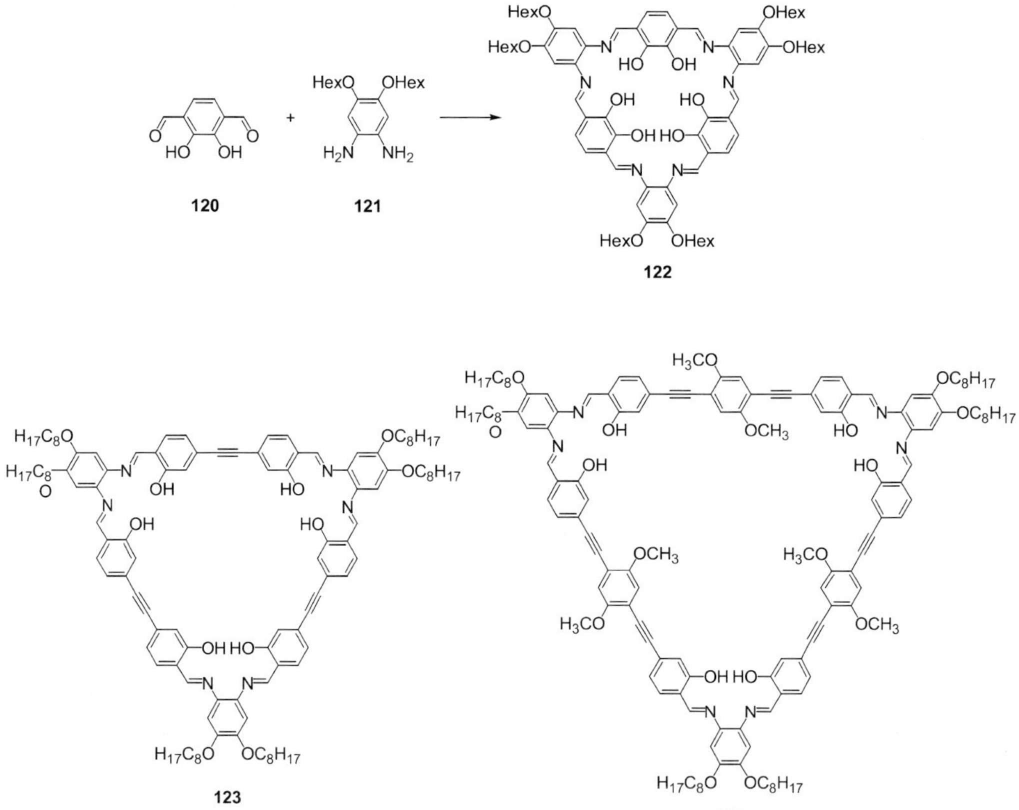

Scheme 6.23 MacLachlan and coworkers' synthesis of shape-persistent macrocycles by reversible imine formation.

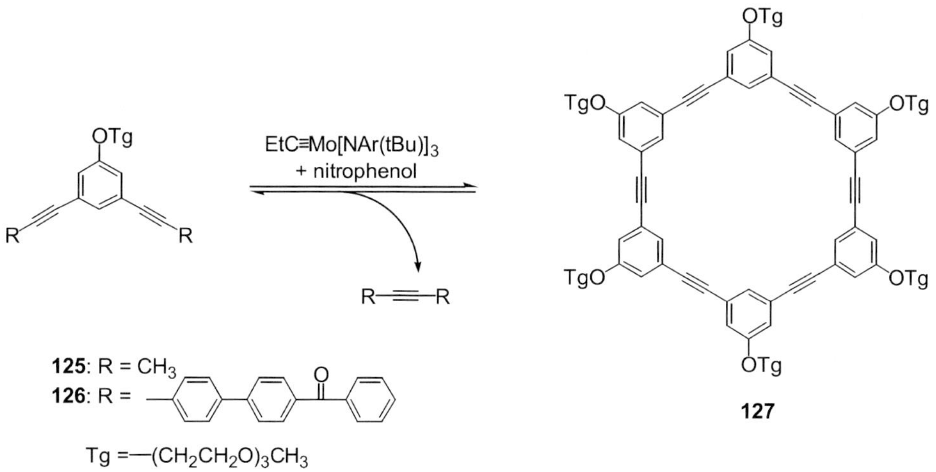

Scheme 6.24 Alkyne metathesis for the preparation of hexameric phenylacetylene macrocycles in high yields.

rigid macrocycles with adjustable diameters and intra-annular binding sites are available in good yield, starting from rather simple building blocks.

Zhang and Moore described a new approach towards arylene–ethynylene macrocycles using the reversibility of the alkyne metathesis reaction [49, 50]. Starting from the simple bisacetylene **125** and a highly active Mo(VI) alkylidyne catalyst, the macrocycle **127** was obtained in high yield (81%) when the reaction was performed on the milligram scale under open driven conditions to remove the byproduct 2-butyne (Scheme 6.24). Several *m*-cyclophanes of this type with other substituents were obtained in comparable yields, thus showing that this concept can be applied fairly generally. However, attempts to increase the scale of the reaction were not successful and mainly oligomeric products were produced. Speculation that the 2-butyne could not be removed fast enough from the reaction mixture if the metathesis is performed on a gram scale or that air might have decomposed the catalysts led to the exploration of an alternative approach for shifting the metathesis equilibrium [51]. Instead of removing a volatile byproduct under vacuum, an insoluble byproduct may separate from the reaction mixture in a closed system. Indeed, under optimized reaction conditions it was possible to prepare in one batch more than 5 g (77%) of **127** when the benzoyl-biphenyl-substituted arylacetylene **126** was cyclized in CCl_4 [52].

The generality of this new strategy was demonstrated with the preparation of the tetrameric carbazole macrocycle **129** that is available in 84% yield from the building block **128** (Scheme 6.25).

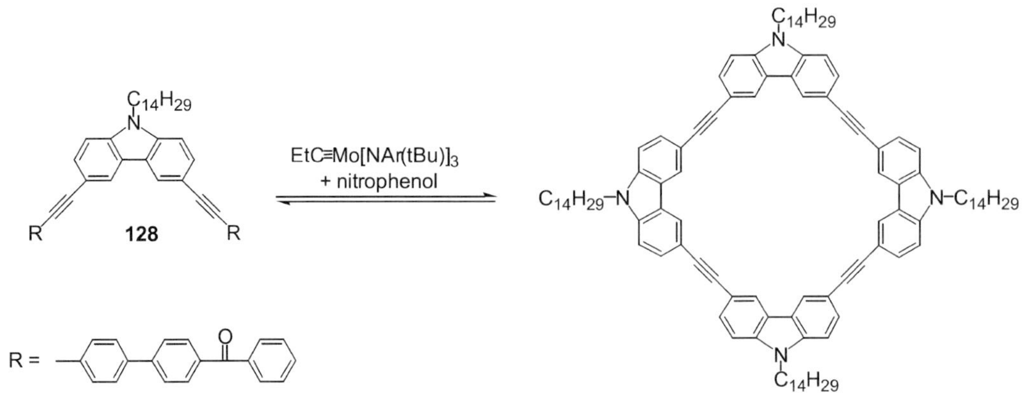

Scheme 6.25 Alkyne metathesis leads in high yields to the formation of tetrameric carbazole–acetylene macrocycles.

6.3
Conclusion

The interest in shape-persistent macrocycles based on the phenylene–acetylene backbone as building blocks in supramolecular chemistry and as functional materials has been growing continuously in the last several years. However, prior to all investigations and possible applications of this class of compounds stands their synthesis. Two general methods are at present commonly used to meet this challenge: the kinetic approach, which starts with the formation of a precursor that is cyclized in the final step, and the thermodynamic approach, in which a self-healing process corrects undesired bond formation and allows the isolation of a product in high yields, if it is the most stable under the given conditions. The advantage of the first approach is that non-symmetrical compounds are also available that need to be investigated in order to obtain a clear picture of the rather complex structure–property relations of the macrocycles. The disadvantage is that high product yields are generally obtained only if a considerable effort is put into the precursor synthesis. Small building blocks give, unless templates (covalent or noncovalent) increase the yield of the desired shape-persistent structure, a complex mixture of macrocycles of different size and oligomers and polymers. The thermodynamic approach, on the other hand, has the advantage of giving high product yields even if small building blocks are used as precursors. However, the structural variability in the macrocycles that can be obtained with the stepwise precursor synthesis has so far not been achieved. Therefore, at present both approaches are used for the creation of new and interesting macrocycles. It is beyond doubt that future investigations on new synthetic approaches towards PAMs will also widen the applications of these compounds in supramolecular chemistry and materials chemistry.

6.4
Experimental Procedures [37]

3-(4-*tert*-Butyl-2,6-diiodophenoxy)propane-1-sulfonic acid chloride (**103**)
(A) Potassium 3-(4-*tert*-butyl-2,6-diiodophenoxy)propanesulfonate: 2,5-diiodo-4-*tert*-butylphenol (30.8 g, 76.6 mmol) was dissolved in *t*-BuOH (100 mL) at 35 °C. KO-*t*-Bu (1 M in *t*-BuOH, 76.6 mL, 76.6 mmol) and then propansultone (6.7 mL, 76.6 mmol) were added. A fine precipitate formed immediately. The mixture slowly became thicker and could not be stirred after 15 min. After an additional hour at 35 °C, the solvent was removed under reduced pressure and the residue was recrystallized from methanol (250 mL) to give potassium 3-(4-*tert*-butyl-2,6-diiodophenoxy) propanesulfonate (37.5 g, 93 %) as a colorless solid, m.p. 262 °C (decomp.). ^{1}H NMR (250 MHz, DMSO-d_6), δ = 7.75 (s, 2 H), 3.87 (t, J = 6.3 Hz, 2 H), 2.65 (m, 2 H), 2.09 (m, 2 H), 1.22 (s, 9 H); ^{13}C NMR (62.5 MHz, DMSO-d_6), δ = 155.23, 150.91, 136.69, 91.67, 72.52, 48.83, 34.01, 31.04, 26.28; MS (FD), m/z = 561.8 (M$^+$). $C_{13}H_{17}I_2KO_4S$ (562.25): calcd. C 27.77, H 3.05; found C 27.48, H 2.95 %.

(B) SO$_2$Cl$_2$ (3.0 g, 24.9 mmol, 1.8 mL) was slowly added to a suspension of the sulfonate described above (14.0 g, 24.9 mmol) in THF–DMF (2:1, 300 mL) at room temperature. The solid dissolved and after the heat evaluation (ca. 1 h) the solution was heated at 60 °C for 2 h. After cooling to room temperature, the mixture was poured slowly on to ice–water (1 L), the product was filtered off and washed three times with ice-cold water and then with petroleum ether. After drying, **103** (10.3 g, 76 %) was obtained as a fawn powder sufficiently pure for the next step. An analytical sample was obtained by chromatography over silica gel, eluting with ethyl acetate (R_F = 0.69), m.p. 94 °C. ^{1}H NMR (250 MHz, CDCl$_3$), δ = 7.71 (s, 2 H), 4.12 (m, 4 H), 2.59 (m, 2 H), 1.25 (s, 9 H); ^{13}C NMR (62.5 MHz, CDCl$_3$), δ = 155.28, 151.09, 136.72, 91.93, 72.11, 48.89, 34.08, 31.12, 26.03; MS (FD), m/z = 542.2 (M$^+$), 1084.1 (2 M$^+$). $C_{13}H_{17}ClI_2O_3S$ (542.60): calcd. C 28.78, H 3.16; found C 28.73, H 2.96 %.

Compound **105**

Triethylamine (0.6 mL, 8.3 mmol) was slowly added to a solution of **103** (2.3 g, 4.3 mmol) and 2,2-bis(4-hydroxyphenyl)propane (**104**) (0.5 g, 2.1 mmol) in THF (10 mL) at 0 °C. The solution was stirred for 1 h at 0 °C and then at room temperature overnight. The precipitate formed was removed by filtration and the filtrate was concentrated under reduced pressure. The yellow, oily residue was purified by chromatography over silica gel, eluting with CH$_2$Cl$_2$–hexane (2:1) (R_F = 0.51), to give **105** (2.2 g, 81 %) as a yellow solid, m.p. 77–80 °C. ^{1}H NMR (250 MHz, CDCl$_3$), δ = 7.70 (s, 4 H), 7.22 (br s, 8 H) 4.07 (t, J = 5.7 Hz, 4 H), 3.67 (m, 4 H), 2.53 (m, 4 H), 1.65 (s, 6 H), 1.25 (s, 18 H); ^{13}C NMR (75.5 MHz, CDCl$_3$), δ = 154.67, 150.87, 138.45, 136.28, 129.62, 127.51, 123.09, 120.91, 90.61, 71.88, 70.07, 47.98, 32.21, 30.54, 24.81; MS (FD), m/z = 1242.5 (M$^+$), 2485.2 (2 M$^+$). $C_{41}H_{48}I_4O_8S_2$ (1240.58): calcd. C 39.70, H 3.90; found C 40.04, H 3.93 %.

Compound 107

Pd(PPh$_3$)$_2$Cl$_2$ (30 mg) and CuI (22 mg) were added to a solution of **105** (0.92 g, 0.74 mmol) and **106** (1.47 g, 3.33 mmol) in triethylamine (12 mL) under argon. The mixture turned dark after a few minutes and was stirred overnight at room temperature and then at 50 °C for 1 h. After cooling to room temperature, diethyl ether (150 mL) and water (150 mL) were added, the organic phase was separated and extracted with water (2 × 75 mL), 10 % acetic acid (5 × 75 mL), water (2 × 75 mL), 10 % aqueous NaOH (3 × 75 mL), water (2 × 75 mL) and brine (75 mL). Drying over MgSO$_4$ and evaporation of the solvent yielded an oily residue, which was purified by chromatography over silica gel, eluting with CH$_2$Cl$_2$–hexane (1:2) (R_F = 0.36), to give **107** (1.68 g, 92 %) as a light-yellow, foamy solid, m.p. 167 °C. ^{1}H NMR (250 MHz, CDCl$_3$), δ = 7.11 (d, J = 8.8 Hz, 4 H), 7.04 (d, J = 8.8 Hz, 4 H), 7.42–7.48 (m, 32 H), 4.44 (t, J = 5.4 Hz, 4 H), 3.69 (m, 4 H), 2.50 (m, 4 H), 1.55 (br s, 6 H), 1.32 (s, 18 H), 1.31 (s, 36 H), 1.13 (s, 28 H); ^{13}C NMR (75.5 MHz, CDCl$_3$), δ = 157.91, 151.75, 149.26, 147.20, 147.16, 133.01, 132.61, 131.93, 131.66, 131.46, 129.32, 129.01, 128.48, 123.73, 123.58, 122.95, 122.89, 121.83, 116.90, 106.87, 93.53, 91.47, 90.88, 89.12, 87.71, 71.47, 48.16, 42.78, 34.90, 34.63, 31.40, 31.31, 30.98, 25.21, 23.55, 18.91, 11.54; MS (FD), m/z = 2483.5 (M$^+$). C$_{165}$H$_{196}$O$_8$S$_2$Si$_2$ (2483.87): calcd. C 79.79, H 7.95; found C 79.69, H 8.00 %.

Compound 108

Tetrabutylammonium fluoride (1 M solution in THF, 9.3 mL, 9.3 mmol) was added to a solution of **107** (1.65 g, 0.66 mmol) in THF–water (39:1, 40 mL). The solution was stirred at room temperature for 4 h. After evaporation of most of the solvent, diethyl ether (300 mL) and water (200 mL) were added and the organic phase was extracted with water (3 × 100 mL) and brine (100 mL). Drying over MgSO$_4$ and evaporation of the solvent yielded an oily residue, which was treated several times with small portions of methanol to give **108** (1.11 g, 90 %) after drying in vacuum as a light-yellow, foamy solid, m.p. 155 °C. ^{1}H NMR (250 MHz, CD$_2$Cl$_2$), δ = 7.12 and 7.05 (d, J = 8.8 Hz, each 4 H), 7.42–7.48 (m, 32 H), 4.42 (t, J = 5.4 Hz, 4 H), 3.67 (m, 4 H), 2.88 (s, 4 H), 2.46 (m, 4 H), 1.55 (br s, 6 H), 1.31 (s, 18 H), 1.30 (s, 36 H); ^{13}C NMR (75.5 MHz, CD$_2$Cl$_2$), δ = 158.32, 152.76, 149.98, 147.93, 147.82, 133.49, 133.00, 132.29, 132.07, 131.87, 129.32, 129.40, 128.85, 124.01, 123.68, 122.91, 122.89, 121.83, 117.40, 106.87, 93.90, 91.66, 90.88, 89.70, 88.35, 77.88, 72.19, 48.72, 43.32, 35.38, 35.09, 31.40, 31.31, 30.98, 25.81, 23.55, 18.91; MS (FD), m/z = 1859.6 (M$^+$), 929.0 (M^{2+}), 619.7 (M^{3+}). C$_{129}$H$_{116}$O$_8$S$_2$ (1858.49): calcd. C 83.37, H 6.29; found C 82.98, H 6.38 %.

Compound 109

A solution of **108** (1.10 g, 0.59 mmol) in pyridine (50 mL) was added to a suspension of CuCl (8.44 g, 85.2 mmol) and CuCl$_2$ (1.68 g, 7.4 mmol) in pyridine (300 mL) over 96 h at room temperature. After completion of the addition, the mixture was stirred for an additional day, then poured into CH$_2$Cl$_2$ (300 mL) and water (200 mL). The organic phase was extracted with water, 25% NH$_3$ solution (in order to remove the copper salts), water, 10% acetic acid, water, 10% aqueous NaOH and brine. After drying over MgSO$_4$, the solvent was evaporated to about 20 mL and the coupling products were precipitated by the addition of methanol (200 mL) and collected by filtration. Recrystallization from ethyl acetate followed by chromatography over silica gel, eluting with CH$_2$Cl$_2$–hexane (1:1) (R_F = 0.74), gave **109** (0.99 g, 90%) as a slightly yellow solid, m.p. >220 °C. ^{1}H NMR (250 MHz, CDCl$_3$), δ = 7.51–7.56 (m, 32 H), 7.28 (d, J = 8.8 Hz, 4 H), 7.20 (d, J = 8.8 Hz, 4 H), 4.42 (t, J = 5.4 Hz, 4 H), 3.75 (m, 4 H), 2.56 (m, 4 H), 1.72 (s, 6 H) 1.34 (s, 18 H), 1.32 (s, 36 H); ^{13}C NMR (75.5 MHz, CD$_2$Cl$_2$), δ = 159.05, 152.98, 150.18, 148.11, 147.41, 133.39, 133.03, 132.19, 132.05, 131.85, 129.21, 128.85, 123.99, 123.94, 123.78, 122.44, 121.83, 117.61, 93.85, 91.46, 89.99, 88.28, 81.97, 74.40, 72.24, 49.34, 43.49, 35.48, 35.13, 31.40, 31.31, 30.98, 25.94; MS (MALDI-TOF), m/z = 1964.3 [M + Ag]$^+$; 1878.1 [M + Na]$^+$. C$_{129}$H$_{112}$O$_8$S$_2$ (1854.45): calcd. C 83.55, H 6.09; found C 83.38, H 5.99%.

Compound **110**
Tetrabutylammonium hydroxide (1.6 mL, 40% in water) was added to a solution of **109** (220 mg, 0.12 mmol) in THF (20 mL). The mixture was stirred overnight at 40 °C and then concentrated to a small volume. Methanol was added to the oily residue (10 mL), the suspension stirred for 4 h and the product collected by filtration to give **110** (190 mg, 85%) as a yellow solid, m.p. > 220 °C. ^{1}H NMR (250 MHz, CDCl$_3$), δ = 7.61–7.44 (m, 32 H), 4.37 (t, J = 5.7 Hz, 4 H), 3.30 (m, 16 H) 3.17 (m, 4 H), 2.48 (m, 4 H), 1.61 (m, 16 H), 1.41 (m, 16 H) 1.32 (s, 18 H), 1.31 (s, 36 H) 0.97 (t, J = 7.2 Hz, 24 H); ^{13}C NMR (75.5 MHz, CDCl$_3$), δ = 158.16, 152.34, 149.56, 148.17, 147.50, 133.48, 132.97, 132.06, 131.94, 131.80, 129.14, 128.76, 123.59, 123.29, 122.49, 121.78, 116.97, 93.47, 91.23, 89.27, 87.93, 81.97, 77.47, 71.77, 48.31, 42.89, 34.95, 34.66, 31.40, 31.31, 25.38, 16.93, 13.08, 11.29; MS (MALDI-TOF), m/z = 1776.3 [M$_{\text{Dinion}}$ + 3 K$^+$]$^+$. C$_{146}$H$_{170}$N$_2$O$_8$S$_2$·4H$_2$O (2217.2): calcd. C 79.08 H 8.11; found C 79.11 H 8.38%.

List of Abbreviations

DCE	1,2-dichloroethane	PAM	phenylacetylene macrocycle
DEAD	diethylazodicarboxylate	PS	polystyrene
GPC	gel permeation chromatography	Py	pyridine
		TBAF	tetrabutylammonium fluoride
MALDI-TOF	matrix-assisted laser desorption/ionization time-of-flight	TBAOH	tetrabutylammonium hydroxide
		THP	tetrahydropyranyl
M.p.	melting-point	TIPS	triisopropylsilyl
NBS	N-bromosuccinimide	TMS	trimethylsilyl

References

1. For recent reviews on shape-persistent macrocycles, see e.g. (a) J. S. Moore, Acc. Chem. Res. 1997, 30, 402–413; (b) S. Höger, J. Polym. Sci. Part A: Polym. Chem. 1999, 37, 2685–2698; (c) M. M. Haley, J. J. Pak, S. C. Brand, Top. Curr. Chem. 1999, 201, 81–130; (d) C. Grave, A. D. Schlüter, Eur. J. Org. Chem. 2002, 3075–3098; (e) D. Zhao, J. S. Moore, Chem. Commun. 2003, 807–818; (f) S. Höger, Chem. Eur. J. 2004, 10, 1320–1329; (g) F. Diederich, P. J. Stang, R. R. Tykwinsky (Eds.), Acetylene Chemistry, Wiley-VCH, Weinheim, 2005.
2. S. Höger, K. Bonrad, A. Mourran, U. Beginn, M. Möller, J. Am. Chem. Soc. 2001, 123, 5651–5659.
3. See e.g. (a) P. Müller, I. Usón, V. Hensel, A. D. Schlüter, G. M. Sheldrick, Helv. Chim. Acta, 2001, 84, 778–784; (b) S. Höger, X. H. Cheng, A.-D. Ramminger, V. Enkelmann, A. Rapp, M. Mondeshiki, I. Schnell, Angew. Chem. 2005, 117, 2862–2866; Angew. Chem. Int. Ed. 2005, 44, 2801–2805.
4. S. Rosselli, A.-D. Ramminger, T. Wagner, B. Silier, S. Wiegand, W. Häussler, G. Lieser, V. Scheumann, S. Höger, Angew. Chem. 2001, 113, 3234–3237; Angew. Chem. Int. Ed. 2001, 40, 3138–3141.
5. D. M. Tiede, R. Zhang, L. X. Cheng, L. Yu, J. S. Lindsey, J. Am. Chem. Soc. 2004, 126, 14054–14062.
6. Although pure aliphatic compounds with a high persistent length are known, as for example [n]staffanes, shape-persistent macrocyclic compounds are dominated by structures with an aromatic backbone.
7. (a) K. Sonogashira, Y. Tohda, N. Hagihara, Tetrahedron Lett. 1975, 4467–4470; (b) K. Sonogashira, in Metal-catalyzed Cross-coupling Reactions, F. Diederich, P. J. Stang (Eds.), Wiley-VCH, Weinheim, 1998, pp. 203–229.
8. (a) T. W. Greene, P. G. M. Wuts, Protective Groups in Organic Synthesis, 3rd edn., Wiley-VCH, Weinheim, 1999, pp. 654–659; (b) S. Höger, K. Bonrad, J. Org. Chem. 2000, 65, 2243–2245.
9. For the concept of intra-annular groups, see e.g. (a) F. Vögtle, P. Neumann, Tetrahedron 1970, 26, 5299–5318; (b) E. Weber, F. Vögtle, Angew. Chem. 1974, 86, 126–127; Angew. Chem. Int. Ed. Engl. 1974, 13, 149–150; (c) E. Weber, F. Vögtle, Chem. Ber. 1976, 109, 1803–1831.
10. For shape-persistent macrocycles with adaptable side-groups, see (a) S. Höger, V. Enkelmann, Angew. Chem. 1995, 107, 2917–2920; Angew. Chem. Int. Ed. Engl. 1995, 34, 2713–2716; (b) D.L. Morrison, S. Höger, J. Chem. Soc., Chem. Commun. 1996, 2313–2314; (c) S. Höger, A.-D. Meckenstock, S. Müller, Chem. Eur. J. 1998, 4, 2421; (d) S. Höger, D. L. Morrison, V. Enkelmann, J. Am. Chem. Soc. 2002, 124, 6734–6736.
11. (a) T. Kawase, H. R. Darabi, M. Oda, Angew. Chem. 1996, 108, 2803–2805; Angew. Chem. Int. Ed. Engl. 1996, 35, 2664–2666; (b) T. Kawase, K. Tanaka, N. Shiono, Y. Seirai, N. Shiono, M. Oda, Angew. Chem. 2004, 116, 1754–1756; Angew. Chem. Int. Ed. 2004, 43, 1722–1724.
12. H. A. Staab, K. Neunhoeffer, Synthesis 1974, 424.
13. (a) P.-H. Ge, W. Fu, W. A. Herrmann, E. Herdtweck, C. Campana, R. D. Adams, U. H. F. Bunz, Angew. Chem. 2000, 112, 3753–3756; Angew. Chem. Int. Ed. 2000, 39, 3607–3610; (b) U. H. F. Bunz, Acc. Chem. Res. 2001, 34, 998–1010.
14. J. S. Moore, J. Zhang, Angew. Chem. 1992, 104, 873–874; Angew. Chem. Int. Ed. Engl. 1992, 31, 922–924.
15. A. S. Shetty, J. Zhang, J. S. Moore, J. Am. Chem. Soc. 1996, 118, 1019–1027.
16. J. Zhang, D. J. Pesak, J. L. Ludwick, J. S. Moore, J. Am. Chem. Soc. 1994, 116, 4227–4239.
17. J. S. Moore, E. J. Winstein, Z. Wu, Tetrahedron Lett. 1991, 32, 2465–2466.
18. H. Staab, F. Binning, Chem. Ber. 1967, 100, 293–305.
19. V. Hensel, K. Lützow, J. Jakob, K. Gessler, W. Saenger, A. D. Schlüter, Angew. Chem. 1997, 109, 2768–2770; Angew. Chem. Int. Ed. Engl. 1997, 36, 2654–

2656; (b) V. Hensel, A. D. Schlüter, Chem. Eur. J. 1999, 5, 421–429.
20. L. T. Scott, M. J. Cooney, D. Johnels, J. Am. Chem. Soc. 1990, 112, 4054–4055.
21. A. M. Boldi, F. Diederich, Angew. Chem. 1994, 106, 482–485; Angew. Chem. Int. Ed. Engl. 1994, 33, 486–489.
22. (a) Y. Tobe, N. Utsumi, K. Kawabata, K. Naemura, Tetrahedron Lett. 1996, 37, 9325–9328; (b) Y. Tobe, N. Utsumi, K. Kawabata, A. Nagano, K. Adachi, S. Araki, M. Sonoda, K. Hirose, K. Naemura, Tetrahedron 2001, 57, 8057–8083; (c) Y. Tobe, N. Utsumi, K. Kawabata, A. Nagano, K. Adachi, S. Araki, M. Sonoda, K. Hirose, K. Naemura, J. Am. Chem. Soc. 2002, 124, 5350–5364.
23. Similar results were also observed for other diethynyl cyclizations, cf. [22a].
24. S. Höger, A.-D. Meckenstock, H. Pellen, J. Org. Chem. 1997, 62, 4556–4557.
25. J. A. Marsden, J. J. Miller, M. M. Haley, Angew. Chem. 2004, 116, 1726–1929; Angew. Chem. Int. Ed. 2004, 43, 1694–1697.
26. (a) N. V. Gerbeleu, V. B. Arion, J. Burgess, Template Synthesis of Macrocyclic Compounds, Wiley-VCH, Weinheim, 1999; (b) F. Diderich, P. J. Stang (Eds.), Templated Organic Synthesis, Wiley-VCH, Weinheim, 2000.
27. D. H. Bush, J. Inclusion Phenom. 1992, 12, 389–395.
28. (a) H. L. Anderson, J. K. M. Sanders, Angew. Chem. 1990, 102, 1478–1480; Angew. Chem. Int. Ed. Engl. 1990, 29, 1400–1403; (b) D. W. J. McCallien, J. K. M. Sanders, J. Am. Chem. Soc. 1995, 117, 6611–6612.
29. J. Li, A. Ambroise, S. I. Yang, J. R. Diers, J. Seth, C. R. Wack, D. F. Bocian, D. Holten, J. S. Lindsey, J. Am. Chem. Soc. 1999, 121, 8927–8940.
30. G. Fuhrmann, T. Debaerdemaeker, P. Bäuerle, Chem. Commun. 2003, 948–949.
31. See e.g. L. R. Cox, S. V. Ley, [26b], pp. 275–395.
32. S. Rosselli, A.-D. Ramminger, T. Wagner, G. Lieser, S. Höger, Chem. Eur. J. 2003, 9, 3481–3491.
33. S. Höger, A.-D. Mechenstock, Tetrahedron Lett. 1999, 39, 1735–1736.
34. The catenation of M by A (solvophobic driven) can be excluded since no influence of the torus volume on the reaction outcome could be observed (82a vs. 82b).
35. S. Höger, A.-D. Ramminger, V. Enkelmann, unpublished; selected crystal data: 85·3.5DCE (T = 120 K), triclinic, P-1, a = 12.9926(7), b = 19.1674(7), c = 23.5859(9) Å, α = 106.5780(14), β = 101.4468(13), γ = 106.4426(15)°, V = 5144.2(4) Å^3, Z = 2, Dx = 1.188 g cm–3, 54 871 independent reflections, 12 063 reflections observed, R = 0.0481, Rw = 0.0555.
36. S. Höger, A.-D. Meckenstock, Chem. Eur. J. 1999, 5, 1686–1691.
37. M. Fischer, S. Höger, Eur. J. Org. Chem. 2003, 441–446.
38. M. Fischer, S. Höger, Tetrahedron 2003, 59, 9441–9446.
39. (a) M. Fischer, G. Lieser, A. Rapp, I. Schnell, W. Mamdouh, S. De Feyter, F. De Schryver, S. Höger, J. Am. Chem. Soc. 2004, 126, 214–222; (b) D. Borissov, A. Ziegler, S. Höger, W. Freyland, Langmuir 2004, 20, 2781–2784.
40. A. Ziegler, W. Mamdouh, A. Ver Heyen, M. Surin, H. Uji-i, M. M. S. Abdel-Motalleb, F. De Schreyver, S. de Feyter, R. Lazarroni, S. Höger, Chem. Mater., 2005, 17, 5670–5683.
41. For a summary, see e.g. S. J. Rowan, S. J. Cantrill, G. R. L. Cousins, J. K. M. Sanders, J. F. Stodart, Angew. Chem. 2002, 114, 938–993; Angew. Chem. Int. Ed. 2002, 41, 898–952.
42. Metal–organic frameworks obtained by self-assembly are topical examples for this strategy, e.g. (a) M. Fujita, K. Umemoto, M. Yoshizawa, N. Fujita, T. Kusokawa, K. Biradha, Chem. Commun. 2001, 509–518; (b) S. R. Seidel, P. J. Stang, Acc. Chem. Res. 2002, 35, 972–983.
43. J. Gawroński, H. Kolbon, M. Kwit, A. Katrusiak, J. Org. Chem. 2000, 65, 5768–5773.
44. One of the most graphical recent examples are Stoddart's Borromean rings: K. S. Chichak, S. J. Cantrill, A. R. Pease, S.-H. Chiu, G. W. V. Cave, J. L. Atwood, J. F. Stoddart, Science 2004, 304, 1308–1312.

45. D. Zhao, J. S. Moore, J. Org. Chem. 2002, 67, 3548–3554. Imine metathesis reactions in different solvents suggest that the formation of assemblies of stacked rings is the thermodynamic driving force for the exclusive cyclodimer formation: D. Zhao, J. S. Moore, Macromolecules 2003, 36, 2712–20.

46. S. Akine, T. Taniguchi, T. Nabeshima, Tetrahedron Lett. 2001, 8861–8864.

47. A. J. Gallant, M. J. MacLachlan, Angew. Chem. 2003, 112, 5465–5468; Angew. Chem. Int. Ed. 2003, 42, 5307–5310.

48. C. Ma, A. Lo, A. Abdolmaleki, M. J. MacLachlan, Org.Lett. 2004, 6, 3841–3844.

49. For a fairly extensive investigation of the alkyne metathesis towards o-dehydrobenzannulenes, see O. S. Miljani, K. P. C. Vollhardt, G. D. Whitener, Synlett 2003, 29–34.

50. W. Zhang, J. S. Moore, J. Am. Chem. Soc. 2004, 126, 12796.

51. Previous experiments by Moore's group have shown that high conversions in the metathesis reaction could be obtained in a closed system if the reaction product is insoluble in the reaction mixture: W. Zhang, S. Kraft, J. S. Moore, J. Am. Chem. Soc. 2004, 126, 329–335.

52. For the large-scale synthesis of 125, the crude reaction product was a mixture of cyclic hexamer and cyclic pentamer (9:1) and the compounds were separated by column chromatography.

7
Functional Materials via Multiple Noncovalent Interactions
Joseph R. Carlisle and Marcus Weck

7.1
Introduction

The concept of self-assembly is ubiquitous in nature. Self-assembly allows for the complexity, variation and differentiation in all life forms in nature [1]. As a result, researchers have been attempting to mimic Nature's self-assembly processes. The concept of self-assembly, however, did not actually begin to gain the enormous popularity it has today until the late 1970s, when Jean-Marie Lehn first introduced the term "supramolecular chemistry" [2]. Soon thereafter, publications on the topic began to grow steadily, from a mere 62 publications in 1980 that concentrated mainly on self-assembly, to 4316 citations in 2004 (Fig. 7.1). The exponential growth of the field of self-assembly and supramolecular chemistry illustrates the enormous impact that it is having on the research community and potentially the world.

Besides attempts at directly mimicking Nature, self-assembly has come to be commonly employed in a large number of other unrelated applications, including electronic and photonic devices [3], liquid crystalline materials [4, 5] biosensors [6], drug delivery [7], molecular electronics [8] and many others [9–16]. Owing to the simple and parallel nature of the process and the low energy demands brought forth through the utilization of self-assembly, this phenomenon is rapidly becoming ubiquitous in the field of materials science also.

Nature rarely, if ever, exhibits a single self-assembly step. Instead, the complexity and superb functionality of natural materials are a result of multiple tiers of numerous hierarchical self-assembly steps that create the elegant order observed in biological systems [1]. Synthetic materials that follow numerous stepwise levels of organization can be recognized as multifunctional materials and comprise perhaps chemistry's closest attempts thus far at following in Nature's footsteps. Such multifunctional materials normally undergo a single self-assembly step, thus creating new materials with different properties than the parent compounds. The emergent new materials, as a result of the noncovalent nature embedded within, are now capable of exhibiting behavior such as secondary self-assembly,

Functional Organic Materials. Syntheses, Strategies, and Applications.
Edited by Thomas J.J. Müller and Uwe H.F. Bunz
Copyright © 2007 WILEY-VCH Verlag GmbH & Co. KGaA, Weinheim
ISBN: 978-3-527-31302-0

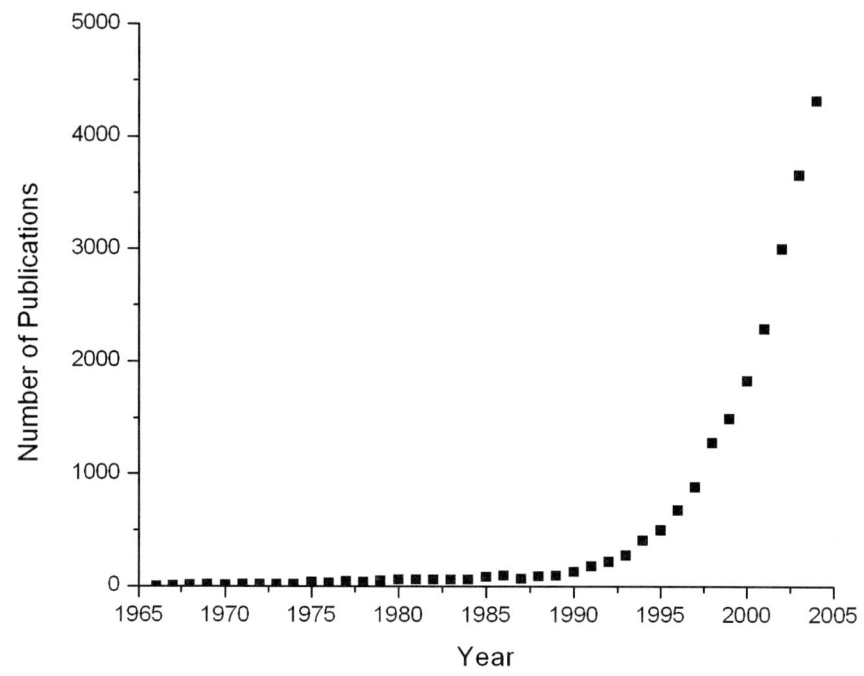

Fig. 7.1 Annual publications focused mainly on self-assembly.

organization or alignment giving the next generation of material a more complex, secondary or tertiary structure. The noncovalent nature of these supramolecules also allows for responsiveness, self-healing and greater adaptability to the surrounding environment, thus giving rise to the term "smart materials" [17–26].

This chapter outlines recent accomplishments in employing self-assembly towards the creation of multifunctional materials. It describes the basic design principles involved and demonstrates these design principles using examples from the recent literature. These examples follow the logical extension from the field starting with bio-inspired materials to small molecule hierarchical assemblies and main-chain and side-chain multifunctional supramolecular polymers, in which scientists have harnessed the power of self-assembly in order to create multifunctional materials.

7.2
Biologically Inspired Materials via Multi-step Self-assembly

The motivation of all scientists to employ self-assembly methods in synthetic chemistry and materials science is the biological self-organization in Nature. Biological self-organization has been defined as the "spontaneous building up of complex structures that takes place under adequate environmental conditions solely on the basis of the respective molecular property, namely, without the effect of external factors, for example, protein folding, formation of lipid double layers, morphogenesis" [27]. Over the years, considerable effort has been placed on attempts to mimic nature, in the hope that we may one day accomplish on a very simple basis what Nature accomplishes every day. Although fulfilling that goal still lies in the unforeseen future, chemists have gained considerable insights through their attempts to mimic the workings of Nature in the laboratory. One concept is clear: the elegant ease with which living organisms exhibit evolution, replication, energy dissipation, self-healing and conversion of food into energy is largely due to a complex web of processes that rely heavily upon the phenomenon of self-assembly.

Peptides lend themselves particularly well to noncovalent interactions and self-assembly, as a result of the possibility of forming a variety of noncovalent interactions including strong hydrogen-bonding interactions, salt bridges and hydrophobic interactions. Making extensive use of these interactions, linear polypeptides self-organize into α-helix, β-sheet and coils followed by the self-assembly of these units into even more complex tertiary and quaternary structures [13, 28–30]. Scientists have tried to mimic this behavior by engineering appropriate amino acid sequences into polypeptides that are capable of self-assembling into a large variety of different structures.

To explain the use of self-assembly in biomimetic materials, we will discuss one example of a synthetic multifunctionalization peptide system in detail that was reported by Mihara and co-workers in 2004. They created nano-fibers with uniform morphologies through the use of self-assembly and the self-organization of designed peptides [28]. By creating a specific peptide sequence based on 10 amino acids per chain, they observed that β-strands organize into β-sheets, forming protofibrils, which in turn self-assemble into ribbons that coil in a left-handed fashion to make a straight fiber (Fig. 7.2).

In order to reduce the unfavorable possibility of aggregation or formation of other types of morphologies which have been observed in previous work [31–33], Mihara and co-workers used a specific alignment of hydrophobic and hydrophilic residues such that the hydrophobic plane and the hydrophilic plane were on opposite sides of the β-strands, with an inversion of planes in the center of the strands. Additionally, they also functionalized both the C- and the N-termini with proline residues. Proline, owing to its cyclic structure, does not have an N–H hydrogen acceptor available for hydrogen bonding and therefore has been known as a β-sheet breaker. By these measures it was observed that the peptides easily self-assembled into various predictable architectures. Furthermore, the resulting macromolecular structures can be controlled and tailored by a number of methods. In the case of Mihara and coworkers' work, it was shown that (1) the order of the amino acid sequence, (2) the choice of N- and C-terminal residues and (3) hydrophobic/hydrophilic effects can be used advantageously in controlling and tailoring macromolecular architecture of the resulting three-dimensional peptides.

A second example where multiple noncovalent interactions have been employed and that is motivated by biological systems is the employment of lipid bi-

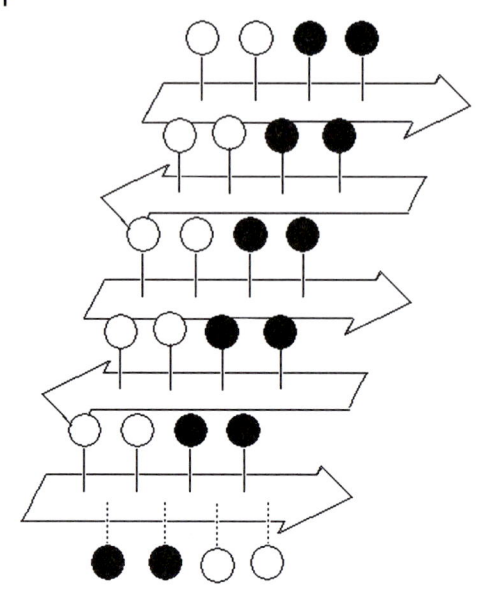

Fig. 7.2 Alignment of peptide 10-mers into three-dimensional architectures.

layers, micelles and vesicles in three-dimensional architectures [34–36]. Nature employs amphiphilic molecules in the assembly of lipids into bilayers. The driving force behind the lipid bilayer formation is the concept of hydrophobicity. In an aqueous environment, lipids containing a polar head group and nonpolar chains can aggregate into three-dimensional structures by minimizing surface tension between the water phase and the hydrophobic chains. Chemists can easily reproduce these types of basic phenomena in the laboratory using various natural or synthetic amphiphiles [37–42]. Depending on the concentration of the amphiphile, several different types of structures result from this hydrophobic self-assembly (Fig. 7.3).

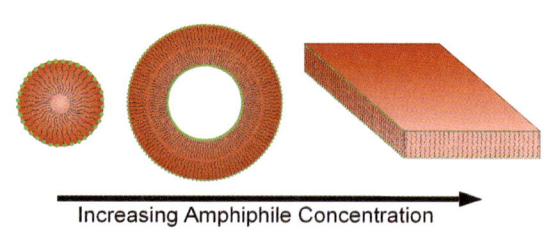

Fig. 7.3 Concentration dependence exhibited by amphiphiles in solution. Micelles form in dilute solution (left); larger suprastructures such as vesicles begin to form at higher concentrations (middle); long extended bilayers form at high concentrations (right). Polar head groups, green; nonpolar tails, black; organic phase, red.

These basic three-dimensional constructs can also be used in the design of three-dimensional architectures. For example, in 2001, Sasaki and co-workers combined two self-assembly techniques to create reversible columns of stacked lipid bilayers [43]. First, vesicles were formed by amphiphilic lipids in water, followed by metal coordination through the polar head groups of certain lipids, resulting in assembly of the vesicles into a three-dimensional layered structure. The metal coordination was found to be strong enough to cause flattening of the vesicles and ultimately rupturing them, resulting in two stacked discs per original vesicle. Sasaki and co-workers reported columns of up to 45 stacked bilayers (Fig. 7.4) that were characterized using electron microscopy.

The vesicles were made from a mixture of two different lipids: 95% distearyl-phosphatidylcholine (DSPC), a zwitterionic lipid, and 5% PSIDA (a Cu^{2+} specific metal-chelating lipid). Once the vesicles were formed, addition of Cu^{2+} caused aggregation, which Sasaki and co-workers found to consist of approximately 15–20% of the material in the columnar formation. This Cu^{2+}-mediated aggregation was fully reversible. On adding EDTA to the reaction mixture, the copper ions could be quantitatively removed from the solution. This combination of reversible self-assembly techniques, the hydrophobic effect and metal coordination, is yet another exciting example of the enormous variety of self-organizational phenomena which are ubiquitously employed by Nature, seemingly effortlessly, in all biological systems.

These are just two examples of bio-inspired materials science where scientists have employed multi-step self-assembly approaches to design and synthesize materials. However, they clearly demonstrate the potential that multi-step self-assembly has in fabricating bio-inspired materials. Biological systems provide a nearly inexhaustible source of inspiration both for discovering new chemical reactions and mimicking currently known biological reactions and processes on continuously greater levels of efficiency and accuracy. Nature very often bases its methods of achieving commonly observed high levels of order on small-molecule self-organization, which was evidenced in the second example above by Sasaki and co-workers. The following section will describe several successful attempts by chemists who have begun to utilize this same small-molecule-based approach to higher levels of hierarchical order using synthetic, non-natural starting materials.

7.3
Small Molecule-based Multi-step Self-assembly

The self-assembly of small molecules into complex material based on a multitude of different noncovalent interactions, such as hydrogen bonding, ionic interactions, hydrophobicity and metal coordination, has been established over the last century [2, 30, 38, 44–49]. In this context, the real power of self-assembly becomes evident when not only multiple noncovalent interactions but also multiple levels of self-assembly occur within the same small molecule system. These multiple tiers of self-organizational hierarchy can yield highly complex structures with sec-

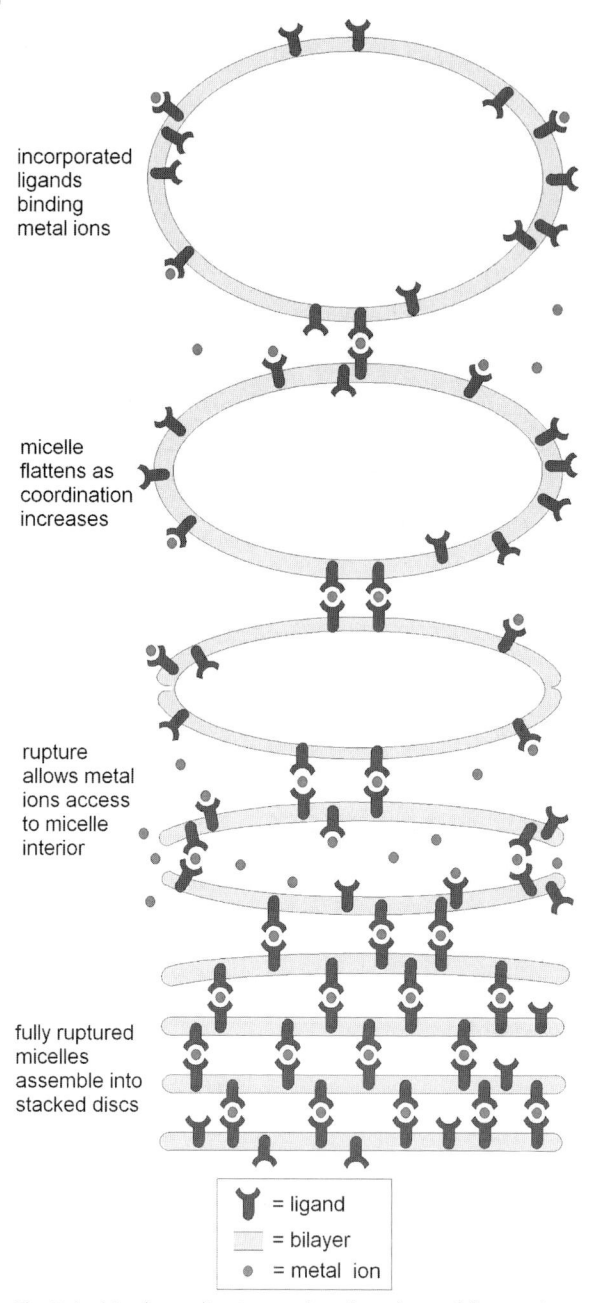

Fig. 7.4 Metal coordination-mediated stacking of flattened, ruptured vesicles.

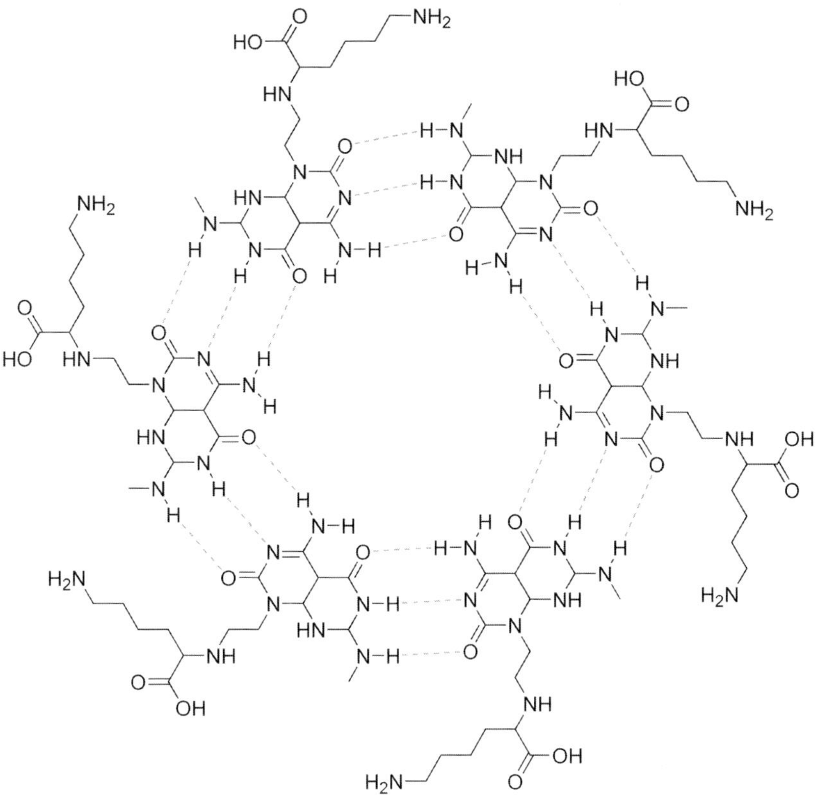

Fig. 7.5 Self-assembled GC units forming a complete rosette structure.

ondary or even tertiary levels of organization. One example is the formation of columnar aggregates resulting from the self-assembly of small molecules into disc-shaped aggregates that then are able to stack face-to-face, forming large cylindrical and rod-like shapes [50]. A recent application of this widely used concept is the employment of this concept in the synthesis of helical rosette nanotubes by Fenniri and co-workers (Fig. 7.5) [46].

In this case, the authors started their material development with small hydrogen-bonding units based on DNA base pairs that have been reported previously to self-assemble into cyclic structures [51]. By functionalizing these hydrogen-bonding units with an amino acid moiety, they imparted to the self-assembled cyclic structure (a) chirality and (b) a highly polar outer portion.

The design of the hydrogen-bonding unit used by Fenniri and co-workers is straightforward. Using a "GC motif"-based molecule, i.e. a molecule that is able to undergo hydrogen bonding through two well-defined recognition sites, (a) a donor–donor–acceptor (DDA) side that resembles guanine and (b) an acceptor–acceptor–donor (AAD) side that resembles cytosine, they designed a molecule that is self-complementary, i.e. one hydrogen-bonding face is the complementary recognition unit to the other hydrogen-bonding face. By fixing the angle between the two faces at approximately 60°, six molecules are required to yield a noncovalent cyclic structure that is known as a rosette structure. Fenniri and co-workers designed three types of hydrogen-bonding building blocks in order to test the effects on the resulting self-assembled superstructures. The first type (**1**) had a single +1 charge and bore one GC base, the second type (**2**) employed a +3 charge and a single GC base and the third type (**3**) had a +3 charge but bore two GC bases. Upon self-assembly, it was observed that all three of these structures formed rosettes; however, only **1** and **3** formed higher ordered aggregates, i.e. stacked nanotubes. Fenniri and coworkers attributed this phenomenon to the greater steric bulk of the rosettes of **2** compared with **1** or **3**, as well as electrostatic repulsion. The hydrophobic character of the hydrogen-bonded interior combined with the chiral, polar exterior of the rosettes created by the amino acid functionalization was sufficient to align these self-assembled structures into the stacked formation characteristic of nanotubes. The chirality directed the stacks into helices, while the void volume in the center of the rosettes extended the entire length of the cylindrical structure, creating nanotubes in a highly efficient manner. The stacking persisted for distances as large as 7.5 μm.

The conformation of these helical rosette nanotubes (HRNs) was found to be highly pH dependent. To demonstrate this phenomenon, HRNs of **3** were assembled at pH values of 4, 7 and 11. It was observed that at pH 4, when all basic sites on each molecule are protonated, giving a net charge of +3 per small molecule, HRNs were significantly shorter (160 ± 111 nm), did not aggregate or associate with one another and showed a bimodal size distribution. At pH 7, where a net charge of +2 is expected per small molecule, HRNs were much longer (approximately 5 μm) and typically the nanotubes aggregated laterally to form ribbons. At pH 11, when there is only a +1 net charge per small molecule, HRNs existed as right-handed superhelices consisting of about 32 nanotubes in width, with lengths up to 7.5 μm. This example of the use of basic hydrogen-bonding self-assembly steps to fabricate nanostructures clearly demonstrates the power of self-assembly. Not only can the nanostructures be fabricated in a fast and very efficient manner, but also the size and other physical properties can be tailored easily. It is hard to imagine that one would have the same control over these characteristics and properties using covalent chemistry.

There are a great many methods by which small molecules can be assembled to form nanotubes and other similar columnar aggregates [50]. Unlike the stacked disc methodology employed by Fenniri's group, Russell and co-workers took a completely different approach, which consisted of nanotubes made by way of rolled-up bilayers [48]. Their study of biocidal nanotubes from a self-assembled diacetylene salt employed a hydrophobically driven bilayer type of self-assembly step. This bilayer formation was based on the alignment of the hydrophobic chains, including aligned π-systems of the acetylenes on the interior of the bilayer and the polar hydrobromide salts of an amine on the exterior of the bilayer (Fig. 7.6).

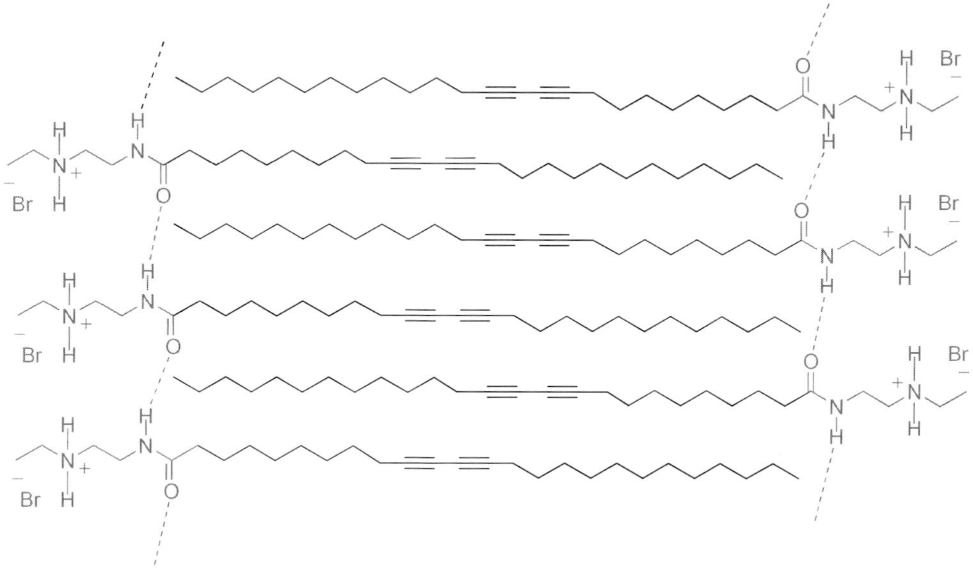

Fig. 7.6 Bilayer formed by self-assembly of amphiphilic hydrobromide diacetylene salts.

In this example, intermolecular hydrogen bonding of the amide proton and carbonyl oxygen is postulated to increase further the stability of the self-assembled structure. Bilayer sheets then bend into a tubular conformation, forming nanotubes, often consisting of several layers of bilayer sheets. Tubes that were five sheets thick were found to be the most stable configuration. Using scanning electron microscopy (SEM), both a very uniform wall thickness of 27 nm and an inner diameter of 35 nm were measured for the nanotubes. The length varied somewhat, but on average was approximately 1 µm. Compared with Fenniri and coworkers' nanotubes, those of Russell and coworkers were approximately an order of magnitude thicker in diameter and shorter by a factor of 5–7, depending on the pH. Russell and coworkers' nanotubes could be photo-cross-linked through UV irradiation, causing polymerization of adjacent diacetylene groups.

Furthermore, the nanotubes were dried onto a glass surface with the addition of a few droplets of chloroform, which were subsequently allowed to evaporate gradually. The resulting material was scraped from the slide and analyzed by SEM, giving some unexpected results: the nanotubes had become unidirectionally aligned in such way as to resemble a carpet and therefore this material was dubbed the "nanocarpet". Russell and coworkers postulated that the chloroform partially melts the surface of the nanotubes into a lamellar structure and then, as it gradually evaporates, the disoriented nanotubes organize themselves. However, the true mechanism of this behavior has not yet been fully elucidated. The "fibers" of the nanocarpet retain the same prealignment length, but exhibit thicker widths on the order of 100 nm, owing to the aggregation of three to four tubes

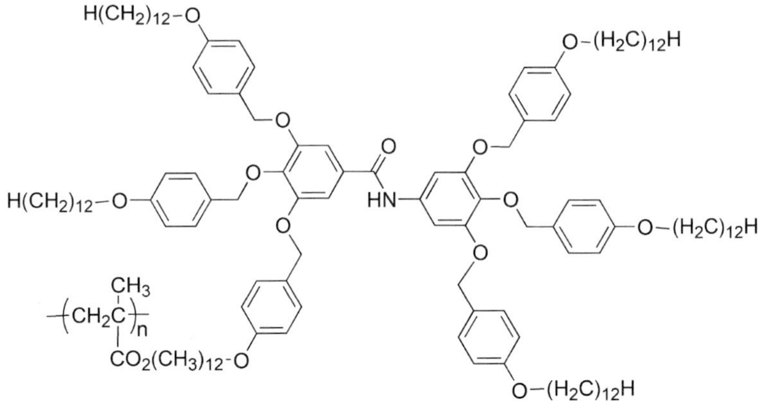

Fig. 7.7 Twin-wedge polymer.

per "fiber" during the alignment process. Russell and co-workers also found that their initial goal of creating a biocidal material from these nanotubes was also met, as the nanotubes proved to be highly effective antimicrobial agents.

Small molecules can undergo self-organization into a wide variety of orientations besides nanotubes. For example, in 2003, Percec designed a system of self-assembled dendritic benzamides based on methyl gallate and trimethoxybenzene that exhibited a liquid crystalline phase both as monomers and as polymers (Fig. 7.7) [52].

Both the twin-wedge complex and the unpolymerized analog exhibited enantiotropic hexagonal columnar LC phases, based on a repeating four-cylinder bundle pattern, as confirmed by X-ray diffraction (Fig. 7.8).

The polymerized complex also exhibits a hexagonal columnar phase, but only shows short-range order that was found to resemble more closely a columnar nematic phase. Percec suggested that the polymer backbone drives the self-organization into a cylindrical, vesicular arrangement that subsequently organizes into a nematic columnar phase with short-range hexagonal order. This is due to "defects" or missing components along the backbone which would be necessary to form the same four-cylinder bundles, that is, the polymer does not provide a dense enough contribution of twin wedges in order to supply the four-fold requirement for the same linear stacking observed in the unpolymerized assemblies.

Interestingly, the ordering of the LC phase of the polymerized twin wedge is highly affected by the addition of an unpolymerized twin wedge and, furthermore, the resulting orientations are dependent on the ratio of polymerized to unpolymerized twin wedges. The ratios tested (polymerized:unpolymerized) ranged from 80:20 to 20:80. At the initial 80:20 ratio, it was observed that significantly more regular ordering was occurring, similar in fashion to the unpolymerized system. Percec postulated that this is in fact due to the monomers acting as guest complexes to the host polymers, occupying the vacancies left along the poly-

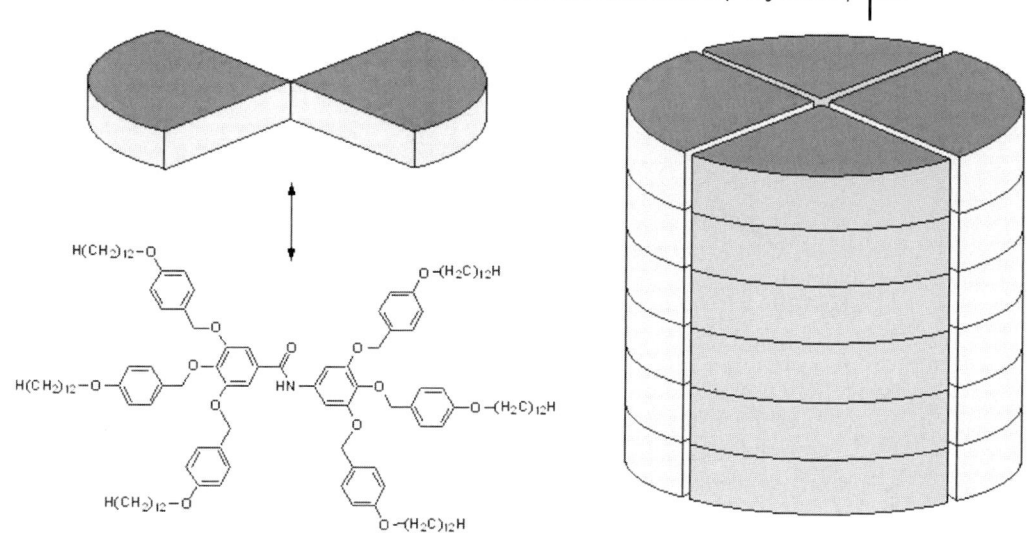

Fig. 7.8 Cartoon of stacked twin-wedge monomers forming a cylinder.

mer backbone and thus resulting in the expected highly ordered four-cylinder bundles.

In a system based largely on self-assembly through mesogenic interactions such as the above example by Percec, it can also be beneficial to incorporate hydrogen bonding as well for an added element of organizational control. For instance, Fréchet and coworkers demonstrated sequential, hierarchical self-assembly of various levels of order in a small molecule-based system which ultimately resulted in a hydrogen-bonded liquid crystalline network (Fig. 7.9) [53].

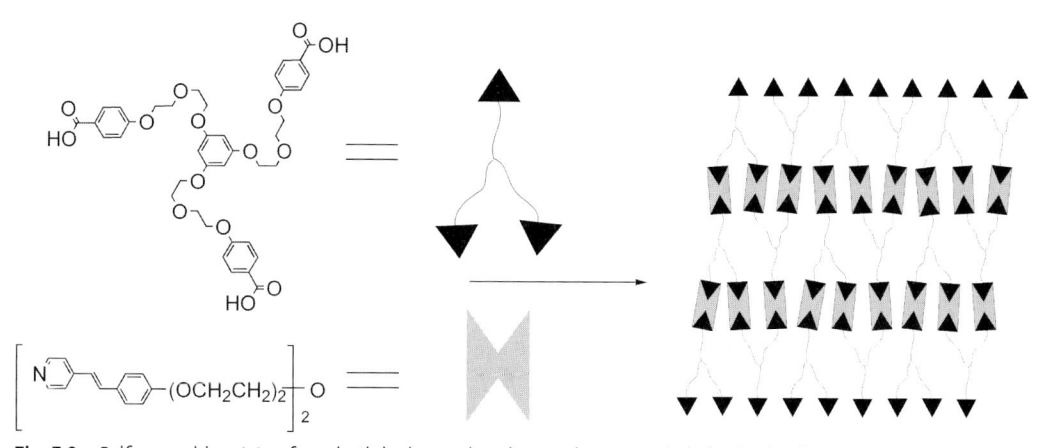

Fig. 7.9 Self-assembly arising from both hydrogen bonding and mesogenic behavior leading to a smectic liquid crystalline network.

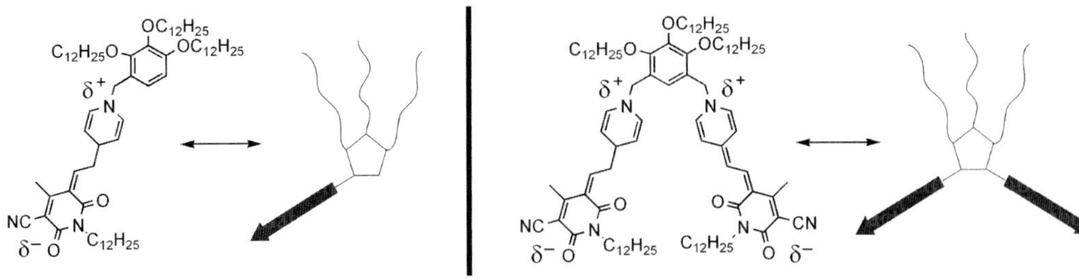

Fig. 7.10 Left, monomerocyanine dye and right, bismerocyanine dye.

The initial organization of the system was driven by the hydrogen-bonding donor and acceptor capabilities of the pyridine- and carboxylic acid-containing small molecules (shown in Fig. 7.9). By including some trifunctional units in the system, supramolecular hydrogen-bonded networks were formed upon combination of the donor and acceptor molecules. These networks subsequently self-assembled into either smectic or nematic liquid crystalline phases, giving higher order to the system. Both of these modes of self-assembly quantitatively and reversibly revert back to the isotropic small-molecule liquid state upon heating. Depending on the architecture of the trifunctional unit, either smectic or nematic liquid crystalline phases could be achieved with a high degree of control and reversibility.

Dipole–dipole interactions are another option for triggering the self-assembly of small molecules. For instance, a strong dipole is created in the conjugated zwitterionic merocyanine form of the photoswitchable molecule spiropyran, which, when in the open form, is a common dye. In 2003, Würthner et al. synthesized a modified merocyanine dye that was capable of forming extended, highly ordered superstructures based on this strong dipole dipole interaction [44]. The highly polar nature of merocyanine dyes in general, combined with the asymmetric addition of nonpolar alkyl chains, results in an amphiphilic structure that is capable of undergoing self-organization on several levels in even the least polar of hydrocarbon solvents. Beginning with the unsubstituted merocyanine molecule, Würthner et al. attached a chloromethyltriiododecyloxybenzyl unit to form a modified dye with a 1:3 ratio of dye molecule to alkyl chain. Additionally, they included a bis(chloromethyl)tridodecyloxybenzyl unit also that is capable of binding two merocyanine molecules, thus forming the dysfunctional bismerocyanine dye in a ratio of 2:3 dye to alkyl chain (Fig. 7.10).

Through strong dipole–dipole interactions (the merocyanine dye has a zwitterionic resonance form that gives this interaction a significant amount of ionic character), possible even in non-polar solvents such as hexane, the monomerocyanine dye forms dimers in solution, whereas the bismerocyanine dye is capable of forming extended structures (Fig. 7.11).

Würthner et al. demonstrated that this supramolecular polymer exists in a helical conformation and thus experiences a second form of higher self-organiza-

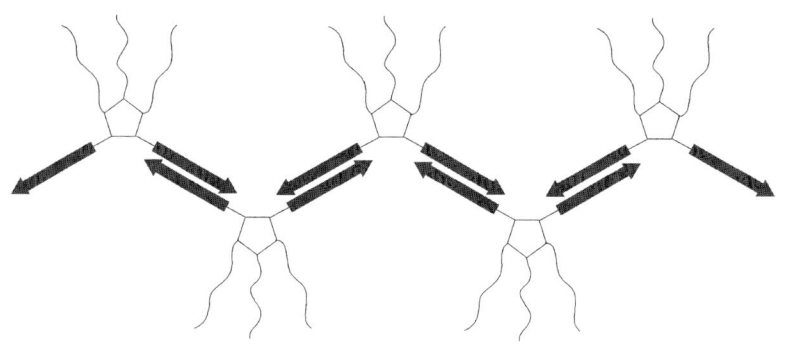

Fig. 7.11 Supramolecular polymerization of bismerocyanine molecules via dipolar aggregation.

Fig. 7.12 Hierarchical self-organization of merocyanine dye molecules into helices (left), followed by alignment into densely packed rods (middle) and finally packing of the rods into a hexagonal arrangement (right).

tion. These supramolecular helices then undergo a third level of organization in that once the helix has formed, six helices preferentially intertwine to form a densely packed rod that then self-organizes yet again into a hexagonal arrangement with other rods (Fig. 7.12).

Metal coordination can also be used as a self-assembly motif, often leading to highly organized and complex superstructures [47]. For example, Lehn and co-workers synthesized several multifunctional ligands, designed for selective complexation of specific metals, and demonstrated the ability of these self-assembled interactions to build both one- and two-dimensional architectures that exhibited magnetic properties [54]. The ligands consisted of a series of several linked N-heterocyclic pyridine-type rings, which resulted in two terpyridine (trpy)-like coordination sites, in addition to two single pyridine coordination sites (Fig. 7.13).

The single pyridine sites were designed with both a *m*- and a *p*-pyridine for varying metal specificity. Upon introduction of an iron(II) salt and subsequent binding of iron into two of the terpyridine-based sites, a square-shaped tetramer results with a metal:ligand ratio of 1:1 (Fig. 7.14).

This can be done for either type of ligand and the resulting square tetramers possess on both the top and bottom faces a new coordination site created by sev-

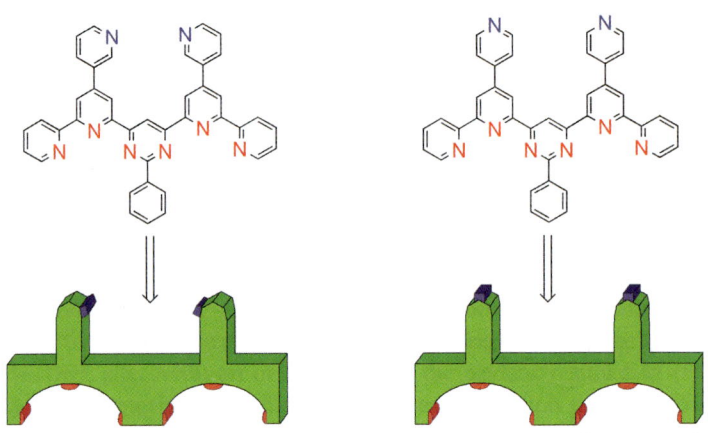

Fig. 7.13 Chemical structures and cartoons of multifunctional ligands containing both pyridine (blue) and terpyridine (red) coordination sites. Left, meta-pyridine building block; right, p-pyridine building block.

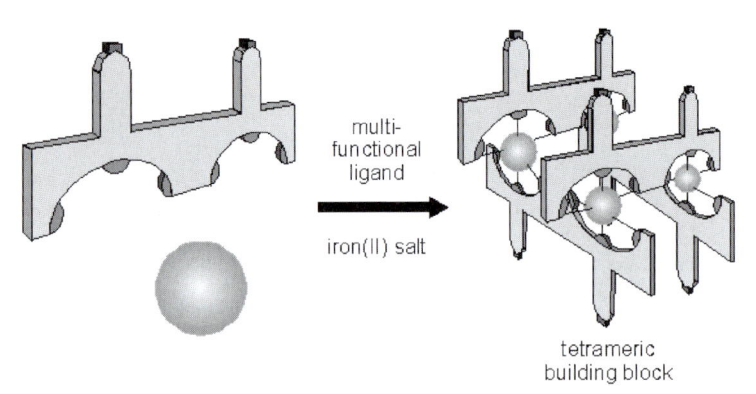

Fig. 7.14 Formation of supramolecular building block via iron(II) coordination.

eral of the single pyridines confined together in close proximity. The tetramers comprise the initial unit of what Lehn and co-workers refer to as a modular approach to supramolecular assembly and architecture. Next, depending on the configuration of the single pyridine portion of the resulting modules, various orientations and structures can be obtained. For instance, addition of a lanthanide salt, such as lanthanum(III) perchlorate, to the module containing the *m*-pyridine sites, creates one-dimensional rods as a result of one lanthanide ion binding to either the entire top or bottom face in a pyridine:lanthanide ratio of 8:1 (Fig. 7.15).

Alternatively, adding a silver salt such as silver tetrafluoroborate to the module based upon the *p*-pyridine-containing ligands creates large two-dimensional networks. These networks arise from a pyridine:silver ratio of 1:2, with the two pyr-

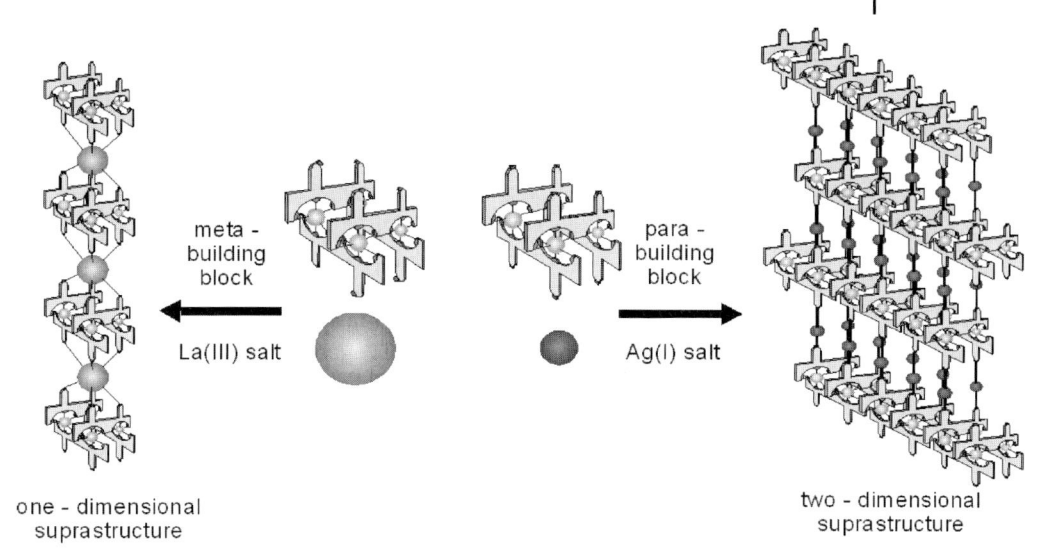

Fig. 7.15 Metal coordination-based suprastructures from combination of *m*- and *p*-iron-based building blocks and a second class of metal salts.

one-dimensional suprastructure (La(III) salt, meta-building block)

two-dimensional suprastructure (Ag(I) salt, para-building block)

idines arranged in a para configuration around each silver ion. Thus, for both the one- and two-dimensional suprastructures formed, a multistep, completely metal coordination-based series of events leads to the desired architectures. Interestingly, these superstructures were found also to exhibit magnetic properties once fully self-assembled.

The above examples illustrate some of the various methods and motifs by which small molecules can undergo self-organization and hierarchical self-assembly to form complex materials with important properties. The self-assembly exhibited by small molecules often gives rise to extended order in one or more directions, over relatively long distances. Polymers also exist as extended structures over long distances and, as such, are an important venue for probing self-assembly on the macromolecular scale.

7.4
Polymer-based Self-assembly

The hierarchical self-organization exhibited routinely by Nature and subsequently mimicked by chemists in the laboratory has opened up the field to a wide variety of potential applications and new methodology in the field of chemistry to which these concepts can be applied. Synthetic polymers can also exhibit several types and hierarchical levels of self-assembly, including (1) main-chain extension based on molecular self-assembly resulting in the formation of high molecular weight polymers, (2) side-chain self-assembly of small molecules onto molecular recognition sites along polymers and (3) macroscopic polymer self-assembly and self-organization/alignment. Whereas the use of self-assembly techniques in small molecules has been utilized and investigated for over a century, the employment of the above-outlined types of self-assembly in non-natural polymer science has been researched extensively only over the last two decades [17, 55–62]. In the following, we will introduce some recent developments in the field by presenting selective examples of each of the three types of supramolecular polymer science in detail.

7.4.1
Main-chain Self-assembly

Multifunctional supramolecular main-chain polymers, i.e. multifunctional polymers that are based on noncovalent interactions and self-assembly, have the advantage of being reversible. This reversibility allows one to tailor molecular weights through a variety of variables, including the strength of the noncovalent interactions between individual monomer units and the incorporation of terminal 'stopper' molecules, i.e. monofunctionalized monomers that terminate a polymer chain. While the incorporation of multiple noncovalent interactions along a polymer main chain is conceptually simple, the realization of this concept has been achieved only very recently. In 2005, Schubert and co-workers reported such a system by combining trpy-based metal coordination with hydrogen bonding. The system is based on poly(-caprolactone) that contains terminal self-assembly groups: on one side is a multidentate trpy ligand that can coordinate to a variety of metals including Ru and Co, while the other end-group contains a ureidopyrimidinone unit (a self-complementary, quadruple hydrogen-bonding unit with a DDAA motif) [63]. The resulting macromonomers can be self-assembled into linear main-chain supramolecular polymers using both hydrogen bonding and metal coordination. The properties of the polymers can be tuned easily through (1) the choice of metal ion, (2) temperature and (3) monomer and/or metal concentration. The trpy ligand is a versatile candidate for this type of supramolecular system owing to the tunability inherent in the very wide range of trpy–metal binding strengths (from 105 kJ/mol for Zn and Cd to 1018 for Ru) [64, 65]. Schubert and coworkers have investigated the potential of the trpy ligand in supramolecular polymers and synthesized a wide variety of polymers for a number of applications, including light-emitting devices [66], molecular switches [67] and solar cells [68]. Furthermore, hydrogen bonding, although typically weaker than metal coordination, is also widely tunable owing to (a) the dramatic increase in K_a as the number of hydrogen-bonding interactions increases, (b) the dependence of the hydrogen-bond strength on the solvent and (c) the dramatic decrease in the hydrogen bond strength at elevated temperatures [69–73]. The quadruple hydrogen-bonding unit employed by Schubert and coworkers, ureidopyrimidinone, has been investigated in detail over the last decade by Meijer and coworkers and has a K_a of 6×10^7 M^{-1} in halogenated solvents [74]. Therefore, the macromole-

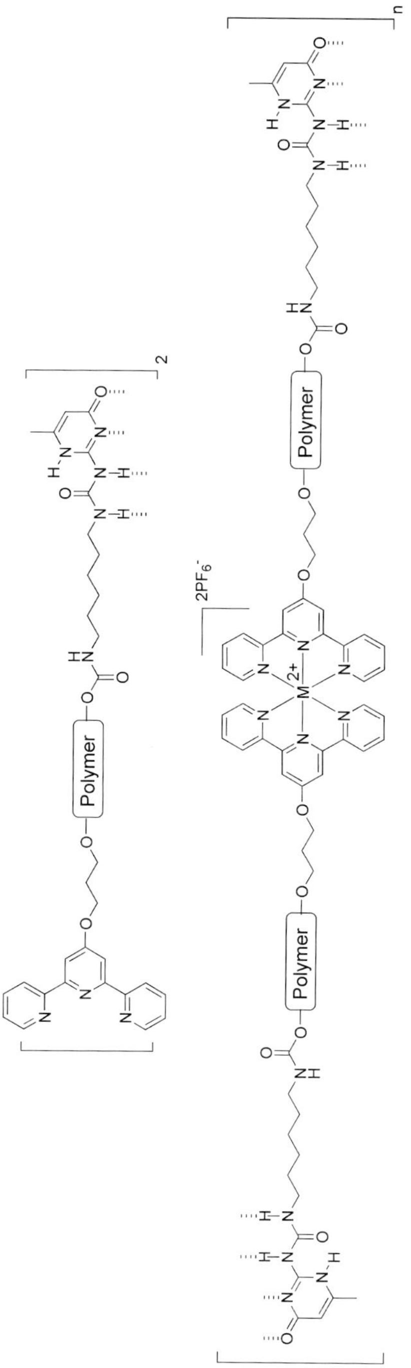

Fig. 7.16 Multi-step main-chain self-assembly to form a flexible, high molecular weight polymer. Step 1: dimerization of bifunctional unit via the self-complimentary ureidopyrimidinone end (top). Step 2: addition of a metal salt such as iron(II) initiates metal coordination-based self-assembly of the trpy-functionalized ends to form extended polymer chains (bottom).

cular building block in the Schubert system exists as a dimer before metal coordination owing to the high association constant for the quadruple hydrogen bonds. Addition of a solution of a metal salt such as $FeCl_2$ or $Zn(OAc)_2$ initiates, via self-assembly, a spontaneous polymerization which afforded, after precipitation with ammonium hexafluorophosphate, a supramolecular polymer with unique physical properties (Fig. 7.16). For example, the hydrogen-bonded dimers are brittle and opaque, whereas the final metal-coordinated polymers are flexible. Furthermore, Schubert and coworkers were able to form transparent thin films of these polymers.

In contrast to covalent-based polymers, characterization of supramolecular polymers is often challenging since basic polymer characterization methods such as gel permeation chromatography for the determination of molecular weights and polydispersities are incompatible with hydrogen bonding and often coordination chemistry (in particular if charged species are involved) [75]. As a result, alternative methods for determining the size of the supramolecules in question are often employed, including small-angle neutron scattering (SANS), small-angle X-ray scattering (SAXS) [76–81] and various light-scattering techniques [21, 82, 83]. The supramolecular polymers reported by Schubert and coworkers were characterized by several methods, including UV–Vis spectroscopy, which confirmed the coordination event, and viscometry, which confirmed the presence of high molecular weight polymers. Interestingly, owing to the reversibility of the self-assembled interactions, an exponential dependence of the viscosity on the polymer concentration was observed, instead of the linear dependence that might be expected from a polymer possessing irreversible linkages throughout. Schubert and coworkers attributed this effect to a ring–chain equilibrium, i.e. the formation of high molecular weight cyclic structures at lower concentrations, which would cause a slower increase in viscosity with increasing concentration. Once a critical concentration is reached, the concentration of cyclic polymers remains constant and the viscosity begins to rise rapidly owing to the formation of high molecular weight linear supramolecular polymers.

The reversibility of the system through metal–trpy decomplexation was also investigated through the addition of the strong transition metal-chelating ligand hydroxyethylethylenediaminetetraacetic acid (HEEDTA) [84]. This ligand has been shown to be capable of opening bis-trpy–iron(II) complexes by abstracting the

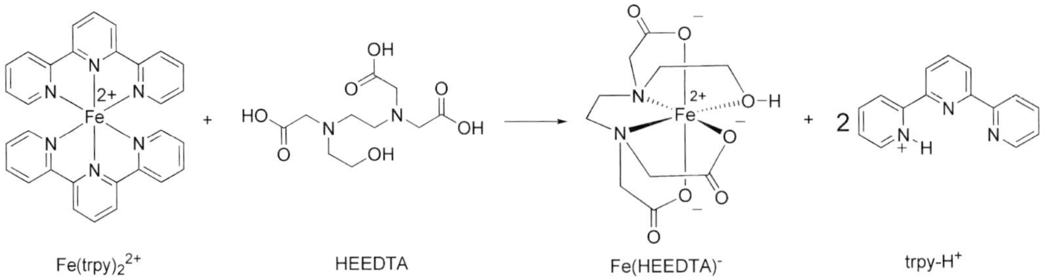

Fig. 7.17 Abstraction of iron(II) ion from the bis-trpy complex with HEEDTA.

iron ion in a ligand-exchange process (Fig. 7.17) [84]. Indeed, addition of HEEDTA to the iron-complexed supramolecular polymer caused full decomplexation within 2 min, as was evident by loss of the characteristic color of the iron–trpy complex, resulting in the depolymerization of the supramolecular polymer into dimers.

Main-chain supramolecular polymers have the advantage of a high degree of control over the primary backbone, thus leading to chain lengthening from oligomers to high polymers, in addition to the capability to depolymerize entirely under the appropriate conditions. However, it is often desirable to be able to functionalize a polymer with small molecule-based functional groups to tailor the properties as well, requiring the self-assembly sites to be relocated from the ends of the main chain to side-chains, leading to another type of supramolecular polymers, side-chain self-assembled polymers.

7.4.2
Side-chain Self-assembly

In addition to main-chain supramolecular polymers, the second major class of supramolecular polymers is side-chain systems where the self-assembly receptors have been positioned on the side-chains of the polymer. The self-assembly event then creates a function along the polymer backbone. This creates the potential for an enormous range of possible modifications to the polymer by rapid, straightforward self-assembly-based methodology. The main goal of this strategy is to generate vast libraries of materials for a wide variety of applications, since side-chain functionalized polymers have been employed or suggested for applications in electro-optics, biomaterials, liquid crystals, etc. [85–90].

Over the past two decades, many workers have employed the basic design strategy of side-chain functionalized supramolecular polymers to synthesize polymers that contain hydrogen bonding and metal coordination receptor molecules in their side-chains [17, 18, 57, 58, 91–102]. Using basic self-assembly strategies, a number of functional polymers such as liquid crystalline polymers have been synthesized [49, 85, 87–89, 103–105]. Although highly successful, the strategies employed do not allow for the controlled multifunctionalization of side-chain supramolecular polymers. Over the past 5 years, it has become clear that controlled multifunctionalization can only be achieved if multiple noncovalent recognition units are engineered into a single polymer backbone and if these multiple noncovalent recognition units can be addressed in an orthogonal fashion. To date, only two basic polymer systems have been reported that follow these design guidelines [58, 95, 96].

In 2004, Weck and co-workers demonstrated that both mono- and multifunctional self-assembly could be employed simultaneously, independently and reversibly on the same side-chain functionalized polymer [96]. A random terpolymer of poly(norbornene) was synthesized consisting of diaminopyridine (DAP) hydrogen-bonding receptors and a palladium-functionalized SCS-type pincer ligand for metal coordination-based self-assembly (Fig. 7.18).

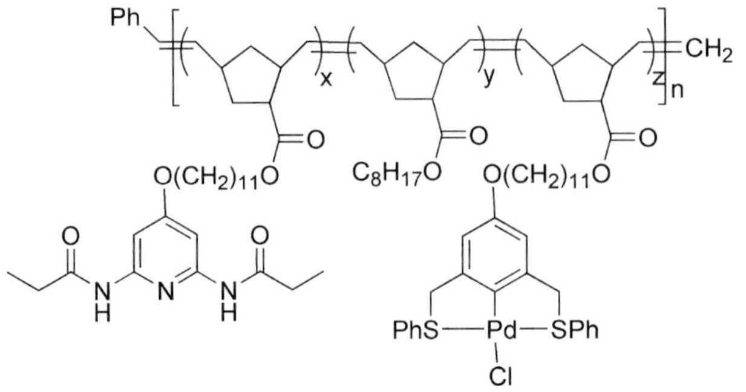

Fig. 7.18 Multifunctional random terpolymer synthesized by Weck and co-workers.

The different unique properties and behaviors of the metal-coordination site versus the hydrogen-bonding sites give the system a high degree of control. For instance, the thermoreversibility of a hydrogen-bonded linkage versus the chemoreversibility of the metal-coordination bond allows for independent control, tunability and optimization of the polymer properties.

Initially, Weck and co-workers demonstrated that controlled, multi-self-assembly in this system can be achieved either in a stepwise fashion or in a one-pot procedure (Fig. 7.19).

Their first report on the system clearly demonstrated that both recognition units function independently of each other and can be addressed selectively. Finally, they demonstrated that the polymer properties of the fully self-assembled polymer are independent of the functionalization route. A variety of requirements are essential in this system. First, the two recognition motifs employed have to be independent of each other; second, the polymerization method for the synthesis of the polymer backbone has to be highly functional group tolerant since most self-assembly receptors have a high functional group and heteroatom content; and finally, the interactions have to be strong enough to allow for controlled polymer functionalization. Since the introduction of their first random copolymer system based on DAP and palladated pincer complexes, Weck and coworkers have extended their concept to other polymer architectures (block copolymers) and self-assembly receptors [95]. Recently, they combined the concepts of polymer architectural control with multifunctionalization [96]. They synthesized a number of poly(norbornene)-based copolymers containing two distinct self-assembly receptor molecules. Again, through simple small-molecule self-assembly, these copolymers can be functionalized in an orthogonal fashion yielding highly functional copolymers in a straightforward fashion (Fig. 7.20).

Finally, Weck's group has also investigated the incorporation of two distinct hydrogen-bonding receptors (Fig. 7.21) along a single polymer backbone (both random and block copolymers were investigated) [93, 95].

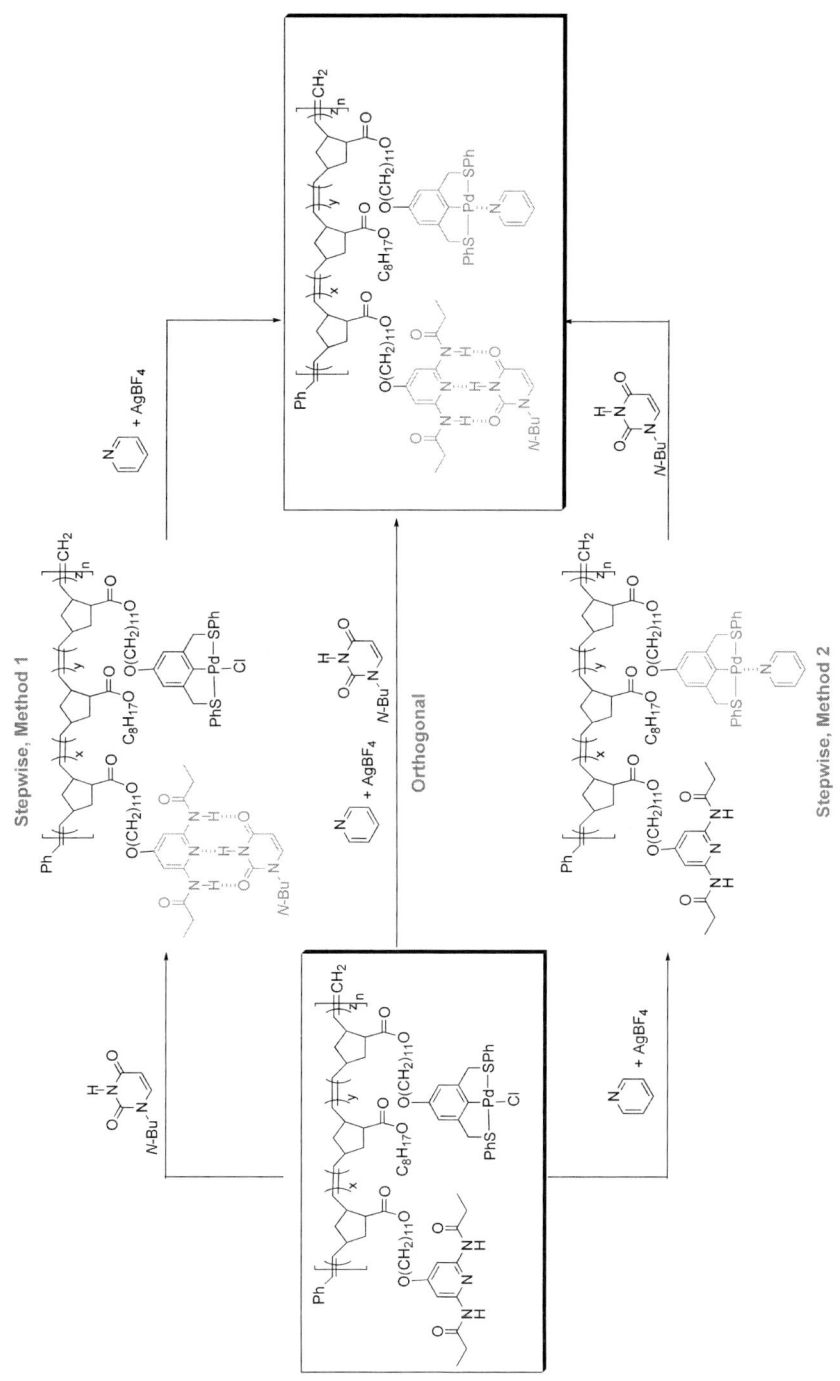

Fig. 7.19 Stepwise and orthogonal routes to functionalization of the universal polymer backbone using DAP–thiamine and pincer–pyridine complementary self-recognition units.

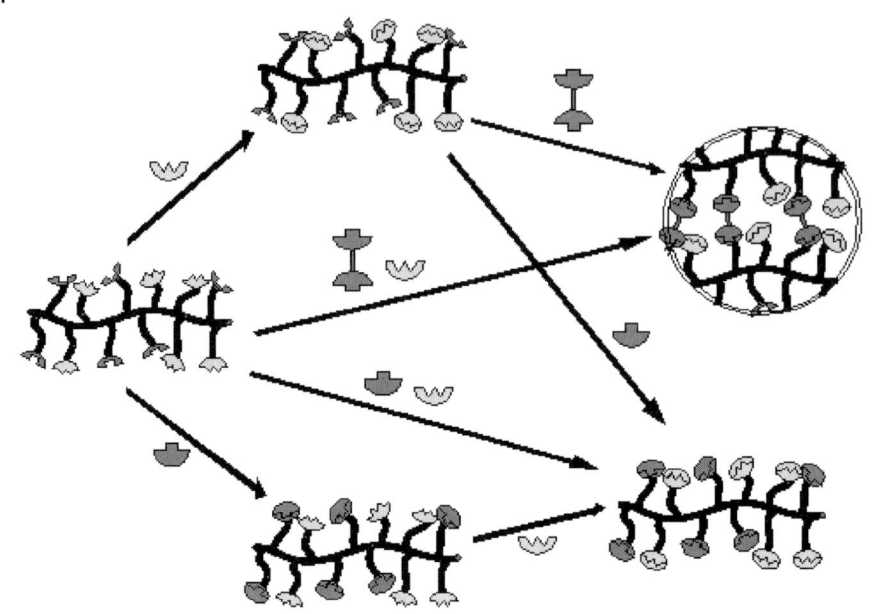

Fig. 7.20 Conceptual depiction of the multiple stepwise and orthogonal self-assembly strategies employed by Weck and co-workers. Clockwise from left: universal polymer backbone (UPB) functionalized with multiple recognition motifs (left); UPB functionalized with single recognition unit (top); cross-linked and fully functionalized UPB system via addition of a bifunctional substrate (top right); UPB fully functionalized with small molecule substrates (lower right); and UPB functionalized with single recognition unit (bottom).

The selective self-assembly of a receptor molecule with its complementary recognition unit in the presence of a competitive recognition unit has been described as self-sorting in the literature. DNA and RNA are the prime examples of this concept. Using the above-described copolymers containing two hydrogen-bonding units, Burd and Weck [95] were able to prove this concept of self-sorting also in synthetic polymer systems (Fig. 7.22).

To demonstrate the practicality and potential importance of their multifunctionalization concept, Weck and coworkers investigated the formation of functional and cross-linked polymeric networks [96]. When employing terpolymers containing monomers with hydrogen-bonding receptor molecules and ones with palladated pincer complexes for metal coordination that are diluted in a matrix of alkyl-based spacer monomers, reversible cross-linking could be achieved through either the hydrogen-bonding unit or the metal-coordination unit by employing either a bisthymine or bisperylene unit to cross-link through the side-chain DAP moieties via hydrogen bonding, as well as a bispyridine molecule for cross-linking through the pincer groups via metal coordination. Extensive cross-linking was observed in all cases. Interestingly, the viscosity increases of the pincer-based cross-linked system were two orders of magnitude higher than those of

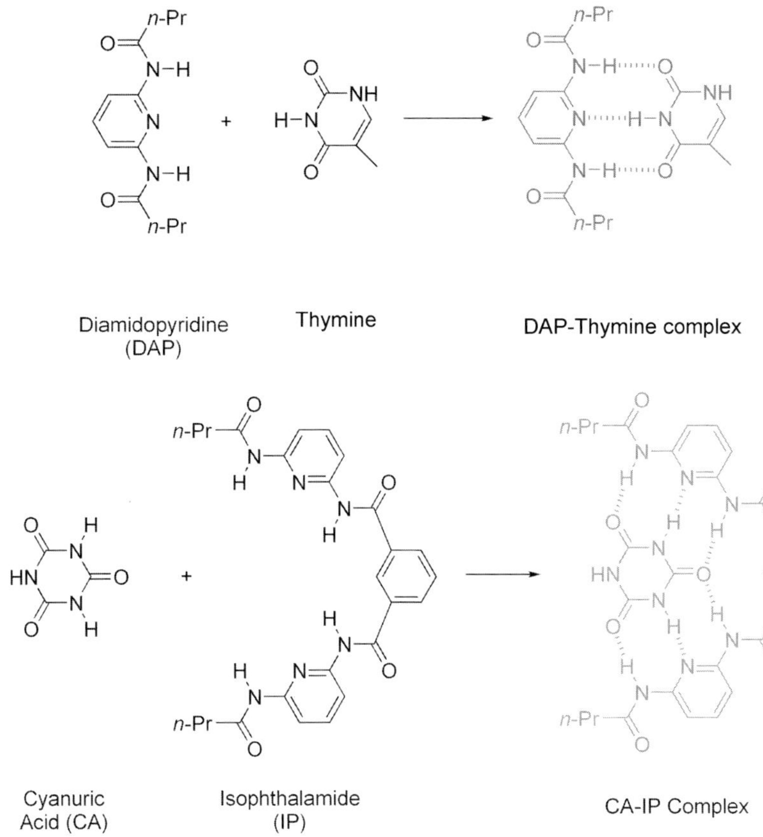

Fig. 7.21 Hydrogen-bonding units designed to exhibit self-sorting when present in the same polymer backbone.

the hydrogen-bonded systems. However, when the side-chain hydrogen-bonding unit attached to the polymer was changed from DAP to cyanuric acid, a stronger interaction with the cross-linker and thus higher solution viscosities were achieved. This is due in large part to the increase in the number of participating hydrogen-bonding interactions involved with the corresponding cross-linker. In the case of the DAP unit, there are three hydrogen bonds involved per self-assembly site, interacting in a DAD fashion, for either of the cross-linkers employed (bisthymine or bisperylene). However, when cyanuric acid is used instead, there are six hydrogen bonds involved at each self-assembly site.

The independent, noninteracting behaviors of these two modes of self-assembly allowed for the creation of a self-assembled, multifunctional, reversible, cross-linked material in one self-assembly step. This simultaneous self-assembly strategy, based on the results outlined above, allows the creation of a fully functionalized, cross-linked systems. Simultaneous addition of a small metal-complexing molecule such as pyridine, along with a hydrogen-bonding cross-linker such as

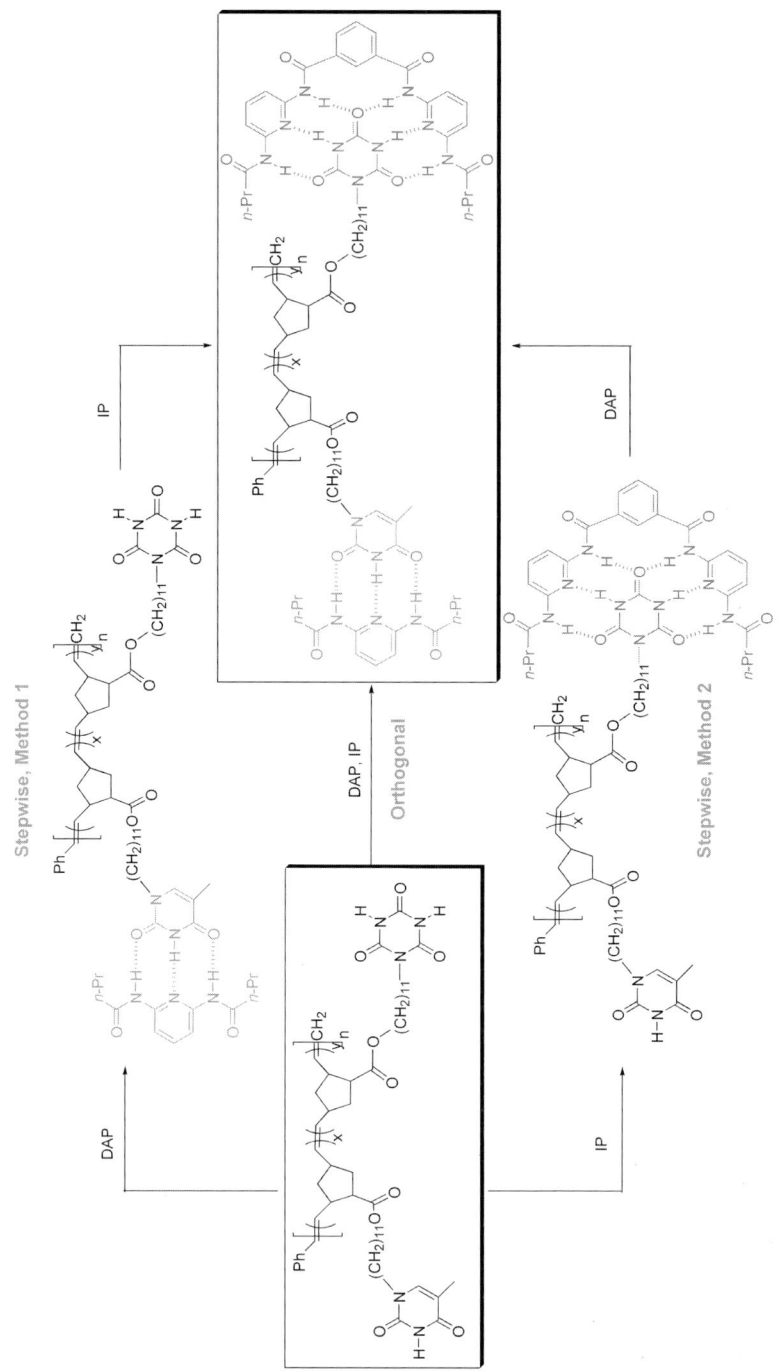

Fig. 7.22 Stepwise and orthogonal routes to functionalization of the universal polymer backbone using thiamine–DAP and iso- phthalamide–cyanuric acid complementary self-recognition units.

the bisthymine unit, results in a fully cross-linked, fully functionalized terpolymer. The converse of this method was equally successful, i.e. addition of a small-molecule hydrogen-bonding unit such as thiamine, along with a bispyridine cross-linker, results again in a fully cross-linked, fully functionalized material. Interestingly, polymers could be fully de-functionalized and de-cross-linked by (1) heating to disrupt the hydrogen bonds and (2) addition of PPh_3 to break the Pd–pyridine dative bond via competitive ligand interaction.

An interesting dimension of metal-coordinated self-assembly that is often ignored, or at least not exploited to its fullest extent, occurs when the resulting coordination complex is a charged species and, as such, in need of a counterion. This counterion itself presents yet another subtle instance of ionic self-assembly, which often is overshadowed by its partner, the coordination complex. The second multi-functional side-chain supramolecular polymer system is based on this simple but important concept [14, 106–111]. In 2003, Ikkala and coworkers reported a study in which they exploited (1) a side-chain functionalized polymer, poly(vinylpyridine), (2) metal-coordination self-assembly via a tridentate Zn^{2+} complex and (3) ionic self-assembly through functionalized counterions, i.e. dodecylbenzenesulfonate ions, to form multiple self-assembled complexes which adopted a cylindrical morphology (Fig. 7.23) [112].

The authors' goal was to synthesize self-assembled polymeric structures closely resembling dendron-modified polymers, but entirely through self-assembly. Ordinarily, during metal coordination-based self-assembly, if a charged species is the result, there will be any of a number of small counterions such as PF_6^-, Cl^-, BF_4^-, etc., to balance the positive residual charge of the metal ion. Ikkala and co-workers' use of dodecylbenzene-functionalized counterions leads to control over solubility, and also contributes significantly to the resulting conformation of the final self-assembled supramolecular structure via steric crowding. Several characterization methods, including (1) SAXS data showing self-organization in the mixture, (2) the lack of macroscopic phase separation and (3) characteristic FT-IR shifts indicative of the coordination event, allowed Ikkala and co-workers to conclude that mutually repulsive moieties are physically bonded within the supramolecular structures [112].

Their work is an impressive demonstration of the overall utility of the self-assembly process and simple alkyl-functionalized counterions constitute one of the most simple counterion modifications, opening the door to more complex and highly functionalized counterion designs. Among the vast possibilities for this concept are mesogen-modified counterions for polymeric LC applications, charge transfer, hole blocking and other useful cooperative combinations for use in emissive devices as well as the use of small chiral counterions which would impart an overall optical activity to the entire polymer system – the list is endless! However, there are still other choices for exploring and utilizing self-assembly in order to obtain multifunctional materials besides the main-chain and side-chain techniques described in this chapter thus far, namely macroscopic self-assembly or self-organization of supermolecular structures into higher ordered materials.

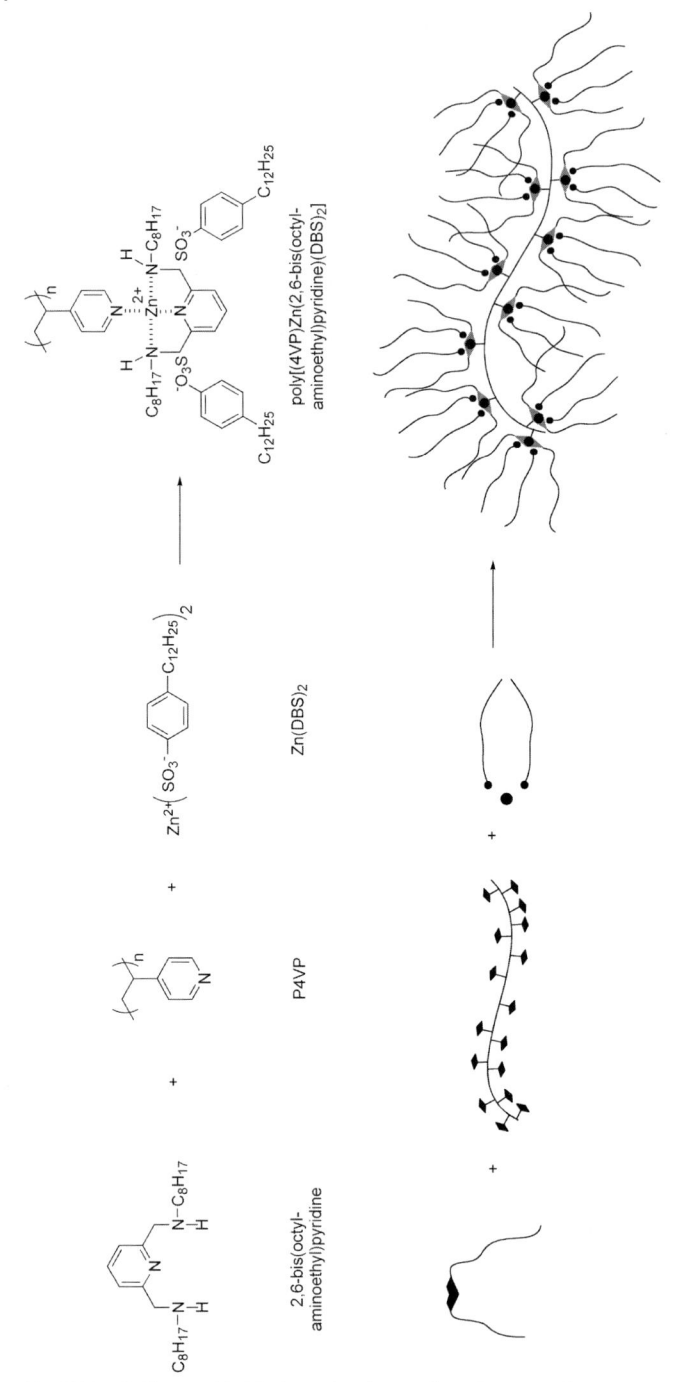

Fig. 7.23 Self-assembled hairy-rod polymers based on metal coordination and ionic interaction.

7.4.3
Macroscopic Self-assembly

In addition to simple small-molecule and polymeric self-assembly of small molecules onto polymers, macromolecules themselves are capable of undergoing self-organization to form higher ordered materials. There are several occasions on which this type of organization has been employed, ranging from the variations in polymer morphology that occur upon changes in composition [113–116] to self-assembly of dendrimers onto functionalized surfaces [117–121].

An interesting example of the latter was accomplished by Reinhoudt and co-workers, who employed cyclodextrin (CD) functionalized self-assembled monolayers (SAMs) to build up layer-by-layer (LBL) thin films via alternating host–guest interactions with functionalized dendrimers and gold nanoparticles [118, 122] The LBL self-assembly onto the CD functionalized surface requires at least two complementary substrates in order that multiple layers can be built up upon repeated alternating applications of the substrates, until a desired thickness has been achieved. Reinhoudt and coworkers chose the self-assembled interaction between CD-modified gold nanoparticles and either adamantyl- or ferrocene-functionalized dendrimers, which in turn were composed either of poly(propylenimine) (PPI) or poly(amidoamine) (PAMAM) [118, 122].

The LBL building up process began with exposure of the CD-functionalized SAMs to an excess of dendrimers that had been functionalized with a complementary guest (either adamantane or ferrocene), thereby depositing a single layer of the dendrimers on the surface. This was followed by exposure of the newly formed surface, which had been covered with guest molecules, with a new layer of CD-modified gold nanoparticles, which in turn was followed by more guest-modified dendrimer and so on (Fig. 7.24).

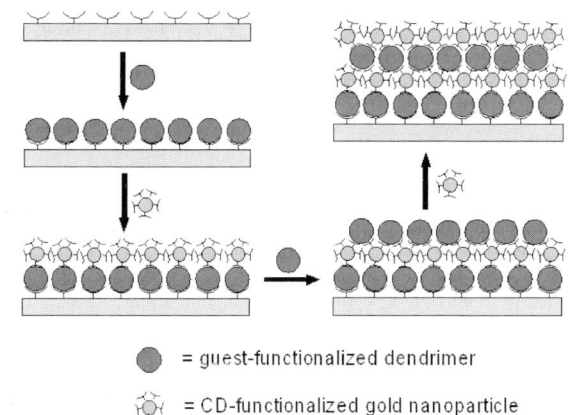

Fig. 7.24 Layer-by layer assembly of dendrimer-based films arising from host–guest interactions between cyclodextrin and small-molecule substrate-functionalized dendrimers.

● = guest-functionalized dendrimer
◯ = CD-functionalized gold nanoparticle
▭ = CD-functionalized SAM

Reinhoudt and co-workers found that the film thickness increased linearly as more layers were added, with the thickness increasing by ca. 2 nm per bilayer. With ferrocene as the guest moiety, it was found that adsorption could be nondestructively reversed via electrochemical means, i.e. conversion of the ferrocenyl-functionalized dendrimers to the corresponding ferrocenium cations. The strength of the binding of the dendrimers to the host-functionalized molecules was found to be the greatest with the highest generation of dendrimer, thereby maximizing the number of guest molecules available for association with the CD portion of the surface, an effect that they attributed to the multivalent properties of the dendrimers.

The self-assembly of the various polymer systems described in the above sections is only a brief summary of the attempts by chemists to create multifunctional materials based on noncovalent interactions. The unique regions present within a polymer including the (1) main-chain/backbone, (2) end groups, (3) side-chains and (4) dendritic periphery, along with the ability to functionalize any of these regions with recognition units, provide chemists with a wide array of self-assembly possibilities with which to build and create multifunctional materials.

7.5
Conclusion and Outlook

The self-assembly strategies presented in this chapter represent only a small selection from the now enormous amount of information being acquired today in this still rapidly expanding arena of multifunctional materials synthesized on the basis of noncovalent interactions. As the field of functional material design and synthesis continues to demand smaller and smaller size ranges for increasingly complex devices and applications, self-assembly will have to emerge as the most important tool available to scientists for the development of these materials. The production of materials in the nanoscale size range becomes much more efficient as the amount of direct manual manipulation of the material on a nanoscale is decreased, i.e. by the creation of "smart materials" that can manipulate themselves on that small scale. The examples described in this chapter demonstrate that scientists are well on their way to this goal and as research continues in this field and scientists continue to take clues from Nature, our understanding and mastery of both materials synthesis and design and also small-scale noncovalent biological processes will continue to improve.

References

1. Voet, D, Voet J. G., Pratt, C. W. *Fundamentals of Biochemistry*, Wiley, New York, **1999**.
2. Lehn, J.-M. *Supramolecular Chemistry*, Wiley-VCH, Weinheim, **1995**.
3. Lehn, J.-M. *Science* **2002**, *295*, 2400–2403.
4. T. Kato, T. Fréchet., J.-M. J. *Macromolecules* **1989**, *22*, 3818.
5. Kumar, U., Kato, T., Fréchet, J.-M. J. *J. Am. Chem. Soc.* **1992**, *114*, 6630–6639.
6. Eiichi, T., Zhi, Z.-L., Yasutaka, M., Quamrul, H. *Methods Mol. Biol.* **2005**, *300*, 369–381.
7. Tu, R. S., Tirrell, M. *Adv. Drug Deliv. Rev.* **2004**, *56*, 1537–1563.
8. Norgaard, K., Bjornholm, T. *Chem. Commun.* **2005**, *14*, 1812–1823.
9. Shenhar, R., Norsten, T. B., Rotello, V. M. *Adv. Mater.* **2005**, *17*, 657–669.
10. Morikawa, M.-A., Yoshihara, M., Endo, T., Kimizuka, N. *Chem. Eur. J.* **2005**, *11*, 1574–1578.
11. Genove, E., Shen, C., Zhang, S., Semino, C. E. *Biomaterials* **2005**, *26*, 3341–3351.
12. Ruotsalainen, T., Turku, J., Heikkilae, P., Ruokolainen, J., Nykaenen, A., Laitinen, T., Torkkeli, M., Serimaa, R., ten Brinke, G., Harlin, A., Ikkala, O. *Adv. Mater.* **2005**, *17*, 1048–1052.
13. Hamley, I. W., Ansari, I. A., Castelletto, V., Nuhn, H., Roesler, A., Klok, H. A. *Biomacromolecules* **2005**, *6*, 1310–1315.
14. Wei, Z., Laitinen, T., Smarsly, B., Ikkala, O., Faul, C. F. J. *Angew. Chem. Int. Ed.* **2005**, *44*, 751–756.
15. Zhao, Y., Mahajan, N., Lu, R., Fang, J. *Proc. Nat. Acad. Sci.USA* **2005**, *102*, 7438–7442.
16. Kim, I.-B., Wilson, J. N., Bunz, U. H. F. *Chem. Commun.* **2005**, 1273–1275.
17. ten Brinke, G., Ikkala, O. *Chem. Rec.* **2004**, *4*, 219–230.
18. Ikkala, O., ten Brinke, G. *Science* **2002**, *295*, 2407–2409.
19. de Jong, J. J. D., Lucas, L. N., Kellogg, R. M., van Esch, J. H., Feringa, B. L. *Science* **2004**, *304*, 278–281.
20. Van Esch, J. H., Feringa, B. L. *Angew. Chem. Int. Ed.* **2000**, *39*, 2263–2266.
21. Xie, P., Zhang, R. *J. Mater. Chem.* **2005**, *15*, 2529–2550.
22. Hofmeier, H., Schubert, U. S. *Chem. Commun.* **2005**, 2423–2432.
23. Andre, X., Zhang, M., Mueller, A. H. E. *Macromol. Rapid Commun.* **2005**, *26*, 558–563.
24. Giuseppone, N., Lehn, J.-M. *J. Am. Chem. Soc.* **2004**, *126*, 11448–11449.
25. Szczubialka, K., Moczek, L., Blaszkiewicz, S., Nowakowska, M. *J. Polym. Sci., Part A: Polym. Chem.* **2004**, *42*, 3879–3886.
26. Barker, I. C., Cowie, J. M. G., Huckerby, T. N., Shaw, D. A., Soutar, I., Swanson, L. *Macromolecules* **2003**, *36*, 7765–7770.
27. Forster, S., Plantenberg, T. *Angew. Chem. Int. Ed.* **2002**, *41*, 688–714.
28. Matsumura, S., Uemura, S., Mihara, H. *Chem. Eur. J.* **2004**, *10*, 2789–2794.
29. Percec, V., Dulcey, A. E., Balagurusamy, V. S. K., Miura, Y., Smidrkal, J., Peterca, M., Nummelin, S., Edlund, U., Hudson, S. D., Heiney, P. A., Duan, H., Magonov, S. N., Vinogradov, S. A. *Nature* **2004**, *430*, 764–768.
30. Vauthey, S., Santoso, S., Gong, H., Watson, N., Zhang, S. *Proc. Natl. Acad. Sci. USA* **2002**, *99*, 5355–5360.
31. Caplan, M. R., Schwartzfarb, E. M., Zhang, S., Kamm, R. D., Lauffenburger, D. A. *Biomaterials* **2002**, *23*, 219–227.
32. Aggeli, A., Bell, M., Boden, N., Carrick, L. M., Strong, A. E. *Angew. Chem. Int. Ed.* **2003**, *42*, 5603–5606.
33. Aggeli, A., Bell, M., Boden, N., Keen, J. N., Knowles, P. F., McLeish, T. C., Pitkeathly, M., Radford, S. E. *Nature* **1997**, *386*, 259.
34. Stephen, M. *Chemical Commun.* **2004**, 1–4.
35. Gonzáles, Y. I., Nakanishi, H., Stjerndahl, M., Kaler, E. W. *J. Phys. Chem. B* **2005**, *109*, 11675–11682.
36. Li, G., Fudickar, W., Skupin, M., Klyszcz, A., Draeger, C., Lauer, M., Fuhrhop, J.-H. *Angew. Chem. Int. Ed.* **2002**, *41*, 1828–1852.
37. Liaw, D.-J., Chen, T.-P., Huang, C.-C. *Macromolecules* **2005**, *38*, 3533–3538.
38. Shimizu, T., Masuda, M., Minamikawa, H. *Chem. Rev.* **2005**, *105*, 1401–1443.

39. Xu, J., Zubarev, E. R. *Angew. Chem. Int. Ed.* **2004**, *43*, 5491–5496.
40. Kurth, D. G., Meister, A., Thuenemann, A. F., Foerster, G. *Langmuir* **2003**, *19*, 4055–4057.
41. Yu, Y.-C., Berndt, P., Tirrell, M., Fields, G. B. *J. Am. Chem. Soc.* **1996**, *118*, 12515–12520.
42. Yang, W.-Y., Ahn, J.-H., Yoo, Y.-S., Oh, N.-K., Lee, M. *Nat. Mater.* **2005**, *4*, 399–402.
43. Waggoner, T. A., Last, J. A., Kotula, P. G., Sasaki, D. Y. *J. Am. Chem. Soc.* **2001**, *123*, 496–497.
44. Würthner, F., Yao, S., Beginn, U. *Angew. Chem. Int. Ed.* **2003**, *42*, 3247.
45. Torkkeli, M., Serimaa, R., Ikkala, O., Linder, M. *Biophys. J.* **2002**, *83*, 2240–2247.
46. Moralez, J. G., Raez, J., Yamazaki, T., Motkuri, R. K., Kovalenko, A., Fenniri, H. *J. Am. Chem. Soc.* **2005**, *127*, 8307–8309.
47. Ruben, M. R., J., Romero-Salguero, F. J., Uppadine, L. H., Lehn, J.-M. *Angew. Chem. Int. Ed.* **2004**, *43*, 3644–3662.
48. Lee, S. B., Koepsel, R., Stolz, D. B., Warriner, H. E., Russell, A. J. *J. Am. Chem. Soc.* **2004**, *126*, 13400–13405.
49. Kato, T., Kihara, H., Uryu, T., Fujishima, A., Fréchet, J. M. J. *Macromolecules* **1992**, *25*, 6836–41.
50. Keizer, H. M., Sijbesma, R. P. *Chem. Soc. Rev.* **2005**, *34*, 226–234.
51. Mascal, M., Hext, N. M., Warmuth, R., Moore, M. H., Turkenburg, J. P. *Angew. Chem. Int. Ed. Engl.* **1996**, *35*, 2204–2206.
52. Percec, V., Bera, T. K., Glodde, M., Fu, Q., Balagurusamy, V. S. K., Heiney, P. A. *Chem. Eur. J.* **2003**, *9*, 921–935.
53. Kihara, H., Kato, T., Uryu, T., Fréchet, J.-M. J. *Chem. Mater.* **1996**, *8*, 961.
54. Ruben, M. Ziener, U., Lehn, J.-M., Ksenofontov, V., Gütlich, P., Vaughan, G. B. M. *Chem. Eur. J.* **2005**, *11*, 94–100.
55. Li, L., Beniash, E., Zubarev, E. R., Xiang, W., Rabatic, B. M., Zhang, G., Stupp, S. I. *Nat. Mater.* **2003**, *2*, 689–694.
56. Klok, H.-A., Lecommandoux, S. *Adv. Mater.* **2001**, *13*, 1217–1229.
57. Gerhardt, W., Crne, M., Weck, M. *Chem. Eur. J.* **2004**, *10*, 6212–6221.
58. Pollino, J. M., Stubbs, L. P., Weck, M. *J. Am. Chem. Soc.* **2004**, *126*, 563–567.
59. Termonia, Y. *Biomacromolecules* **2004**, *5*, 2404–2407.
60. Bladon, P., Griffin, A. C. *Macromolecules* **1993**, *26*, 6604–6610.
61. Bronich, T. K., Ouyang, M., Kabanov, V. A., Eisenberg, A., Szoka, F. C., Jr., Kabanov, A. V. *J. Am. Chem. Soc.* **2002**, *124*, 11872–11873.
62. Percec, V., De, B. B., Cho, W.-D., Singer, K. D., Zhang, J. *Polym. Mater. Sci. Eng.* **1999**, *80*, 262–263.
63. Hofmeier, H., Hoogenboom, R., Wouters, M. E. L., Schubert, U. S. *J. Am. Chem. Soc.* **2005**, *127*, 2913–2921.
64. Holyer, R. H., Hubbard, C. D., Kettle, S. F. A., Wilkins, R. G. *Inorg. Chem.* **1966**, *5*, 622–625.
65. Hogg, R., Wilkins, R. G. *J. Chem. Soc.* **1962**, 341–350.
66. Tekin, E., Holder, E., Marin, V., De Gans, B.-J., Schubert, U. S. *Macromol. Rapid Commun.* **2005**, *26*, 293–297.
67. Hofmeier, H., Schubert, U. S. *Chem. Soc. Rev.* **2004**, *33*, 373–399.
68. Andres, P. R., Schubert, U. S. *Adv. Mater.* **2004**, *16*, 1043–1068.
69. Beta, I. A., Sorensen, C. M. *J. Phys. Chem. A* **2005**, *109*, 7850–7853.
70. Bian, L. *J. Phys. Chem. A* **2003**, *107*, 11517–11524.
71. Grabowski, S. J. *J. Phys. Chem. A* **2001**, *105*, 10739–10746.
72. Remer, L. C., Jensen, J. H. *J. Phys. Chem. A* **2000**, *104*, 9266–9275.
73. Silverstein, K. A. T., Haymet, A. D. J., Dill, K. A. *J. Am. Chem. Soc.* **2000**, *122*, 8037–8041.
74. El-ghayoury, A., Peeters, E., Schenning, A. P. H. J., Meijer, E. W. *Chem. Commun.* **2000**, 1969–1970.
75. Meier, M. A. R., Lohmeijer, B. G. G., Schubert, U. S. *Macromol. Rapid Commun.* **2003**, *24*, 852.
76. Tiitu, M., Volk, N., Torkkeli, M., Serimaa, R., ten Brinke, G., Ikkala, O. *Macromolecules* **2004**, *37*, 7364–7370.
77. Ikkala, O., ten Brinke, G. *Mater. Res. Soc. Symp. Proc.* **2003**, *775*, 213–223.
78. Kosonen, H., Valkama, S., Ruokolainen, J., Knaapila, M., Torkkeli, M., Serimaa, R., Monkman, A. P., ten Brinke, G.,

Ikkala, O. *Synth. Met.* **2003**, *137*, 881–882.
79. Valkama, S., Ruotsalainen, T., Kosonen, H., Ruokolainen, J., Torkkeli, M., Serimaa, R., ten Brinke, G., Ikkala, O. *Macromolecules* **2003**, *36*, 3986–3991.
80. Knaapila, M., Torkkeli, M., Makela, T., Horsburgh, L., Lindfors, K., Serimaa, R., Kaivola, M., Monkman, A. P., ten Brinke, G., Ikkala, O. *Mater. Res. Soc. Symp. Proc.* **2001**, *660*, JJ5 21/1–JJ5 21/6.
81. Ruokolainen, J., Tanner, J., ten Brinke, G., Ikkala, O., Torkkeli, M., Serimaa, R. *Macromolecules* **1995**, *28*, 7779–7784.
82. Gohy, J.-F., Lohmeijer, B. G. G., Alexeev, A., Wang, X.-S., Manners, I., Winnik, M. A., Schubert, U. S. *Chem. Eur. J.* **2004**, *10*, 4315–4323.
83. Gohy, J.-F., Lohmeijer, B. G. G., Schubert, U. S. *Macromol. Rapid Commun.* **2002**, *23*, 555–560.
84. Schmatloch, S., Fernandez-González, M., Schubert, U. S. *Macromol. Rapid Commun.* **2002**, *23*, 957.
85. Keith, C., Reddy, R. A., Tschierske, C. *Chem. Commun.* **2005**, 871–873.
86. Hagen, R., Bieringer, T. *Adv. Mater.* **2001**, *13*, 1805–1810.
87. Lee, J.-H., Lee, S.-D., Choi, D. H. *Mol. Crys. Liq. Cryst.* **1999**, *329*, 933–940.
88. Demikhov, E. I., Kozlovsky, M. V. *Liq. Cryst.* **1995**, *18*, 911–914.
89. Helgee, B., Hjertberg, T., Skarp, K., Andersson, G., Gouda, F. *Liq. Cryst.* **1995**, *18*, 871–878.
90. Pollino, J. M., Weck, M. *Chem. Soc. Rev.* **2005**, *34*, 193–207.
91. Carlise, J. R., Weck, M. *J. Polym. Sci. Part A: Polym. Chem.* **2004**, *42*, 2973–2984.
92. Kosonen, H., Valkama, S., Ruokolainen, J., Torkkeli, M., Serimaa, R., ten Brinke, G., Ikkala, O. *Eur. Phys. J. E: Soft Matter* **2003**, *10*, 69–75.
93. Stubbs, L. P., Weck, M. *Chem. Eur. J.* **2003**, *9*, 992–999.
94. Wang, X.-Y., Weck, M. *Macromolecules* **2005**, *38*, 7219–7224.
95. Burd, C., Weck, M. *Macromolecules* **2005**, *38*, 7225–7230.
96. Pollino, J. M., Nair, K. P., Stubbs, L. P., Adams, J., Weck, M. *Tetrahedron* **2004**, *60*, 7205–7215.
97. Meyers, A., South, C., Weck, M. *Chem. Commun.* **2004**, 1176–1177.
98. Meyers, A., Weck, M. *Chem. Mater.* **2004**, *16*, 1183–1188.
99. Pollino, J. M., Stubbs, L. P., Weck, M. *Macromolecules* **2003**, *36*, 2230–2234.
100. Meyers, A., Weck, M. *Macromolecules* **2003**, *36*, 1766–1768.
101. Pollino, J. M., Weck, M. *Synthesis* **2002**, 1277–1285.
102. Pollino, J. M., Weck, M. *Org. Lett.* **2002**, *4*, 753–756.
103. Ishii, T., Hirayama, T., Murakami, K., Tashiro, H., Thiemann, T., Kubo, K., Mori, A., Yamasaki, S., Akao, T., Tsuboyama, A., Mukaide, T., Ueno, K., Mataka, S. *Langmuir* **2005**, *21*, 1261–1268.
104. Kato, T., Kihara, H., Ujiie, S., Uryu, T., Fréchet, J. M. J. *Macromolecules* **1996**, *29*, 8734–8739.
105. Kumar, U., Kato, T., Fréchet, J. M. J. *J. Am. Chem. Soc.* **1992**, *114*, 6630–6639.
106. Guan, Y., Yu, S.-H., Antonietti, M., Boettcher, C., Faul, C. F. J. *Chem. Eur. J.* **2005**, *11*, 1305–1311.
107. Zhang, T., Spitz, C., Antonietti, M., Faul, C. F. J. *Chem. Eur. J.* **2005**, *11*, 1001–1009.
108. Wang, Z., Medforth, C. J., Shelnutt, J. A. *J. Am. Chem. Soc.* **2004**, *126*, 15954–15955.
109. Zakrevskyy, Y., Faul, C. F. J., Guan, Y., Stumpe, J. *Adv. Funct. Mater.* **2004**, *14*, 835–841.
110. Kadam, J., Faul, C. F. J., Scherf, U. *Chem. Mater.* **2004**, *16*, 3867–3871.
111. Faul, C. F. J., Antonietti, M. *Adv. Mater.* **2003**, *15*, 673–683.
112. Valkama, S., Lehtonen, O., Lappalainen, K., Kosonen, H., Castro, P., Repo, T., Torkkeli, M., Serimaa, R., ten Brinke, G., Leskelae, M., Ikkala, O. *Macromol. Rapid Commun.* **2003**, *24*, 556–560.
113. Okada, M., Inoue, G., Ikegami, T., Kimura, K., Furukawa, H. *Polymer* **2004**, *45*, 4315–4321.
114. Yoshie, N., Saito, M., Inoue, Y. *Polymer* **2004**, *45*, 1903–1911.
115. Jana, S. C., Sau, M. *Polymer* **2004**, *45*, 1665–1678.
116. Kontopoulou, M., Wang, W., Gopakumar, T. G., Cheung, C. *Polymer* **2003**, *44*, 7495–7504.

117. Boubbou, K. H., Ghaddar, T. H. *Langmuir* **2005**, *21*, 8844–8851.
118. Nijhuis, C. A., Yu, F., Knoll, W., Huskens, J., Reinhoudt, D. N. *Langmuir* **2005**, *21*, 7866–7876.
119. Lee, S. R., Yoon, D. K., Park, S.-H., Lee, E. H., Kim, Y. H., Stenger, P., Zasadzinski, J. A., Jung, H.-T. *Langmuir* **2005**, *21*, 4989–4995.
120. Nijhuis, C. A., Huskens, J., Reinhoudt, D. N. *J. Am. Chem. Soc.* **2004**, *126*, 12266–12267.
121. Mark, S. S., Sandhyarani, N., Zhu, C., Campagnolo, C., Batt, C. A. *Langmuir* **2004**, *20*, 6808–6817.
122. Crespo-Biel, O., Dordi, B., Reinhoudt, D. N., Huskens, J. *J. Am. Chem. Soc.* **2005**, *127*, 7594–7600.

Part III
Molecular Muscles, Switches and Electronics

8
Molecular Motors and Muscles
Sourav Saha and J. Fraser Stoddart

8.1
Introduction

Richard Feynman [1], in his seminal lecture in 1959, noted that "I can hardly doubt that when we have some control of the arrangement of the things on a molecular scale, we will get an enormously greater range of possible properties that substances can have and of different things we can do." Since then, rapid advances in the physical and biological sciences have shed an enormous amount of light on natural and biological machines [2], enabling us to comprehend them down to the molecular level. Being amongst the most accomplished imitators of Nature, scientists have devoted a lot of effort to the development of artificial molecular machines [3, 4], trying to mimic functions which are reminiscent of those in biology. A class of molecules that executes specific mechanical movements (outputs), when triggered by appropriate external stimuli (inputs), can be defined [5] as molecular machines. Artificial molecular machines can be classified [6] into several categories based on the following considerations and criteria: (1) the types of the energy inputs that power and control the machines – chemical [7], electrochemical [8] and photochemical [9] inputs; (2) the molecular structures [5–9] of the machines – bistable catenanes and rotaxanes, chiral molecular motors and molecular ratchets; (3) the types of molecular actuations displayed – linear or rotary translational motions; (4) the fidelity – reversibility to establish a cyclic process and its repetition; (5) the means of detection of the molecular actuations; and (6) the time-scale required for the machines to cycle.

The feasibility and limitations of molecular machines can hardly be emphasized any better than by Feynman's mixed message [1], namely that "An internal combustion engine of molecular scale is impossible. Other chemical reactions, liberating energy when cold, can be used instead." Nanoscale machines, like their macroscopic counterparts, require power supplies of appropriate kinds and magnitudes for their functions. While macroscopic machines enjoy the simplicity of distinct active (ON) and inactive (OFF) states in the presence and absence of power supplies, respectively, molecular machines are in perpetual Brow-

Functional Organic Materials. Syntheses, Strategies, and Applications.
Edited by Thomas J.J. Müller and Uwe H.F. Bunz
Copyright © 2007 WILEY-VCH Verlag GmbH & Co. KGaA, Weinheim
ISBN: 978-3-527-31302-0

nian motion at ambient temperature. Therefore, in order to make a molecular-level machine do work, an energy input is required to trigger a specific and directional motion within the molecule, one which would otherwise be largely inaccessible amidst random molecular motions. Since the functions of the molecular machines originate from chemical changes within molecules, the archetypical power supplies employed to induce molecular movements are chemical stimuli [7] – for example, acid–base reactions (changes in pH), ion-exchange and chemical redox processes. The use of chemical stimuli, however, leads [8g] to the accumulation of chemical waste in the machines' working environments – an issue which is detrimental to the machines' functions after only a few cycles. Moreover, removal of the chemical waste is a daunting task and is seldom a successful venture. In this regard, electrochemical and photochemical inputs are much more desirable when it comes to driving molecular machines. Employing either of these inputs, molecular actuations can be powered and controlled [8g] by introducing electrons (or holes) or photons in processes that are much cleaner and more sophisticated.

An important issue associated with molecular machines is the detection of actuations on the nanoscale level. When a chemical stimulus induces movement in a machine, several spectroscopic techniques, such as nuclear magnetic resonance (NMR) spectroscopy, UV–Vis spectroscopy, emission spectroscopy and X-ray photoelectron spectroscopy (XPS) can be used to detect their outputs. More intriguingly, electrochemical and photochemical inputs often provide [6, 8g] a two-fold advantage by inducing the mechanical movements and detecting them. Additionally, the dual actions of the these two types of stimuli can be exploited when the time-scale of the molecular actuations, which ranges from picoseconds to seconds, falls within the detection time-scale of the apparatus.

The very first generation of molecules that were claimed to be nanomachines involved [6] configurational isomerization, for example, the cis–trans isomerization of C=C, C=N or N=N bonds in response to photo and/or thermal stimuli. The movements of the molecular components during these isomerizations were so small in magnitude, however, that they satisfied no more than a chemist's dream when it came to molecular machines. Nevertheless, they laid the foundation for more advanced nanoscale molecular machines that are now gradually becoming more promising in the realm of nanotechnology. The nature of a molecular actuation depends largely on a molecule's structure and superstructure, that is, whether covalent or noncovalent bonding interactions can be used to address and control nanomechanical movements. There exist [4] various classical molecular motors and ratchets whose stimuli-induced rotational motions are based on specific covalent bond modifications. In recent years, however, with the advent of supramolecular chemistry, the paradigm has shifted [6] largely towards mechanically interlocked molecules. The dynamic properties of mechanically interlocked molecules originate [10] from inherent intramolecular noncovalent bonding interactions and molecular recognition between their components. In this chapter, we will focus on bistable rotaxanes which can be stimulated externally to perform large-amplitude molecular motions.

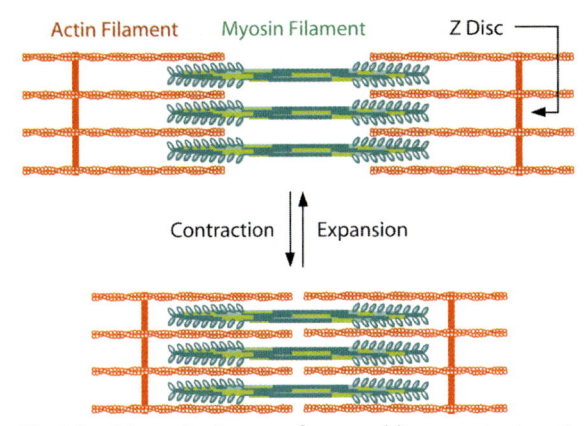

Fig. 8.1 Schematic diagram of a natural linear motor in action.

As we learn more about biomolecules with mechanical functions, we begin to appreciate that the miniaturization of artificial molecular machines benefits [6] more from the "bottom-up" approach [11] than from the intrinsically limited "top-down" approach used in modern macroscale manufacturing. It can be argued that the numerous natural molecular machines which orchestrate movements in the living world are built up of the self-assembly and self-organization of proteins. Furthermore, the dynamic properties of these motor-proteins arise from changes in molecular recognition within their subunits. One example of such a natural machine [6, 12] is the actin–myosin-based linear motor (Fig. 8.1) present in skeletal muscles. Myosin is a long, linear protein composed of two head groups bearing bottlebrush-like bristles at its ends. Actin is a channel-shaped, thin filamentous protein, which acts as a ladder to allow the myosin heads to migrate along its channels. The insertion of the myosin thick filaments into the actin channels results in the contraction of the muscle – a process that is powered by the hydrolysis of adenosine triphosphate (ATP). It is a highly reversible and repeatable process, which produces five strokes per second with displacements of 10 nm during each stroke. The translational motions displayed by the artificial linear motor molecules that will be discussed in the following sections are reminiscent of the workings of skeletal muscle.

8.2
Mechanically Interlocked Molecules as Artificial Molecular Machines

Mechanically interlocked molecules, such as bistable catenanes [13] and [2]rotaxanes [14], constitute some of the most appropriate candidates to serve as nanoscale switches and machines in the rapidly developing fields of nanoelectronics [15] and nanoelectromechanical systems (NEMS) [16]. The advantages of using mechanically interlocked molecules in the fields of molecular electronics and machines stem from the abilities of their components to undergo controllable relative mechanical motions in response to external stimuli. In a bistable rotaxane, for example, the competitive binding affinities of two recognition sites on its dumbbell component with its ring component allow the latter to be translated between the two sites. Thus a large-amplitude molecular mechanical motion can be produced within a mechanically interlocked molecule without damage to its molecular structure. The necessary conditions for generating these highly controllable molecular actuations is a large difference in the binding affinities of the ring component for two sites on the dumbbell component and their complete reversal in two different states of the molecule. A free energy difference greater than 1.2 kcal mol^{-1} between the two translational isomers of a bistable rotaxane in its ground state ensures [8g] that, at room temperature, 90% of the ring resides around one of the two recognition sites in preference to the other. In order to induce motion of the ring component within such skewed bistable rotaxanes, a variety of external stimuli, such as chemical, electrochemical and photoelectrochemical inputs, have been employed with considerable success [7–9].

In this chapter, we will address how bistable rotaxanes [17], equipped with a variety of different noncovalent bonding interactions, can be induced to undergo linear translational movements in precise and controllable ways. Another family of mechanically interlocked molecules – bistable catenanes [17], in which circumrotatory motions can be generated – is beyond the scope of this chapter. A bistable [2]rotaxane is comprised of a ring component that encircles a dumbbell-shaped component. The interaction between the ring and dumbbell components is such that the ring can choose between one of two recognition sites along the rod section of the dumbbell component (see above). If two recognition sites with very different binding affinities can be grafted into the rod section of the dumbbell component, a bistable [2]rotaxane with switchable properties can be obtained. The movement of the ring between two recognition sites along the dumbbell component gives rise to two translational isomers [6, 18]: (1) a ground-state co-conformation (GSCC) when the macrocyclic ring encircles the recognition site which provides the stronger stabilizing interaction and (2) a metastable-state co-conformation (MSCC) which arises when the ring component finds itself positioned around the second and less desirable recognition site after the ground-state interactions have been turned OFF externally by some stimulus. Provided the binding affinities of the recognition sites can be modulated reversibly by some means or other, for example, by chemical, electrochemical or photochemical inputs, the ring's mechanical movement can be activated and controlled on demand. Thus, the recognition sites in such bistable rotaxanes act essentially as controllable switches. The detection of these molecular switches' stimuli-induced motions in their working environments, which can range from the solution phases through condensed phases to the device settings, while non-trivial, can be detected and understood in some detail. Several different experiments are required to establish the mechanism of their operation. The stimuli-response behavior of the dumbbell component, devoid of the ring component, is often employed as

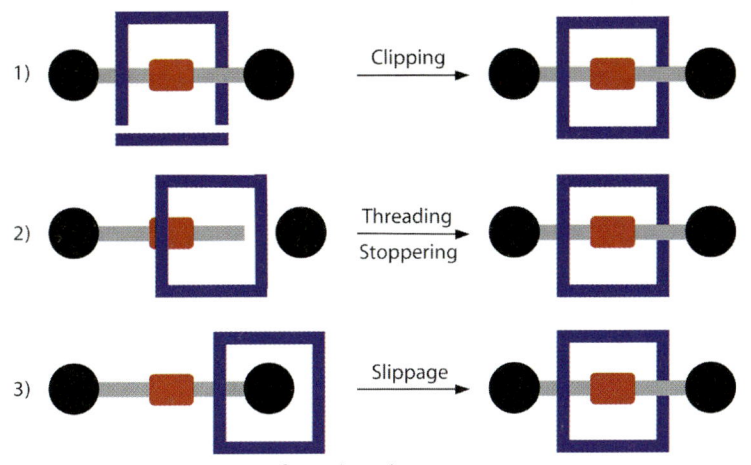

Fig. 8.2 Synthetic strategies for making the [2]rotaxanes.

a control in order to detect the changes in a bistable rotaxane's properties and related parameters when its components undergo relative mechanical movements.

The first step towards developing a molecular machine is its assembly by synthesis using chemical reactions. The rotaxanes that will be discussed in this chapter were assembled [6] by a variety of different template-directed [10, 19] synthetic strategies (Fig. 8.2): (1) clipping of a ring around a recognition site on a preformed dumbbell component, (2) threading of a half-dumbbell-shaped component into the cavity of a preformed ring by exploiting the mutual recognition they have for each other, followed by stoppering of the open end of the pseudorotaxane, and (3) slippage of a preformed ring over one of the terminal stoppers onto a preformed dumbbell at elevated temperatures.

In the following sections, bistable rotaxanes will be discussed and classified based on the types of the energy inputs utilized to power their mechanical motions. Since many of these compounds have already made their way into the device world, the prospects for their further development in an applied sense will also be addressed.

8.3
Chemically Induced Switching of the Bistable Rotaxanes

A chemical stimulus [7] is one of the external stimuli that can be used to trigger mechanical movements in bistable rotaxanes. Depending on the nature of the noncovalent bonding interactions involved within the rotaxane molecules, several types of chemical stimuli have been employed to induce relative molecular motion in ring and dumbbell components, for example, (1) if a bistable rotaxane involves hydrogen bonding (H-bonding) interactions, an acid–base (pH change) stimulus can sometimes be employed [20]; (2) if a bistable rotaxane is capable of binding metal ions by its coordination sites selectively, the relative movements of its components can be stimulated [21] by exchanging metal ions; and (3) chemical oxidants/reductants are often employed [22] as stimuli for bistable rotaxanes that comprise redox-active recognition sites. Chemically induced switching in bistable rotaxanes can be monitored by, for example, NMR spectroscopy, absorption and emission spectroscopy or XPS.

8.3.1
A Bistable [2]Rotaxane Driven by Acid–Base Chemistry

A prototype of linear motor molecule (Fig. 8.3) has been developed [7b, 20a] in the form of a bistable [2]rotaxane $1\text{-}H^{3+}$ that can be switched by simple acid–base chemistry. The [2]rotaxane $1\text{-}H^{3+}$ is composed of a π-electron-rich dibenzo[24]-crown-8 (DB24C8) macrocycle interlocked with a dumbbell component, which carries a pH-sensitive dialkylammonium ion ($-NH_2^+-$) center as its primary recognition site and a 4,4′-bipyridinium ($BIPY^{2+}$) unit as its secondary recognition site. In the GSCC of the rotaxane $[1\text{-}H]^{3+}$, the crown ether ring encircles predominantly the $-NH_2^+-$ center on account of hydrogen bonding (H-bonding) interactions. Two different kinds of bulky aromatic stoppers – 9-anthracenylmethyl and 3,5-di-*tert*-butylbenzyl – have been attached to the dialkylammonium ion center. In the protonated form of the rotaxane $[1\text{-}H]^{3+}$, the DB24C8 ring encircles exclusively the $-NH_2^+-$ center on account of the H-bonding interactions with the $-NH_2^+-$ center. Addition of $(i\text{-}Pr)_2NEt$ deprotonates the $-NH_2^+-$ center to produce a secondary amine 1^{2+} – a situation that turns OFF its H-bond-donating ability to the DB24C8 ring and triggers the mechanical movement of the ring component

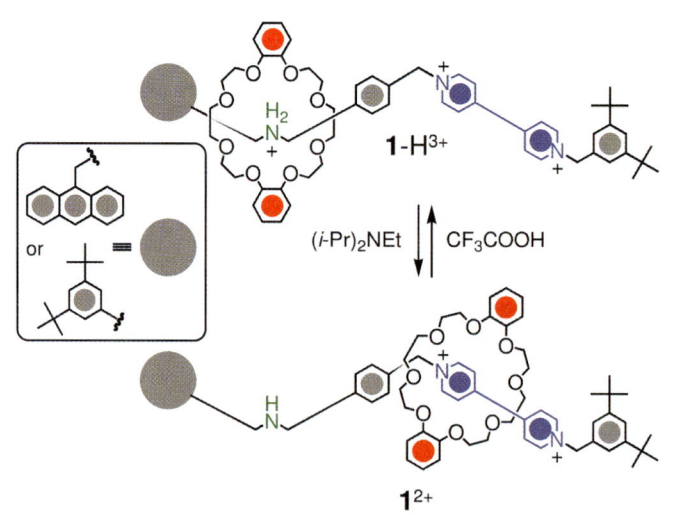

Fig. 8.3 Chemically controlled switching of a bistable rotaxane $1\text{-}H^{3+}/1^{2+}$.

to encircle the π-electron-deficient BIPY^{2+} unit exclusively. Addition of trifluoroacetic acid (TFA) reprotonates the amine center, regenerating **1-H**$^{3+}$, and the DB24C8 ring moves back to the original $-NH_2^+-$ center, completing a full cycle of the linear motor-like motion. This bistable [2]rotaxane lays the foundations for a more intriguing molecular machine in the form of a molecular elevator, in which three of these switching elements are integrated in one mechanically interlocked trivalent [2]rotaxane.

8.3.2
A pH-driven Molecular Elevator

The molecular elevator [20b] **2-H**$_3^{9+}$ comprises (Fig. 8.4) a tripodal rig-like component bearing two different recognition sites on each of its three legs positioned at two different levels. The bistable tripod-shaped component is triply interlocked with a platform component, which is composed of three macrocyclic rings fused together with a trigonal core, forming the molecular elevator. The platform could be switched chemically between two different recognition sites on each of the three legs of the rig-like component. Each of the three legs furnishes two dissimilar electron-deficient stations: an $-NH_2^+-$ center on the upper level and a BIPY^{2+} unit on the lower level. Each leg of the rig component is interlocked with a DB24C8 ring, all three of which are fused together to constitute the symmetric trigonal core of the platform component. The DB24C8 ring can form H-bonds with the $-NH_2^+-$ center and a π-complex with the BIPY^{2+} center. Bulky 3,5-di-*tert*-butylaryl stoppers at the bottom end of each leg prevent the DB24C8 platform from slipping off the rig component. In the ground state of the molecular elevator, stabilizing interactions, arising from strong [N$^+$–H···O] hydrogen bonding and weak [C–H···O] interactions between the DB24C8 rings and the $-NH_2^+-$ centers, set the platform component on the upper level. Addition of 3.4 equiv. of a strong, non-nucleophilic phosphazene base to the CD$_3$CN solution of the elevator deprotonates all three $-NH_2^+-$ centers and thereby erases their H-bonding ability with the crown ether rings. Such chemical modification drives the electron-rich DB24C8 platform to the π-electron-accepting BIPY^{2+} stations, located at the lower level of the tripod. Addition of a stoichiometric amount of TFA regenerates the $-NH_2^+-$ centers on all three legs and the DB24C8 platform moves back to the upper level. The pH-driven movements of the molecular elevator can be monitored by absorption spectroscopy, and also by the electrochemistry of the BIPY^{2+} stations. In the deprotonated form **2**$^{6+}$, where the DB24C8 rings encircle the BIPY^{2+} stations, the diagnostic 310-nm absorption band shows a significantly higher intensity than its original **2-H**$_3^{9+}$ state. The intensity of the 310-nm band increases and decreases reversibly upon respective additions of stoichiometric amounts of base and acid. as shown in Fig. 8.4. Cyclic voltammetry (CV) shows that the BIPY^{2+} units exhibit more negative reduction potentials in the deprotonated **2**$^{6+}$ form than for those in the original **2-H**$_3^{9+}$ state. In the deprotonated form **2**$^{6+}$, the reduction of the BIPY^{2+} station occurs at a more negative potential than does the protonated form **2-H**$_3^{9+}$ – a phenomenon that is attributed

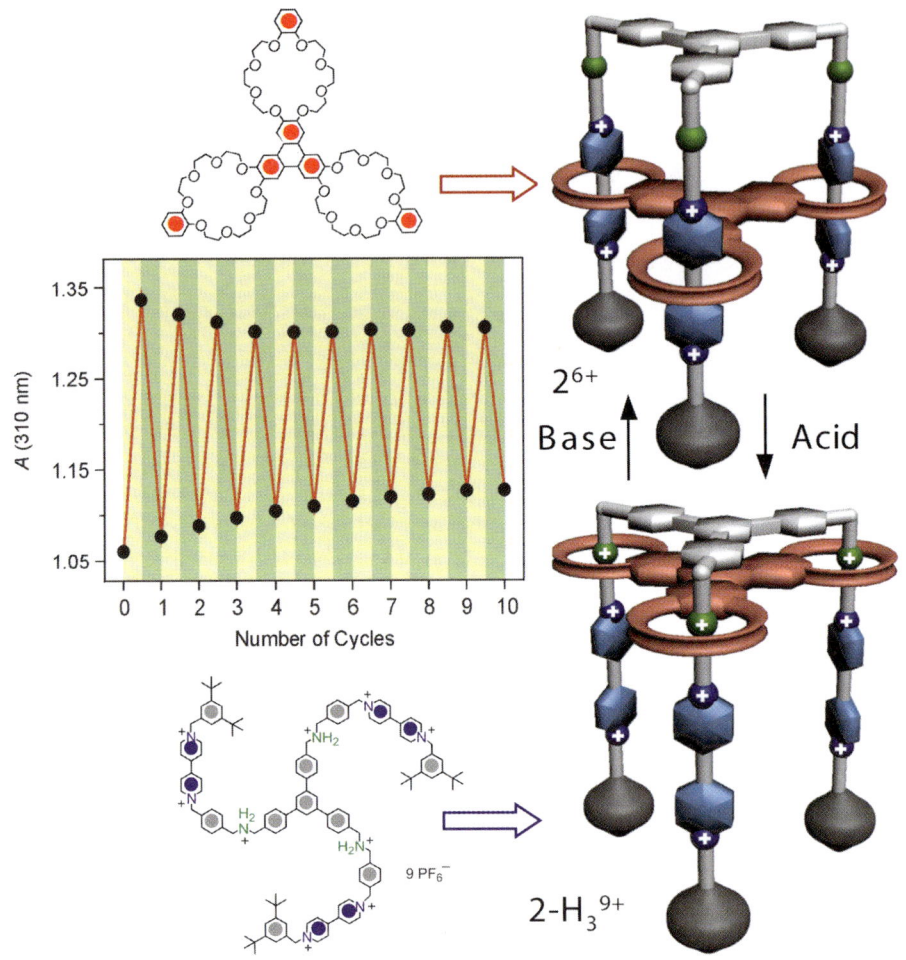

Fig. 8.4 A pH-driven molecular elevator, 2-$H_3^{9+}/2^{6+}$. The red saw-tooth trace shows the increase in the absorption intensity on addition of a base (yellow region) and its decrease on addition of an acid (green region). The elevator was cycled reversibly 10 times.

to the π-electron-rich DB24C8 ring's location around the $BIPY^{2+}$ in 2^{2+}. The increase and decrease in the reduction potentials of the $BIPY^{2+}$ units could also be controlled reversibly by adding acid and base, respectively. The reversible spectroscopic changes, along with the changes in the reduction potential of the $BIPY^{2+}$ units in the molecular elevator in response to the pH changes, confirm the movement of the platform component between two recognition sites on each legs of the rig component.

In response to the chemical stimuli, the platform component travels a distance of 0.7 nm from the upper to the lower level of the rig, generating a force up to 200 pN, a value which is one order of magnitude higher than that exerted by the natural motors, such as myosin and kinesin. Furthermore, the base-induced "power stroke" of the DB24C8 platform from the upper ($-NH_2^+-$) to the lower ($BIPY^{2+}$) level opens up a cavity (1.5 × 0.8 nm), which can potentially act as a host for a variety of guest molecules. Therefore, the pH-driven molecular elevator may be used in principle for the controlled binding and release of appropriate guest molecules by its two distinct forms.

8.3.3
A Molecular Muscle Powered by Metal Ion Exchange

The molecular muscle 3^{2+} is (Fig. 8.5) essentially a two-component rotaxane dimer [21] composed of two self-complementary, identical ring and rod components. Each unit contains a ring that incorporates a bidentate 1,10-phenanthroline (phen) ligand and a rod that incorporates a second phen ligand close to the ring and a remote terdentate 2,2′:6′,2′′-terpyridine (terpy) ligand and a bulky tetraarylmethane stopper at the end prevents the dethreading of the rod from the ring of the complementary unit. The rotaxane dimer 3^{2+} was synthesized by exploiting the templating ability of the four-coordinating copper(I) (Cu^I) ions, which favor [21] a tetrahedral coordination geometry. The tetrahedral geometry of the Cu^I ion is satisfied only when two bidentate phen ligands – one in the ring component and another on the rod component – chelate the Cu^I ion to afford the dark-red extended isomer 3^{2+}. A complete demetalation, however, of two Cu^I ions by potassium cyanide (KCN), followed by remetalation with zinc(II) [Zn^{II}] ions, furnishes the colorless contracted isomer 4^{4+}. In the contracted form, the bidentate phen ligand in the ring and the remote terdentate terpy ligand

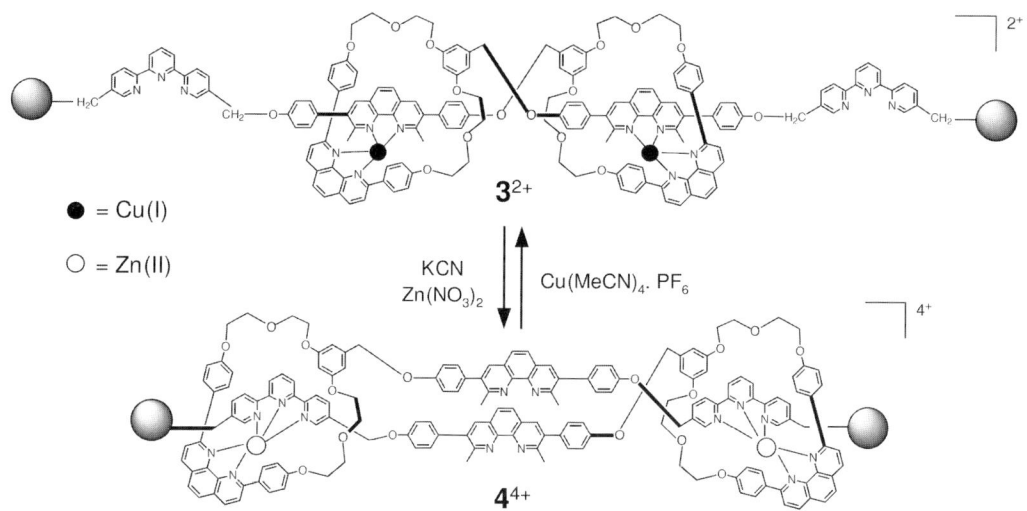

Fig. 8.5 Molecular muscle based on a self-complementary [2]rotaxane dimer Cu(I)-3^{2+}–Zn(II)-4^{4+}. Cu(I) chelation provides its extended form 3^{2+}, whereas Zn(II) complexation generates the contracted form 4^{4+}.

on the complementary rod chelate the pentacoordinating Zn^{II} ion to satisfy its five-coordination geometry. The extended isomer can be regenerated by exchanging two Zn^{II} ions with two Cu^{I} ions in the presence of excess of $[Cu(MeCN)_4][PF_6]$ at room temperature. Corey–Pauling–Kultun (CPK) space-filling molecular models show that the length of the muscle molecule shrinks from 8.3 nm in the extended form to 6.5 nm in the contracted form, a change that is roughly of the same magnitude (~27%) as that observed in the natural actin/myosin-based muscles.

8.3.4
Redox and Chemically Controlled Molecular Switches and Muscles

Another class of bistable rotaxanes was developed [22] based on two nonequivalent π-electron-donating recognition sites grafted on to a dumbbell component and a π-electron-accepting ring that can switch between two recognition sites when triggered by a chemical or electrochemical impetus. A tetrathiafulvalene (TTF) unit – an efficient π-electron donor with reasonably low oxidation potentials (+350 and +710 mV vs. SCE) for its radical cation and dication, respectively – and a 1,5-dioxynaphthalene (DNP) unit – a second, weaker π-electron donor with a much higher oxidation potential (+1300 mV vs. SCE) – are located along the dumbbell components, which, therefore, serve as templates for the formation of the ring components. The π-electron-deficient tetracationic cyclobis(paraquat-p-phenylene) ($CBPQT^{4+}$) cyclophane [23] was clipped around the π-electron-donating TTF recognition sites by template-directed [10, 19] synthesis. A variety of bulky stoppers have been incorporated into this class of TTF/DNP-based bistable rotaxanes in order to tune their polarities. The spacer units between the TTF and DNP recognition sites were also varied so as to control the distance between them, and also to provide rigidity or flexibility to the dumbbell component. In general, polyethylene glycol units, which also serve as flexible spacers, are required on at least one side of each of the TTF and DNP sites to provide stronger binding affinities and hence enhanced templating abilities to the recognition sites. A one order of magnitude difference in binding interactions with the $CBPQT^{4+}$ ring in favor of the TTF site ($K_a = 8000$ M^{-1} in MeCN) [24] over the DNP station ($K_a = 800$ M^{-1} in MeCN) [25] provides an emerald green GSCC, which bears the ring almost exclusively around the TTF site on account of strong π–π charge-transfer (CT) interactions [26] and weak [C–H···O] interactions. Two sets of oxidants – either $Fe(ClO_4)_3$ or tris(p-bromophenyl)iminium hexachloroantimonate – were introduced to turn OFF the stabilizing interactions in the ground state by oxidizing the π-electron-rich TTF unit to $TTF^{+\cdot}$ or TTF^{2+} units, neither of which are π-electron donors. In this mechanically excited state, electrostatic repulsion between the tetracationic ring and the oxidized $TTF^{+\cdot}$ unit and the attractive interaction offered by the alternate π-electron-donating DNP site induces the ring to move to the DNP site. This translational motion is triggered by the change in the binding affinities of the recognition sites and is assisted by thermal energy from the surrounding environment. Reduction of the previously oxidized $TTF^{+\cdot}$

or TTF^{2+} units to its neutral TTF form by chemical reductants, such as ascorbic acid or Zn^0 dust, produces translational isomers of these [2]rotaxanes with their $CBPQT^{4+}$ ring encircling a DNP site. These translational isomers constitute metastable states and correspond to a geometry where an MSCC is adopted. At this point, the ambient thermal energy or a two-electron reduction of the $CBPQT^{4+}$ ring destabilizes the MSCC and allows the ring to return to its original TTF station, thus completing the machine's full cycle.

8.3.4.1 Solution-phase Switching

In one such TTF/DNP-based bistable [2]rotaxane 5^{4+} (Fig. 8.6a), a long, rigid *p*-terphenyl spacer was employed between the TTF and DNP recognition sites and two tetraarylmethane stoppers were employed [22a] on both ends of the dumbbell component for the solution-phase switching studies. UV–Vis spectroscopy (Fig. 8.6b), which reveals the precise location of the $CBPQT^{4+}$ ring on the dumbbell

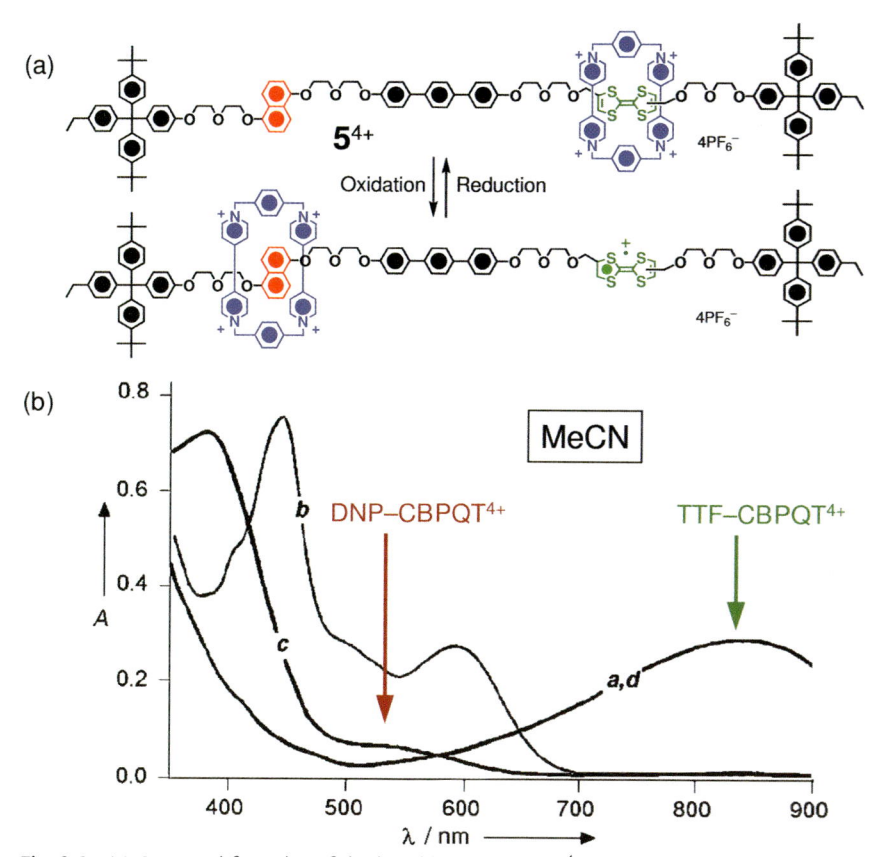

Fig. 8.6 (a) Structural formulas of the bistable [2]rotaxane 5^{4+} and its oxidized derivative. (b) Chemical switching of the bistable [2]rotaxane 5^{4+} monitored by UV–Vis spectroelectrochemistry.

component, was employed to monitor the movement of the ring component in response to oxidants and reductants. The TTF/DNP-based [2]rotaxanes show a characteristic charge-transfer (CT) band at ca. 846 nm (Fig. 8.6b, curve a), corresponding to the TTF–CBPQT^{4+} CT interaction in the GSCC. The absence of the characteristic DNP–CBPQT^{4+} CT band at 500–600 nm demonstrates the structural integrity of the GSCC. Addition of an 1.0 equiv. of Fe(ClO$_4$)$_3$ leads (Fig. 8.6b, curve b) to the attenuation of the 846-nm CT band intensity and the appearance of bands at 450 and 600 nm, which correspond to the TTF$^{+\cdot}$ radical cation's absorptions. A new absorption band at 515 nm was also observed, revealing the presence of a DNP–CBPQT^{4+} CT interaction – a phenomenon which arises from the ring's translational movement along the dumbbell from the TTF to the DNP site. Addition of 2.0 equiv. of the oxidant resulted in (Fig. 8.6b, curve c) the disappearance of the TTF$^{+\cdot}$ species' absorption peaks at 450 and 600 nm and the appearance of a peak centered on 375 nm corresponding to the TTF^{2+} dication's absorption. The DNP–CBPQT^{4+} CT band centered at ca. 530 nm became even more evident at this point. Addition of 2.0 equiv. of ascorbic acid regenerated the original spectrum (Fig. 8.6b, curve d) – an occurrence that established the reversibility of the redox-driven cycle of the linear molecular motor. CPK space-filling molecular models showed that, in the idealized extended orientation of the dumbbell component, the ring travels a distance of 3.7 nm in each direction along the rod section of the dumbbell. However, in the solution-phase, flexible polyethylene glycol spacers between the TTF and DNP recognition sites allow the molecules to fold freely, leading to random geometry where the distances between the recognition sites are not fixed.

8.3.4.2 Condensed-phase Switching

In order to be able to incorporate molecular motors into devices, it is imperative that the mechanical dynamic properties of the bistable rotaxanes are preserved in condensed phases, where the mechanical movements of the molecules can be harnessed in a cooperative and coherent fashion. In order to demonstrate that the TTF/DNP-based rotaxanes maintain their solution-phase switching characteristics in condensed phases, the tetraarylmethane stopper in [2]rotaxane **5**$^{4+}$ was replaced with a hydrophilic stopper at the TTF end of the dumbbell. This modular structural modification afforded (Fig. 8.7) an amphiphilic bistable [2]rotaxane **6**$^{4+}$, which can form [22b] Langmuir monolayers at the air/water interfaces and can then be transferred, using the Langmuir–Blodgett technique, on to solid substrates. The nitrogen 1s emission from the CBPQT^{4+} ring was chosen as the diagnostic XPS peak to monitor the redox chemically induced molecular movement in the condensed phase, where the amphiphilic [2]rotaxane molecules are present in a highly organized and oriented fashion. When the oxidant Fe(ClO$_4$)$_3$ was added to the aqueous subphase in highly packed Langmuir monolayers, XPS of the film displayed (Fig. 8.7) a much higher N 1s peak intensity for the switched molecules compared with that for the starting state of the amphiphilic bistable [2]rotaxane monolayer. The orientation of the [2]rotaxane **6**$^{4+}$ in the Langmuir

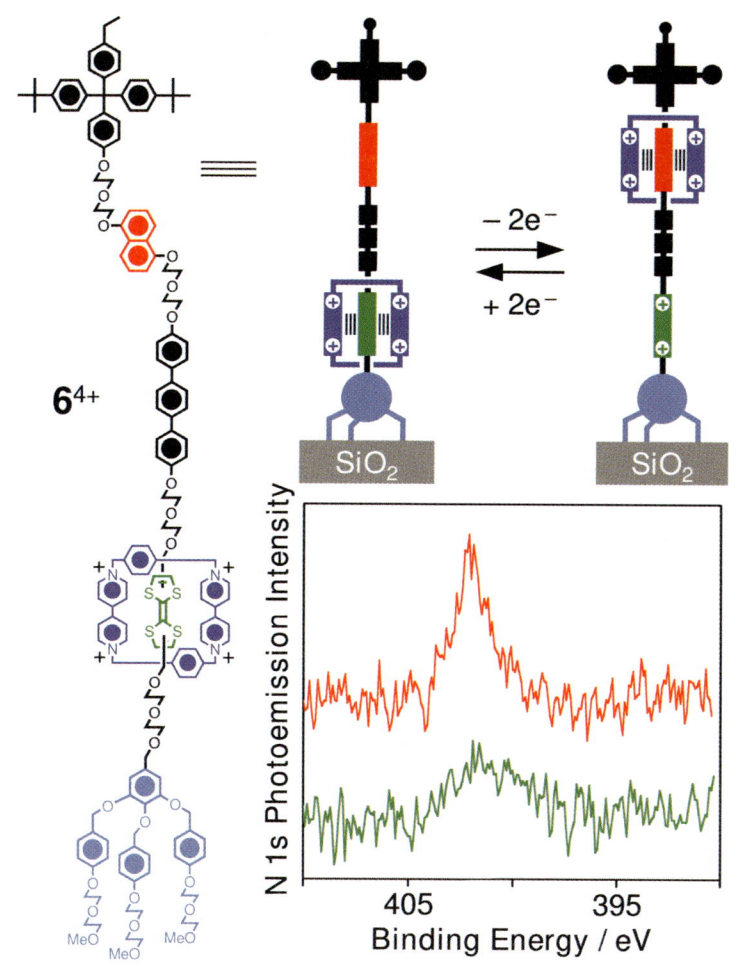

Fig. 8.7 Chemical switching of an amphiphilic bistable [2]rotaxane 6^{4+} in a condensed phase, monitored by X-ray photoelectron spectroscopy.

monolayer is such that, in the original state (GSCC), the CBPQT^{4+} rings are embedded deep in the monolayer as they encircle the TTF sites, whereas, upon oxidation of the TTF units, the rings move to the DNP sites which are located much closer to the film's surface. This explanation accounts for the much higher intensity of the N 1s peak in the oxidized sate than that in the initial state. As expected, in the control experiment, using the corresponding amphiphilic dumbbell compound which lacks the CBPQT^{4+} ring, no N 1s pcak is observed. Detailed ellipsometric analyses showed that the ring travels 1.9 nm in the 4.4-nm thick film, generating a 42% of displacement – a movement which is in excellent agreement with the 46% displacement estimated by molecular modeling. These studies demonstrate that the stimuli-induced, mechanical motion of the linear molecular motors can be preserved in condensed phases, making bistable rotaxane-based nanoelectromechanical systems (NEMS) viable devices in the near future.

8.3.4.3 A Solid-state Nanomechanical Device

The concept of a TTF/DNP-based bistable [2]rotaxane has been extended [22c] into a doubly bistable palindromic [3]rotaxane 7^{8+} to develop (Fig. 8.8) a molecular muscle. These molecular muscles, when self-assembled on microcantilever beams (500 × 100 × 1 µm), are capable of bending and stretching the beams when appropriate redox reagents are injected into the device environment in a microfluidic cell. The palindromic bistable [3]rotaxane 7^{8+} is composed of a symmetrical dumbbell component with two $CBPQT^{4+}$ rings interlocked on to the dumbbell. The symmetrical dumbbell component incorporates two TTF recognition sites close to its ends, two DNP sites bridged by a rigid dialkyne spacer at its center and two 2,6-diisopropylphenyl stoppers at both termini. Each of the $CBPQT^{4+}$ rings carries a pendant disulfide to allow the self-assembly on to gold surfaces of the [3]rotaxane through its two rings. As a result, when the GSCC, having both $CBPQT^{4+}$ rings around the remote TTF sites, is self-assembled on to gold-coated thin microcantilever beams, the positions of the $CBPQT^{4+}$ rings are fixed on the gold surface. Chemical oxidation of both TTF units to TTF^{2+} dications by $Fe(ClO_4)_3$ induces an electrostatic charge repulsion between the TTF^{2+} dications and $CBPQT^{4+}$ rings, which drives them both towards the DNP sites near the center – a process that bends the underlying cantilever beams upwards by ca. 35 nm to their apparent saturation point. Covalent attachments of both $CBPQT^{4+}$ rings to the cantilever beams via Au–S bond formation restrict their free translation from the remote TTF sites to the internal DNP sites along the cantilever surface. However, the $CBPQT^{4+}$ rings can still encircle the DNP sites near the center by bending the beam in the upward direction, as demonstrated in Fig. 8.8. Addition of a reductant (ascorbic acid) in the medium returns the beam to its starting position by bending it downwards. This process could be reproduced on a set of four cantilever beams for 25 cycles, monitored by an optical lever. Chemical and/or physical passivation of the self-assembled monolayers (SAMs) of the [3]rotaxane 7^{8+} might account for the gradual attenuation of the beam deflection in successive cycles. In summary, the chemically induced mechanical force generated by a doubly bistable palindromic [3]rotaxane could be utilized to harness nanomechanical bending of Au-coated silicon cantilever beams. It also demonstrates strongly that molecular nanomachines, in the true sense, are viable targets.

It should be noted that, in all these cases, appropriate chemical reagents need to be added to destabilize the ground-state interactions between the ring and the recognition site, producing a force that ultimately propels the ring's movement to the second recognition site with the assistance of ambient thermal energy. Similarly, a second chemical reagent is required to reverse the influence of the first one and hence restore the ground state from the metastable state. Consequently, chemical reagents keep being accumulated in the medium of the molecular ma-

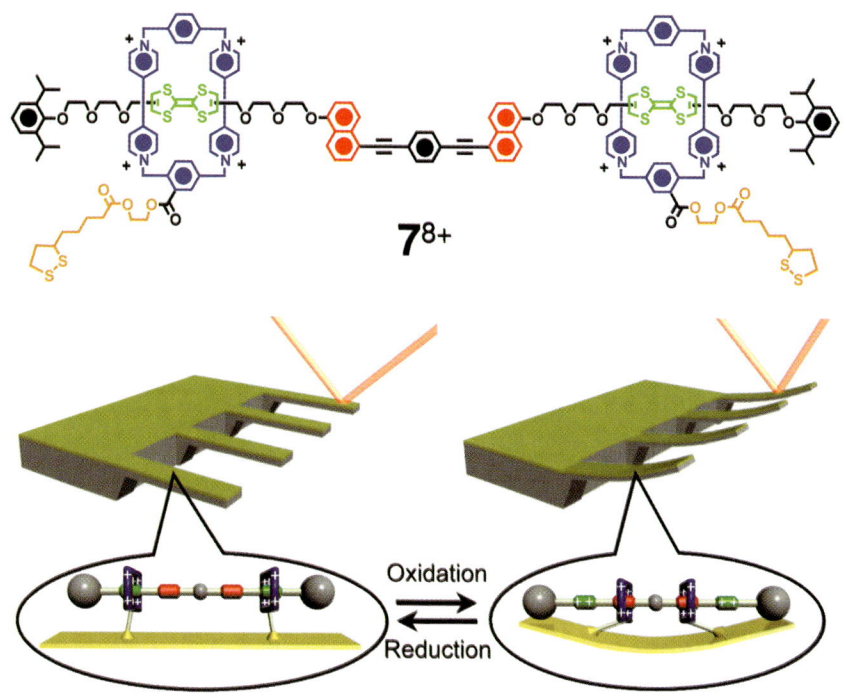

Fig. 8.8 Bending and stretching microcantilever beams by the SAMs of a chemically controlled doubly bistable [3]rotaxane 7^{8+}.

chines, producing a substantial amount of chemical waste after only a few cycles. This situation becomes detrimental to the performance of the molecular machines in later cycles. Nevertheless, these examples demonstrate that, when appropriate stimuli are employed, bistable rotaxanes can execute piston-like translational motions. It also provided a launching pad for the development of the bistable linear molecular motors that can be driven by electrochemical and photochemical inputs in waste-free manners.

8.4
Electrochemically Controllable Bistable Rotaxanes

An electrochemical stimulus provides a two-fold advantage since it can stimulate molecular actuations through clean redox reactions and also detects simultaneously the resulting molecular movements by sensing the changes in the electromechanical properties of a bistable molecule. CV is often employed to both write and read the electromechanical movements in the redox-active bistable rotaxanes.

8.4.1
A Benzidine/Biphenol-based Molecular Switch

The [2]rotaxane 8^{4+} is one (Fig. 8.9) of the earlier examples [14a] of a redox-active molecular switch, composed of a π-electron-deficient CBPQT^{4+} ring, which is interlocked on to a dumbbell component comprising a good π-electron-donating benzidine unit and a weaker π-electron-donating biphenol unit. The ^{1}H NMR spectrum (CDCN$_3$) of [2]rotaxane 8^{4+} at 229 K shows the coexistence of both possible co-conformations in favor (84:16) of the one having the stronger π-donor benzidine station encircled by the CBPQT^{4+} ring. The lack of a more pronounced presence of the thermodynamically more stable isomer in the ground state may be attributed to the relatively small difference in the binding affinities of the two recognition sites. The CV of the [2]rotaxane 8^{4+} shows that the benzidine unit undergoes its first oxidation at a more positive potential compared with a model benzidine compound devoid of a CBPQT^{4+} ring around it. However, the second oxidation of the benzidine unit occurs essentially at the same potential as that observed for the uncomplexed model compound. These observations indicate that the π-electron-deficient CBPQT^{4+} ring around the benzidine station makes its first oxidation more difficult in the major GSCC. The first oxidation of the benzidine, however, destabilizes the CT interaction with the CBPQT^{4+} ring, allowing it to move under thermal assistance to the weaker π-electron-donat-

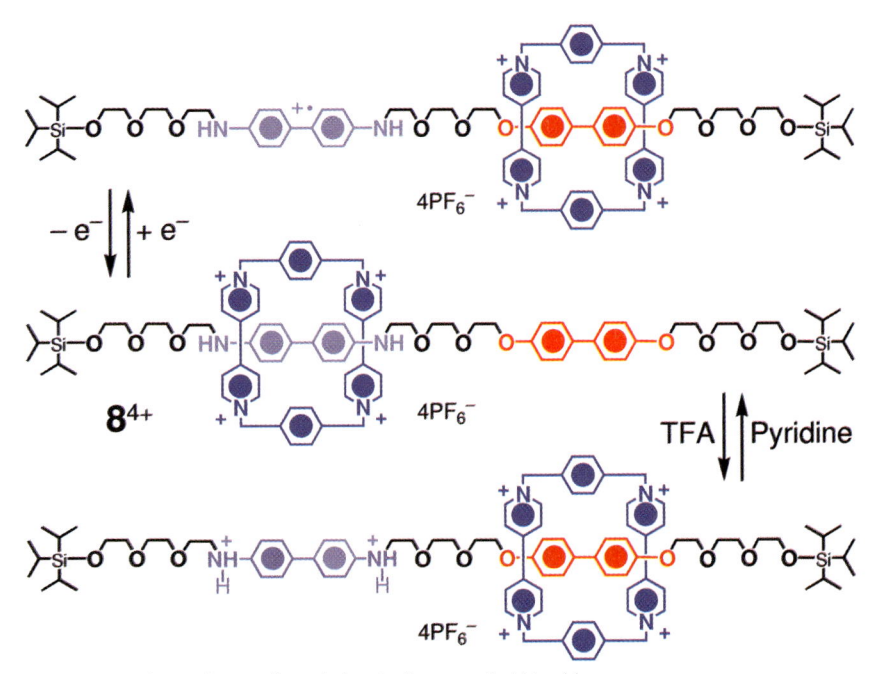

Fig. 8.9 An electrochemically and chemically controlled bistable [2]rotaxane 8^{4+}.

ing biphenol unit. In this oxidized state, the CBPQT^{4+} ring does not have any influence on the second oxidation of the bare benzidine radical cation. Reduction of the benzidine radical cation produces the MSCC with CBPQT^{4+} ring around the biphenol site, which relaxes thermally back to the GSCC.

It is worth mentioning that movement of the CBPQT^{4+} ring from the benzidine unit to the biphenol unit could also be effected by protonation of the benzidine nitrogen atoms by trifluoroacetic acid, making the unit a much weaker π-electron donor. The recovery stroke of the CBPQT^{4+} ring could be effected by the addition of a stoichiometric amount of pyridine, which deprotonates the protonated benzidine unit.

8.4.2
Electrochemically Controlled Switching of TTF/DNP-based [2]Rotaxanes

8.4.2.1 Solution-phase Switching

Movement of the CBPQT^{4+} ring between the TTF and DNP units in the bistable [2]rotaxanes 5^{4+} and 6^{4+} can also be triggered [14e] by electrochemical inputs. The CV of these bistable [2]rotaxanes revealed that a more positive potential (200–400 mV) is required for the first one-electron oxidation of TTF in both [2]rotaxanes than the standard oxidation potential of +350 mV for a free TTF unit. This large shift indicates a strong CT interaction between π-electron-rich TTF unit and the surrounding π-electron-deficient CBPQT^{4+} ring – a situation which makes the oxidation of the TTF unit to TTF$^{+\cdot}$ radical cation more difficult. However, the second one-electron oxidation to the TTF^{2+} dication takes place at a similar potential (+800 mV) to that observed for the bare TTF unit in the corresponding dumbbell compounds. Moreover, the oxidation potential of the DNP unit in the [2]rotaxanes shifts towards more positive potentials (ca. +1600 mV) from those (ca. +1300 mV) for the dumbbell components. All of these observations lead to the conclusion that, once the TTF unit is oxidized to a TTF$^{+\cdot}$ radical cation, the electrostatic repulsion moves the CBPQT^{4+} ring to the second π-electron-donating DNP station, where the ring has no influence on its second oxidation. However, the CBPQT^{4+} ring encircling the DNP unit makes its oxidation more difficult. The reduction of the TTF^{2+} dication back to its neutral form brings the CBPQT^{4+} ring back to its original TTF unit. In the solution phase, the CBPQT^{4+} ring relaxes back to its original TTF unit so quickly that capturing the MSCC at room temperature becomes virtually impossible.

8.4.2.2 Metastability of a Redox-driven [2]Rotaxane SAM on Gold Surfaces

A TTF/DNP-based [2]rotaxane [15e] 9^{4+} was functionalized (Fig. 8.10a) with a disulfide-based anchoring group at the DNP end of the dumbbell component in order to allow its self-assembly on to gold surfaces. The CV of the SAM/Au was recorded at a scan rate of 300 mV s^{-1} at 288 K. The first CV cycle (Fig. 8.10b, green trace) displays a higher positive potential (+490 mV as opposed to +290 mV for the dumbbell) for the first one-electron oxidation of the TTF unit

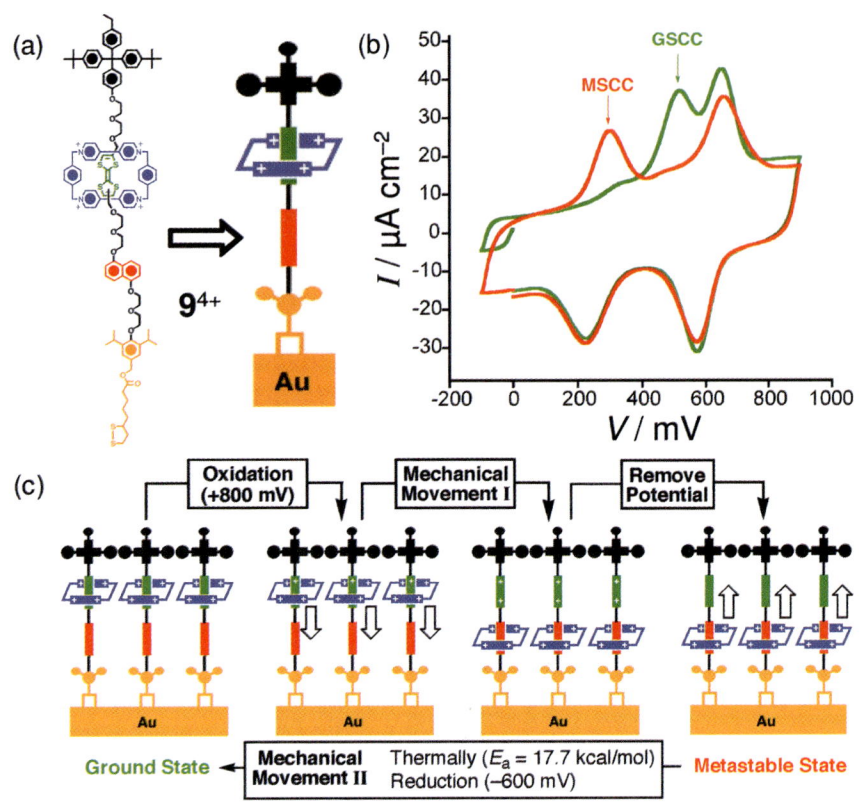

Fig. 8.10 (a) Structural formula of the bistable [2]rotaxane 9^{4+} self-assembled on gold surfaces. (b) The green CV trace indicates the GSCC and the red CV trace displays the presence of an MSCC at a low temperature. (c) Electrochemical switching of the SAMs.

to the TTF$^{+\cdot}$ radical cation on account of the TTF–CBPQT^{4+} CT interaction in the GSCC. The second and all subsequent CV cycles (Fig. 8.10b, red trace) did not display the +490-mV peak. A peak, however, corresponding to the first oxidation (+280 mV) of the free TTF unit appeared in the CV trace. These observations provide direct evidence of the metastable state during the second and subsequent CV cycles. The first oxidation of the TTF unit in the GSCC induces the CBPQT^{4+} ring to move (Fig. 8.10c) to the DNP unit, generating a bare TTF unit, which can be captured in the subsequent cycles at a lower temperature and higher scan rates. The MSCC could be replaced completely with the GSCC by sweeping the CV cycle through an anodic potential (–600 mV), which reduces the CBPQT^{4+} ring to the CBPQT$^{2\cdot/2+}$ diradical, destabilizing the MSCC. It allows the relaxation of the MSCC more promptly than the thermally excited relaxation.

8.4.2.3 A TTF/DNP [2]Rotaxane-based Electrochromic Device

One of the most appealing properties of the TTF/DNP-based [2]rotaxanes originates from the unique colors of the GSCC and MSCC. The GSCC of the TTF/DNP-based bistable rotaxanes displays an emerald green color for the TTF/CBPQT^{4+} CT complex, while the MSCC shows a reddish purple color on account of the DNP/CBPQT^{4+} CT interaction. When the [2]rotaxane [15f] **10**$^{4+}$ is admixed (Fig. 8.11a) with an optically transparent polymer matrix, its redox-induced mechanical movements are slowed significantly. This situation allows both the GSCC and the oxidatively generated MSCC to live long enough to display their own distinct colors. Initially, the polymer film containing **10**$^{4+}$ displayed (Fig. 8.11b) the green color of the GSCC, which turned (Fig. 8.11c) to the reddish purple color of the MSCC at an applied potential of +1.0 V. The green color of the GSCC could be regenerated by removing the bias and allowing the MSCC to relax thermally.

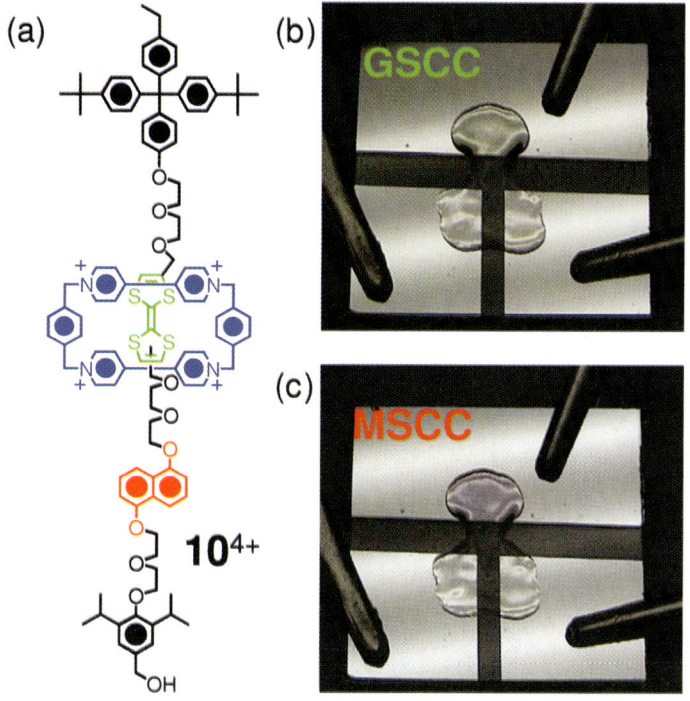

Fig. 8.11 (a) Structural formula of the bistable [2]rotaxane **10**$^{4+}$ introduced into a polymer matrix. The electrochromic responses of the bistable [2]rotaxane **10**$^{4+}$ trapped inside a polymer matrix in the GSCC (b) and in the oxidized state (c).

8.4.2.4 A Redox-driven [2]Rotaxane-based Molecular Switch Tunnel Junctions (MSTJs) Device

The redox-controllable switching behavior of the TTF/DNP-based bistable [2]rotaxanes has been exploited [15c] in the fabrication of two-dimensional molecular electronic crossbar circuits which can be made to function as memory and logic devices. Langmuir–Blodgett (L–B) monolayers of the amphiphilic [2]rotaxane 11^{4+} (Fig. 8.12a) and several other catenanes and rotaxanes containing the same switching gears were transferred [15c] to a highly doped n-type polysilicon-based bottom electrode. The monolayer was then covered with a 5–15-nm titanium layer followed by the deposition of a 100-nm thick layer of aluminum or aluminum plus nickel, to form the top electrodes. It was demonstrated that the resulting crossbar electrical circuits of micrometer and nanometer dimensions could be switched between open (OFF) or closed (ON) states by applying threshold voltages. A remnant molecular signature (RMS) can be recorded [15c] by sweeping a write potential between −2.0 and +2.0 V in 40-mV steps. Between each of the 40-mV steps, a small read voltage of −0.2 or +0.2 V is applied. In each case, the Ti/Al top electrode was grounded while the voltages were applied to the bottom Si electrodes. The hysteresis loop in the RMS shows (Fig. 8.12b, black trace) that the MSTJ is closed at −2.0 V and opened at +2.0 V. The current flow (∼40 pA) between the electrodes through the monolayer of the bistable rotaxane molecules in the switch-open state is about four times larger than that (∼10 pA) in the switch-closed state. A control system, involving electronically inactive eicosanoic acid, did not show (Fig. 8.12b, red trace) any such switching.

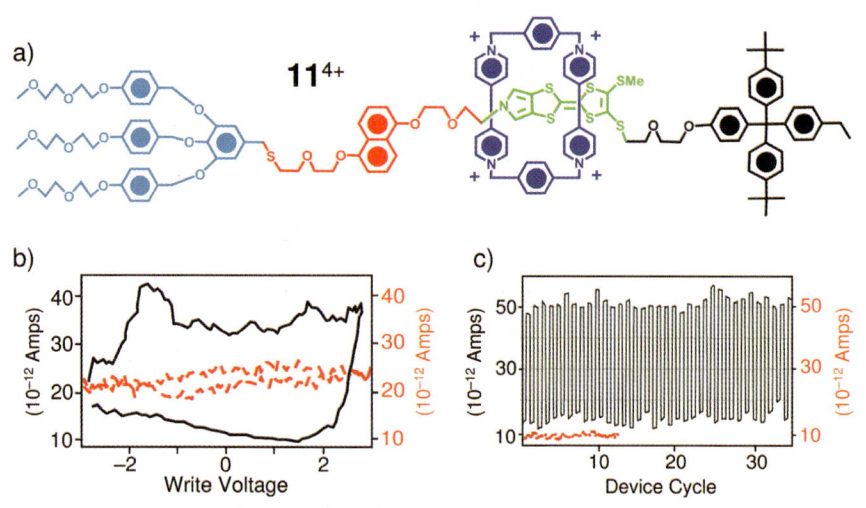

Fig. 8.12 (a) Structural formula of a redox-driven amphiphilic bistable [2]rotaxane, 11^{4+}, which is incorporated in a crossbar MSTJ. (b) Hysteresis loop displayed by 11^{4+} (black trace) and an electrochemically inactive eicosanoic acid (red trace). (c) Switch opening and closing by 11^{4+} at +2.0 and −2.0 V, respectively (black trace). No such response by eicosanoic acid.

Furthermore, a nanometer-sized MSTJ fabricated with the bistable rotaxane 11^{4+} could be switched ON/OFF (Fig. 8.12c, black trace) for 35 cycles by applying +2.0 and −2.0 V, respectively, while degenerate, non-bistable controls do not switch (Fig. 8.12c, red trace). It is believed that the GSCC of the bistable [2]rotaxane 11^{4+} with the CBPQT^{4+} ring around the monopyrrolo-TTF (MP-TTF) is responsible for the switch closed state, whereas the MSCC with the ring around the DNP unit accounts for the switch open state. This hypothesis is supported by the fact that the voltages required to activate the switch opening and closing states match roughly with the voltages required for the switching of the bistable [2]rotaxane 11^{4+}. Computational studies based on first principles [26] to obtain the transmission functions reflect [3n] that the switch ON and OFF states correspond to the smaller and larger energy difference between the rotaxane's highest occupied and lowest occupied molecular orbitals (HOMO, LUMO), respectively. The simulation studies show [3n] that the HOMO–LUMO gap in 11^{4+} is larger when the CBPQT^{4+} ring encircles the TTF site compared with that when it encircles the DNP site – findings that are extremely consistent with the ON/OFF states observed for MSTJs. Similar behavior was observed for MSTJs based on bistable [2]catenanes. These MSTJ devices have been used [15c] to build AND and XOR gates.

8.4.3
A Redox and Chemically Controllable Bistable Neutral [2]Rotaxane

8.4.3.1 Electrochemical Switching
The bistable, neutral [2]rotaxane 12 (Fig. 8.13) is composed [14g] of a π-electron-rich 1,5-dinaphtho[38]crown-10 (1/5DNP38C10; crown ether hereafter) and a dumbbell component that provides two π-electron-deficient recognition sites – a stronger π-electron-accepting 1,4,5,8-naphthalenetetracarboxylate diimide (NpI) unit and a second weaker π-electron-accepting pyromellitic diimide (PmI) unit, separated within the rod section by a pentamethylene chain. The size of the tetraarylmethane stoppers located at both ends of the dumbbell is such that they can allow the crown ether to slip on to the rod section of the dumbbell component at elevated temperatures to form the [2]rotaxane. The UV–Vis absorption spectrum of the [2]rotaxane 12 displays only one CT band centered on 490 nm; it is characteristic of the NpI–crown ether CT interaction. This band, along with the absence of a PmI–crown ether CT interaction (435-nm band), confirms that, in the GSCC of the rotaxane 12, the crown ether encircles the NpI unit. The first one-electron reduction of the NpI unit, encircled by the crown ether in the GSCC of the [2]rotaxane, occurs at a more negative potential (−740 mV) than that (−650 mV) observed for the corresponding dumbbell component. This reductive process causes movement of the crown ether to the second π-electron-accepting PmI unit, making the first one-electron reduction of the PmI unit to happen at a more negative potential (−1060 mV) than its free form (−900 mV). Interestingly, the second reduction the PmI$^-$ unit in the rotaxane 12$^{··}$ takes place at exactly the same potential (−1470 mV) as that of the bare PmI$^-$ unit, indicating the disengagement of the

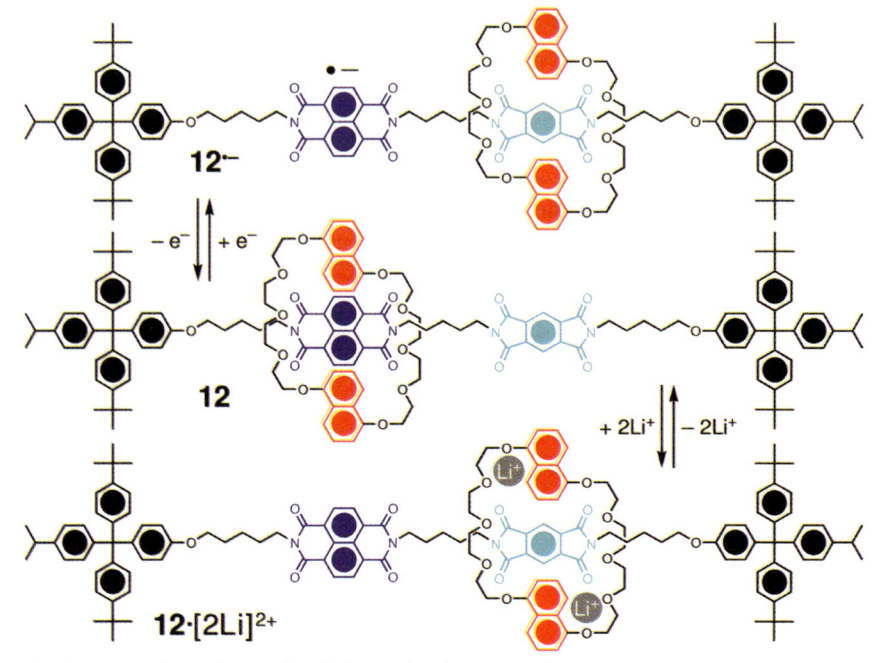

Fig. 8.13 An electrochemical and chemical (Li$^+$ ion) switching in a neutral bistable [2]rotaxane, **12**.

PmI$^-$ unit from the crown ether. Regeneration of the neutral NpI unit by an oxidative process restores the bistable, neutral [2]rotaxane to its ground state.

8.4.3.2 Chemical Switching Induced by Lithium Ion (Li$^+$)

In the neutral [2]rotaxane **12**, the movement of the crown ether from the original NpI to the PmI unit can also be induced (Fig. 8.13) chemically [7g] by introducing Li$^+$ ions, which help the PmI unit to bind the 1/5DNP38C10 ring more strongly than it does the NpI unit to generate **12**·[2Li]$^{2+}$. Sequestering of the Li$^+$ ions by 12-crown-4 obliges the 1/5DNP38C10 ring to return to the NpI station.

8.5
Photochemically Powered Molecular Switches

So far, several example of the chemically and electrochemically controlled switching of bistable linear molecular machines have been presented. The final section of this chapter will be dedicated to illustrating how such molecular switches and motors, when designed ingeniously, can also be powered by nature's most abundant and powerful energy source – light.

8.5 Photochemically Powered Molecular Switches

Not only can photo-active molecular motors be driven in a waste-free manner [6], but also their eventual introduction into molecular electronics and NEMS will have a tremendous impact because of their properties relating to photo-to-mechanical energy transduction. Photochemically controlled molecular switches are among the most advanced and fascinating class of artificial nanomachines. Just like electrons, photons can induce (write) as well as detect (read) molecular actuations. In addition to the basic structural units of a bistable [2]rotaxane, a photo-driven molecular abacus should have a photoactive unit capable of harnessing light energy to generate chemical changes in a process that ultimately culminates in some mechanical movements. Two classes of photo-driven molecular abacuses are known [9, 27, 29] to date: (1) one [9i] involving photoisomerization (cis–trans) of double bonds (C=C, C=N and N=N), which, in turn, modifies the binding abilities of the recognition sites, and (2) the other [9b,c, 28] stimulated by photo-induced electron transfer (PET) processes. Two strategies have been adopted in the design of the light-driven molecular abacuses where the selective noncovalent binding abilities of the recognition sites are altered photochemically: (1) one introduces an external reagent that can undergo PET with the recognition sites in the presence of light and (2) the other incorporates a built-in light-harvesting unit which can undergo an intramolecular PET to the recognition sites. In some cases, sacrificial chemical reagents are required in order to quench the unproductive pathways and to facilitate the molecular motions.

8.5.1
Molecular Switching Caused by Photoisomerization

The simplest example of photoactive molecular switches relies on light-activated cis–trans isomerization to effect switching. The [2]rotaxane (E/Z)-**13**-H_2^{2+} (Fig. 8.14) incorporates [9i] a fumaramide/maleamide unit –a photoisomerizable alkene-diamide that also plays the role a strong H-bond-acceptor in its E-form (fumaramide) to the surrounding H-bond donating tetramide macrocycle in (E)-**13**-H_2^{2+}. A second H-bond-accepting glycylglycine (Gly-Gly) unit and a nearby anthracene unit to act as a fluorescent label are also present in the dumbbell component of the rotaxane. The tetramide-based macrocycle provides four H-bond donating centers. Two pyridinium units in the macrocycle are known to quench the anthracene-based fluorescence through electron transfer when they are in close proximity. Irradiation of the (E)-**13**-H_2^{2+} isomer with 312-nm light converts the (E)-fumaramide into its Z-isomer, maleamide, which has a significantly lower H-bond-accepting ability. This change in the binding affinity allows the macrocycle to move to the intermediate H-bond accepting station Gly-Gly, generating the (Z)-**13**-H_2^{2+} translational isomer. Heating of the (Z)-**13**-H_2^{2+} isomer at 115 °C reproduces the fumaramide unit and the macrocycle returns to its original binding site in (E)-**13**-H_2^{2+}. The longer distance between the anthracene fluorescent tag and the macrocycle containing the two pyridinium units in the (E)-**13**-H_2^{2+} isomer allows this isomer to fluoresce (λ_{exc} = 365 nm), whereas the anthracene fluorescence in the (Z)-**13**-H_2^{2+} rotaxane is quenched by the close proximity of the pyridinium ions.

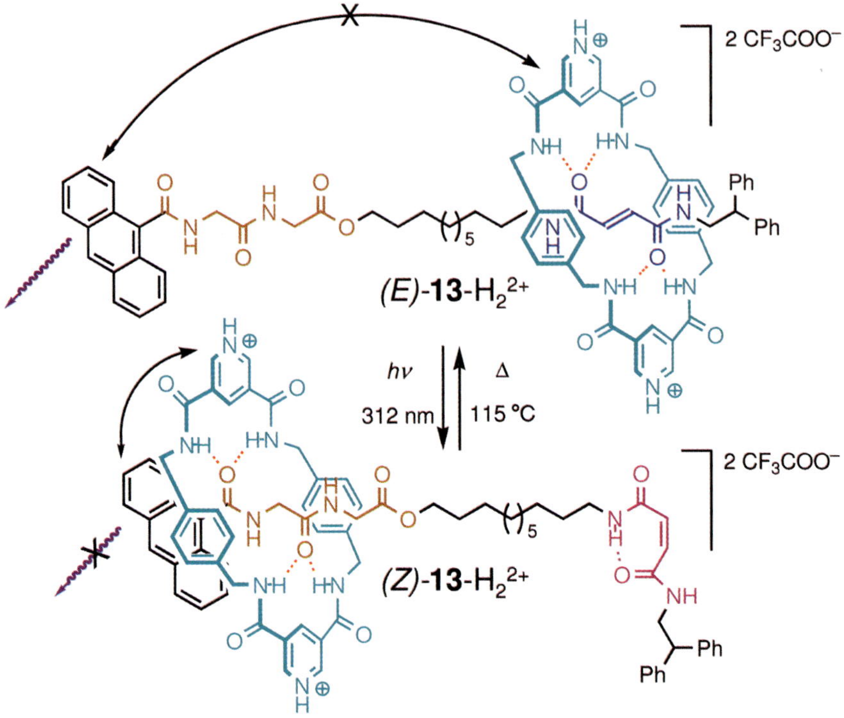

Fig. 8.14 Photochemically and thermally induced switching in the [2]rotaxane (E/Z)-**13**.

8.5.2
PET-induced Switching of an H-bonded Molecular Motor

The [2]rotaxane **14** (Fig. 8.15) is composed [9c] of an H-bond-donating benzylic amide macrocycle, mechanically interlocked with a dumbbell component, which is equipped with a strong H-bond-accepting unit, succinamide (*succ*), and a poor H-bond-accepting unit, 3,6-di-*tert*-butyl-1,8-naphthalimide (*ni*). The chair-shaped macrocycle encircles the *succ* unit in the neutral GSCC of the rotaxane. Photoreduction of the *ni* station to *ni*$^{\cdot-}$ radical anion by irradiation at 355 nm in the presence of an electron donor (D), 1,4-diazabicyclo[2.2.2]octane (DABCO), makes it a stronger H-bond acceptor than the *succ* unit. This modification in the H-bonding abilities of the recognition sites provides a driving force for the ring to move to the *ni*$^{\cdot-}$ radical anion along the C_{12} alkyl spacer. Charge recombination (CR) between the *ni*$^{\cdot-}$ radical anion and the DABCO$^{+\cdot}$ radical cation generates the neutral MSCC, which thermally relaxes back to the GSCC as the ring returns to the original *succ* unit. Hence the light-induced movement of the macrocycle resembles a "power stroke" and the CR-induced return of the ring is similar to the

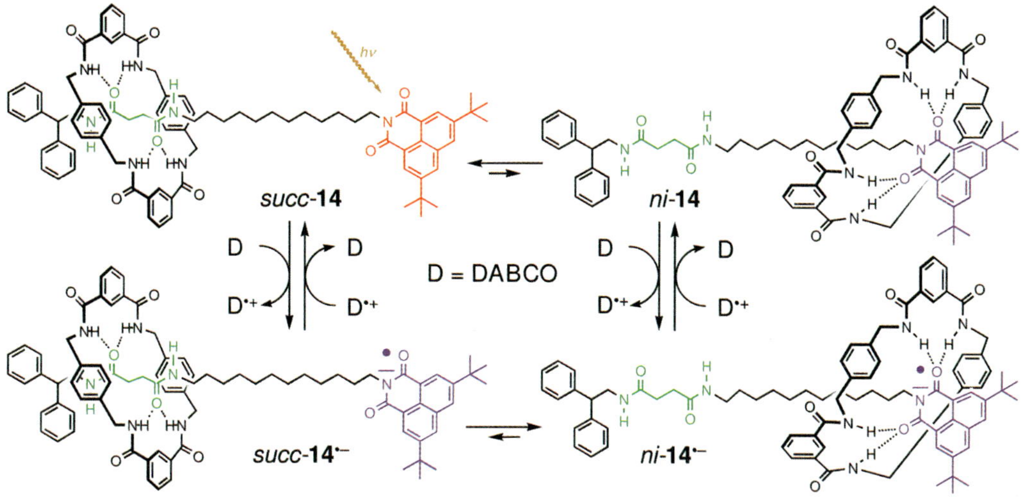

Fig. 8.15 Photo-induced switching in the [2]rotaxane **14**.

"recovery stroke" of a piston's mechanism. This reversible photo-induced movement of components in the [2]rotaxane **14** generates a mechanical power of ca. 10^{-15} W molecule^{-1}.

The [2]rotaxane **14** could also be powered by an electrochemical stimulus [8g], in which case, the "power stroke" is generated by a direct reduction of the *ni* unit to *ni*$^{-}$ and the "recovery stroke" is induced by its neutralization.

8.5.3
MLCT-induced Switching of a Metal Ion-based Molecular Motor

The transition metal-based [2]rotaxane **15**$^{+/2+}$ is composed (Fig. 8.16) of a 30-membered macrocycle which furnishes [28] a bidentate phen ligand and a two-station dumbbell bearing another disubstituted phen ligand and a terdentate terpy ligand. The interlocked ring can be switched between the phen and terpy ligands based on the preferred coordination environments of the oxidation states of the complexed Cu^{I} or Cu^{II} ions. The Cu$^{I}_{(4)}$ state favors tetrahedral geometry, a requirement which can be satisfied when the bidentate phen ligands from both the ring and the dumbbell are involved in chelation. By contrast, the pentacoordination of the Cu$^{II}_{(5)}$ state can be satisfied by the involvement of the phen ligand from the ring and the terpy ligand from the dumbbell component. Thus, a photochemical or electrochemical oxidation of the complexed Cu$^{I}_{(4)}$ to Cu$^{II}_{(5)}$ drives the ring from the phen site to the terpy site along the rod section of the dumbbell and vice versa. Irradiation of the [2]rotaxane **15**$^{+}$-Cu^{I} with 464-nm light in the presence of an electron acceptor (*p*-nitrobenzyl bromide) generates **15**$^{2+}$-Cu$^{II}_{(4)}$, which relaxes to the more stable **15**$^{2+}$-Cu$^{II}_{(5)}$ when the ring moves to the terpy site. Addition of a reductant (ascorbic acid) reduces the Cu$^{II}_{(5)}$ to the Cu$^{I}_{(4)}$ state, which, in turn, induces

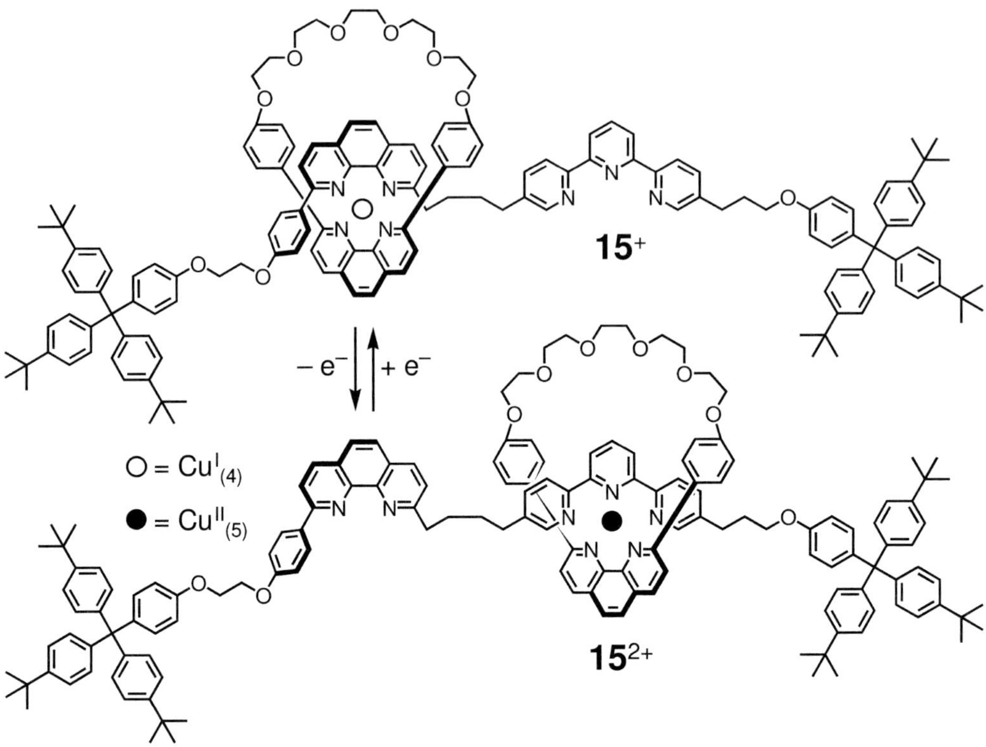

Fig. 8.16 Photochemically and electrochemically induced switching of the [2]rotaxane $15^+/15^{2+}$.

the return of the ring to its original phen site on the dumbbell. As a result, the movement of the ring can be powered by hybrid photochemical–chemical stimuli. A direct electrochemical redox process involving the $Cu^I_{(4)}/Cu^{II}_{(5)}$ couple also triggers the same translational motion in the bistable [2]rotaxane.

8.5.4
A Photo-driven Molecular Abacus

The multicomponent [2]rotaxane 16^{6+} has been designed (Fig. 8.17) to work [9b] as a visible light-driven molecular abacus. It is composed of a π-electron-rich macrocycle, bis-p-phenylene-34-crown-10 (BPP34C10), interlocked with a dumbbell component, which comprises an $Ru^{II}(bipy)_3$ (P)-based light-harvesting unit, two π-electron-accepting sites – a highly efficient π-acceptor, 4,4′-bipyridinium, unit (A_1) and a second less efficient π-electron acceptor, 3,3′-dimethyl–4,4′-bipyridinium unit (A_2) – and a rigid terphenyl spacer between the light-harvesting unit and the recognition sites. In the GSCC, the crown ether encircles the stronger π-accepting A_1 unit. Photosensitization of the $Ru(bipy)^{2+}$ complex with 532-nm light

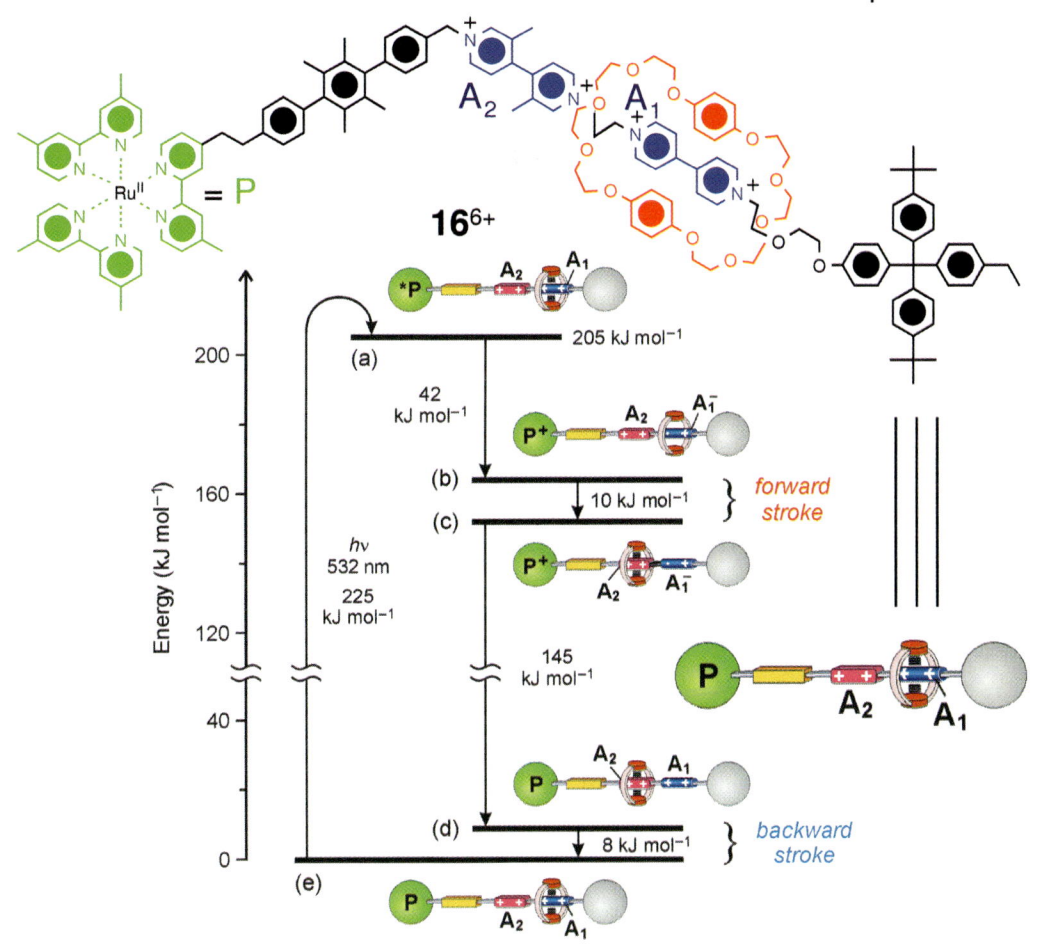

Fig. 8.17 Light-driven switching in the [2]rotaxane 16^{6+}.

generates its metal-to-ligand charge-transfer (MLCT) excited state, which undergoes a PET to the stronger π-electron-accepting A_1 unit to produce the A_1^- entity. In the presence of an external electron donor (D), phenothiazine [29], which directly reduces the photo-oxidized P^+ unit to prevent the back electron transfer (BET) process, the A_1^- entity enjoys a sufficient lifetime to repel the π-electron-rich BPP34C10 ring to the weaker π-electron-accepting A_2 unit. An intermolecular electron transfer between the A_1^- entity and the oxidized phenothiazine unit regenerates the neutral A_1 unit and generates the MSCC, which is represented by the BPP34C10 ring's location around the A_2 unit. Finally, Brownian motion of the molecules, assisted by the ambient thermal energy, moves the BPP34C10 ring back to its original A_1 unit. This movement demonstrates that an autonomous photo-driven molecular abacus can be developed in the presence of a dual-action electron donor–acceptor reagent. Previously, switching of the [2]rotaxane **16**$^{6+}$ was achieved[9b] by introducing individual sacrificial donor and acceptor reagents to restrict the detrimental charge recombination processes.

8.6
Conclusions

Supramolecular chemistry has empowered chemists to develop nanoscale molecular machinery based on interlocked molecules in a "bottom-up" approach. The intrinsic dynamic nature of such molecules, which relies on self-assembly and molecular recognition for their structural integrity, makes them one of the best candidates to qualify as nanomechanical machines. Mechanical movements in molecular machines can be triggered and controlled on the nanoscale level by a variety of external stimuli. To establish the proof-of-principle behind molecular machines, most of them have been demonstrated to perform in the solution phase. Their prospects as NEMS depend heavily on their ability to reproduce the same mechanical movements on surfaces and at interfaces and in condensed phases. Some bistable rotaxanes and catenanes display stimuli-induced reversible movements in situations, such as in SAMs, in polymer matrices and in Langmuir monolayers. Several miniature solid-state devices – molecular switch tunnel junctions (MSTJs), molecular logic gates, molecular muscles, molecular nanovalves – have already been developed, based on controllable, switchable, interlocked molecules. It is already evident that a successful marriage between the "top-down" strategies of the device world and the "bottom-up" approach of the molecular world will expedite the development of nanotechnology.

Acknowledgments

We thank all researchers, including our collaborators, colleagues and coworkers, whose work has been presented in this chapter. We thank especially Dr Amar H. Flood for his valuable inputs and critical discussions. The research was supported by the Defense Advanced Research Projects Agency (DARPA), the National Science Foundation (NSF) and the Office of Naval Research (ONR).

References

1. R. P. Feynman, *Eng. Sci.* **1960**, *23*, 22–36.
2. (a) M. Schliwa (Ed.), *Molecular Motors*, Wiley-VCH, Weinheim, **2003**; (b) Goodsell, D. S., *Bionanotechnology: Lessons from Nature*, Wiley, New York, **2004**; (c) G. Oster, H. Wang, *Trends Cell Biol.*, **2004**, *13*, 114–121.
3. (a) J. F. Stoddart, *Chem. Aust.* **1992**, *59*, 576–577, 581; (b) M. Gómez-López, J. A. Preece, J. F. Stoddart, *Nanotechnology* **1996**, *7*, 183–192; (c) V. Balzani, M. Gómez-López, J. F. Stoddart, *Acc. Chem. Res.* **1998**, *31*, 405–414; (d) V. Balzani, A. Credi, F. M. Raymo, J. F. Stoddart, *Angew. Chem. Int. Ed.* **2000**, *39*, 3348–3391; (e) A. Harada, *Acc. Chem. Res.* **2001**, *34*, 456–464; (f) C. A. Schalley, K. Beizai, F. Vögtle, *Acc. Chem. Res.* **2001**, *34*, 465–476; (g) J.-P. Collin, C. Dietrich-Buchecker, P. Gaviña, M. C. Jiménez-Molero, J.-P. Sauvage, *Acc. Chem. Res.* **2001**, *34*, 477–487; (h) R. Ballardini, V. Balzani, A. Credi, M. T. Gandolfi, M. Venturi, *Struct. Bonding* **2001**, *99*, 163–188; (i) L. Raehm, J.-P. Sauvage, *Struct. Bonding* **2001**, *99*, 55–78; (j) C. A. Stainer, S. J. Alderman, T. D. W. Claridge, H. L. Anderson, *Angew. Chem. Int. Ed.* **2002**, *41*, 1769–1772; (k) V. Balzani, A. Credi, M. Venturi, *Chem. Eur. J.* **2002**, *8*, 5524–5532; (l) H.-R. Tseng, J. F. Stoddart, in *Modern Arene Chemistry*, D. Astruc (Ed.), Wiley-VCH, Weinheim, **2002**, pp. 574–599; (m) V. Balzani, A. Credi, M. Venturi, *Molecular Devices and Machines – A Journey into the Nano World*, Wiley-VCH, Weinheim, **2003**; (n) A. H. Flood, R. J. A. Ramirez, W.-Q. Deng, R. P. Muller, W. A. Goddard, III, J. F. Stoddart, *Aust. J. Chem.* **2004**, *57*, 301–322; (o) P. M. Mendes, A. H. Flood, J. F. Stoddart, *Appl. Phys. A* **2005**, *80*, 1197–1209.
4. (a) T. R. Kelly, H. D. Silva, R. A. Silva, *Nature* **1999**, *401*, 150–152; (b) N. Koumura, R. W. Zijlstra, R. A. van Delden, H. Harada, B. L. Feringa, *Nature* **1999**, *401*, 152–155; (c) Y. Yokoyama, *Chem. Rev.* **2000**, *100*, 1717–1740; (d) G. Berkovic, V. Krongauz, V. Weiss, *Chem. Rev.* **2000**, *100*, 1741–1754; (e) B. L. Feringa, R. A. van Delden, N. Koumura, E. M. Geertsema, *Chem. Rev.* **2000**, *100*, 1789–1816; (f) T. R. Kelly, *Acc. Chem. Res.* **2001**, *34*, 514–522; (g) S. Shinkai, M. Ikeda, A. Sugasaki, M. Takeuchi, *Acc. Chem. Res.* **2001**, *34*, 494–503; (h) K. Oh, K.-S. Jeong, J. S. Moore, *Nature* **2001**, *414*, 889–893; (i) N. Koumura, E. M. Geertsema, M. B. van Gelder, A. Meetsma, B. L. Feringa, *J. Am. Chem. Soc.* **2002**, *124*, 5037–5051; (j) F. Hawthorne, J. I. Zink, J. M. Skelton, M. J. Bayer, C. Liu, E. Livshits, R. Baer, D. Neuhauser, *Science* **2004**, *303*, 1849–1851; (k) J. J. D. de Jong, L. N. Lucas, R. M. Kellogg, J. H. van Esch, B. L. Feringa, *Science* **2004**, *304*, 278–281; (l) A. Koçer, M. Walco, W, Meijberg, B. L. Feringa, *Science* **2005**, *309*, 755–758; (m) S. P. Fletcher, F. Dumur, M. M. Pollard, B. L. Feringa, *Science* **2005**, *310*, 80–82; (n) J. Berná, D. A. Leigh, M. Lubomska, S. M. Mendoza, E. M. Pérez, P. Rudolf, G. Teobaldi, F. Zerbetto, *Nat. Mater.* **2005**, *4*, 704–710.
5. S. Shinkai, O. Manabe, *Top. Curr. Chem.* **1984**, *121*, 76–104.
6. V. Balzani, A. Credi, F. M. Raymo, J. F. Stoddart, *Angew. Chem. Int. Ed.* **2000**, *39*, 3348–3391.
7. (a) A. S. Lane, D. A. Leigh, A. Murphy, *J. Am. Chem. Soc.* **1997**, *119*, 11092–11093; (b) P. R. Ashton, R. Ballardini, V. Balzani, I. Baxter, A. Credi, M. C. T. Fyfe, M. T. Gandolfi, M. Gómez-López, M.-V. Martínez-Díaz, A. Piersanti, N. Spencer, J. F. Stoddart, M. Venturi, A. J. P. White, D. J. Williams, *J. Am. Chem. Soc.* **1998**, *120*, 11932–11942; (c) J. W. Lee, K. Kim, K. Kim, *Chem. Commun.* **2001**, 1042–1043; (d) A. M. Elizarov, H.-S. Chiu, J. F. Stoddart, *J. Org. Chem.* **2002**, *67*, 9175–9181; (e) G. Kaiser, T. Jarrosson, S. Otto, Y.-F. Ng, A. D. Bond, J. K. M. Sanders *Angew. Chem. Int. Ed.* **2004**, *43*, 1959–1962; (f) Y. Liu, A. H. Flood, J. F. Stoddart, *J. Am. Chem. Soc.* **2004**, *126*, 9150–9151; (g) S. A. Vignon, T. Jarrosson, T. Iijima, H.-R. Tseng, J. K. M. Sanders, J. F. Stoddart, *J. Am. Chem. Soc.* **2004**, *126*, 9884–9885.

8. (a) L. Raehm, J. M. Kern, J.-P. Sauvage, *Chem. Eur. J.* **1999**, *5*, 3310–3317; (b) V. Bermudez, N. Capron, T. Gase, F. G. Gatti, F. Kajzar, D. A. Leigh, F. Zerbetto, S. Zhang, *Nature* **2000**, *406*, 608–611; (c) J.-M. Kern, L. Raehm, J.-P. Sauvage, B. Divisia-Blohorn, P.-L. Vida, *Inorg. Chem.* **2000**, *39*, 1555–1560; (d) R. Ballardini, V. Balzani, W. Dehaen, A. E. Dell'Erba, F. M. Raymo, J. F. Stoddart, M. Venturi, *Eur. J. Org. Chem.* **2000**, *65*, 591–602; (e) V. Balzani, A. Credi, G. Mattersteig, O. A. Matthews, F. M. Raymo, J. F. Stoddart, M. Venturi, A. J. P. White, D. J. Williams, *J. Org. Chem.* **2000**, *65*, 1924–1936; (f) J.-P. Collin, J.-M. Kern, L. Raehm, J.-P. Sauvage, in *Molecular Switches*, B. L. Feringa (Ed.), Wiley-VCH, Weinheim, 2000, pp. 249–280; (g) A. Altieri, F. G. Gatti, E. R. Kay, D. A. Leigh, F. Paolucci, A. M. S. Slawin, J. K. Y. Wong, *J. Am. Chem. Soc.* **2003**, *125*, 8644–8654; (h) I. Poleschak, J.-M. Kern, J.-P. Suavage, *Chem. Commun.* **2004**, 474–476.

9. (a) R. Ballardini, V. Balzani, M. T. Gandolfi, L. Prodi, M. Venturi, D. Philp, H. G. Ricketts, J. F. Stoddart, *Angew. Chem. Int. Ed. Engl.* **1993**, *32*, 1301–1303; (b) P. R. Ashton, R. Ballardini, V. Balzani, A. Credi, R. Dress, E. Ishow, O. Kocian, J. A. Preece, N. Spencer, J. F. Stoddart, M. Venturi, S. Wenger, *Chem. Eur. J.* **2000**, *6*, 3558–3574; (c) A. M. Brower, C. Frochot, F. G. Gatti, D. A. Leigh, L. Mottier, F. Paolucci, S. Roffia, G. W. H. Wurpel, *Science* **2001**, *291*, 2124–2128; (d) J.-P. Collin, A.-C. Laemmel, J.-P. Sauvage, *New J. Chem.* **2001**, *25*, 22–24; (e) G. Bottari, D. A. Leigh, E. M. Pérez, *J. Am. Chem. Soc.* **2003**, *125*, 1360–1361; (f) F. G. Gatti, S. Len, J. K. Y. Wong, G. Bottari, A. Altieri, M. A. F. Morales, S. J. Teat, C. Frochot, D. A. Leigh, A. M. Brower, F. Zerbetto, *Proc. Natl. Acad. Sci. USA* **2003**, *100*, 10–14; (g) A. Altieri, G. Bottari, F. Dehez, D. A. Leigh, J. K. Y. Wong, F. Zerbetto, *Angew. Chem. Int. Ed.* **2003**, *42*, 2296–2300; (h) A. M. Brower, S. M. Fazio, C. Frochot, F. G. Gatti, D. A. Leigh, J. K. Y. Wong, G. W. H. Wurpel, *J. Pure Appl. Chem.* **2003**, *75*, 1055–1060; (i) E. M. Pérez, D. T. F. Dryden, D. A. Leigh, G. Teobaldi, F. Zerbetto, *J. Am. Chem. Soc.* **2004**, *126*, 12210–12211.

10. (a) D. H. Busch, N. A. Stephenson, *Coord. Chem. Rev.* **1990**, *100*, 119–154; (b) J. S. Lindsey, *New J. Chem.* **1991**, *15*, 153–180; (c) D. Philp, J. F. Stoddart, *Synlett* **1991**, 445–458; (d) C. A. Hunter, *J. Am. Chem. Soc.* **1992**, *114*, 5303–5311; (e) S. Anderson, H. L. Anderson, J. K. M. Sanders, *Acc. Chem. Res.* **1993**, *26*, 469–475; (f) A. P. Bisson, F. J. Carver, C. A. Hunter, J. P. Waltho, *J. Am. Chem. Soc.* **1994**, *116*, 10292–10293; (g) C. A. Hunter, *Angew. Chem. Int. Ed. Engl.* **1995**, *34*, 1079–1081; (h) J. P. Schneider, J. W. Kelly, *Chem. Rev.* **1995**, *95*, 2169–2187; (i) D. Philp, J. F. Stoddart, *Angew. Chem. Int. Ed. Engl.* **1996**, *35*, 1154–1196; (j) A. G. Kolchinski, N. W. Alcock, R. A. Roesner, D. H. Busch, *Chem. Commun.* **1998**, 1437–1438; (k) M. Fujita, *Acc. Chem. Res.* **1999**, *32*, 53–61; (l) J. Rebek, Jr., *Acc. Chem. Res.* **1999**, *32*, 278–286; (m) F. Diederich, P. J. Stang (Eds.), *Templated Organic Synthesis*, Wiley-VCH, Weinheim, 1999; (n) M. Nakash, Z. Clyde-Watson, N. Feeder, S. J. Teat, J. K. M. Sanders, *Chem. Eur. J.* **2000**, *6*, 2112–2119; (o) J. K. M. Sanders, *Pure Appl. Chem.* **2000**, *72*, 2265–2274; (p) D. T. Bong, T. D. Clark, J. R. Granja, M. R. Ghadiri, *Angew. Chem.* **2001**, *113*, 1016–1041; *Angew. Chem. Int. Ed.* **2001**, *40*, 988–1011; (q) L. J. Prins, D. N. Reinhoudt, P. Timmerman, *Angew. Chem. Int. Ed.* **2001**, *40*, 2382–2426; (r) S. R. Seidel, P. J. Stang, *Acc. Chem. Res.* **2002**, *35*, 972–983; (s) J. F. Stoddart, H.-R. Tseng, *Proc. Natl. Acad. Sci. USA* **2002**, *99*, 4797–4800; (t) R. L. E. Furlan, S. Otto, J. K. M. Sanders, *Proc. Natl. Acad. Sci. USA* **2002**, *99*, 4801–4804; (u) D. Joester, E. Walter, M. Losson, R. Pugin, H. P. Merkle, F. Diederich, *Angew. Chem. Int. Ed.* **2003**, *42*, 1486–1490; (v) L. Hogg, D. A. Leigh, P. J. Lusby, A. Morelli, S. Parsons, J. K. Y. Wong, *Angew. Chem. Int. Ed.* **2004**, *43*, 1218–1221; (w) X. Zheng, M. E. Mulcahy, D. Horinek, F. Galeotti, T. F. Magnera, J. Michl, *J. Am. Chem. Soc.* **2004**, *126*, 4540–4542; (x) F. Aricó, J. D. Badjic, S. J. Cantrill, A. H. Flood,

K. C.-F. Leung, Y. Liu, J. F. Stoddart, *Top. Curr. Chem.* **2005**, *249*, 203–259.

11. (a) F. L. Carter (Ed.), *Molecular Electronic Devices*, Marcel Dekker, New York, **1982**; (b) F. L. Carter (Ed.), *Molecular Electronic Devices*, Marcel Dekker, New York, **1987**; (c) F. L. Carter, R. E. Siatkowski, H. Wohltjie (Eds.), *Molecular Electronic Devices*, Elsevier, Amsterdam, **1988**; (d) A. Aviram (Ed.), *Molecular Electronics – Science and Technology*, Engineering Foundation, New York, **1989**; (e) P. I. Lazarev (Ed.), *Molecular Electronics: Materials and Methods*, Kluwer, Dordrecht, **1991**; (f) A. Aviram (Ed.), *Molecular Electronics – Science and Technology*, American Institute of Physics, Washington, DC, **1992**; (g) G. J. Ashwell (Ed.), *Molecular Electronics*, Wiley, New York, **1992**; (h) K. Sienicki (Ed.), *Molecular Electronics and Molecular Electronic Devices*, CRC Press, Boca Raton, FL, **1993**; (i) R. R. Birge (Ed.), *Molecular and Biomolecular Electronics*, American Chemical Society, Washington, DC, **1994**; (j) M. C. Petty, M. R. Bryce, D. Bloor (Eds.), *Introduction to Molecular Electronic Devices*, Oxford University Press, New York, **1995**; (k) J. Jortner, M. Ratner (Eds.), *Molecular Electronics*, Blackwell Science, Oxford, **1997**; (l) A. Aviram, M. Ratner (Eds.), *Molecular Electronics: Science and Technology*, Engineering Foundation, New York, **1998**; (m) *Chem. Rev.* **1999**, *99*, 1641–1990 (Special Issue on Nanoscale Materials); (n) *Acc. Chem. Res.* **1999**, *32*, 387–454 (Special Issue on Nanoscale Materials); (o) H. S. Nalwa (Ed.), *Handbook of Nanostructured Materials and Nanotechnology*, Academic Press, New York, **1999**.

12. (a) D. W. Urry, *Angew. Chem. Int. Ed. Engl.* **1993**, *32*, 819–841; (b) J. Howard, *Nature* **1997**, *389*, 561–567.

13. P. R. Ashton, R. Ballardini, V. Balzani, S. E. Boyd, A. Credi, M. T. Gandolfi, M. Gómez-López, S. Iqbal, D. Philp, J. A. Preece, L. Prodi, H. G. Ricketts, J. F. Stoddart, M. S. Tolley, M. Venturi, A. J. P. White, D. J. Williams, *Chem. Eur. J.* **1997**, *3*, 152–170.

14. (a) R. A. Bissell, E. Córdova, A. E. Kaifer, J. F. Stoddart, *Nature* **1994**, *369*, 133–137; (b) P.-L. Anelli, M. Asakawa, P. R. Ashton, R. A. Bissell, G. Clavier, R. Górski, A. E. Kaifer, S. J. Langford, G. Mattersteig, S. Menzer, D. Philp, A. M. Z. Slawin, N. Spencer, J. F. Stoddart, M. S. Tolley, D. J. Williams, *Chem. Eur. J.* **1997**, *3*, 1136–1150; (c) J. O. Jeppesen, K. A. Nielsen, J. Perkins, S. A. Vignon, A. Di Fabio, R. Ballardini, M. T. Gandolfi, M. Venturi, V. Balzani, J. Becher, J. F. Stoddart, *Chem. Eur. J.* **2003**, *9*, 2982–3007; (d) J. O. Jeppesen, S. A. Vignon, J. F. Stoddart, *Chem. Eur. J.* **2003**, *9*, 4611–4625; (e) H.-R. Tseng, S. A. Vignon, P. C. Celestre, J. Perkins, J. O. Jeppesen, A. Di Fabio, R. Ballardini, M. T. Gandolfi, M. Venturi, V. Balzani, J. F. Stoddart, *Chem. Eur. J.* **2004**, *10*, 155–172; (f) B. W. Laursen, S. Nygaard, J. O. Jeppesen, J. F. Stoddart, *Org. Lett.* **2004**, *6*, 4167–4170; (g) T. Iijima, S. A. Vignon, H.-R. Tseng, T. Jarrosson, J. K. M. Sanders, F. Marchioni, M. Venturi, E. Apostoli, V. Balzani, J. F. Stoddart, *Chem. Eur. J.* **2004**, *10*, 6375–6392; (h) J. O. Jeppesen, S. Nygaard, S. A. Vignon, J. F. Stoddart, *Eur. J. Org. Chem.* **2005**, 196–220.

15. Bistable catenanes and bistable rotaxanes have been employed as switches in a range of devices and across different environments, see (a) C. P. Collier, G. Mattersteig, E. W. Wong, Y. Luo, K. Beverly, J. Sampaio, F. M. Raymo, J. F. Stoddart, J. R. Heath, *Science* **2000**, *289*, 1172–1175; (b) A. R. Pease, J. O. Jeppesen, J. F. Stoddart, Y. Luo, C. P. Collier, J. R. Heath, *Acc. Chem. Res.* **2001**, *34*, 434–444; (c) Y. Luo, C. P. Collier, J. O. Jeppesen, K. A. Neilsen, E. DeIonno, G. Ho, J. Perkins, H.-R. Tseng, T. Yamamoto, J. F. Stoddart, J. R. Heath, *ChemPhysChem* **2002**, *3*, 519–525; (d) M. R. Diehl, D. W. Steuerman, H.-R. Tseng, S. A. Vignon, A. Star, P. C. Celestre, J. F. Stoddart, J. R. Heath, *ChemPhysChem* **2003**, *4*, 1335–1339; (e) H.-R. Tseng, D. Wu, N. Fang, X. Zhang, J. F. Stoddart, *ChemPhysChem* **2004**, *5*, 111–116; (f) D. W. Steuerman, H.-R. Tseng, A. J. Peters, A. H. Flood, J. O. Jeppesen, K. A. Nielsen, J. F. Stoddart, J. R. Heath, *Angew. Chem. Int. Ed.* **2004**, *43*, 6486–6491; (g) A. H. Flood, A. J. Peters, S. A. Vignon, D. W. Steuerman,

H.-R. Tseng, S. Kang, J. R. Heath, J. F. Stoddart, *Chem. Eur. J.* **2004**, *10*, 6558–6464; (h) A. H. Flood, J. F. Stoddart, D. W. Steuerman, J. R. Heath, *Science* **2004**, *306*, 2055–2056; (i) S. S. Jang, Y. H. Kim, W. A. Goddard, III, A. H. Flood, B. W. Laursen, H.-R. Tseng, J. F. Stoddart, J. O. Jeppesen, J. W. Choi, D. W. Steuerman, E. DeIonno, J. R. Heath, *J. Am. Chem. Soc.* **2005**, *127*, 1563–1575; (j) S. S. Jang, Y. H. Jang, Y.-H. Kim, W. A. Goddard, III, J. W. Choi, J. R. Heath, B. W. Laursen, A. H. Flood, J. F. Stoddart, K. Nørgaard, T. Bjørnholm, *J. Am. Chem. Soc.* **2005**, *127*, 14804–14816.

16. (a) S. Chia, J. Cao, J. F. Stoddart, J. I. Zink, *Angew. Chem. Int. Ed.* **2001**, *40*, 2447–2451; (b) K. Kim, W. S. Jeon, J.-K. Kang, J. W. Lee, S. Y. Jon, T. Kim, K. Kim, *Angew. Chem. Int. Ed.* **2003**, *42*, 2293–2296; (c) B. Long, K. Nikitin, D. Fitzmaurice, *J. Am. Chem. Soc.* **2003**, *125*, 15490–15498; (d) R. Hernandez, H.-R. Tseng, J. W. Wong, J. F. Stoddart, J. I. Zink, *J. Am. Chem. Soc.* **2004**, *126*, 3370–3371; (e) I. C. Lee, C. W. Frank, T. Yamamoto, H.-R. Tseng, A. H. Flood, J. F. Stoddart, J. O. Jeppesen, *Langmuir* **2004**, *20*, 5809–5818; (f) E. Katz, L. Sheeney-Haj-Ichia, I. Willner, *Angew. Chem. Int. Ed.* **2004**, *43*, 3292–3300; (g) O. Lioubashevski, V. I. Chegel, F. Patolsky, E. Katz, I. Willner, *J. Am. Chem. Soc.* **2004**, *126*, 7133–7143; (h) T. Nguyen, H.-R. Tseng, P. C. Celestre, A. H. Flood, Y. Liu, J. I. Zink, J. F. Stoddart, *Proc. Natl. Acad. Sci. USA* **2005**, *102*, 10029–10034.

17. (a) G. Schill, *Catenenes, Rotaxanes and Knots*, Academic Press, New York, **1971**; (b) J.-P. Sauvage, C. O. Dietrich-Buchecker (Ed.), *Molecular Catenanes, Rotaxanes and Knots*, Wiley-VCH, Weinheim, **1999**.

18. According to the classical stereochemical definition, the term "conformation" describes "the different spatial arrangements of atoms in molecules that result solely from torsions (rotations) around single and/or partial double bonds." However, in the domain of supramolecular chemistry, molecular motions arise from the competitive noncovalent interactions between the components. Such molecular motions are not restricted to the atoms, but part of an interlocked molecule can move from one recognition site to another, depending on the binding affinities of the recognition units. Therefore, the term "coconformation" was introduced to describe the translational isomers of mechanically interlocked molecules. In a wider sense, it illustrates the spatial location of one mechanically interlocked component, which can move around or along the other "stationary" component. See M. C. T. Fyfe, P. T. Glink, S. Menzer, J. F. Stoddart, A. J. P. White, D. J. Williams, *Angew. Chem. Int. Ed. Engl.* **1997**, *36*, 2068–2070.

19. (a) D. H. Busch, N. A. Stephenson, *J. Inclusion Phenom. Mol. Recognit. Chem.* **1992**, *12*, 389–395; (b) R. Cacciapaglia, L. Madolini, *Chem. Soc. Rev.* **1994**, *106*, 389–398; (c) R. Hoss, F. Vögtle, *Angew. Chem. Int. Ed. Engl.* **1994**, *33*, 375–384; (d) F. M. Raymo, J. F. Stoddart, *Pure Appl. Chem.* **1996**, *68*, 313–322; (h) T. J. Hubin, A. G. Kolchinski, A. L. Vance, D. L. Busch, *Adv. Supramol. Chem.* **1999**, *5*, 237–357.

20. (a) M. Martínez-Díaz, N. Spencer, J. F. Stoddart, *Angew. Chem. Int. Ed. Engl.* **1997**, *36*, 1904–1907; (b) J. D. Badjic, V. Balzani, A. Credi, S. Silvi, J. F. Stoddart, *Science* **2004**, *303*, 1845–1849; (c) J. D. Badjic, C. M. Ronconi, A. H. Flood, J. F. Stoddart, S. Silvi, A. Credi, V. Balzani, *J. Am. Chem. Soc.* **2006**, *128*, 1489–1499.

21. (a) M. Consuelo-Jiménez, C. Dietrich-Buchecker, J.-P. Sauvage, *Angew. Chem. Int. Ed.* **2000**, *39*, 3284–3287; (b) M. Consuelo-Jiménez, C. Dietrich-Buchecker, J.-P. Sauvage, *Chem. Eur. J.* **2002**, *39*, 1456–1466; (c) C. Dietrich-Buchecker, M. Consuelo-Jiménez, V. Sartor, J.-P. Sauvage, *Pure Appl. Chem.* **2003**, *75*, 1383–1390.

22. (a) H.-R. Tseng, S. A. Vignon, J. F. Stoddart, *Angew. Chem. Int. Ed.* **2003**, *42*, 1491–1495; (b) T. J. Huang, H.-R. Tseng, L. Sha, W. Lu, B. Brough, A. H. Flood, B.-D. Yu, P. C. Celestre, J. P. Chang, J. F. Stoddart, C.-M. Ho, *Nano Lett.* **2004**, *4*, 2065–2071; (c) T. J. Huang, B. Brough, C.-M. Ho, Y. Liu, A. H. Flood, P. A. Bonvallet, H.-R. Tseng, J. F.

Stoddart, M. Baller, S. Magonov, *Appl. Phys. Lett.* **2004**, *85*, 5391–5393; (d) Y. Liu, A. H. Flood, P. A. Bonvallet, S. A. Vignon, H.-R. Tseng, B. Brough, M. Baller, S. Magonov, S. Solares, W. A. Goddard, III, C.-M. Ho, J. F. Stoddart, *J. Am. Chem. Soc.* **2005**, *127*, 127, 9745–9759.

23. A. Asakawa, W. Dahaen, G. L'abbé, S. Menzer, J. Nouwen, F. M. Raymo, J. F. Stoddart, D. J. Williams, *J. Org. Chem.* **1996**, *61*, 9591–9595.

24. P.-L. Anelli, M. Asakawa, P. R. Ashton, R. A. Bissell, G. Clavier, R. Górski, A. E. Kaifer, S. J. Langford, M. Mattersteig, S. Menzer, D. Philp, A. M. Z. Slawin, N. Spencer, J. F. Stoddart, M. S. Tolley, D. J. Williams, *Chem. Eur. J.* **1997**, *3*, 1113–1135.

25. R. Castro, K. R. Nixon, J. D. Evanseck, A. E. Kaifer, *J. Org. Chem.* **1996**, *61*, 7289–7303.

26. (a) W.-Q. Deng, R. P. Muller, W. A. Goddard, III, *J. Am. Chem. Soc.* **2004**, *126*, 13562–13563; (b) also see a report on "computational nanotechnology" by E. K. Wilson, *Chem. Eng. News* **2003**, 28 April, 27–29.

27. (a) C. A. Hunter, J. K. M. Sanders, *J. Am. Chem. Soc.* **1990**, *112*, 5525–5534; (b) M. H. Schwartz, *J. Inclusion Phenom. Macrocycl. Chem.* **1990**, *9*, 1–35; (c) J. H. Williams, *Acc. Chem. Res.* **1993**, *26*, 539–598; (d) C. A. Hunter, *Angew. Chem. Int. Ed. Engl.* **1993**, *32*, 1584–1586; (e) C. A. Hunter, *J. Mol. Biol.* **1993**, *230*, 1025–1054; (f) T. Dahl, *Acta. Chem. Scand.* **1994**, *48*, 95–116; (g) F. Cozzi, J. S. Siegel, *Pure Appl. Chem.* **1995**, *67*, 683–689; (h) C. G. Claessens, J. F. Stoddart, *J. Phys. Org. Chem.* **1997**, *10*, 254–272; (i) G. Chessari, C. A. Hunter, C. M. R. Low, M. J. Packer, J. G. Vinter, C. Zonta, *Chem. Eur. J.* **2002**, *8*, 2860–2867.

28. (a) N. Armaroli, V. Balzani, J.-P. Collin, P. Gaviña, J.-P. Sauvage, B. Ventura, *J. Am. Chem. Soc.* **1999**, *121*, 4397–4408; (b) J.-P. Collin, C. Dietrich-Buchecker, P. Gaviña, M. C. Jiminez-Molero, J.-P. Sauvage, *J. Am. Chem. Soc.* **2001**, *34*, 477–487.

29. V. Balzani, M. Clemente-León, A. Credi, B. Ferrer, M. Venturi, A. H. Flood, J. F. Stoddart, *Proc. Natl. Acad. Sci. USA*, **2006**, *103*, 1178–1183.

9
Diarylethene as a Photoswitching Unit of Intramolecular Magnetic Interaction

Kenji Matsuda and Masahiro Irie

9.1
Introduction

When two unpaired electrons are placed in proximity, the exchange interaction operates between the two electrons [1–4]. The exchange interaction results in the separation of the energy of the singlet state (the spin quantum number equals zero) and that of the triplet state (the spin quantum number equals one). In the singlet state, the two spins align antiparallel, while the two spins align parallel in the triplet state. The exchange interaction is essential in magnetism. From the Heisenberg Hamiltonian [Eq. (1)], the singlet-triplet energy gap ΔE_{S-T} can be expressed by the exchange interaction J as in Eq. (2).

$$\hat{H} = -2J\mathbf{S}_1 \cdot \mathbf{S}_2 \qquad (1)$$

$$\Delta E_{S-T} = 2J \qquad (2)$$

When J is positive, the ground state is triplet; when J is negative, the ground state is singlet.

When an unpaired electron is placed at each end of a π-conjugated system, the two spins of the unpaired electrons interact magnetically through the π-system effectively. The π-conjugated system can be regarded as a "spin coupler" [5]. The direction of the spin alignment and the sign of the exchange interaction are controlled by the Ovchinnikov rule [6]. The magnitude of the interaction becomes weaker exponentially with increase in the spacer length. Figure 9.1 shows several examples of bis(nitronyl nitroxide)s and the exchange interactions between two spins. When molecules have even numbers of sp^2 carbon atoms between the two radicals, as in **1** [7, 8], **3** [9], **5** [10] and **6** [10], the interaction is antiferromagnetic ($J < 0$). On the other hand, the interaction becomes ferromagnetic ($J > 0$) when molecules have odd numbers of sp^2 carbon atoms between two radicals as in the case of **2** [11] and **4** [9]. Compound **6** has a weaker interaction than **5** because **6** has a longer π-conjugated chain length. Thiophene is superior to phen-

Functional Organic Materials. Syntheses, Strategies, and Applications.
Edited by Thomas J.J. Müller and Uwe H.F. Bunz
Copyright © 2007 WILEY-VCH Verlag GmbH & Co. KGaA, Weinheim
ISBN: 978-3-527-31302-0

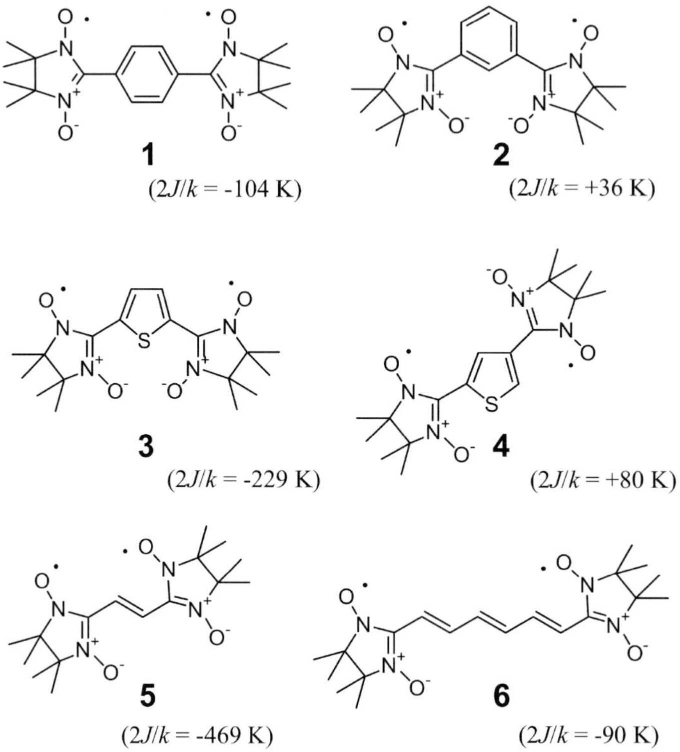

Fig. 9.1 Bis(nitronyl nitroxide)s and the exchange interactions between two spins.

Compound **1**: ($2J/k = -104$ K)
Compound **2**: ($2J/k = +36$ K)
Compound **3**: ($2J/k = -229$ K)
Compound **4**: ($2J/k = +80$ K)
Compound **5**: ($2J/k = -469$ K)
Compound **6**: ($2J/k = -90$ K)

ylene as a spin coupler. The magnetic exchange interaction is dependent on the nature of the π-system placed between the spins.

There are several measurement methods to determine the magnetic interactions. The temperature dependence of the magnetic susceptibility is the most commonly used method. When the two electron spins ($S = \frac{1}{2}$) are coupled with exchange interaction J, the magnetic susceptibility values χ are expressed as a function of temperature T according to Eq. (3) [12].

$$\chi = \frac{2Ng^2\mu_B^2}{kT[3 + \exp(-2J/kT)]} \tag{3}$$

where g is the Landé g-factor, μ_B is the Bohr magneton and k is the Boltzmann constant. When $J < 0$, χT decreases as the temperature decreases and when $J > 0$, χT increases as the temperature decreases. This method can be used to determine fairly strong interactions ($|J/k| > \sim 10$ K), but cannot be applied to measure weak interactions.

A convenient method to estimate weak interactions is to measure ESR spectra in solution [13, 14]. The hyperfine couplings in the ESR spectra give information concerning the magnetic interactions. For example, nitronyl nitroxides themselves have two identical nitrogen atoms; the molecules give five-line ESR spectra with relative intensities 1:2:3:2:1 and a 7.5 G spacing. When two nitronyl nitroxides are magnetically coupled via an exchange interaction, the biradical gives a nine-line ESR spectrum with relative intensities 1:4:10:16:19:16:10:4:1 and a 3.7 G spacing. If the exchange interaction is weaker than the hyperfine coupling in the biradical, the two nitroxide radicals are magnetically independent and give the same spectrum as the independent monoradical. This method is applicable to measure fairly weak interactions ($|J/k| \approx 0.01$ K). The weak interaction can also be detected by measuring the field dependence of the magnetization [15, 16].

In this chapter, several attempts to control the magnetic interactions between two separated spins by photoirradiation will be described.

9.2
Photochromic Spin Coupler

Photochromism is a reversible phototransformation of a chemical species between two forms having different absorption spectra [17–20]. Photochromic compounds reversibly change not only the absorption spectra but also their geometric and electronic structures. The geometric and electronic structural changes induce some changes in physical properties, such as fluorescence, refractive index, polarizability and electric conductivity. When the photochromic compounds are used as "spin couplers", the magnetic interaction can be controlled by photoirradiation.

Diarylethenes with heterocyclic aryl groups are well known as thermally irreversible, highly sensitive and fatigue-resistant photochromic compounds [21, 22]. The photochromic reaction is based on a reversible transformation between an open-ring isomer with a hexatriene structure and a closed-ring isomer with a cyclohexadiene structure according to the Woodward–Hoffmann rule, as shown in Scheme 9.1. While the open-ring isomer **7a** is colorless in most cases, the closed-ring isomer **7b** can be yellow, red or blue, depending on the molecular structure.

In the open-ring isomer, free rotation is possible between the ethene moiety and the aryl group. Therefore, the open-ring isomer is non-planar and the π-electrons are localized in the two aryl groups. On the other hand, the closed-ring isomer has a bond-alternative polyene structure and the π-electrons are delocalized throughout the molecule. These geometric and electronic structural differences result in some differences in their physical properties. For example, the closed-

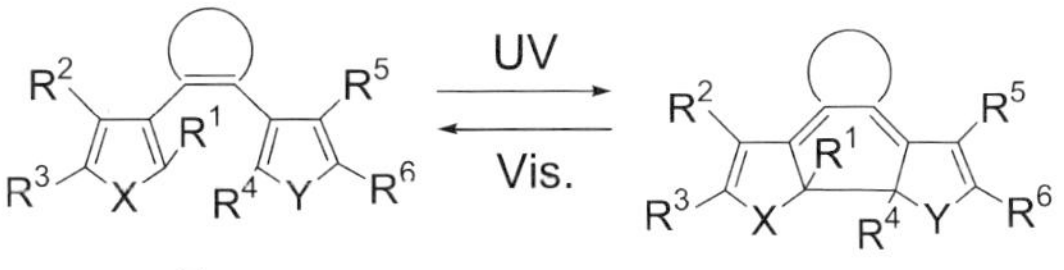

Scheme 9.1 Photochromism of diarylethene.

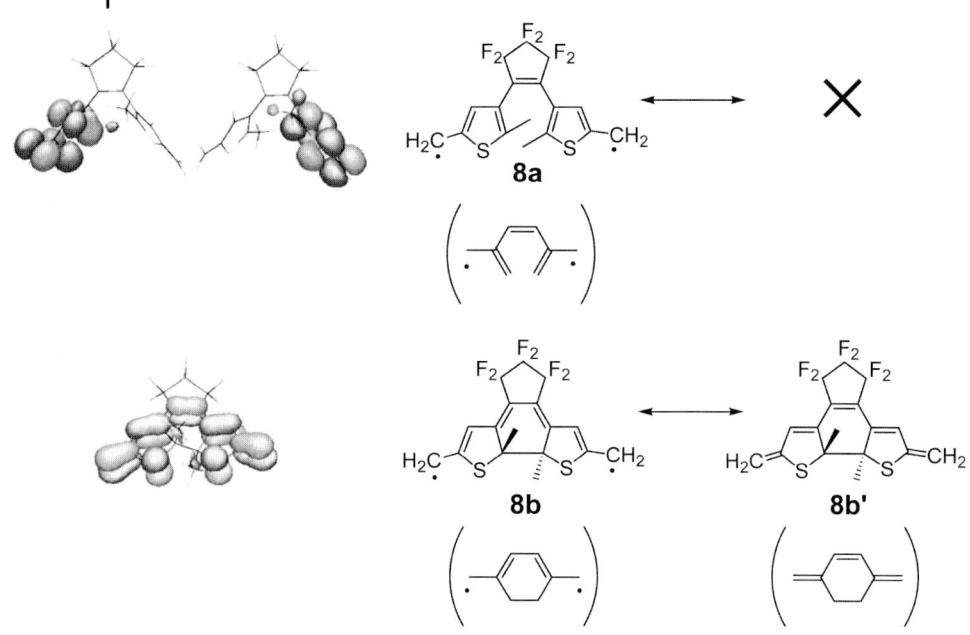

Fig. 9.2 The open-ring isomer **8a** and the closed-ring isomer **8b** of the radical-substituted diarylethene. Two SOMOs of **8a** and HOMO of **8b'** are also shown.

ring isomer has a higher polarizability, because the closed-ring isomer has more delocalized π-electrons [23, 24]. Not only the polarizability, but also fluorescence [25–33] and electronic conduction [34–36] can be switched.

There is a characteristic feature in the electronic structural changes between the two isomers. Figure 9.2 shows model structures of a radical-substituted diarylethene, the open-ring isomer **8a** and closed-ring isomer **8b**. Whereas there is no resonant closed-shell structure for **8a**, there exists **8b'** as the resonant quinoid-type closed-shell structure for **8b**. Structure **8a** is a non-Kekulé biradical and **8b** is a normal Kekulé molecule. In other words, **8a** has two unpaired electrons, whereas **8b** has no unpaired electrons. Moreover, the calculated shapes of two SOMOs of **8a** are separated in the molecule and there is no overlap [37]. Therefore, the configuration of biradical **8a** is classified as a disjoint biradical, in which the intramolecular radical–radical interaction is weak [15, 16]. On the other hand, the closed-ring isomer **8b'** is a normal Kekulé molecule. In this case, the ground electronic state has no unpaired electrons. In the singlet ground state, the magnetic interaction is strongly antiferromagnetic.

The electronic structural change of radical-substituted diarylethenes accompanying the photoisomerization is the transformation between a disjoint non-Kekulé structure and a closed-shell Kekulé structure. One may infer from the above consideration that the interaction between spins in the open-ring isomer of diarylethene is weak, whereas significant antiferromagnetic interaction takes place in the closed-ring isomer. In other words, the open-ring isomer is the "OFF" state and the closed-ring isomer is the "ON" state.

By using this principle, the photoswitching of magnetism has been planned and performed. Molecular magnetism can also be photocontrolled by using the spin crossover phenomena, which are light-induced excited spin state trapping (LIESST), light-induced thermal hysteresis (LITH) and ligand-driven light-induced spin change (LD-LISC) [38–47]. In addition to spin crossover systems, several other systems using photochromic units have also been reported [48–61]. Here, we focus on the photoswitching of the intramolecular magnetic interaction based on the photochromism of diarylethenes.

9.3
Synthesis of Diarylethene Biradicals

To realize the above-mentioned systems, we carefully chose suitable switching units and radical moieties. As an initial attempt, we employed 1,2-bis(2-methyl-1-benzothiophen-3-yl)perfluorocyclopentene (**9a**) as a photochromic spin coupler (Scheme 9.2). Compound **9a** is one of the most fatigue-resistant diarylethenes [21]. Nitronyl nitroxide was chosen for the spin source, because this radical is π-conjugative. Thus, we designed molecule **10a**, which is an embodiment of the simplified model **8a** [37, 62].

The synthesis was performed according to Scheme 9.3. 1,2-Bis(2-methyl-1-benzothiophen-3-yl)perfluorocyclopentene (**9a**) was formylated with dichloromethyl methyl ether to give diformyl compound **14**, which treated with 2,3-bis(hydroxyamino)-2,3-dimethylbutane sulfate followed by sodium periodate. Compound **10a** was obtained and purified by column chromatography and gel permeation chromatography (GPC). Recrystallization from hexane–CH$_2$Cl$_2$ gave dark-blue plate crystals of **10a**.

Using this switching unit, several biradical molecules were examined (see below). In the syntheses of these compounds, diiodo derivative **15** was used as

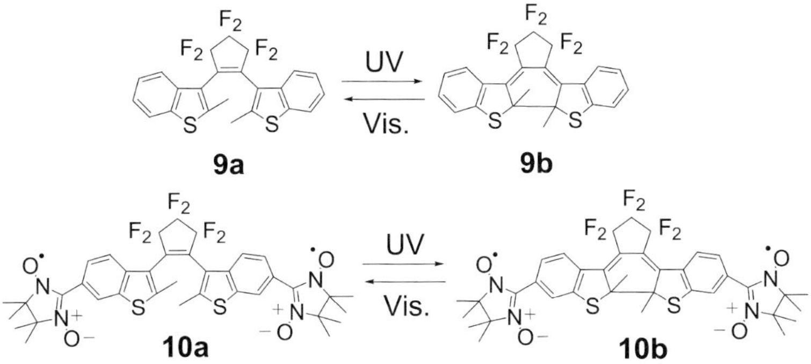

Scheme 9.2 Photochromism of the switching unit, 1,2-bis(2-methyl-1-benzothiophen-3-yl)perfluorocyclopentene (**9a**) and its biradical derivative **10a**.

Scheme 9.3 Synthetic scheme for diarylethene biradical **10a**.

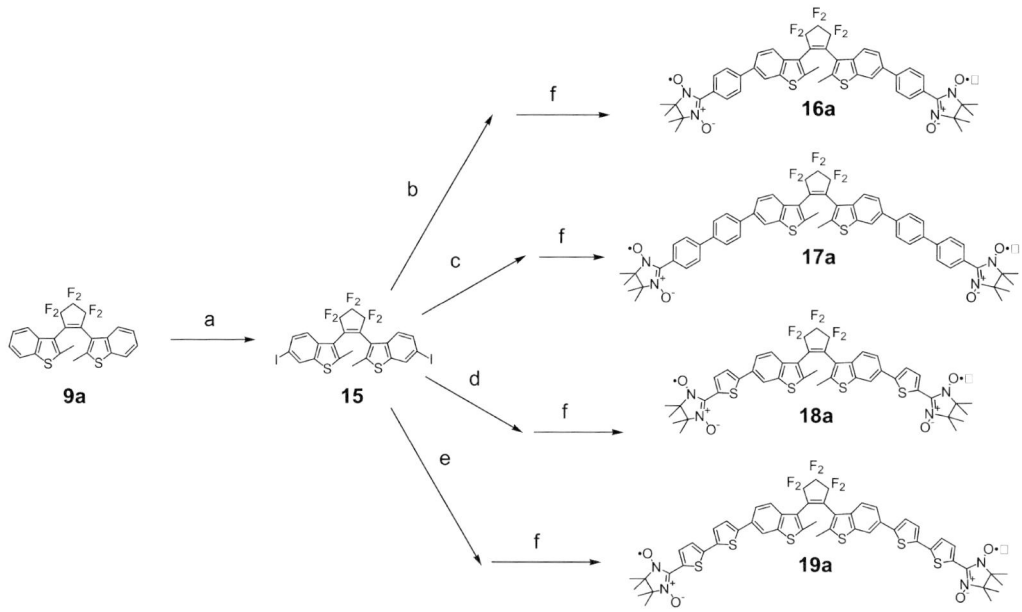

Scheme 9.4 Syntheses of diarylethene derivatives using the common diiodo intermediate. Reagent and conditions: (a) I₂, H₅IO₆, H₂SO₄, AcOH, H₂O, 76%; (b) Pd(PPh₃)₄, 4-formylphenylboronic acid, Na₂CO₃, THF, H₂O, 49%; (c) n-BuLi, B(OBu)₃ then Pd(PPh₃)₄, 4-formyl–4′-iodobiphenyl, Na₂CO₃, THF, H₂O, 60%; (d) Pd(PPh₃)₄, 2-formylthiophene-5-boronic acid, Na₂CO₃, THF, H₂O, 50%; (e) n-BuLi, B(OBu)₃ then Pd(PPh₃)₄, 5-iodo–5′-formyl-2,2′-bithiophene, Na₂CO₃, THF, H₂O, 48%; (f) 2,3-dimethyl-2,3-bis(hydroxyamino)butane sulfate, methanol then NaIO₄, CH₂Cl₂, 2–15%.

9.4
Photoswitching Using Bis(3-thienyl)ethene

Photoswitching behavior was examined for photochromic biradical **10a**. A schematic representation is shown in Fig. 9.3.

In solution, **10a** showed typical photochromic behavior on irradiation with UV and visible light (Fig. 9.4). Although the radical moiety has a weak absorption in the region from 550 to 700 nm, this did not prevent the photochromic reaction. Almost 100% photochemical conversions were observed in both the cyclization from the open-ring isomer **10a** to the closed-ring isomer **10b** and the cycloreversion from **10b** to **10a**. For the practical use of photochromic devices, high conversion is one of the most important characteristics.

Magnetic susceptibilities of **10a** and **10b** were measured on a SQUID susceptometer in microcrystalline form. χT–T plots are shown in Fig. 9.5. The data were analyzed in terms of a modified singlet–triplet two-spin model (the Bleaney–Bowers–type), in which two spins ($S = \frac{1}{2}$) couple antiferromagnetically within a biradical molecule by exchange interaction J. The best-fit parameters obtained by means of a least-squares method were $2J/k_B = -2.2 \pm 0.04$ K for **10a** and -11.6 ± 0.4 K for **10b**. Although the interaction ($2J/k_B = -2.2$ K) between the two spins in the open-ring isomer **10a** was weak, the spins of **10b** showed a remarkable antiferromagnetic interaction ($2J/k_B = -11.6$ K).

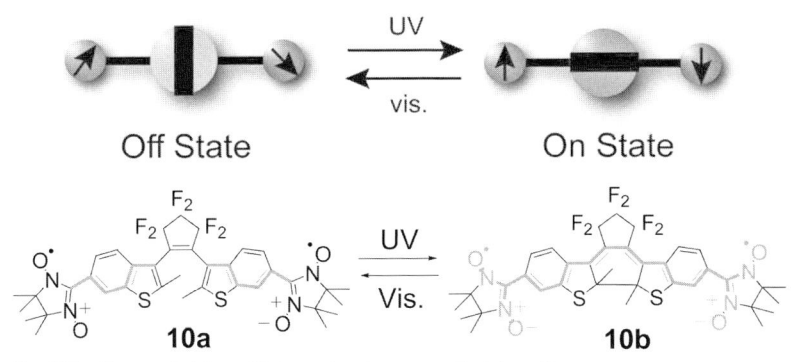

Fig. 9.3 Photoswitching of intramolecular magnetic interaction. Top, schematic representation; bottom, synthesized molecule. The thick lines represent the pathway of π-conjugation.

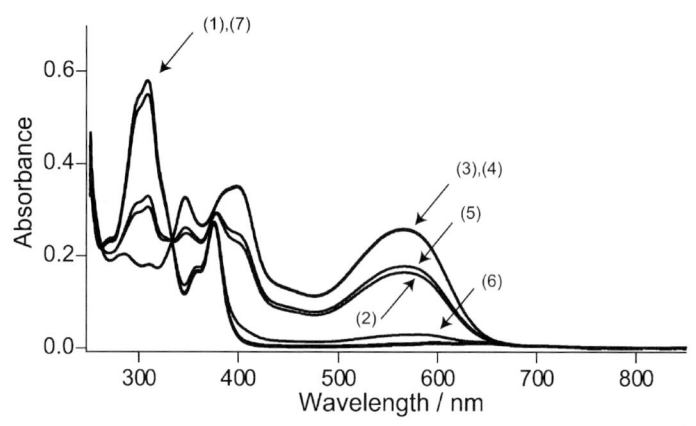

Fig. 9.4 Photochromic reaction of **10** (in ethyl acetate, 1.7×10^{-5} M): (1) the open-ring isomer; (2) irradiated with 313-nm light for 1 min; (3) for 5 min; (4) for 10 min; (5) irradiated with 578-nm light for 5 min; (6) for 30 min; (7) for 60 min. (Reprinted by permission from *Chemistry Letters* **29**, 16, 2000. Copyright Chemical Society of Japan.)

The open-ring isomer **10a** has a twisted molecular structure and a disjoint electronic configuration. On the other hand, the closed-ring isomer **10b** has a planar molecular structure and a non-disjoint electronic configuration. The photoinduced change in magnetism agrees well with the prediction that the open-ring isomer has an "OFF" state and the closed-ring isomer has an "ON" state.

Although the switching of exchange interaction was detected by a susceptibility measurement of biradical **10**, both open- and closed-ring isomers **10a** and **10b** had nine-line ESR spectra because the exchange interaction between the two radicals was much stronger than the hyperfine coupling constant in both isomers. For one

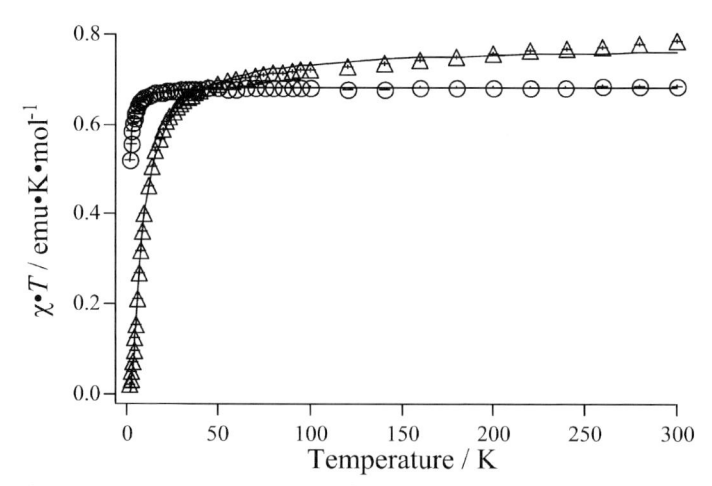

Fig. 9.5 Temperature dependence of the magnetic susceptibility of **10a** (○) and **10b** (△) (χT–T plot). (Reprinted by permission from *Chemistry Letters* **29**, 16, 2000. Copyright Chemical Society of Japan.)

Scheme 9.5

to detect the change of the exchange interaction by ESR spectroscopy, the value of the interaction should be comparable to the hyperfine coupling constant. Therefore, biradicals **16** and **17**, in which *p*-phenylene spacers are introduced to control the strength of the exchange interaction, were designed and synthesized (Scheme 9.5) [63, 64].

As described earlier, when two nitronyl nitroxides are magnetically coupled via an exchange interaction, the biradical gives a nine-line ESR spectrum. If the exchange interaction is weaker than the hyperfine coupling, the two nitroxide radicals are magnetically independent and give a five-line spectrum. In intermediate situations, the spectrum becomes complex.

Diarylethenes **16** and **17** underwent reversible photochromic reactions by alternate irradiation with UV and visible light. A typical absorption spectral change for **16** is shown in Fig. 9.6. The changes in the ESR spectra accompanying the photochromic reaction were examined for diarylethenes **16** and **17**. Figure 9.7 shows the ESR spectra at different stages of the photochromic reaction of **16a**. The ESR spectrum of **16a** showed 15 complex lines. This suggests that the two

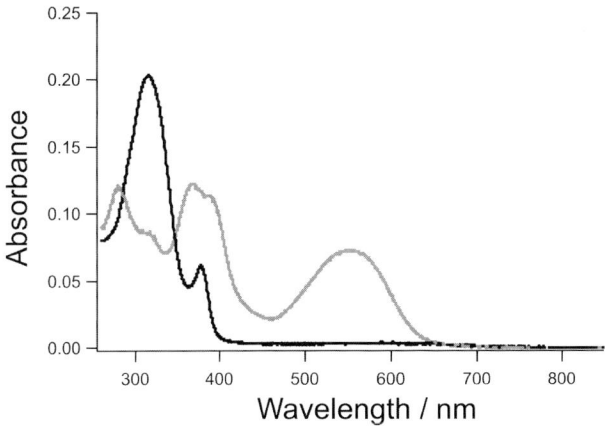

Fig. 9.6 Absorption spectrum of **16** along with photochromism (ethyl acetate solution, 2.8×10^{-6} M). Bottom (dark) line, open-ring isomer **16a**; top (pale) line, closed-ring isomer **16b**; dashed line (coincident with the line for **16b**), in the photostationary state under irradiation with 313-nm light.

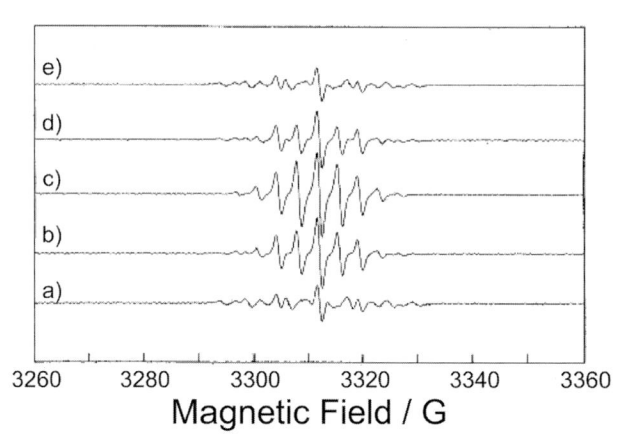

Fig. 9.7 ESR spectral change of **16a** along with photochromism (benzene solution, 1.1×10^{-4} M). (a) initial; (b) irradiation with 366-nm light for 1 min; (c) 4 min; (d) irradiation with >520 nm light for 20 min; (e) 50 min. (Reprinted by permission from Journal of the American Chemical Society **22**, 8309, 2000. Copyright American Chemical Society.)

spins of nitronyl nitroxide radicals are coupled by an exchange interaction that is comparable to the hyperfine coupling constant. Upon irradiation with 366-nm light, the spectrum converted completely to a nine-line spectrum, corresponding to the closed-ring isomer **16b**. The nine-line spectrum indicates that the exchange interaction between the two spins in **16b** is much stronger than the hyperfine coupling constant.

The ESR spectral change was also observed for **17a**. The open-ring isomer **17a** showed a five-line spectrum, whereas the closed-ring isomer **17b** had a distorted nine-line spectrum. A simulation of the ESR spectra was performed. The exchange interaction decreases with increase in the π-conjugated chain length, as shown in Table 9.1. The exchange interaction change in **16** and **17** upon photoirradiation was more than 30-fold. This result shows that the bisbenzothienylethene unit is a favorable spin coupler for photoswitching. Although the absolute value of the exchange interaction is small, the information of the spins can be clearly transmitted through the closed-ring isomer and the switching can be detected by ESR spectroscopy.

Oligothiophenes are good candidates for conductive molecular wires. The thiophene-2,5-diyl moiety has been used as a molecular wire unit for energy and elec-

18a (n=1)
19a (n=2)

18b (n=1)
19b (n=2)

Scheme 9.6

Table 9.1 Magnetic interaction between two nitronyl nitroxide connected by diarylethene photoswitches.

Compound	Open-ring isomer		Closed-ring isomer	
	ESR line shape	|2J/k_B| (K)	ESR line shape	|2J/k_B| (K)
10	9 lines	2.2	9 lines	11.6
16	15 lines	$\begin{cases} 1.2 \times 10^{-3} \\ <3 \times 10^{-4} \end{cases}$	9 lines	>0.04
17	5 lines	$<3 \times 10^{-4}$	Distorted 9 lines	0.010
18	13 lines	$\begin{cases} 5.6 \times 10^{-3} \\ <3 \times 10^{-4} \end{cases}$	9 lines	>0.04
19	5 lines	$<3 \times 10^{-4}$	9 lines	>0.04

tron transfer and can serve as a stronger magnetic coupler than p-phenylene [65, 66]. Therefore, diarylethenes **18** and **19** having one nitronyl nitroxide radical at each end of a molecule containing oligothiophene spacers were synthesized and their photo- and magnetochemical properties were studied (Scheme 9.6).

Photochromic reactions and an ESR spectral change were also observed for **18** and **19** as shown in Fig. 9.8. Table 9.1 lists the exchange interaction between the two diarylethene-bridged nitronyl nitroxide radicals. For all five biradicals, the closed-ring isomers have stronger interactions than the open-ring isomers. The exchange interactions through oligothiophene spacers were stronger than the corresponding biradicals with oligophenylene spacers. The efficient π-conjugation in

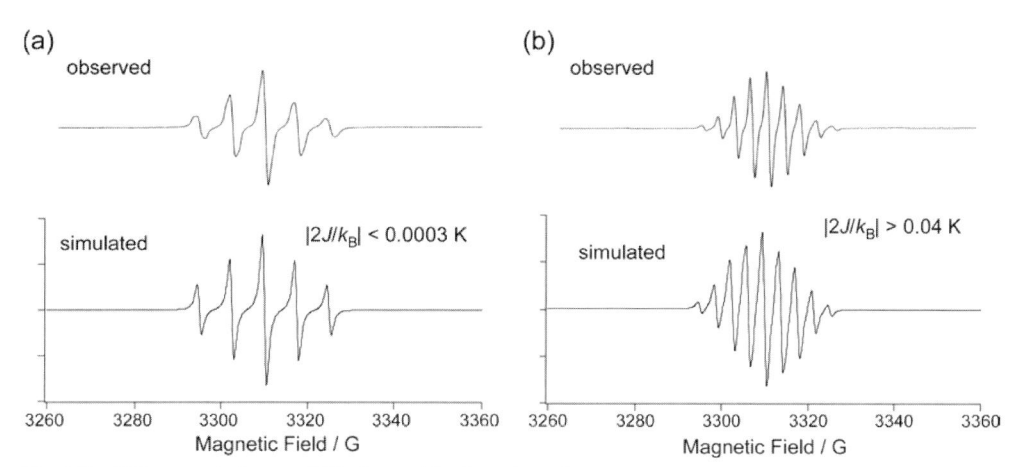

Fig. 9.8 ESR spectral change of **19** along with photochromism (benzene solution). (a) Open-ring isomer **19a**; (b) closed-ring isomer **19b**.

thiophene spacers results in strong exchange interactions between the two nitronyl nitroxide radicals. In the case of bithiophene spacers, the exchange interaction difference between open- and closed-ring isomers was estimated to be more than 150-fold.

9.5
Reversed Photoswitching Using Bis(2-thienyl)ethene

Photoswitching using bis(2-thienyl)ethene was performed. Bis(2-thienyl)ethene isomers have two 2-thienyl groups connected to the perfluorocyclopentene ring (Scheme 9.7). In the case of regular bis(3-thienyl)ethene, the bond alternation is discontinued in the open-ring isomer and the π-electron is delocalized throughout the molecule in the closed-ring isomer. Therefore, the open-ring isomer is in the "OFF" state due to the disconnection of the π-system and the closed-ring isomer is in the "ON" state due to the delocalization of the π-conjugated system. The situation is reversed when thiophene rings are substituted to the ethene moiety at the 2-position. The bond alternation is continued throughout the molecule in the open-ring isomer, whereas in the closed-ring isomer two aryl rings are separated by the sp^3 carbon and sulfur atoms. The magnetic interaction between the two unpaired electrons at 5-positions in the closed-ring isomer is expected to become much weaker than that in the open-ring isomer. In other words, the open-ring isomer is in the "ON" state and the closed-ring isomer in the "OFF" state.

Photoswitching using bis(2-thienyl)ethene **20** was examined (Scheme 9.8) [67]. Unfortunately, the synthesized **20** did not undergo photochromism, but the precursor underwent the photochromic reaction. The closed-ring isomer of the precursor was isolated and the closed-ring isomer was converted to the desired molecule. The closed-ring isomer **20b** obtained was converted to the open-ring isomer **20a** by irradiation with visible light. The ESR spectrum of the closed-ring isomer **20b** was a five-line spectrum, but that of the open-ring isomer **20a** was a distorted nine-line spectrum. This suggests that the magnetic interaction is stronger in the open-ring isomer than in the closed-ring isomer. The switching direction is reversed by using bis(2-thienyl)ethene.

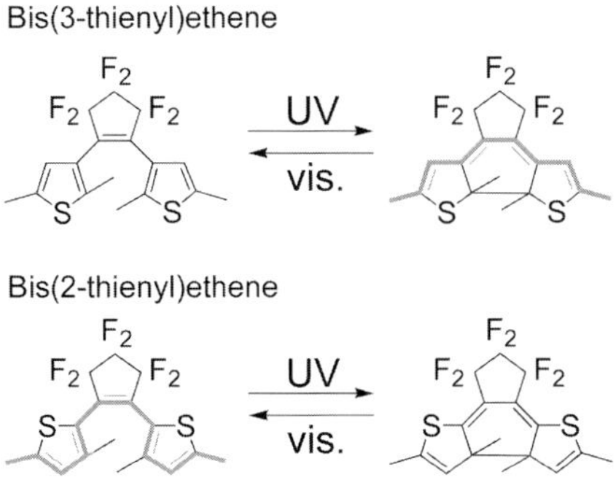

Scheme 9.7 Photochromism of bis(3-thienyl)ethene and bis(2-thienyl)ethene. The thick lines represent the pathway of π-conjugation.

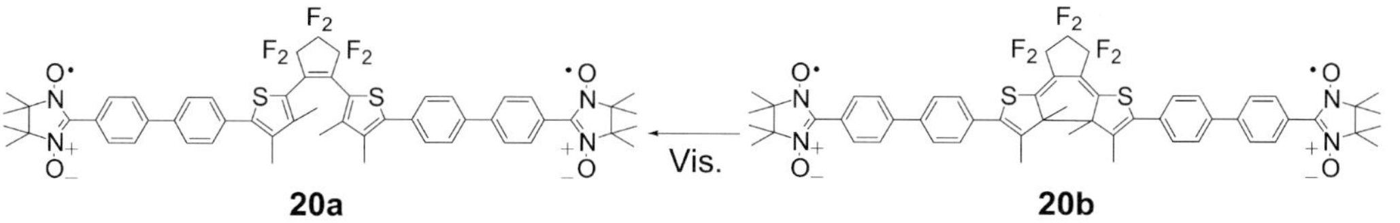

Scheme 9.8

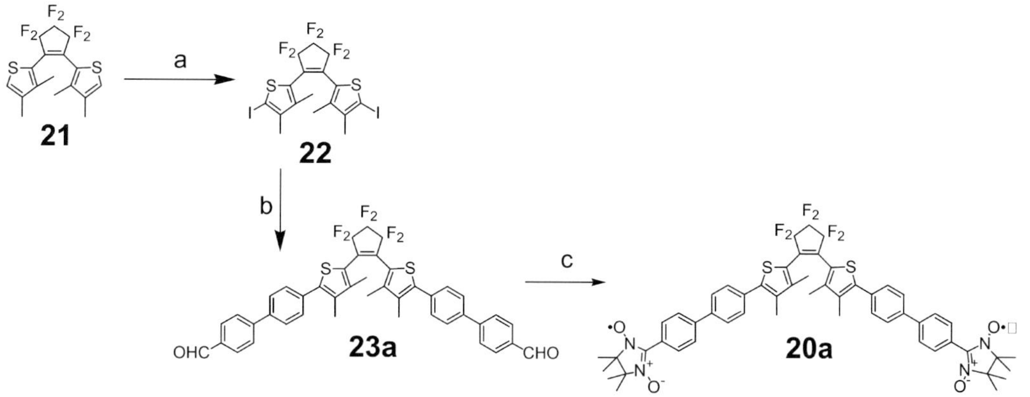

Scheme 9.9 Reagents and conditions: (a) I₂, H₅IO₆, H₂SO₄, AcOH, H₂O, 71 %; (b) n-BuLi, B(OBu)₃ then Pd(PPh₃)₄, 4-formyl-4′-iodobi- phenyl, Na₂CO₃, THF, H₂O, 26 %; (c) 2,3-di-methyl-2,3-bis(hydroxyamino)butane sulfate, methanol then NaIO₄, CH₂Cl₂, 21 %.

The synthesis was performed according to Scheme 9.9. Compound **20a** was synthesized from 1,2-bis(3,4-dimethyl-2-thienyl)hexafluorocyclopentene (**21**). After iodination, the transformation to diboronic acid followed by Suzuki coupling with 4-formyl-4′-iodobiphenyl gave diformyl compound **23a**. The conventional treatment was employed for formyl compounds **23a** to convert nitronyl nitroxide radical **20a**.

9.6
Photoswitching Using an Array of Photochromic Molecules

In the previous section, it was demonstrated that the exchange interaction between two nitronyl nitroxide radicals located at each end of a diarylethene was photoswitched reversibly by alternate irradiation with ultraviolet and visible light. The difference in the exchange interaction between the two switching states was more than 150-fold. ESR spectra can be used as a good tool for detecting small magnetic interactions in molecular systems. In this section, photoswitching of an intramolecular magnetic interaction using a diarylethene dimer is described [68].

342 | *9 Diarylethene as a Photoswitching Unit of Intramolecular Magnetic Interaction*

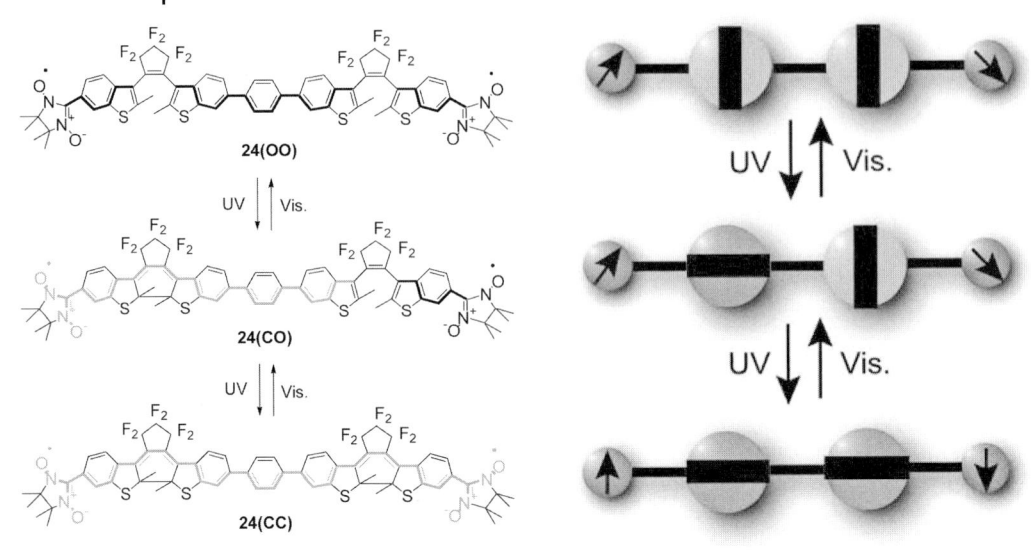

Fig. 9.9 Photochromic reaction and schematic illustration of diarylethene **24**. The thick gray lines in the bottom two structures represent the pathway of π-conjugation.

When a diarylethene dimer is used as a switching unit, there are three kinds of photochromic states: open–open (OO), closed–open (CO) and closed–closed (CC). From the analogy of an electric circuit, one may infer that the dimer has two switching units in series. Diarylethene dimer **24**, which has 28 carbon atoms between two nitronyl nitroxide radicals, was synthesized as shown in Fig. 9.9. When the two radicals are separated by 28 conjugated carbon atoms, the five- and nine-line spectra were clearly distinguishable upon irradiation. A *p*-phenylene spacer is introduced so that the cyclization reaction can occur at both diarylethene moieties. Bond alternation is discontinued at the open-ring moieties of **24(OO)** and **24(CO)**. As a result, the spin at each end of **24(OO)** and **24(CO)** cannot interact with each other. On the other hand, the π-system of **24(CC)** is delocalized throughout the molecule and the exchange interaction between the two radicals is expected to occur.

Compound **24(OO)** was synthesized according to Scheme 9.10. The key step was the Suzuki coupling of iodo-substituted diarylethene and *p*-phenylenediboronic acid.

Compound **24(OO)** underwent photochromic reaction by alternate irradiation with UV and visible light. Upon irradiation of the ethyl acetate solution of **24(OO)** with 313-nm light, an absorption at 560 nm appeared, as shown in Fig. 9.10. This absorption grew and shifted and the system reached the photostationary state after 120 min. The color of the solution changed from pale blue to red–purple and then to blue–purple. Such a red spectral shift suggests the formation of **24(CC)**. The isosbestic point was maintained at an initial stage of irradiation,

Scheme 9.10 Reagents and conditions: (a) n-BuLi, THF then DMF; (b) p-phenylenediboronic acid, Pd(PPh$_3$)$_4$, Na$_2$CO$_3$, THF, H$_2$O; (c) 2,3-dimethyl-2,3-bis(hydroxyamino)butane sulfate, methanol then NaIO$_4$, CH$_2$Cl$_2$, (4.7%, in three steps).

but it later deviated. The blue–purple solution was completely bleached by irradiation with 578-nm light. Compounds **24(CO)** and **24(CC)** were isolated from the blue–purple solution by HPLC. The spectra of **24(OO)**, **24(CO)** and **24(CC)** are also shown in Fig. 9.10. Compound **24(CC)** has an absorption maximum at 576 nm, which is red shifted by as much as 16 nm in comparison with its location in **24(CO)**.

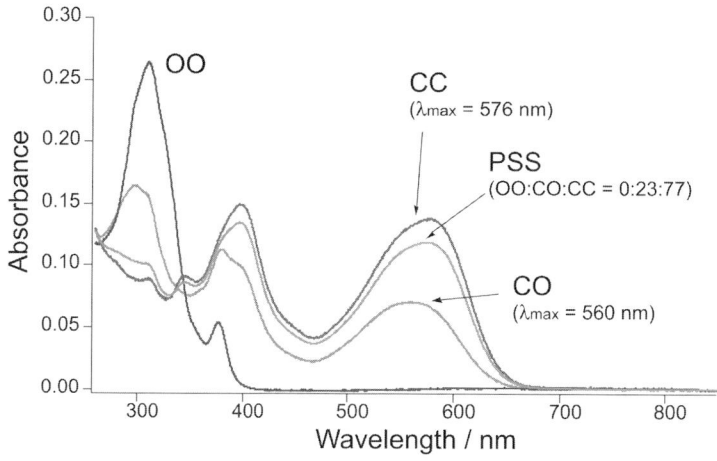

Fig. 9.10 Absorption spectra of **24(OO)**, **24(CO)**, **24(CC)** and sample in the photostationary state under irradiation with 313-nm light. (Reprinted by permission from *Journal of the American Chemical Society* **123**, 9896, 2001. Copyright American Chemical Society.)

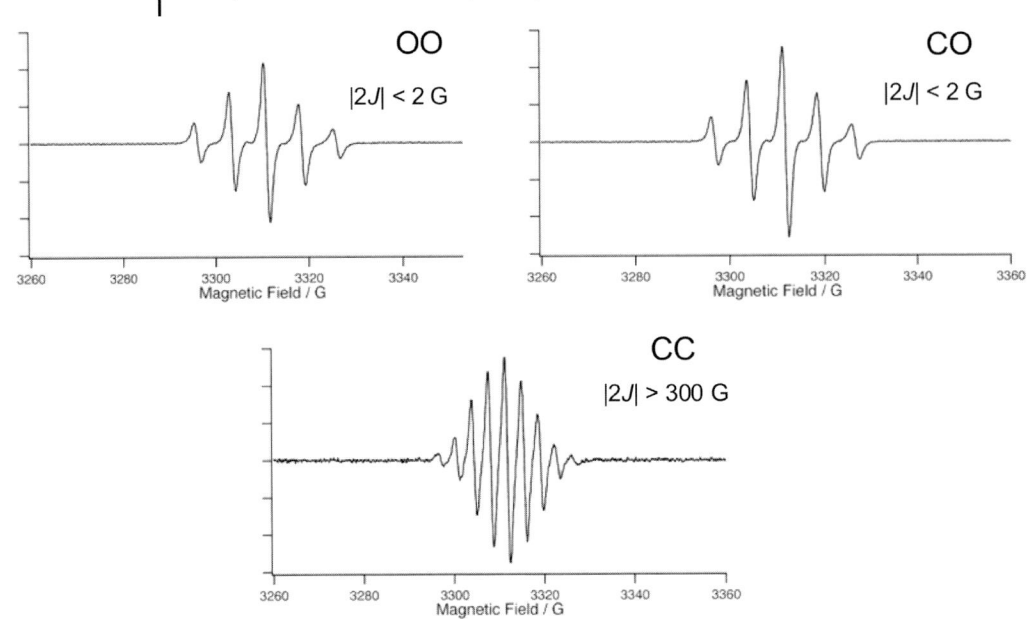

Fig. 9.11 ESR spectra of (a) **24(OO)**, (b) **24(CO)** and (c) **24(CC)**. (Reprinted by permission from *Journal of the American Chemical Society* **123**, 9896, 2001. Copyright American Chemical Society.)

ESR spectra of isolated **24(OO)**, **24(CO)** and **24(CC)** were measured in benzene at room temperature as shown in Fig. 9.11. The spectra of **24(OO)** and **24(CO)** are five-line spectra, suggesting that the exchange interaction between the two nitronyl nitroxide radicals is much weaker than the hyperfine coupling constant ($2J/k_B < 3 \times 10^{-3}$ K).

However, the spectrum of **24(CC)** has a clear nine-line spectrum, indicating that the exchange interaction between the two spins is much stronger than the hyperfine coupling constant ($2J/k_B > 0.04$ K). The result indicates that each diarylethene chromophore serves as a switching unit to control the magnetic interaction. The magnetic interaction between terminal nitronyl nitroxide radicals was controlled by the switching units in series.

9.7
Development of a New Switching Unit

In the previous photoswitches, a radical unit is placed at each side of the diarylethene photoswitching unit and separated by an extended π-conjugated chain. When the π-conjugated chain length between the radical becomes longer, both photocyclization and cycloreversion reactivities are reduced. This is attributed to the reduced excitation density at the central diarylethene unit [69]. The excitation

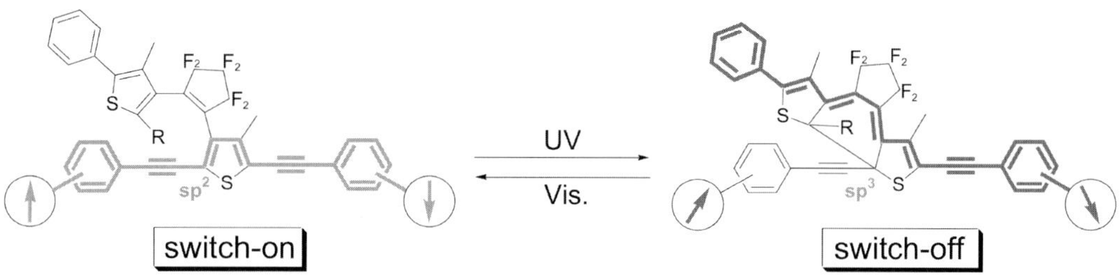

9.7 Development of a New Switching Unit

Fig. 9.12 Photochromism of diarylethene having a 2,5-arylethynyl-3-thienyl unit. The thick lines represent the pathway of π-conjugation.

density localized at the center of both sides of the π-conjugated aryl unit. To solve the problem, it is necessary to develop new switching systems, in which the excitation density at the switching unit is not strongly reduced. The proposed new switching molecule has its switching unit located in the middle of the π-conjugated chain.

In this section, we propose a new switching unit, in which two radicals are placed in the same aryl unit and π-conjugated chain is extended from 2- and 5-positions of the thiophene ring in one aryl unit of the diarylethene [70]. The photochromic reactivity and magnetic switching of the new diarylethene derivatives will be discussed.

Figure 9.12 shows the photochromic behavior of the newly developed diarylethenes. The photocyclization reaction of the diarylethene unit breaks the π-conjugation in 2,5-bis(arylethynyl)-3-thienyl unit due to the change of the orbital hybridization from sp^2 to sp^3 at the 2-position of the thiophene ring. One of such molecules is shown in Scheme 9.11. An imino nitroxide radical is introduced at each end of the 2,5-bis(arylethynyl)thiophene π-conjugated chain and m-phenylene was chosen as a spacer between the radicals and reactive center [71]. A methoxy group was introduced at the reactive carbon [72]. The magnetic interaction between the two radicals via the π-conjugated chain can be altered by the photocyclization. The open-ring isomer represents the "ON" state because the π-conjugated system is delocalized between two radicals, whereas the closed-ring isomer represents the "OFF" state because the π-conjugated system is disconnected at the 2-position of the thiophene ring.

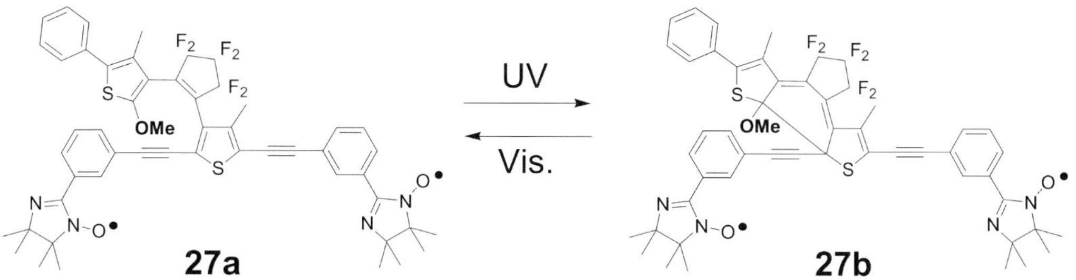

Scheme 9.11

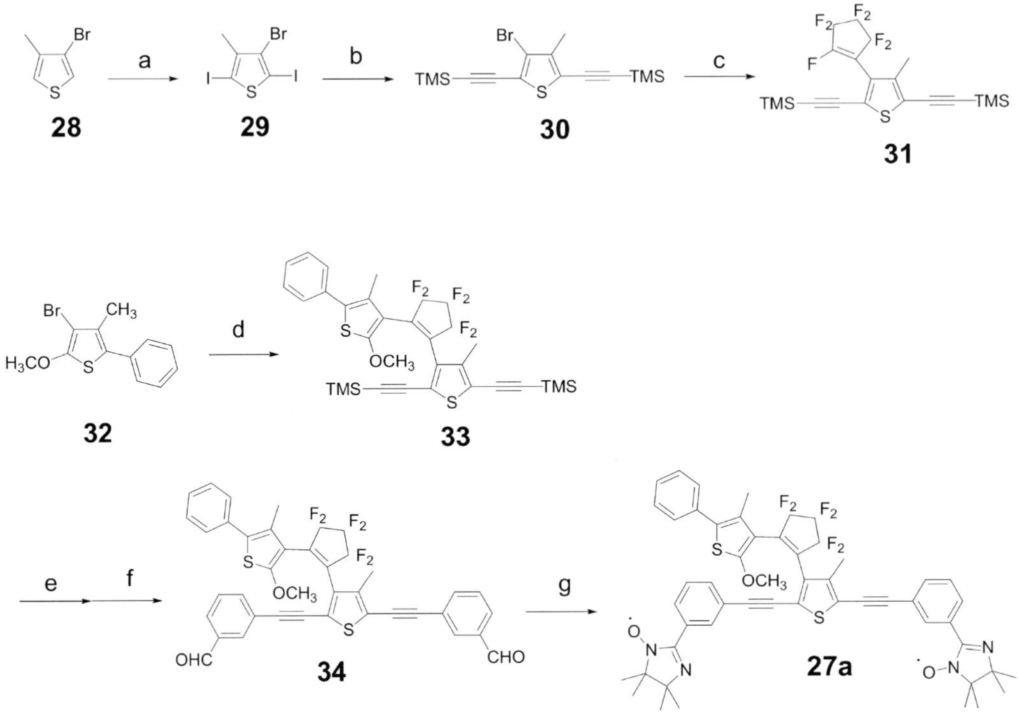

Scheme 9.12 Reagents and conditions:
(a) I_2, H_5IO_6, CH_3COOH, H_2SO_4, H_2O, 96%;
(b) $Pd(PPh_3)_2Cl_2$, CuI, trimethylsilylacetylene, Et_3N, 40%; (c) n-BuLi, THF then perfluorocyclopentene, 81%; (d) THF, n-BuLi then **9**, 90%; (e) 20% aq. KOH, MeOH, THF, 86%;
(f) $Pd(PPh_3)_2Cl_2$, CuI, 3-bromobenzaldehyde, diisopropylamine, 54%; (g) 2,3-dimethyl-2,3-bis(hydroxyamino)butane sulfate and methanol then $NaIO_4$ and CH_2Cl_2, 22%.

Scheme 9.12 shows the synthetic route. [4-Methyl-2,5-bis(trimethylsilylethynyl)-3-thienyl]heptafluorocyclopentene (**31**) was prepared from 3-bromo-4-methylthiophene (**28**) in three steps. 2′-methoxy derivative **33** was synthesized by the coupling of **31** with lithiated **32**. After desilylation with KOH, Sonogashira coupling with bromoformyl compounds gave bisformylated diarylethenes **34**. Formyl derivative **34** was converted into nitroxide radical **27a**.

Compounds **27a** underwent a photochromic reaction upon alternate irradiation with UV and visible light. ESR spectral change of the toluene solution containing **27** was followed, keeping the sample in the ESR cavity during irradiation with UV and visible light (Fig. 9.13). The photoreaction was started from the isolated closed-ring isomer **27b**. The closed-ring isomer **27b** showed a seven-line spectrum, which is a spectrum of an isolated imino nitroxide ($|2J/k| < 3 \times 10^{-4}$ K). Upon irradiation with 578-nm light, the closed-ring isomer converted to the open-ring isomer. At the same time, a 13-line spectrum with a ratio of 1:2:5:6:10:10:13:10:10:6:5:2:1 appeared. Upon further irradiation, the spectrum

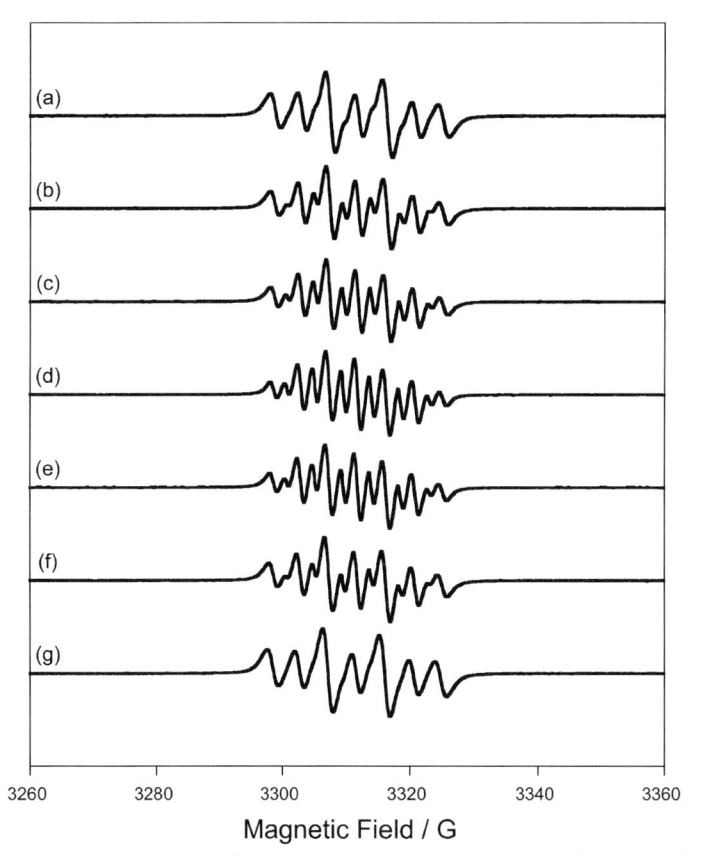

Fig. 9.13 ESR spectra of biradical **27** along with photochromic reaction measured in toluene at room temperature (9.32 GHz): (a) the closed-ring isomer **27b**; (b) after irradiation with 578-nm light for 5 min; (c) for 7 min; (d) for 10 min, which is identical with the spectrum of **27a**; (e) after irradiation with 365-nm light for 1 min; (f) for 3 min; (g) for 5 min, which is identical with the spectrum of **27b**. (Reprinted by permission from *Journal of the American Chemical Society* **127**, 13344, 2005. Copyright American Chemical Society.)

was completely converted to the 13-line spectrum. The 13-line spectrum indicates that the exchange interaction takes place between the two radicals ($|2J/k| > 0.04$ K). Subsequent irradiation with 365-nm light regenerated the seven-line spectrum along with the regeneration of the closed-ring isomer **27b**. The difference in the exchange interaction between the open- and the closed-ring isomers was estimated to be >150-fold. The difference in the exchange interaction is attributed to the change in the hybridization of the carbon atom from sp^2 to sp^3 atom at the reaction center.

9.8
Conclusions

Attempts to photoswitch the intramolecular magnetic interaction based on photochromism of diarylethenes have been overviewed. The switching between the disjoint and non-disjoint structures caused a change in the interaction between the separated spins. Photochromic diarylethene is one of the most favorable photoswitching units for magnetic interactions. This system has the possibility to be applied to the molecular-scale information processing system [73–77].

9.9
Experimental Procedures

As a typical example of diarylethene derivatives, the procedure for the synthesis of diarylethene biradical **10a** [37] is given below.

1,2-Bis(6-formyl-2-methyl-1-benzothiophen-3-yl)hexafluorocyclopentene (**14**)
To a stirred solution of 1,2-bis(2-methyl-1-benzothiophen-3-yl)hexafluorocyclopentene (**9a**) (2.35 g, 5.02 mmol) in nitrobenzene (25 mL) was added dichloromethyl methyl ether (7.5 mL) and anhydrous aluminum chloride (2.8 g) at 0 °C and stirred for 15 h at room temperature under an argon atmosphere. Water was poured into the reaction mixture and then extracted with ethyl acetate, washed with water, dried with magnesium sulfate and concentrated. After evaporation of nitrobenzene in vacuum, column chromatography [hexane–Et$_2$O (3:1–1:1)] gave diformyl compound **14** (2.16 g, 82%) as a white powder, m.p. 183.0–184.0 °C. ^{1}H NMR (CDCl$_3$, 250 MHz), δ 2.31 and 2.56 (s × 2, 6 H), 7.68–8.20 (m, 6 H), 9.95 and 10.05 (s × 2, 2 H). Anal. Calcd for $C_{25}H_{14}F_6O_2S_2$: C, 57.25; H, 2.69. Found: C, 57.22; H, 2.70 %.

1,2-Bis[6-(1-oxyl-3-oxide-4,4,5,5-tetramethylimidazolin-2-yl)-2-methyl-1-benzothiophen-3-yl]hexafluorocyclopentene (**10a**)
A solution of **14** (2.0 g, 3.8 mmol), 2,3-bis(hydroxyamino)-2,3-dimethylbutane sulfate (4.75 g, 19.3 mmol) and potassium carbonate (2.85 g, 20.6 mmol) in methanol (40 mL) was refluxed for 24 h. The reaction mixture was poured into water, extracted with ethyl acetate, washed with water, dried over magnesium sulfate and concentrated to give tetrahydroxylamine as a yellow oil. Purification was not performed.

To a solution of tetrahydroxylamine in dichloromethane (300 mL) was added a solution of sodium periodate (2.45 g, 11.5 mmol) in water (500 mL) and the mixture was stirred for 1 h in the open air. The organic layer was separated, washed with water, dried over magnesium sulfate and concentrated. Purification was performed by column chromatography [silica, hexane–Et$_2$O (1:1–0:1)] followed by GPC and then recrystallized from CH$_2$Cl$_2$–hexane by a diffusion method in the dark. Compound **10a** was obtained as a dark-blue microcrystalline solid

(470 mg, 16 %), m.p. 231.0–232.0 °C (decomp.). UV–Vis (AcOEt), λ_{max} (ε) 300 (sh), 309 (3.4 × 10^4), 359 (sh), 377 (1.6 × 10^4), 553 (sh), 598 (6.3 × 10^2), 646 (6.3 × 10^2), 706 (sh); ESR (benzene), 1:4:10:16:19:16:10:4:1, nine lines, g = 2.007, $|a_N|$ = 3.7 G; FAB HRMS (m/z): [M + H]$^+$ calcd for C$_{37}$H$_{37}$F$_6$N$_4$O$_4$S$_2$, 779.2160; found, 779.2173.

Corresponding closed-ring isomer **10b**
This was isolated as a black microcrystalline solid. UV–Vis (AcOEt), λ_{max} (ε) 280 (1.7 × 10^4), 348 (1.9 × 10^4), 385 (sh), 400 (2.0 × 10^4), 565 (1.5 × 10^4); ESR (benzene), 1:4:10:16:19:16:10:4:1, nine lines, g = 2.007, $|a_N|$ = 3.7 G.

Photochemical measurements
Absorption spectra were measured on a Hitachi U-3500 spectrophotometer. Photoirradiation was carried out by using a USHIO 500-W super high-pressure mercury lamp. Mercury lines of 313, 365, 517 and 578 nm were isolated by passing the light through a combination of a Toshiba band-pass filter (UV-D33S) or sharp cut filter (Y-48, Y-52) and monochromator.

Magnetic measurements
Fine crystalline and polymer samples were mounted in a capsule and measured on a Quantum Design MPMS-5S SQUID susceptometer at 5000 G. Corrections for the diamagnetic contribution were made using Pascal's constants.

ESR spectroscopy
A Bruker ESP 300E spectrometer was used to obtain X-band ESR spectra. The sample was dissolved in solvent (ca. 1 mM) and degassed by bubbling argon for 5 min.

Acknowledgments

The authors acknowledge coworkers as cited in the references for their key contributions to this research. The authors are grateful for financial support from the Japan Science and Technology Agency and the Ministry of Education, Culture, Sports, Science and Technology, Japan.

References

1. O. Kahn, *Molecular Magnetism*, VCH, New York, **1993**.
2. J. S. Miller, A. J. Epstein, *Angew. Chem., Int. Ed. Engl.* **1994**, *33*, 385.
3. C. Benelli, D. Gatteschi, *Chem. Rev.* **2002**, *102*, 2369.
4. J. S. Miller, M. Drillon (Eds.), *Magnetism: Molecules to Materials*, Wiley-VCH, Weinheim, **2001**.
5. H. Iwamura, N. Koga, *Acc. Chem. Res.* **1993**, *26*, 346.
6. A. A. Ovchinnikov, *Theor. Chim. Acta*, **1978**, *47*, 297.
7. L. Catala, J. Le Moigne, N. Kyritsakas, P. Rey, J. J. Novoa, P. Turek, *Chem. Eur. J.* **2001**, *7*, 2466.
8. A. Izuoka, M. Fukada, T. Sugawara, M. Sakai, S. Bandow, *Chem. Lett.*, **1992**, 1627.
9. T. Mitsumori, K. Inoue, N. Koga, H. Iwamura, *J. Am. Chem. Soc.*, **1995**, *117*, 2467.
10. C. Stroh, P. Turek, R. Ziessel, *Chem. Commun.*, **1998**, 2337.
11. A. Caneschi, P. Chiesi, L. David, F. Ferraro, D. Gatteschi, R. Sessoli, *Inorg. Chem.* **1993**, *32*, 1445.
12. B. Bleany, K. D. Bowers, *Proc. R. Soc. London A* **1952**, *214*, 451.
13. R.Brière, R.-M. Dupeyre, H. Lemaire, C. Morat, A. Rassat, P. Rey, *Bull. Soc. Chim. Fr.* **1965**, *11*, 3290.
14. S. H. Glarum, J. H. Marshall, *J. Chem. Phys.* **1967**, *47*, 1374.
15. K. Matsuda, H. Iwamura, *J. Am. Chem. Soc.* **1997**, *119*, 7412.
16. K. Matsuda, H. Iwamura, *J. Chem. Soc., Perkin Trans. 2* **1998**, 1023.
17. B. L. Feringa (Ed.), *Molecular Switches*, Wiley-VCH, Weinheim, **2001**.
18. G. H. Brown, *Photochromism*, Wiley-Interscience, New York, **1971**.
19. H. Dürr, H. Bouas-Laurent (Eds.), *Photochromism: Molecules and Systems*, Elsevier, Amsterdam, **2003**.
20. H. Bouas-Laurent, H. Dürr, *Pure Appl. Chem.* **2001**, *73*, 639.
21. M. Irie, *Chem. Rev.* **2000**, *100* 1685 (2000).
22. M. Irie, K. Uchida, *Bull. Chem. Soc. Jpn.* **1998**, *71*, 985.
23. T. Kawai, N. Fukuda, D. Dröschl, S. Kobatake, M. Irie, *Jpn. J. Appl. Phys.* **1999**, *38*, L1194.
24. J. Chauvin, T. Kawai, M. Irie, *Jpn. J. Appl. Phys.* **2001**, *40*, 2518.
25. G. M. Tsivgoulis, J.-M. Lehn, *Angew. Chem., Int. Ed. Engl.* **1995**, *34*, 1119.
26. G. M. Tsivgoulis, J.-M. Lehn, *Chem. Eur. J.* **1996**, *2* 1399.
27. M. Takeshita, M. Irie, *Chem. Lett.* **1998** 1123.
28. K. Yagi, C. F. Soong, M. Irie, *J. Org. Chem.*, **2001**, *66*, 5419.
29. A. Osuka, D. Fujikane, H. Shinmori, S. Kobatake, M. Irie, *J. Org. Chem.* **2001**, *66*, 3913.
30. T. Kawai, T. Sasaki, M. Irie, *Chem. Commun.* **2001**, 711.
31. T. B. Norsten, N. R. Branda, *Adv. Mater.* **2001**, *13*, 347.
32. A. Fernández-Acebes, J.-M. Lehn, *Chem. Eur. J.*, **1999**, *5*, 3285.
33. M.-S. Kim, T. Kawai, M. Irie, *Chem. Lett.*, **2001** 702.
34. S. L. Gilat, S. H. Kawai, J.-M. Lehn, *Chem. Eur. J.* **1995**, *1*, 275.
35. S. L. Gilat, S. H. Kawai, J.-M. Lehn, *J. Chem. Soc., Chem. Commun.* **1993**, 1439.
36. T. Kawai, T. Kunitake, M. Irie, *Chem. Lett.* **1999**, 905.
37. K. Matsuda, M. Irie, *J. Am. Chem. Soc.* **2000**, *122*, 7195.
38. S. Decurtins, P. Gütlich, C. P. Köhler, H. Spiering, A. Hauser, *Chem. Phys. Lett.* **1984**, *105*, 1.
39. A. Hauser, *Chem. Phys. Lett.* **1986**, *124*, 543.
40. T. Buchen, P. Gütlich, H. A. Goodwin, *Inorg. Chem.* **1994**, *33*, 4573.
41. M.-L. Boillot, C. Roux, J.-P. Audière, A. Dausse, J. Zarembowitch, *Inorg. Chem.* **1996**, *35*, 3975.
42. O. Sato, T. Iyoda, A. Fujishima, K. Hashimoto, *Science* **1996**, *272*, 704.
43. J.-F. Létard, P.Guionneau, L. Rabardel, J. A. K. Howard, A. E. Goeta, D. Chasseau, O. Kahn, *Inorg Chem.* **1998**, *37*, 4432.
44. S. Hayami, Z.-Z. Gu, M. Shiro, Y. Einaga, A. Fujishima, O. Sato, *J. Am. Chem. Soc.* **2000**, *122*, 7126.

45. E. Breuning, M. Ruben, J.-M. Lehn, F. Renz, Y. Garcia, V. Ksenofontov, P. Gütlich, E. K. Wegelius, K. Rissanen, *Angew. Chem. Int. Ed.* **2000**, *39*, 2504.
46. V. A. Money, J. S. Costa, S, Marcén, G. Chastanet, J. Elhaïk, M. A. Halcrow, J. A. K. Howard, J.-F. Létard, *Chem. Phys. Lett.* **2004**, *391*, 273.
47. J.-F. Létard, P. Guionneau, O. Nguyen, J. S. Costa, S. Marcén, G. Chastanet, M. Marchivie, L. Goux-Capes, *Chem. Eur. J.* **2005**, *11*, 4582.
48. W. Fujita, K. Awaga, *J. Am. Chem. Soc.* **1997**, *119*, 4563.
49. K. Hamachi, K. Matsuda, T. Itoh, H. Iwamura, *Bull. Chem. Soc. Jpn.* **1998**, *71*, 2937.
50. H. Kurata, T. Tanaka, M. Oda, *Chem. Lett.* **1999**, 749.
51. K. Tanaka, F. Toda, *J. Chem. Soc., Perkin Trans. 1* **2000**, 873.
52. S. Nakatsuji, Y. Ogawa, S. Takeuchi, H. Akutsu, J.-i. Yamada, A. Naito, K. Sudo, N. Yasuoka, *J. Chem. Soc., Perkin Trans. 2* **2000**, 1969.
53. J. Abe, T. Sano, M. Kawano, Y. Ohashi, M. M. Matsushita, T. Iyoda, *Angew. Chem. Int. Ed.* **2001**, *40*, 580.
54. Y. Teki, S. Miyamoto, M. Nakatsuji, Y. Miura, *J. Am. Chem. Soc.* **2001**, *123*, 294.
55. S. Bénard, E. Rivière, P. Yu, K. Nakatani, J. F. Delouis, *Chem. Mater.* **2001**, *13*, 159.
56. S. Bénard, A. Léaustic, E. Rivière, P. Yu, R. Clément, *Chem. Mater.* **2001**, *13*, 3709.
57. S. Nakatsuji, T. Ojima, H. Akutsu, J.-i. Yamada, *J. Org. Chem.* **2002**, *67*, 916.
58. T. Kaneko, H. Akutsu, J.-i. Yamada, S. Nakatsuji, *Org. Lett.* **2003**, *5*, 2127.
59. L. Xu, T. Sugiyama, H. Huang, Z. Song, J. Meng, T. Matsuura, *Chem. Commun.* **2003**, 2328.
60. H. Kurata, Y. Takehara, T. Kawase, M. Oda, *Chem. Lett.* **2003**, *32*, 538.
61. I. Ratera, D. Ruiz-Molina, J. Vidal-Gancedo, J. J. Novoa, K. Wurst, J.-F. Létard, C. Rovira, J. Veciana, *Chem. Eur. J.* **2004**, *10*, 603.
62. K. Matsuda, M. Irie, *Chem. Lett.*, **2000**, 16.
63. K. Matsuda, M. Irie, *J. Am. Chem. Soc.* **2000**, *122*, 8309.
64. K. Matsuda, M. Irie, *Chem. Eur. J.* **2001**, *7*, 3466.
65. K. Matsuda, M. Matsuo, M. Irie, *Chem. Lett.*, **2001**, 436.
66. K. Matsuda, M. Matsuo, M. Irie, *J. Org. Chem.* **2001**, *66*, 8799.
67. K. Matauda, M. Matsuo, S. Mizoguti, K. Higashiguchi, M. Irie, *J. Phys. Chem. B*, **2002**, *106*, 11218.
68. K. Matsuda, M. Irie, *J. Am. Chem. Soc.* **2001**, *123*, 9896.
69. A. T. Bens, D. Frewert, K. Kodatis, C. Kryschi, H.-D. Martin, H. P. Trommsdorf, *Eur. J. Org. Chem.* **1998**, 2333.
70. N. Tanifuji, M. Irie, K. Matsuda, *J. Am. Chem. Soc.* **2005**, *127*, 13344.
71. N. Tanifuji, K. Matsuda, M. Irie, *Org. Lett.* **2005**, *7*, 3773.
72. K. Shibata, S. Kobatake, M. Irie, *Chem. Lett.* **2001**, 618.
73. C. Joachim, J. K. Gimzewski and A. Aviram, *Nature* **2000**, *408*, 541.
74. R. F. Service, *Science* **2001**, *293*, 782.
75. J. M. Tour, *Acc. Chem. Res.* **2000**, *33*, 791.
76. B. Xu and N. J. Tao, *Science* **2003**, *301*, 1221.
77. T. Nakamura, T. Matsumoto, H. Tada, K.-I. Sugiura (Eds.), *Chemistry of Nanomolecular Systems: Towards the Realization of Nanomolecular Devices*, Springer, Heidelberg, 2003.

10
Thiol End-capped Molecules for Molecular Electronics: Synthetic Methods, Molecular Junctions and Structure–Property Relationships

Kasper Nørgaard, Mogens Brøndsted Nielsen and Thomas Bjørnholm

10.1
Introduction

Silicon based microelectronics has undergone relentless miniaturization during the past three decades, resulting in substantial improvements in information storage capacity and processing power. This development has been driven by numerous scientific and technological innovations, in addition to considerable financial awareness. If this miniaturization trend is to continue, the size of electronic components will eventually reach the scale of atoms or small molecules – a size-regime that is beyond the reach of current photolithographic fabrication techniques [1]. The development of novel manufacturing schemes for small-scale electronics is therefore an essential requirement for future advances within this field. An attractive alternative approach is to explore the electronic capabilities of organic materials at the single-molecule level. In particular, the study of π-conjugated molecules could potentially overcome some of the limitations inherent to the current "top-down" techniques. Through organic synthesis, it is possible to produce large quantities ($\sim 10^{23}$) of identical molecules. Rational design of the molecular structure provides a way to tune the electronic properties of the molecules and thereby possibly mimic the function of electronic components, such as wires, rectifiers, switches and transistors. This ambition has been a major driving force in the evolution of *molecular electronics* [2].

The use of intrinsic molecular properties in the construction of electronic devices was first proposed in 1974 by Aviram and Ratner [3], who suggested that a single molecule with a donor–spacer–acceptor structure would behave as a (uni)molecular rectifier when connected to two electrodes. Experimental work along these lines was pioneered in the early 1970s by Mann and Kuhn [4], who measured conduction through monomolecular layers (Langmuir–Blodgett monolayers) of different fatty acids, sandwiched between two metal electrodes. At that time, however, no techniques were available for making electronic contact to individual molecules and feasible ways to attach molecules chemically to metals were lacking. The demonstration by Nuzzo and Allara [5] that sulfur-contain-

Functional Organic Materials. Syntheses, Strategies, and Applications.
Edited by Thomas J.J. Müller and Uwe H.F. Bunz
Copyright © 2007 WILEY-VCH Verlag GmbH & Co. KGaA, Weinheim
ISBN: 978-3-527-31302-0

ing molecules could form well-ordered self-assembled monolayers (SAMs) [6, 7], when deposited on flat gold surfaces, provided a means of chemically attaching organic molecules to macroscopic electrodes.

It was not until the late 1990s, however – and after the dawn of nano-scale technologies – that the attachment of single molecules to external electrodes became technological feasible. This breakthrough [8] launched the field of experimental molecular electronics and paved the road for fundamental investigations of the relationship between molecular and electronic structure at the single molecule level.

This chapter describes different synthetic approaches towards the fabrication of candidate molecules for use in molecular electronics applications with an emphasis on thiol end-capped π-conjugated molecules, followed by a survey of the electronic transmission properties in two- and three-terminal devices.

10.2
Synthetic Procedures

10.2.1
Protecting Groups for Arylthiols

The synthesis of thiol-end-capped π-conjugated wire compounds relies on suitable thiol protecting groups that can be removed *in situ*. The acetyl group provides one such group and is removed by aqueous base [9]. The lability of the acetyl protecting group during synthetic transformations has stimulated, however, the exploitation of other protecting groups that can withstand a larger selection of synthetic manipulations and that in a final step can be converted to the acetyl group. One successful approach is to introduce the thiol functionality as the *tert*-butyl sulfide at the beginning of the synthetic sequence and to maintain it through the following steps owing to the resistance of *t*-BuS-Ar to both strongly basic and acidic conditions [10]. In a final step, the *t*-BuS group can be converted into the AcS moiety by means of AcCl–BBr$_3$ or by AcCl–cat. Br$_2$ [10, 11]. The deprotection and conversion between a large selection of ArSH protecting groups is illustrated in Scheme 10.1 [9–12].

10.2.1.1 Synthesis of Arylthiol "Alligator Clips"
Compounds **1–8**, containing either acetyl or *tert*-butyl protecting groups, have found wide applicability as thiol-based alligator clips [10, 13], i.e. where the thiol end-group facilitates adhesion of the molecular compound to external electrodes. In particular, these compounds (**1–8**) allow the construction of molecular wire compounds containing oligophenylene–ethynylene (OPE) or oligophenylene–vinylene (OPV) backbones. Thus, aryl halides **1–3** and terminal alkynes **4** and **5** can be subjected to Sonogashira cross-coupling conditions, whereas alde-

Scheme 10.1

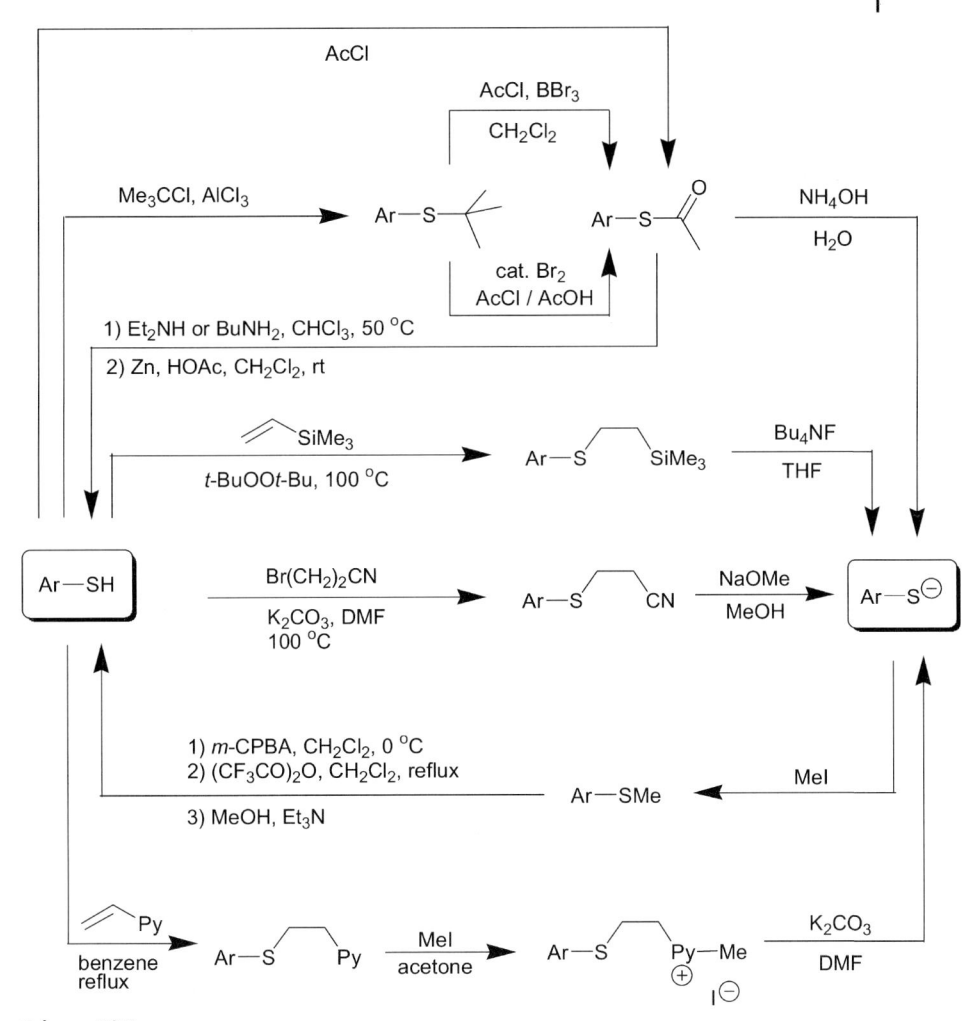

hydes **6** and **7** and phosphonate **8** can be employed for Horner–Wadsworth–Emmons (HWE) reactions.

Scheme 10.2 shows a protocol for synthesizing **1** and its further transformation into **4** [13a,d]. Monolithiation of the diiodide **9** followed by treatment with S_8 and AcCl generated **1**. Next, a Sonogashira cross-coupling reaction with trimethylsilylacetylene afforded **10** in high yield. Desilylation was obtained upon treatment with n-Bu$_4$NF, which provided the terminal alkyne **4**.

The synthesis of **1** as described in Scheme 10.2 relies on monolithiation of **9**, which is not always easy to reproduce. In order to avoid this problematic step, Bryce and coworkers [13f] developed an alternative synthesis based on four high-yielding steps. The synthesis starts from 4-iodophenol **11**, which was reacted

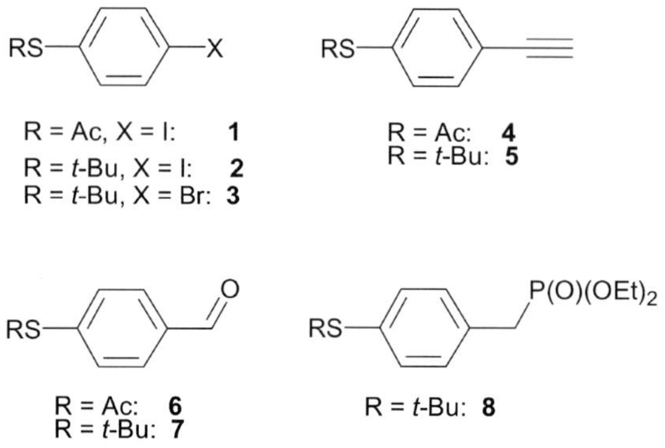

Chart. 10.1

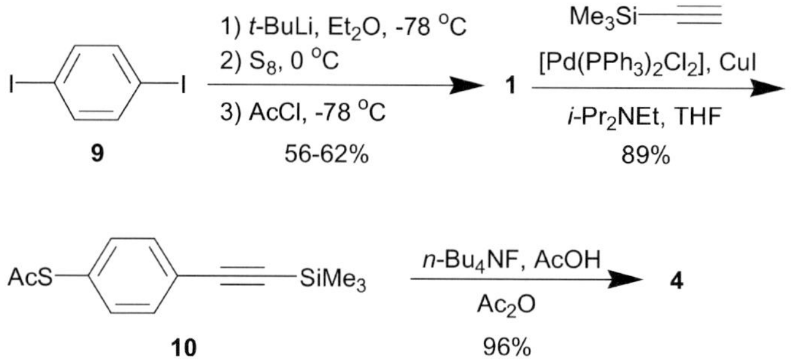

Scheme 10.2

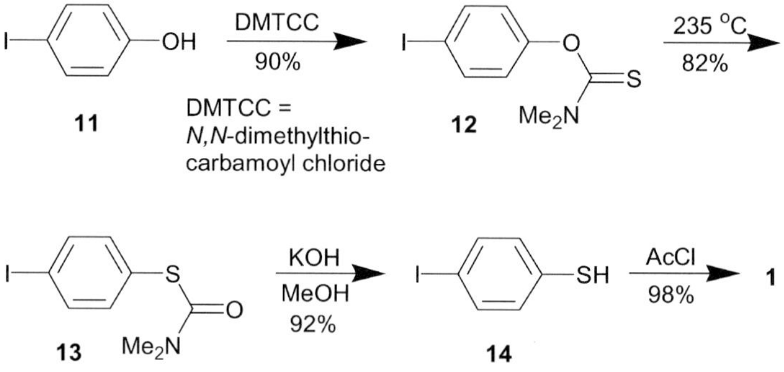

Scheme 10.3

with N,N-dimethylthiocarbamoyl chloride (DMTCC) to afford the O-ester **12**. This ester was converted to the S-ester **13** at 235 °C. Subsequent hydrolysis gave **14** and, after acetylation, **1** was finally obtained.

An efficient synthesis of the aldehyde **6** was reported by Lindsey and coworkers (Scheme 10.4) [13b]. First, the aldehyde **15** was converted into the acetal **16**. Sub-

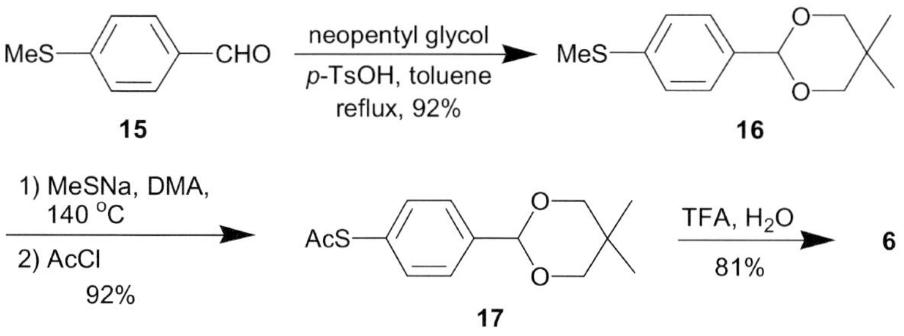

Scheme 10.4

DMA = N,N-dimethylacetamide
TFA = trifluoroacetic acid

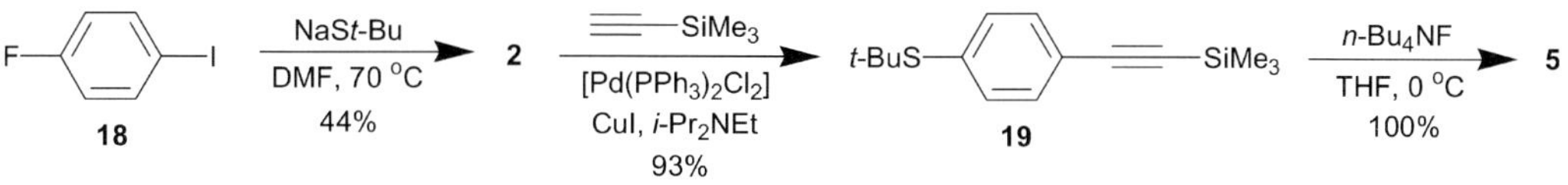

Scheme 10.5

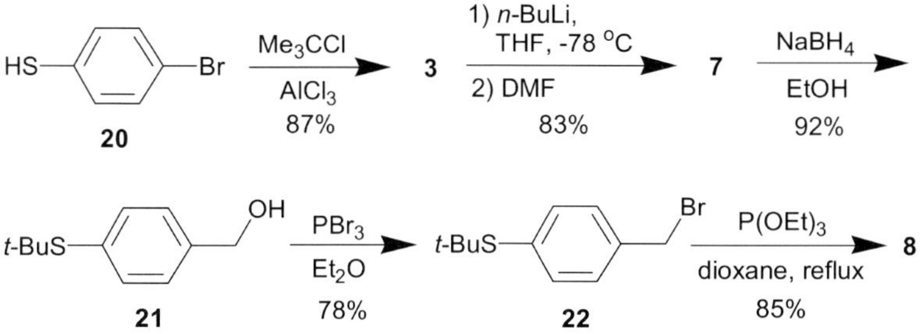

Scheme 10.6

sequent transformation of the SMe group into SAc (**17**) followed by selective TFA cleavage of the acetal group afforded **6** in good yield.

Mayor et al. [13g] prepared the iodide **2** via substitution of the fluoride **18** by using sodium *tert*-butylthiolate (Scheme 10.5). A cross-coupling reaction with trimethylsilylacetylene generated the alkyne **19** that was desilylated by tetrabutylammonium fluoride finally to provide **5**.

The bromide **3** was prepared by Stuhr-Hansen et al. [10] from the thiol **20** upon treatment with 2-chloro-2-methylpropane in the presence of catalytic amount of aluminum chloride (Scheme 10.6). Lithiation followed by reaction with DMF afforded the aldehyde **7**. This aldehyde can be converted to the diethyl phosphonate **8** in three high-yielding steps, via alcohol (**21**) and bromide (**22**) intermediates.

10.2.2
One-terminal Wires

Scheme 10.7 depicts the synthesis of an OPE3 one-terminal wire compound by repeated Sonogashira cross-couplings [13c]. The synthesis takes advantage of the more reactive iodide as compared with the bromide, which allows for stepwise scaffolding. Thus, two concomitant Pd-catalyzed cross-couplings of **23** produced **24**. The terminal alkyne obtained after desilylation was subsequently cross-coupled with **1** to afford **25**.

The donor–acceptor substituted OPE **26** is particularly interesting since it shows a high negative differential resistance (see below). The original synthesis reported by Tour et al. [13c] was in some respects inconvenient and therefore Bryce and coworkers [13f] developed an improved procedure. The details are given in the original literature. Scheme 10.8 show the final step in the synthesis of **26**, the coupling between alkyne **27** and iodide **1** [13f].

Scheme 10.9 shows the synthesis of a one-terminal OPV by two subsequent Heck coupling reactions of the unsymmetrical dihalide **23**, again exploiting the higher reactivity of an aryl iodide relative to an aryl bromide [13e]. Thus, a selective coupling reaction between **23** and **28** gave **29**. This bromide was then reacted with the alkene **30**. Treatment of the resulting compound **31** with fluoride ions liberated the thiolate via an elimination reaction. This thiolate was then subjected to acetylation to provide OPV3 **32**.

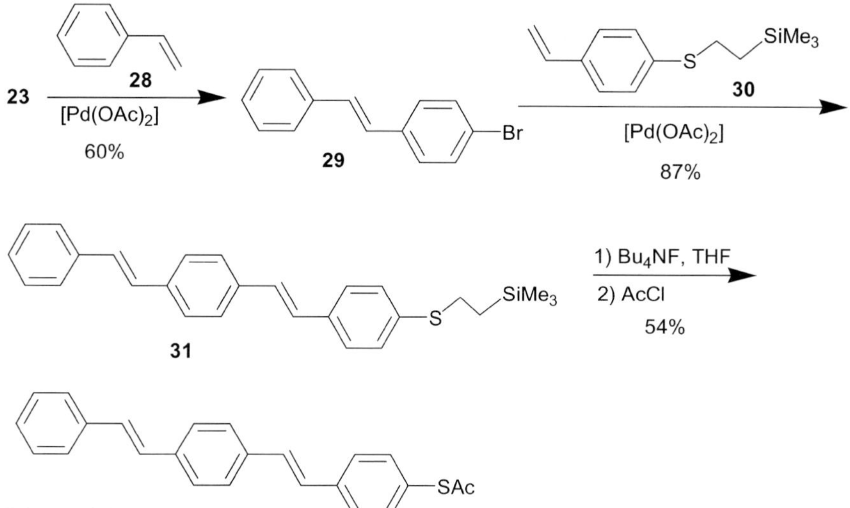

Scheme 10.9

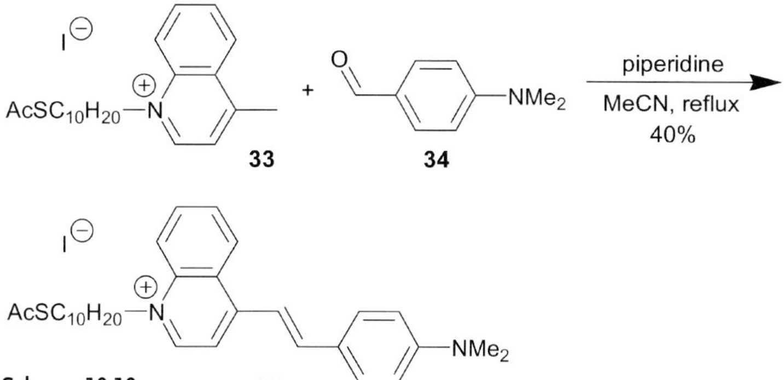

Scheme 10.10

Donor–π-acceptor dyads can be employed as molecular rectifiers. For this purpose, a quinolinium hemicyanine was prepared by Ashwell et al. [14]. Reaction between **33** and **34** in the presence of piperidine base gave the dyad **35** (Scheme 10.10).

10.2.3
Two-terminal Wires

A cross-coupling reaction between diiodide **9** and 2 equiv. of alkyne **4** provided directly the two-terminal OPE3 **36** (Scheme 10.11) [13c].

Two-terminal OPV compounds were efficiently prepared by Bjørnholm and coworkers from the aldehyde **6** and suitable phosphonates via HWE reactions [10a].

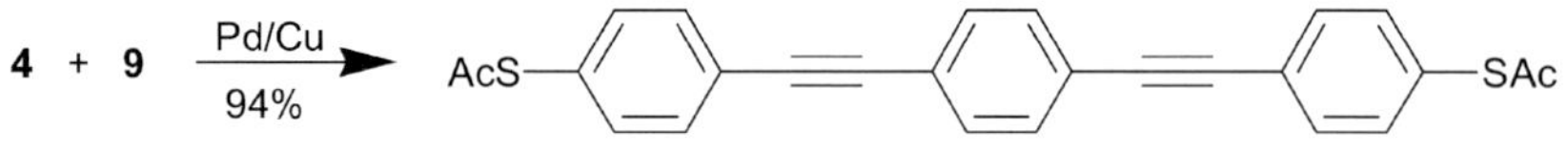

Scheme 10.11

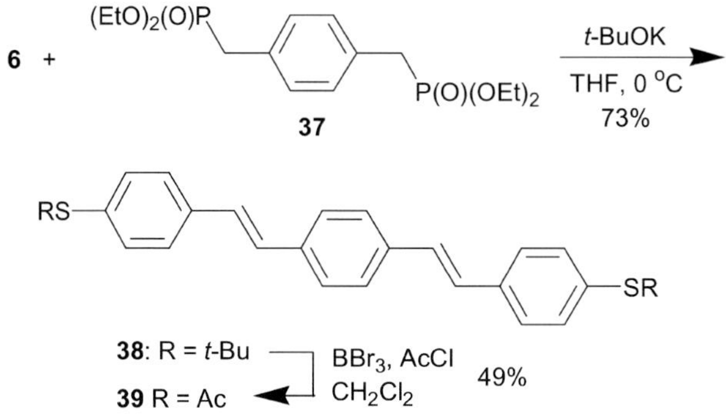

4 + 9 →[Pd/Cu, 94%] AcS—⟨⟩—≡—⟨⟩—≡—⟨⟩—SAc
36

Scheme 10.12

6 + (EtO)₂(O)P—CH₂—⟨⟩—CH₂—P(O)(OEt)₂ →[t-BuOK, THF, 0 °C, 73%]
37

RS—⟨⟩—CH=CH—⟨⟩—CH=CH—⟨⟩—SR

38: R = t-Bu
39 R = Ac

BBr₃, AcCl / CH₂Cl₂, 49%

One example is illustrated in Scheme 10.12. In the synthetic approach adopted here, the arylthiol functionality was introduced as the *tert*-butyl sulfide at the beginning of the synthetic sequence. A two-fold HWE reaction between aldehyde **6** and diphosphonate **37** gave OPV3 **38**. The *t*-BuS group of **38** was then converted into the AcS moiety by means of AcCl–BBr₃.

Mayor et al. [13g] prepared anthracene-spaced wire compounds by two-fold Sonogashira cross-couplings of the dibromide **40** (Scheme 10.13). Treatment of **40** with the alkyne **5** gave **41** that was subsequently converted to the SAc end-capped compound **42** under the standard conditions (see above). Treating instead **40** with alkyne **43** gave **44**, albeit in low yield. Similarly, **42** was produced in very low yield (5 %) on reacting instead dibromide **40** and the SAc-containing alkyne **4**. Compounds **42** and **44** differ in the position of the thiol anchor groups, being at para and meta positions, respectively. The lack of conjugation in the meta position and therefore the reduced electronic communication compared with that present in the para isomer results in reduced conductance of **44** (see below).

Tour et al. [13e] also replaced the phenyl or the biphenyl core of OPEs with one or two pyridine moieties. The aim was to lower the LUMO energy and hereby to produce a better match with the Fermi level of the metal contact and higher current through the device. An example is shown in Scheme 10.14. After desilylation, **45** was subjected to a two-fold cross-coupling with **1** to provide **46**.

Synthetic routes to mesitylthio-capped oligothiophenes are outlined in Schemes 15 and 16 [12c]. Lithiation of thiophene **47** followed by treatment with mesitylenesulfenyl chloride gave **48**. Monolithiation of **49** with BuLi and reaction with MesSCl generated **50**. Another lithium–halogen exchange followed by an oxida-

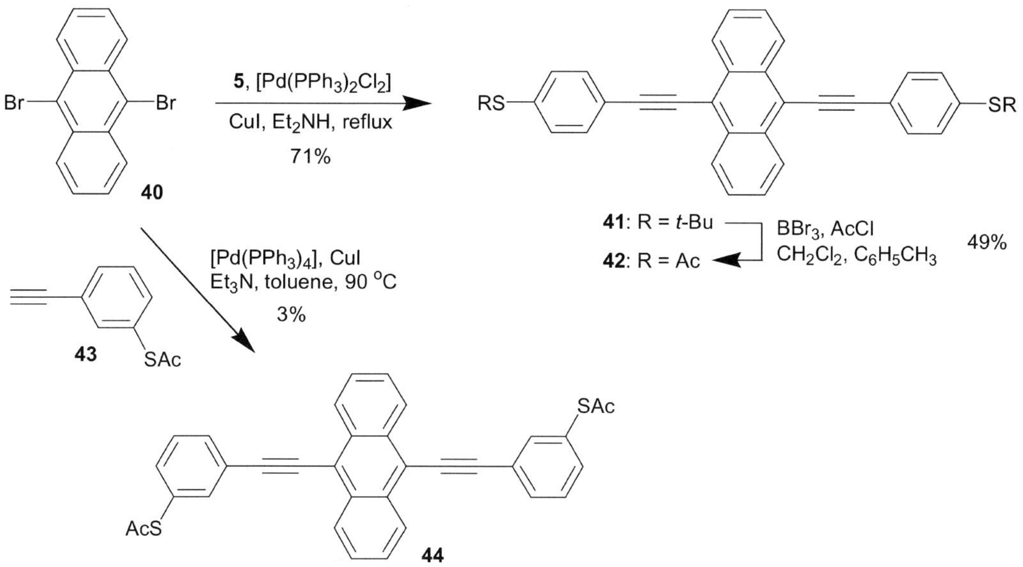

Scheme 10.13

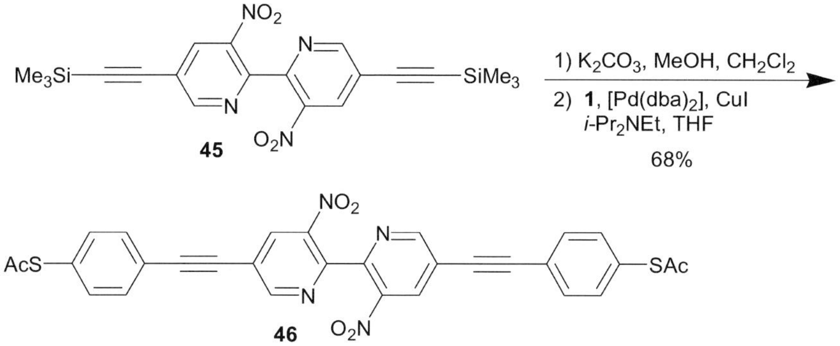

Scheme 10.14

tive coupling employing CuCl$_2$ produced the bithiophene **51**. A Stille coupling between bromide **50** and the bisstannyl thiophene **52** gave the terthiophene **53**.

Two-fold Menschutkin reactions on 4,4′-bipyridine provide a route to compounds containing a redox-active paraquat unit [15]. Thus, reacting **54** with 6-tosyloxyhexyl thioacetate afforded **55** (Scheme 10.17). Perylene tetracarboxylic diimide presents another electron acceptor that has been functionalized with thiol end-groups. Thus, heating a mixture of the dianhydride **56** and 4-aminothiophenol (**57**) gave the product **58** (Scheme 10.18) [16].

Maskus and Abruña [17] functionalized terpyridine (tpy) ligands with thiol groups. These ligands can be employed for Co^{2+}-bridged molecular wires. The synthesis proceeds according to Scheme 10.19. After deprotonation, **59** was alky-

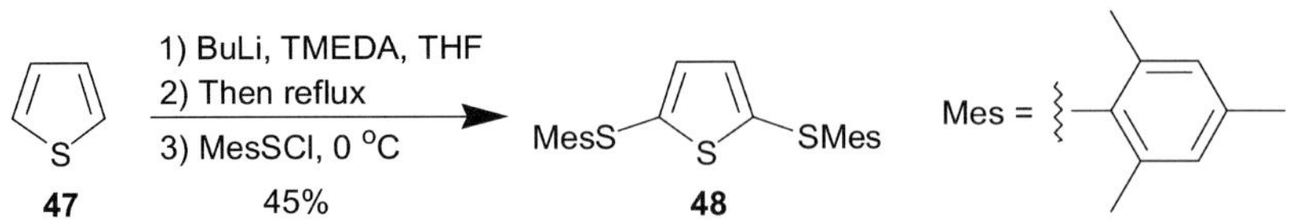

Scheme 10.15

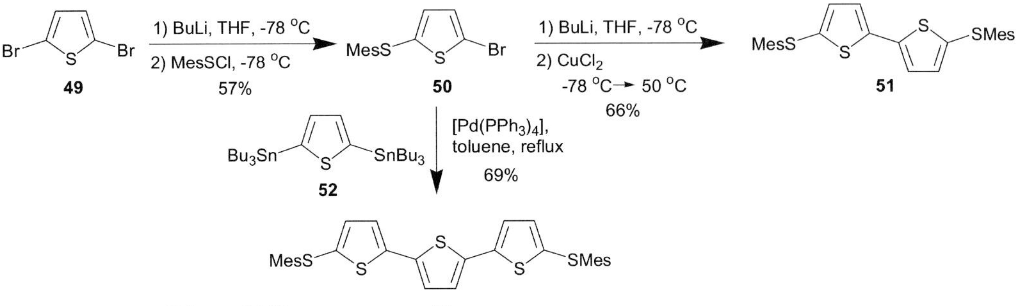

Scheme 10.16

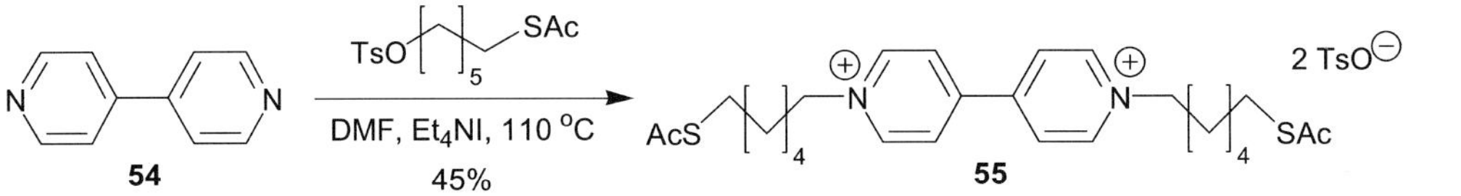

Scheme 10.17

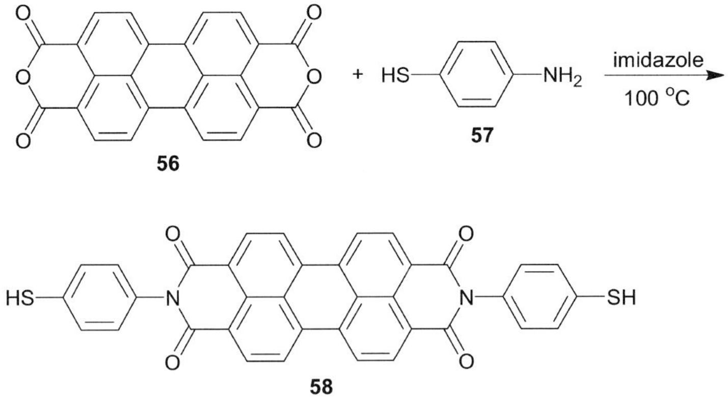

Scheme 10.18

lated with 1-bromo-4-chlorobutane, which afforded the product **60**. A Finkelstein reaction with NaI followed by reaction with thiourea gave a stable thiouronium salt that was finally hydrolyzed to afford the thiol **61**. This ligand was treated with [Co(tpy)Cl$_2$], which yielded the complex **62**.

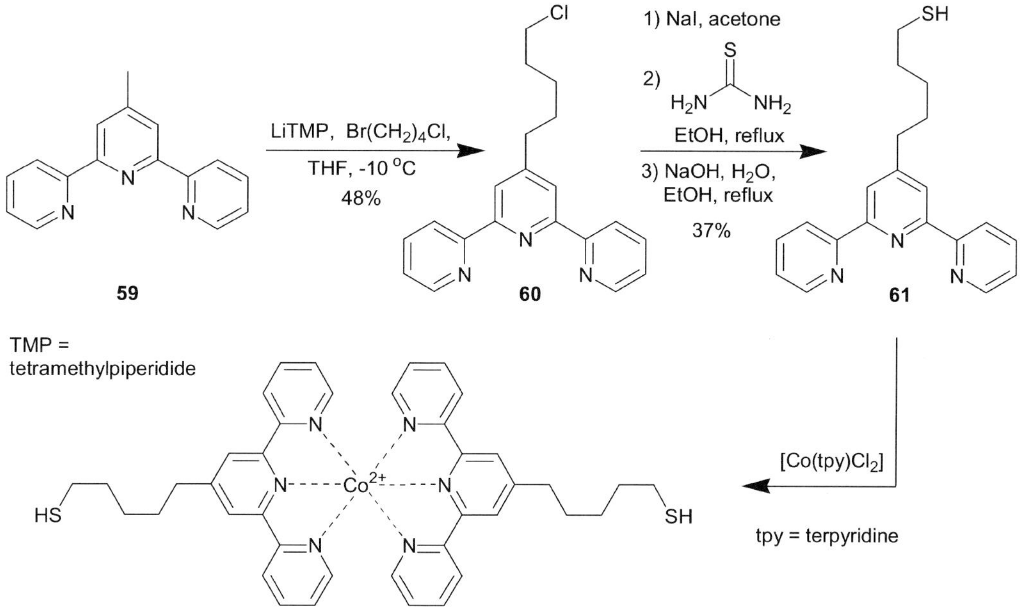

Scheme 10.19

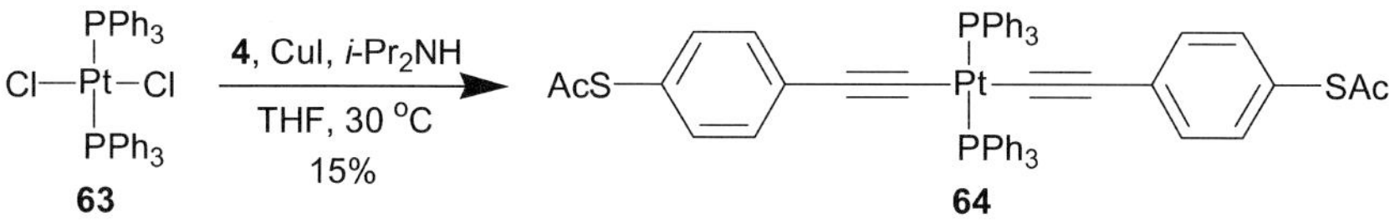

TMP = tetramethylpiperidide

Scheme 10.20

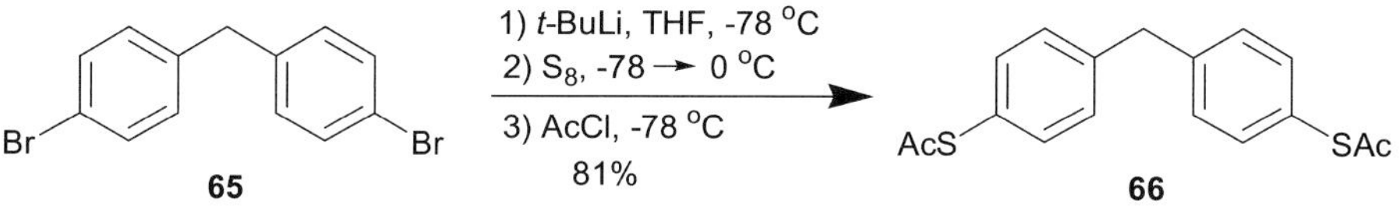

Scheme 10.21

A Pt(II)-bridged wire compound was prepared by Mayor et al. [18] by treating trans-bis(triphenylphosphane)platinum(II) chloride (**63**) with the alkyne **4** and CuI as catalyst (Scheme 10.20). The trans configuration of the product **64** was confirmed by X-ray crystallographic analysis.

Several wire compounds in which the conjugation is broken by a methylene group within the molecule have been reported by Tour et al. [13c]. The synthesis of a two-terminal compound is presented in Scheme 10.21. Lithiation of the dibromide **65** followed by treatment with S_8 and AcCl provided **66**. Such molecules are interesting for the study of eventual tunnel effects.

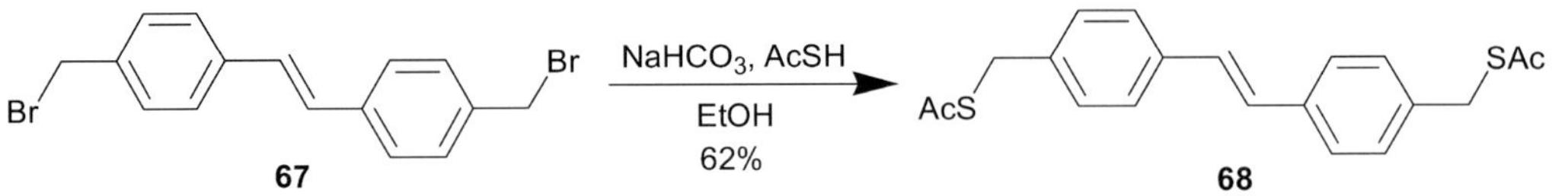

364 | *10 Thiol End-capped Molecules for Molecular Electronics*

Scheme 10.22

Treatment of the dibromide **67** with AcSH gave **68** in which the conjugation is instead disrupted between the conjugated moiety and the thiol end-caps (Scheme 10.22) [19].

10.2.4
Three-terminal Wires

1,3,5-Triiodobenzene (**69**) serves as a suitable core for the synthesis of three-terminal OPEs. Thus, a three-fold cross-coupling with the alkyne **70** produced **71** (Scheme 10.23) [13c].

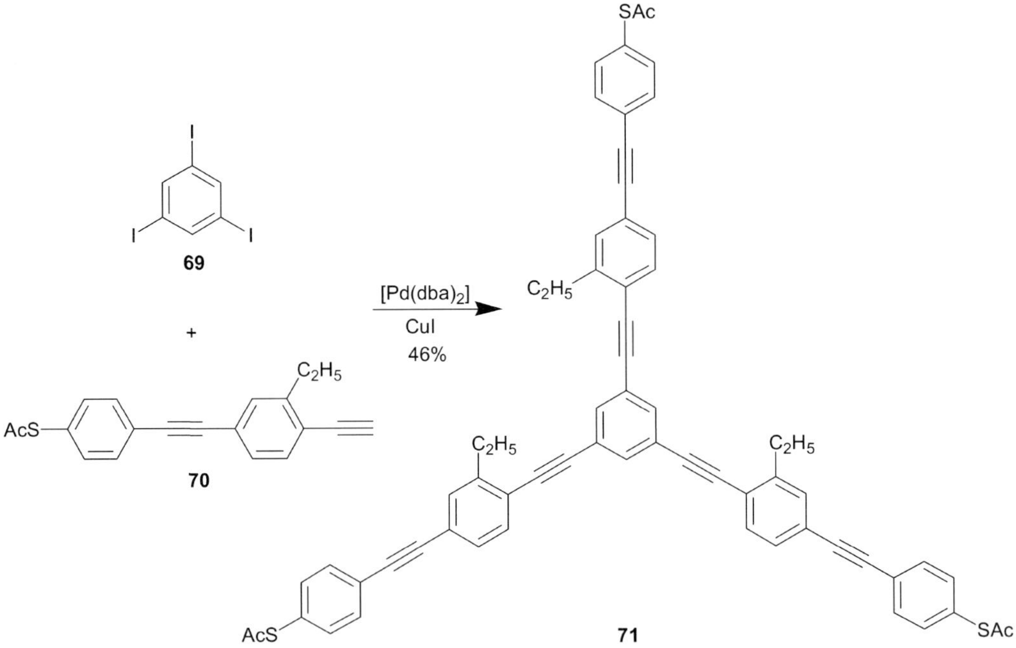

Scheme 10.23

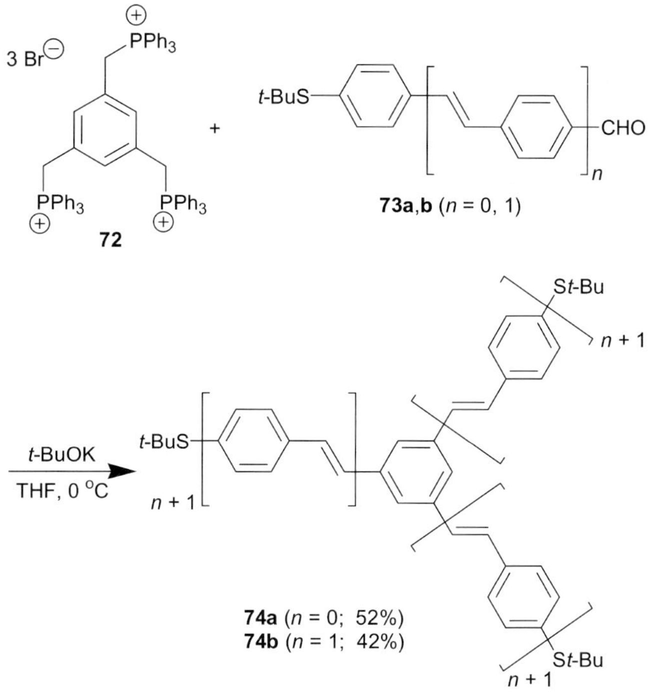

Scheme 10.24

A three-fold Wittig reaction of the trisphosphonium salt **72** with suitable aldehydes **73a** and **b** represents an efficient route to three-terminal OPVs **74a** and **b**, respectively (Scheme 10.24) [10a].

10.2.5
Four-terminal Wires

Two routes for the synthesis of a four-terminal wire compound based on a central porphyrin unit are presented in Scheme 10.25 [13b]. In one route, the tetrabromide **75** was treated with potassium thioacetate, which resulted in the substitution product **76**. Alternatively, reaction of aldehyde **77** with pyrrole **78** at room temperature using mixed acid catalysis conditions provided **76** after oxidation with DDQ.

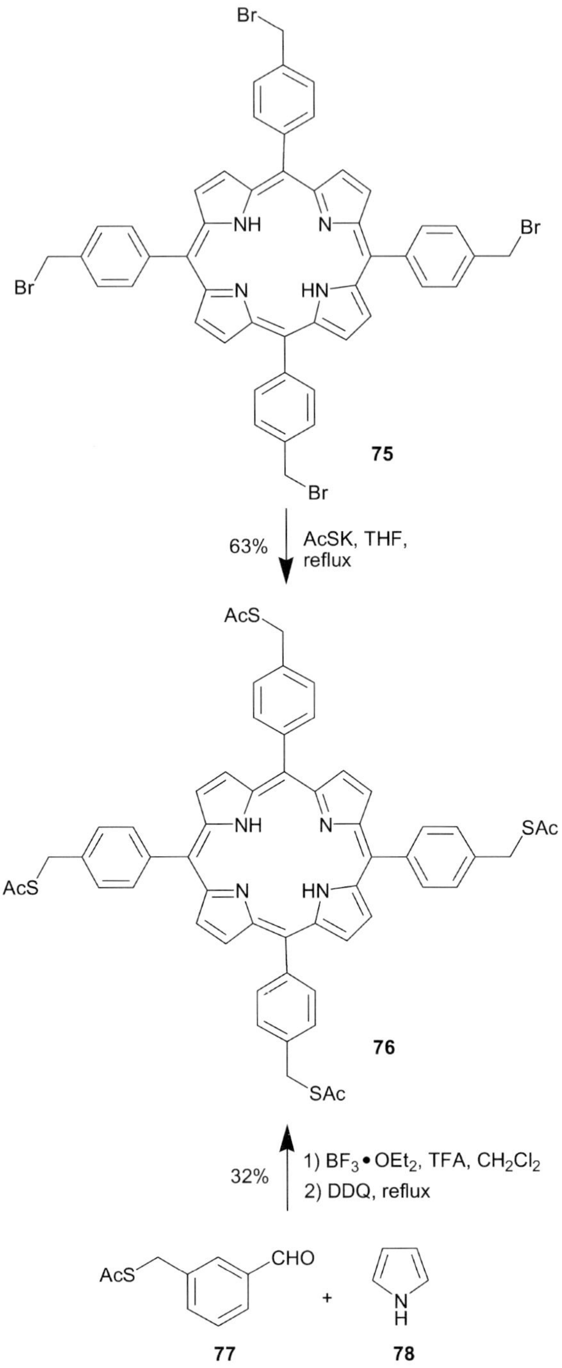
Br
NH
N
HN
Br
Br
Br
75
63%
AcSK, THF,
reflux
AcS
NH
N
HN
SAc
AcS
SAc
76
32%
1) BF3 • OEt2, TFA, CH2Cl2
2) DDQ, reflux
AcS
CHO
+
N
H
77
78

Scheme 10.25

10.2.6
Caltrops

A classical caltrop is a tetrahedrally arranged four-pointed device such that three of the prongs naturally form a tripod on the ground while the fourth prong projects upward when the caltrop is deployed. Based on a tetrahedral Si atom (**79**), the caltrop **80** was prepared (Scheme 10.26) [20]. This caltrop contains three legs bearing thiol-tipped feet for adhesion to Au.

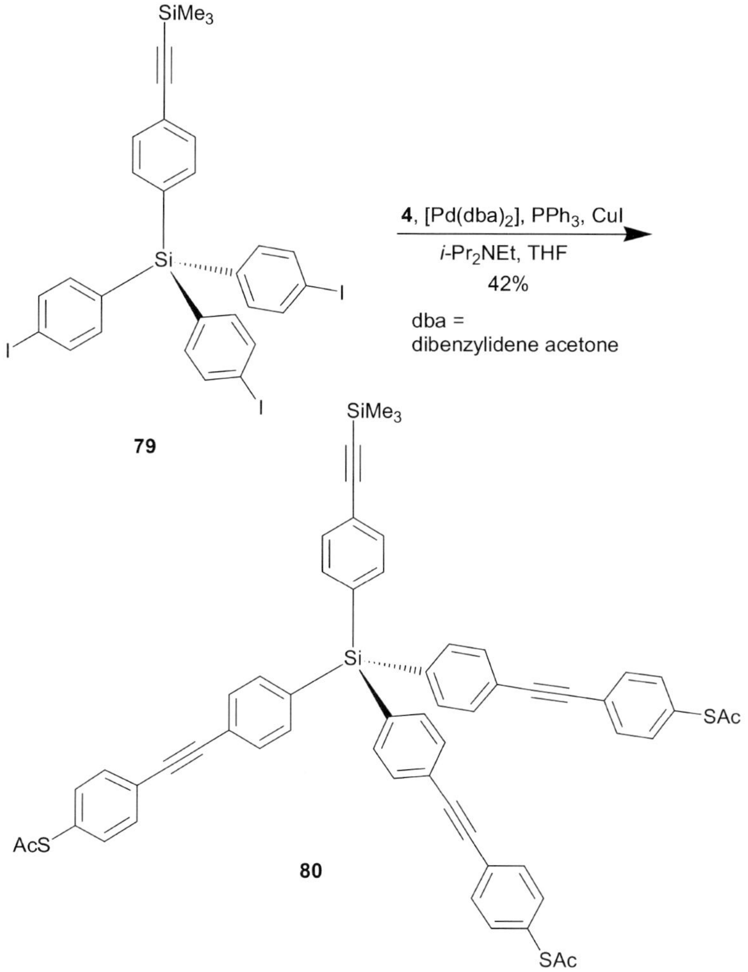

Scheme 10.26

10.3
Electron Transport in Two- and Three-terminal Molecular Devices

As described in the previous section, synthetic chemistry can produce a wide variety of π-conjugated organic molecules upon which future electronic devices may be based. In the context of molecular-scale electronics, a related (and equally important) challenge concerns the insertion of these organic molecules into actual functional devices and integrated circuits. Unlike the field of synthetic organic chemistry, which has developed and matured for more than a century, experimental work on single-molecule electronics is still very much in its infancy, and developing convenient methods for contacting and characterizing individual molecules electronically is currently a very active field of research. Central to this work is the construction of *molecular junctions* in which the organic molecule(s) is sandwiched between two (or more) conducting electrodes. A schematic illustration of a molecular junction is shown in Fig. 10.1. The molecular junction facilitates a measurement of the current versus voltage, $I(V)$, characteristics of the molecule when sandwiched between the electrodes.

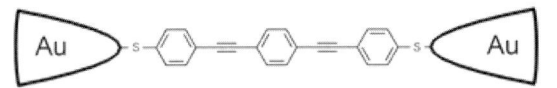

Fig. 10.1 Schematic illustration of molecular junction.

Electron transport through a molecule can, to a first approximation, be considered as classical or coherent tunneling (ballistic transport), although incoherent tunneling and other types of charge transport could significantly alter this picture. The conductance, G, in the coherent tunneling regime can be described by the Landauer equation [56]:

$$G = \frac{2e^2}{h} T \tag{1}$$

where T is a transmission factor that reflects the probability of electron transmission through the junction. This factor can be divided into contributions from the contacts and the molecule itself, according to

$$T = T_1 T_2 T_{mol} \tag{2}$$

where T_1 and T_2 represent the charge transport efficiency across contacts 1 and 2 and T_{mol} represents the charge transport through the molecule. Electron tunneling through a molecule can – in the coherent tunneling regime – be approximated by

$$T_{mol} = \exp(-\beta l) \tag{3}$$

where l is the length of the molecule and β is called the tunneling decay parameter, which is given by the expression

$$\beta = 2\sqrt{\frac{2m\alpha\left(\phi - \frac{eV}{2}\right)}{\hbar^2}} \tag{4}$$

where ϕ represents the tunnel barrier height for tunneling trough the HOMO or the LUMO, i.e. $\phi = (E_{HOMO} - E_F)$ and $\phi = (E_F - E_{LUMO})$, respectively, where E_F is the Fermi energy, m is the electron mass, V is the applied bias and α represents the asymmetry of the junction potential profile ($\alpha = 1$ for symmetric junctions).

When molecules couple to metallic electrodes, the discrete energy levels in the molecule mix with the continuous electronic states of the electrodes. Therefore, when interpreting the current–voltage response of a molecular junction, it is necessary to view the entire electrode-molecule–electrode setup as one electronic system. A factor that is likely to influence conductance through a molecular junction include, in addition to the molecular structure, also the coupling of the molecule to the electrodes. In addition to classical tunneling behavior, a number of nonlinear effects have been observed in $I(V)$ characteristics of molecular junctions (see below). Examples of such phenomena include negative differential resistance (NDR), conductance switching and rectification of current. NDR is a situation where $dV/dI < 0$ and thus occurs if the slope of the $I(V)$ curve becomes negative. A central question is whether these nonlinear effects are inherent to the molecules themselves or related to the molecule-metal contact. Table 10.1 provides a survey of measurements performed on different molecular wires.

Table 10.1 Survey of molecules investigated at different junctions: N = number of molecules; β = tunneling decay parameter; R = resistance; compound numbers in parentheses refer to protected thiol precursors.

Junction type	Molecular structure		N	β (Å^{-1})	R (GΩ)	Ref.
STM on SAM		n = 0: **81a**	1			21
		n = 1: **81b**	1			21
		n = 2: **81c**	1			21
		82	1			22, 23
		n = 1: **83a**	1			22, 23
		n = 2: **83b**	1			21
		n = 3: **83c**	1			21
		R^1 = H, R^2 = Et: **84a**	1			24
		R^1 = H, R^2 = NO$_2$: **84b**	1	0.45		24
		R^1 = NO$_2$, R^2 = NO$_2$: **84c**	1	0.34		24
		R^1 = NH$_2$, R^2 = NO$_2$: **84d** ("**26**")	1			24
		85	1			24
		86	1	0.15		24
		87	1	0.50		24
		88 ("**35**")	1			25
		89 ("**46**")	1			26

Table 10.1 (continued).

Junction type	Molecular structure		N	β (Å^{-1})	R (GΩ)	Ref.
STM in matrix	[structure with R¹, R²]	R¹ = H, R² = H: **84e** ("**25**")	1	0.94		27
		R¹ = H, R² = Et: **84a**	1			28
		R¹ = H, R² = NO$_2$: **84b**	1	0.99		27
		R¹ = NH$_2$, R² = NO$_2$: **84d** ("**26**")	1			27
	[structure]	90	1	0.65		19
	[structure with n]	n = 1: **91a**	1	0.53		19
		n = 2: **91b**	1	0.63		19
	[structure with NO$_2$]	92	1	0.78		19
	[structure with NO$_2$]	93	1	0.80		19
	[structure with O$_2$N, NO$_2$]	94	1	0.72		19
	[structure]	95 ("**36**")	1		1.7	29
	[structure with OBu, BuO]	96	1		0.4	29
	[structure with NO$_2$, O$_2$N]	97	1			26

Table 10.1 (continued).

Junction type	Molecular structure		N	β (Å^{-1})	R (GΩ)	Ref.
CP-AFM		n = 0: **98a**	10^2	0.42	0.05	30
		n = 1: **98b**	10^2	0.42	0.4	30
		n = 2: **98c**	10^2	0.42	1.5	30
CP-AFM (AuNP)		R^1 = H: **95** ("25") 1–5			51	31
		R^1 = NO$_2$: **99** 1–5			< 2	31
		100	1		4.9	32
STM (liq.)		**101** ("55")	1			33
		102	1			34
STM/BJ (liq.)		**103** ("58")	1		1	16
		104	1		0.0013	35
		83a	1		0.0012	36
		82	1		0.021	36
MCB		**83a**	1			37
		105 ("53")	1		~0.1	38
		106 ("42")	1		0.003	39

10.3 Electron Transport in Two- and Three-terminal Molecular Devices | 373

Table 10.1 (continued).

Junction type	Molecular structure	N	β (Å^{-1})	R (GΩ)	Ref.
MCB		107 ("44")	1		13g
		108	1		39
		109	1	5–50	18
Crossed wires		84e ("25")	10^3	5	40
		95 ("36")	10^3	1.7	40, 41
		110	10^3	0.5	40
		111	10^3		42
		112 ("46")	10^3		26
Nanopore		R^1 = H, R^2 = NO$_2$: **84b**	10^3		43
		R^1 = NH$_2$, R^2 = NO$_2$: **84d** ("26")	10^3		44
		81b	10^3		45

Table 10.1 (continued).

Junction type	Molecular structure	N	β (Å^{-1})	R (GΩ)	Ref.
Square tip		n = 2: **105a**			46
		n = 3: **105b**			46
		n = 4: **105c**			46
		113			46
Hg drop		n = 0: **81a**	10^{11}–10^{13}	0.61	47
		n = 1: **81b**	10^{11}–10^{13}	0.61	48
		n = 2: **81c**	10^{11}–10^{13}	0.61	47
		n = 0: **98a**	10^{11}–10^{13}	0.67	47
		n = 1: **98b**	10^{11}–10^{13}	0.67	48
		n = 2: **98c**	10^{11}–10^{13}	0.67	47
		114	10^{11}–10^{13}		48
		115	10^{11}–10^{13}		48
Particle junction		R^1 = H: **95** ("36")			49
		R^1 = NO$_2$: **84b**			49
		91b ("39")			50
Nanowire		**95** ("36")	2500		51
		96	1000		51

Table 10.1 (continued).

Junction type	Molecular structure	N	β (Å⁻¹)	R (GΩ)	Ref.
3-Terminal		116 ("62")	1		52
		117	1		53
		118	1		54
		91b ("39")	1		55

10.3.1
Molecular Junctions

Owing to the technological difficulties associated with contacting a minute number of organic molecules with two (or more) macroscopic electrodes, a variety of different approaches are currently being pursued – each with distinct advantages and disadvantages relative to the others. Molecular junctions can, for example, be categorized according to fabrication methodology (scanning-probe methods, break junctions, cross-wire, etc.), structure (two- or three-terminal devices with single or multiple molecules) or the nature of the linker chemistry (SAMs, LB monolayers, covalent linking, etc.).

10.3.1.1 Scanning Tunneling-based Molecular Junctions

Since the invention of the scanning tunneling microscope (STM) in 1982 [57], it has had a tremendous impact on the developing field of nanoscale sciences. In addition to the general nanoscale imaging capabilities, the STM introduced a con-

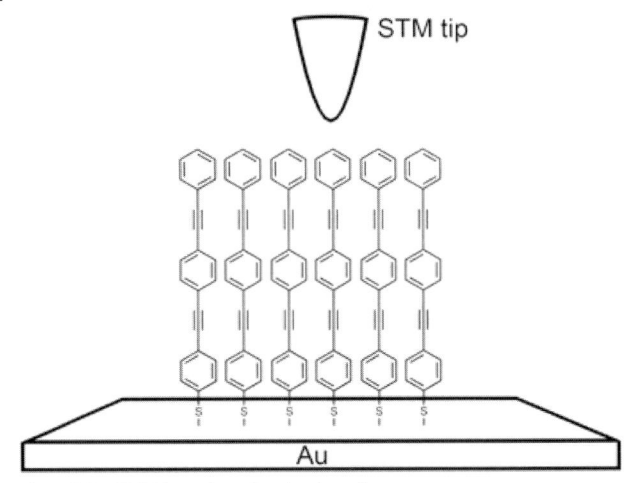

Fig. 10.2 STM-based molecular junction.

venient method to characterize electronic transmission through molecules [58]. In an STM experiment, the tunnel current is measured across the gap between the STM tip and a conducting substrate. The insertion of one or more molecules into this gap produces a molecular junction, which can then be characterized electronically by varying the bias voltage between the tip and the substrate.

The fabrication of STM-based molecular junctions have relied – in most reported cases –on the deposition of thiol end-capped molecules as self-assembled monolayers on a flat gold substrate. By using an exceedingly small tip, the measurement of no more than a single or a few molecules becomes possible. A schematic illustration of an STM-based molecular junction is shown in Fig. 10.2.

Kubiak and coworkers used STM in an ultra-high vacuum environment to study the conductance through self-assembled monolayers of simple π-conjugated, thiol-terminated molecules on gold [21–23, 59] Early work, which focused on the *p*-xylyldithiol **82** monolayers [22, 23], was later expanded to include also short oligophenylenes (OPs) (**81** and **83**) that were terminated at either one or both ends by thiol or isocyanide groups [21]. Based on this work, a relatively simple model was presented to account for the observed conductance spectra of the junctions [22].

A similar experimental approach was employed by Bard and colleagues [24, 60] to study SAMs of a series of thiol end-capped OPEs and OPs with different substitution patterns (**85–87**). Owing to the highly conjugated, linear and rigid backbone of OPE molecules, these are considered prime candidates for use as molecular wires in molecular-scale applications. Using a tuning fork-based STM [60b], a series of sharp peaks were observed in the $I(V)$ curves in the OPE junctions, which were attributed to NDR. Another observation based on this work reported the ability of nitro-substituted OPEs to trap charges inside the molecular monolayer.

Fig. 10.3 Molecular junction of donor–π-acceptor molecule **88**.

SAMs of a molecular rectifier candidate – donor–π-acceptor molecule **88** – were investigated with an alkanethiolate-covered STM probe by Ashwell et al. [25]. The molecular junction shown in Fig. 10.3 demonstrated highly asymmetric $I(V)$ characteristics, which was taken as evidence for rectification of the monolayer. Depending on the experimental details, rectification ratios up to around 18 was observed. Reversible protonation and deprotonation of the donor moiety, $C_6H_4N(CH_3)_2$, in acidic or basic media, respectively, switched the rectification capability of the monolayer OFF and ON. This observation strongly indicates that the rectification behavior is an inherent property of the molecule.

Weiss and coworkers [27, 28, 61] used an STM probe to study the conductance of mixed SAMs of n-alkanethiols and thiol end-capped OPEs. By using a large excess of the alkanethiol, single OPE molecules **84a,b,d** and **e** could be isolated in a two-dimensional matrix of "non-conducting" alkanethiolates. An illustration of these mixed monolayers is shown in Fig. 10.4. Enhanced tunneling current, relative to the alkanethiolate background, was observed at sites that corresponded to the position of an OPE molecule [28]. This observation clearly supports the notion that π-conjugated molecules are better conductors than saturated molecules of similar length.

A surprising conductance switching between an ON and an OFF state was observed over time for single OPE molecules in the alkanethiolate matrix [27]. This switching behavior was first attributed to conformational changes in the OPE

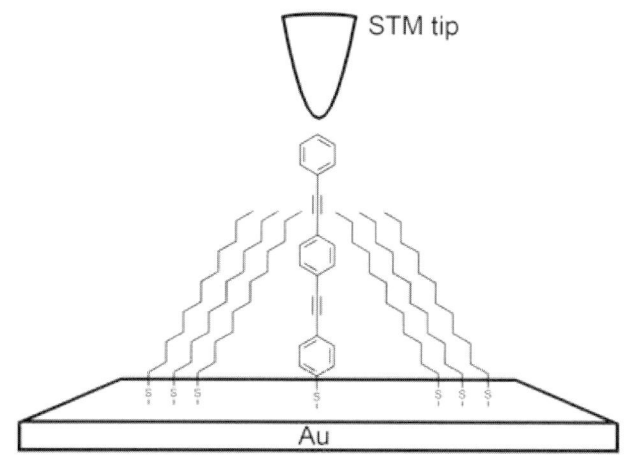

Fig. 10.4 Mixed monolayer of OPE and alkanethiol.

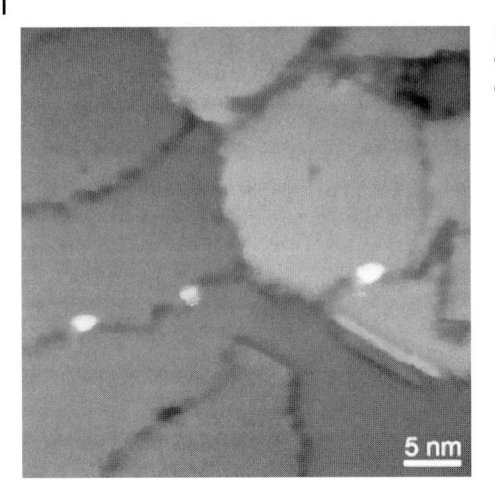

Fig. 10.5. STM micrograph of individual OPV molecules **91b** inserted in a SAM of dodecanethiol. Adapted from [19].

backbone. Later work, by Lindsay and coworkers [62], indicated that such conductance switching behavior was caused – at least in part –by a stochastic breaking and reformation of the Au–S bond in the SAMs.

The concept of making hybrid monolayers to isolate single molecules was later adopted by Blum and coworkers [29, 63] to compare the conductivity of an isolated OPE molecule (**95**) to that of crystalline SAMs of identical OPE molecules. The reported gap resistances were almost identical, which indicated that electron transport in the molecular junction occurred through individual molecules, even in crystalline self-assembled monolayers, as opposed to through intermolecular electron hopping.

A comparative study of the conductance through an OPV dithiol (**96**), an OPE dithiol (**95**) and a dodecanedithiol was also described. The observed order of conductance in these three molecules decreased in the order OPV > OPE > dodecanedithiol. A similar study performed by Bjørnholm and coworkers [19] involved a series of OPVs (**90–94**) with different lengths, functionalizations and substitution patterns embedded in a matrix of alkanethiol. Figure 10.5 shows an STM micrograph of individual OPV molecules **91b** inserted in a SAM of dodecanethiol.

The study showed that the insertion of a methylene spacer between the sulfur atom and the π-conjugated system (**90**) clearly reduces charge transport through the molecule compared with **91a**. Nitro substitution in molecule **92** also reduced the conductance considerably compared with the unsubstituted OPV **91b**. However, the difference in conductance between OPVs that were para or meta coupled through the central benzene in **92** and **93** was surprisingly small.

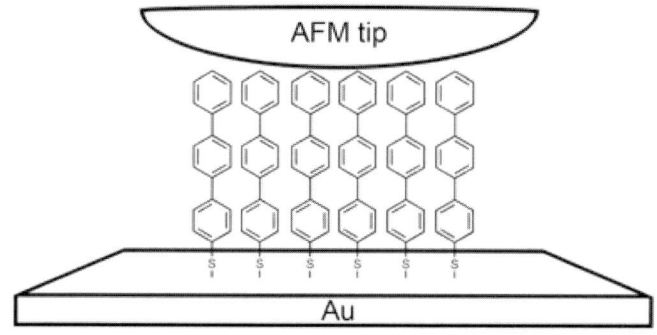

Fig. 10.6 CP-AFM-based molecular junction.

10.3.1.2 Conducting-probe Atomic Force Microscopy

Two groups from the universities in Minnesota [30, 64] and Arizona [31, 32, 62, 65] used conducting-probe atomic force microscopy (CP-AFM) to study the electronic transport properties of molecular SAMs on gold. CP-AFM (Fig. 10.6) utilizes a gold-coated AFM tip to make direct contact with the monolayer while simultaneously measuring the $I(V)$ characteristics of the tip-molecule(s)–substrate junction. The studies encompassed several alkanethiols, simple OPs **98a–c** and OPEs **95** and **99** as well as a thiol end-capped carotene molecule (**100**) on gold substrates. Despite having similar thicknesses, junctions that were based on benzylthiol **98a** showed resistances that were about 10 times lower than observed for junctions based on hexanethiols.

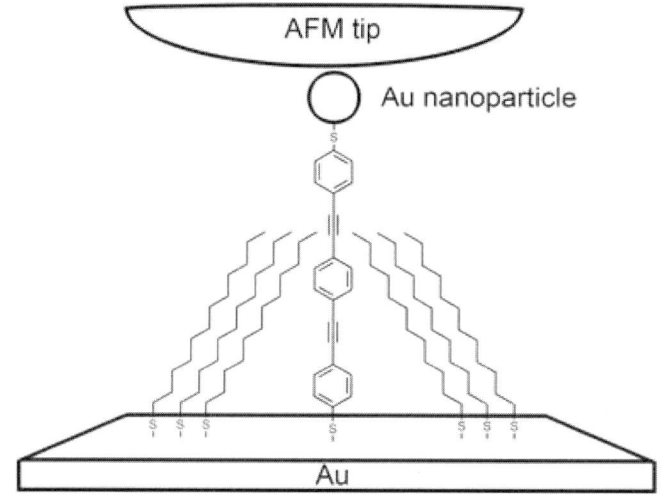

Fig. 10.7 Isolated substrate–molecule–particle junction.

A general challenge for these types of measurements relates to determining the number of molecules in a single junction. This issue was addressed by Lindsay and coworkers [65b], who attached small Au nanoparticles to the free end of dithiol molecules embedded in a SAM of monothiol molecules. This approach, illustrated in Fig. 10.7, produced isolated substrate-molecule-particle junctions that could be addressed with the conducting AFM probe. This procedure ensured that only a small – and countable – number of molecules were addressed at a time. Covalently attaching both ends of the molecule to a gold contact also generally results in a significantly higher conductance of the molecular junction compared with the situation where only one end of the molecule is chemically bound to an electrode.

10.3.1.3 Solution-phase Molecular STM Junctions

STM measurements on molecular junctions in solution have been realized by groups in Miami [66], Lyngby [34] and Liverpool [33, 67]. Performing the STM experiments in a liquid environment provides a way to combine well-established electrochemical techniques with *in situ* electronic STM characterization of a single or a small number of molecules. One particular advantage of this approach is the option to use an electrochemical reference electrode as a third electrode (gate) in the setup in addition to the tip and substrate electrodes (source and drain). Such a setup closely resembles a (three-terminal) transistor setup, as the third electrode can be used to manipulate the transmission properties of the junction molecules by applying a potential between the substrate and the reference electrode.

Schiffrin and coworkers [33] trapped a single redox-active molecule bearing two thiol end-groups (**101**) between a gold STM tip and a gold substrate in an aqueous solution. The redox-active part of the molecule (a viologen) could be reduced electrochemically by varying the potential, which resulted in a noticeable change (by a factor of ~5) of the conductivity through the junction.

Xu and coworkers [35, 36, 68] used a combination of STM and break junction approaches to form a statistically significant number of molecular junctions (>1000) for each molecule. A gold STM tip was repeatedly moved in and out of contact with a gold substrate in a solution containing either alkanedithiols, bipyridine (**104**) or dithiols (**103, 83a** and **82**). As the tip was moved slowly away from the gold surface, a chain of Au atoms would form between the tip and the substrate. This chain was broken as the tip was pulled further away and a molecule from the solution could bridge the gap while a break junction was formed. The junction conductance was monitored throughout the withdrawal. A conductance histogram showed discrete steps in conductance related to (1) the formation of the Au chain, followed by (2) the formation of the molecular junction and finally (3) an open gap.

Recently, this method has been modified to accommodate also an electrochemical gate [68] in the experimental setup – a situation that resembles a three-terminal device. This setup was used to study the electronic transmission properties of the redox-active molecule **103** – a perylene tetracarboxylic diimide (PTCDI) – as a function of gate voltage. A large enhancement of the conductance was observed at increased gate bias. This effect was not observed for non-redox molecules, such as alkanethiols or bipyridine (**104**).

10.3.1.4 Break Junctions

A break junction is formed by breaking a thin metallic wire to produce a narrow gap between two conductors. Bridging this gap by a single or a few molecules creates a metal–molecule–metal junction, as illustrated in Fig. 10.8. The metallic wire can be broken by mechanical deformation (mechanically controlled break junctions, MCBs) or by electromigration.

The first molecular measurement using MCBs was reported in 1997 by Reed and coworkers [37], who measured the conductance of a single benzene-1,4-dithiol molecule (**83a**). This work was followed in 1999 by measurements on oligothiophene **105** by Bourgoin and coworkers [38].

One advantage of the MCB technique, as opposed to the STM-based techniques, is the possibility of forming almost completely symmetrical molecular junctions since both electrodes are identical. A group in Karlsruhe [13g, 18, 39, 69] used MBCs to measure the conductance of molecules **106** and **108** both at room temperature [39] and at low temperature (~30 K) [69c]. $I(V)$ curves for the symmetric molecule **106** were symmetric with respect to voltage inversion, whereas the corresponding $I(V)$ curves for the asymmetric molecule **108** were clearly nonsymmetrical. Molecule **107** with the thiol anchor groups in the meta position was also measured in a MCB setup [13g] and the results were compared with those of the para-coupled molecule **106**. The currents measured for **107** were almost two orders of magnitude smaller than values recorded for **106**. Hence the lack of a fully conjugated path through molecule **107** noticeably lowers the junction conductance.

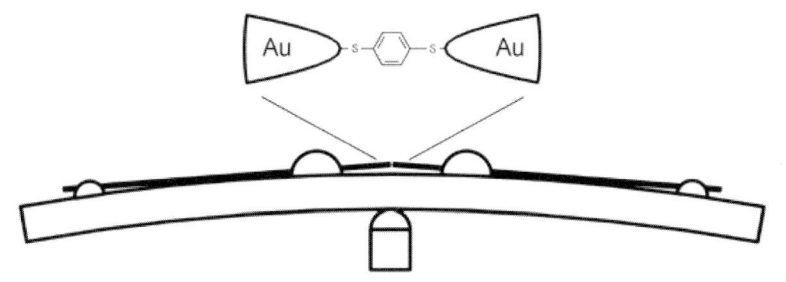

Fig. 10.8 Metal–molecule–metal junction.

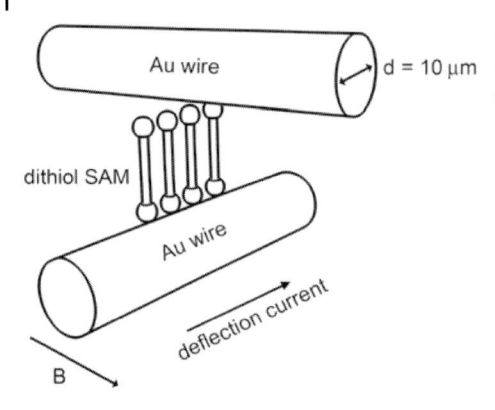

Fig. 10.9 SAM-covered wire in the vicinity of a second wire in a crossed-wire geometry.

10.3.1.5 Crossed Wires

Kushmerick and coworkers [40–42, 70] developed an elegant way to sandwich around 1000 molecules in a molecular junction. An Au wire with a diameter of 10 μm was modified with a SAM of the molecules of interest. The SAM-covered wire was brought into the vicinity of a second 10 μm diameter wire in a crossed-wire geometry (Fig. 10.9). A molecular junction was formed by applying a small d.c. current across the first wire in a magnetic field, which caused the wire to deflect. Upon touching the other wire, the $I(V)$ characteristics of the molecular junction could be measured.

The crossed-wire junction method was used to measure charge transport through OPE **95** and OPV **110**. The order of conductance was again found to follow the series OPV > OPE > dodecanedithiol [41], despite disruption of conjugation in **110** by the two CH_2 units. Oriented SAMs of molecule **112** displayed a slight rectification of the tunnel current when measured in a crossed-wire geometry [42], thus reflecting the asymmetry of the molecule.

10.3.1.6 Nanopore Junctions

Nanopore junctions, shown in Fig. 10.10, were investigated by Reed and coworkers [43–45, 71]. A small hole, 30–50 nm in diameter, was etched in a silicon nitride membrane by e-beam lithography. One side of the hole was filled with evaporated gold and a SAM (consisting of ~1000 molecules) was formed inside the hole when placed in a solution. A second Au electrode was then deposited on the other side of the hole by evaporation.

A sharp peak at ~2 V, demonstrating a clear NDR effect, was observed in the $I(V)$ characteristics of a nanopore junction containing substituted OPE molecule **84d** when measured at 60 K [44]. Later studies, involving nitro- and amino-substituted OPE-based nanopores, also demonstrated a controllable memory effect of the junction [71b]. This effect was based on the switching between a high and a low conductivity state of the OPE, which could be controlled by the bias voltage.

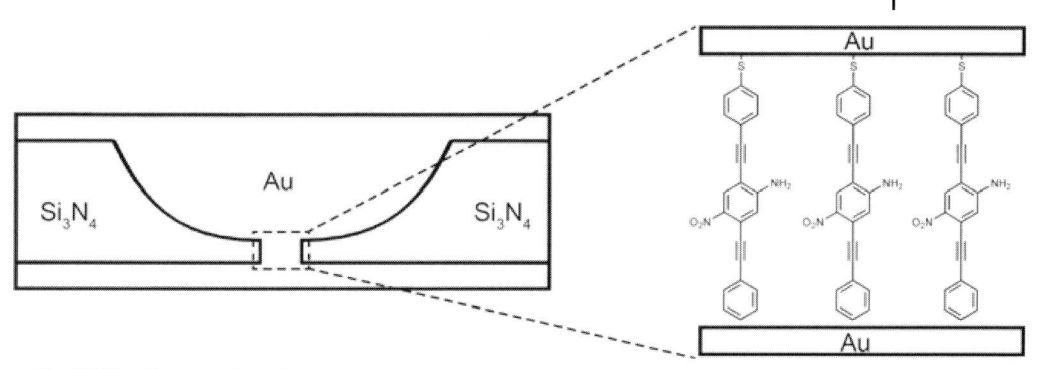

Fig. 10.10 Nanopore junction.

An early nanopore study focused on an asymmetric Au–molecule–Ti junction based on thiol end-capped biphenyl **81b** molecules [45]. The asymmetry of the structure led to the observation of a prominent rectifying behavior with larger current when the Ti electrode was negatively biased. Recent work by Bao and coworkers [72] has shown that vapor deposition of Ti on SAMs results in penetration of the monolayer, thus destroying it. Similar observations were made using Au and Al deposition. However, destruction of the monolayer could in this case be prevented if SAMs of dithiols were used, since the Au or Al would react with the free thiol end.

10.3.1.7 Square-tip Junctions

Scientists at Bell Laboratories [46, 73] used shadow angle evaporation to deposit metallic electrodes on different faces of a square quartz tip, with a diameter of 20–30 nm, as illustrated in Fig. 10.11. The square shape of the tip is thought to reduce the risk of shorts produced by the evaporation process. The fabrication method was divided into several steps. First, an Al gate was evaporated onto one

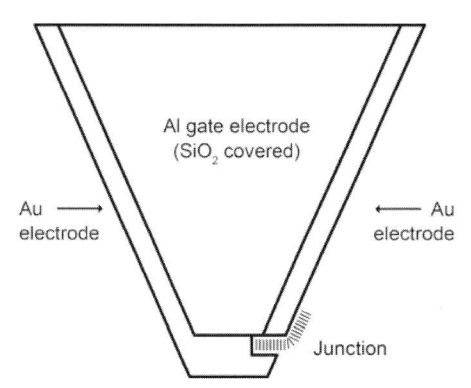

Fig. 10.11 Metallic electrodes deposited on different faces of a square quartz tip.

front face of the tip and isolated with a layer of SiO_2. This process was followed by the addition of contact pads and external wiring to the sides. An Au drain-electrode was then evaporated on to one side of the tip and a self-assembled monolayer was deposited on this electrode. Finally the Au source-electrode was evaporated on to the opposite side of the tip and on to the end of the tip until it contacted the SAM on the drain electrode.

Thiol end-capped oligothiophenes **105a–c** were used to form the SAMs between the electrodes. A series of distinct periodic steps in the conductance was observed for all samples at low temperature (<100 K). These features were suggested to originate from vibrational modes in the molecules. A (weakly coupled) gate potential could be applied to the molecular junction, which shifted the step position in the $I(V)$ curves but not the step widths. This observation was taken as an indication that only a single molecule was electrically active in the molecular junction.

10.3.1.8 Mercury Drop Junctions

The groups of Majda [74] and Whitesides [47, 48, 75] have both developed metal–SAM/SAM–metal junctions using Hg drops as one or both electrodes (Fig. 10.12). In a mercury drop experiment, both electrodes are covered with a SAM and immersed in a solution. The contact area between the two SAMs is fairly large – up to 1 mm in diameter, which corresponds to ca. 10^{11}–10^{13} molecules. Yet, despite producing nowhere near single molecular junctions, the ease of fabrication and versatility of this method make it an excellent candidate as a standard laboratory test-bed and β-values for several π-conjugated molecules have been determined in this way.

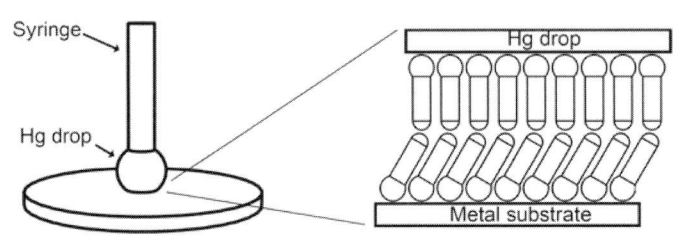

Fig. 10.12 Mercury drop junction.

10.3.1.9 Particle Junctions

Amlani et al. [49] combined conventional photolithographic techniques with self-assembly aspects to form a metal–SAM–metal–SAM–metal junction. Au-covered electrodes with a separation of 40–100 nm were covered with SAMs of OPEs or alkanethiols and an Au nanoparticle (d = 40–100 nm) was trapped in the gap between the electrodes by applying an alternating-current bias. An illustration of the system is shown in Fig. 10.13.

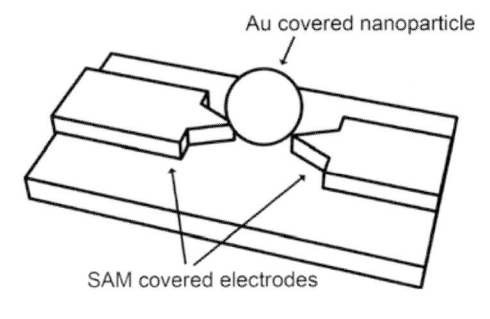

Fig. 10.13 Metal–SAM–metal–SAM–metal junction.

$I(V)$ measurements on nitro-substituted OPE **84b** revealed two NDR peaks at ~1.3 and ~1.7 V, as might be expected for two molecular junctions in a series. Analogous measurements on the unsubstituted OPE **95** showed no NDR peaks.

A conceptually similar approach was recently developed by Kushmerick and coworkers [50], who used Au-covered magnetic particles to effect directed assembly into lithographically defined magnetic arrays that were functionalized with SAMs of molecule **91b**. The magnetic particles were made by depositing first Ni and then Au on a silica microsphere.

10.3.1.10 Nanowire Junctions

A method to form metal–SAM–metal nanowires with a diameter < 40 nm was developed by Mallouk and coworkers [51, 76]. The nanowires were produced by electrodeposition of Au or Pd into the nanopores of a polycarbonate membrane. A SAM was formed at the end of the wire and a second metal contact (Au, Ag or Pd) was deposited on top of this. The polycarbonate was subsequently dissolved in dichloromethane, which released a large quantity (10^{11} cm^{-2}) of nanowires that could be aligned individually between pairs of lithographically fabricated metal electrodes. A schematic illustration of the nanowire molecular junctions is shown in Fig. 10.14.

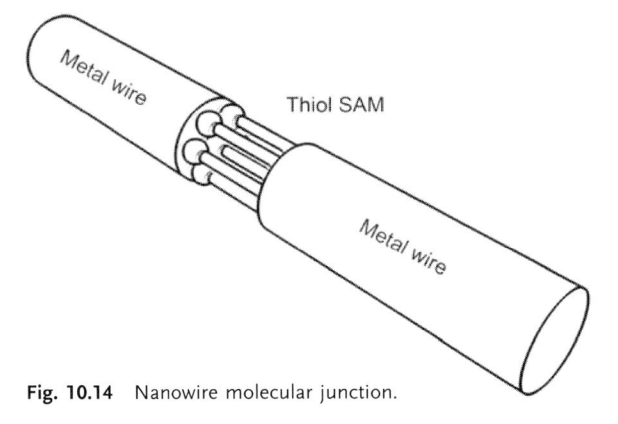

Fig. 10.14 Nanowire molecular junction.

Using the nanowire junction, a comparative study of OPV **96**, OPE **95** and dodecanedithiol was performed [51]. The conductances of the OPV and OPE molecules were 2–3 orders of magnitude higher than for the dodecanedithiol. Using different combinations of metals for the top and bottom electrodes showed that a symmetrical Pd–SAM–Pd junction produced the highest conductance, presumably owing to better metal–molecule coupling. In a related study, nanowire junctions incorporating the nitro-substituted OPE molecule **84b** demonstrated NDR at room-temperature [76c].

10.3.1.11 Three-terminal Single-molecule Transistors

The addition of a third (gate) electrode to a molecular junction provides a way to adjust the electronic levels of the molecules relative to the Fermi energies of the electrodes independently from the bias voltage. A gate can be used to oxidize and reduce molecules or to tune the molecules between conducting and nonconducting states. This ability is central for a thorough study of molecular conductance, including an identification of the transport mechanism and for the integration of molecular devices in circuits [77].

Current fabrication schemes for solid-state three-terminal molecular junctions typically rely on electromigration [78] to produce a nanoscale gap between two metallic electrodes which rests on top of a gate electrode, separated by a thin oxide layer. A schematic representation of a three-terminal device is shown in Fig. 10.15.

Experimental studies of single-molecule transistors have not been reported until relatively recently and, owing to the considerable technical challenges associated with the fabrication of such devices, only a moderate number of scientific publications have emerged in this field so far. Early measurements using this method focused on the Kondo effect [52, 79] and observations of single-electron phenomena, such as Coulomb blockades [80]. In Coulomb blockade behavior, the low-bias conductance is suppressed owing to capacitive charging of a molecule (or small atomic island), whereas the Kondo effect is related to a change in conductivity affected by the spin state of an impurity (i.e. the molecule). Both of these effects could be tuned by the gate voltage.

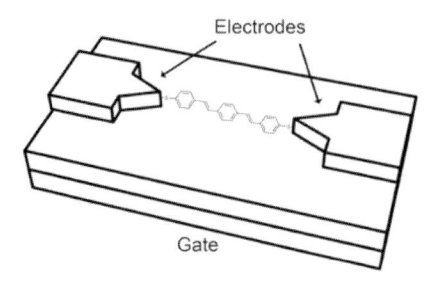

Fig. 10.15 Three-terminal device.

Later work by Kubatkin et al. [53] focused on the OPV molecule **117**, which was physisorbed on the electrodes in a single-electron transistor, leading only to weak van der Waals contact between the single molecule and the device. By varying the gate voltage, eight distinct transmitting redox states of the molecule could be probed. However, the energies of these states were highly perturbed compared with the corresponding energies measured in solution. This perturbation was explained as an effect of the metal–molecule contact, where image charges in the metal electrodes strongly altered the electronic levels in the molecule [53b], and illustrates that molecular properties cannot be decoupled from the properties of the electrodes in a molecular junction.

10.4
Summary and Outlook

A central issue in molecular-scale electronics is to relate the electronic behavior of the system to the structure of the molecules being probed. This involves systematically varying the properties by changing the molecular structure (i.e. length, conjugation, substituents, number of thiol groups, etc.). The field of molecular electronics is in this respect strongly fuelled by organic synthesis and relies crucially on the ability to produce well-defined molecular species. However, in attempting to rationalize what influence the molecular structure has on the properties of metal–molecule–metal junctions, one is confronted with the evident lack of statistical material. Furthermore, as most of the reported molecular junctions are made one at a time, it can be difficult to rationalize how consistent their behavior is. Continuous investigations of further molecules will be necessary to provide larger libraries for determining the fundamental principles governing electronic transport in single molecules. These issues aside, from the data collected in Table 10.1 one can nevertheless extract some important structure–property rela-

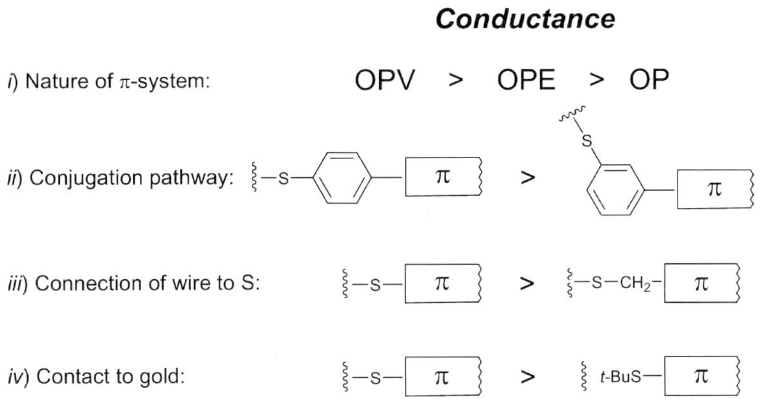

Fig. 10.16 Schematic illustration of how the conductance qualitatively changes as a function of molecular structure.

tionships as illustrated schematically in Fig. 10.16. Thus, the conductance increases: (i) along the progression OP, OPE, OPV, (ii) when sulfur is situated para to the π-system rather than meta,(iii) in the absence of insulating methylene bridges and (iv) when sulfur is directly adhered to gold.

Once the basic concepts governing the behavior of single-molecule electronics have been successfully harnessed, a next step should target the incorporation of these systems into large integrated circuit structures. One possible strategy towards this objective includes the "bottom-up" self-assembly of supramolecular architectures consisting of functional single-molecule building-blocks [2h]. The interconnections between each functional unit could, for example, be facilitated by self-assembled metallic nanowires [81] or arrays of metal islands [82] connected to macroscopic electrodes in hybrid semiconductor–nanodevice circuits. Another strategy could be the integration of several individual molecular functions into the same molecule [77, 83] in a true monomolecular electronic device.

10.5
Experimental

General procedure for conversion of ArSt-Bu to ArSAc
To a solution of the ArSt-Bu (2 mmol) and AcCl (1 mL) in CH_2Cl_2 (20 mL) was added BBr_3 (1.0 M solution in CH_2Cl_2, 2.2 mL, 2.2 mmol). After stirring for 2 h at room temperature, the mixture was poured into ice and worked up in the usual manner [10].

Synthesis of 4-bromophenyl tert-butyl sulfide (**3**)
To a slurry of 4-bromothiophenol (100 g, 0.53 mmol) in tert-butyl chloride (400 mL) was added $AlCl_3$ (3.5 g, 0.03 mol) in small portions. The reaction became vigorously foaming and evolved HCl that was led through solid NaOH. Stirring at room temperature was maintained for 30 min, whereupon the reaction mixture was poured into H_2O (700 mL) and extracted with pentane (3 × 150 mL). The combined organic phase was washed with H_2O (40 mL), dried ($MgSO_4$), filtered and concentrated in vacuo. The residual oil was purified by distillation (b.p. 93–95 °C/12 Pa) [10b].

General procedure for coupling of a terminal alkyne with an aryl halide
The reaction is performed under an atmosphere of nitrogen. The aryl halide, the Pd catalyst, e.g. [$PdCl_2(PPh_3)_2$] (3–5 mol% per halide) and CuI (6–10 mol% per halide) were dissolved in THF and/or benzene and/or CH_2Cl_2. Then NEt_3 or $HN(i-Pr)_2$ was added and subsequently the terminal alkyne (1–1.5 mol% per halide), whereupon the reaction was heated until complete, then it was quenched with H_2O and a saturated aqueous solution of NH_4Cl. The organic layer was subjected to the usual extraction followed by column chromatography [13].

Synthesis of OPV3 **38** by the Horner–Wadsworth–Emmons reaction

To a solution of aldehyde **6** (7.77 g, 40 mmol) and diphosphonate **37** (7.57 g, 20 mmol) in THF (200 mL) cooled in an ice-bath was added *t*-BuOK (4.94 g, 44 mmol) in small portions during 10 min. After stirring for 6 h at room temperature under a nitrogen atmosphere, the mixture was poured into H_2O (300 mL). The yellow precipitate was filtered off, washed with H_2O and dried. The product was dissolved in a minimum amount of boiling THF containing I_2 (0.1 mM). The mixture was refluxed for 12 h and then slowly cooled to room temperature. The pure *trans*-stilbene **38** crystallized [10a].

References

1. (a) M. Schulz, *Nature* **1999**, *399*, 729–730; (b) T. Ito, S. Okazaki, *Nature* **2000**, *406*, 1027–1031.
2. (a) W. A. Reinerth, L. Jones, II, T. P. Burgin, C. Zhou, C. J. Muller, M. R. Deshpande, M. A. Reed, J. M. Tour, *Nanotechnology* **1988**, *9*, 246–250; (b) J. M. Tour *Acc. Chem. Res.* **2000**, *33*, 791–804; (c) R. L. Carroll, C. B. Gorman, *Angew. Chem. Int. Ed.* **2002**, *41*, 4378–4400; (d) N. Robertson, C. A. McGowan, *Chem. Soc. Rev.* **2003**, *32*, 96–103; (e) A. M. Rawlett, T. J. Hopson, I. Amlani, R. Zhang, J. Tresek, L. A. Nagahara, R. K. Tsui, H. Goronkin, *Nanotechnology* **2003**, *14*, 377–384; (f) G. Marruccio, R. Cingolani, R. Rinaldi, *J. Mater. Chem.* **2004**, *14*, 542–554; (g) A. H. Flood, J. F. Stoddart, D. W. Steuerman, J. R. Heath, *Science* **2004**, *306*, 2055–2056; (h) K. Nørgaard, T. Bjørnholm, *Chem. Commun.* **2005**, 1812–1823.
3. A. Aviram, M. Ratner, *Chem. Phys. Lett.* **1974**, *29*, 277.
4. B. Mann, H. Kuhn, *J. Appl. Phys.* **1971**, *42*, 4398.
5. R. G. Nuzzo, D. L. Allara, *J. Am. Chem. Soc.* **1983**, *105*, 4481.
6. A. Ulman, *Chem. Rev.* **1996**, *96*, 1533–1554.
7. J. C. Love, L. A. Estroff, J. K. Kriebel, R. G. Nuzzo, G. M. Whitesides, *Chem. Rev.* **2005**, *105*, 1103–1169.
8. R. F. Service, *Science*, **2001**, *294*, 2442–2443.
9. (a) R. P. Hsung, J. R. Babcock, C. E. D. Chidsey, L. R. Sita, *Tetrahedron Lett.* **1995**, *36*, 4525–4528; (b) Z. J. Donhauser, B. A. Mantooth, K. F. Kelly, L. A. Bumm, J. D. Monnell, J. J. Stapleton, D. W. Price, Jr., A. M. Rawlett, D. L. Allara, J. M. Tour, P. S. Weiss, *Science* **2001**, *292*, 2303–2307.
10. (a) N. Stuhr-Hansen, J. B. Christensen, N. Harrit, T. Bjørnholm, *J. Org. Chem.* **2003**, *68*, 1275–1282; (b) N. Stuhr-Hansen, *Synth. Commun.* **2003**, *33*, 641–646.
11. A. Blaszczyk, M. Elbing, M. Mayor, *Org. Biomol. Chem.* **2004**, *2*, 2722–2724.
12. (a) R. N. Young, J. Y. Gauthier, W. Coombs, *Tetrahedron Lett.* **1984**, *25*, 1753–1756; (b) C. J. Yu, Y. Chong, J. F. Kayyem, M. Gozin, *J. Org. Chem.* **1999**, *64*, 2070–2079; (c) R. C. Hicks, M. B. Nodwell, *J. Am. Chem. Soc.* **2000**, *122*, 6746–6753; (d) C. Wang, A. S. Batsanov, M. R. Bryce, I. Sage, *Org. Lett.* **2004**, *6*, 2181–2184.
13. (a) D. L. Pearson, J. M. Tour, *J. Org. Chem.* **1997**, *62*, 1376–1387; (b) D. T. Gryko, C. Clausen, J. S. Lindsey, *J. Org. Chem.* **1999**, *64*, 8635–8647; (c) J. M. Tour, A. M. Rawlett, M. Kozaki, Y. Yao, R. C. Jagessar, S. M. Dirk, D. W. Price, M. A. Reed, C.-W. Zhou, J. Chen, W. Wang, I. Campbell, *Chem. Eur. J.* **2001**, *7*, 5118–5134; (d) C. Hortholary, C. Coudret, *J. Org. Chem.* **2003**, *68*, 2167–2174; (e) A. K. Flatt, S. M. Dirk, J. S. Henderson, D. E. Shen, J. Su, M. E. Reed, J. M. Tour, *Tetrahedron* **2003**, *59*, 8555–8570; (f) C. Wang, A. S. Batsanov, M. R. Bryce, I. Sage, *Synthesis* **2003**, 2089–2095; (g) M. Mayor, H. B. Weber, J. Reichert, M. Elbing, C. von Hänisch,

D. Beckmann, M. Fischer, *Angew. Chem. Int. Ed.* **2003**, *42*, 5834–5838; (h) C. Xue, F.-T. Luo, *Tetrahedron* **2004**, *60*, 6285–6294.

14. G. J. Ashwell, W. D. Tyrrell, A. J. Whittam, *J. Am. Chem. Soc.* **2004**, *126*, 7102–7110.

15. (a) D. I. Gittins, D. Bethell, R. J. Nichols, D. J. Schiffrin, *Adv. Mater.* **1999**, *11*, 737–740; (b) W. Haiss, R. J. Nichols, S. J. Higgins, D. Bethell, H. Höbenreich, D. J. Schiffrin, *Faraday Discuss.* **2004**, *125*, 179–194.

16. B. Xu, X. Xiao, X. Yang, L. Zang, N. Tao, *J. Am. Chem. Soc.* **2005**, *127*, 2386–2387.

17. M. Maskus, H. Abruña, *Langmuir* **1996**, *12*, 4455–4462.

18. M. Mayor, C. von Hänisch, H. B. Weber, J. Reichert, D. Beckmann, *Angew. Chem. Int. Ed.* **2002**, *41*, 1183–1186.

19. K. Moth-Poulsen, L. Patrone, N. Stuhr-Hansen, J. B. Christensen, J.-P. Bourgoin, T. Bjørnholm, *Nano Lett.* **2005**, *5*, 783–785.

20. Y. Yao, J. M. Tour, *J. Org. Chem.* **1999**, *64*, 1968–1971.

21. S. Hong, R. Reifenberger, W. Tian, S. Datta, J. I. Henderson, C. P. Kubiak, *Superlattices Microstruct.* **2000**, *28*, 289–303.

22. W. D. Tian, S. Datta, S. H. Hong, R. Reifenberger, J. I. Henderson, C. P. Kubiak, *J. Chem. Phys.* **1998**, *109*, 2874–2882.

23. R. P. Andres, T. Bein, M. Dorogi, S. Feng, J. I. Henderson, C. P. Kubiak, W. Mahoney, R. G. Osifchin, R. Reifenberger, *Science* **1996**, *272*, 1323–1325.

24. F. R. F. Fan, J. P. Yang, L. T. Cai, D. W. Price, S. M. Dirk, D. V. Kosynkin, Y. X. Yao, A. M. Rawlett, J. M. Tour, A. J. Bard, *J. Am. Chem. Soc.* **2002**, *124*, 5550–5560.

25. G. J. Ashwell, W. D. Tyrrell, A. J. Whittam, *J. Am. Chem. Soc.* **2004**, *126*, 7102–7110.

26. A. S. Blum, J. G. Kushmerick, D. P. Long, C. H. Patterson, J. C. Yang, J. C. Henderson, Y. Yao, J. M. Tour, R. Shashidhar, B. Ratna, *Nat. Mater.* **2005**, *4*, 167–172.

27. Z. J. Donhauser, B. A. Mantooth, K. F. Kelly, L. A. Bumm, J. D. Monnell, J. J. Stapleton, D. W. Price, A. M. Rawlett, D. L. Allara, J. M. Tour, P. S. Weiss, *Science* **2001**, *292*, 2303–2307.

28. L. A. Bumm, J. J. Arnold, M. T. Cygan, T. D. Dunbar, T. P. Burgin, L. Jones, D. L. Allara, J. M. Tour, P. S. Weiss, *Science* **1996**, *271*, 1705–1707.

29. A. S. Blum, J. C. Yang, R. Shashidhar, B. Ratna, *Appl. Phys. Lett.* **2003**, *82*, 3322–3324.

30. D. J. Wold, R. Haag, M. A. Rampi, C. D. Frisbie, *J. Phys. Chem. B* **2002**, *106*, 2813–2816.

31. A. M. Rawlett, T. J. Hopson, L. A. Nagahara, R. K. Tsui, G. K. Ramachandran, S. M. Lindsay, *Appl. Phys. Lett.* **2002**, *81*, 3043–3045.

32. G. K. Ramachandran, J. K. Tomfohr, J. Li, O. F. Sankey, X. Zarate, A. Primak, Y. Terazono, T. A. Moore, A. L. Moore, D. Gust, L. A. Nagahara, S. M. Lindsay, *J. Phys. Chem. B* **2003**, *107*, 6162–6169.

33. (a) W. Haiss, H. van Zalinge, S. J. Higgins, D. Bethell, H. Höbenreich, D. J. Schiffrin, R. J. Nichols, *J. Am. Chem. Soc.* **2003**, *125*, 15294–15295; (b) W. Haiss, R. J. Nichols, H. van Zalinge, S. J. Higgins, D. Bethell, D. J. Schiffrin, *Phys. Chem. Chem. Phys.* **2004**, *6*, 4330–4337.

34. (a) T. Albrecht, K. Moth-Poulsen, J. B. Christensen, A. Guckian, T. Bjørnholm, J. G. Vos, J. Ulstrup, *Faraday Discuss.* **2005**, submitted. **2006**, *131*, 265–279; (b) T. Albrecht, K. Moth-Poulsen, J. B. Christensen, J. Hjelm, T. Bjørnholm, J. Ulstrup, *J. Am. Chem. Soc.* **2006**, *128*, 6545–6575.

35. B. Q. Xu, N. J. J. Tao, *Science* **2003**, *301*, 1221–1223.

36. X. Y. Xiao, B. Q. Xu, N. J. Tao, *Nano Lett.* **2004**, *4*, 267–271.

37. M. A. Reed, C. Zhou, C. J. Muller, T. P. Burgin, J. M. Tour, *Science* **1997**, *278*, 252–254.

38. C. Kergueris, J. P. Bourgoin, S. Palacin, D. Esteve, C. Urbina, M. Magoga, C. Joachim, *Phys. Rev. B* **1999**, *59*, 12505–12513.

39. J. Reichert, R. Ochs, D. Beckmann, H. B. Weber, M. Mayor, H. von Lohneysen, *Phys. Rev. Lett.* **2002**, *88*, 176804.

40. J. G. Kushmerick, D. B. Holt, J. C. Yang, J. Naciri, M. H. Moore, R. Shashidhar, *Phys. Rev. Lett.* **2002**, *89*, 086802.
41. J. G. Kushmerick, D. B. Holt, S. K. Pollack, M. A. Ratner, J. C. Yang, T. L. Schull, J. Naciri, M. H. Moore, R. Shashidhar, *J. Am. Chem. Soc.* **2002**, *124*, 10654–10655.
42. S. K. Pollack, J. Naciri, J. Mastrangelo, C. H. Patterson, J. Torres, M. Moore, R. Shashidhar, J. G. Kushmerick, *Langmuir* **2004**, *20*, 1838–1842.
43. J. Chen, W. Wang, M. A. Reed, A. M. Rawlett, D. W. Price, J. M. Tour, *Appl. Phys. Lett.* **2000**, *77*, 1224–1226.
44. J. Chen, M. A. Reed, A. M. Rawlett, J. M. Tour, *Science* **1999**, *286*, 1550–1552.
45. C. Zhou, M. R. Deshpande, M. A. Reed, L. Jones, J. M. Tour, *Appl. Phys. Lett.* **1997**, *71*, 611–613.
46. N. B. Zhitenev, H. Meng, Z. Bao, *Phys. Rev. Lett.* **2002**, *88*, 226801.
47. R. E. Holmlin, R. Haag, M. L. Chabinyc, R. F. Ismagilov, A. E. Cohen, A. Terfort, M. A. Rampi, G. M. Whitesides, *J. Am. Chem. Soc.* **2001**, *123*, 5075–5085.
48. R. Haag, M. A. Rampi, R. E. Holmlin, G. M. Whitesides, *J. Am. Chem. Soc.* **1999**, *121*, 7895–7906.
49. I. Amlani, A. M. Rawlett, L. A. Nagahara, R. K. Tsui, *Appl. Phys. Lett.* **2002**, *80*, 2761–2763.
50. D. P. Long, C. H. Patterson, M. H. Moore, D. S. Seferos, G. C. Bazan, J. G. Kushmerick, *Appl. Phys. Lett.* **2005**, *86*, 153105.
51. L. T. Cai, H. Skulason, J. G. Kushmerick, S. K. Pollack, J. Naciri, R. Shashidhar, D. L. Allara, T. E. Mallouk, T. S. Mayer, *J. Phys. Chem. B* **2004**, *108*, 2827–2832.
52. J. Park, A. N. Pasupathy, J. I. Goldsmith, C. Chang, Y. Yaish, J. R. Petta, M. Rinkoski, J. P. Sethna, H. D. Abruna, P. L. Mceuen, D. C. Ralph, *Nature* **2002**, *417*, 722–725.
53. (a) S. Kubatkin, A. Danilov, M. Hjort, J. Cornil, J. L. Brédas, N. Stuhr-Hansen, P. Hedegard, T. Bjørnholm, *Nature* **2003**, *425*, 698–701; (b) P. Hedegård, T. Bjørnholm, *Chem. Phys.* **2005**, *319*, 350–359; (c) A. Danilov, S. Kubatkin, S. G. Kafanov, K. Flensberg, T. Bjørnholm, *Nano Letters*, **2006**, in press.
54. L. H. Yu, Z. K. Keane, J. W. Ciszek, L. Cheng, M. P. Stewart, J. M. Tour, D. Natelson, *Phys. Rev. Lett.* **2004**, *93*, 266802.
55. H. S. J. van der Zant, Y.-V. Kervennic, M. Poot, K. O'Neill, Z. de Groot, J. M. Thijssen, H. B. Heershe, N. Stuhr-Hansen, T. Bjørnholm, D. Vanmaekelbergh, C. A. van Walree, L. W. Jenneskens, *Faraday Discuss.*, **2006**, *131*, 347–356.
56. S. Datta, *Electronic Transport in Mesoscopic Systems*, Cambridge University Press, Cambridge, **2001**.
57. G. Binnig, H, Rohrer, *Helv. Phys. Acta* **1982**, *55*, 726–735.
58. P. Samori, J. Rabe, *J. Phys.: Condens. Matter* **2002**, *14*, 9955–9973.
59. (a) S. Datta, W. D. Tian, S. H. Hong, R. Reifenberger, J. I. Henderson, C. P. Kubiak, *Phys. Rev. Lett.* **1997**, *79*, 2530–2533; (b) Y. Q. Xue, S. Datta, S. Hong, R. Reifenberger, J. I. Henderson, C. P. Kubiak, *Phys. Rev. B* **1999**, *59*, R7852–R7855.
60. (a) F. R. F. Fan, J. P. Yang, S. M. Dirk, D. W. Price, D. Kosynkin, J. M. Tour, A. J. Bard, *J. Am. Chem. Soc.* **2001**, *123*, 2454–2455; (b) F. R. F. Fan, R. Y. Lai, J. Cornil, Y. Karzazi, J. L. Bredas, L. T. Cai, L. Cheng, Y. X. Yao, D. W. Price, S. M. Dirk, J. M. Tour, A. J. Bard, *J. Am. Chem. Soc.* **2004**, *126*, 2568–2573.
61. (a) M. T. Cygan, T. D. Dunbar, J. J. Arnold, L. A. Bumm, N. F. Shedlock, T. P. Burgin, L. Jones, D. L. Allara, J. M. Tour, P. S. Weiss, *J. Am. Chem. Soc.* **1998**, *120*, 2721–2732; (b) L. A. Bumm, J. J. Arnold, T. D. Dunbar, D. L. Allara, P. S. Weiss, *J. Phys. Chem. B* **1999**, *103*, 8122–8127; (c) G. S. McCarty, P. S. Weiss, *Chem. Rev.* **1999**, *99*, 1983–1990.
62. G. K. Ramachandran, T. J. Hopson, A. M. Rawlett, L. A. Nagahara, A. Primak, S. M. Lindsay, *Science* **2003**, *300*, 1413–1416.
63. A. S. Blum, J. G. Kushmerick, S. K. Pollack, J. C. Yang, M. Moore, J. Naciri, R. Shashidhar, B. R. Ratna, *J. Phys. Chem. B* **2004**, *108*, 18124–18128.
64. (a) D. J. Wold, C. D. Frisbie, *J. Am. Chem. Soc.* **2000**, *122*, 2970–2971; (b) D. J. Wold, C. D. Frisbie, *J. Am. Chem. Soc.* **2001**, *123*, 5549–5556; (c) J. M. Beebe,

V. B. Engelkes, L. L. Miller, C. D. Frisbie, *J. Am. Chem. Soc.* **2002**, *124*, 11268–11269.

65. (a) G. Leatherman, E. N. Durantini, D. Gust, T. A. Moore, A. L. Moore, S. Stone, Z. Zhou, P. Rez, Y. Z. Liu, S. M. Lindsay, *J. Phys. Chem. B* **1999**, *103*, 4006–4010; (b) X. D. Cui, A. Primak, X. Zarate, J. Tomfohr, O. F. Sankey, A. L. Moore, T. A. Moore, D. Gust, G. Harris, S. M. Lindsay, *Science* **2001**, *294*, 571–574; (c) X. D. Cui, A. Primak, X. Zarate, J. Tomfohr, O. F. Sankey, A. L. Moore, T. A. Moore, D. Gust, L. A. Nagahara, S. M. Lindsay, *J. Phys. Chem. B* **2002**, *106*, 8609–8614.
66. N. J. Tao, *Phys. Rev. Lett.* **1996**, *76*, 4066–4069.
67. D. I. Gittins, D. Bethell, D. J. Schiffrin, R. J. Nichols, *Nature* **2000**, *408*, 67–69.
68. B. Q. Xu, X. Y. Xiao, X. M. Yang, L. Zang, N. J. Tao, *J. Am. Chem. Soc.* **2005**, *127*, 2386–2387.
69. (a) H. B. Weber, J. Reichert, F. Weigend, R. Ochs, D. Beckmann, M. Mayor, R. Ahlrichs, H. von Lohneysen, *Chem. Phys.* **2002**, *281*, 113–125; (b) H. B. Weber, J. Reichert, R. Ochs, D. Beckmann, M. Mayor, H. von Lohneysen, *Physica E* **2003**, *18*, 231–232; (c) J. Reichert, H. B. Weber, M. Mayor, H. v. Löhneysen, *Appl. Phys. Lett.*, **2003**, *82*, 4137–4139.
70. J. G. Kushmerick, J. Lazorcik, C. H. Patterson, R. Shashidhar, D. S. Seferos, G. C. Bazan, *Nano Lett.* **2004**, *4*, 639–642.
71. (a) J. Chen, L. C. Calvet, M. A. Reed, D. W. Carr, D. S. Grubisha, D. W. Bennett, *Chem. Phys. Lett.* **1999**, *313*, 741–748; (b) M. A. Reed, J. Chen, A. W. Rawlett, D. W. Price, J. M. Tour, *Appl. Phys. Lett.* **2001**, *78*, 3735–3737.
72. B. de Boer, M. M. Frank, Y. J. Chabal, W. Jiang, E. Garfunkel, Z. Bao, *Langmuir* **2004**, *20*, 1539–1542.
73. N. B. Zhitenev, A. Erbe, Z. Bao, *Phys. Rev. Lett.* **2004**, *92*, 186805.
74. (a) K. Slowinski, H. K. Y. Fong, M. Majda, *J. Am. Chem. Soc.* **1999**, *121*, 7257–7261; (b) K. Slowinski, M. Majda, *J. Electroanal. Chem.* **2000**, *491*, 139–147.
75. M. A. Rampi, G. M. Whitesides, *Chem. Phys.* **2002**, *281*, 373–391.
76. (a) N. I. Kovtyukhova, B. R. Martin, J. K. N. Mbindyo, P. A. Smith, B. Razavi, T. S. Mayer, T. E. Mallouk, *J. Phys. Chem. B* **2001**, *105*, 8762–8769; (b) J. K. N. Mbindyo, T. E. Mallouk, J. B. Mattzela, I. Kratochvilova, B. Razavi, T. N. Jackson, T. S. Mayer, *J. Am. Chem. Soc.* **2002**, *124*, 4020–4026; (c) I. Kratochvilova, M. Kocirik, A. Zambova, J. Mbindyo, T. E. Mallouk, T. S. Mayer, *J. Mater. Chem.* **2002**, *12*, 2927–2930.
77. C. Joachim, J. K. Gimzewski, A. Aviram, *Nature* **2000**, *408*, 541–548.
78. H. Park, A. K. L. Lim, A. P. Alivisatos, J. Park, P. L. McEuen, *Appl. Phys. Lett.* **1999**, *75*, 301–303.
79. W. J. Liang, M. P. Shores, M. Bockrath, J. R. Long, H. Park, *Nature* **2002**, *417*, 725–729.
80. H. Park, J. Park, A. K. L. Lim, E. H. Anderson, A. P. Alivisatos, P. L. McEuen, *Nature* **2000**, *407*, 57–60.
81. (a) T. Hassenkam, K. Nørgaard, L. Iversen, C. J. Kiely, M. Brust, T. Bjørnholm, *Adv. Mater.* **2002**, *14*, 1126–1130. (b) T. Hassenkam, K. Moth-Poulsen, N. Stuhr-Hansen, K. Nørgaard, M. S. Kabir, T. Bjørnholm, *Nano Lett.* **2004**, *4*, 19–22.
82. J. M. Tour, L. Cheng, D. P. Nackashi, Y. Yao, A. K. Flatt, S. K. St. Angelo, T. E. Mallouk, P. D. Franzon, *J. Am. Chem. Soc.* **2003**, *125*, 13279–13283.
83. (a) C. Joachim, *Superlattices Microstruct.* **2000**, *28*, 305–315; (b) C. Joachim, *Nanotechnology* **2002**, *13*, R1-R7.

11
Nonlinear Optical Properties of Organic Materials
Stephen Barlow and Seth R. Marder

11.1
Introduction to Nonlinear Optics

11.1.1
Introduction

In emerging photonic technologies for use in areas such as telecommunications, sensors and information technology, light is used as the carrier of information; optical waveguides including optical fibers are used to modulate and transport this light. For many applications, "passive" materials are used to transport light; the properties of these materials are not altered by the light, nor, to a large extent, does the material alter the properties of the light. However, to modulate the propagation characteristics of light it is necessary to use "active" nonlinear optical (NLO) materials in which the optical properties of the material can be modified by application of an electric field (such as the electric field associated with light). NLO phenomena include a range of interactions that light can have with a material. These include sum and difference frequency generation (including frequency doubling and tripling), the electrooptic (EO) effect (in which the refractive index of the material depends on an electric field) and self-focusing and defocusing effects (in which the refractive index of the material depends on the light intensity). These effects may be used to control the phase, polarization state and frequency of light beams, with a variety of applications in telecommunications and other photonic technologies.

An additional class of nonlinear optical effects is that of multi-photon absorption processes. Using these process, one can create excited states (and, therefore, their associated physical and chemical properties) with a high degree of three-dimensional (3D) spatial confinement, at depth in absorbing media. There are potential applications of multi-photon absorbing materials in 3D fluorescence imaging, photodynamic therapy, nonlinear optical transmission and 3D microfabrication.

This chapter gives an overview of the origin of NLO effects and focuses on the design of chromophores for two current applications of technological interest: poled polymer materials for electrooptic switching and two-photon absorption.

NLO effects arise from the interaction of light with loosely held electrons, typically, in the case of organic materials, those associated with conjugated π-systems. In the following section we examine the details of this interaction.

11.1.2
Linear and Nonlinear Polarization

Nonlinear optical properties are associated with the ability of a material to undergo nonlinear polarization under the influence of electric fields (either static fields or the oscillating fields associated with electromagnetic radiation). To understand nonlinear polarization, it is helpful first to consider linear polarization of a material. Application of an electric field E to a system of charges will result in a charge separation; in the example of a conjugated organic molecule, this charge separation will principally take place within the more polarizable π-system. If we consider only linear polarization, then the polarization induced by an electric field E, i.e. the induced dipole moment, μ, is proportional to the field:

$$(\text{polarization})_i = \mu_i = \alpha_{ij} E_j \tag{1}$$

where i and j refer to components in the molecular frame. The constant of proportionality, α_{ij}, is the relevant component of the linear polarizability, α, which is a second-rank tensor. For example, the component of the dipole moment induced in the x direction by an electric field oriented along the x direction is determined by α_{xx}, whereas the component induced along the y-axis is determined by α_{yx} and so on. In a bulk material composed of these molecules, the linear polarization per unit volume is given by an analogous equation:

$$P_i = \chi_{ij} E_j \tag{2}$$

where χ_{ij} is the linear susceptibility tensor of the material and, in the absence of significant intermolecular interactions, is related to the sum of all the individual polarizabilities, α_{ij}. Equations (1) and (2) are reasonable approximations at small electric fields (as shown in Fig. 11.1). However, in general, polarization is not a linear function of field and nonlinearities become more significant at increasingly intense fields, such as the intense fields associated with laser light; indeed, the observation of NLO effects has been greatly aided by the availability of lasers. The generalized nonlinear dependence of dipole moment on field can be expressed as a Taylor expansion series:

$$\mu_i(E) = \mu_i(0) + \frac{\partial \mu_i}{\partial E_j}\bigg|_0 E_j + \frac{1}{2!}\frac{\partial^2 \mu_i}{\partial E_j \partial E_k}\bigg|_0 E_j E_k + \frac{1}{3!}\frac{\partial^3 \mu_i}{\partial E_j \partial E_k \partial E_l}\bigg|_0 E_j E_k E_l + \ldots \tag{3}$$

where the 0 subscripts indicate the values of the differentials at $E = 0$ and i, j, ... refer to components in the molecular frame. The first term, $\mu_i(0)$, is the i component of the dipole moment of the molecule in the absence of any electric field. The second term is the linear polarization, equivalent to the linearly induced dipole moment of Eq. (1), while the remaining terms describe nonlinear polarization. The value of the first differential of dipole moment with respect to field is the linear polarizability, α. The first hyperpolarizability, β, is a third-rank tensor and can be equated to the evaluated second differential of dipole moment with respect to field, while the second hyperpolarizability, γ, a fourth-rank tensor, can be defined as the evaluated third differential. Thus, Eq. (3) can be rewritten as

$$\mu_i(E) = \mu_i(0) + \alpha_{ij} E_j + \frac{1}{2!} \beta_{ijk} E_j E_k + \frac{1}{3!} \gamma_{ijkl} E_j E_k E_l + \ldots \quad (4)$$

Unfortunately, however, there are several alternative definitions of β and γ (and higher hyperpolarizabilities) and many studies do not explicitly state which convention is being followed. Therefore, one should be careful in comparing values from different studies. These are discussed in detail, along with conversion factors, in Ref. [1]. For example, the $(n - 1)$th-order hyperpolarizability is commonly equated to $(1/n!) \times (\partial^n \mu / \partial E^n)$, rather than simply to the differential.

The nonlinear bulk polarization density is given by an expression analogous to Eq. (4):

$$P_i(E) = P_i(0) + \chi^{(1)}_{ij} E_j + \chi^{(2)}_{ijk} E_j E_k + \chi^{(3)}_{ijkl} E_j E_k E_l \ldots \quad (5)$$

where the $\chi^{(n)}$ are linear ($n = 1$) and nonlinear ($n > 1$) susceptibilities and $P(0)$ is the intrinsic static dipole-moment density of the sample. The nonlinear nature of the polarization is shown in Fig. 11.1 for cases where $\chi^{(2)}$ or $\chi^{(3)}$ are non-zero.

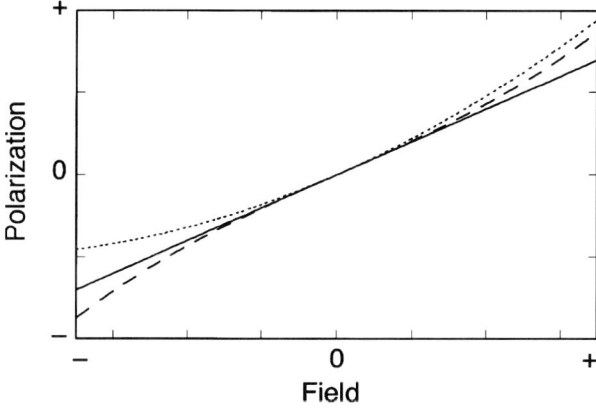

Fig. 11.1 Schematic showing how induced polarization varies with electric field for a materials with only linear polarizability [$\chi^{(1)} \neq 0$, $\chi^{(2)} = 0$, $\chi^{(3)} = 0$; solid line], linear and quadratic polarizabilities [$\chi^{(1)} \neq 0$, $\chi^{(2)} \neq 0$, $\chi^{(3)} = 0$; dotted line] and linear and cubic polarizabilities [$\chi^{(1)} \neq 0$, $\chi^{(2)} = 0$, $\chi^{(3)} \neq 0$; dashed line].

Before we examine how second- and third-order NLO effects are related to nonlinear polarization, we briefly examine an important symmetry restriction on second-order NLO properties. From Eq. (5), we can see that $P(E) = P(0) + \chi^{(1)}E + \chi^{(2)}E^2 + \chi^{(3)}E^3+...$ and $P(-E) = P(0) - \chi^{(1)}E + \chi^{(2)}E^2 - \chi^{(3)}E^3+....$; we can also see from Fig. 11.1 that $P(E) \neq P(-E)$ if $\chi^{(2)} \neq 0$. In a centrosymmetric material, $P(E)$ is necessarily equal to $P(-E)$ and, therefore, $P(0)$, $\chi^{(2)}$ and other even-order terms must be zero. Therefore, for second-order effects to be observed in a molecule or material, the molecule or material must be non-centrosymmetric. However, no such requirement applies to odd-order processes, such as third-order effects [Fig. 11.1 shows $P(E) = P(-E)$ for a material with only linear and cubic susceptibilities non-zero].

11.1.3
Second-order Nonlinear Optical Effects

Second-order NLO effects can be considered as the interaction of the polarizable electrons of the NLO material with two electric fields, E_1 and E_2, these fields potentially having different polarizations and potentially oscillating with frequencies ω_1 and ω_2, respectively. For example, consider the interaction of the material with two laser beams of different frequencies. The second-order term of Eq. (5) becomes

$$\chi^{(2)}E_1\cos(\omega_1 t)E_2\cos(\omega_2 t) \qquad (6)$$

which, from trigonometry, is equivalent to

$$\tfrac{1}{2}\chi^{(2)}E_1 E_2\cos[(\omega_1 + \omega_2)t] + \tfrac{1}{2}\chi^{(2)}E_1 E_2\cos[(\omega_1 - \omega_2)t] \qquad (7)$$

Hence it can be seen that when two light beams of frequencies ω_1 and ω_2 interact in an NLO material, nonlinear polarization occurs at sum $(\omega_1 + \omega_2)$ and difference $(\omega_1 - \omega_2)$ frequencies. This oscillating polarization can be regarded as a classical oscillating dipole that emits radiation at its oscillation frequencies. Hence light will be emitted at the sum and difference frequencies: this process is called sum (or difference) frequency generation (SFG). In the special case where $\omega_1 = \omega_2$, the sum is $2\omega_1$, the second harmonic of ω_1, and the difference is a d.c. electric field; the sum and difference generation in this case are referred to second harmonic generation (SHG) and optical rectification, respectively. SHG is a form of three-wave mixing, since two photons with frequency ω combine to generate a single photon with frequency 2ω. Since the oscillating dipole re-emits at all of its polarization frequencies, one observes light at both ω and 2ω. In a similar manner, this idea can be extended to third- and higher order nonlinear terms; by analogy, third-order processes involve interaction of three fields in a molecule to produce a fourth field (four-wave mixing).

In another special case, one of the fields in Eq. (7) is a d.c. electric field, E_2, applied to the material, i.e. $\omega_2 = 0$. The optical frequency polarization arising from the second-order susceptibility is

$$P^{(2)}_{opt} = \chi^{(2)} E_1 E_2 (\cos \omega_1 t) \tag{8}$$

and the total optical polarization [ignoring $\chi^{(3)}$ and higher terms] is

$$P_{opt} = \chi^{(1)} E_1 (\cos \omega_1 t) + \chi^{(2)} E_1 E_2 (\cos \omega_1 t) = \left(\chi^{(1)} + \chi^{(2)} E_2\right) E_1 (\cos \omega_1 t) \tag{9}$$

Thus, the applied field, E_2, changes the *effective* linear susceptibility (i.e. the dependence of the polarization on the light field, E_1). Since the linear susceptibility is related to the refractive index, the refractive index of the material is also changed by the applied field. This is known as the linear electrooptic (EO) or Pockels effect and can be used to modulate the polarization or phase of light by changing the applied voltage.

11.1.4
Measurement Techniques for Second-order Properties, β and $\chi^{(2)}$

Second-order organic NLO materials are non-centrosymmetric assemblies of second-order NLO chromophores – molecular fragments exhibiting high β. These assemblies may be either acentric crystals composed of chromophores or systems in which the chromphore is doped into a passive guest material, such as an NLO-inactive polymer. In general, for a chromophore to find useful application, a wide range of properties must be optimized, some of which are discussed in more detail in later sections. However, it is clearly important that the chromophore exhibits a high first hyperpolarizability, β, and that bulk materials (crystals, poled-polymer films) incorporating this chromophore can exhibit high $\chi^{(2)}$. Here we briefly describe how these second-order properties are measured.

Many early measurements of the second-order properties concentrated on determining the SHG efficiencies of microcrystalline samples, using the Kurtz–Perry powder technique [2]. These efficiencies have typically been reported relative to those of well-known acentric crystals, such as urea or KH_2PO_4 (KDP). Complications can arise from SHG due to the acentric nature of crystal surfaces, even those of crystals in centrosymmetric space groups; for example, SHG similar to that of urea has been measured for centrosymmetric crystals of an organoruthenium species (as mentioned in Section 11.1.2, second-order NLO effects are only possible in non-centrosymmetric media) [3]. Moreover, SHG gives us only limited insight into molecular structure–property relations, since it is dependent on the vagaries of the crystal packing. For example, many high-β chromophores have rather high dipole moments and these tend to align antiparallel with one another leading to zero $\chi^{(2)}$.

SHG can also be used to determine $\chi^{(2)}$ (or the related quantity d_{33}, which specifically relates to a mixing of fields of optical frequencies) for poled chromophore-doped polymer films, i.e. films in which the acentricity necessary for second-order NLO activity is imparted by aligning the guest chromophores using an electric field above the glass-transition temperature of the host polymer (discussed in more detail in Section 11.2). Electrooptic coefficients, r_{33}, can also be measured on these films. If one can make a reasonable estimate of degree of alignment of the molecules in these materials, one can extract estimates of the molecular hyperpolarizability, β, from these bulk quantities.

Molecular first hyperpolarizabilities, β, can also be measured in solution using either electric-field-induced second-harmonic generation (EFISH) or hyper-Rayleigh scattering (HRS). The former technique [4, 5] involves measuring second-harmonic efficiencies from a solution of a neutral dipolar molecule in a strong electric field, which can induce the acentric order necessary for the observation of second-order properties. The quantity measured is the scalar product of the vectorial part of β with the dipole moment, μ. An advantage of this technique is that $\mu\beta$ is a useful figure-of-merit for poled chromophore-doped polymer systems, (except where the chromophore has a high dipole moment and is present at high concentration, in which case dipole–dipole repulsions can become important; see below). A disadvantage is that one only obtains the component of β along the dipole moment axis, although in many dipolar more or less linear π-systems, the principal component of β often does lie more or less along the axis of μ. Moreover, EFISH is, in general, also unsuitable for ionic and non-dipolar chromophores (as is electric-field poling). In contrast, HRS is suitable for all types of chromophores; however, HRS relies on incoherent frequency doubling arising from the locally asymmetric environments of molecules in isotropic solution and so the signals measured are typically rather weak, requiring great care to be taken in correcting for a variety of other processes, especially two photon-excited fluorescence and resonance effects [6–9].

Measurements using any of the techniques discussed above will be complicated if the fundamental frequency of the light used in the measurement and, in frequency doubling-based measurements, its second harmonic, are absorbed by the chromophore. However, even when the light involved in the measurement is not absorbed (i.e., when one is in the so-called "non-resonant" régime), the values of $\chi^{(2)}$, r_{33}, β, etc., obtained will depend on the frequency of light: they are often said to exhibit a *dispersion* with frequency. Only when the frequencies of light involved are far from any absorption will these values approach their so-called zero-frequency "static" values, which are typically denoted with a "0"; for example, the static value of β is referred to as β_0. In the case of systems that are well described by the two-state model (see Section 11.2.1), the dispersion of $\chi^{(2)}$, r_{33}, β, etc., can be well described, at least where the frequencies do not approach the absorption maxima too closely, by simple equations (see Section 11.2.1); in these cases, estimates of static values or of values at a second frequency of interest can be made from measured values. If one is interested in, say, an application involving the frequency doubling of 532-nm light, the most useful values of $\chi^{(2)}$ or β will be those determined by frequency doubling-based techniques determined using a 532-nm fundamental. On the other hand, if one is interested in understanding structure–property relationships, it is important to be able to compare values from different studies (potentially at different frequencies) and to establish the extent to which trends are due to underlying trends in the static hyperpolarizability, rather than to differing dispersion effects; here it is desirable that good estimates of β_0 be made. In general, these estimates will be most reliable when the fundamental employed in the measurements has as long a wavelength as possible.

A variety of quantum-chemical computational methods have also been used to compute β. Most commonly, semi-empirical methods [such as the Zerner Intermediate Neglect of Differential Overlap (ZINDO) method] are used to evaluate β using either finite field (FF) or sum-over-states (SOS) calculations (for a more detailed discussion of semi-empirical computational techniques for second-order NLO, see Ref. [10]). In the FF method [11–13], the ground-state dipole moment is calculated in the presence of a static electric field and α_0, β_0 and γ_0 are obtained from the first, second and third derivatives, respectively, of dipole moment with respect to field [according to Eqs. (3) and (4)]. More recently, related approaches have been developed using density functional theory [14, 15]. The alternative SOS approach [16] is based on perturbation theory (see Section 11.2.2), which relates β to simple few-state expressions [such as Eqs. (16–18)] in terms of parameters – energies, transition dipoles and changes in static dipole moment – associated with electronic transitions. These parameters are calculated for a large number of low-lying excited states and β (at the frequency of interest) is obtained as the sum of these few-level terms. Advantages of the SOS method are that the frequency dependence of β can be predicted and that chemical insight can often be gained into the origin of trends in β in a series of molecules through analysis of the changes in the parameters entering the few-state expressions. It is worth noting that comparison between experimental and computational methods should be made with caution: the latter typically apply to the gas phase where it has been shown that β can be significantly smaller than in the condensed phases [17, 18]. Moreover, confusion may also arise from the choice of conventions defining the experimental and computational hyperpolarizabilities (see Section 11.1.2).

11.1.5
Third-order Nonlinear Optical Effects

Just as second-order effects involve the interaction of two electric fields with the electrons of a material, third-order effects involve three electric fields, E_1, E_2 and E_3. In the special case where all these three fields have the same frequency and where $\chi^{(2)}$ is zero (which is necessarily the case in centrosymmetric materials – see Section 11.1.2), we see that

$$P_{opt} = \chi^{(1)} E \cos(\omega t) + \chi^{(3)} E^3 \cos^3(\omega t) \dots \quad (10)$$

which, using the trigonometric identity

$$\cos^3(\omega t) = \frac{3}{4}\cos(\omega t) + \frac{1}{4}\cos(3\omega t) \qquad (11)$$

can be rewritten as

$$P_{opt} = \chi^{(1)} E \cos(\omega t) + \chi^{(3)} E^3 \frac{3}{4}\cos(\omega t) + \chi^{(3)} E^3 \frac{1}{4}\cos(3\omega t)\ldots \qquad (12)$$

or

$$P_{opt} = \left(\chi^{(1)} + \frac{3\chi^{(3)} E^2}{4}\right) E\cos(\omega t) + \frac{\chi^{(3)} E^3}{4}\cos(3\omega t) \qquad (13)$$

Hence the interaction of light with third-order NLO molecules creates a polarization component at its third harmonic. In addition, there is a component at the fundamental and we note that the $[\chi^{(1)} + 3\chi^{(3)}E^2/4]$ term of Eq. (13) is similar to the term responsible for the linear electrooptic effect. Similarly, it can be shown that the application of an intense light field or voltage will also induce a refractive index change in a third-order NLO material; these effects are known as the optical and the d.c. Kerr effects, respectively. The sign of $\chi^{(3)}$ will determine if the third-order contribution to the refractive index is positive or negative in sign. For positive $\chi^{(3)}$, increased intensity leads to increased refractive index and, hence, reduced light velocity; such a the material is said to be *self-focusing* since the differences in intensity, and hence in light velocity, between the center and periphery of a laser beam propagating in the material will lead to distortion of the wavefronts such that the wavevectors converge to a focus. Conversely, materials with negative $\chi^{(3)}$ are self-defocusing.

It should be noted that γ and $\chi^{(3)}$ are frequency-dependent complex quantities; properties such as self-focusing depend on the real part of the refractive index, which in turn depends on the real part of the susceptibility, Re($\chi^{(3)}$), whereas frequency tripling depends on the modulus of the susceptibility $|\chi^{(3)}|$. One of the factors that can give rise to positive imaginary contributions to the hyperpolarizability and susceptibility, Im(γ) and Im($\chi^{(3)}$), respectively, is two-photon absorption (2PA) at the frequency in question. In this chapter we shall use 2PA to refer to the "simultaneous" absorption of two photons, i.e. the situation where the second photon is absorbed within the lifetime of the "virtual state" formed by the first photon, rather than to sequential absorption, i.e. absorption of one photon to afford an intermediate excited state which is metastable with respect to the ground state and which then absorbs a second photon. The energy of the excited state formed is equal to the sum of the energies of the two photons absorbed (these energies are usually equal for potential applications and also for many measurements). The molecular cross-section for 2PA, δ (which should not be confused with the third hyperpolarizability!), is proportional to the imaginary part of γ:

$$\delta(\omega) \propto \text{Im}[\gamma(-\omega;\omega,\omega,-\omega)] \qquad (14)$$

The molecular 2PA cross-section is sometimes denoted by σ_2 or σ_2' rather than δ, while the corresponding bulk 2PA coefficient is, rather confusingly, typically represented by β.

It is interesting to note that for many years 2PA was regarded as a parasitic effect by many in the nonlinear optics community. For example, in the case of all-optical switching using $\chi^{(3)}$ materials, 2PA can lead to strong attenuation of the signals and lead to sample damage through severe heating. In this context, therefore, there was no desire to maximize 2PA; indeed, efforts were made to prevent or minimize its appearance. However, as is often the case, what can be regarded as highly detrimental in one field can be exploited in other fields. 2PA is advantageous when one wishes to excite molecules with a high degree of 3D spatial confinement and/or in media that are strongly absorbing at the appropriate one-photon wavelength. The ability of 2PA to excite molecules with this 3D spatial resolution arises because 2PA is proportional to the square of the light intensity at the relevant wavelength, whereas one-photon absorption (1PA) is proportional to the intensity. The intensity, I, of a laser beam falls away as the square of the distance, z, from the focus. Thus, the number of excited states formed by 2PA is proportional to z^{-4}, whereas the number of excited states formed by one-photon absorption is proportional to z^{-2}. The much stronger distance dependence of 2PA means that excitation can be essentially confined to a volume of λ^3, where λ is the wavelength of excitation light. The I^2 and z^{-4} dependence of 2PA also means that a material can be essentially transparent away from the focus of a laser beam and, therefore, that excitation can be carried out at depth in the material. Moreover, since the photon energy required for 2PA is lower than (potentially as low as half) that required for 1PA, in many cases the photons are less likely to be absorbed by other one-photon chromophores present (for example, in biological media which are considerably less absorbing in the near-infrared than the visible/ultraviolet). The applications of 2PA are discussed in more detail in Section 11.3.3.

11.1.6
Measurement Techniques for 2PA Cross-section, δ

Since this chapter is focused on electrooptic materials and two-photon absorption, we will not consider measurement techniques for γ and $\chi^{(3)}$ in detail, but refer the interested reader to Ref. [19]. In this section, we will briefly describe some of the techniques that have been used for the measurement of the two-photon cross-section, δ. We first recall that the 2PA behavior of a molecule is characterized by a spectrum entirely analogous to a one-photon absorption (1PA) spectrum. Although, from the point of view of an application at a particular wavelength, measurements of δ at only that wavelength may contribute to a figure-of-merit for that application, more generally, to understand structure–property relationships, it is helpful to acquire 2PA data over as wide a frequency range as permitted by the lasers available and by the onset of 1PA.

Nonlinear transmission (NLT) measurements have frequently been used for single wavelength measurements (although some studies have exploited frequency-dependent NLT, for example, see Ref. [20]); here a transmitted light intensity, I_t, is simply measured as a function of incident light intensity, I_0. In the case of a linearly absorbing material (i.e. 1PA only), a linear relationship is obtained, i.e. the transmission, T, defined as the ratio I_t/I_0, is a constant. If one uses a wavelength at which there is no 1PA, but simultaneous 2PA, a nonlinear relationship is expected and δ can be obtained from

$$1/T = 1 + CI_0\delta \qquad (15)$$

where C is a constant composed of factors taking account of the sample path length, the coherence of the laser pulses and the number density of chromophores. However, depending on the laser pulse length employed, measurements can be complicated by excited-state absorption. For example, if NLT data for **1** (Fig. 11.2) acquired in toluene with nanosecond pulses are fitted to a 2PA model, an apparent δ of 9300 GM (see the Appendix for a definition of the units of 2PA) is obtained at 600 nm [21], whereas a value of only 210 GM is obtained from two-photon fluorescence measurements in the same solvent [22, 23]; this discrepancy is due to strong excited-state absorption at 600 nm in the NLT measurements subsequent to 2PA excitation and relaxation to the first singlet excited state.

Two-photon fluorescence (2PF) [24] is advantageous in that data acquired with nanosecond and picosecond laser pulses have been shown to give consistent values for δ in many cases. The fluorescence intensity of a sample is recorded as a function of excitation wavelength and intensity, using wavelengths at which no linear absorption occurs. The intensity dependence allows one to verify that the process responsible is indeed 2PA; the wavelength dependence allows one to construct the 2PA spectrum. Indeed, regardless of the measurement technique employed, it is useful to perform the measurements over a range of intensities and demonstrate that the observed effects are strictly quadratic with input intensity; deviations from quadratic behavior are often signs of linear absorption or higher order nonlinear optical processes contributing to the observed signal. An important assumption in deriving δ from the data is that the fluorescence quantum yield is independent of whether the molecule is excited using 1PA or 2PA; however, this assumption should be valid for molecules obeying Kasha's rule. An obvious drawback of the method is that it is limited to fluorescent molecules. Typically, δ values are determined using a reference material with a known δ spectrum in the region of interest; hence it is also a requirement that standards are available with well-characterized 2PA spectra in the appropriate wavelength region.

Third-order nonlinear optical phenomena, such as nonlinear absorption, are only significant close to the focus of a laser beam. The Z-scan technique exploits this distance dependence to probe both real and imaginary parts of $\chi^{(3)}$ [25, 26]. The sample is translated along the axis of a focused laser beam ("z") and changes in the transmitted beam are detected. In the "closed aperture" mode the nature of the signal as a function of z indicates whether the solution is self-focusing or defocusing, i.e. whether it has positive or negative $\text{Re}(\chi^{(3)})$. With an open aperture, the Z-scan of a 2PA sample shows a negative peak at $z = 0$, corresponding to the situation where the focus of the laser beam is centered with the sample and 2PA

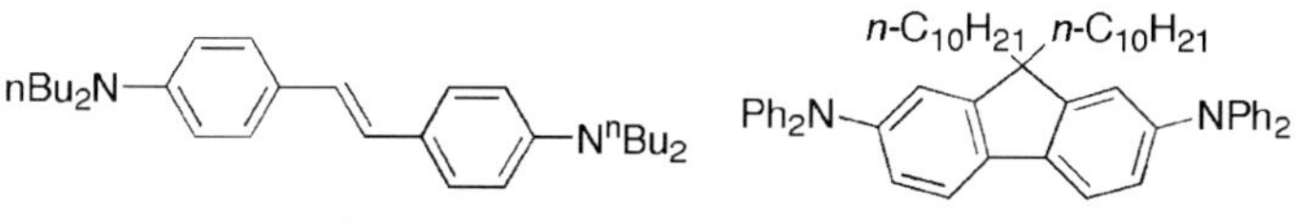

Fig. 11.2 Structures of compounds referred to in Section 11.1.6.

is at a maximum and, therefore, transmitted light is at a minimum. As with NLT measurements, complications can arise from excited-state absorption at long pulse lengths; however, the details of the fit of the Z-scan data help in distinguishing this possibility. Degenerate four-wave mixing (DFWM) is another technique that can be used to obtain both real and imaginary parts of $\chi^{(3)}$ [27], but has been little used in measurement of δ (but, for an example, see Ref. [28]).

All the above techniques rely upon *degenerate* 2PA, i.e. 2PA in which the two photons have the same energies. However, a recently developed technique relies upon *non-degenerate* 2PA; a strong pump beam, of frequency insufficient for 1PA, is combined with a weak white-light continuum (WLC) probe beam [29]. The technique allows for rapid acquisition of 2PA spectra as a function of probe frequency; 2PA occurs when the sum of pump and probe energies equals the energy of a 2PA-allowed state. A limitation of the technique is that one cannot acquire as wide a spectral range as is possible with degenerate techniques, since 1PA of either pump or probe light will set in at lower state energy. In addition, it should be noted that values of δ for non-degenerate 2PA can be much larger than those for degenerate 2PA, this difference depending on the detuning (energy difference) between the pump and probe energies and the various resonant states. This has clearly been shown in the case of **2** (Fig. 11.2); the peak δ varies from 89 GM for degenerate 2PA at 2 eV to 427 GM when pump and probe energies are 1 and 3 eV respectively [30]. This behavior can be rationalized using the more complex essential-states expressions derived for δ in the non-degenerate case [30] than those described in Section 11.3.1 for the degenerate case. In the case of dipolar systems, where two-level terms, rather than three-level terms, are important (Section 11.3.1), less severe divergence between non-degenerate and degenerate δ is found.

A variety of quantum-chemical calculations have also been used to predict 2PA spectra for molecules. Most frequently, semi-empirical SOS calculations are employed using few-state terms such as those discussed in Section 11.3.1 [Eqs. (20) and (21)], with damping terms chosen to afford linewidths for the transitions in accordance with known experimental data. To describe 2PA adequately using the SOS methodology, it is typically necessary to take account of extensive configuration interaction.

11.2
Second-order Chromophores for Electrooptic Applications

11.2.1
Design of Second-order Chromophores: the Two-level Model

An expansion of the Stark energies of a molecule can be derived from perturbation theory. In this so-called sum-over-states approach, hyperpolarizabilities are related to the transition moments, dipole-moment changes and energy separations between the ground and excited states of the molecule and between various excited states of the molecule. This approach can greatly aid the understanding and optimization of nonlinear optical properties through the consideration of excited-state properties. In the case of β, transition moments, dipole-moment changes and energy terms enter into the sum-over-states expression in several different ways. We have already seen (Section 11.1.2) that molecules with non-zero β should be non-centrosymmetric; the most widely investigated class of non-centrosymmetric molecules is that of dipolar "donor–acceptor" species and in dipolar species the most important contributions to β are so-called "two-level" contributions. In simple donor–acceptor compounds, β can be well described as arising principally from the contributions of the ground state, g, and one low-lying charge-transfer state, e, strongly coupled to the ground state (often this is the first excited state) [31, 32]. Using this two-state approximation, the static (zero-frequency) limit of β is given by

$$\beta_0 \propto \frac{\mu_{ge}^2 (\Delta\mu_{ge})}{\omega_{ge}^2} \tag{16}$$

where μ_{ge} is the transition dipole moment linking ground and excited state, $\Delta\mu_{ge}$ is the change in static dipole moment between ground and excited state and ω_{ge} is the frequency corresponding to the excited-state energy. The constants of proportionality employed depend on the precise definition of β used (see Section 11.1.2 and Ref. [1]). The two-level model also describes the dispersion of β with frequency, ω. For β, as sampled by frequency-doubling phenomena, this is given by

$$\beta(-2\omega;\omega,\omega) = \beta_0 \frac{\omega_{ge}^4}{\left(\omega_{ge}^2 - (2\omega)^2\right)\left(\omega_{ge}^2 - \omega^2\right)} \tag{17}$$

where ω is the fundamental frequency and ω_{ge} is the absorption maximum of the molecule. An analogous expression relates $\chi^{(2)}(-2\omega;\omega,\omega)$ to $\chi^{(2)}_0$. This equation has been widely used to estimate static β_0 from single-frequency measurements of β. However, this expression is really only applicable:

1. where $\omega_{ge} \gg 2\omega$, i.e. in the so-called non-resonant régime;
2. where one excited state dominates the nonlinearity; and
3. when one tensor component dominates the NLO response.

The third condition is more or less met in many dipolar chromophores, but Eq. (17) has often been applied to estimate β_0 from experimental β data where the first two conditions are not met, i.e. where ω_{ge} is close to 2ω [damping terms should be included in Eq. (17) as $2\omega \to \omega_{ge}$ to avoid $\beta_0 \to 0$] or where more than one excited state contributes significantly to ω (as is the case for some classes of organometallic compounds, notably those based on ferrocene donors [33]). In some cases, however, there is some evidence that, owing to the cancellation of opposing effects, reasonable estimates can sometimes be obtained from this equation even when not all the conditions are met [34]. In any case, values of β_0 given in this chapter are those given in the original publications and inevitably vary in reliability from compound to compound. Another two-level expression relates to β (or r_{33}), as sampled by electrooptic measurements [35, 36]:

$$\beta(-\omega;\omega,0) = \beta_0 \frac{\left(\omega_{ge}^2 - (\omega^2/3)\right)\omega_{ge}^2}{\left(\omega_{ge}^2 - \omega^2\right)^2} \qquad (18)$$

It should be noted that although β at a given frequency is different for different NLO phenomena, both tend to the same zero-frequency limit, β_0. Figure 11.3 compares the predictions of the two-level model for the frequency dispersion of the electrooptic effect with that for frequency doubling. While dispersion effects complicate comparison of the "intrinsic nonlinearity" of various chromophores

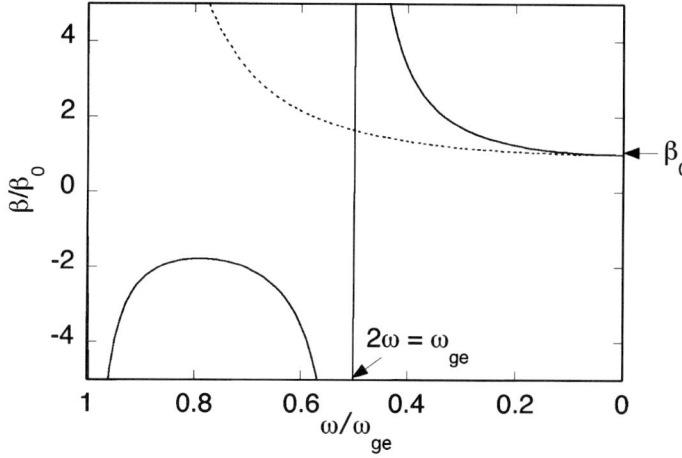

Fig. 11.3 Variation of β for frequency doubling (solid lines) and the electrooptic effect (dotted line) with incident light frequency, ω, according to the two-level Eqs. (17) and (18), respectively. β_0 is the static hyperpolarizability and ω_{ge} is the transition energy from the ground state to the excited state responsible for β. Note that for frequency doubling there is an additional resonance seen when $\omega = \omega_{ge}/2$, i.e. when the frequency doubled light is resonant with ω_{ge}, in addition to the resonance seen when the incident light is resonant with ω_{ge}. It should also be noted that these two-level expressions are only useful for real systems under certain conditions (see text); in particular, no damping is included and so the expressions are not useful close to the resonances.

and, thus, hinder the deciphering of basic structure–property relationships, one can exploit these effects for applications by choosing a molecule with a large dispersion-enhanced hyperpolarizability at the frequency one wishes to use. At the same time, if one is interested in processing a light signal, one does not want to lose any of this light (or, in SHG applications, its second harmonic) through electronic absorption by the chromophore; it is generally desirable to use chromophores with ω_{ge} as close to the frequency (frequencies) of interest as possible without incurring significant loss through absorption.

Returning to Eq. (16), it can clearly been seen that molecules with linear spectra showing intense (i.e. large μ_{ge}) low-energy (i.e. low ω_{ge}) charge-transfer (i.e. large $|\Delta\mu_{ge}|$, as indicated by solvatochromism or by Stark spectroscopy) transitions are good candidates for second-order NLO investigations. The first hyperpolarizability, β, is determined by the interplay of these three parameters, each of which varies in a different way with the strengths of donor and acceptor, with the nature of the bridge and with the nature of the environment (e.g. solvent polarity). As examples of dipolar molecules that are well described by the two-state model, consider a π-donor group linked by a polyene bridge to a π-acceptor. Here both the ground and the first excited states can be viewed as linear combinations of the neutral and charge-separated (zwitterionic) limiting-resonance forms, shown in Fig. 11.4 for three examples of such compounds (3–5). The ground- and excited-state structures are determined by the relative contribution of each limiting resonance form. Thus, if the ground-state is dominated by the neutral structure, the excited state will be dominated by the zwitterionic structure. Hence a large increase in dipole moment is expected on excitation, in the same direction as the ground-state dipole moment ($\mu_g \Delta\mu_{ge} > 0$). Conversely, if the charge-separated form is the main contributor to the ground state, then the neutral form dominates the excited state so that the excited-state dipole moment is smaller than

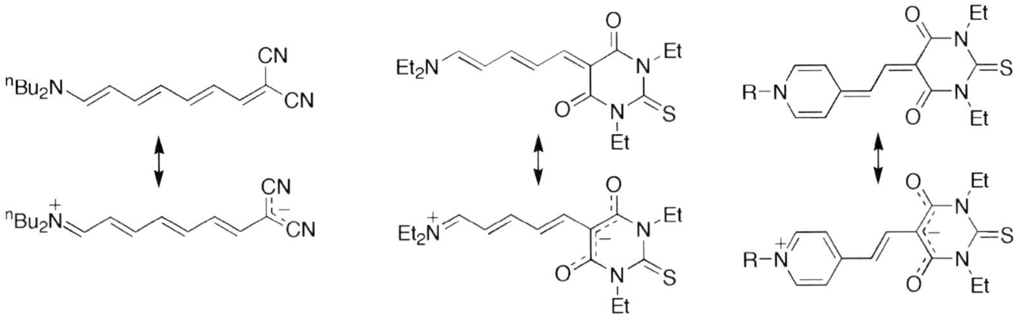

Fig. 11.4 Neutral (upper) and zwitterionic (lower) resonance structures for three examples of donor–acceptor polyenes studied in Ref. [37] (R = 3-methylbutyl) and found to belong to BOA régimes B (neutral form dominates ground state; zwitterionic form dominates excited state), C (neutral and zwitterionic forms make comparable contributions to both ground and excited states) and D/E (zwitterionic form dominates ground state; neutral form dominates excited state) respectively (see Fig. 11.5).

the ground-state dipole ($\mu_g \Delta\mu_{ge} < 0$). Therefore, a positive β_0 indicates a predominance of the neutral form in the ground state. On the other hand, a negative β_0 implies that the zwitterionic form prevails in the ground state. If the neutral and zwitterionic resonance structures make equal contributions to the ground state, these contributions will also be equal in the excited state and $\Delta\mu_{ge}$ and, therefore, β will be zero. The contributions of the two resonance structures can be gauged by the bond-length alternation (BLA), i.e. the alternation between adjacent C–C bonds in the bridge or by the related bond-order alternation (BOA). When the neutral resonance structure dominates the ground state, the BLA (BOA) will have the opposite sign (defined as negative) to that found when the zwitterionic structure dominates (defined as positive). In the "cyanine" limit, where neutral and zwitterionic resonance structures make equal contributions to the ground state, the BLA (BOA) is zero.

This has been confirmed by semi-empirical calculations performed on donor–acceptor polyenes in which β_0 (and γ_0) were correlated with BLA and BOA [37–39].

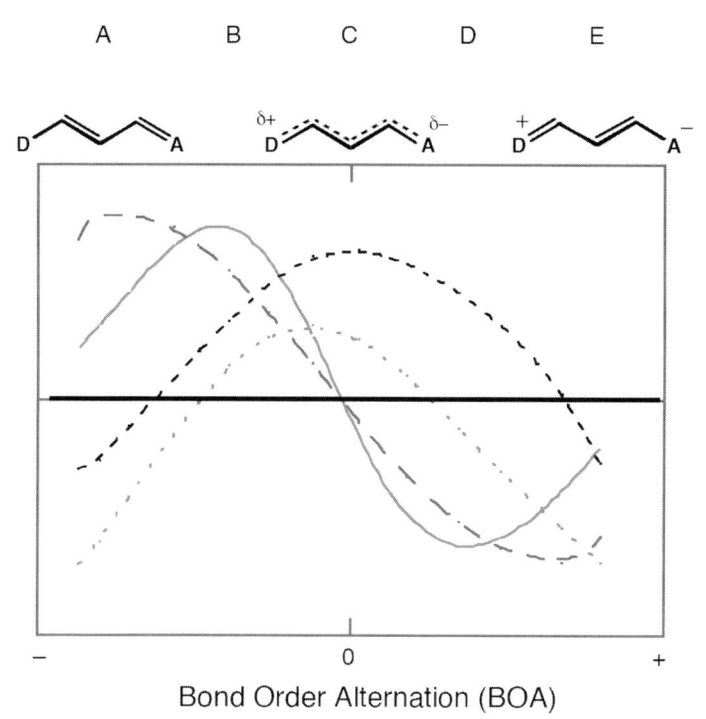

Fig. 11.5 Schematic figure showing how β_0 (solid line) varies with bond-order alternation (BOA) for a donor–acceptor polyene from a neutral ground-state structure (left) to a fully zwitterionic ground-state structure (right), along with the variation of the constituent terms according to the two-state Eq. (16): μ_{ge}^2 (medium dashed line), $\Delta\mu_{ge}$ (long dashed line) and $1/\omega_{ge}^2$ (short dashed line). Values of β_0 and $\Delta\mu_{ge}$ are positive above the black horizontal line and negative below; values of the remaining quantities are always positive. The labels A–E refer to régimes of "BOA space" referred to in the text.

The ground-state structure of gas-phase molecules was artificially tuned between neutral to zwitterionic extreme structures, by varying an external static electric field parallel to the acceptor–donor axis. It was shown that there is positive peak in α_0, β_0 vanishes and γ_0 peaks negatively at zero BOA independently of the nature of the push–pull molecule [40]. Figure 11.5 shows the positions of peaks in β_0 – positive at intermediate negative BOA (in what we have labeled the "B" régime of "BOA space") and negative at intermediate positive BOA ("D" régime) – along with the variation of the contributing parameters involved in the two-state expression [Eq. (16)] for β_0. These results illustrate that relative contributions of the two resonance forms must be carefully balanced to optimize β (or γ) in polyene-bridged molecules.

Other factors besides the donor and acceptor strength can influence the degree of BOA in a dipolar chromophore, most notably the nature of the bridge. Thus, replacement of a non-aromatic segment of the bridge by a ring that would be aromatic in the neutral representation of the structure and that gives the same overall conjugation length (say, replacement of a butadienylene bridge with a p-phenylene bridge) will shift the BOA towards to left-hand side of Fig. 11.5 [41, 42]. On the other hand, inclusion of a ring that becomes aromatic in the charge-separated structure will move the structure towards the right on the BOA plot. For example, while the neutral resonance structures of compounds **4** and **5** both have three double bonds between an amine donor and a thiobarbituric acid acceptor group, the zwitterionic ionic form of **5** is aromatic, whereas that of **4** is not; consequently, whereas **4** falls in the "C" region of "BOA space", **5** falls in the "D/E" region. We also note in passing here that NLO data for donor–acceptor stilbenes [43] and tolanes [44] indicates that ethynylene bridges are less effective mediators of donor–acceptor π interactions than vinylene groups.

Another factor is the polarity of the medium; a more polar medium favors greater contributions from the charge-separated resonance structure than a less polar medium. This phenomenon has been demonstrated experimentally: as a function of increasing solvent polarity, β for **3** (Fig. 11.4) increases to a maximum and then decreases (régime "B" of Fig. 11.5), β for **4** can be tuned from positive to negative ("C") and β for **5** can be tuned through a maximum negative value ("D" to "E") [37]. Indeed, it is this sensitivity of β to solvent polarity that allows the confident assignment of positions on the BOA curve to **3–5**.

One can also use Eq. (16) to predict that, at least for a chromophore with significant $\Delta\mu_{ge}$ (i.e. excluding chromophores from the "C" régime of BOA), β should increase as the length of the conjugated bridge between donor and acceptor is increased; intuitively we would expect μ_{ge} and $\Delta\mu_{ge}$ to increase and ω_{ge} to decrease as the chain length increases (longer donor–acceptor compounds tend to have stronger, lower energy, more solvatochromic transitions in their absorption spectra than their shorter analogs). An increase in β with chain length is indeed typically observed for donor–acceptor polyenes [45].

11.2.2
Other Chromophore Designs

In addition to the many studies of all-organic dipolar second-order chromophores of the general type discussed in Section 11.2.1, β has also been measured for many dipolar chromophores incorporating inorganic coordination complexes and organometallic groups [10, 46–49]. The ferrocenyl donor has been particularly widely studied [50], partly owing to its well-known electron-donor properties, its reasonable stability, at least by the standards of much of organometallic chemistry, and its well-developed functionalization chemistry; moreover, at the time of its report in 1987, the SHG efficiency for microcrystalline (Z)-p-nitrostyrylferrocene, **6** (Fig. 11.6), was one of the largest values known for a molecular compound [51]. Subsequently, very large values of β and $\mu\beta$ have been obtained in extended chromophores with stronger acceptor groups, for example, **7** (Fig. 11.6) [52]. In some cases the $\mu\beta$ values of these longer species approach those of analogous compounds with amine donors and can have comparable transparency [45].

A wide range of other transition metal centers have also been found to be capable of acting as strong donors and acceptors in extended conjugated molecules. For example, half-sandwich group 8 phosphine complexes [53–57] can act as stronger donors and give higher nonlinearities than the 4-(dimethylamino)phenyl group, as illustrated by the comparison of **8** and **9**, respectively, in Fig. 11.7 [44, 55]; unfortunately, these donors have not been combined with the strongest acceptors or most effective bridging groups. Despite the large body of work on transition metal chromophores, no second-order material based on a metal-containing molecular chromophore has yet been demonstrated to be competitive with the best all-organic materials in devices. Moreover, our understanding of other important factors for practical device applications, such as thermal and photostability (see Section 11.2.3), is generally much less well developed for metal–organic species that for all-organic molecules. One often-cited potential advantage of metal-containing species is that nonlinearities might be switchable through reversible redox chemistry at the metal center; although electrochemical and chemical redox switching of second-order nonlinearities has been demonstrated for several metal–organic chromophores [58–60], this switching has yet to be exploited in a

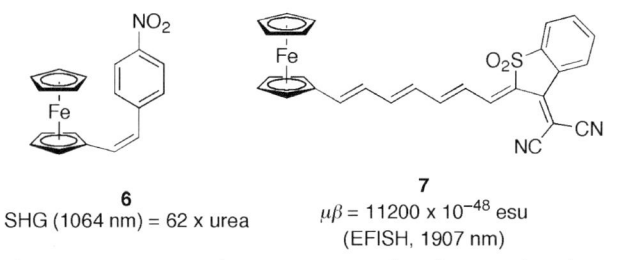

6
SHG (1064 nm) = 62 × urea

7
$\mu\beta = 11200 \times 10^{-48}$ esu
(EFISH, 1907 nm)

Fig. 11.6 Structures and NLO properties of two ferrocene-based dipolar chromophores [51, 52].

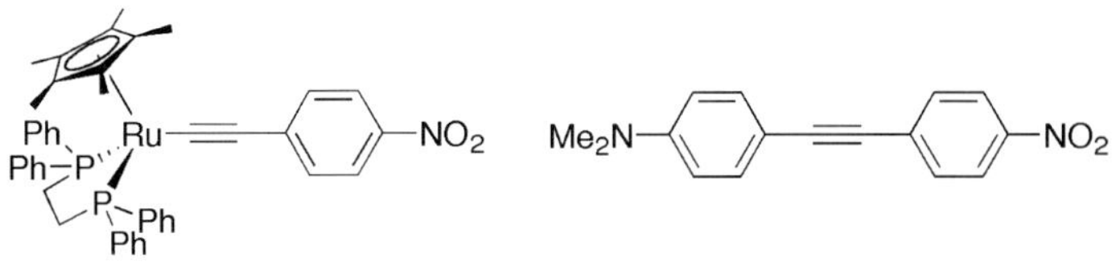

8
$\mu\beta$ ($\mu\beta_0$) = 2100 (1155) × 10^{-48} esu
β (β_0) = 169 (93) × 10^{-30} esu

9
$\mu\beta$ = 280 × 10^{-48} esu
β = 46 × 10^{-30} esu

Fig. 11.7 Structures and nonlinear optical data (EFISH, 1907 nm) for an organoruthenium chromophore [55] and an all-organic analog [44].

functional device. Moreover, redox switching should also be possible for some organic species, albeit at different potentials.

Most chromophores studied for second-order applications have a dipolar donor–acceptor character; however, non-zero β is possible for any non-centrosymmetric molecule. Octupolar organic, inorganic and organometallic molecules have accordingly attracted considerable interest [61–63]. Owing to the lack of dipole moments in these species, it is anticipated that it may be easier to achieve non-centrosymmetry in crystals, although most studies thus far have usually focused on determining β in solution using HRS [64]. In purely octupolar species, no two-level dipolar contributions to β [i.e. those described by Eq. (16)] are expected since μ_g and μ_e are both zero by symmetry (although two-level contributions have been reported in some systems, where, despite a non-polar ground state, the excited state breaks symmetry to acquire a dipole moment [65]), but rather multi-level octupolar contributions to β. As an example, Eq. (19) shows the three-level expression that has been derived for molecules with D_3 symmetry; this involves the ground state and two degenerate excited states, e and e'. An excited-state/excited-state transition dipole moment, $\mu_{ee'}$, enters the expression; chemists typically have relatively little intuition regarding this type of quantity and, unlike μ_{ge}, ω_{ge} and $\Delta\mu_{ge}$, insight cannot be gained from absorption spectra, meaning that this equation is less readily translated into design guidelines than Eq. (16).

$$\beta_0 \propto \frac{\mu_{ge}^2 \mu_{ee'}}{\omega_{ge}^2} \tag{19}$$

Many octupolar chromophores have threefold symmetry based around 1,3,5-substituted benzene or triazine cores; an example [66] is shown in Fig. 11.8 and this type of chromphore has more recently also been used in extended dendrimeric structures [67]. However, other non-centrosymmetric cores have been used, for example, four-coordinate tetrahedral tin centers [68].

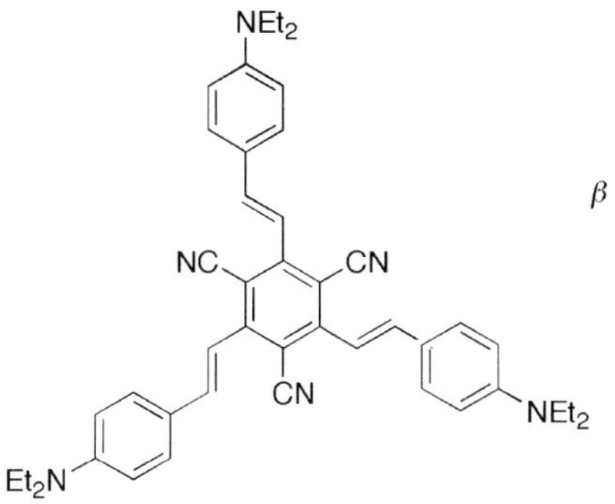

Fig. 11.8 Structures and NLO data for an example of an octupolar chromophore [66].

Compound **10**: $\beta\ (\beta_0) = 121\ (65) \times 10^{-30}$ esu (HRS, 1560 nm); SHG = ca. 45 × urea.

11.2.3
Other Considerations

In Section 11.2.1, we have seen how β can be optimized in donor–acceptor compounds, by balancing of donor and acceptor strengths and by varying the length and nature of the bridge and how it varies with the frequency of light employed. We also noted the importance of transparency as the frequency (frequencies) of interest. However, although large β and good transparency are necessary requirements for an organic material to be useful for applications, for example, in electrooptic devices, they are not sufficient; many additional factors must be taken into consideration.

As discussed above (Section 11.1.2), bulk materials, as well as the molecules themselves, must be non-centrosymmetric for second-order effects to be observed. Molecules that crystallize in non-centrosymmetric space groups provide one possible solution to the problem of achieving the required asymmetry. Indeed, the inorganic materials employed for second-order NLO applications, such as $LiNbO_3$ and KH_2PO_4 (KDP), are extended solids with non-centrosymmetric crystalline structures. However, prediction and control of the crystal structures of organics are far from straightforward and the dipolar chromophores with the highest β values typically also possess large dipole moments that favor structures in which the dipoles, and thus the principal components of β, align antiparallel to one another. Crystals are also difficult to integrate with other device components, relative to approaches based on solution-processed films. The Langmuir–Blodgett technique [69] and other self-assembly processes based on successive chemisorption processes [70–73] can be used to achieve acentric thin films. However, by far the most general and the most widely used technique for achieving non-centrosymmetry for device applications is electric-field poling.

In electric-field poling, a guest dipolar chromophore and a chemically and optically passive host polymer (the chromophore can also be covalently tethered to the polymer [74] or incorporated into the main chain of the polymer) is heated to above the glass transition temperature, T_g, of the mixture. A large electric field (typically ca. 10^6 V^{-1} cm^{-1}) is applied; under the influence of this field the chromophores can reorient so that their dipoles partially align with the electric field. The material is then cooled back below the T_g, still in the electric field. When the field is turned off, the orientational order is "frozen in" owing to the limited motional freedom below T_g. The second-order susceptibility expected for a poled polymer film (in the absence of significant chromophore–chromophore interactions) is proportional to the number density of chromophores multiplied by $\mu\beta$, the scalar product of the vectorial component of β with the ground-state dipole moment, μ, which readers may recall from Section 11.1.4 is the same quantity sampled by solution EFISH experiments. Hence it is generally $\mu\beta$, rather than β, that should be maximized for poled-polymer applications, although in the case of molecules with very large μ, dipole–dipole interactions between chromophores can create strong internal forces that oppose the poling field (see Section 11.2.4).

Most poled-polymer films are prepared by spin-casting and, therefore, it is necessary that the chromophore and the polymer host both have good solubility in some solvent. It is also necessary chromophores do not phase-segregate from the polymer host either during spin-coating or during poling; this can decrease $\chi^{(2)}$ if the chromophores form centrosymmetric crystallites and can lead to optical losses through scattering. One widely used solution to these problems is to functionalize the chromophore with long alkyl chains. Another common strategy is to attach the chromophore covalently to the host polymer; this can also lead to improved orientational stability.

The orientational order induced by poling tends to decrease slowly with time. In general, the orientational stability increases as the difference between T_g and the operating temperature of the device is increased, hence one should use host polymers with high T_g. It should be noted that chromophores in host–guest systems often act as plasticizers towards the host polymer and T_g for the composite can be significantly lower than for the pure polymer, especially at high chromophore loading. However, if a high-T_g polymer system is used, the chromophore should have good short-term stability at the poling temperature in order to survive the device-fabrication process. Post-poling cross-linking can also increase orientational stability, although again one must take care that the chromophore is stable under the conditions required for cross-linking. Also, the chromophore should have good long-term thermal stability at the operating temperature. In addition, the material should have good photostability, at least at the frequencies to which the material is exposed during operation.

To address issues of chromophore photochemical and thermal stability, several strategies have been adopted. Although donor–acceptor polyenes are well known to exhibit $\mu\beta$ values exceeding those of systems of comparable conjugation length in which aromatic rings are interposed between donor and acceptor [75], extended polyenes often exhibit poor photostability and are typically not sufficiently stable to withstand poling at temperatures greater than 200 °C [76, 77]. To address these issues of chromophore stability, several strategies have been adopted. Several groups have worked on reducing the possibility for cis–trans isomerization and

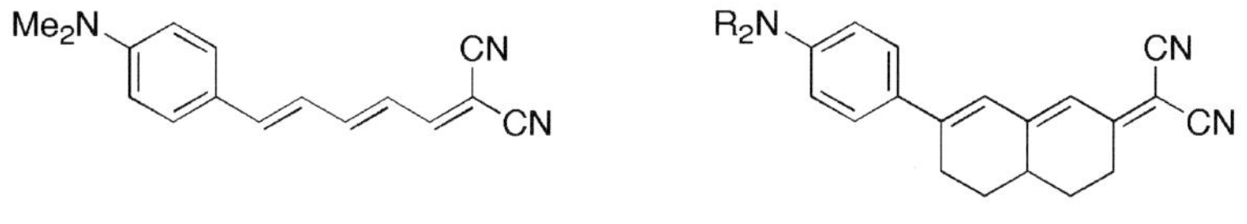

11
$T_d = 209\ °C$

12 ($R = {}^nBu$)
$T_d = 280\ °C$

13 ($R = 4\text{-}{}^nBuC_6H_4$)
$T_d = 330\ °C$

Fig. 11.9 Thermal stability data for three dipolar chromophores illustrating the results of conformationally locking polyenes and of replacing dialkylamino groups with diarylamino groups. T_d is the onset of weight loss according to thermogravimetric analysis [80]. Comparable $\mu\beta$ (EFISH, 1907 nm) values of 1470×10^{-48} and 1800×10^{-48} esu have been reported for **11** [92] and **13** [80], respectively.

increasing steric protection of the π-system by locking the conformations of some or all of the double bonds in a polyene chains by their incorporation into ring systems (e.g. Fig. 11.9). Several comparisons of ring-locked and non-ring-locked polyenes demonstrate greater thermal stability for the former class of compounds, whilst comparable $\mu\beta$ values are attainable in ring-locked systems to those in non-locked polyenes [75, 78–80]. These ring-locked systems also potentially act to reduce dipole–dipole interactions between chromophores (see Fig. 11.10, Section 11.2.4). Typically, rather unstable structures are obtained when dialkylamino (or diarylamino) donors are directly attached to polyene bridges; thus, in most candidate chromophores for applications, 4-(dialkylamino)phenyl or 4-(diarylamino)phenyl donors are used. One compromise that has had some success is the use of 5-(dialkylamino)-2-thienyl or 5-(diarylamino)-2-thienyl donors [81–87]; here the reduced aromaticity of the heterocycle leads to greater nonlinearities than in the phenylene species, but the amine–heterocycle link is still considerably more stable than the direct amino–polyene link. Thiophene and thiazole have also been incorporated into the conjugation pathway in the center of the bridge or adjacent to the acceptor, again as a tradeoff between stability and nonlinearity [88–91].

It has also been found that chromophores based on 4-(diarylamino)phenyl donors exhibit enhanced thermal stability relative to 4-(dialkylamino)phenyl analogs, presumably because of their lack of α-hydrogens adjacent to the nitrogen (Fig. 11.9) [78, 80, 93, 94].

11.2.4
High-performance Electooptic Poled-polymer Systems

In this section, we will highlight how recent advances in chromophore and polymer design have led to polymer systems with advantages over inorganic crystals, such as $LiNbO_3$, for electrooptic (EO) applications in the so-called "telecommunications" wavelength range (1.30–1.55 μm); we will also briefly describe how similar systems have already been found to be advantageous in the exciting new area of generating and detecting terahertz (THz) radiation. For telecommunications applications, advantages of organic materials include: the ability to achieve electrooptic coefficients (see below) considerably larger than that of LiNbO$_3$ (30 pm V^{-1} at 1.3 µm), which leads to the possibility of devices with low drive voltages; ease of fabrication and processing; and low density (advantageous for space applications where minimizing weight is critical) [95]. The potential for high bandwidths is another strong point of organic systems for other applications where light of multiple frequencies must be processed. Drawbacks of organic materials relative to LiNbO$_3$ that one must consider include relatively poor thermal and photochemical stability and potentially higher optical losses. A range of efficient EO devices based on poled polymers has now been demonstrated. The basis for many of these devices, and one of the simplest EO devices to understand, is the EO Mach–Zehnder interferometer, a device whereby a light beam is split and each half passes through an EO material along a different path, before being recombined. If the two path lengths are the same, the two light beams will recombine constructively. However, if one of the paths is subjected to an electric field, the refractive index along that path will be modified, inducing a phase shift of the light beam following that path relative to the other. Hence the electric field can be used to switch the light beams from a constructively interfering to destructively interfering situation, thus enabling one to convert an electrical signal to an optical signal.

Firstly, we briefly survey some of the high-$\mu\beta$ organic chromophores that have been developed, starting from the design guidelines discussed in Section 11.2.1. Table 11.1 compares a number of these species with the well-known commercial chromophore Disperse Red (**14**) [96]; $\mu\beta$ values many times greater than that of Disperse Red can be obtained by using longer conjugation pathways and stronger heterocyclic acceptors (these more efficient chromophores would be generally located slightly to the left of the positive maximum in β on the BOA diagram of Fig. 11.5). However, many high-$\mu\beta$ chromophores have relatively high molecular weights; hence we include the ratio of $\mu\beta$ to molecular weight to give some idea of the nonlinearity that might be expected at a given percentage loading into a polymer host. The chromophores included in Table 11.1 exploit many of the structural features discussed in Section 11.2.3, i.e. the incorporation of five-membered aromatic heterocycles into the conjugation path (**15, 16, 18, 19**), use of diarylamino donor groups (**16**) and partial locking of polyene chains using saturated ring systems (**20**). It is also worth noting that the organic polyene **17** exhibits a slightly higher nonlinearity than its close ferrocene analog **7** (Fig. 11.6), despite having a shorter polyene bridge between donor and acceptor. The highest nonlinearities have been obtained with strong heterocyclic acceptors (**17–20**). In particular, the so-called TCF (tricyanofuran) acceptor of **19** and **20** has attracted increasing attention owing to its powerful acceptor strength and high stability. Moreover, TCF derivatives have been developed that are even stronger acceptors (as in chromophore **26**, Fig. 11.12) and that incorporate additional groups for solubility or for attachment to polymerizable or cross-linkable groups. In addition, many of these TCF derivatives are readily accessible through a recently described microwave synthesis [97].

Table 11.1. Some representative NLO chromophores with solution EFISH $\mu\beta$ data measured at 1907 nm and electrooptic data measured at 1330 nm. Adapted from a more extensive table in Ref. [96].

Chromophore		$\mu\beta$ (10^{-48} esu)	$(\mu\beta/M_w)$[a] (10^{-48} esu mol g^{-1})	r_{33} (pm V^{-1})[b]
14	Me₂N–⟨⟩–N=N–⟨⟩–NO₂	580	2.1	13 (30%)
15	Et₂N–⟨⟩–CH=CH–thiophene–CH=C(CN)₂	1300	3.9	
16	Ph₂N–thiophene–CH=CH–thiophene–C(CN)=C(CN)₂	10200	22.1	
17	^{n}Bu₂N–⟨⟩–(CH=CH)₂–benzo[b]thiophene-SO₂–C=C(CN)₂	13500	27.1	55 (20%)
18	^{n}Bu₂N–⟨⟩–CH=CH–thiophene–benzo[b]thiophene-SO₂–C=C(CN)(CN)	15000	27.1	
19	(AcO-CH₂-CH₂)₂N–⟨⟩–CH=CH–thiophene–CH=C-furan-C(CN)=C(CN)₂	18000	25.9	
20	(^{t}BuMe₂SiO-CH₂-CH₂)₂N–⟨⟩–CH=CH–cyclohexenyl(^{n}Bu,^{n}Bu)–CH=CH–furan-C(CN)=C(CN)(CN)	35000	45.7	>60 (30%)

[a] Ratio of $\mu\beta$ to molecular weight of the chromophore.
[b] Electrooptic coefficient measured for poled host–guest systems with the chromophore weight percentage in parentheses.

We mentioned in the previous section that the bulk NLO properties of a poled-polymer host–guest system might be expected to be proportional to the chromophore $\mu\beta$ multiplied by the number density of chromophores. However, for high loadings of molecules with large μ, the d_{33} and r_{33} coefficients do not continue to increase linearly with chromophore loading, but exhibit maxima; these are attributed to competition between the interaction of the chromophores and the poling

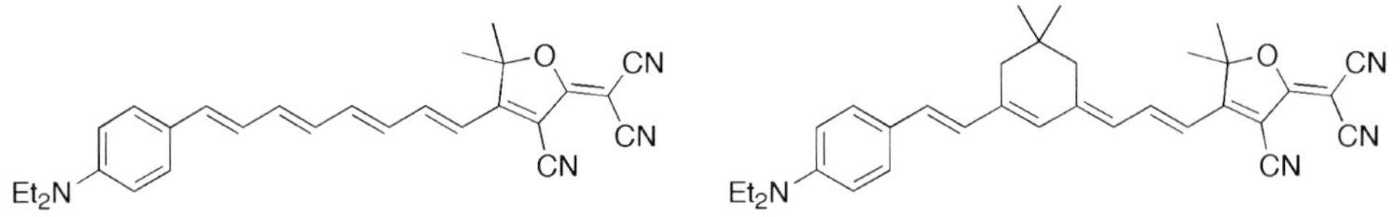

Fig. 11.10 A pair of chromophores with almost identical electronic structures but different shapes and poling behavior [102].

field and significant chromophore–chromophore electrostatic interactions at high chromophore loadings [98, 99]. It is predicted that these poling difficulties should be greatly reduced in the case of chromophores with spherical or, even better, oblate or discoid, shapes relative to the case of "traditionally" shaped chromophores in which the dipole moment lies along the long axis of the molecule [100, 101]. It has been shown experimentally that even the changes in chromophore shape between **21** and **22** (Fig. 11.10) can increase the achievable r_{33} by ca. 20% when loaded at comparable concentrations into a polymer host [102].

More dramatic strategies have been employed to modify the chromophore shape, including several based on dendrimers [96, 103–107]. For example, a poled film of a polycarbonate containing the dendronized chromophore **24** exhibits three times the electrooptic activity of a film containing the same concentration of the non-dendronized chromophore **23** [105]. An additional feature of **24** that deserves comment is the use of pentafluorophenyl groups. We have already alluded to the importance of minimizing optical loss due to chromophores' electronic absorptions and to scattering; however, at telecommunications wavelengths, loss can also occur due to absorption by overtones of C–H vibrational modes. While this issue is particularly important for passive optical materials (for an example of a fluorinated polymer designed for passive applications, see Ref. [108]), through which the light may need to travel for substantial distances, and while it is generally impractical to synthesize active EO materials without C–H bonds, it is desirable to minimize the number of these bonds, for example, by replacing C–H bonds of chromophores by C–F (or even C–D) bonds [109] or by using fluorinated host polymers [110]. The low polarizability of the perfluoroaryl environment of **24** also leads to a blue shift of the absorption maximum relative to **23**, and, thus, to better transparency. The dendronized material is also somewhat more thermally stable. Dendronized chromophores, including highly nonlinear chromophores similar to **20**, have also been cross-linked during the poling process and have been incorporated as side-chains in polymers, with and without cross-linking groups. The same basic chromophore as **23** has also been used as arms of a dendritic-type structure, with trifluorovinyl ether groups as cross-linking groups (**25**, Fig. 11.11) [103], which can also be poled very effectively. Trifluorovinyl ethers undergo efficient [2 + 2]-cycloaddition reactions at moderate temperatures (ca. 140 °C for **25**) and so are suitable for cross-linking EO materials without significant damage to the chromophores [103, 111, 112]. After thermal

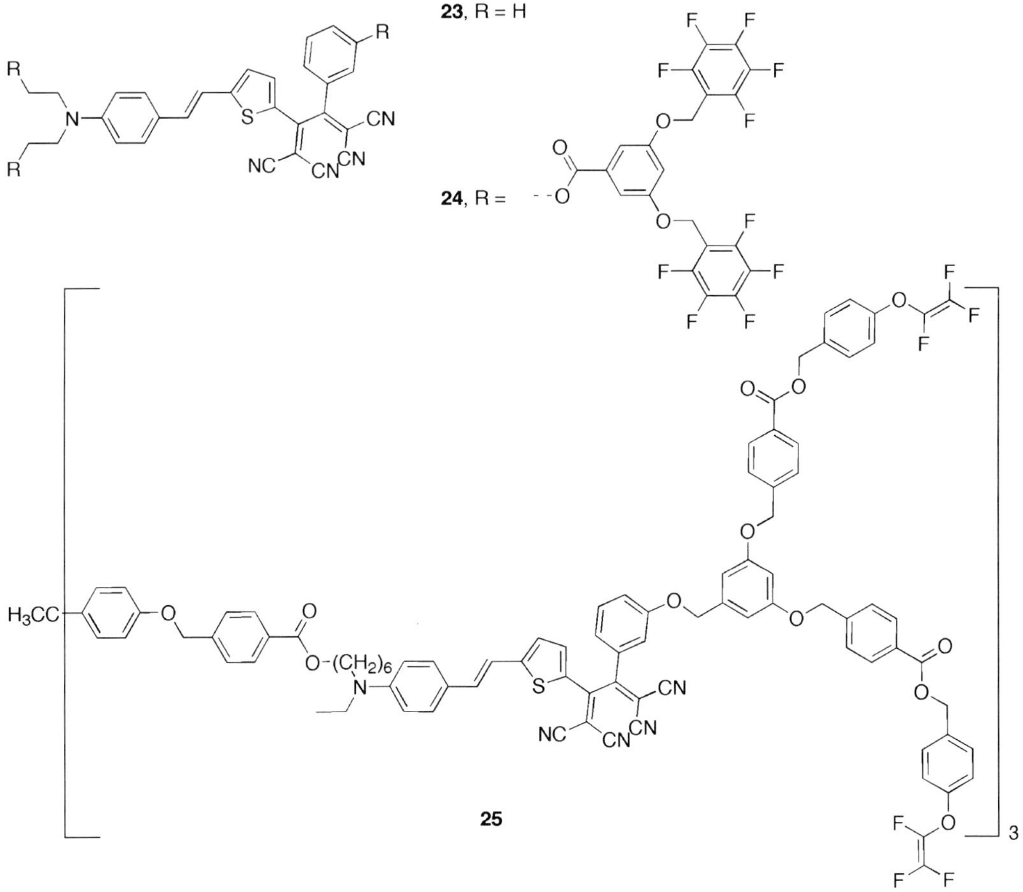

Fig. 11.11 An EO chromophore and two dendronized derivatives.

cross-linking during poling, **25** shows excellent orientational stability with r_{33} retaining 90% of its initial value after 1000 h at 85 °C. Other cross-linking groups that have recently been used in EO materials include benzocyclobutenones [107], which undergo ring-opening above 200 °C to give reactive vinylketenes and maleimide–furan Diels–Alder systems [113]. The latter system is particularly interesting, in that the cross-linking can be thermally reversible or irreversible depending on the substitution of the furan.

Finally, we note that recently EO polymers have been shown to be effective media for the generation and detection of terahertz (THz) radiation [114–117]. This is likely to be an area of significant interest for organic NLO materials owing to its potential applications, ranging from medical [118] and near-field imaging [119] to spectroscopy of biological and chemical agents [120]. THz radiation can be generated in both inorganic (ZnTe is a standard material) and organic materials by pumping the material with short (femtosecond) optical laser pulses,

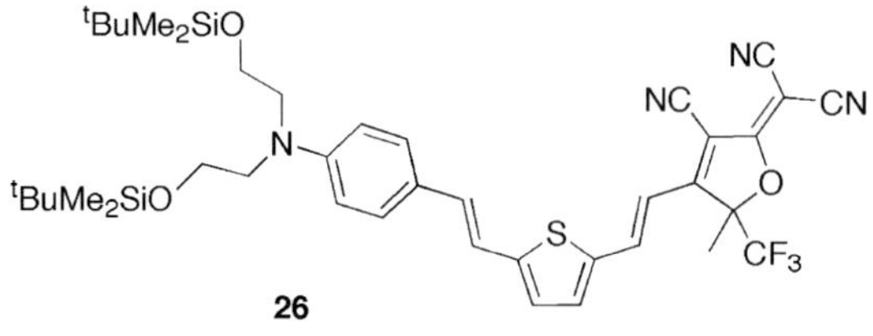

Fig. 11.12 Chromophore used for effective THz generation.

which inevitably, owing to considerations arising from the uncertainty principle, consist of a range of frequencies; difference mixing of these frequencies within a $\chi^{(2)}$ material gives broadband THz radiation [121]. Detection takes place through exploiting the linear electrooptic effect of the electric field associated with incident THz radiation upon polarization of an optical probe beam. Organic films have been shown to exhibit THz generation efficiencies considerably larger than comparable thicknesses of ZnTe in the non-resonant régime, but even more so if one uses a pump radiation close to the absorption maximum of the chromophore; a 3.1-μm thick poled film of **26** (Fig. 11.12, λ_{max} = 710 nm, 20% composite with a polycarbonate) shows a larger THz amplitude than a 1-mm thick sample of ZnTe when pumped at 800 nm and r_{33} is estimated to be over 1250 pm V^{-1} at 800 nm [117]. So far, the organic materials used for THz generation have been limited to those used for other EO applications.

In summary, considerable recent advances have been made in EO poled-polymer systems, in developing highly nonlinear chromophores, in increasing poling efficiency and in increasing the temporal orientational stability of poled systems. It has been predicted that it should be possible to realize EO coefficients an order of magnitude greater than that of LiNbO$_3$ [95]. As researchers continue to improve the materials properties of organic EO materials, they are likely to find real applications, particularly in space applications where the low weight of organics and their resistance to high-energy radiation are advantageous over traditional materials. Meanwhile, the new area of exploiting organic EO materials for THz generation and detection is likely to expand, and other completely new applications may emerge.

11.3
Design and Application of Two-photon Absorbing Chromophores

11.3.1
Essential-state Models for Two-photon Cross-section

As with β and γ, few-state models are useful for understanding structure–property relationships for the two-photon absorption (2PA) cross-section, δ. In dipolar molecules, the peak cross-sections associated with 2PA into an excited state e from the ground state, g, will have a two-level contribution according to

$$\delta_{2-\text{state}} \propto \frac{\mu_{ge}^2 \Delta\mu_{ge}^2}{\Gamma} \tag{20}$$

where μ_{ge} and $\Delta\mu_{ge}$ are the transition dipole moment and change in static dipole moment, respectively, associated with the transition and Γ is a damping term, related to the width of the 2PA absorption band. Of course μ_{ge} also dictates the strength of one-photon absorption into e (the peak absorptivity is proportional to the product of μ_{ge}^2 and ω_{ge}, again divided by a damping term); thus Eq. (20) tells us that states with charge-transfer character (i.e. exhibiting strong solvatochromism and a strong Stark effect) and showing strong one-photon absorption (1PA) are likely to show large δ. These guidelines are reminiscent of those derived from the two-level model for maximizing the static first hyperpolarizability, β_0 (Section 11.2.1). However, the differences in Eqs. (16) and (20) mean that $\delta_{2\text{-state}}$ should be maximized in structures with a balance of neutral and zwitterionic resonance contributions slightly different to that required for maximizing β_0; in terms of the bond-order alternation (BOA) of Fig. 11.5, the maxima in $\delta_{2\text{-state}}$ occur at more negative (more positive) BOA than the maximum (minimum) in β_0.

For centrosymmetric molecules, however, $\Delta\mu_{ge} = 0$ and accordingly $\delta_{2\text{-state}} = 0$. Moreover, 2PA into one-photon-allowed states is forbidden according to the parity selection rule; whereas one-photon transitions in centrosymmetric systems are accompanied by a change in state parity (g→u or u→g), 2PA is only possible between states of the same parity (g→g or u→u). Indeed, this feature has long been exploited in spectroscopy to obtain information complementary to that accessible from 1PA; for example, see Refs. [122] and [123].

In centrosymmetric systems, it is only three-level terms (which can also contribute in dipolar systems, their significance depending on the relative magnitude of two- and three-level terms for the state in question) that are responsible for δ. For absorption into a state, e′:

$$\delta_{3-\text{state}} \propto \left(\frac{\omega_{ge'}}{2}\right)^2 \frac{\mu_{ge}^2 \mu_{ee'}^2}{[\omega_{ge} - (\omega_{ge'}/2)]^2 \Gamma} \tag{21}$$

where $\omega_{ge'}$ is the frequency of the transition and so $\omega_{ge'}/2$ is the photon frequency, μ_{ge} and $\mu_{ee'}$ are transition dipole moments, ω_{ge} is the frequency corresponding to the energy of an intermediate state above the ground state and Γ is a damping term. It should be noted that both Eqs. (20) and (21) apply only in the case of degenerate 2PA; when the two photons have different energies (as in the pump–probe WLC method discussed in Section 11.1.6), considerably more complex two- and three-level expressions are obtained (see Ref. [30]). In many cases, a single three-level model can be used to account for a given 2PA peak; however, in general, multiple alternative "pathways" involving alternative intermediate states, e, between the ground state, g, and final 2PA excited state, e′, may contribute to the cross-section for g to e′ transition.

The dependence of δ upon μ_{ge} and $\mu_{ee'}$ tells us that absorption from the ground state to the intermediate state should be strongly one-photon allowed (which one can, of course, readily establish from linear spectra) and that absorption from the intermediate to two-photon state should also be strongly allowed (which is less easily probed experimentally). The term $\omega_{ge} - (\omega_{ge'}/2)$ is known as the "detuning term" and indicates that 2PA will be strongest when the intermediate state lies half way in energy between the ground state and the 2PA state; this situation is known as "double resonance". However, at or near double resonance, significant 1PA is expected at the same photon frequency and, hence, (i) δ will be rather difficult to measure reliably and (ii) the advantages of 2PA for potential applications will be lost. Ideally, to maximize δ one would like the intermediate state to be as close in energy to the double resonance situation as possible, whilst retaining sufficient detuning that there is no one-photon absorption at $\omega_{ge'}/2$.

11.3.2
Chromophore Designs

Our own interest in 2PA was kindled by the observation by Perry's group of strong blue fluorescence when a toluene solution of (E)-4,4'-bis(di-n-butyl)aminostilbene (**1**) (Figs. 11.2 and 11.13), was irradiated with 605 nm 5-ns laser pulses, a wavelength at which this species (λ_{max} = 385 nm) shows no linear absorption. Moreover, the fluorescence intensity was proportional to the square of the laser light intensity. This two-photon-excited fluorescence spectrum was essentially identical

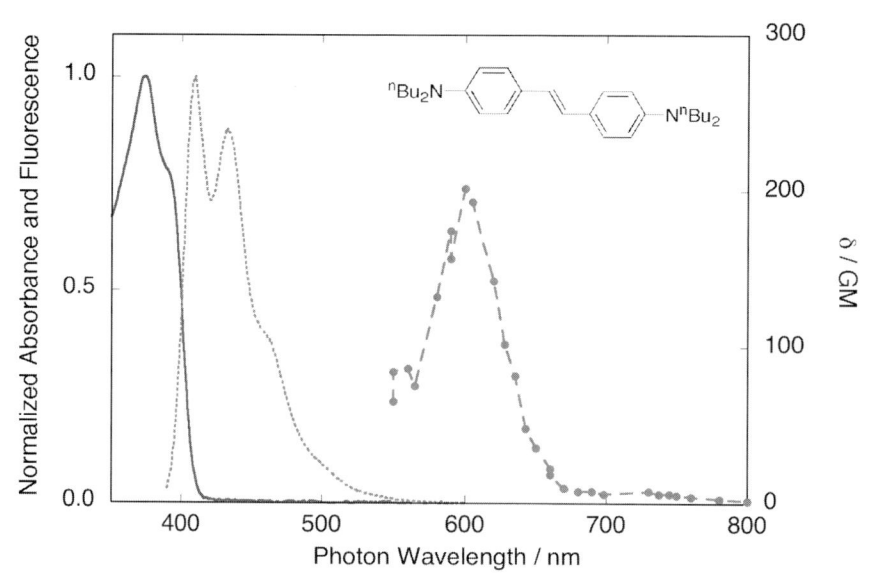

Fig. 11.13 1PA (solid line) and fluorescence (dotted line) spectra of (E)-4,4'-bis(di-n-butylamino)stilbene (**1**) (Fig. 11.2), compared with the 2PA spectrum determined by the 2PF method (data points and broken line) [23]. This plot against photon wavelength emphasizes that blue fluorescence (peaking at 410 nm) can be obtained from up-conversion of 605-nm incident light.

with that excited by 1PA, consistent with Kasha's rule and suggesting that there was rapid relaxation of the 2PA state to the same state accessed by 1PA and subsequent fluorescence from that state. Measurement of the two-photon excitation cross-section for **1** using the 2PF method gave $\delta_{max} = 210$ GM (see the Appendix, Section 11.4, for a discussion of the units of 2PA) at an excitation wavelength of 605 nm, which is almost 20 times greater than the peak cross-section for *trans*-stilbene and was, at the time, among the largest values of δ reported for organic compounds [22, 124].

To gain insight into the different δ_{max} values for **1** and *trans*-stilbene, quantum-chemical calculations were performed [22, 125]. The 2PA spectra were obtained from the dispersion of $\text{Im}[\gamma(-\omega;\omega,\omega,-\omega)]$ [Eq. (14)] using the sum-over-states expression and reproduced the experimentally observed order-of-magnitude enhancement in δ_{max} on substitution of *trans*-stilbene with terminal dimethylamino groups. From the calculations, it was shown that the lowest singlet excited state is a one-photon allowed state with B_u symmetry, whereas the second singlet excited state has A_g symmetry and corresponds to the observed 2PA peak. Figure 11.14 displays the calculated terms entering the three-level expression for δ [Eq. (21)], with the B_u state as the intermediate 1PA state. The changes in these calculated values of all three of these parameters contribute to the observed increase in δ on bis(donor) substitution of *trans*-stilbene; however, although μ_{ge} increases and the detuning term $[\omega_{ge} - (\omega_{ge'}/2)]$ decreases, it is the increase in $\mu_{ee'}$ that is most dramatic. The calculations also show that the g→e excitation is accompanied by a substantial charge transfer from the amino groups to the central vinylene group, leading to a large change in quadrupole moment upon excitation; a similar sense and magnitude of charge transfer are calculated for the g→e' transition. This pronounced redistribution of the π-electronic density is correlated with an increase of electron delocalization in the first excited state and results in a significant increase in the e→e' transition dipole moment, which is the major contri-

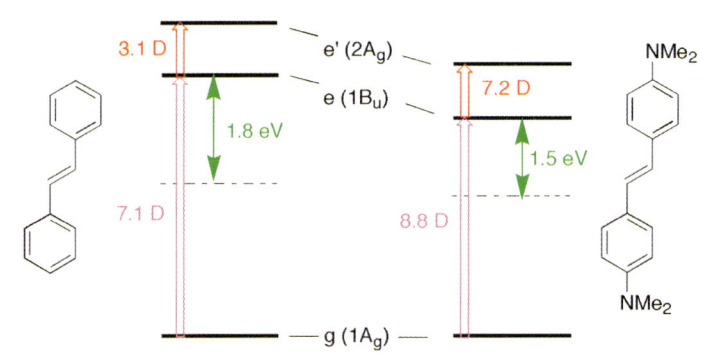

Fig. 11.14 Energy level diagram showing the lowest energy 1PA- and 2PA-allowed states (e and e', respectively) for *trans*-stilbene (left) and (E)-4,4'-bis(dimethylamino)stilbene (right) according to the quantum-chemical calculations results of Ref. [22]. The dotted lines represent the virtual states midway between g and e'. Calculated transition dipoles μ_{ge} and $\mu_{ee'}$ are shown in purple and red respectively; calculated detuning energies are indicated in green.

butor to the enhanced value of **1** with respect to that of *trans*-stilbene. Another consequence of the terminal substitution with electron donors is a shift of the position of the two-photon resonance to lower energy.

These results suggested several strategies to enhance δ_{max} and to tune the wavelength of the 2PA peak for π-conjugated organic molecules. Because the symmetric charge transfer and change in quadrupole moment appear to be important for molecules with small ground-state mesomeric quadrupole moments, we reasoned that structural features that could further enhance the change in quadrupole moment upon excitation might be beneficial for enhancing the corresponding transition dipole moments and the magnitude of δ. Consequently, we examined systems in which the π-conjugation length between the two-donor groups ("D–π–D" chromophores) is increased [22, 23] and in which acceptor groups were attached to the center of the π-conjugated bridge ("D–A–D" chromophores) [22]. These strategies are indeed effective in increasing δ: Fig. 11.15 shows an example, **27**, where both the conjugation length is increased and acceptors are introduced on the bridge. Systems in which the sense of the symmetric charge transfer is reversed ("A–π–A" and "A–D–A" chromophores) were also shown to exhibit large δ [22].

A large number of "D–π–D" and "D–A–D" chromophores with amine donors have subsequently been investigated [20, 126–132]. These include compounds with shorter conjugation pathways than that in **1** to blue shift the accessible

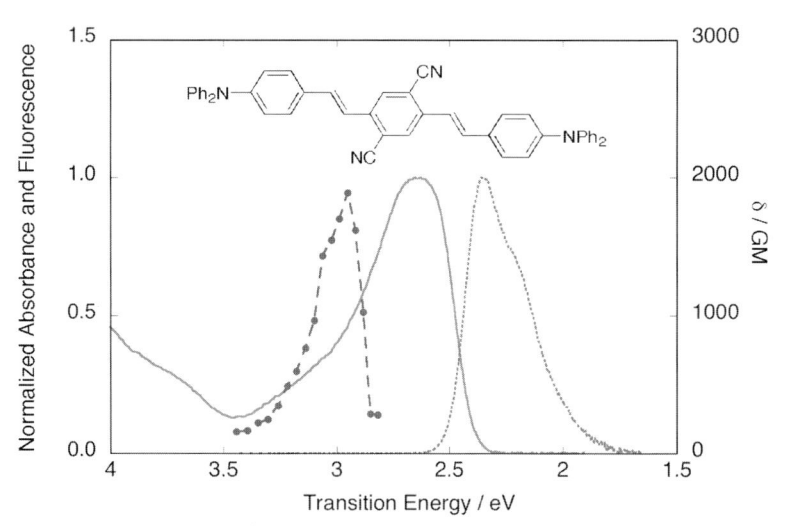

Fig. 11.15 Comparison of the 2PA spectrum (data points and broken line) of a D–A–D chromophore, **27**, determined by the 2PF technique with the 1PA spectrum (solid line) and fluorescence spectrum (dotted line) [22, 126]. Plotting the data versus transition energy (rather than photon energy or wavelength, as in Fig. 11.13) emphasizes how 1PA and 2PA access different states, as expected from the parity selection rules applicable in centrosymmetric systems. Note that δ_{max} is approximately an order of magnitude larger than in the shorter D–π–D molecule, **1**.

Table 11.2 2PA data for some D–π–D and D–A–D chromophores: all data from 2PF measurements [22, 23, 126, 132].

Chromophore		$\lambda^{(2)}_{max}$ [a] (nm)	δ_{max} [a] (GM)
1		605	210
28		640	260
29		730	1300
30		730	900
31		830	1750
32		840	1420
33		970	5300

[a] Photon wavelength and cross-section corresponding to the peak in the 2PA spectrum.

2PA peak [30], those with much longer bridges to red shift the maximum and increase the cross-section for the 2PA peak [131, 132] and those in which phenylene bridging groups are replaced by other aromatic units [131] or in which vinylene bridges are replaced with ethynylene bridges [130]. Table 11.2 summarizes 2PA data from the Perry and Marder groups for a range of bis(dialkylamino)phenyl derivatives with polyene and phenylene-vinylene bridges.

"A–π–A" and "A–D–A" chromophores have been less widely investigated [22, 133–136], but cross-sections as high as 8000 GM have recently been measured using Z-scan for perylene derivatives such as **34** (Fig. 11.16); these large values are principally due to small detuning energies, rather than especially remarkable transition dipole moments, since the highest values were measured close to double resonance [137].

Another class of more or less symmetric chromophores exhibiting strong 2PA are porphyrin oligomers and polymers [28, 138–143]. Materials from Anderson's group show large real and imaginary components of $\chi^{(3)}$ at 1064 nm [28, 138–140], whereas other porphyrin oligomers have been shown to have large $\chi^{(3)}$ at 800 nm [144]. The large imaginary values can be attributed to strong 2PA, whereas the magnitude of the real values is enhanced significantly by two-photon

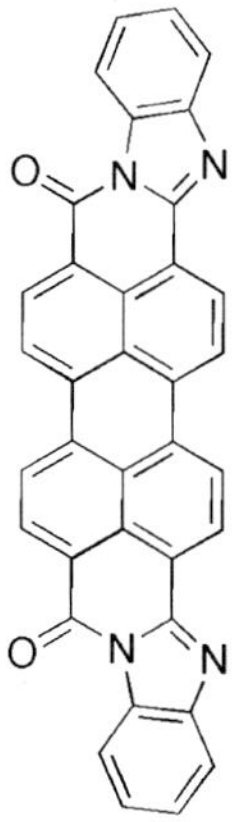

Fig. 11.16 Some chromophores with high 2PA cross-sections discussed in the text.

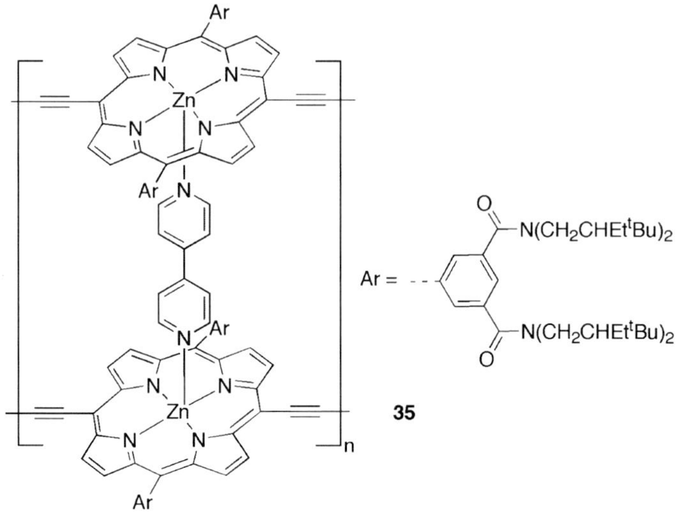

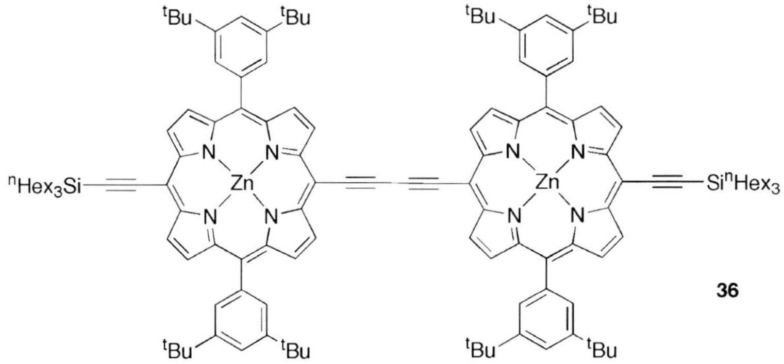

resonance. For the double-strand polymer **35** (Fig. 11.16), DFWM data show that Im($\chi^{(3)}$) is considerably larger than for single-stranded analogs and δ_{1064} is estimated to be 50 000 GM per porphyrin unit [28]. More recently, 2PA spectra have been acquired for a number of examples using the 2PF method. Whereas a monomeric porphyrin model showed δ_{max} = 20 GM at a photon wavelength of ca. 850 nm, dimers with a variety of conjugated bridges showed huge values in this range; **36** (Fig. 11.16) showed δ_{max} = 11 000 GM at a photon wavelength of ca. 880 nm [141, 142]. An important factor contributing to the dimers is that 2PA state appears at an energy only slightly detuned from resonance with the 1PA state; as with the perylene-based materials (see above), these species have reasonably sharp cut-offs on the low-energy side of the lowest energy one-photon transition, thus permitting one to approach the doubly resonant situation relatively closely without interference from 1PA.

Fig. 11.17 Structures of dipolar 2PA chromophores.

Dipolar donor–acceptor (D–π–A) chromophores have also been investigated [20, 129, 145, 146]. In quadrupolar chromophores the selection rules are mutually exclusive for 1PA and 2PA and often the lowest excited state is 1PA allowed; thus, 2PA takes place at a photon energy greater than half that of the 1PA maximum. In dipolar systems, 2PA can take place at half the 1PA energy, this expanding the potential energy range over which 2PA can be measured and potentially exploited. In addition, dipolar chromophores often have low-energy charge-transfer-type transitions and are therefore good candidates for achieving 2PA well into the near-IR region. For example, the lowest lying 2PA state of **37** (Fig. 11.17) shows a δ_{max} of 200 GM according to femtosecond Z-scan data [146], similar to the 210 GM determined using 2PF for the structurally closely related D–π–D chromophore, **1** (see above). However, the 2PA absorption maximum for the D–π–A corresponds to a photon energy of ca. 900 nm, considerably red shifted relative to the 605 nm seen for its D–π–D analog. For **38** (Fig. 11.17), δ_{max} was measured using the WLC pump–probe method (Section 11.1.6) as 1300 GM at probe and pump wavelengths of 670 and 1210 nm, respectively (this corresponds to degenerate 2PA of photons of ca. 860 nm, for which δ is expected to be somewhat lower) [145].

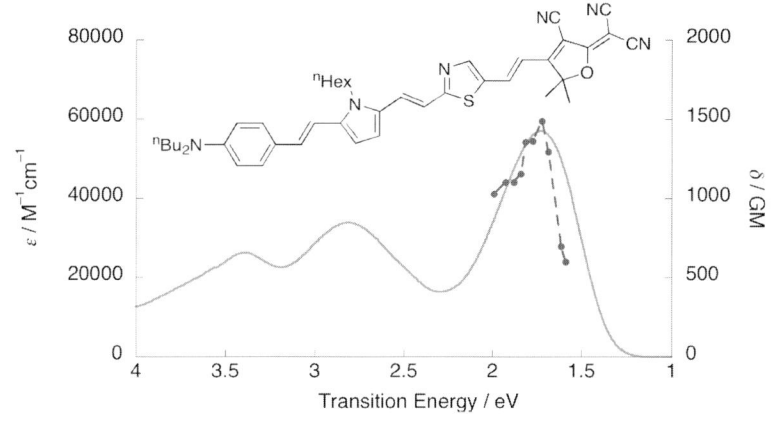

Fig. 11.18 Structure of a dipolar chromophore, **39**, along with its 1PA (solid line) and 2PA (dashed line) spectra, both plotted as a function of transition energy, emphasizing that 1PA and 2PA access the same state [147]. The 2PA spectrum was obtained by the WLC pump–probe method using a pump wavelength of 1800 nm.

Recently, we have reported a non-degenerate pump–probe value of δ = ca. 1500 GM for **39** (Fig. 11.18) in the telecommunications wavelength range (1.3–1.55 µm) [147]. Quantum-chemical calculations suggest that the value for degenerate 2PA is significantly (ca. 40%) lower, but still large. The transition energy corresponding to two 1.44-µm photons and, as expected for a strongly dipolar chromophore, this corresponds to the same state accessed by 1PA.

In addition to the quadrupolar and dipolar chromophores discussed above, several other chromophore designs have been investigated, including octupolar systems and other branched and linked molecules, including dendrimers and species linked through cyclophane bridges [148–159]; in some cases these can be regarded as composed of interacting dipolar or quadrupolar chromophores, whereas in others the constituent chromophores cannot or do not interact significantly. In an example of a system with cooperative interactions between the constituent chromophores, Fig. 11.19 shows how δ_{max} for (E)-4,4'-bis(diarylamino)stilbene-based oligomers increases in a superlinear fashion with chromophore size [155]. However, in higher generation dendrimers, the ratio of δ_{max} to the number of constituent chromophores seems to saturate [152, 155].

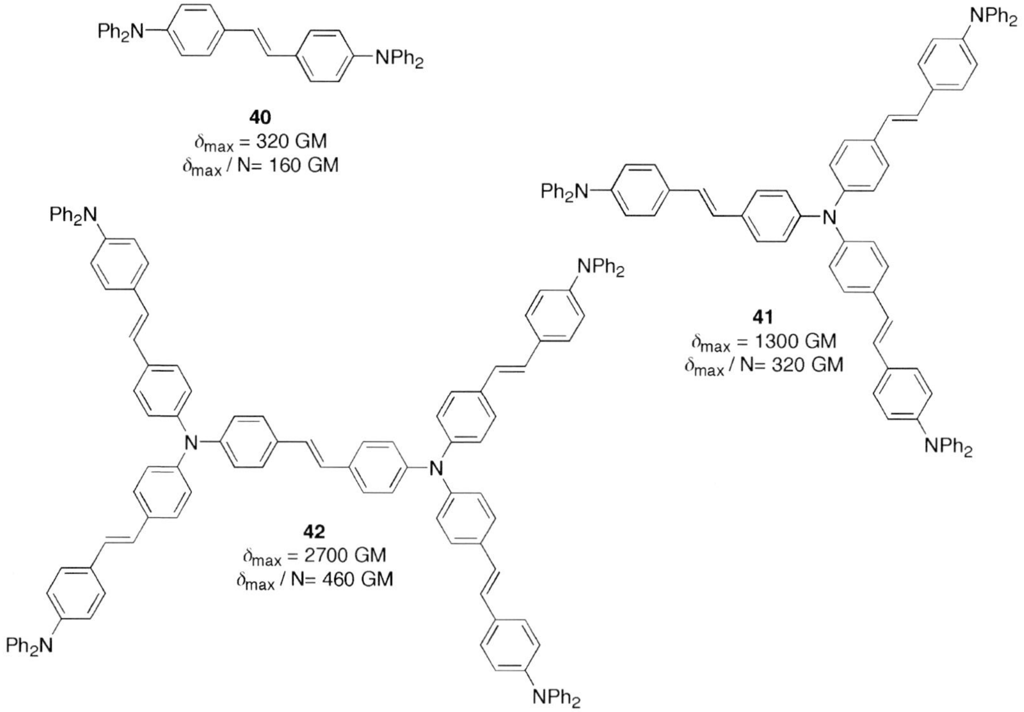

40
δ_{max} = 320 GM
δ_{max}/N = 160 GM

41
δ_{max} = 1300 GM
δ_{max}/N = 320 GM

42
δ_{max} = 2700 GM
δ_{max}/N = 460 GM

Fig. 11.19 Structures and data showing how δ_{max} (2PF method) for dendritic-type bis(diarylamino)stilbene chromophores increase in a superlinear fashion with the number of triarylamine groups, N [155]. In all three cases, 2PA peaks fall in the range $\lambda^{(2)}_{max}$ = 670–694 nm.

11.3.3
Applications of Two-photon Absorption

In Section 11.1.5 we discussed how 2PA allows for the creation of excited states with high 3D resolution at depth in absorbing media. In this section, we will discuss work focused on exploiting these properties and other properties of 2PA. Many of these applications were proposed or demonstrated using traditional 1PA chromophores with low 2PA cross-sections; however, their effectiveness, and consequently interest in these applications, have greatly increased with the recent developments in 2PA chromophores described in Section 11.3.2.

2PA-induced photopolymerization is one such example that was first demonstrated using direct photoexcitation of monomers or through the use of standard UV photoinitiators irradiated with visible light [160–162]. Most common photoinitiators have rather small cross-sections [163]; hence, recently developed high-δ chromophores such as **1** and **30** are considerably more effective as 2PA initiators at 600 nm of radical polymerization of acrylates than standard UV initiators [164]. The polymerization of acrylates by chromophores such as **1** and **30** is believed to be initiated by electron transfer from the excited state of the dye to the acrylate monomer and subsequently initiation of polymerization by the acrylate radical anion. Negative-tone resist materials based on acrylates have been used for 2PA microfabrication. The pattern of interest is traced out using a focused laser beam in a resin composed of a cross-linkable acrylate monomer, a 2PA chromophore. Cross-linking of the acrylates to form an insoluble solid takes place along and close to the path traced by the beam. Finally, unreacted monomer and the binder are removed with solvent, leaving the insoluble cross-linked structure. This type of microfabrication has been exploited by a number of groups to make structures with potential use as photonic bandgaps, waveguides, photoresponsive microstructures and microelectromechanical systems [164–173]. An example of a cross-linked acrylate structure obtained in this way is shown in Fig. 11.20.

Fig. 11.20 A photonic crystal lattice fabricated by two-photonic lithography.

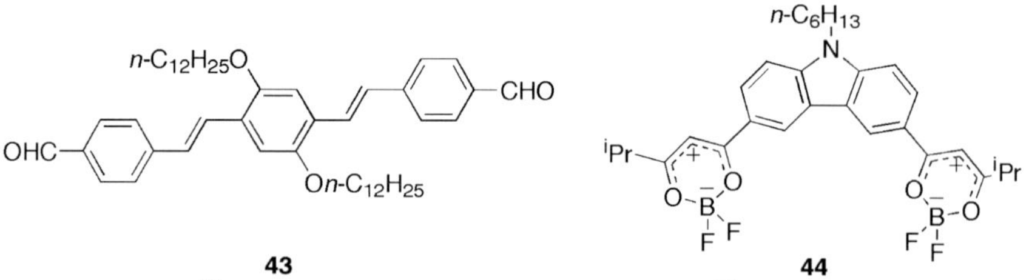

Fig. 11.21 Structures and properties of 2PA dyes used for the deposition of metallic silver.

Direct 2PA-induced electron-transfer reactions have also been used to deposit metallic silver wires. In this case, D–π–D dyes such as **1** (and the other species of Table 11.2) cannot be used as initiators since the ground-state molecule, in addition to the excited state, is capable of reducing Ag^+. However, A–D–A dyes such as **43** and **44** (Fig. 11.21), which are much less easily oxidized than **1**, have been used to write Ag lines from a resin composed of 2PA dye, thiol-coated Ag nanoparticles, $AgBF_4$, poly(N-vinylcarbazole) and N-ethylcarbazole [133, 174]. Compound **43** has also been used to deposit Au metal [174].

Although direct excited-state electron transfer from 2PA dyes to monomer is successful for polymerizing acrylates and depositing silver, few other materials can be patterned in the same way; for effective initiation, the reduction potential for the monomer, $E_{1/2}(M/M^{-\bullet})$, needs to be greater, i.e. less negative, than the excited-state oxidation potential for the initiator, $E_{1/2}(M^+/M^*)$, which can be estimated from

$$E_{1/2}(M^+/M^*) = E_{1/2}(M^+/M) - E_{0,0} \tag{22}$$

where $E_{0,0}$ is the excited-state energy of the 2PA initiator, for example, as estimated from the intersection of normalized absorption and fluorescence spectra. Moreover, this method is unsuitable for patterning positive tone resists (materials that become more soluble on exposure). In addition, polymerization of acrylates is typically accompanied by shrinkage, which can lead to poor fidelity of the 2PA-writing process. As a step towards addressing these issues, a 2PA photoacid has been developed that shows a high 2PA cross-section and high quantum yield for acid generation (Fig. 11.22) [175–177]. The design strategy for this material is also based on excited-state electron transfer, in this case intramolecular electron transfer from the central D–π–D chromophore to one of the terminal sulfonium groups, where subsequent scission of the S–Me bond ultimately leads to proton generation. This photoacid has been used to pattern negative tone resists based on epoxides. and positive tone resists where a polymer is solubilized by the acid-induced cleavage of tetrahydropyran moieties.

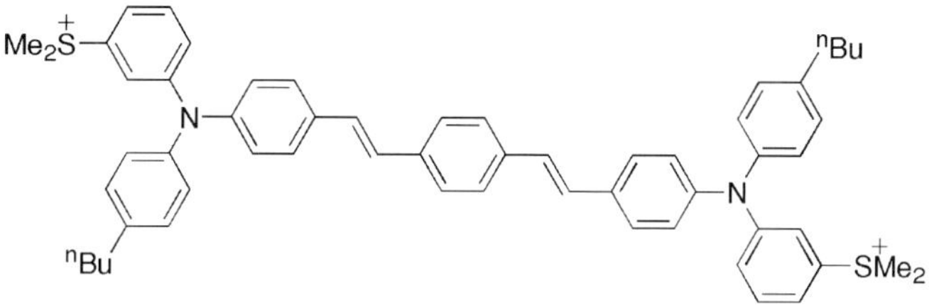

Fig. 11.22 Structure and properties of a 2PA photoacid [175–177].

$\lambda^{(2)}_{max}$ = 710 nm
δ_{max} = 690 GM
$\Phi(H^+)$ = 0.5

2PA-induced acrylate polymerization has also been used as a way of writing bits for high-density read-only memory devices. In early work, acrylates were polymerized directly, with the information readable through the different refractive indices of acrylate monomers and polymerized acrylates; however, high-power excitation was required due to the low cross-sections of the materials employed [160]. An alternative scheme involves resins incorporating effective 2PA dyes similar to **1**, but functionalized with covalently attached acrylate groups. In unexposed regions, the fluorescence of the chromophore is largely quenched by electron transfer to the acrylates, whereas in exposed regions where bits are written, the fluorescence is "turned on" on by polymerization of the acrylates [164, 178]. Several alternative 2PA memory schemes, based both on conventional 1PA chromophores and on newer 2PA chromophores, have also been reported.

Another class of potential applications centers around the use of 2PA dyes for biological and medical applications. The two-photon microscope enables one to obtain highly resolved 3D images at depth in biological media [179]. Generally, fluorophores with specific binding characteristics, often conferred by binding groups covalently attached to the basic chromophore, are used to label the structures or chemical species of interest and the two-photon microscope is used to create a 3D map of fluorescence intensity. Much of the work done with two-photon microscopy has employed standard chromophores originally optimized for one-photon excitation and generally having small cross-sections. Some progress has been made in making the high-δ chromophores described in Section 11.3.2 compatible with aqueous and biological systems and in investigating their potential for membrane labeling [128, 180, 181]. High-δ chromophores have been covalently linked to crown ethers, which can bind metal ions (albeit not in aqueous media) with changes in the 2PA response [182], to nuclei acids [183] and to an anti-cancer drug [184]. Recent reports, however, suggest that water-soluble inorganic quantum dots may have some advantages over organic chromophores for biological 2PA applications [185]. In addition to imaging and labeling applications, 2PA chemistry may potentially be used to release or sensitize pharmacologically active chemical species with high 3D resolution. Already it has been shown that $^1O_2$ [186–189] and NO [190] can be generated using systems incorporating 2PA chromophores.

In addition, some applications of 2PA do *not* specifically exploit the spatial characteristics. The use of 2PA for optical pulse suppression simply takes advantage of the nonlinear transmission of 2PA materials, i.e. that the transmission of a sample decreases with incident light intensity, as shown in Eq. (15) (Section 11.1.6). Comparison of 2PA with other mechanisms for obtaining strong nonlinear transmission has been covered elsewhere in a review [191]. Here, we merely note that efficient optical pulse suppression can be observed using 2PA processes, especially when 2PA is followed by excited-state absorption [136, 192–194]. Moreover, for certain chromophores or mixtures of chromophores, a broadband response across the visible–near-IR region has been demonstrated [21, 195, 196].

Finally, we note that 2PA, and more generally simultaneous multi-photon absorption processes, has been used as a mechanism for pumping organic lasers using light available from semiconductor lasers [197–203].

11.4
Appendix: Units in NLO

Although the bulk SHG and electrooptic coefficients, d_{33} and r_{33}, are frequently quoted in pm V^{-1}, many NLO quantities (and most of those discussed in this chapter) are typically expressed in c.g.s. units. These are not usually stated explicitly, but "esu" is used to denote that the units are the appropriate "electrostatic units" for that quantity in the c.g.s. system. We have followed this practice in this chapter. Some important conversions from esu to SI units are given in Table 11.3. In addition, we have followed the practice of quoting 2PA cross-sections in Goeppert-Mayer (GM) units, which are named after the scientist who first predicted 2PA processes [204, 205], and defined as 1 GM = 10^{-50} cm^4 s (photon)$^{-1}$. In addition, readers should note that cross-sections are sometimes quoted in cm^4 GW^{-1}; cross-sections in these units can be converted to cm^4 s (photon)$^{-1}$ by multiplying by the photon energy.

Table 11.3 Conversions for some properties important in NLO[a].

Quantity	SI units	Conversion from esu[b]
Dipole moment, μ	C m	$\mu_{SI} = (1/3) \times 10^{-11} \times \mu_{esu}$[c]
β	C m^3 V^{-2}	$\beta_{SI} = (1/3)^3 \times 10^{-19} \times \beta_{esu}$
$\mu\beta$	C^2 m^4 V^{-2}	$\mu\beta_{SI} = (1/3)^4 \times 10^{-30} \times \mu\beta_{esu}$
$\chi^{(2)}$	m V^{-1}	$\chi^{(2)}_{SI} = (4\pi/3) \times 10^{-4} \times \chi^{(2)}_{esu}$
γ	C^2 m^4 V^{-3}	$\gamma_{SI} = (1/3)^4 \times 10^{-23} \times \gamma_{esu}$
$\chi^{(3)}$	m^2 V^{-2}	$\chi^{(3)}_{SI} = (4\pi/3)^2 \times 10^{-8} \times \chi^{(3)}_{esu}$

[a] Adapted from a more extensive table in Ref. [10].
[b] The subscripts "SI" and "esu" denote the quantity in question expressed in those unit conventions.
[c] 1 D = 10^{-18} esu.

Acknowledgments

We would like to thank Mariacristina Rumi and Joe Perry for the data plotted in Figs. 11.13 and 11.15, Jie Fie, David Hagan and Eric Van Stryland for the data plotted in Fig. 11.18 and Joe Perry for Fig. 11.20.

References

1. A. Willetts, J. E. Rice, D. M. Burland, D. P. Shelton, *J. Chem. Phys.* **1992**, *97*, 7590.
2. S. K. Kurtz, T. T. Perry, *J. Appl. Phys.* **1968**, *39*, 3798.
3. I. R. Whittall, M. P. Cifuentes, M. J. Costigan, M. G. Humphrey, S. C. Goh, B. W. Skelton, A. H. White, *J. Organomet. Chem.* **1994**, *471*, 193.
4. D. S. Chemla, J. L. Oudar, J. Jerphagnon, *Phys. Rev. B* **1975**, *12*, 4534.
5. C. G. Bethea, *Appl. Opt.* **1975**, *14*, 1447.
6. E. Hendrickx, K. Clays, A. Persoons, *Acc. Chem. Res.* **1988**, *31*, 675.
7. K. Clays, A. Persoons, *Phys. Rev. Lett.* **1991**, *66*, 2980.
8. I. D. Morrison, R. G. Denning, W. M. Laidlaw, M. Stammers, *Rev. Sci. Instrum.* **1996**, *67*, 1445.
9. K. Clays, K. Wostyn, A. Persoons, *Adv. Funct. Mater.* **2002**, *12*, 557.
10. I. R. Whittall, A. M. McDonagh, M. G. Humphrey, M. Samoc, *Adv. Organomet. Chem.* **1998**, *42*, 291.
11. J. Zyss, *J. Chem. Phys.* **1979**, *71*, 909.
12. M. J. S. Dewar, J. J. P. Stewart, *Chem. Phys. Lett.* **1984**, *111*, 416.
13. M. J. S. Dewar, C. H. Reynolds, *J. Comput. Chem.* **1986**, *7*, 140.
14. S. M. Colwell, C. W. Murray, N. C. Handy, R. D. Amos, *Chem. Phys. Lett.* **1993**, *210*, 261.
15. B. Delley, in *Modern Density Functional Theory: a Tool for Chemistry*, J. M. Seminario, P. Politzer (Eds.), Elsevier, Amsterdam, **1995**, 221.
16. D. Pugh, J. O. Morley, in *Nonlinear Optical Properties of Organic Molecules and Crystals*, D. S. Chemla, J. Zyss (Eds.), Academic, New York, **1987**, 193.
17. P. Kaatz, D. P. Shelton, *J. Chem. Phys.* **1996**, *105*, 3918.
18. P. Kaatz, E. A. Donley, D. P. Shelton, *J. Chem. Phys.* **1998**, *108*, 849.
19. I. R. Whittall, A. M. McDonagh, M. G. Humphrey, M. Samoc, *Adv. Organomet. Chem.* **1998**, *43*, 349.
20. X. Wang, D. Wang, G. Y. Zhou, W. Yu, Y. Zhou, Q. Fang, M. Jiang, *J. Mater. Chem.* **2001**, *11*, 1600.
21. J. E. Ehrlich, X. L. Wu, L.-Y.S. Lee, Z.-Y. Hu, H. Röckel, S. R. Marder, J. W. Perry, *Opt. Lett.* **1997**, *22*, 1843.
22. M. Albota, D. Beljonne, J.-L. Brédas, J. E. Ehrlich, J.-Y. Fu, A. A. Heikal, S. E. Hess, T. Kogej, M. D. Levin, S. R. Marder, D. McCord-Maughon, J. W. Perry, H. Röckel, M. Rumi, G. Subramanian, W. W. Webb, X.-L. Wu, C. Xu, *Science* **1998**, *281*, 1653.
23. M. Rumi, J. E. Ehrlich, A. A. Heikal, J. W. Perry, S. Barlow, Z. Hu, D. McCord-Maughon, T. C. Parker, H. Röckel, S. Thayumanavan, S. R. Marder, D. Beljonne, J.-L. Brédas, *J. Am. Chem. Soc.* **2000**, *122*, 9500.
24. C. Xu, W. W. Webb, *J. Opt. Soc. Am. B* **1996**, *13*, 481.
25. M. Sheik-bahae, A. A. Said, E. W. Van Stryland, *Opt. Lett.* **1989**, *14*, 955.
26. M. Sheik-bahae, A. A. Said, T.-H. Wei, D. J. Hagan, E. W. Van Stryland, *IEEE J. Quantum Electron.* **1990**, *26*, 760.
27. S. R. Friberg, P. W. Smith, *IEEE J. Quantum Electron.* **1987**, *23*, 2089.
28. T. E. O. Screen, J. R. G. Thorne, R. G. Denning, D. G. Bucknall, H. L. Anderson, *J. Am. Chem. Soc.* **2002**, *124*, 9712.
29. R. A. Negres, J. M. Hales, A. Kobyakov, D. J. Hagan, E. W. Van Stryland, *Opt. Lett.* **2002**, *27*, 270.
30. J. M. Hales, D. J. Hagan, E. W. V. Stryland, K. J. Schafer, A. R. Morales, K. D. Belfield, P. Pacher, O. Kwon, E. Zojer, J. L. Bredas, *J. Chem. Phys.* **2004**, *121*, 3152.
31. J. L. Oudar, *J. Chem. Phys.* **1977**, *67*, 446.

32. J. L. Oudar, D. S. Chemla, *J. Chem. Phys.* **1977**, *66*, 2664.
33. S. Barlow, H. E. Bunting, C. Ringham, J. C. Green, G. U. Bublitz, S. G. Boxer, J. W. Perry, S. R. Marder, *J. Am. Chem. Soc.* **1999**, *121*, 3715.
34. S. Di Bella, *New J. Chem.* **2002**, *26*, 495.
35. K. D. Singer, S. J. Lalama, J. E. Sohn, R. D. Small, in *Nonlinear Optical Properties of Organic Molecules and Crystals*, D. S. Chemla, J. Zyss (Eds.), Academic Press, New York, **1987**, 437.
36. K. D. Singer, M. G. Kuzyk, J. E. Sohn, *J. Opt. Soc. Am. B* **1987**, *4*, 968.
37. S. R. Marder, C. B. Gorman, F. Meyers, J. W. Perry, G. Bourhill, J.-L. Brédas, B. M. Pierce, *Science* **1994**, *265*, 632.
38. S. R. Marder, J. W. Perry, G. Bourhill, C. B. Gorman, B. G. Tiemann, K. Mansour, *Science* **1993**, *261*, 186.
39. C. B. Gorman, S. R. Marder, *Proc. Natl. Acad. Sci. USA* **1993**, *90*, 11297.
40. In addition to the strong negative peak at BOA = 0, γ_0 shows positive peaks at intermediate positive, intermediate negative BOA; this more complex pattern can be understood in the light of the two-level expression for γ_0 (M. G. Kuzyk, C. W. Dirk, *Phys. Rev. B* **1990**, *41*, 5098) in which there is a more complex interplay of the same parameters as those appearing in the expression for β_0:

$$\gamma_0 \propto \frac{\mu_{ge}^2 (\Delta\mu_{ge})^2 - \mu_{ge}^4}{\omega_{ge}^2}$$

41. S. R. Marder, L.-T. Cheng, B. G. Tiemann, A. C. Friedli, M. Blanchard-Desce, J. W. Perry, J. Skindhøj, *Science* **1994**, *263*, 511.
42. B. G. Tiemann, L. T. Cheng, S. R. Marder, *J. Chem. Soc., Chem. Commun.* **1993**, 735.
43. L.-T. Cheng, W. Tam, S. H. Stevenson, G. R. Meredith, G. Rikken, S. R. Marder, *J. Phys. Chem.* **1991**, *95*, 10631.
44. A. E. Stiegman, E. M. Graham, K. J. Perry, L. R. Khundkar, L.-T. Cheng, J. W. Perry, *J. Am. Chem. Soc.* **1991**, *113*, 7658.
45. M. Blanchard-Desce, C. Runser, A. Fort, M. Barzoukas, J.-M. Lehn, V. Bloy, V. Alain, *Chem. Phys.* **1995**, *199*, 253.
46. N. J. Long, *Angew. Chem. Int. Ed. Engl.* **1995**, *34*, 21, and references cited therein.
47. S. R. Marder, in *Inorganic Materials*, D. W. Bruce, D. O'Hare (Eds.), Wiley, Chichester, **1996**, 121.
48. B. J. Coe, in *Comprehensive Coordination Chemistry II*, Vol. 9, M. D. Ward (Ed.), Elsevier, Oxford, **2004**, 621.
49. M. E. Thompson, P. E. Djurovich, S. Barlow, S. R. Marder, in *Comprehensive Organometallic Chemistry III*, Vol. 12, D O'Hare (Ed.), Elsevier, Amsterdam, in press.
50. S. Barlow, S. R. Marder, *Chem. Commun.* **2000**, 1555.
51. M. L. H. Green, S. R. Marder, M. E. Thompson, J. A. Bandy, D. Bloor, P. V. Kolinsky, R. J. Jones, *Nature* **1987**, *330*, 360.
52. V. Alain, M. Blanchard-Desce, C. T. Chen, S. R. Marder, A. Fort, M. Barzoukas, *Synth. Met.* **1996**, *81*, 133.
53. M. H. Garcia, M. P. Robalo, A. R. Dias, M. Teresa Duarte, W. Wenseleers, G. Aerts, E. Goovaerts, M. P. Cifuentes, S. Hurst, M. G. Humphrey, M. Samoc, B. Luther-Davies, *Organometallics* **2002**, *21*, 2107.
54. C. E. Powell, M. P. Cifuentes, A. M. McDonagh, S. K. Hurst, N. T. Lucas, C. D. Delfs, R. Stranger, M. G. Humphrey, S. Houbrechts, I. Asselberghs, A. Persoons, D. C. R. Hockless, *Inorg. Chim. Acta* **2003**, *352*, 9.
55. F. Paul, K. Costuas, I. Ledoux, S. Deveau, J. Zyss, J.-F. Halet, C. Lapinte, *Organometallics* **2002**, *21*, 5229.
56. S. Houbrechts, K. Clays, A. Persoons, V. Cadierno, M. P. Gamasa, J. Gimeno, *Organometallics* **1996**, *15*, 5266.
57. I. R. Whittall, M. P. Cifuentes, M. G. Humphrey, B. Luther-Davies, M. Samoc, S. Houbrechts, A. Persoons, G. A. Heath, D. C. R. Hockless, *J. Organomet. Chem.* **1997**, *549*, 127.
58. B. J. Coe, *Chem. Eur. J.* **1999**, *5*, 2464.
59. T. Weyland, I. Ledoux, S. Brasselet, J. Zyss, C. Lapinte, *Organometallics* **2000**, *19*, 5235.
60. I. Asselberghs, K. Clays, A. Persoons, A. M. McDonagh, M. D. Ward, J. A. McCleverty, *Chem. Phys. Lett.* **2003**, *368*, 408.

61. J. Zyss, I. Ledoux, *Chem. Rev.* **1994**, *94*, 77.
62. T. Verbiest, S. Houbrechts, M. Kauranen, K. Clays, A. Persoons, *J. Mater. Chem.* **1997**, *7*, 2175, and references cited therein.
63. C. Andraud, T. Zabulon, A. Collet, J. Zyss, *Chem. Phys.* **1999**, *245*, 243.
64. J. Zyss, C. Dhenaut, T. Chauvan, I. Ledoux, *Chem. Phys. Lett.* **1993**, *206*, 409.
65. F. W. Vance, R. D. Williams, J. T. Hupp, *Int. Rev. Phys. Chem.* **1998**, *17*, 302.
66. B. R. Cho, S. B. Park, S. J. Lee, K. H. Son, S. H. Lee, M.-J. Lee, J. Yoo, Y. K. Lee, G. J. Lee, T. I. Kang, M. Cho, S.-J. Jeon, *J. Am. Chem. Soc.* **2001**, *123*, 6421.
67. H. C. Jeong, M. J. Piao, S. H. Lee, M.-Y. Jeong, K. M. Kang, G. Park, S.-J. Jeon, B. R. Cho, *Adv. Funct. Mater.* **2004**, *14*, 64.
68. M. Lequan, C. Branger, J. Simon, T. Thami, E. Chauchard, A. Persoons, *Chem. Phys. Lett.* **1994**, *229*, 101.
69. M. Blanchard-Desce, S. R. Marder, M. Barzoukas, in *Comprehensive Supramolecular Chemistry*, Vol. 10, D. N. Reinhoudt (Ed.), Elsevier, Oxford, **1996**, 833.
70. D. Li, M. A. Ratner, T. J. Marks, C. Zhang, J. Yang, G. K. Wong, *J. Am. Chem. Soc.* **1990**, *112*, 7389.
71. H. E. Katz, G. Scheller, T. M. Putvinski, M. L. Schilling, W. L. Wilson, C. E. D. Chidsey, *Science* **1991**, *254*, 1485.
72. W. Lin, W. Lin, G. K. Wong, T. J. Marks, *J. Am. Chem. Soc.* **1996**, *118*, 8034.
73. H. Kang, P. Zhu, Y. Yang, A. Facchetti, T. J. Marks, *J. Am. Chem. Soc.* **2004**, *126*, 15974.
74. K. Chane-Ching, M. Lequan, R. M. Lequan, F. Kajzar, *Chem. Phys. Lett.* **1995**, *242*, 598.
75. C.-F. Shu, W. J. Tsai, J.-Y. Chen, A. K.-Y. Jen, Y. Zhang, T.-A. Chen, *Chem. Commun.* **1996**, 2279.
76. S. Gilmour, R. A. Montgomery, S. R. Marder, L.-T. Cheng, A. K.-Y. Jen, Y. Cai, J. W. Perry, L. R. Dalton, *Chem. Mater.* **1994**, *6*, 1603.
77. V. P. Rao, K. Y. Wong, A. K.-Y. Jen, K. J. Drost, *Chem. Mater.* **1994**, *6*, 2210.
78. S. Ermer, S. M. Lovejoy, D. S. Leung, H. Warren, C. R. Moylan, R. J. Twieg, *Chem. Mater.* **1997**, *9*, 1437.
79. Y.-C. Shu, Z.-H. Gao, C.-F. Shu, E. M. Breitung, R. J. McMahon, G.-H. Lee, A. K.-Y. Jen, *Chem. Mater.* **1999**, *11*, 1628.
80. K. Staub, G. A. Levina, S. Barlow, T. C. Kowalczyk, H. S. Lackritz, M. Barzoukas, A. Fort, S. R. Marder, *J. Mater. Chem.* **2003**, *13*, 825.
81. P. V. Bedworth, Y. M. Cai, A. Jen, S. R. Marder, *J. Org. Chem.* **1996**, *61*, 2242.
82. S.-S. P. Chou, D.-J. Sun, J.-Y. Huang, P.-K. Yang, H.-C. Lin, *Tetrahedron Lett.* **1996**, *37*, 7279.
83. F. Steybe, F. Effenberger, U. Gubler, C. Bosshard, P. Gunter, *Tetrahedron* **1998**, *54*, 8469.
84. E. M. Breitung, C.-F. Shu, R. J. McMahon, *J. Am. Chem. Soc.* **2000**, *122*, 1154.
85. B. R. Cho, S. H. Lee, J. C. Lim, T. I. Kang, S.-J. Jeon, *Mol. Cryst. Liq. Cryst.* **2001**, *370*, 77.
86. Z.-Y. Hu, A. Fort, M. Barzoukas, A. K.-Y. Jen, S. Barlow, S. R. Marder, *J. Phys. Chem. B* **2004**, *108*, 8626.
87. O. Maury, L. Viau, K. Senechal, B. Corre, J.-P. Guegan, T. Renouard, I. Ledoux, J. Zyss, H. le Bozec, *Chem. Eur. J.* **2004**, *10*, 4454.
88. P. R. Varanasi, A. K.-Y. Jen, J. Chandrasekhar, I. N. N. Nemboothiri, A. Rathna, *J. Am. Chem. Soc.* **1996**, *118*, 12443.
89. I. D. L. Albert, T. J. Marks, M. A. Ratner, *J. Am. Chem. Soc.* **1997**, *119*, 6575.
90. C.-F. Shu, Y.-K. Wang, *J. Mater. Chem.* **1998**, *8*, 833.
91. S. Thayumanavan, J. Mendez, S. R. Marder, *J. Org. Chem.* **1999**, *64*, 4289.
92. L.-T. Cheng, W. Tam, S. R. Marder, A. E. Stiegman, G. Rikken, C. W. Spangler, *J. Phys. Chem.* **1991**, *95*, 10643.
93. C. R. Moylan, R. J. Twieg, V. Y. Lee, S. A. Swanson, K. M. Betterton, R. D. Miller, *J. Am. Chem. Soc.* **1993**, *115*, 12599.
94. M. Bösch, C. Fisher, C. Cai, I. Liakatas, C. Bosshard, P. Günter, *Synth. Met.* **2001**, *2001*, 241.
95. L. R. Dalton, *J. Phys: Condens. Matter* **2003**, *15*, R897.
96. H. Ma, S. Liu, J. Luo, S. Suresh, L. Liu, S. H. Kang, M. Haller, T. Sassa, L. R. Dalton, A. K.-Y. Jen, *Adv. Funct. Mater.* **2002**, *12*, 565.

97. S. Liu, A. Haller, H. Ma, L. R. Dalton, S.-H. Jang, A. K.-Y. Jen, *Adv. Mater.* **2003**, *15*, 605.
98. L. R. Dalton, A. W. Harper, B. H. Robinson, *Proc. Natl. Acad. Sci. USA* **1997**, *94*, 4842.
99. A. W. Harper, S. Sun, L. R. Dalton, S. M. Garner, A. Chen, S. Kalluri, W. H. Steier, B. H. Robinson, *J. Opt. Soc. Am. B* **1998**, *15*, 329.
100. B. H. Robinson, L. R. Dalton, *J. Phys. Chem. A* **2000**, *104*, 4785.
101. Y. V. Pereverzev, O. V. Prezhdo, L. R. Dalton, *ChemPhysChem.* **2004**, *5*, 1821.
102. Y. Shi, C. Zhang, H. Zhang, J. H. Bechtel, L. R. Dalton, B. H. Robinson, W. H. Steier, *Science* **2000**, *288*, 119.
103. H. Ma, B. Chen, T. Sassa, L. R. Dalton, A. K.-Y. Jen, *J. Am. Chem. Soc.* **2001**, *123*, 986.
104. Y. V. Pereverzev, O. V. Prezhdo, L. R. Dalton, *Chem. Phys. Lett.* **2003**, *373*, 207.
105. J. Luo, M. Haller, H. Ma, S. Liu, T.-D. Kim, Y. Tian, B. Chen, S.-H. Jang, L. R. Dalton, A. K.-Y. Jen, *J. Phys. Chem. B* **2004**, *108*, 8523.
106. T. Gray, R. M. Overney, M. Haller, J. Luo, A. K.-Y. Jen, *Appl. Phys. Lett.* **2005**, *86*, 211908.
107. Y. Bai, N. Song, J. P. Gao, X. Sun, X. Wang, G. Yu, Z. Y. Wang, *J. Am. Chem. Soc.* **2005**, *127*, 2060.
108. S. H. Kang, J. Luo, H. Ma, R. R. Barto, C. W. Frank, L. R. Dalton, A. K.-Y. Jen, *Macromolecules* **2003**, *36*, 4355.
109. A. K.-Y. Jen, Q. Yang, S. R. Marder, L. R. Dalton, C. F. Shu, *Mater. Res. Soc., Symp. Proc.* **1998**, *2527*, 150.
110. H. Ma, B. Q. Chen, L. R. Dalton, A. K.-Y. Jen, *Polym. Mater. Sci. Eng.* **2000**, *83*, 165.
111. D. Babb, B. R. Ezzell, K. S. Clement, W. F. Richey, A. P. Kennedy, *J. Polym. Sci., Part A: Polym. Chem.* **1993**, *31*, 3465.
112. D. W. Smith, H. W. Boone, R. Traiphol, H. Shah, D. Perahia, *Macromolecules* **2000**, *33*, 1126.
113. M. Haller, J. Luo, H. Li, T.-D. Kim, Y. Liao, B. H. Robinson, L. R. Dalton, A. K.-Y. Jen, *Macromolecules* **2004**, *37*, 688.
114. A. M. Sinyukov, L. M. Hayden, *Opt. Lett.* **2002**, *27*, 55.
115. L. M. Hayden, A. M. Sinyukov, M. R. Leahy, J. French, P. Lindahl, W. N. Herman, R. J. Twieg, M. He, *J. Polym. Sci., Part B: Polym. Phys.* **2003**, *41*, 2493.
116. A. M. Sinyukov, L. M. Hayden, *J. Phys. Chem. B* **2004**, *108*, 8515.
117. A. M. Sinyukov, M. R. Leahy, L. M. Hayden, M. Haller, J. Luo, A. K.-Y. Jen, L. R. Dalton, *Appl. Phys. Lett.* **2004**, *85*, 5827.
118. A. J. Fitzgerald, E. Berry, N. N. Zinovev, G. C. Walker, M. A. Smith, J. M. Chamberlain, *Phys. Med. Biol.* **2002**, *47*, R67.
119. Q. Chen, Z. Jiang, G. X. Xu, X. C. Zhang, *Opt. Lett.* **2000**, *25*, 1122.
120. B. Fisher, M. Walther, P. U. Jepsen, *Phys. Med. Biol.* **2002**, *47*, 3807.
121. X.-C. Zhang, Y. Jin, X. F. Ma, *Appl. Phys. Lett.* **1992**, *61*, 2764.
122. R. G. Denning, *Eur. J. Solid State Inorg. Chem.* **1991**, *28*, 33.
123. K. Pachucki, D. Leibfried, M. Weitz, A. Huber, W. Koenig, T. W. Haensch, *J. Phys. B* **1996**, *29*, 177.
124. A much larger value of δ_{max} of ca. 5800 GM (also corresponding to a photon wavelength of ca. 600 nm), obtained by NLT measurements, has been reported by Wang et al. for a similar molecule, presumably reflecting contributions from excited-state absorption: X.-M. Wang, Y.-F. Zhou, W.-T. Yu, C. Wang, Q. Fang, M.-H. Jiang, H. Lei, H.-Z. Wang, *J. Mater. Chem.* **2000**, *10*, 2698.
125. E. Zojer, D. Beljonne, T. Kogej, H. Vogel, S. R. Marder, J. W. Perry, J.-L. Brédas, *J. Chem. Phys.* **2002**, *116*, 3646.
126. S. J. K. Pond, M. Rumi, M. D. Levin, T. C. Parker, D. Beljonne, M. W. Day, J. L. Bredas, S. R. Marder, J. W. Perry, *J. Phys. Chem. A* **2002**, *106*, 11470.
127. L. Ventelon, M. Blanchard-Desce, L. Moreaux, J. Mertz, *Chem. Commun.* **1999**, 2055.
128. L. Ventelon, S. Charier, L. Moreaux, J. Mertz, M. Blanchard-Desce, *Angew. Chem. Int. Ed.* **2001**, *40*, 2098.
129. B. Strehmel, A. M. Sarker, H. Detert, *ChemPhysChem* **2003**, *4*, 249.
130. B. Strehmel, S. Amthor, J. Schelter, C. Lambert, *ChemPhysChem* **2005**, *6*, 893.
131. S. K. Lee, W. J. Yang, J. J. Choi, C. H. Kim, S.-J. Jeon, B. R. Cho, *Org. Lett.* **2005**, *7*, 323.

132. S.-J. Chung, M. Rumi, V. Alain, S. Barlow, J. W. Perry, S. R. Marder, *J. Am. Chem. Soc.* **2005**, *127*, 10844.
133. M. Halik, W. Wenseleers, C. Grasso, F. Stellacci, E. Zojer, S. Barlow, J.-L. Brédas, J. W. Perry, S. R. Marder, *Chem. Commun.* **2003**, 1490.
134. E. Zojer, W. Wenseleers, M. Halik, C. Grasso, S. Barlow, J. W. Perry, S. R. Marder, J.-L. Brédas, *ChemPhysChem.* **2004**, *5*, 982.
135. E. Zojer, W. Wenseleers, P. Pacher, S. Barlow, M. Halik, C. Grasso, J. W. Perry, S. R. Marder, J.-L. Brédas, *J. Phys. Chem. B* **2004**, *108*, 8641.
136. A. Abbotto, L. Beverina, R. Bozio, A. Facchetti, C. Ferrante, G. A. Pagani, D. Pedron, R. Signorini, *Org. Lett.* **2002**, *4*, 1495.
137. S. L. Oliveira, D. S. Corrêa, L. Misoguti, C. J. L. Constantino, R. F. Aroca, S. C. Zilio, C. R. Mendonça, *Adv. Mater.* **2005**, *17*, 1890.
138. J. R. G. Thorne, S. M. Kuebler, R. G. Denning, I. M. Blake, P. N. Taylor, H. L. Anderson, *Chem. Phys.* **1999**, *248*, 181.
139. S. M. Kuebler, R. G. Denning, H. L. Anderson, *J. Am. Chem. Soc.* **2000**, *122*, 339.
140. T. E. O. Screen, K. B. Lawton, G. S. Wilson, N. Dolney, R. Ispasoiu, T. Goodson, S. J. Martin, D. D. C. Bradley, H. L. Anderson, *J. Mater. Chem.* **2001**, *11*, 312.
141. M. Drobizhev, Y. Stepanenko, Y. Dzenis, A. Karotki, A. Rebane, P. N. Taylor, H. L. Anderson, *J. Am. Chem. Soc.* **2004**, *126*, 15352.
142. M. Drobizhev, Y. Stephanenko, Y. Dzenis, A. Karotki, A. Rebane, P. N. Taylor, H. L. Anderson, *J. Phys. Chem. B* **2005**, *109*, 7223.
143. Y. Inokuma, N. Ono, H. Uno, D. Y. Kim, S. B. Noh, D. Kim, A. Osuka, *Chem. Commun.* **2005**, 3782.
144. K. Ogawa, T. Zhang, K. Yoshihara, Y. Kobuke, *J. Am. Chem. Soc.* **2002**, *124*, 22.
145. K. D. Belfield, D. J. Hagan, E. W. Van Stryland, K. J. Schaefer, R. A. Negres, *Org. Lett.* **1999**, *1*, 1575.
146. L. Antonov, K. Kamada, K. Ohta, F. S. Kamounah, *Phys. Chem. Chem. Phys.* **2003**, *5*, 1193.
147. L. Beverina, J. Fu, A. Leclercq, E. Zojer, P. Pacher, S. Barlow, E. W. Van Stryland, D. J. Hagan, J.-L. Brédas, S. R. Marder, *J. Am. Chem. Soc.* **2005**, *127*, 7282.
148. S.-J. Chung, K.-S. Kim, T.-C. Lin, G. S. He, J. Swiatkiewicz, P. N. Prasad, *J. Phys. Chem. B* **1999**, *103*, 10741.
149. A. Adronov, J. M. J. Fréchet, G. S. He, K.-S. Kim, S.-J. Chung, J. Swiatkiewicz, P. N. Prasad, *Chem. Mater.* **2000**, *12*, 2838.
150. W.-H. Lee, H. Lee, J.-A. Kim, J.-H. Choi, M. Cho, S.-J. Jeon, B. R. Cho, *J. Am. Chem. Soc.* **2001**, *123*, 10658.
151. B. R. Cho, K. H. Son, S. H. Lee, Y.-S. Song, Y.-K. Lee, S.-J. Jeon, J. H. Choi, H. Lee, M. Cho, *J. Am. Chem. Soc.* **2001**, *123*, 10039.
152. M. Drobizhev, A. Karotki, A. Rebane, C. W. Spangler, *Opt. Lett.* **2001**, *26*, 1081.
153. B. R. Cho, M. J. Piao, K. H. Son, S. H. Lee, S. J. Yoon, S.-J. Jeon, M. Cho, *Chem. Eur. J.* **2002**, *8*, 3907.
154. D. Beljonne, W. Wenseleers, E. Zojer, Z. Shuai, H. Vogel, S. J. K. Pond, J. W. Perry, S. R. Marder, J.-L. Brédas, *Adv. Funct. Mater.* **2002**, *12*, 631.
155. M. Drobizhev, A. Karotki, Y. Dzenis, A. Rebane, Z. Suo, C. W. Spangler, *J. Phys. Chem. B* **2003**, *107*, 7540.
156. B. J. Zhang, S.-J. Jeon, *Chem. Phys. Lett.* **2003**, *377*, 210.
157. J. Yoo, S. K. Yang, M.-Y. Jeong, H. C. Ahn, S.-J. Jeon, B. R. Cho, *Org. Lett.* **2003**, *5*, 645.
158. Y. Wang, G. S. He, P. N. Prasad, T. Goodson, *J. Am. Chem. Soc.* **2005**, *127*, 10129.
159. G. P. Bartholomew, M. Rumi, S. J. K. Pond, J. W. Perry, S. Tretiak, G. C. Bazan, *J. Am. Chem. Soc.* **2004**, *126*, 11529.
160. J. H. Strickler, W. W. Webb, *Opt. Lett.* **1991**, *16*, 1780.
161. E. S. Webb, J. H. Strickler, W. R. Harrell, W. W. Webb, *Proc. SPIE* **1992**, *1674*, 776.
162. S. Maruo, O. Nakamura, S. Kawata, *Opt. Lett.* **1997**, *22*, 132.
163. K. J. Schafer, J. M. Hales, M. Balu, K. D. Belfield, E. W. Van Stryland, D. J. Hagan, *J. Chem. Photochem. Photobiol. A.* **2004**, *162*, 497.
164. B. H. Cumpston, S. P. Ananthavel, S. Barlow, D. L. Dyer, J. E. Ehrlich, L. L.

Erskine, A. A. Heikal, S. M. Kuebler, I.-Y. S. Lee, D. McCord-Maughon, J. Qin, H. Röckel, M. Rumi, X.-L. Wu, S. R. Marder, J. W. Perry, *Nature* **1999**, *398*, 51.
165. M. P. Joshi, H. E. Pudavar, J. Swiatkiewicz, P. N. Prasad, *Appl. Phys. Lett.* **1999**, *74*, 170.
166. K. D. Belfield, X. Ren, E. W. Van Stryland, D. J. Hagan, V. Dubikovsky, E. J. Miesak, *J. Am. Chem. Soc.* **2000**, *122*, 1217.
167. S. M. Kuebler, M. Rumi, T. Watanabe, K. Braun, B. H. Cumpston, A. A. Heikal, L. L. Erskine, S. Thayumanavan, S. Barlow, S. R. Marder, J. W. Perry, *J. Photopolym. Sci. Technol.* **2001**, *14*, 657.
168. T. Watanabe, M. Akiyama, K. Totani, S. M. Kuebler, F. Stellacci, W. Wenseleers, K. L. Braun, S. R. Marder, J. W. Perry, *Adv. Funct. Mater.* **2002**, *12*, 611.
169. C. Martineau, R. Anemian, C. Andraud, I. Wang, M. Bouriau, P. L. Baldeck, *Chem. Phys. Lett.* **2002**, *362*, 291.
170. I. Wang, M. Bouriau, P. L. Baldeck, C. Martineau, C. Andraud, *Opt. Lett.* **2002**, *27*, 1348.
171. H.-B. Sun, H. Maeda, K. Takada, J. W. M. Chon, M. Gu, S. Kawata, *Appl. Phys. Lett.* **2003**, *83*, 819.
172. Y. Lu, F. Hasegawa, T. Goto, S. Ohkuma, S. Fukuhara, Y. Kawazu, K. Totani, T. Yamashita, T. Watanabe, *J. Mater. Chem.* **2004**, *14*, 75.
173. Y. Lu, F. Hasegawa, S. Ohkuma, T. Goto, S. Fukuhara, Y. Kawazu, K. Totani, T. Yamashita, T. Watanabe, *J. Mater. Chem.* **2004**, *14*, 1391.
174. F. Stellacci, C. A. Bauer, T. Meyer-Friedrichsen, W. Wenseleers, V. Alain, S. M. Kuebler, S. J. K. Pond, Y. Zhang, S. R. Marder, J. W. Perry, *Adv. Mater.* **2002**, *14*, 194.
175. W. Zhou, S. M. Kuebler, K. L. Braun, T. Yu, J. K. Cammack, C. K. Ober, J. W. Perry, S. R. Marder, *Science* **2002**, *296*, 1106.
176. S. M. Kuebler, K.L. Braun, W. Zhou, J. K. Cammack, T. Yu, C. K. Ober, S. R. Marder, J. W. Perry, *J. Photchem. Photobiol. A: Chem.* **2003**, *158*, 163.
177. T. Yu, C. K. Ober, S. M. Kuebler, W. Zhou, S. R. Marder, J. W. Perry, *Adv. Mater.* **2003**, *15*, 517.
178. D. J. Dyer, B. H. Cumpston, D. McCord-Maughon, S. Thayumanavan, S. Barlow, J. W. Perry, S. R. Marder, *Nonlinear Opt.* **2004**, *31*, 175.
179. W. Denk, J. H. Strickler, W. W. Webb, *Science* **1990**, *248*, 73.
180. L. Moreaux, O. Sandre, M. Blanchard-Desce, J. Mertz, *Opt. Lett.* **2000**, *25*, 320.
181. H. Y. Woo, J. W. Hong, B. Liu, A. Mikhailovky, D. Kortystov, G. C. Bazan, *J. Am. Chem. Soc.* **2005**, *127*, 820.
182. S. J. K. Pond, O. Tsutsumi, M. Rumi, O. Kwon, E. Zojer, J.-L. Brédas, S. R. Marder, J. W. Perry, *J. Am. Chem. Soc.* **2004**, *126*, 9291.
183. T. Y. Ohulchanskyy, H. E. Pudavar, S. M. Yarmoluk, V. M. Yashchuk, E. J. Bergey, P. N. Prasad, *Photochem. Photobiol.* **2003**, *77*, 138.
184. X. Wang, L. J. Krebs, M. Al-Nuri, H. H. Pudavar, S. Ghosal, C. Liebow, A. A. Nagy, A. V. Schally, P. N. Prasad, *Proc. Natl. Acad. Sci. USA* **1999**, *96*, 11081.
185. D. R. Larson, W. R. Zipfel, R. M. Williams, S. W. Clark, M. P. Bruchez, F. W. Wise, W. W. Webb, *Science* **2003**, *5624*, 1434.
186. J. D. Bhawalkar, N. D. Kumar, C.-F. Zhao, P. N. Prasad, *J. Clin. Laser Med. Surg.* **1997**, *15*, 201.
187. P. K. Frederiksen, M. Jørgensen, P. R. Ogilby, *J. Am. Chem. Soc.* **2001**, *123*, 1215.
188. A. Karotki, M. Kruk, M. Drobizhev, A. Rebane, E. Nickel, C. W. Spangler, *IEEE J. Quantum Electron.* **2001**, *7*, 971.
189. W. R. Dichtel, J. M. Serin, C. Edder, J. M. J. Fréchet, M. Matuszewski, L.-S. Tan, T. Y. Ohulchanskyy, P. N. Prasad, *J. Am. Chem. Soc.* **2004**, *126*, 5380.
190. S. Wecksler, A. Mikhailovsky, P. C. Ford, *J. Am. Chem. Soc.* **2004**, *126*, 13566.
191. C. W. Spangler, *J. Mater. Chem.* **1999**, *9*, 2013.
192. G. S. He, G. C. Xu, P. N. Prasad, B. A. Reinhardt, J. C. Bhatt, A. G. Dillard, *Opt. Lett.* **1995**, *20*, 435.
193. G. S. He, J. D. Bhawalkar, C. F. Zhao, P. N. Prasad, *Appl. Phys. Lett.* **1995**, *67*, 24333.
194. M. G. Silly, L. Porrès, O. Mongin, P.-A. Chollet, M. Blanchard-Desce, *Chem. Phys. Lett.* **2003**, *379*, 74.

195. J. Zhang, Y. Cui, M. Wang, J. Liu, *Chem. Commun.* **2002**, 2526.
196. C. Li, C. Liu, Q. Li, Q. Gong, *Chem. Phys. Lett.* **2004**, *400*, 569.
197. G. S. He, J. D. Bhawalkar, C. F. Zhao, C.-K. Park, P. N. Prasad, *Opt. Lett.* **1995**, *20*, 2393.
198. G. S. He, J. D. Bhawalkar, C. F. Zhao, P. N. Prasad, *Appl. Phys. Lett.* **1995**, *67*, 3703.
199. G. S. He, J. D. Bhawalkar, C. F. Zhao, P. N. Prasad, *IEEE J. Quantum Electron.* **1996**, *32*, 749.
200. J. D. Bhawalkar, G. S. He, C.-K. Park, C. F. Zhao, G. Ruland, P. N. Prasad, *Opt. Commun.* **1996**, *124*, 33.
201. G. S. He, P. P. Markowicz, T.-C. Lin, P. N. Prasad, *Nature* **2002**, *415*, 767.
202. C. Bauer, B. Schnabel, E.-B. Kley, U. Scherf, H. Giessen, R. F. Mahrt, *Adv. Mater.* **2002**, *14*, 673.
203. G. S. He, R. Helgeson, T.-C. Tin, Q. Zhang, F. Wudl, P. N. Prasad, *IEEE J. Quantum Electron.* **2003**, *39*, 1003.
204. M. Goeppert-Mayer, *Ann. Phys.* **1931**, *9*, 273.
205. http://nobelprize.org/physics/laureates/1963/mayer-bio.html.

Part IV
Electronic Interaction and Structure

12
Photoinduced Electron Transfer Processes in Synthetically Modified DNA

Hans-Achim Wagenknecht

12.1
DNA as a Bioorganic Material for Electron Transport

With the introduction of π-conjugated systems in electronic devices and the challenge to achieve molecular electronics based on these systems, the detailed understanding of the supramolecular interactions between the individual π-conjugated molecules has become the subject of very intense research. The central problem for advances in nanoelectronics is the miniaturization of integrated circuits, which drives the search for efficient molecular devices at the nanometer scale. The basic idea of "molecular electronics" is to use individual molecules as wires, switches, rectifiers and memories [1]. Conceptually, this can be realized by using the so-called bottom-up approach in which the system is composed of small building blocks with recognition, structuring and, most importantly, self-assembling properties. With respect to this strategy, DNA plays an important role as a bioorganic material since its unique properties provide a suitable basis for using duplex DNA or DNA-like architectures as a structural scaffold for molecular electronics:

1. *Self-assembly:* Two oligonucleotides spontaneously organize themselves in duplex structures as encoded by the DNA base sequence. Even complex supramolecular arrangements can be realized, e.g. polyhedra or four-way junctions [2].
2. *Regular helical structure:* Duplex DNA forms regular helical structures with defined dimensions. In the B-DNA conformation, the base-pair distance along the helical axis is 3.4 Å, providing the structural basis for electron transport.
3. *Molecular recognition:* Double-helical DNA is able to interact with ligands, e.g. DNA-binding proteins, often in a very site-selective way, providing the basis for the modulation of electron transport.
4. *Synthesis:* The automated oligonucleotide chemistry makes DNA available in any desired base sequence and also in large

Functional Organic Materials. Syntheses, Strategies, and Applications.
Edited by Thomas J.J. Müller and Uwe H.F. Bunz
Copyright © 2007 WILEY-VCH Verlag GmbH & Co. KGaA, Weinheim
ISBN: 978-3-527-31302-0

quantities. Moreover, the building block strategy can be applied for synthetic modifications on the way to new DNA-inspired structures and architectures.

The central question for the application of DNA as a π-conjugated material for functional π-systems is how well DNA transport charges over nanometer distances. The planar heterocyclic aromatic systems of the DNA bases are π-stacked at a distance of 3.4 Å, which initially brought up the idea that DNA could conduct electrons like a molecular wire. The first remark about charge migration in DNA was published by Eley and Spivey over 40 years ago [3]. Especially in the 1990s, DNA was considered as a conducting biopolymer for long-range charge migration, which was discussed highly controversially. Remarkably, DNA was described to be either a molecular wire, semiconductor or insulator [4]. At the beginning of those investigations, most research groups interpreted their results according to the Marcus theory for non-adiabatic electron transfer [5]: mainly three different variables influence the rate of the electron-transfer process (k_{ET}) – the electronic coupling, the driving force for the electron-transfer process and the reorganization energy. The full understanding of DNA-mediated charge transfer requires a knowledge of how these three variables are influenced by the medium DNA as the bridge between donor (**Do**) and acceptor (**Ac**) and in most synthetic DNA systems this is not very clear. The β-value that describes the exponential distance dependence of k_{ET} has been highly overrated and overinterpreted within the controversy. A simplified picture on the relative energies shows (Fig. 12.1) that the level of the bridge in relation to the energetic levels of **Do** and **Ac** decides on molecular wire-like behavior or charge transfer via the superexchange mechanism. In the case of a molecular wire, the bridge states are energetically comparable to the level of **Do** and the electron could be injected into the bridge. Upon injection, the electron is localized within the bridge and moves incoherently to **Ac**. By now it is clear that DNA is not a molecular wire. Alternatively, in the case of the superexchange mechanism, the bridge states lie above the level of **Do**. Consequently, the electron is transferred in one coherent jump and is never localized

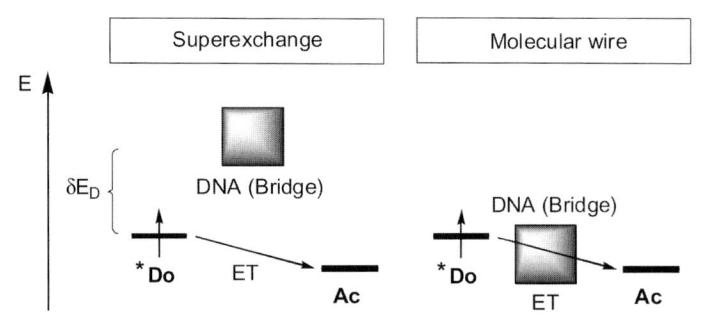

Fig. 12.1 Comparison of charge transfer via the superexchange mechanism and via a molecular wire. **Do** = donor; **Ac** = acceptor; ET = electron transfer.

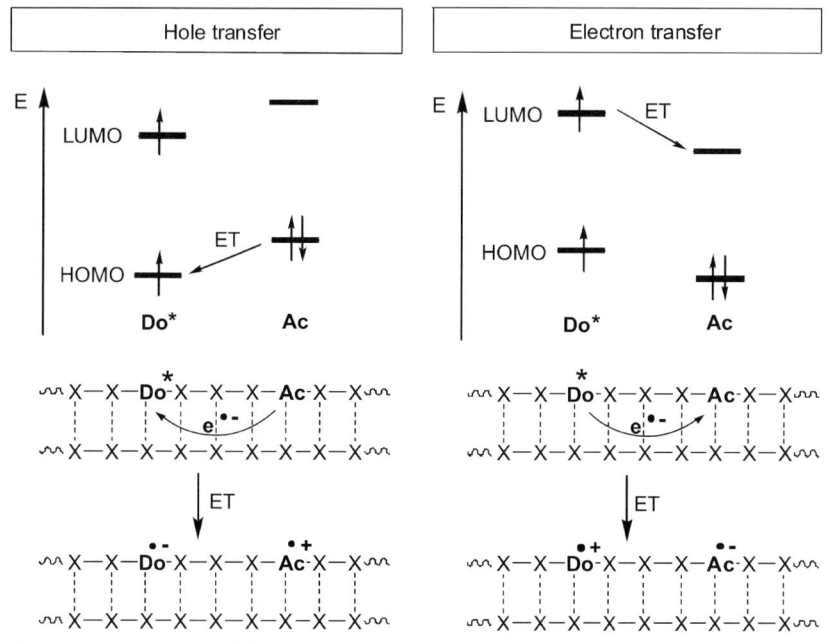

Fig. 12.2 Comparison of photoinduced hole transfer (HOMO control) and electron transfer (LUMO control) in DNA. **Do** = donor; **Ac** = acceptor; ET = electron transfer.

within the bridge. For the superexchange mechanism, the distance dependence behavior of k_{ET} is clearly exponential and the β-value applies.

Barton and coworkers pushed this research with astonishing and controversial contributions about photoinduced charge-transfer chemistry in DNA [6]. At least in part, the research was motivated by the search for the primary chemical processes during DNA damage [7]. Today, DNA-mediated charge-transfer processes are categorized into oxidative processes including hole transfer/transport/migration and reductive types, which are excess electron transfer/transport/migration (Fig. 12.2) [8, 9]. It is important to point out that both types of processes are in fact electron-transfer reactions, but with different orbital control. During the hole transfer an electron is transferred from the DNA or the final acceptor (**Ac**) to the photoexcited charge donor (**Do**); this is a HOMO-controlled process. In the case of an electron transfer, the photoexcited electron of **Do** is injected into the DNA or transferred to the final electron acceptor (**Ac**); this process is LUMO-controlled. Hence this categorization is not just formalism about the different directions of the electron which is transferred.

In contrast to the initial controversy, a fairly clear picture about charge-transfer processes in DNA has now been obtained. It is known that charge-transfer processes may follow different mechanistic descriptions, mainly the superexchange and the hopping mechanisms [10]. It became evident that DNA-mediated charge transfer can occur on an ultrafast time-scale and can result in reactions over long distances [11, 12]. Especially the latter observation shows that the charge-transfer chemistry in DNA is biologically relevant in the formation of oxidative damage to the DNA, leading to mutagenesis, apoptosis or carcinogenesis [7]. More recent publications discuss the role of DNA-mediated charge-transfer processes during the sensing of base damage by DNA-binding proteins followed by DNA repair [13]. Additionally, charge transfer plays a growing role in the development of DNA chips and microarrays detecting single base mutations or various DNA lesions by electrochemical readout methods [14]. Moreover, the knowledge about charge-transfer processes in DNA can be used for nanotechnological devices based on DNA [15]. Despite these research efforts, a complete and clear description of the electronic properties inside the DNA base stack is still lacking. Hence DNA still represents a very special and exciting medium in terms of energy and charge-transfer processes.

12.2
Mechanism of Hole Transfer and Hole Hopping in DNA

With respect to the biological relevance for oxidative DNA damage, most research groups initially focused their work on the photochemically induced hole transfer and the mobility of the positive charge created in DNA. Despite some initial controversies [4], it turned out that DNA as a medium for charge transfer is *not* a molecular wire. Accordingly, the DNA-mediated hole-transfer processes over short distances are described best by using a superexchange mechanism [10]. The charge tunnels in one coherent step from **Do** to **Ac** and never resides on the DNA bridge between (Fig. 12.1), since the bridge states are energetically higher than the photoexcited donor state (**Do***). According to the Marcus equation [5], the rate k_{ET} of such a single-step process depends exponentially on the distance R between **Do** and **Ac** and β represents the parameter for this distance dependence. Four important observations are important to point out here:

1. The hole transfer via the superexchange mechanism is limited to short distances (< 10 Å).
2. Short-range hole transfer reactions occur on a very fast time-scale ($k_{ET} = 10^9$–10^{12} s^{-1}).
3. The typical β-value of DNA-mediated hole transfer is 0.6–0.8 Å^{-1}.
4. The intercalation of the charge donor and acceptor is crucial for a fast and efficient hole-transfer process

Values of β for charge transfer in proteins were determined experimentally in the range 1.0–1.4 Å^{-1} [16]. In comparison, apparent β values for DNA-mediated hole transfer were found in a wide range from $\beta < 0.1$ up to 1.5 Å^{-1} [11, 12]. For the alternative mechanistic description of DNA-mediated charge-transport phenomena over long distances which typically exhibit a very shallow distance depen-

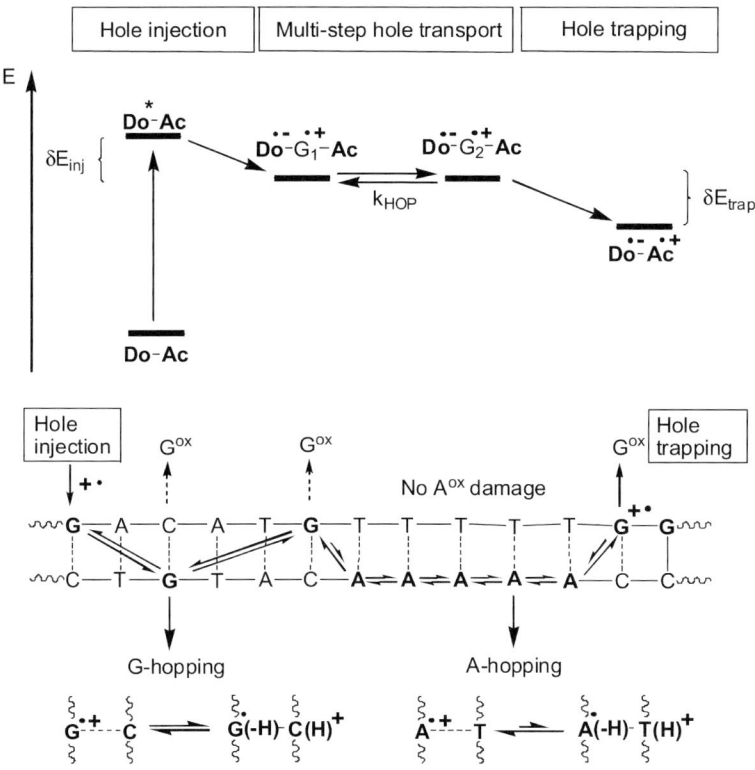

Fig. 12.3 Photoinduced hole transport via hopping. **Do** = donor; **Ac** = acceptor; **B** = base; inj = injection; HOP = hopping; trap = trapping.

dence ($\beta < 0.1$ Å^{-1}), the hopping model for conducting polymers was applied (Fig. 12.3) [10]. It is important to note that for this process DNA is considered as a functional π-conjugated biopolymer. Among the four different DNA bases, guanine (G) is most easily oxidized [17, 18] and the G radical cation serves as the intermediate charge carrier during the hole hopping in DNA. In contrast to the superexchange mechanism, the level of the photoexcited donor (**Do***) has to be higher than or at least similar to the localized bridge states in order to inject a hole thermally into the DNA base stack. Subsequently, the positive charge hops from G to G and is finally trapped at a distant guanine or any other suitable charge acceptor. The dynamics of hole hopping depends on the number of hopping steps N [10]. The rate for a single hopping step from G to GG was determined to be $k_{HOP} = 10^6$–10^8 s^{-1} [19]. Using the site-specific binding of methyltransferase HhaI to DNA, a lower limit for hole hopping in DNA $k_{ET} > 10^6$ s^{-1} was measured over 50 Å through the base stack [20]. Based on the absence of a significant distance dependence, it was concluded that hole hopping through the DNA is not a rate-limiting step.

More recently, it became evident that adenines can also play the role of intermediate carriers during hole hopping in DNA (Fig. 12.3) [21]. Such A-hopping can occur if G is not present within the sequential context, as this is the case in stretches with at least four subsequent A–T base pairs between the guanines. The oxidation of A by $G^{+\bullet}$ is a slightly endothermic reaction. However, once $A^{+\bullet}$ has been generated, the A-hopping proceeds fast. In fact, the rate of A-hopping has been determined to be $k_{ET} = 10^{10}$ s^{-1} [22]. Moreover, it is remarkable that hole transport over eight A–T base pairs is nearly as efficient as the hole transport over only two A–T base pairs [23].. In comparison with G-hopping, A-hopping proceeds faster, more efficiently and almost without any distance dependence. Taken together, hole transport via the hopping mechanism seems to be a highly sequence-dependent process.

12.3
Reductive Electron Transfer and Excess Electron Transport in DNA

12.3.1
Strategies for the Synthesis of DNA Donor–Acceptor Systems

For the investigation of photoinduced electron-transfer processes through DNA, it is necessary to modify oligonucleotides with suitable chromophores or fluorophores. The whole range of different methods for the building-block strategy via phosphoramidite chemistry and protocols for post-synthetic oligonucleotide modifications have been applied, developed and improved dramatically during the last 10–15 years in order to prepare such structurally well-defined DNA systems [24, 25]. A broad variety of organic or inorganic intercalators, sugar modifications and natural or modified DNA bases have been used as charge donors and charge acceptors in order to study charge-transfer phenomena in DNA. Nearly all existing DNA assays can be categorized by their characteristic structural features (Fig. 12.4) [25]:

    A. DNA duplexes with unnatural or artificial DNA bases;
    B. DNA duplexes with DNA base modifications pointing into the major or minor groove;
    C. capped DNA hairpins with a duplex stem;
    D. DNA duplexes with intercalators covalently attached via a flexible linker.

In cases A, B and C, the DNA base or sugar modifications can be introduced via automated solid-phase synthetic methods using suitable DNA building blocks. As an alternative route, DNA modifications can be introduced by solid-phase methods which are applied during or after the complete automated solid-phase synthesis, as is the case for the preparation of the DNA assays B and D.

If the natural DNA bases are used as charge acceptors in such DNA systems, it is difficult to track the charge transfer within the context of a complex DNA base

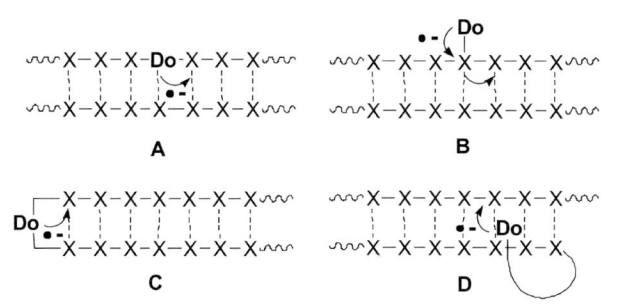

Fig. 12.4 Principle structures A–D of DNA donor–acceptor assays for the investigation of charge-transfer processes in DNA (Do = donor).

sequence. Hence, in many cases, it is suitable to incorporate a second oligonucleotide modification as an artificial charge acceptor which allows the charge-transfer process to be probed spectroscopically, chemically or electrochemically. As a result, synthetic DNA systems can be obtained bearing charge donor and charge acceptor covalently bound at a distinct distance on the oligonucleotide strands.

The major part of recent photochemical assays for electron transport in DNA has focused on the chemical trapping of the excess electrons. The quantification of the resulting DNA strand cleavages which are analyzed chemically by either gel electrophoresis or HPLC separation allow the comparison of the electron-transport efficiency within a set of samples. Mainly, two different chemical electron traps have been applied which are (i) a special T–T dimer lacking the phosphodiester bridge (T^T) [26, 27] and (ii) 5-bromo-2′-deoxyuridine (Br-dU) [28–30]. Both radical clocks yield a strand cleavage at the site of electron trapping (Br-dU only after piperidine treatment at elevated temperature) but the kinetic regime of the radical intermediate is significantly different (Fig. 12.5). The exact dynamics have been determined with the isolated nucleoside monomers: the radical anion of Br-dU loses its bromide at a rate of 7 ns^{-1} [31] whereas the radical anion of the T^T dimer splits with a much slower rate of 556 ns^{-1} [32].

This striking difference has important consequences for the elucidation of the distance dependence and DNA base sequence dependence of the excess electron-transport efficiency since the hopping rates are expected to occur on the fast nanosecond time-scale. Hence it is not surprising that in the assay of Carell and co-workers the amount of T^T dimer cleavage depended rather weakly on the distance to the electron donor [26]. In their studies a photoexcited and deprotonated flavine (Fl) derivative was used as the electron source. On the other hand, when Br-dU was applied as the electron trap, a significant dependence of the strand cleavage efficiency on the intervening DNA base sequence was observed by Ito and Rokita [28] and our group [30]. This is even more remarkable since the electron donors, the naphthalenediamine (Nd) incorporated into an abasic site of oligonucleotides and our phenothiazine-modified uridine (Pz-dU) represent signifi-

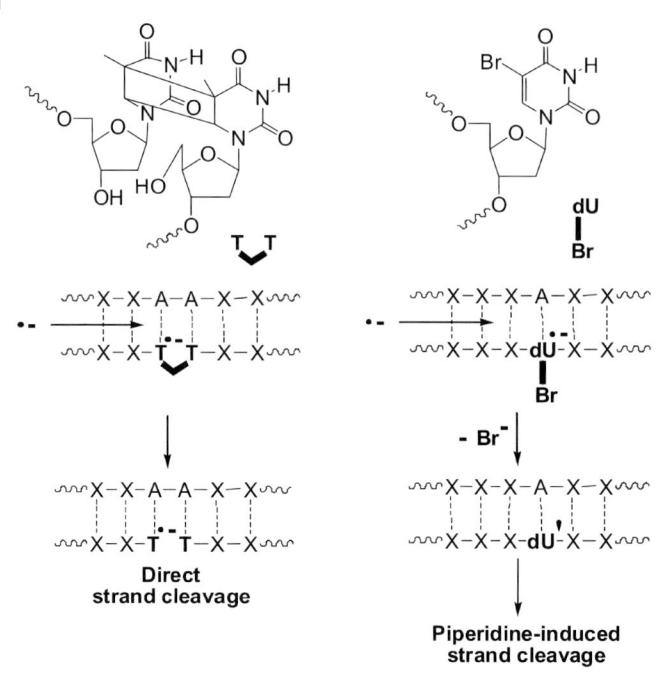

Fig. 12.5 Comparison between the T–T dimer and Br-dU as kinetic electron traps.

cantly different photoexcitable electron donors with respect to their reduction potential and their interaction with the DNA. Hence Br-dU seems to be more suitable as a kinetic electron trap since the time resolution is better for the exploration of details of an assumingly fast electron-transport process. It is important to point out that in contrast to Br-dU where the trapped electron is consumed by the loss of the bromide anion, the cleavage of the T^T dimer is redox-neutral. That means that subsequent to the T^T dimer cleavage, the excess electron could be transported further away. In fact, Giese et al. showed recently that a single injected electron can cleave more than one T^T dimer in the same DNA duplex [27]. Accordingly, they called it a "catalytic electron".

12.3.2
Chromophore Functionalization of DNA Bases via Synthesis of DNA Building Blocks

In our group, DNA bases are typically functionalized with chromophores and fluorophores using the following general synthetic strategy (Fig. 12.6) [29, 30, 33]. The nucleosides are halogenated using either N-bromosuccinimide or N-iodosuccinimide. In the case of the pyrimidine bases C and dU, halogenation occurs mainly at the 5-position and in the case of the purine bases G and A at the 8-position. These halogenated DNA bases represent the starting material for the subsequent Pd-catalyzed Suzuki–Miyaura-type cross-coupling reactions using the cor-

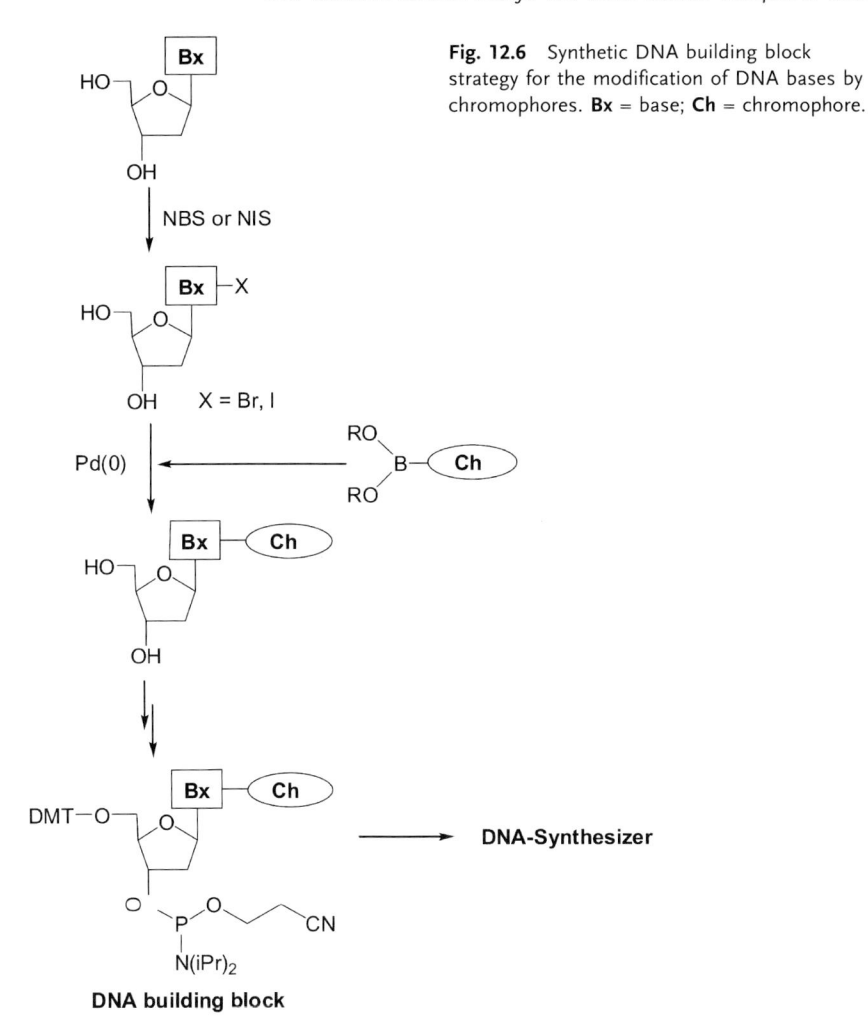

Fig. 12.6 Synthetic DNA building block strategy for the modification of DNA bases by chromophores. **Bx** = base; **Ch** = chromophore.

responding boronic acid or boronic acid esters of the chromophores. Subsequently, the 4,4-dimethoxytrityl protecting group is attached to the 5′-hydroxy group of the chromophore-modified nucleosides and the 3′-hydroxy group is converted to the phosphoramidite group yielding the final DNA building block. In principle, the last two steps represent standard reactions for the synthesis of oligonucleotides. Nevertheless, in the case of the chromophore-modified nucleosides the reaction conditions often have to be varied, mainly with respect to solvents, concentrations and reaction duration. Finally, the phosphoramidites are used as DNA building blocks for the automated synthesis of oligonucleotides on solid phases. Extended coupling times have to be applied in many cases owing to the enhanced steric hindrance of the chromophore-modified nucleoside phosphoramidite.

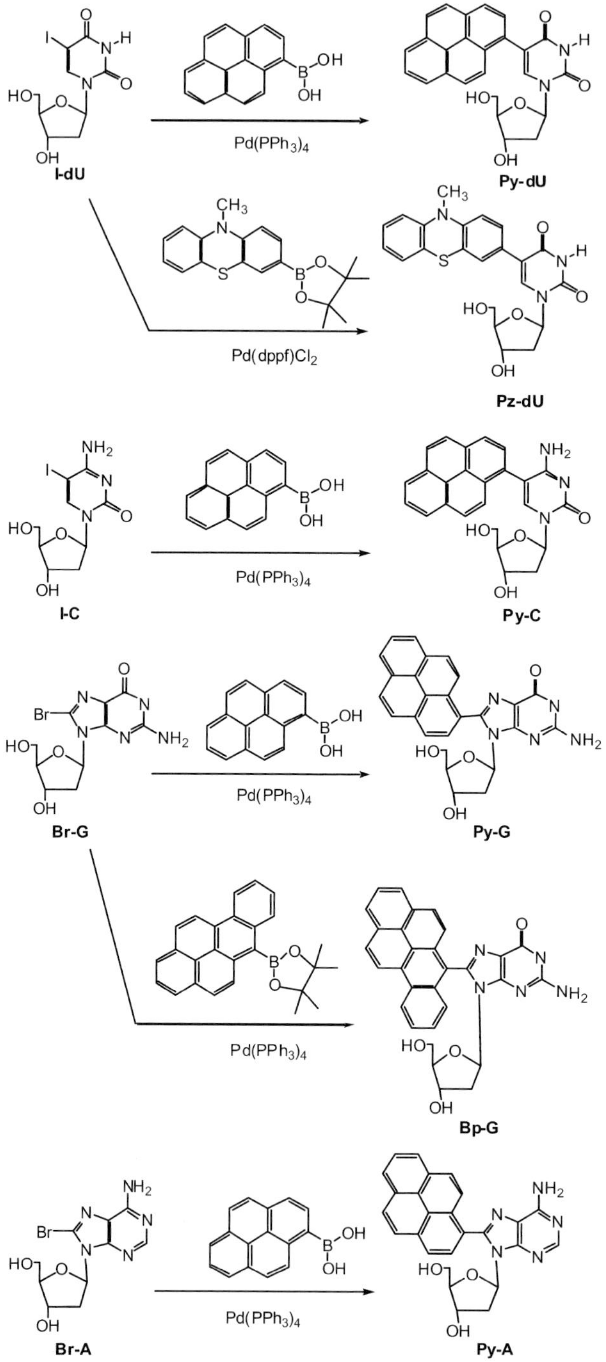

Fig. 12.7 Examples of Suzuki–Miyaura cross-couplings yielding chromophore-modified DNA bases. THF–MeOH–H$_2$O was used as the solvent and NaOH as the base.

We applied this synthetic protocol in order to attach pyrene [30, 33], phenothiazine [29] and benzpyrene[34] covalently as charge donors to the DNA bases dU, C, G and A (Fig. 12.7). Especially the 5-position of the pyrimidine DNA bases dU (and C) seems to be highly suitable for the attachment of electron donors since these base modifications can be regarded as unnatural and functionalized derivatives of the natural DNA base T. It is important to note that the design of this type of modification allows principally that the chromophores should point into the major groove of the DNA duplex while maintaining the natural Watson–Crick base pairing of the dU moiety to A as the counterbase in the complementary oligonucleotide strand. Hence this functionalization by the chromophores introduces only a local perturbation of the normal B-DNA conformation.

Accordingly, the pyrene and the phenothiazine group were attached to either 5-iodo-2′-deoxyuridine (I-dU) or 5-iodo-2′-deoxycytidine (I-C) via Pd-catalyzed Suzuki-type cross-couplings. 1-Pyrenylboronic acid or 10-methyl-3-(4,4,5,5-tetramethyl[1,3,2]dioxaborolan-2-yl)phenothiazine have been used as the chromophore starting materials.[29, 30, 33] In general, Suzuki–Miyaura-type couplings can be conducted in moist or even aqueous solutions and they tolerate the presence of unprotected functional groups [35]. Hence no protecting groups were applied for the hydroxy functions of the 2′-deoxyuridine moiety.

A critical structural issue with the subsequent incorporation of this type of chromophore-modified nucleosides into DNA duplexes is the assumption that it forms Watson–Crick base pairs with A as the counterbase. Our concern was that the large pyrene moiety may force the nucleosides into a syn conformation. This conformation would place the chromophore residue into a stacked situation within the adjacent DNA bases while replacing the counterbase of the complementary strand. To obtain more structural information, NOESY experiments were performed on the pyrene-modified nucleosides. The spectra of the modified pyrimidines Py-dU and Py-C clearly showed a significant NOESY cross peak between H-6 of the dU part or the C part, respectively and H-2′ of the corresponding 2′-deoxyribose moiety. The NOESY cross peak was comparably as strong as the cross peaks between H-2′ and H-1′ or H-3′, respectively. These NMR results can only be explained with the preferred anti conformation of these nucleosides. In contrast, the spectra of modified purines Py-G and Py-A showed a significant NOE between H-1 of the pyrene part and H-1′ of the 2′-deoxyribose moiety. The NOESY cross peak was comparably as strong as the cross peaks between H-2′ and H-1′. These NMR results can only be explained with the preferred syn conformation of the nucleosides. Conclusively, it is expected that the chromophore-modified purines induce a more drastic perturbation into the B-DNA conformation than the modified pyrimidines.

12.3.3
DNA Base Modifications via a Solid-phase Synthetic Strategy

Alternatively to the previously described phosphoramidite synthesis, organic or inorganic intercalators, which have been applied as charge donors by various research groups, were incorporated into oligonucleotides by different solid-phase methods. In most cases the corresponding modification protocols follow the automated synthesis of oligonucleotides when it is still attached to the solid phase [25]. After complete modification, the DNA conjugate is cleaved off the beads and purified by HPLC, gel electrophoresis or capillary electrophoresis. The major advantages of this modification strategy are as follows:

1. The preparation of the corresponding water-sensitive phosphoramidite which represents in a lot of cases a synthetic bottleneck can be avoided.
2. The modification protocol is versatile and can be applied to oligonucleotides of different lengths and/or different base sequence compositions.
3. Photolabile or redox-active intercalators or cofactors can be attached at the very end of the oligonucleotide synthesis.

Hence the post-synthetic modification of oligonucleotides represents a very fast and efficient way to obtain a variety of different DNA–charge-donor conjugates. Most of the applied methods have molecular probes attached to the 5′-terminal hydroxy group and rely on an amide bond formation between the carboxylate group as part of the charge donor and the amino group as part of a linker or a modified DNA base in the oligonucleotide. For example, Barton and coworkers tethered ethidium as an intercalating charge donor via an amide bond and a flexible alkyl linker to the 5′-terminus of oligonucleotides [36].

In our group, a fast and versatile synthetic approach is applied for the preparation of modified oligonucleotides in which the chromophore is attached to the DNA base via an acetylene bridge (Fig. 12.8) [37–39]. Three structural features of these chromophore-modified DNA are important to point out: (i) a clear steric separation of the chromophore moiety from the DNA base stack via the rigid ethynyl group, (ii) strong electronic coupling between the chromophore and the base moiety provided by the acetylene bridge and (iii) partial stacking of the base moiety as part of the delocalized chromophore–base conjugate. The incorporation of chromophores via the acetylene bridge is based on Pd-catalyzed Sonogashira-type cross-couplings [40] which work at room temperature. Thus, in contrast to the Suzuki-type cross-couplings as described in the previous section, these Sonogashira-type modifications can be performed by a solid-phase strategy as follows (Fig. 12.8) [37–39, 41]. (i) The oligonucleotide is synthesized following standard protocols on a DNA synthesizer up to the position of the modified DNA base. (ii) A halogenated nucleoside is inserted automatically without the final deprotection of the terminal 5′-hydroxy group. (iii) Subsequently, the CPG vials are removed from the synthesizer and a Sonogashira coupling reagent mixture containing

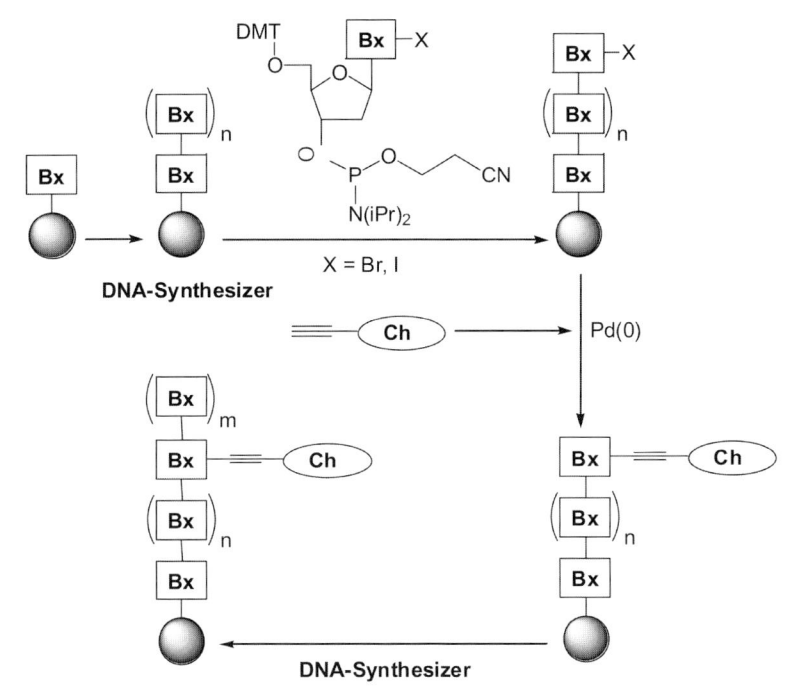

Fig. 12.8 Synthetic semi-automated solid-phase strategy for the modification of DNA bases by chromophores via an acetylene bridge and Sonogashira-type cross-couplings.

Pd(PPh$_3$)$_4$, the 1-ethynyl chromophore and CuI in DMF–Et$_3$N (3.5:1.5) is added to the CPG vials under dry conditions via syringes. After a coupling time of 3 h at room temperature, the CPGs were washed using different solvents and dried. (iv) The CPG vials were attached to the DNA synthesizer and the synthesis continued automatically. Modification of the standard procedures for deprotection and cleavage of the oligonucleotides from the solid phase or during workup is not necessary in most cases.

This synthetic protocol exhibits very high versatility. In principle, the chromophore can be attached to any of the four DNA bases A, C, G or dU (T). The site of attachment can be designed by choosing the right phosphoramidite of the halogenated nucleosides as a DNA building block during the procedure, either 8-bromo-2′-deoxyadenosine, 2′-deoxy-5-iodocytidine, 8-bromo-2′-deoxyguanosine or 2′-deoxy-5-iodouridine. Two of these modifications can be introduced simultaneously [40]. In our group, the synthetic protocol has been applied for the incorporation of 1-ethynylpyrene into oligonucleotides. However, in principle, this synthetic strategy should be applicable to any desired acetylene-linked chromophore.

12.3.4
Chromophores as Artificial DNA Base Substitutes

Alternatively to the DNA modifications in the previous two sections where the chromophore was attached to one of the four DNA bases, chromophores can be incorporated as artificial DNA bases substituting a natural base or even a whole base-pair. There is a large number of recently reported syntheses of chromophores as DNA base surrogates, e.g. flavine derivatives [26] and thiazole orange derivatives [42]. Additionally, a variety of phosphoramidites as DNA building blocks for the introduction of fluorophores into DNA are commercially available, e.g. acridine derivatives. Clearly, the synthetic protocols for this kind of DNA modification do not follow a principle strategy which can be applied in a versatile fashion, as is the case for the DNA base modifications mentioned in the previous sections. It is important to point out that in many cases it turned out to be useful to replace the 2'-deoxyribose moiety with acyclic linker systems. This was also the case during our attempts to synthesize ethidium-modified DNA, which will be described here briefly.

3,8-Diamino-5-ethyl-6-phenylphenanthridinium ("ethidium") has been widely used in fluorescence assays with nucleic acids [43]. The site-specific intercalation of ethidium in DNA is crucial for a detailed study of the binding interactions and the charge donor properties. Hence, we incorporated the phenanthridinium heterocycle of ethidium as an artificial base at specific sites in duplex DNA (Fig. 12.9) [44–46]. The amino group in position C-3 of ethidium is deactivated in comparison with the C-8 amino group owing to the mesomeric stabilization of the positive charge. Accordingly, we synthesized 8-(2'-deoxy-D-ribofuranosyl)-3-acetamido-5-ethyl-6-phenylphenanthridinium via a regioselective reaction of the C-8 amino group [47]. This nucleoside contains the ethidium moiety glycosidically linked to 2'-deoxyribofuranose. Although significant mesomeric stabilization exists within the ethidium aromatic system, this nucleoside does not exhibit stability towards basic hydrolysis which is high enough to guarantee the subsequent successful incorporation into oligonucleotides. Therefore, we decided to replace the 2'-deoxyribofuranoside moiety by an acyclic linker system which is tethered to the N-5 position of the phenanthridinium system [44–46]. Avoiding the labile glycosidic bond, the ethidium nucleoside analog should be suitable for the preparation of DNA building blocks and phenanthridinium–DNA conjugates via automated phosphoramidite chemistry.

The synthesis of the corresponding DNA building block [45] starts with the protection of the two exocyclic amino functions of 3,8-diamino-6-phenylphenanthridine by chloroallyl formate. Subsequently, the bis-allyloxycarbonyl-protected phenanthridine derivative is alkylated with 1,3-diiodopropane and linked to DMT-protected 3-amino-1,3-propanediol using the typical conditions for a nucleophilic substitution. Subsequently, the alloc protecting groups had to be changed to trifluoroacetyl groups. This procedure is necessary since trifluoroacetyl groups are not stable enough for the previous alkylation at N-5 of the phenanthridine heterocycle but can be cleaved under typical DNA workup conditions. Additionally, this

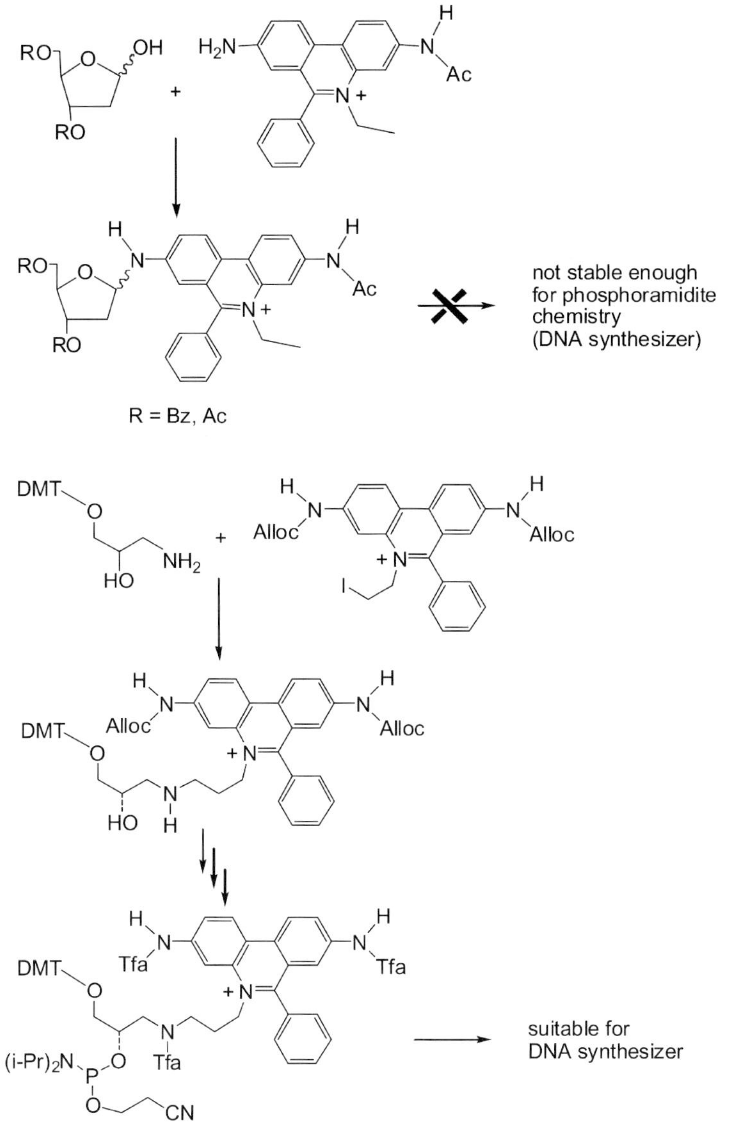

Fig. 12.9 Synthesis of DNA bearing ethidium as an artificial DNA base (alloc = allyloxycarbonyl, Tfa = trifluoroacetyl).

protecting group strategy has the advantage that the secondary amino function of the alkyl linker also gets protected. The preparation of the phosphoramidite was finished using standard procedures and subsequently used for the automated preparation of phenanthridinium-modified oligonucleotides. An extended coupling time (1 h instead of 1.5 min. for standard couplings), a higher phosphoramidite concentration (0.2 instead of 0.067 M) and three coupling cycles interrupted by washing steps were necessary to achieve nearly quantitative coupling.

It is important to point out that this synthetic route has been optimized for the incorporation of ethidium as an artificial DNA base and therefore cannot simply be transferred for the preparation of other chromophores as DNA base surrogates. This makes it clear that this type of DNA–chromophore modification represents the most time-consuming one and much synthetic research effort needs to be invested in order to obtain a reliable synthetic procedure for the routine synthesis of chromophore-modified DNA.

12.4
Results from the Electron Transfer Studies

A significant amount of knowledge about reductive electron transport in DNA comes from γ-pulse radiolysis studies using DNA samples doped with intercalated and randomly spaced electron traps [48]. The major disadvantage of this experimental setup is that the electron injection and the electron trapping do not occur regioselectively. Nevertheless, it became clear that above 170 K the electron-transport mechanism follows a thermally activated process. Synthetic photochemical assays have the advantage that the electron injection can be initiated site selectively by the covalently attached chromophore.[9, 24, 25] In most assays the electron transport was probed chemically: (i) Carell and coworkers showed that the amount of T–T dimer cleavage depends rather weakly on the distance [26]; (ii) Giese et al. showed that a single injected electron can cleave more than one T–T dimer [27]; and (iii) Ito and Rokita detected a significant base sequence dependence of the electron-transport efficiency [28]. The proposal of a thermally activated electron hopping mechanism involving C$^-$• and T$^-$• as intermediates [8] was supported by all studies. So far, only Lewis et al. [49], Netzel and coworkers [50] and our group [30, 51] have focused on the dynamics of ET processes.

We use pyrene (Py), phenothiazine (Pz) and ethidium (Et) as photoinducible electron donors. Interestingly, Py in the excited state (Py*) is a significantly weaker electron donor than calculated and therefore expected. Combining the potentials $E(Py^{+}•/Py^*) = -1.8$ V (vs. NHE) [52] and $E(dU/dU^-•) = -1.1$ V [53], the driving force ΔG for the ET process in Py-dU could be a maximum of -0.6 eV. However, our studies revealed a driving force $\Delta G \approx 0$ eV [51], which requires the potential $E(dU/dU^-•)$ to be ca. -1.8 V. In this context, the measured value $E(dU/dU^-•) = -1.1$ V provided by Steenken et al. [53] is difficult to understand and could reflect the result of a proton-coupled electron transport [54]. Hence it is likely that the -1.1 V potential corresponds to $E[dU/dU(H)•]$. Pz is a stronger electron donor than Py since the reduction potential of Pz in the excited state, $E(Pz^{+}•/Pz^*)$, is -2.0 V [52]. In order to use Py and Pz as electron donors in DNA, we synthesized Py-dU and Pz-dU as described in the previous sections. By this synthetic approach, we are able to photoinitiate exclusively a reductive electron transfer since the intramolecular electron transfer in the Py-dU and the Pz-dU moieties can be regarded

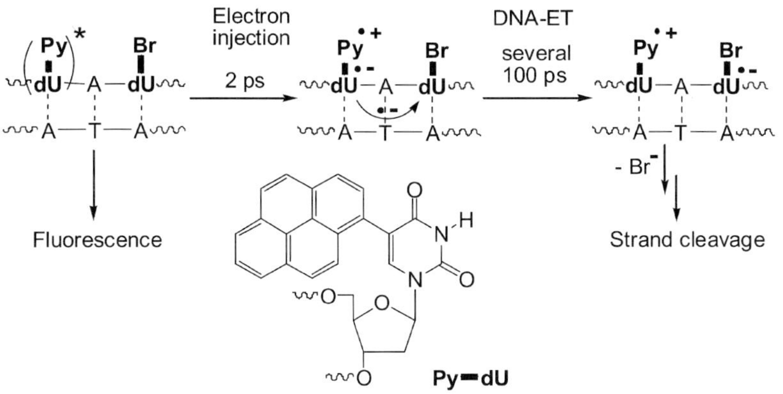

Fig. 12.10 Photochemical assay for the spectroscopic and chemical investigation of electron transport in Py-dU/Br-dU-modified DNA.

as an electron injection into the DNA preceding the electron hopping (Figs. 12.10 and 12.11).

We studied the electron-transfer processes using a combination of different techniques, comprising mainly steady-state fluorescence spectroscopy, time-resolved laser spectroscopy and chemical probing using Br-dU (Figs. 12.10 and 12.11). It is known that Br-dU undergoes a chemical modification after its one-electron reduction which can be analyzed by piperidine-induced strand cleavage and has been applied to quantify the efficiency of DNA-mediated ET processes.[28, 29, 55] It is important to point out that, based on reduction potentials, Br-dU is not a significantly better electron acceptor [49] and hence Br-dU represents a kinetic electron trap.

In recent years, it has become apparent that electron-transfer phenomena in DNA cannot be understood without considering the manifold of conformational states present in DNA [56]. Since ET rates depend strongly on the microscopic environment, one might not observe single kinetic rate constants for DNA-mediated electron transport but rather a distribution of rates. From recent studies of oxidative hole transport in DNA, it is known that DNA has to be considered as a structurally flexible medium and DNA-mediated electron transfer cannot be reduced to a static donor–bridge–acceptor situation. In particular, there is strong experimental evidence for an involvement of base stacking fluctuations and hydrogen bonding interactions (not necessarily proton transfer) inside the DNA helix which we observed in the Py-dU-modified DNA samples (Fig. 12.10) [30]. The charge-separated state Py$^+$•-dU$^-$• which is formed after excitation and subsequent electron injection exhibits a strong kinetic dispersion in its lifetimes which is consistent with multi-conformational DNA. Interestingly, the electron injection process in our functionalized DNA shows only minor variations due to structural inhomogeneity because the injection process occurs between the covalently connected Py and dU moieties. To answer the question of whether subsequent electron shift into the

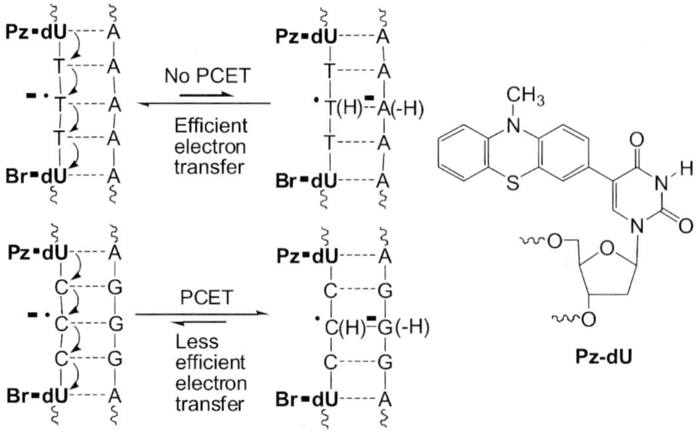

Fig. 12.11 DNA-mediated electron transport in Pz-dU/Br-dU-modified DNA via cytosine and thymine hopping.

base stack competes with charge recombination, we had to combine these time-resolved measurements with chemical probing using the Br-dU radical clock. The results indicate that the electron shift occurs on the same time-scale (several hundred picoseconds) as charge recombination to the ground state. It is reasonable to assume that additional electron migration steps will be faster since the Coulomb interaction between the electron and Py^{+}• decreases drastically with distance. These results therefore provide a lower limit for excess electron-transfer rates through DNA. Our results demonstrate the importance of probing both the early time events and the product states for obtaining conclusive mechanistic insights. Since DNA-mediated electron transport is a multi-step process that occurs on various time-scales, the measured electron injection rates may not necessarily correlate with the observed strand cleavage efficiencies as the chemical result of DNA-mediated electron transport.

As mentioned above, it was postulated that electron hopping in DNA involves all base pairs (T–A and C–G), meaning that both pyrimidine radical anions, C¯• and T¯•, play the role of intermediate charge carriers [8], although it was known that C¯• exhibits a significantly stronger basicity than T¯• [51, 54]. Remarkably, from the strand cleavage experiments with the Pz-dU/Br-dU DNA duplexes, it becomes clear that T–A base pairs transport electrons more efficiently than C–G base pairs (Fig. 12.11) [29]. This implies that C¯• is not likely to play a major role as an intermediate electron carrier. This observation is supported by a number of recent publications: (i) the presence of C–G base pairs lowered significantly the ET efficiency in the DNA assay of Ito and Rokita et al. [28]; (ii) Cai and Sevilla showed that proton transfer can slow excess electron transfer but does not stop it [48]; (iii) we showed using pyrene-modified nucleosides as models for electron transport in DNA that the non-protonated radical anion of C could not be observed in aqueous solution, suggesting that the protonation of C¯• by the complementary DNA base G or the surrounding water molecules will occur rapidly [51]. During this time, the hydrogen bond interface can readjust and stabilize the excess electron on the base by separating its spin from its charge. Although these processes must be microscopically reversible, they may ultimately terminate electron migration in DNA. In summary, it is clear now that proton transfer interferes with electron transport. Thus electron hopping in DNA is highly sequence dependent and occurs faster and more efficiently over T–A base pairs than over C–G base pairs.

Ethidium (Et) plays an important role as a charge donor in studies of photoinduced processes through DNA [57]. This is remarkable since, based on relative redox potentials, ethidium in the photoexcited state (Et*$^+$) is not able to oxidize [E(Et*$^+$/Et•) = −0.5 V] [58] or reduce [E(Et^{2+}•/Et*$^+$) = 1.2 V] [59] DNA in order to initiate hole or electron hopping, respectively. Hence a suitable charge acceptor has to be provided for each of the two different charge-transfer processes. Real-time measurements were carried out on Et covalently bound to DNA by a flexible molecular tether which permitted intercalation only at specific DNA sites [57]. In these systems, Et was the oxidative charge donor and 7-deazaguanine (Zg) as the hole acceptor quenching the emission of photoexcited Et* as a result of a charge transfer [E(Zg$^+$•/Zg = 1.0 V] [59]. For the reductive electron transfer, 5-nitroindole (Ni) with a reduction potential E(Ni/Ni$^-$•) = −0.3 V [60] should be suitable as an electron acceptor. The corresponding phosphoramidite is commercially available.

As described in the previous section, we incorporated Et as an artificial DNA base at specific sites in duplex DNA (Fig. 12.12) [44–46]. Temperature-dependent absorption and steady-state fluorescence measurements prove the intercalation of the Et moiety in duplex DNA [44]. The intercalation properties of the Et moiety do not depend significantly on the local duplex environment. Interestingly, the optical properties of Et seem not to interfere with the presence of the different counterbases T, G, C or A [45]. This result is remarkable with respect to the steric demand of the Et heterocycle and indicates a bulged position of the counterbase.

The charge donor properties of Et as an artificial DNA base were elucidated using either Zg as the hole acceptor or Ni as the electron acceptor (Fig. 12.12). Those charge acceptors were separated from the charge donor Et by up to three intervening base pairs. As expected, the fluorescence quenching is highest in the DNA duplexes with Et and Zg (or Ni) closest together. In the latter arrangement the charge-transfer process is most efficient. Over three base pairs, the fluorescence quenching drops to less than 10%. These observations stand in agreement with a Marcus type of charge transfer which occurs via a superexchange mechanism and are typically limited to a distance of 3–4 base pairs. It is important to point out that using this DNA assay it is possible to compare oxidative hole transfer (Et–DNA–Zg) and reductive electron transfer (Et–DNA–Ni) in a structurally well-defined and very similar donor–acceptor-system.

The DNA–charge-transfer system consisting of Et and Zg as a distinct donor–acceptor couple was applied in order to detect DNA base mismatches and the abasic site as typical DNA lesion [46]. By using charge-transfer processes additionally to the emission properties of Et*, the detection of base mismatches does not rely solely on the small differences in the hybridization energies between matched

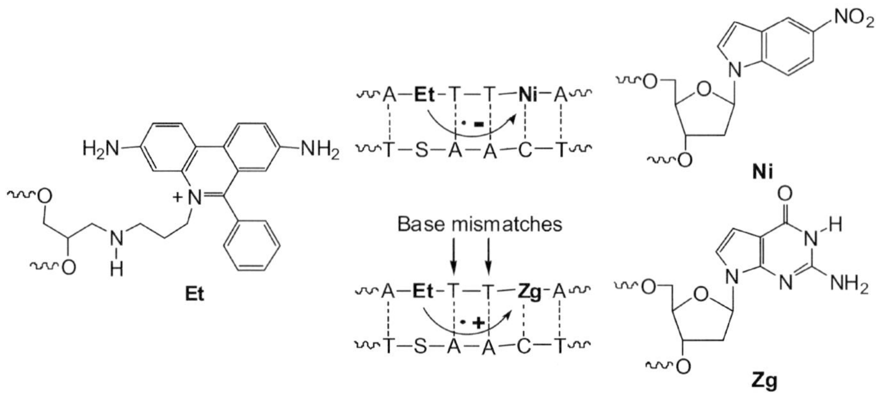

Fig. 12.12 Hole transfer in Et/Zg-modified DNA in comparison with electron transfer in Et/Ni-modified DNA. S = abasic site analog.

and mismatched duplexes. Our assay is based on the measurement of the fluorescence intensities of Et-modified DNA bearing a base mismatch or abasic site (S) either one base pair away or two base pairs away from the Et chromophore. First, we measured the steady-state fluorescence of the DNA control duplexes lacking Zg as the charge acceptor. As expected, the Et chromophore exhibits no fluorescence sensitivity towards adjacent base mismatches without the charge-transfer process. In contrast, the fluorescence spectra of the DNA duplexes bearing Zg as the charge acceptor show remarkable differences of the emission quantities. In fact, the presence of a base mismatch or abasic site (S) shows an enhanced fluorescence quenching compared with the matched duplexes. From these studies, it is evident that combining the fluorescence properties of Et as an artificial DNA base together with DNA-mediated charge-transfer processes represents a new method for the homogeneous detection of DNA base mismatches and abasic sites. The method has the advantage that the changes in fluorescence are not limited to the directly adjacent bases and allow the scanning of a sequence of two base pairs. Existing DNA arrays which are based on charge-transfer processes require the attachment of DNA duplexes as self-assembled monolayers on gold electrodes combined with electrochemical measurements as the readout [14]. Our results could lead to new DNA microarrays which are based on charge-transfer processes and can be analyzed by commonly used fluorescence readout techniques.

12.5
Outlook: Towards Synthetic Nanostructures Based on DNA-like Architecture

As already described in Section 12.1, the construction of nanoscale electronic devices from conducting biopolymers remains a challenging task. DNA represents a superior material for this research by moving away from the natural biological

12.5 Outlook: Towards Synthetic Nanostructures Based on DNA-like Architecture

role and considering it as a supramolecular architecture which features important structural aspects such as the regular linear structure. Moreover, the most remarkable aspect of oligonucleotides is the highly selective recognition ability with their complementary parts as encoded by the DNA base sequence. DNA-based molecular wires are expected to represent an important nanomaterial with a high potential for new electronic devices.

In principle, both types of charge transport (oxidative and reductive) could be applied for molecular electronics since they potentially occur both on long distances. However, one of the major drawbacks of the oxidative hole transport with respect to electronic devices is the oxidative DNA damage which occurs as a side product of the guanine radical cation (see Fig. 12.4). This hole trapping represents a serious disadvantage. For this reason, Saito's group developed an N-phenylguanosine, PG (Fig. 12.13), as a substitute for G in duplex DNA. PG represents the an improved G substitute as an intermediate charge carrier for hole transport by avoiding oxidative degradation [61]. Another way to improve the hole-transport efficiency is the extension of the π–π-stacking interactions. Benzo-fused DNA bases like BA (Fig. 12.13) have such an expanded aromatic surface and are able to increase the hole-transport efficiency due the enhanced stacking interactions inside the DNA double helix [62].

On the other hand, several proposals support strongly the idea that the excess electron transport has a significantly higher potential for the application in electronic devices on the molecular level. First of all, all base pairs (C–G and T–A) are involved as intermediate charge carriers during electron hopping (see Fig. 12.12). In contrast, during hole hopping, domains of G-hopping and A-hopping occur (see Fig. 12.3). Moreover, the proposed lack of DNA damage as the result of reductive electron transport is ideal for the development of DNA-nanowires. A few measurements of the electronic conductivity of DNA have been performed [63]. It is important to point out that in many cases the DNA secondary structure which is crucial for fast and efficient electron transport remained undetermined owing to the absence of water in the experiments. Nevertheless, it was shown that it is possible to transport electrons along the DNA molecules, but the conductivity is rather poor. There are several attempts to use metallized DNA in order to improve the conductivity [64]. Metal-mediated base pairing represents an important way to enhance the genetic code, and more importantly, to gain access to a structurally well-defined row of potentially interacting metal ions [65]. As a prominent recent example, Shionoya's group prepared a DNA duplex bearing five metal-chelating base pairs (Fig. 12.13). However, the general disadvantage of this approach for metalation of DNA is that the self-complementary and specific base-pairing as encoded by the DNA base sequence is lost. Hence current research is focused additionally on the sequence-specific metalation of the DNA duplex in the major or minor groove. These are important steps to nanoelectronic tools and, combined with chromophores as photoinducible electron donors, the construction of complex photoelectronic devices can be envisioned.

Taken together with the results from the photoinduced electron-transport experiments in which diffusive thermal electron hopping can occur over a few

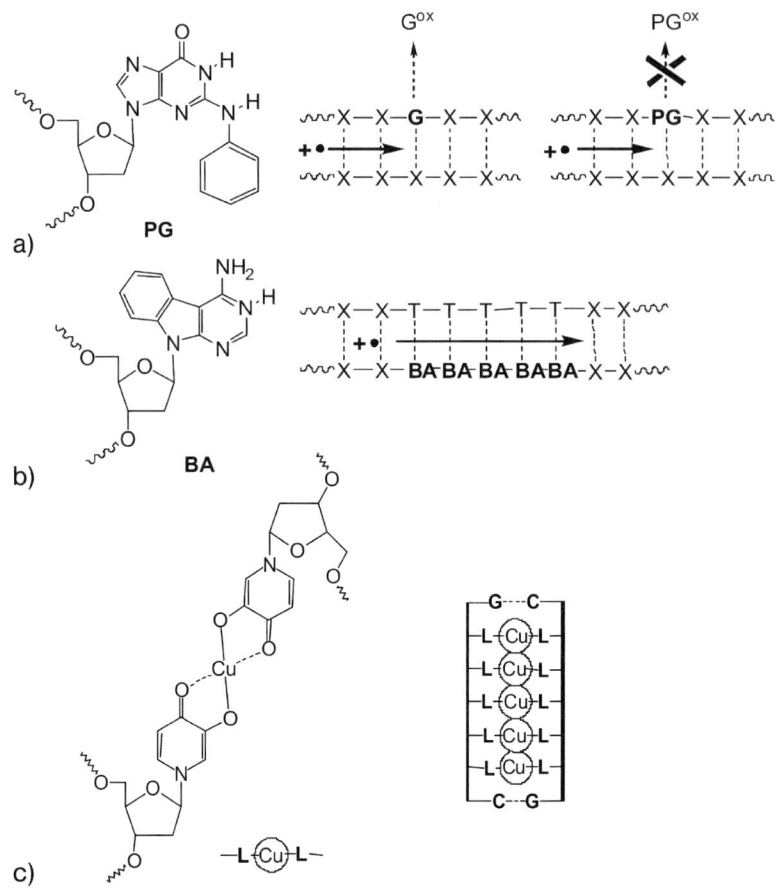

Fig. 12.13 Modified bases for the development of electronic devices based on DNA-mediated charge transport: PG (a), BA (b) and Cu-mediated base-pairing of L (c).

nanometers, it becomes clear that electron transport through longer single DNA molecules (>20 nm) requires more research effort. Hence, in order to use DNA as a molecular wire, it is necessary to develop DNA-inspired materials which contain the DNA-typical structural features (as described in Section 12.1) but exhibit improved electron-transport capabilities. It will be interesting to see how this research develops over the next few years.

References

1. See reviews: (a) A. Aviram, M. A. Ratner, *Chem. Phys. Lett.* **1974**, *29*, 277; (b) R. M. Metzger, *Acc. Chem. Res.* **1999**, *32*, 950–957; (c) J. M. Tour, *Acc. Chem. Res.* **2000**, *33*, 791–804; (d) C. Joachim, J. K. Gimzewski, A. Aviram, *Nature* **2000**, *408*, 541–548; (e) V. Mujica, M. A. Ratner, A, Nitzan, *Chem. Phys.* **2002**, *281*, 147–150; (f) M. Mayor, H. B. Weber, in *Nanoelectronics and Information Technology*, R. Waser (Ed.), Wiley-VCH, Weinheim, **2003**, pp. 501–525; (g) R. A. Wassel, C. B. Gorman, *Angew. Chem. Int. Ed.* **2004**, *43*, 5120–5123; (h) P. Bäuerle, *Nachr. Chem.* **2004**, *52*, 19–24.
2. See review: N. C. Seeman, *Chem. Biol.* **2003**, *10*, 1151–1159.
3. D. D. Eley, D. I. Spivey, *Trans. Faraday Soc.* **1962**, *58*, 411–415.
4. (a) S. Priyadarshy, S. M. Risser, D. N. Beratan, *J. Phys. Chem.* **1996**, *100*, 17678–17682; (b) N. J. Turro, J. K. Barton, *J. Biol. Inorg. Chem.* **1998**, *3*, 201–109; (c) Y. A. Berlin, A. L. Burin, M. A. Ratner, *Superlattices Microstruct.* **2000**, *28*, 241–252.
5. R. A. Marcus, N. Sutin, *Biochim. Biophys. Acta* **1985**, *811*, 265–322.
6. C. J. Murphy, M. R. Arkin, Y. Jenkins, N. D. Ghatlia, S. H. Bossmann, N. J. Turro, J. K. Barton, *Science* **1993**, *262*, 1025–1029.
7. See reviews: (a) P. O'Neill, E. M. Frieden, *Adv. Radiat. Biol.* **1993**, *17*, 53–120; (c) C. J. Burrows, J. G. Muller, *Chem. Rev.* **1998**, *98*, 1109–1151; (c) D. Wang, D. A. Kreutzer, J. M. Essigmann, *Mutat. Res.* **1998**, *400*, 99–115; (e) S. Kawanashi, Y. Hiraku, S. Oikawa, *Mutat. Res.* **2001**, *488*, 65–76.
8. See review: B. Giese, *Annu. Rev. Biochem.* **2002**, *71*, 51–70.
9. See review: H.-A. Wagenknecht, *Angew. Chem. Int. Ed.* **2003**, *42*, 2454–2460.
10. J. Jortner, M. Bixon, T. Langenbacher, M. E. Michel-Beyerle, *Proc. Natl. Acad. Sci. USA* **1998**, *95*, 12759–12765.
11. See reviews: *Top. Curr. Chem.* **2004**, *236* and *237*; whole issues.
12. H.-A. Wagenknecht, *Charge Transfer in DNA: From Mechanism to Application*, Wiley-VCH, Weinheim, **2005**.
13. E. Yavin, A. K. Boal, E. D. A. Stemp, E. M. Boon, A. L. Livingston, V. L. O'Shea, S. S. David, J. K. Barton, *Proc. Natl. Acad. Sci. USA* **2005**, *102*, 3546–3551.
14. See review: T. G. Drummond, M. G. Hill, J. K. Barton, *Nat. Biotechnol.* **2003**, *21*, 1192–1199.
15. See review: D. Porath, G. Cuniberti, R. Di Felice, *Top. Curr. Chem.* **2004**, *237*, 183–227.
16. (a) J. R. Winkler, H. B. Gray, *Chem. Rev.* **1992**, *92*, 369–379; (b) M. R. Wasielewski, *Chem. Rev.* **1992**, *92*, 435–461.
17. S. Steenken, S. V. Jovanovic, *J. Am. Chem. Soc.* **1997**, *119*, 617–618.
18. C. A. M. Seidel, A. Schulz, M. H. M. Sauer, *J. Phys. Chem.* **1996**, *100*, 5541–5553.
19. F. D. Lewis, X. Liu, J. Liu, S. E. Miller, R. T. Hayes, M. R. Wasielewski, *Nature* **2000**, *406*, 51–53.
20. H.-A. Wagenknecht, S. R. Rajski, M. Pascaly, E. D. A. Stemp, J. K. Barton, *J. Am. Chem. Soc.* **2001**, *123*, 4400.
21. B. Giese, M. Spichty, *ChemPhysChem* **2000**, *1*, 195–198; (b) B. Giese, J. Amaudrut, A.-K. Köhler, M. Spormann, S. Wessely, *Nature* **2001**, *412*, 318–320.
22. T. Takada, K. Kawai, X. Cai, A. Sugimoto, M. Fujitsuka, T. Majima, *J. Am. Chem. Soc.* **2004**, *126*, 1125–1129.
23. B. Giese, T. Kendrick, *Chem. Commun.* **2002**, 2016–2017.
24. See review: M. W. Grinstaff, *Angew. Chem. Int. Ed.* **1999**, *38*, 3629–3635.
25. See review: H.-A. Wagenknecht, *Curr. Org. Chem.* **2004**, *8*, 251–266.
26. (a) C. Behrens, L. T. Burgdorf, A. Schwögler, T. Carell, *Angew. Chem. Int. Ed.* **2002**, *41*, 1763–1766; (b) C. Haas, K. Kräling, M. Cichon, N. Rahe, T. Carell, *Angew. Chem. Int. Ed.* **2004**, *43*, 1842–1844.
27. B. Giese, B. Carl, T. Carl, T. Carell, C. Behrens, U. Hennecke, O. Schiemann, E. Feresin, *Angew. Chem. Int. Ed.* **2004**, *43*, 1848–1851.
28. (a) T. Ito, S. E. Rokita, *J. Am. Chem. Soc.* **2003**, *125*, 11480–11481; (b) T. Ito, S. E. Rokita, *Angew. Chem. Int. Ed.* **2004**, *43*, 1839–1842.

29. C. Wagner, H.-A. Wagenknecht, *Chem. Eur. J.* **2005**, *22*, 1871–1876.
30. (a) N. Amann, E. Pandurski, T. Fiebig, H.-A. Wagenknecht, *Chem. Eur. J.* **2002**, *8*, 4877–4883; (b) P. Kaden, E. Mayer, A. Trifonov, T. Fiebig, H.-A. Wagenknecht, *Angew. Chem. Int. Ed.* **2005**, *44*, 1636–1639.
31. E. Rivera, R. H. Schuler, *J. Am. Chem. Soc.* **1983**, *87*, 3966–3971.
32. S.-R. Yeh, D. E. Falvey, *J. Am. Chem. Soc.* **1997**, *113*, 8557–8558.
33. E. Mayer, L. Valis, R. Huber, N. Amann, H.-A. Wagenknecht, *Synthesis* **2003**, 2335–2340.
34. L. Valis, H.-A. Wagenknecht, *Synlett* **2005**, 2281–2284.
35. See review: N. Miyaura, A. Suzuki, *Chem. Rev.* **1995**, *95*, 2457–2483.
36. S. O. Kelley, R. E. Holmlin, E. D. A. Stemp, J. K. Barton, *J. Am. Chem. Soc.* **1997**, *119*, 9861–9870.
37. M. Rist, N. Amann, H.-A. Wagenknecht, *Eur. J. Org. Chem.* **2003**, 2498–2504.
38. E. Mayer, L. Valis, C. Wagner, M. Rist, N. Amann, H.-A. Wagenknecht, *ChemBioChem* **2004**, *5*, 865–868.
39. C. Wagner, M. Rist, E. Mayer-Enthart, H.-A. Wagenknecht, *Org. Biomol. Chem.* **2005**, *3*, 2062–2063.
40. See review: K. Sonogashira, *J. Organomet. Chem.* **2002**, *653*, 46–49.
41. S. I. Khan, M. W. Grinstaff, *J. Am. Chem. Soc.* **1999**, *121*, 4704–4705.
42. D. V. Jarokite, O. Köhler, E. Socher, O. Seitz, *Eur. J. Org. Chem.* **2005**, 3187–3195.
43. (a) A. R. Morgan, J. S. Lee, D. E. Pulleyblank, N. L. Murray, D. H. Evans, *Nucleic Acids Res.* **1979**, *7*, 547–569; (b) A. R. Morgan, J. S. Lee, D. E. Pulleyblank, N. L. Murray, D. H. Evans, *Nucleic Acids Res.* **1979**, *7*, 571–594.
44. N. Amann, R. Huber, H.-A. Wagenknecht, *Angew. Chem. Int. Ed.* **2004**, *43*, 1845–1847.
45. R. Huber, N. Amann, H.-A. Wagenknecht, *J. Org. Chem.* **2004**, *69*, 744–751.
46. L. Valis, N. Amann, H.-A. Wagenknecht, *Org. Biomol. Chem.* **2005**, *3*, 36–38.
47. N. Amann, H.-A. Wagenknecht, *Tetrahedron Lett.* **2003**, *44*, 1685–1690.
48. See review: Z. Cai, M. D. Sevilla, *Top. Curr. Chem.* **2004**, *237*, 103–128.
49. F. D. Lewis, X. Liu, S. E. Miller, R. T. Hayes, M. R. Wasielewski, *J. Am. Chem. Soc.* **2002**, *124*, 11280–11281.
50. S. T. Gaballah, J. D. Vaught, B. E. Eaton, T. L. Netzel, *J. Phys. Chem. B* **2005**, *109*, 5927–5934.
51. (a) R. Huber, T. Fiebig, H.-A. Wagenknecht, *Chem. Commun.* **2003**, 1878–1879; (b) M. Raytchev, E. Mayer, N. Amann, H.-A. Wagenknecht, T. Fiebig, *ChemPhysChem* **2004**, *5*, 706–712.
52. T. Kubota, K. Kano, T. Konse, *Bull. Chem. Soc. Jpn.* **1987**, *60*, 3865–3877.
53. S. Steenken, J. P. Telo, H. M. Novais, L. P. Candeias, *J. Am. Chem. Soc.* **1992**, *114*, 4701–4709.
54. (a) S. Steenken, *Biol. Chem.* **1997**, *378*, 1293–1297; (b) S. Steenken, *Free Rad. Res. Commun.* **1992**, *16*, 349–379.
55. E. Mayer-Enthart, P. Kaden, A. Wagenknecht, *Biochemistry* **2005**, *44*, 11749–11757.
56. (a) M. O'Neill, J. K. Barton, *J. Am. Chem. Soc.* **2004**, *126*, 11471–11483; (b) M. O'Neill, J. K. Barton, *J. Am. Chem. Soc.* **2004**, *126*, 13234–13235.
57. See e.g.: C. Wan, T. Fiebig T., S. O. Kelley, C. R. Treadway, J. K. Barton, A. H. Zewail, *Proc. Natl. Acad. Sci. USA* **1999**, *96*, 6014–6019.
58. D. A. Dunn, V. H. Lin, I. E. Kochevar, *Biochemistry* **1992**, *31*, 11620–11625.
59. S. O. Kelley, J. K. Barton, *Chem. Biol.* **1998**, *5*, 413–425.
60. G. Kokkinidis, A. Kelaidopoulou., *J. Electronanal. Chem.* **1996**, *414*, 197–208.
61. A. Okamoto, K. Tanaka, I. Saito, *J. Am. Chem. Soc.* **2003**, *125*, 5066–5071.
62. K. Nakatani, C. Dohno, I. Saito, *J. Am. Chem. Soc.* **2002**, *124*, 6802–6803.
63. See review: D. Porath, G. Cuniberti, R. Di Felice, *Top. Curr. Chem.* **2004**, *237*, 183–227.
64. See reviews: (a) I. Willner, *Science* **2002**, *298*, 2407–208; (b) T. Carell, C. Behrens, J. Gierlich, *Org. Biomol. Chem.* **2003**, *1*, 2221–2228; (c) J. Richter, *Physica E* **2003**, *16*, 157–173.
65. See review: M. Shionoy, K. Tanaka, *Curr. Opin. Chem. Biol.* **2004**, *8*, 592–597.

13
Electron Transfer of π-Functional Systems and Applications
Shunichi Fukuzumi

13.1
Introduction

Light energy from sunlight is vital to our sustenance, because all the food we eat and all the fossil fuel we use are actually products of photosynthesis, which is the process that converts energy in sunlight to chemical forms of energy used by biological systems. The overwhelming part of our energy supply comes from the chemical energy which has been stored in fossil fuels over several billion years. The world's energy consumption rate of fossil fuels is expected to increase further in the next decades owing to increased demand in the developing countries. The rapid consumption of fossil fuel has created unacceptable environmental problems such as greenhouse effects, which could lead to disastrous climatic consequences. Hence renewable and clean energy resources are definitely required in order to maintain the quality of human life and the environment. One of the most attractive strategies is the development of systems that mimic natural photosynthesis in the conversion and storage of solar energy.

Energy from sunlight is captured by photosynthetic π-pigments (primarily chlorophylls and carotenoids) which cover the wide spectral range of the solar irradiation [1–3]. There are many different types of bacteriochlorophyll, carotenoids and phycobilins, differing from each other in their chemical structure. π-Pigments generally are bound to proteins, which provide the pigment molecules with the appropriate orientation and positioning with respect to each other. Light energy is absorbed by individual π-pigments, but is not used immediately by these π-pigments for energy conversion. Instead, the light energy is transferred to chlorophylls that are in a special protein environment where the actual energy conversion event starts via electron-transfer processes. Pigments and protein involved in this actual primary electron- transfer event together are called the reaction center. The three-dimensional X-ray crystal structures of reaction centers of *Rhodobacter* (*Rb.*) *sphaeroides* [4] and other purple bacteria including *Rhodopseudomonas* (*Rh.*) *viridis* [5] have provided valuable mechanistic insight into the photosynthesis. A large number of π-pigment molecules (100–5000), collectively re-

Functional Organic Materials. Syntheses, Strategies, and Applications.
Edited by Thomas J.J. Müller and Uwe H.F. Bunz
Copyright © 2007 WILEY-VCH Verlag GmbH & Co. KGaA, Weinheim
ISBN: 978-3-527-31302-0

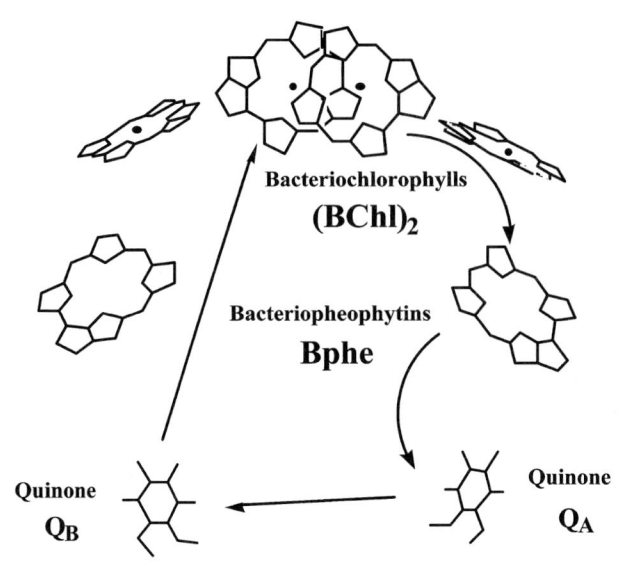

Fig. 13.1 Structure of the reaction center of Rb. spaeroides by X-ray analysis.

ferred to as antenna, "harvest" light and transfer the light energy to the same reaction center. The purpose of such antenna molecules is to maintain a high rate of electron transfer in the reaction center, even at lower light intensities. The structure of the photosynthetic reaction center of purple bacteria is shown in Fig. 13.1, where the central part is referred to as the special pair [(BChl)$_2$], while the other two bacteriochlorophylls (BChl) are referred to as "accessory" bacteriochlorophylls. There are also two bacteriopheophytins (BPhe) and two ubiquinones (Q$_A$ and Q$_B$), which, together with the special pair, are placed at appropriate positions in the protein matrix. Our knowledge of the mechanisms in natural photosynthesis has made great progress in the past two decades owing to the developments in spectroscopic analysis techniques. A number of modern physical techniques including fast transient optical spectroscopy and magnetic resonance have been used to determine both the thermodynamics and kinetics of photoinduced electron transfer within protein matrix [6, 7]. Photoinduced electron transfer occurs from the singlet excited state of the special pair [(BChl)$_2$] to the neighboring pigments, leading to a radical cation [(BChl)$_2^{+\bullet}$] and a radical anion (Bphe$^{-\bullet}$). The subsequent electron-transfer process is found to occur very rapidly from the special pair toward the quinones to produce the final charge-separated state that has a surprisingly long lifetime (ca. 1 s) with nearly 100 % quantum yield [1–3]. Ultimately the product of this cascade is conversion of light into usable chemical energy [1–3]. The structure and function of the bacterial photosynthetic reaction center (Fig. 13.1) has provided valuable incentives to construct the artificial model systems.

In order to develop efficient photoenergy conversion systems such as photosynthesis, it is of primary importance to understand the electron-transfer processes of π-systems involved in photosynthesis. The functional π-compounds to be used in artificial photosynthesis are not necessarily the same as those employed in natural photosynthesis. The first promising candidates are porphyrins, which contain an extensively conjugated two-dimensional π-system and are thereby suitable for efficient electron transfer, because the uptake or release of electrons results in minimal structural and solvation change upon electron transfer, resulting in a small reorganization energy of electron transfer [9]. In addition, rich and extensive absorption features of porphyrinoid systems guarantees increased absorption cross-sections and an efficient use of the solar spectrum [9].

In contrast with the two-dimensional porphyrin π-system, fullerenes contain an extensively conjugated three-dimensional π-system [10]. Buckminsterfullerene (C_{60}), for example, is described as having a closed-shell configuration consisting of 30 bonding molecular orbitals with 60 π-electrons, which is also suitable for efficient electron transfer. In particular, a set of 60 molecular orbitals split into 30 bonding/30 antibonding π-molecular orbitals and thereby accommodating 60 π-electrons. The resulting electronic configuration discloses for C_{60} a five-fold degenerate HOMO (h_u) and a three-fold degenerate LUMO (t_{1u}), which are separated by an energy gap of 1.8 eV [11]. The three isoenergetic LUMOs have consequences that in electrochemical experiments, six equally spaced reduction waves, with the first reduction step resembling that of quinones, were registered for C_{60} [11]. The separation between any two successive reduction steps is ca. 450 ± 50 mV. This is regarded as a clear manifestation of conditions that guarantee the optimal delocalization of charges (i.e. electrons). In other words, even in a highly reduced fullerene state, electrons, as they are subsequently added to the fullerene's π-system, experience little, if any, repulsive forces. Hence an electron transfer to C_{60} is expected to be highly efficient because of the minimal changes in structure and solvation associated with the electron-transfer reduction [12]. The specific objective of this chapter is therefore to review recent developments in electron-transfer systems of functional π-compounds (porphyrins, fullerenes and other π-chromophores) and organization of π-compounds to construct efficient light energy conversion systems such as photovoltaic devices.

13.2
Efficient Electron-transfer Properties of Zinc Porphyrins

Since Ward and Weissman determined the rates of fast electron self-exchange reaction between naphthalene radical anion and naphthalene from the linewidth broadening in the ESR spectra [13], a number of fast electron self-exchange reactions have been determined for radical cations and radical anions [14, 15]. Rates of self-exchange electron transfer between porphyrins and the porphyrin radical cations have also been determined from the linewidth broadening in the ESR spectra, providing interesting mechanistic insight into the electron-transfer mechanism (see below) [16, 17].

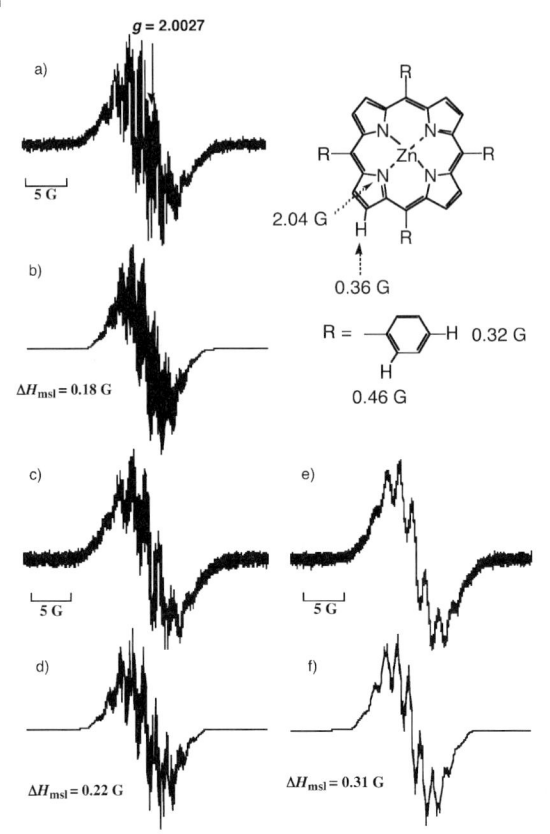

Fig. 13.2 (a) ESR spectrum of ZnTPP$^{+\bullet}$ (5.0 × 10^{-4} M) in MeCN at 233 K and (b) the computer simulation spectrum (ΔH_{msl} = 0.18 G) [16, 17]. (c) ESR spectrum of ZnTPP$^{+\bullet}$ (5.0 × 10^{-4} M) in the presence of ZnTPP (5.0 × 10^{-4} M) in toluene at 233 K and (d) the computer simulation spectrum (ΔH_{msl} = 0.22 G) [16, 17]. (e) ESR spectrum of ZnTPP$^{+\bullet}$ (5.0 × 10^{-4} M) in the presence of ZnTPP (5.0 × 10^{-4} M) in toluene at 313 K and (f) the computer simulation spectrum (ΔH_{msl} = 0.31 G) [16, 17].

When ZnTPP (TPP^{2-} = teteraphenylporphyrin dianion: 5.0 × 10^{-4} M) is oxidized by exactly 1 equiv. of Ru(bpy)$_3^{3+}$ (bpy = 2,2′-bipyridine: 5.0 × 10^{-4} M), ZnTPP$^{+\bullet}$ is produced without leaving any neutral species. In such a case, no self-exchange electron transfer of ZnTPP$^{+\bullet}$ with ZnTPP occurs when the ESR spectrum of ZnTPP$^{+\bullet}$ detected at 233 K exhibits well-resolved hyperfine structures as shown in Fig. 13.2a. The hyperfine coupling constants (*hfc*) are determined by comparison of the observed spectrum with the computer simulation spectrum as shown in Fig. 13.2b.

When excess of ZnTPP (1.0 × 10^{-3} M) is used for the oxidation with Ru(bpy)$_3^{3+}$ (5.0 × 10^{-4} M) at 233 K, the ESR linewidth becomes broader than the spectrum obtained in the equimolar condition (Fig. 13.2c). The computer simulation spectrum with a broader linewidth is shown in Fig. 13.2d. The line broadening results from the self-exchange electron transfer between ZnTPP$^{+\bullet}$ and ZnTPP which is left in solution [Eq. (1)]. When the temperature is raised from 233 to 313 K, the linewidth becomes broader because of the faster electron-transfer rate at a higher reaction temperature as shown in Fig. 13.2e, together with the computer simulation spectrum in Fig. 13.2f. The $\Delta H°_{msl}$ value in the absence of excess porphyrin varies slightly depending on temperature and solvent [16, 17].

The rate constants (k_{ex}) of the electron exchange reactions between ZnTPP$^{+\bullet}$ and ZnTPP [Eq. (1)] were determined using Eq. (2), where ΔH_{msl} and $\Delta H°_{msl}$ are the maximum slope linewidths of the ESR spectra in the presence and absence of ZnTPP$^{+\bullet}$, respectively, and P_i is a statistical factor [14]. From the linear plots of ($\Delta H_{msl} - \Delta H°_{msl}$) and [ZnTPP] at various temperatures are obtained the self-exchange electron-transfer rate constant (k_{ex}). The Arrhenius plots are shown in Fig. 13.3 together with the observed activation enthalpies ($\Delta H_{obs}^{\neq}$), where the effect of diffusion (k_{diff}) is taken into account. The $\Delta H_{obs}^{\neq}$ values are all positive and decrease in order: toluene > MeCN > CH$_2$Cl$_2$ [16].

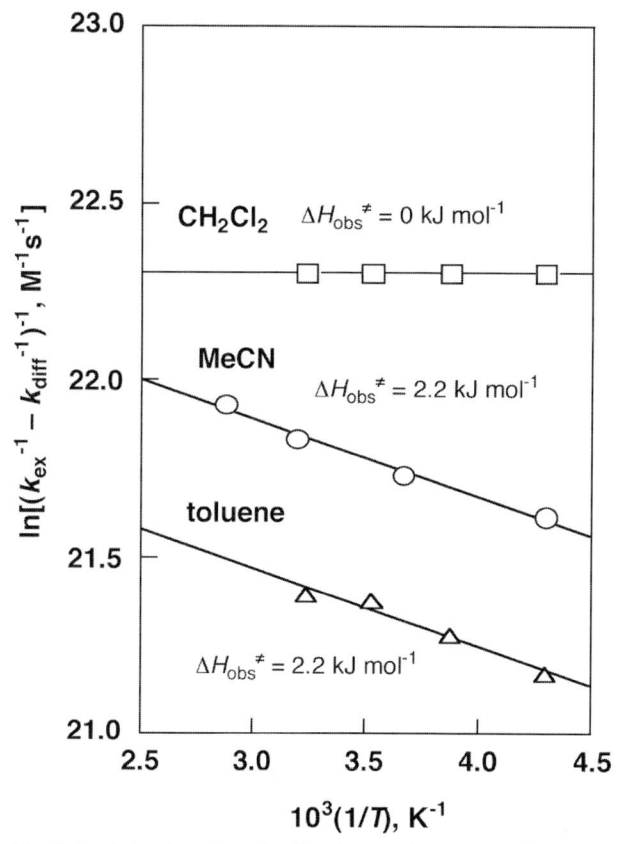

Fig. 13.3 Arrhenius plots of self-exchange electron transfer between ZnTPP$^{+\bullet}$ and ZnTPP in different solvents [16].

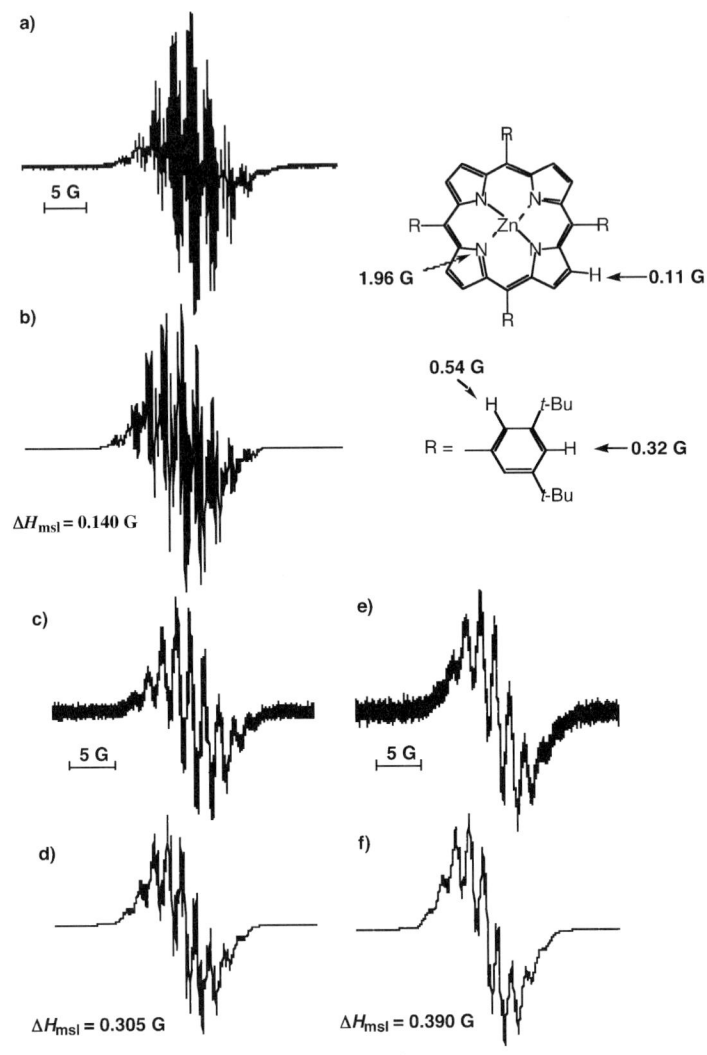

Fig. 13.4 (a) ESR spectra of ZnT(t-Bu)PP$^{+\bullet}$ formed by the oxidation of ZnT(t-Bu)PP (5.0 × 10^{-4} M) by Ru(bpy)$_3$(PF$_6$)$_3$ (5.0 × 10^{-4} M) in MeCN at 298 K and (b) the computer simulation spectrum [16]. (c) ESR spectrum of ZnT(t-Bu)PP$^{+\bullet}$ (5.0 × 10^{-4} M) in the presence of ZnT(t-Bu)PP (5.0 × 10^{-4} M) in toluene at 313 K and (d) the computer simulated spectrum (ΔH_{msl} = 0.305 G) [16]. (e) ESR spectrum of ZnT(t-Bu)PP$^{+\bullet}$ (5.0 × 10^{-4} M) in the presence of ZnT(t-Bu)PP (5.0 × 10^{-4} M) in toluene at 233 K and (f) the computer simulated spectrum (ΔH_{msl} = 0.390 G) [16].

$$\text{ZnTPP}^{\bullet+} + \text{ZnTPP} \underset{}{\overset{k_{ex}}{\rightleftarrows}} \text{ZnTPP} + \text{ZnTPP}^{\bullet+} \quad (1)$$

$$k_{ex} = (1.52 \times 10^7)(\Delta H_{msl} - \Delta H^\circ{}_{msl})/\{(1 - P_i)[\text{ZnTPP}]\} \quad (2)$$

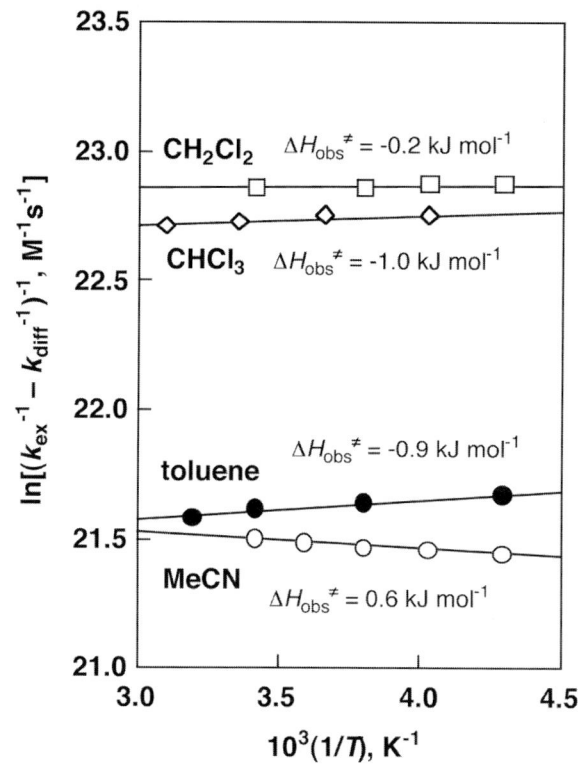

Fig. 13.5 Arrhenius plots of electron self-exchange reaction between ZnT(t-Bu)PP$^{+\bullet}$ and ZnT(t-Bu)PP in different solvents [16].

The ESR spectrum of ZnT(t-Bu)PP$^{+\bullet}$ [T(t-Bu)PP^{2-} = 5,10,15,20-tetrakis(3,5-di-*tert*-butylphenyl)porphyrin dianion] is also obtained by the one-electron oxidation of ZnT(t-Bu)PP (5.0 × 10^{-4} M) by Ru(bpy)$_3$$^{3+}$ (5.0 × 10^{-4} M) in toluene as shown in Fig. 13.4a [16].

The hyperfine coupling constants (*hfc*) are determined by comparison of the observed spectrum with the computer simulated spectrum as shown in Fig. 13.4b. As in the case of ZnTPP$^{+\bullet}$ (Fig. 13.2), the ESR linewidth is broadened in the presence of excess ZnT(t-Bu)PP due to the electron self-exchange between ZnT(t-Bu)PP$^{+\bullet}$ and ZnT(t-Bu)PP (Fig. 13.4a). In contrast to the case of ZnTPP$^{+\bullet}$, however, the linewidth of the ESR signal of ZnT(t-Bu)PP$^{+\bullet}$ becomes sharper as the temperature is raised from 233 K (Fig. 13.4e) to 313 K (Fig. 13.4c). This indicates that the electron self-exchange becomes slower at higher temperature. From the slopes of the linear plots of ΔH_{msl} and [ZnT(t-Bu)PP] at various temperature are obtained the self-exchange electron-transfer rate constants (k_{ex}). The Arrhenius plots are shown in Fig. 13.5 together with the observed activation enthalpies ($\Delta H_{obs}^{\neq}$) [16]. The $\Delta H_{obs}^{\neq}$ value is positive in MeCN but the values are negative in CH$_2$Cl$_2$, CHCl$_3$ and toluene.

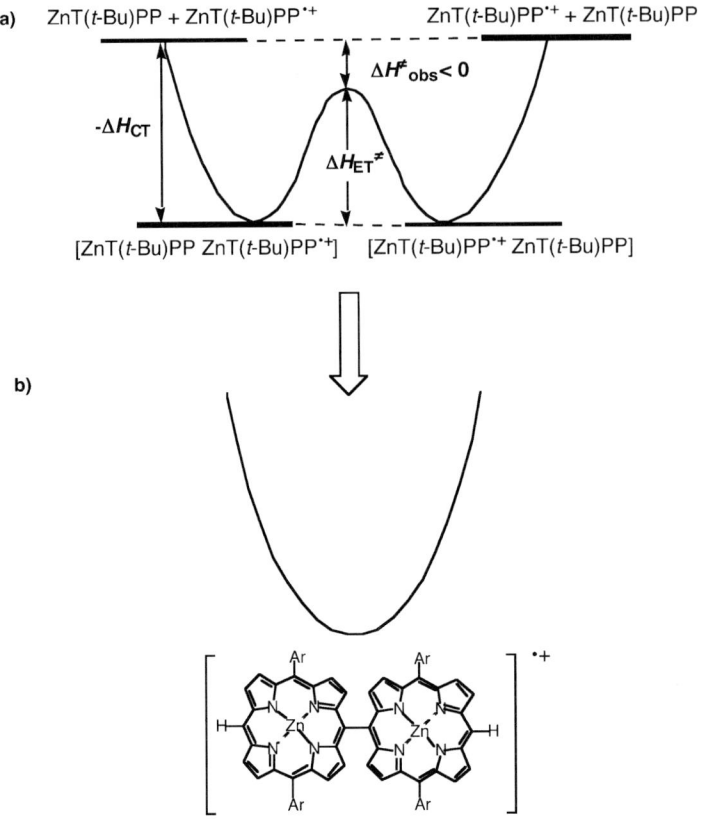

Fig. 13.6 (a) Potential energy diagram for self-exchange electron transfer between ZnT(t-Bu)PP$^{+\bullet}$ and ZnT(t-Bu)PP. (b) Potential energy diagram for (ZnP)$_2^{+\bullet}$.

The observation of a negative activation enthalpy indicates that self-exchange electron transfer proceeds via an intermediate that is a charge-transfer π-complex formed between ZnT(t-Bu)PP$^{+\bullet}$ and ZnT(t-Bu)PP. The energy of the π-complex is lower than that of the reactant pair and the energy difference between the reactant pair and the intermediate ($-\Delta H_{CT}$) is larger than the activation energy from the intermediate ($\Delta H_{ET}^{\neq}$) as show in Fig. 13.6a. In general, an electron-transfer reaction proceeds via a precursor complex formed between an electron donor and an acceptor [16, 19]. In the case of electron self-exchange, an electron donor is ZnT(t-Bu)PP, which can form a strong charge-transfer π-complex with an electron acceptor, ZnT(t-Bu)PP$^{+\bullet}$. In the case of electron transfer between a neutral species and a cationic species, there is no net change in solvation before and after the electron transfer. In such a case, the solvent reorganization energy becomes smaller as the solvent polarity decreases [18]. This results in a decrease in the λ value of ZnT(t-Bu)PP$^{+\bullet}$. This may be the reason why a negative activation energy is ob-

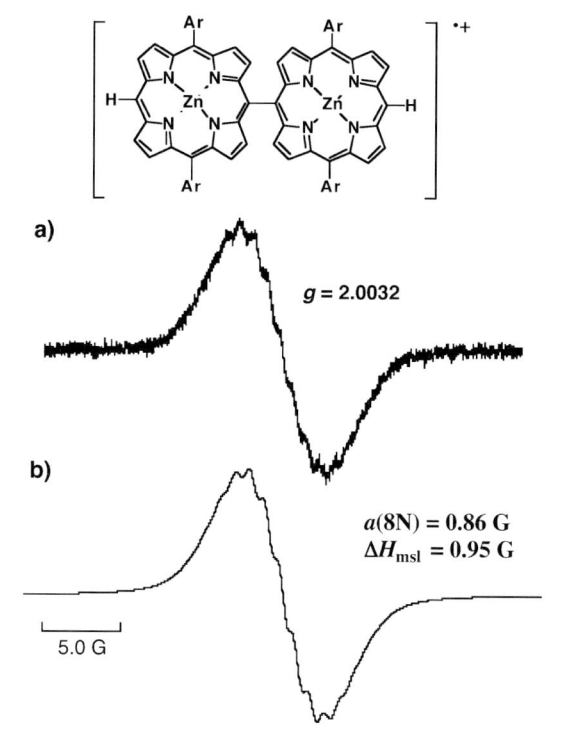

Fig. 13.7 (a) ESR spectra of $(ZnP)_2^{+\bullet}$ formed by the oxidation of $(ZnP)_2$ (1.0×10^{-3} M) by $Fe(bpy)_3(PF_6)_3$ (1.0×10^{-3} M) in PhCN at 298 K and (b) the computer simulation spectrum [17].

a)

$g = 2.0032$

b)

$a(8N) = 0.86$ G
$\Delta H_{msl} = 0.95$ G

5.0 G

served in less polar solvents such as CH_2Cl_2, $CHCl_3$ and toluene as compared with MeCN (Fig. 13.5).

When two zinc porphyrin moieties are linked to each other by a covalent bond, intramolecular self-exchange electron transfer should take place in the radical cation of the zinc porphyrin dimer. With decreasing linkage distance, the self-exchange electron-transfer rate is expected to increase. The ESR spectrum of the radical cation of the meso-linked bis-substituted zinc porphyrin dimer {(ZnP)₂: 5,5′-bis[10,20-bis(3,5-di-*tert*-butylphenyl)]porphyrinatozinc(II)} exhibits a hyperfine structure due to at least four nitrogen atoms as shown in Fig. 13.7a [17]. The computer simulation spectrum with eight equivalent hyperfine coupling constant due to nitrogen atoms (Fig. 13.7b) agrees with the observed spectrum in Fig. 13.7a. This indicates that the unpaired electron is fully delocalized on two porphyrin rings. The delocalization of unpaired electron is confirmed by the DFT calculation of the radical cation of the free base porphyrin dimer as shown in Fig. 13.8 [17]. The torsional angle between neighboring π-planes of two porphyrin rings is calculated to be 75°, which allows orbital overlap between two porphyrin rings leading to the full delocalization of the unpaired electron over two porphyrin rings. Such orbital overlap results in a decrease in the activation barrier for the intramolecular electron transfer, leading to a single potential energy well for the radical cation of the zinc porphyrin dimer as shown in Fig. 13.6b. This is

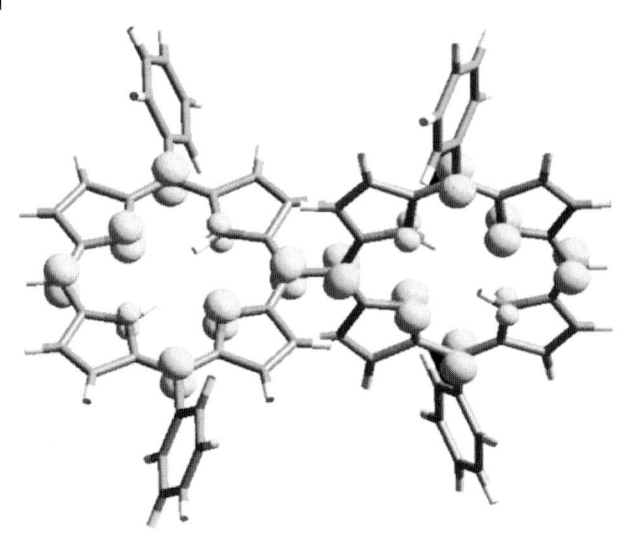

Fig. 13.8 The spin distribution of meso-linked free base porphyrin dimer obtained by DFT calculation (B3LYP/6–31G* basis set) with the optimized structure; 3,5-di-*tert*-butylphenyl groups are replaced by phenyl groups [17].

regarded as an extreme case of the fast self-exchange electron transfer between zinc porphyrin radical cation and neutral zinc porphyrin. In conclusion, the porphyrin π-systems are well suited as components of efficient electron-transfer systems.

13.3
Efficient Electron-transfer Properties of Fullerenes

The efficient electron-transfer properties of fullerenes are characterized by the driving force dependence of the rate constants (k_{et}) for intermolecular electron transfer from fullerenes (C_{60}, C_{76} and C_{78}) to a series of arene π-radical cations in Fig. 13.9, which reveals a striking parabolic dependence including the Marcus *inverted* region [8, 18]. This is the most important prediction of the Marcus theory of electron transfer in the strongly exergonic region where the electron-transfer rate is expected to decrease as the driving force of electron transfer ($-\Delta G°_{et}$) increases provided that the $-\Delta G°_{et}$ value is larger than the reorganization energy of electron transfer (λ) [8, 19–21]. The reorganization energy of electron transfer (λ) is the energy required to reorganize the donor, acceptor and their solvation spheres structurally upon electron transfer [8]. It should be noted that the reorganization energy (λ) consists of two constituents [8]: a purely structural component (λ_i), including vibrations of the molecules, etc., and contributions stemming from the polarization changes in the solvent environment (λ_S) [8]. For intermolecular electron-transfer reactions, definitive evidence for the inverted region has been

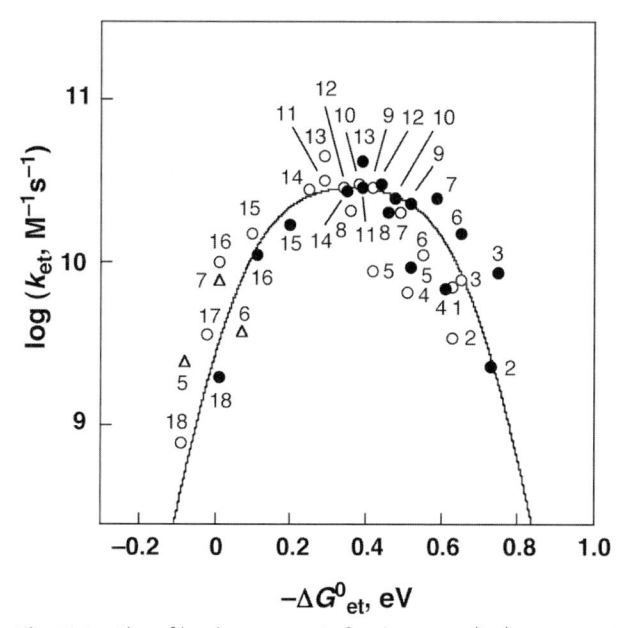

Fig. 13.9 Plot of log k_{et} vs. $-\Delta G°_{et}$ for electron transfer from C_{60} (open triangle), C_{76} (open circle) and C_{78} (closed circle) to arene radical cations in CH_2Cl_2 [18]. The solid line is drawn based on the Marcus theory of electron transfer [Eq. (3)].

very rare. Normally, these intermolecular ET reactions follow the Rehm–Weller behavior such that the ET rate increases with an increase in driving force, reaches a diffusion limit and remains unchanged no matter how exergonic ET might become [22–24]. There have been considerable discussions on the reasons to explain the non-observance of the Marcus inverted region in the bimolecular electron-transfer reactions [25]. In Fig. 13.9, the log k_{et} value increases with increasing driving force to reach a diffusion-limited value and then decreases with further increase in the driving force [18]. The k_{et} value (2.3×10^9 M^{-1} s^{-1}) for a highly exergonic electron transfer from C_{78} to mesitylene radical cation ($-\Delta G°_{et} = 0.73$ eV) is about 20 times smaller than the value (4.5×10^{10} M^{-1} s^{-1}) for a much less exergonic electron transfer from C_{76} to, for example, chrysene radical cation ($-\Delta G°_{et} = 0.32$ eV) [18]. Such a pronounced decrease towards the highly exothermic region represents the first definitive confirmation of the existence of the Marcus *inverted* region in a truly bimolecular electron transfer.

The plateau in Fig. 13.9 corresponds to the diffusion-limited region where the rate of electron transfer is faster than the rate of diffusion. According to the Marcus theory of electron transfer, the observed rate constant of an intermolecular electron transfer is given by [23]

$$\frac{1}{k_{et}} = \frac{1}{k_{diff}} + \frac{1}{Z \exp[-(\lambda/4)(1+G°_{et}/\lambda)^2/k_B T]} \quad (3)$$

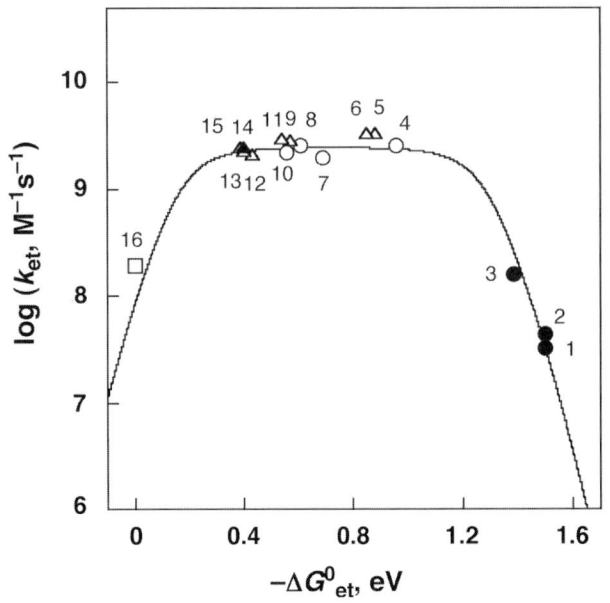

Fig. 13.10 Plot of logk_{et} vs. $-\Delta G°_{et}$ for electron transfer from anthracene radical anion to C_{60}, C_{70} and 1,4-t-Bu(PhCH$_2$)C_{60} in PhCN (closed circle) [18], electron transfer from $C_{60}{}^{\bullet-}$ to α-chloranil in PhCN (closed triangle) [28], electron transfer from NADH analogs to [3]$C_{60}{}^*$ and [3]$C_{70}{}^*$ in PhCN (open triangle) [29, 30], electron self-exchange between t-BuC$_{60}{}^{\bullet}$ and t-BuC$_{60}{}^-$ in PhCN–toluene (open rectangle) [31] and electron transfer from metalloporphyrin π-radical anions to C_{60} (open circle) in propanol–toluene–acetone [32]. The solid line is drawn based on the Marcus theory of electron transfer [Eq. (6)].

where k_{diff} is the diffusion rate constant, Z is the collision frequency which is taken as 1×10^{11} M^{-1} s^{-1}, λ is the reorganization energy of electron transfer and k_B is the Boltzmann constant. By fitting the data in Fig. 13.9 with the Marcus equation for bimolecular ET reactions [Eq. (3)], an experimental λ value of 0.36 eV is deduced for the reorganization energy for electron-transfer oxidation of fullerenes (C_{60}, C_{76} and C_{78}) in CH$_2$Cl$_2$. The small λ value results from the charge-shift type electron-transfer reactions of the large three dimensional π-systems in CH$_2$Cl$_2$, which is less polar than typical polar solvents such as benzonitrile (PhCN) and acetonitrile (MeCN). Electron-transfer reactions are normally performed in polar solvents, in which the product ions of the electron transfer are subject to stabilization via strong solvation. When cationic electron acceptors such as arene radical cations are employed together with a neutral electron donor (e.g. fullerene), the solvation before and after the electron transfer may be largely canceled out [26, 27], especially when the free energy change of electron transfer is expected to be independent of the solvent polarity. In addition, the solvent reorganization energy for the charge-shift type electron-transfer reaction is known to decrease with decreasing solvent polarity [27]. In addition, the choice of the fullerenes as electron donors is important. The cationic charge of the radical cations is highly delocalized in the large, three-dimensional π-system. Such delocalization of a positive charge results in minimal reorganization of bonds and solvation upon electron transfer. Arene radical cations also have small reorganization energies upon electron transfer due to the delocalized charge [15]. It is important to note that the deduced reorganization energy from the driving force dependence of k_{et} is that of the average of the reorganization energies between arene radical cations and fullerenes. Thus, uniformly small reorganization energies for electron-transfer reactions from fullerenes to arene radical cations, regardless of the large variation of the redox potentials, make it possible for the first time to observe the Marcus inverted region clearly for intermolecular electron-transfer reactions, which are normally masked by the diffusion process.

A parabolic driving force dependence of $\log k_{et}$ is also observed for electron-transfer reduction of fullerenes in PhCN, as shown in Fig. 13.10 [18, 28–32]. By fitting the data in Fig. 13.10 with the Marcus equation for intermolecular ET reactions [Eq. (3)], the average λ value for electron-transfer reduction of fullerenes is determined as 0.72 eV [18]. This λ value agrees well with the reported λ value (0.73 eV) for electron transfer from $C_{60}^{-\bullet}$ to various electron acceptors in PhCN [31]. The larger λ value for electron-transfer reduction of C_{60} in PhCN than the λ value for electron-transfer oxidation of fullerenes in CH_2Cl_2 (0.36 eV) may result from the larger solvent reorganization in the more polar solvent. In any case, the small reorganization energies for electron-transfer reactions of fullerenes compared with those of other smaller π-systems have been demonstrated to be essential in order to observe the Marcus inverted region for bimolecular electron-transfer reactions.

13.4
Photoinduced Electron Transfer in Electron Donor-Acceptor Linked Molecules Mimicking the Photosynthetic Reaction Center

According to the Marcus theory of electron transfer [8], the rate constant of non-adiabatic intramolecular electron transfer (k_{ET}) is given by

$$k_{ET} = \left(\frac{4\pi^3}{h^2 \lambda k_B T}\right)^{\frac{1}{2}} V^2 \exp\left[-\frac{(\Delta G_{ET} + \lambda)^2}{4\lambda k_B T}\right] \qquad (4)$$

where V is the electronic coupling matrix element, h is Planck's constant, T is the absolute temperature, ΔG_{ET} is the free energy change of electron transfer and λ is the reorganization energy of electron transfer. According to Eq. (4), the logarithm of the rate constant ($\log k_{ET}$) of electron transfer in an electron donor–acceptor dyad (D–A) is related parabolically to the driving force of electron transfer from electron donors to acceptors ($-\Delta G_{ET}^{\circ}$) and to the reorganization energy (λ) of electron transfer, that is, the energy required to reorganize the donor, acceptor and their solvation spheres structurally upon electron transfer as shown in Fig. 13.11. The parabolic dependence of $\log k_{ET}$ on $-\Delta G_{ET}^{\circ}$ is determined by two factors:

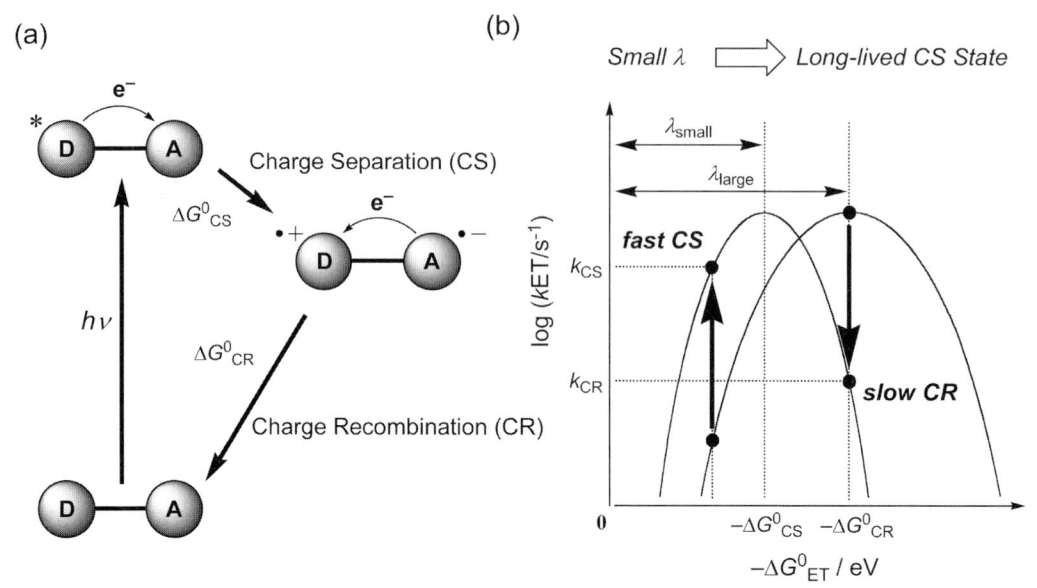

Fig. 13.11 (a) Energy diagram of photoinduced electron transfer of an electron donor–acceptor dyad with photoexcitation of the donor moiety. (b) Driving force dependence of $\log k_{ET}$ of the CS and CR processes with different λ values.

the electronic coupling matrix element V, which determines the maximum k_{ET} value, and the reorganization energy (λ). When the magnitude of the driving force of electron transfer becomes the same as the reorganization energy ($-\Delta G_{ET} \approx \lambda$), the k_{ET} value reaches a maximum and is basically controlled by the magnitude of electronic coupling (V) between the donor and acceptor moieties (Fig. 13.11b). Upon passing this thermodynamic maximum, the highly exothermic region of the parabola ($-\Delta G_{ET} > \lambda$) is entered, in which an additional increase in the driving force results in an actual slowing down of the electron-transfer rate, due to an increasingly poor vibrational overlap of the product and reactant wavefunctions. This highly exergonic range is generally referred to as the Marcus inverted region (see above) [8, 19–21]. In such a case, the magnitude of the reorganization energy is the key parameter to control the electron-transfer process. The smaller the reorganization energy, the faster is the forward photoinduced charge-separation (CS) process of an electron donor–acceptor dyad (D–A), but the charge-recombination (CR) process becomes slower when the driving force for back electron transfer ($-\Delta G_{ET}$) is larger than the reorganization energy (λ) of electron transfer (Fig. 13.11b).

As described above, both porphyrins and fullerenes have highly delocalized π-systems suitable for efficient electron transfer, because the uptake or release of electrons results in minimal structural and solvation changes upon electron transfer to afford small reorganization energies of electron transfer. Therefore, extensive efforts have been devoted to study intramolecular electron transfer in elec-

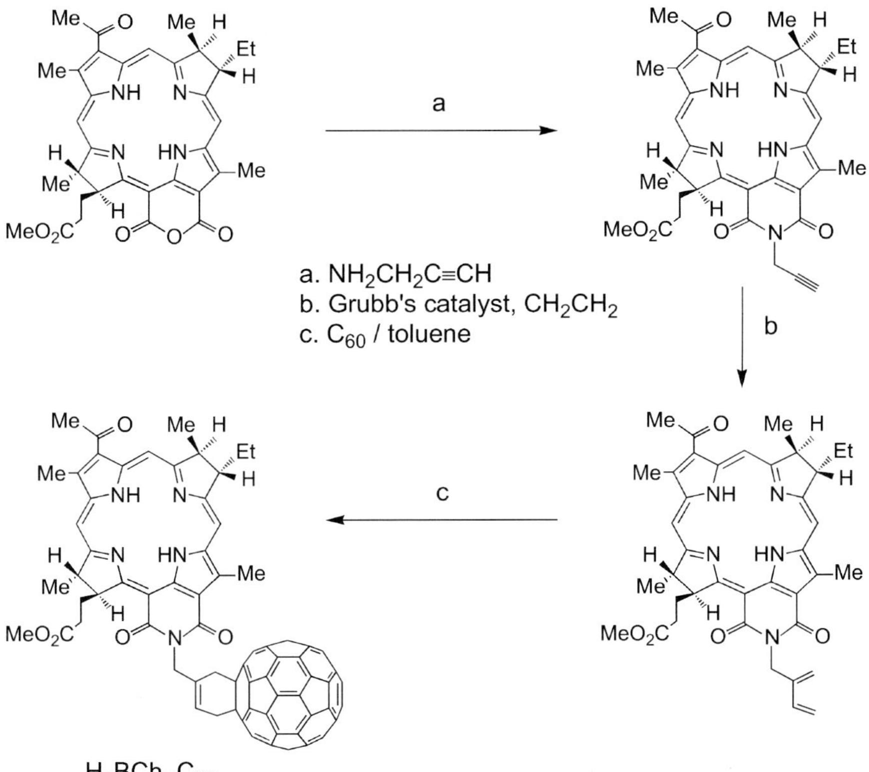

Scheme 13.1 Synthesis of H$_2$BCh–C$_{60}$.

a. NH$_2$CH$_2$C≡CH
b. Grubb's catalyst, CH$_2$CH$_2$
c. C$_{60}$ / toluene

tron donor–acceptor ensembles containing porphyrins and fullerenes, which is started by photoexcitation of porphyrins used as antenna molecules for efficient light capture in the visible region of the spectrum [35–38]. Natural photosynthesis utilizes chlorophylls rather than porphyrins as antenna molecules. The number of reduced double bonds in the pyrrole rings is zero in the case of porphyrins, one in the case of chlorins and two in the case of bacteriochlorins.

A series of chlorophyll-like donor (a chlorin) linked having C$_{60}$ (chlorin–C$_{60}$) or porphyrin–C$_{60}$ dyads with the same short spacer have been synthesized as shown in Schemes 13.1 and 13.2 [39, 40]. The photoinduced electron-transfer dynamics have been reported [39, 40]. A deoxygenated PhCN solution containing ZnCh–C$_{60}$ gives rise upon a 388-nm laser pulse to a transient absorption maximum at 460 nm due to the singlet excited state of ZnCh [39]. The decay rate constant was determined as 1.0×10^{11} s^{-1}, which agrees with the value determined from fluorescence lifetime measurements [39]. This indicates that electron transfer from ^{1}ZnCh* to C$_{60}$ occurs rapidly to form the CS state, ZnCh$^{\bullet+}$–C$_{60}$$^{\bullet-}$. The CS state has absorption maxima at 790 and 1000 nm due ZnCh$^{\bullet+}$ and C$_{60}$$^{\bullet-}$,

480 | *13 Electron Transfer of π-Functional Systems and Applications*

Scheme 13.2 Synthesis of ZnCh–C_{60} and ZnCh–ref.

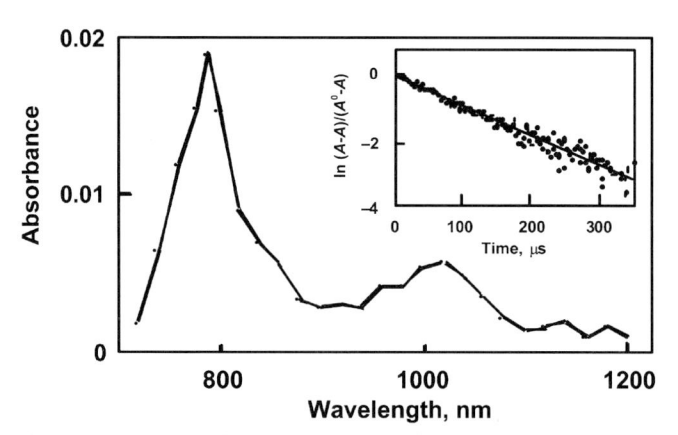

Fig. 13.12 Transient absorption spectrum of ZnCh–C_{60} in deaerated PhCN at 298 K at 0.1 μs after laser excitation at 355 nm [39]. Inset: first-order plot of the decay at 790 nm [39].

M = Zn (**1**, ZnCh–C$_{60}$)
= H$_2$ (**3**, H$_2$Ch–C$_{60}$)

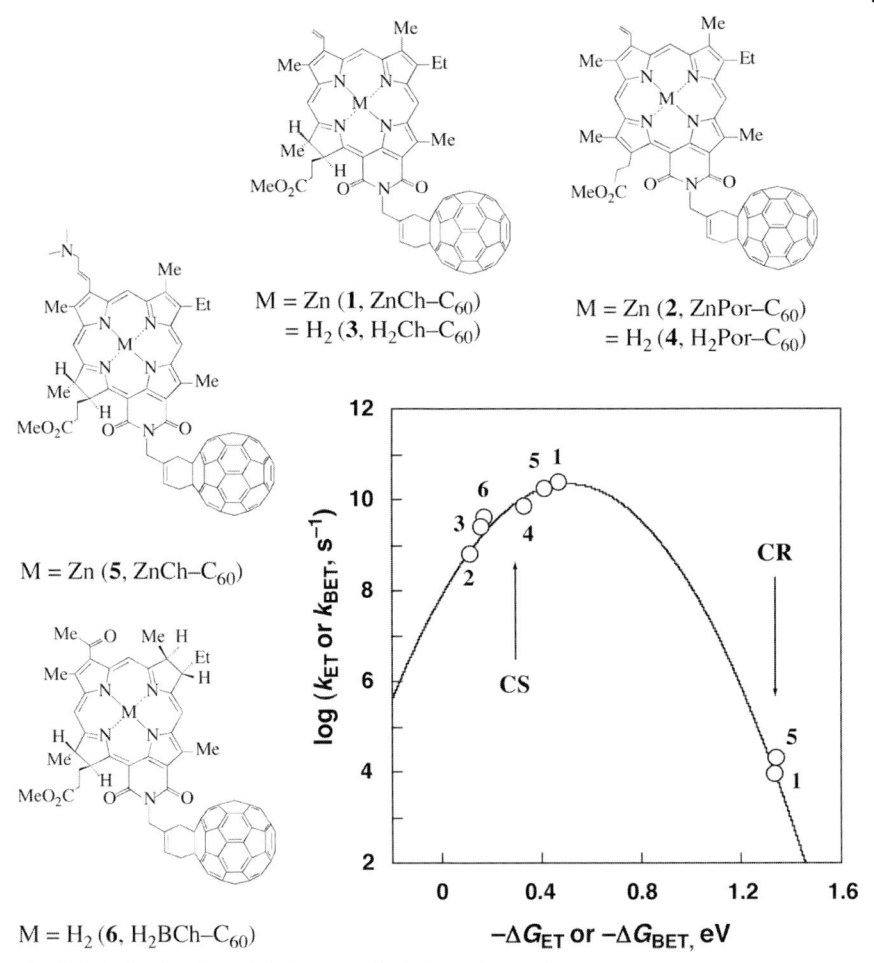

M = Zn (**2**, ZnPor–C$_{60}$)
= H$_2$ (**4**, H$_2$Por–C$_{60}$)

M = Zn (**5**, ZnCh–C$_{60}$)

M = H$_2$ (**6**, H$_2$BCh–C$_{60}$)

Fig. 13.13 Driving force ($-\Delta G_{ET}$ or $-\Delta G_{BET}$) dependence of intramolecular ET rate constants (k_{ET} or k_{BET}) in C$_{60}$-linked dyads in PhCN [39, 40].

respectively, as shown in Fig. 13.12. The CS state decays, obeying first-order kinetics, to afford the lifetime of 110 μs [39].

The photoexcitation of the free-base bacteriochlorin–C$_{60}$ dyad with the same short linkage also leads to the formation of the radical ion pair, but it decays quickly to the triplet excited state of the bacteriochlorin moiety, because the CS energy is higher than the triplet energy [40].

Figure 13.13 shows the driving force dependence of the CS rate constants (logk_{ET}) and the CR rate constants (or k_{BET}) for the C$_{60}$-containing dyads and the best fit of Eq. (4) provides λ = 0.51 eV and V = 7.8 cm^{-1} [39, 40]. The photoinduced ET processes in the dyads are located in the normal region of the Marcus

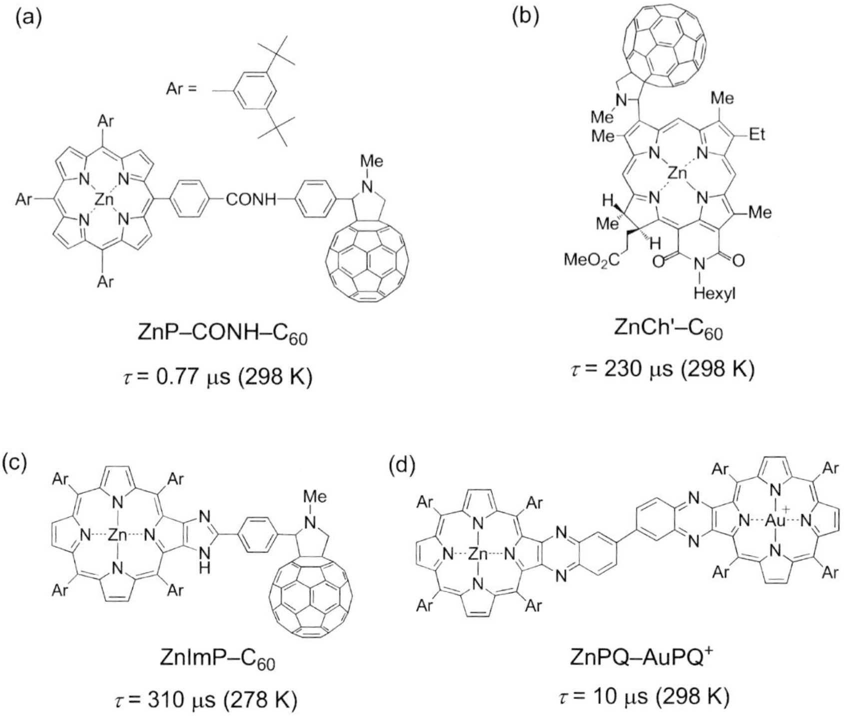

Fig. 13.14 Electron donor–acceptor dyads affording long-lived CS states [41–44].

parabola ($-\Delta G_{ET} < \lambda$) whereas the CR process from $C_{60}^{-\bullet}$ to ZnCh$^{+\bullet}$ is in the Marcus inverted region ($-\Delta G_{ET} > \lambda$). A similar parabolic dependence of logk_{ET} has been reported for zinc porphyrin-linked fullerene dyads with a longer amide linkage (ZnP–CONH–C$_{60}$) [41]. The longer edge-to-edge distance (R_{ee} = 11.9 Å) of ZnP–CONH–C$_{60}$ than ZnCh–C$_{60}$ (R_{ee} = 5.9 Å) results in a smaller V value (3.9 cm^{-1}) and a larger λ value (0.66 eV) as expected from the Marcus theory of electron transfer. Thus, a short edge-to-edge distance is required to attain the small reorganization energy of electron transfer in electron donor–acceptor dyads, affording long-lived CS states.

Along this line, simple donor–acceptor dyads have been designed and synthesized to attain long-lived CS states, where the donor and acceptor molecules are linked with a short spacer to minimize the solvent reorganization energy as shown in Fig. 13.14. A zinc chlorin–C$_{60}$ dyad with a shorter linkage between zinc chlorin and C$_{60}$ (ZnCh'–C$_{60}$) in Fig. 13.14b affords a CS lifetime of 230 μs [42], which is significantly longer than that of ZnP–CONH–C$_{60}$ (0.77 μs) in Fig. 13.14a [41]. A zinc imidazoporphyrin–C$_{60}$ dyad (ZnImP–C$_{60}$) with a short linkage (Fig. 13.14c) also affords a long-lived CS state with a lifetime of 330 μs at 278 K, which is much longer than those of conventional donor–acceptor dyads with longer spacers [35–38].

An Au(III) porphyrin can also be used as an electron acceptor moiety linked with Zn(II) porphyrin with a short spacer (ZnPQ–AuPQ$^+$ in Fig. 13.14d) [44]. The introduction of quinoxaline to the porphyrin ring results in a lowering of the electron-transfer state energy, which becomes lower than the energies of the triplet excited states of ZnPQ (1.32 eV) and AuPQ$^+$ (1.64 eV) [44]. In this case, photoinduced electron transfer occurs from the singlet excited state of the ZnPQ (^{1}ZnPQ*) to the metal center of the AuPQ$^+$ moiety to produce ZnPQ$^{+•}$–Au^{II}PQ, which has the lifetime of 10 μs in a nonpolar solvent such as cyclohexane [44]. The observed long lifetime of ZnPQ$^{+•}$–Au^{II}PQ results from a small reorganization energy for the *metal-centered* electron transfer of AuPQ$^+$ in nonpolar solvents due to the small change in solvation upon the electron transfer as compared with that in polar solvents. In a polar solvent such as PhCN, no CS state was observed, but instead only the triplet–triplet absorption due to ^{3}ZnPQ*–AuPQ$^+$ was observed [44]. The absence of an observable CS state in PhCN is ascribed to the much slower photoinduced electron transfer due to the large reorganization energy compared with that in nonpolar solvents, allowing an efficient intersystem crossing process in the ZnPQ–AuPQ$^+$ dyad to produce the triplet excited state ^{3}ZnPQ*–AuPQ$^+$.

In the natural photosynthetic reaction center, ubiquinones (Q_A and Q_B), which are organized in the protein matrix, are used as electron acceptors. Thus, covalently and non-covalently linked porphyrin–quinone dyads constitute one of the most extensively investigated photosynthetic models, in which the fast photoinduced electron transfer from the porphyrin singlet excited state to the quinone occurs to produce the CS state, mimicking well the photosynthetic electron transfer [45–47]. However, the CR rates of the CS state of porphyrin–quinone dyads are also fast and the CS lifetimes are mostly of the order of picoseconds or sub-nanoseconds in solution [45–47]. A three-dimensional π-compound, C_{60}, is super-

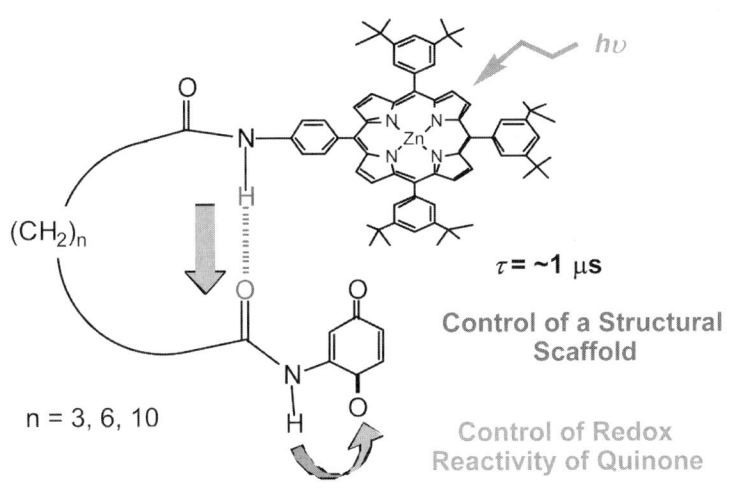

Fig. 13.15 Zinc porphyrin–quinone linked dyads (ZnP–n-Q; n = 3, 6, 10) with hydrogen bonds [49].

ior to a two-dimensional quinone in terms of the smaller reorganization of electron transfer of C_{60} as compared with quinone (see above) to attain the long-lived CS state [12, 48]. In the natural photosynthetic reaction center, the electron transfer to quinones is finely controlled by hydrogen bonding. Thus, a long lifetime of up to a 1 μs has been attained when the geometry between a porphyrin ring and quinone is optimized by using hydrogen bonds, which can also control the redox potentials of quinones (see above) [49]. In a series of ZnP–n-Q ($n = 3, 6, 10$) in Fig. 13.15, the hydrogen bond between two amide groups provides a structural scaffold to assemble the donor (ZnP) and the acceptor (Q) moiety, allowing one to attain the long-lived CS state [49].

In Fig. 13.13, the CR rates in the Marcus inverted region are much slower than the CS rates from both the singlet and triplet excited states in the Marcus normal region. This allows a subsequent electron transfer from an additional electron donor such as ferrocene (Fc) to $ZnP^{+\bullet}$ in the triad molecule (Fc–$ZnP^{+\bullet}$–$C_{60}^{-\bullet}$) to produce the final CS state, Fc^+–ZnP–$C_{60}^{-\bullet}$, in competition with the back electron transfer in the initial CS states [41]. Such multi-step electron-transfer processes are expanded to the tetrad molecule (Fc–ZnP–H_2P–C_{60}) as shown in Fig. 13.16a [50]. In the final CS state, Fc^+–ZnP–H_2P–$C_{60}^{-\bullet}$, charges are separated at

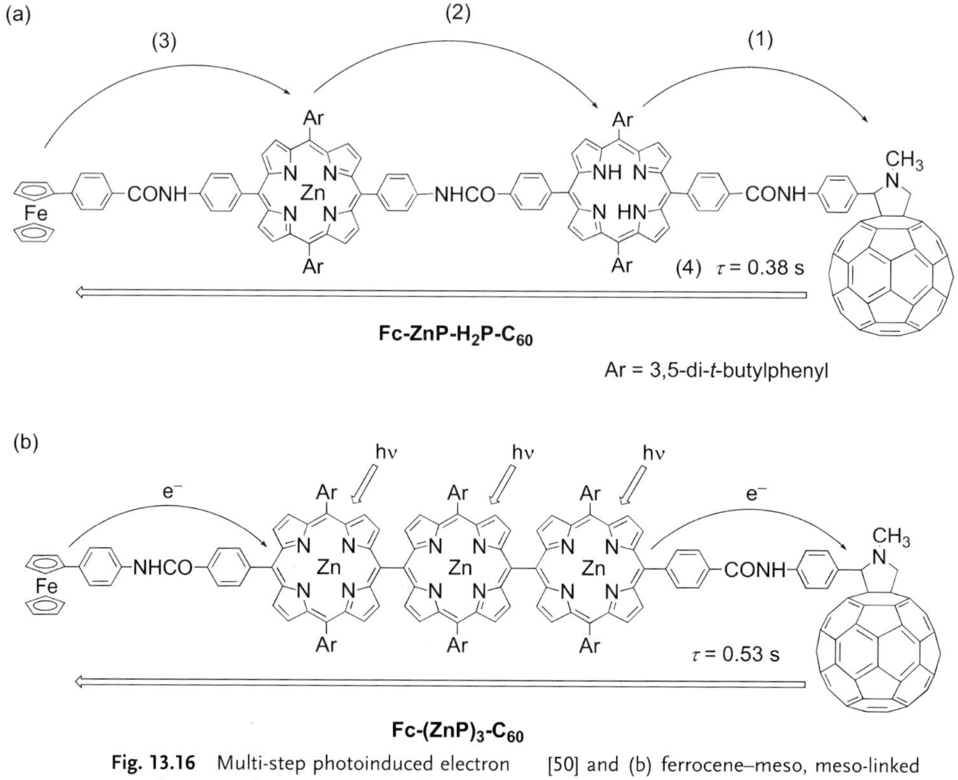

Fig. 13.16 Multi-step photoinduced electron transfer in (a) a ferrocene–zinc porphyrin–free base porphyrin–C_{60} tetrad (Fc–ZnP–H_2P–C_{60}) [50] and (b) ferrocene–meso, meso-linked porphyrin trimer–fullerene pentad [Fc–$(ZnP)_3$–C_{60}]; Ar = 3,5-$Bu^t_2C_6H_3$ [51].

a long distance (R_{ee} = 48.9 Å) [50]. The lifetime of the resulting CS state at such a long distance in a frozen PhCN has been determined to be as long as 0.38 s [50]. It should be noted that the CS lifetime is temperature independent, since the CR process is at the Marcus top region in Fig. 13.11b [50]. However, such an extremely long CS lifetime could only be determined in frozen media, since in condensed media bimolecular back electron transfer between two Fc^+–ZnP–H_2P–$C_{60}^{-\bullet}$ is much faster than the unimolecular CR process [50].

The best molecule mimicking multi-step electron-transfer processes in the photosynthetic reaction center so far reported is a ferrocene–meso, meso-linked porphyrin trimer–fullerene pentad [Fc–$(ZnP)_3$–C_{60}] in Fig. 13.16b, where the C_{60} and the ferrocene (Fc) are tethered at both the ends of $(ZnP)_3$ (R_{ee} = 46.9 Å) [51]. The lifetime of the final CS state (0.53 s at 163 K) has been attained without lowering the CS efficiency (Φ = 0.83) [51].

13.5
An Orthogonal π-Donor-Acceptor Dyad Affording an Infinite CS Lifetime

As described above, donor–acceptor dyads containing porphyrins and C_{60} linked with a short spacer afford long-lived CS states. As long as porphyrins and C_{60} are used as components of donor–acceptor dyads, however, the low-lying triplet energies of porphyrins and C_{60} have precluded attempts to attain long-lived CS states with a higher energy than the triplet energies. As shown in Fig. 13.11b, the CS lifetime becomes longer with increasing driving force of the CR process. In such a case, it is highly desirable to find a π-chromophore which has a high triplet energy and a small λ value of electron transfer. Acridinium ion can be regarded as the best candidate for such a purpose, since the λ value for the electron self-exchange between the acridinium ion and the corresponding one-electron reduced radical is the smallest (0.3 eV) among the redox-active organic compounds [27] and the triplet excited energy is significantly higher than those of porphyrins and C_{60} [52, 53]. Thus, an electron donor moiety (mesityl group) is directly connected at the 9-position of the acridinium ion to yield 9-mesityl-10-methylacridinium ion (Acr^+–Mes), in which the solvent reorganization of electron transfer is minimized because of the short linkage between the donor and acceptor moieties [54]. The X-ray crystal structure of Acr^+–Mes is shown in Fig. 13.17a [54]. The dihedral angle made by aromatic ring planes is perpendicular. Hence there is no π-conjugation between the donor and acceptor moieties as indicated by the absorption and fluorescence spectra of Acr^+–Mes, which are superpositions of the spectra of each component, i.e. mesitylene and 10-methylacridinium ion [54]. The HOMO and LUMO orbitals of Acr^+–Mes calculated by a DFT method with Gaussian 98 (B3LYP/6–31G* basis set) are localized on mesitylene and acridinium moieties (Fig. 13.17b and c), respectively. The energy of the electron-transfer state ($Acr^{\bullet}$–$Mes^{+\bullet}$) is determined by the redox potentials of each component of Acr^+–Mes as 2.37 eV, which is significantly higher than the CS states of electron donor–acceptor dyads containing porphyrins and C_{60} (see above) [54].

486 | *13 Electron Transfer of π-Functional Systems and Applications*

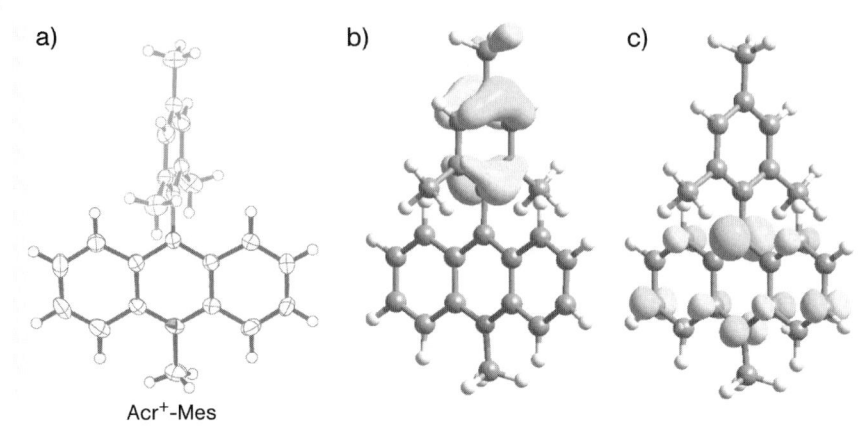

a) b) c)

Acr⁺-Mes

Fig. 13.17 (a) X-ray crystal structure of 9-mesityl-10-methylacridinium ion (Acr⁺–Mes) [54]. (b) HOMO and (c) LUMO orbitals calculated by a DFT method with Gaussian 98 (B3LYP/6–31G* basis set) [54].

Photoexcitation of a deaerated PhCN solution of Acr⁺–Mes by a nanosecond laser light flash at 430 nm results in the formation of Acr•–Mes⁺• with a quantum yield close to unity (98%) via photoinduced electron transfer from the mesitylene moiety to the singlet excited state of the acridinium ion moiety (^{1}Acr⁺*–Mes) [54]. The decay of Acr•–Mes⁺• obeyed second- rather than first-order kinetics at ambient temperature as observed in the case of Fc⁺–ZnP–H$_2$P–C$_{60}$⁻•, when the bimolecular back electron transfer predominates owing to the slow intramolecular back electron transfer (see above) [50]. In contrast, the decay of Acr•–Mes⁺• obeys first-order kinetics in PhCN at high temperatures (e.g. 373 K). This indicates that the rate of the intramolecular back electron transfer of Acr•–Mes⁺• becomes much faster than the rate of the intermolecular back electron transfer at higher tempera-

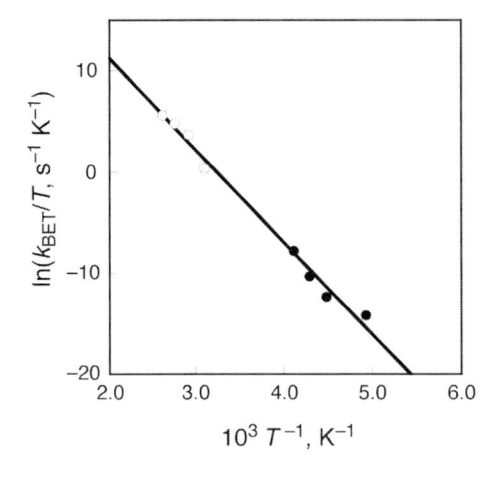

Fig. 13.18 Plot of k_{BET}/T vs. T^{-1} for intramolecular back electron transfer from the Acr• moiety to the Mes⁺• moiety in Acr•–Mes⁺• [54].

tures, because the activation energy of the former is higher than that of the latter. A large temperature dependence of the rate constant of intramolecular back electron transfer (k_{BET}) in Fig. 13.18 agrees with the Marcus equation with a large driving force for the back electron-transfer process in the deeply inverted region [Eq. (4)]. In such a case, the lifetime of the electron-transfer state becomes surprisingly longer with decreasing temperature, to approach a virtually infinite value at 77 K [54]. Such a simple molecular dyad capable of fast charge separation but extremely slow charge recombination has been recognized as a clear advantage with regard to synthetic feasibility and the applications [55].

The large temperature dependence of the rate constant of back electron transfer (k_{BET}) in Fig. 13.18 and the virtually infinite lifetime come from the high energy of the electron-transfer state, which is still lower than the energy of the triplet excited state. However, Benniston et al. have recently reported that the triplet excitation energy of Acr$^+$–Mes is 1.96 eV based on the phosphorescence spectrum, which is lower than the energy of the electron-transfer state (2.37 eV), and that the triplet excited state of the acridinium ion moiety (^{3}Acr^{+*}–Mes) might be formed rather than the electron-transfer state (Acr$^{\bullet}$–Mes$^{+\bullet}$) [56]. If this value were correct, the one-electron oxidation potential (E_{ox}) of ^{3}Acr^{+*}–Mes would be −0.08 V vs. SCE, which is determined from the one-electron oxidation potential of the Mes moiety (1.88 V) [49] and the triplet excitation energy (1.96 V). In such a case, electron transfer from the triplet excited state of Acr$^+$–Mes to N,N-dihexylnaphthalenediimide (NIm: E_{red} = −0.46 V vs. SCE) would be energetically impossible judging from the positive free energy change of electron transfer (0.38 eV). However, the addition of NIm (1.0 × 10^{-3} M) to a PhCN solution of Acr$^+$–Mes and the laser photoexcitation results in the formation of NIm$^{-\bullet}$ as detected by the well-known absorption bands at 480 and 720 nm [57, 58], accompanied by the decay of transient absorption at 510 nm due to the Acr$^{\bullet}$ moiety of the ET state as shown in Fig. 13.19a [59]. Similarly, the addition of aniline (3.0 × 10^{-5} M) to a PhCN solution of Acr$^+$–Mes results in formation of aniline radical cation (λ_{max} = 430 nm) [60], accompanied by decay of the Mes$^{+\bullet}$ moiety at 500 nm as shown in Fig. 13.19b [59]. The formation rate constant of aniline radical cation is determined as 5.6 × 10^9 M^{-1} s^{-1}, which is close to the diffusion rate constant in PhCN.

Thus, the photogenerated state of Acr$^+$–Mes has both the reducing and the oxidizing ability: to reduce NIm and to oxidize aniline, respectively. Only the electron-transfer state (Acr$^{\bullet}$–Mes$^{+\bullet}$) has such a dual ability. However, this conclusion is contradictory to the triplet energy (1.96 eV), which was reported to be lower than the energy of the electron-transfer state by Benniston et al (see above) [56]. This contradiction comes from acridine contained as an impurity in the preparation of Acr$^+$–Mes by Benniston et al., who synthesized the compound via methylation of the corresponding acridine [56]. It was confirmed that the phosphorescence maximum of 9-phenylacridine in glassy 2-MeTHF at 77 K afforded a similar spectrum to that reported by Benniston et al. [59]. In contrast to the synthetic method that Benniston et al. employed [59], we prepared Acr$^+$–Mes by the Grignard reaction of 10-methyl-9(10H)-acridone with 2-mesitylmagnesium bro-

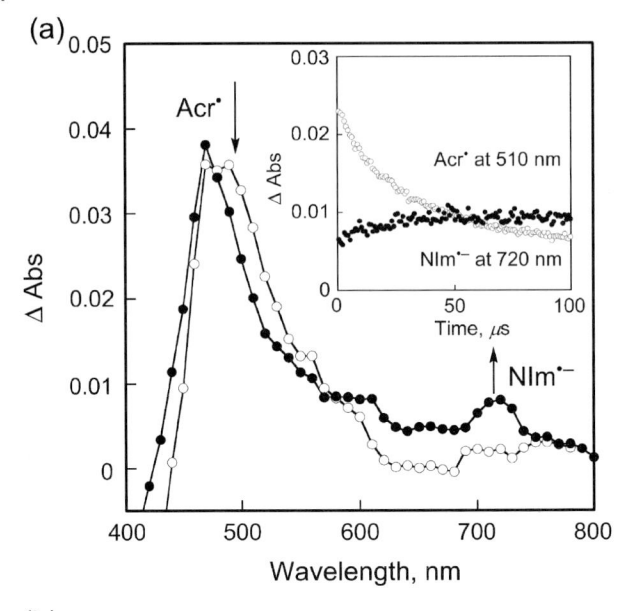

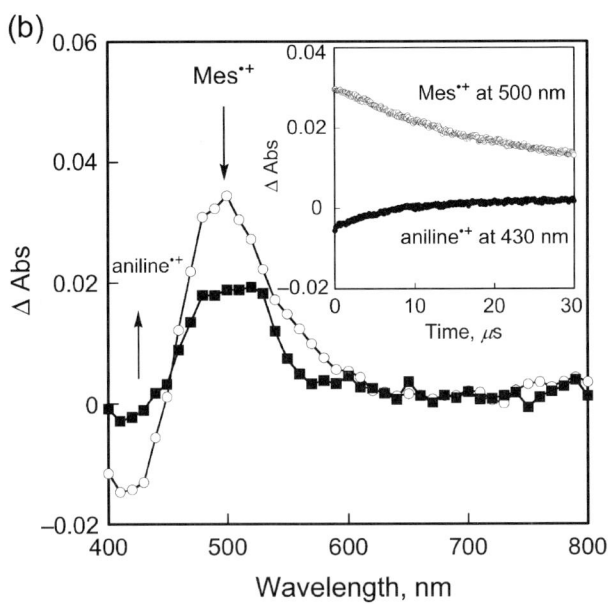

Fig. 13.19 Transient absorption spectra of Acr$^+$–Mes (5.0 × 10^{-5} M) in deaerated MeCN at 298 K taken at 2 and 20 μs after laser excitation at 430 nm in the presence of (a) N,N-dihexylnaphthalenediimide (1.0 × 10^{-3} M) or (b) aniline (3.0 × 10^{-5} M) [59]. Inset: time profiles of the absorbance decay at 510 nm and the rise at 720 nm and (b) the decay at 500 nm and the rise at 430 nm [59].

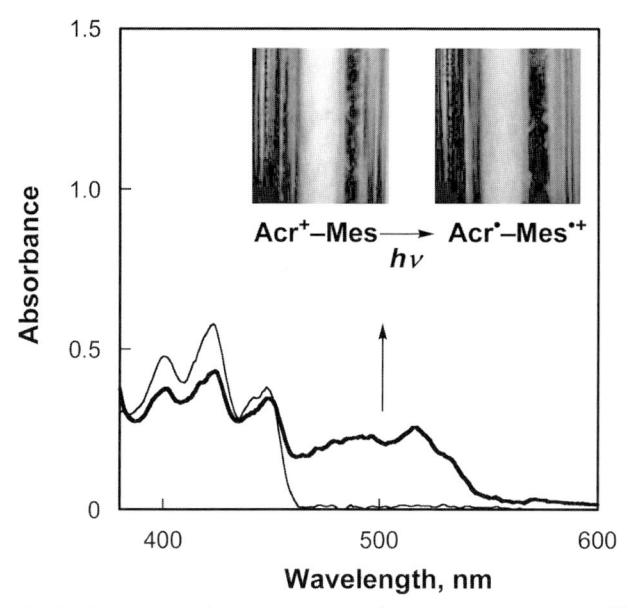

Fig. 13.20 UV–Vis absorption spectra obtained by photoirradiation with a high-pressure mercury lamp of deaerated 2-MeTHF glasses of Acr$^+$–Mes at 77 K [59]. Inset: picture images of frozen PhCN solutions of Acr$^+$–Mes before and after photoirradiation at low temperatures and taken at 77 K [59].

mide without involving methylation [54]. In this case there is no acridine impurity in Acr$^+$–Mes.

Benninston et al. also reported that photoirradiation of a PhCN solution of Acr$^+$–Mes results in the formation of the acridinyl radical (Acr$^\bullet$–Mes) [56]. They implied that this stable radical species could be mistaken for a long-lived electron-transfer state [56]. When PhCN is purified, however, no change in the absorption spectrum is observed [54, 59]. The formation of Acr$^\bullet$–Mes results from electron transfer from an donor impurity contained in unpurified PhCN (e.g. aniline) to the Mes$^{\bullet+}$ moiety of Acr$^\bullet$–Mes$^{\bullet+}$. In fact, even an extremely small amount (5.0×10^{-5} M) of aniline is enough to react with Acr$^\bullet$–Mes$^{+\bullet}$ to produce Acr$^\bullet$–Mes, which is stable owing to the bulky Mes substituent, because the lifetime of Acr$^\bullet$–Mes$^{+\bullet}$ is long enough to react with such a small concentration of an electron donor. Thus, misleading effects of impurities indeed result from the long-lived electron-transfer state that has both the oxidizing and reducing ability.

In contrast to the photoirradiation of a purified PhCN solution of Acr$^+$–Mes at 298 K, when the photoirradiation of the same solution was performed at low temperatures (213–243 K) with a 1000-W high-pressure mercury lamp through the UV light-cutting filter (>390 nm) and the sample was cooled to 77 K, the color of the frozen sample at 77 K was clearly changed from green to brownish, as shown in the inset in Fig. 13.20. When a glassy 2-methyltetrahydrofuran (2-MeTHF) is employed for the photoirradiation of Acr$^+$–Mes at low temperature, the resulting glassy solution measured at 77 K affords the absorption spectrum due to the electron-transfer state, which consists of the absorption bands of the Acr• moiety (500 nm) and the Mes•+ moiety (470 nm), as shown in Fig. 13.20. No decay of the absorption due to the electron-transfer state in Fig. 13.20 was observed at 77 K until liquid nitrogen runs out [59]. It is obvious that any excited state would have such a nearly infinite lifetime at 77 K.

13.6
A Long-lived ET State Acting as an Efficient ET Photocatalyst

As Harriman has recently pointed out, the CS state is now sufficiently long-lived and energetic to do something useful [55]. Thus, the long-lived electron-transfer state (Acr•–Mes+•), which has both high oxidizing and reducing ability (Fig. 13.19), has been utilized as an efficient photocatalyst for radical coupling reactions between radical cations and radical anions, which can be produced by the electron-transfer oxidation and reduction of external electron donors and acceptors, respectively (see above).

Visible light irradiation ($\lambda > 430$ nm) of the absorption band of Acr+–Mes in an O_2-saturated MeCN solution containing anthracene derivatives results in the formation of the oxygenation products, i.e. epidioxyanthracenes [Eq. (5)] [61]. The photocatalytic oxygenation of 9,10-dimethylanthracene (Me_2An) with O_2 on a preparative scale (100 mg, 5.0×10^{-4} mol) with Acr+–Mes (8.2 mg, 2.0×10^{-5} mol) resulted in the isolation of dimethylepidioxyanthracene (Me_2An-O_2) in 80% yield [61].

Addition of Me_2An to an MeCN solution of Acr+–Mes and laser photoexcitation result in the formation of Me_2An radical cation ($Me_2An^{+\bullet}$: $\lambda_{max} = 660$ nm) [62] with a concomitant decrease in the absorption band due to the Mes+• moiety [61]. The rate constant (k_{et}) of electron transfer from Me_2An to the Mes+• moiety of Acr•–Mes+• is determined as 1.4×10^{10} M^{-1} s^{-1}, which is close to the diffusion-limited value as expected from the exergonic electron transfer [61]. The absorption band due to the Acr• moiety remains virtually the same in the absence of O_2 during the time-scale for the electron-transfer oxidation of Me_2An with Mes+• moiety of Acr•–Mes+•. In the presence of O_2, the absorption band due to the Acr• moiety decays by electron transfer from the Acr• moiety to O_2. The formation of $O_2^{-\bullet}$ was confirmed by the ESR spectrum measured at 233 K with the typical anisotropic g values due to $O_2^{-\bullet}$ ($g_{//} = 2.1050$ and $g_{\perp} = 2.0032$) [61]. The rate constant of electron transfer from the Acr• moiety to O_2 (k'_{et}) was determined as 6.8×10^8 M^{-1} s^{-1}. The absorbance at 660 nm due to $Me_2An^{+\bullet}$ in the presence of O_2 decays, obeying second-order kinetics, by the bimolecular reaction between $Me_2An^{+\bullet}$ and $O_2^{-\bullet}$. The second-order rate constant was determined as 1.7×10^{10} M^{-1} s^{-1}, which is close to the diffusion-limited value in MeCN. The quantum yields of the formation of anthracene radical cations in the presence of O_2 are significantly larger than those of epidioxyanthracene formation [61]. This indicates that the back electron transfer from $O_2^{-\bullet}$ to anthracene radical cations to regenerate the reactant

Scheme 13.3 Photocatalytic oxygenation of anthracene with O_2 using Acr^+–Mes [61].

pair is also involved in the second-order decay of anthracene radical cations in addition to the radical coupling between anthracene radical cations and $O_2^{-\bullet}$ to afford the corresponding epidioxyanthracenes (Scheme 13.3). It is important to note that An-O_2 is formed exclusively by the radical coupling between anthracene radical cation and $O_2^{-\bullet}$ (k_c) rather than the reaction of anthracene and $^1O_2$ [61].

In the case of anthracene, An-O_2 is converted to 10-hydroxyanthrone, which is further oxidized to yield the final six-electron oxidation product, i.e. anthraquinone, accompanied by generation of H_2O_2 with the further photoirradiation ($\lambda > 430$ nm) of an O_2-saturated CD_3CN solution of An-O_2 and Acr^+–Mes [61]. When the reaction is started from the isolated An-O_2, no photochemical reaction has occurred without Acr^+–Mes or no thermal reaction has taken place with Acr^+–Mes [61]. Under photoirradiation in the presence of Acr^+–Mes, electron transfer from An-O_2 to the $Mes^{\bullet+}$ moiety of $Acr^{\bullet}$–$Mes^{\bullet+}$ results in O–O bond cleavage of An-$O_2^{+\bullet}$, followed by facile intramolecular hydrogen transfer to produce the 10-hydroxyanthrone radical cation as shown in Scheme 13.4 [61]. The back electron transfer from the $Acr^{\bullet}$ moiety to 10-hydroxyanthrone radical cation affords 10-hy-

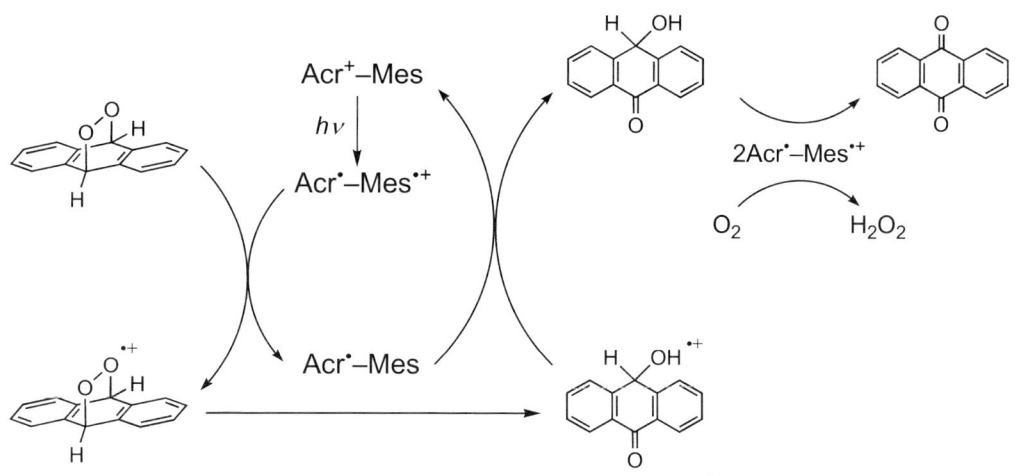

Scheme 13.4 Photocatalytic oxygenation of AnO_2 to anthraquinone with $Acr^{+\bullet}$–Mes [61].

droxyanthrone, accompanied by regeneration of Acr[+]–Mes. The electron-transfer oxidation of 10-hydroxyanthrone by the Mes[+•] moiety of Acr[•]–Mes[+•] results in the further two-electron oxidation to yield anthraquinone by releasing two protons, whereas the two-electron reduction of O_2 by the Ac[•] moiety of Acr[•]–Mes[+•] with two protons yields H_2O_2 (Scheme 13.4).

The radical coupling reaction between anthracene radical cation and $O_2^{-•}$ to produce An-O_2 in Scheme 13.3 has been expanded to the dioxetane formation from alkenes [63]. 1,2-Dioxetanes have attracted considerable interest because of the key roles in chemiluminescence and bioluminescence [64, 65], which have a broad range of biological, chemical and medical applications [66, 67]. 1,2-Dioxetanes have been commonly prepared by the formal [2 + 2]-cycloaddition of singlet oxygen ($^1O_2$) to electron-rich alkenes [68, 69]. If alkenes are too electron poor to react with $^1O_2$, however, no oxygenated products were obtained. For example, it was reported that no products were formed in an oxygen-saturated MeCN solution of tetraphenylethylene (TPE) in the presence of $^1O_2$ sensitizers under photoirradiation [70]. No oxygenation occurred in an O_2-saturated solution of TPE containing C_{60} or tetraphenylporphyrin as $^1O_2$ sensitizers, since the reaction of TPE with $^1O_2$ has been reported to be very slow [71]. In contrast, when Acr[+]–Mes is used as a photocatalyst, oxygenation of TPE with oxygen occurs efficiently via electron-transfer reactions of TPE and oxygen with photogenerated electron-transfer state of Acr[+]–Mes, followed by the radical coupling reaction between TPE radical cation and $O_2^{-•}$ to produce 1,2-dioxetane selectively (Scheme 13.5) [63]. The photocatalytic oxygenation of TPE with O_2 in a preparative scale (60 mg, 1.8×10^{-4} mol) with Acr[+]–Mes (3.8 mg, 8.7×10^{-6} mol) in $CHCl_3$ (2.0 mL) afforded the corresponding 1,2-dioxetane (27% isolated yield) after 4 h of photoirradiation at 278 K [63]. The second-order rate constant (k_{et}) of electron

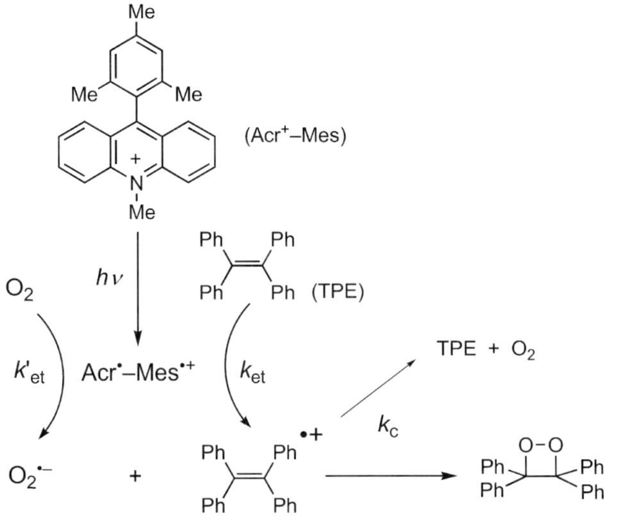

Scheme 13.5 Photocatalytic oxygenation of tetraphenylethylene (TPE) with O_2 using Acr[+]–Mes [63].

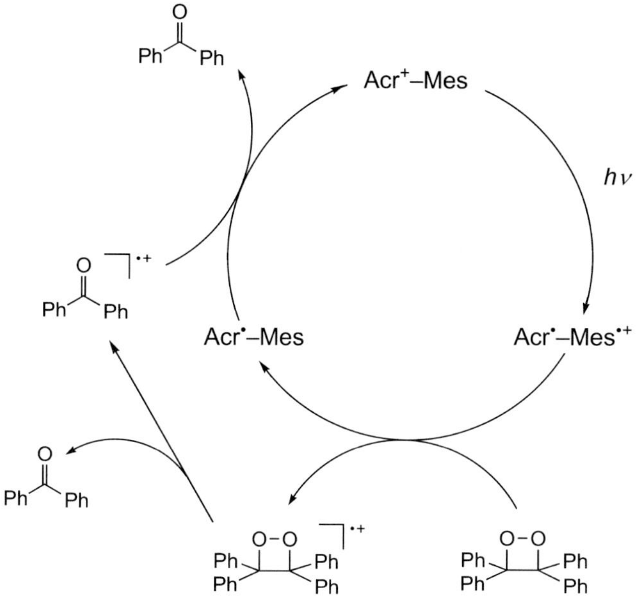

13.6 A Long-lived ET State Acting as an Efficient ET Photocatalyst

Scheme 13.6 Photocatalytic decomposition of TPE dioxetane with Acr⁺–Mes [63].

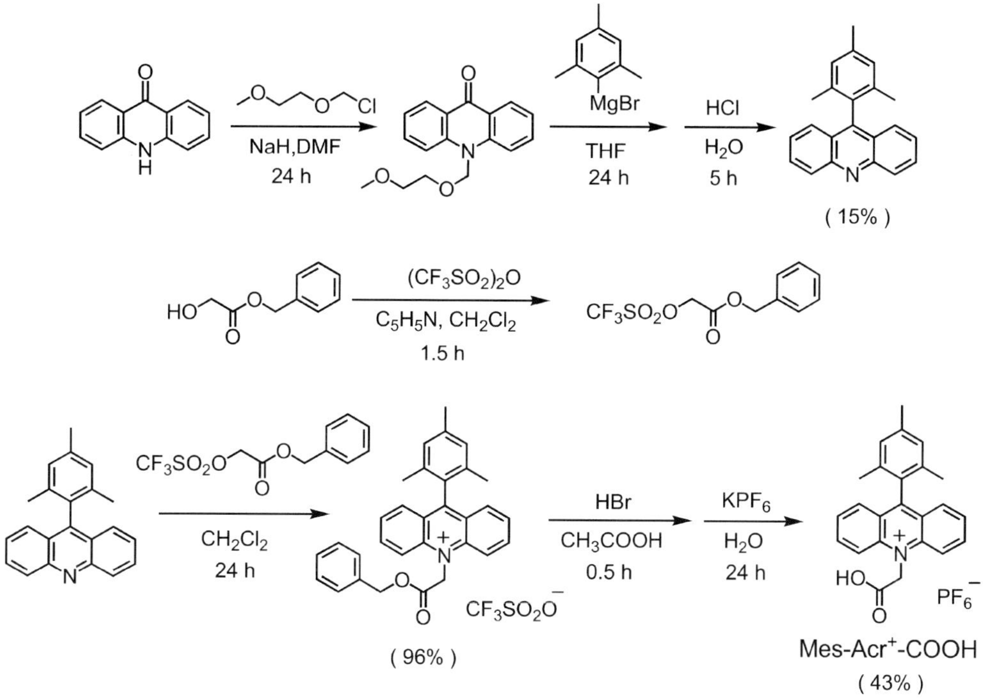

Scheme 13.7 Synthesitc procedure for Mes–Acr⁺–COOH [77].

transfer from TPE to Acr•–Mes+• is determined as 2.5×10^9 M^{-1} s^{-1} in CHCl$_3$, which is close to be the diffusion-limited value as expected from the exergonic electron transfer [63]. The second-order rate constant of electron-transfer reduction of O$_2$ (k'_{et}) by the Acr• moiety was also determined as 3.8×10^8 M^{-1} s^{-1} in CHCl$_3$ [63].

The dioxetane formed by the radical coupling reaction between TPE$^{+•}$ and O$_2^{-•}$ is further oxidized by Acr•–Mes$^{+•}$ rather than by being reduced to produce the dioxetane radical cation, which undergoes O–O bond homolysis to produce benzophenone and the radical cation as shown in Scheme 13.6 [63]. The benzophenone radical cation is reduced by Acr•–Mes to produce another benzophenone molecule, accompanied by regeneration of Acr$^+$–Mes (Scheme 13.6). The formation of the dioxetane radical cation was confirmed by electron spin resonance (ESR) measurements under photoirradiation at low temperature [63]. The resulting ESR spectrum observed at 143 K exhibits anisotropic signals at $g_{//} = 2.020$ and $g_\perp = 2.004$. The isotropic g value (g_{iso}) is determined as 2.009 ± 0.001, which agrees with the reported value of a dioxetane radical cation (2.0099) [72].

As demonstrated above, the use of Acr$^+$–Mes as an electron-transfer photocatalyst in the presence of O$_2$ provides a convenient methodology to produce radical cations of electron donors of substrates and O$_2^{-•}$, which can combine together to yield the oxygenated products selectively in a preparative scale.

13.7
Organic Solar Cells Using Simple Donor-Acceptor Dyads

The requirement to develop inexpensive renewable energy sources has stimulated new approaches for the production of efficient, low-cost organic photovoltaic devices [73–76]. A simple molecular dyad (Acr$^+$–Mes) capable of fast charge separation but extremely slow charge recombination has allowed us to develop a unique organic photovoltaic cell using the molecular dyad. In order to assemble the dyad on an optically transparent electrode (OTE) of nanostructured SnO$_2$ (OTE/SnO$_2$), a carboxyl group was introduced to the dyad to prepare 9-mesityl-10-carboxymethylacridinium ion (Mes–Acr$^+$–COOH) as shown in Scheme 13.7 [77]. An organic photovoltaic cell composed of Mes–Acr$^+$–COOH and fullerene nanoclusters has been constructed as shown in Scheme 13.8 [77]. Fullerene (C$_{60}$) clusters prepared from C$_{60}$ suspension in acetonitrile–toluene (3:1) have been deposited electrophoretically on the OTE/SnO$_2$/Mes–Acr$^+$–COOH electrode [denoted OTE/SnO$_2$/Mes–Acr$^+$–COOH + (C$_{60}$)$_n$] in order to improve the light-harvesting and electron transport efficiency [77].

Photoelectrochemical measurements were performed using a standard two-electrode system consisting of a working electrode and a Pt wire gauze electrode in air-saturated acetonitrile containing 0.5 M NaI and 0.01 M I$_2$ [77]. The incident photon-to-photocurrent efficiency (IPCE) values were calculated by normalizing the photocurrent values for incident light energy and intensity and using the equation [78]

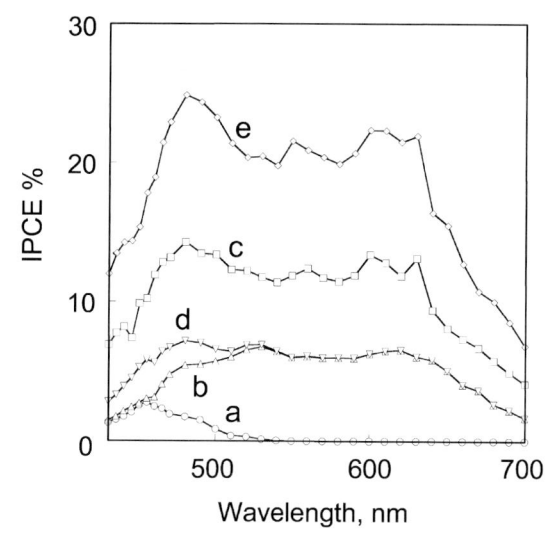

Fig. 13.21 Comparison of plots of IPCE vs. wavelength of (a) OTE/SnO$_2$/Mes–Acr$^+$–COOH electrode ([C$_{60}$] = 0.62 mM), (b) OTE/SnO$_2$/(C$_{60}$)$_n$ electrode ([C$_{60}$] = 0.62 mM), (c) OTE/SnO$_2$/Mes-Acr$^+$-COOH+(C$_{60}$)$_n$ electrode ([C$_{60}$] = 0.62 mM), (d) the sum of the IPCE response of OTE/SnO$_2$/Mes–Acr$^+$–COOH and OTE/SnO$_2$/(C$_{60}$)$_n$ electrodes ([C$_{60}$] = 0.62 mM) and (e) OTE/SnO$_2$/Mes–Acr$^+$–COOH + (C$_{60}$)$_n$ electrode ([C$_{60}$] = 0.62 mM) with applied bias potential: 0.2 V vs. SCE [77].

$$\text{IPCE } (\%) = 100 \times 1240 \times i_{sc}/(I_{inc} \times \lambda) \tag{6}$$

where i_{sc} is the short-circuit photocurrent (A cm^{-2}), I_{inc} is the incident light intensity (W cm^{-2}) and λ is the wavelength (nm). The maximum IPCE value of OTE/SnO$_2$/Acr$^+$–Mes–COOH (spectrum a in Fig. 13.21) is only 2% (445 nm), whereas the IPCE value of OTE/SnO$_2$/Acr$^+$–Mes–COOH + (C$_{60}$)$_n$ (spectrum c in Fig. 13.21) reaches 15% (480 nm) [77]. The IPCE value of OTE/SnO$_2$/Acr$^+$–Mes–COOH + (C$_{60}$)$_n$ is much higher than the sum of the two individual IPCE values of the individual systems [OTE/SnO$_2$/Acr$^+$–Mes–COOH and OTE/SnO$_2$/(C$_{60}$)$_n$; spectrum d in Fig. 13.21] in the visible region [77].

The charge separation in the OTE/SnO$_2$/Mes–Acr$^+$–COOH + (C$_{60}$)$_n$ electrode can be further modulated by controlling the applied potential in a standard three-compartment cell with a working electrode along with a Pt wire gauze counter electrode and a saturated calomel reference electrode (SCE) [77]. The maximum IPCE value is obtained as 25% at an applied potential of 0.2 V vs. SCE (spectrum e in Fig. 13.21) [77]. Such a high IPCE value indicates that photocurrent is initiated via electron transfer between excited Mes Acr$^+$ COOH and C$_{60}$ clusters, followed by the charge transport to the collective surface of an OTE/SnO$_2$ electrode. The charge transport is significantly improved under the influence of an applied bias.

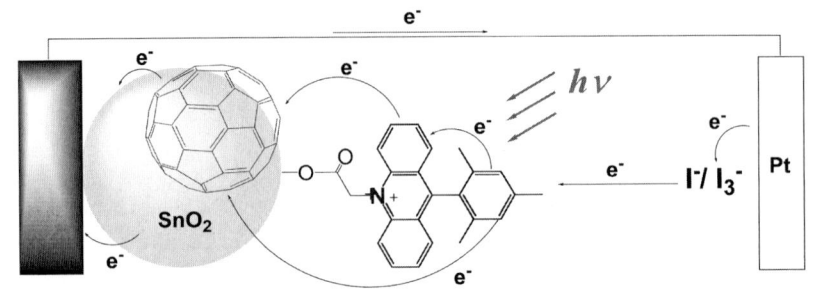

Scheme 13.8 The photovoltaic cell composed of fullerene nanoclusters and 9-mesityl-10-carboxymethylacridinium ion [77].

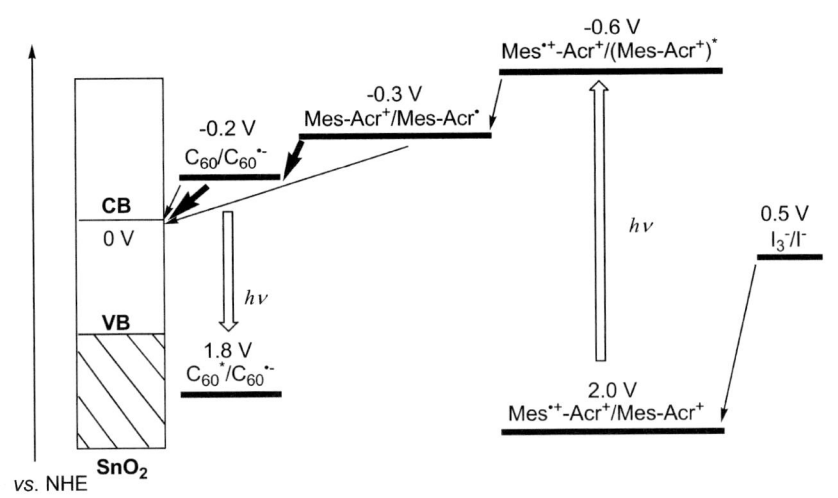

Scheme 13.9 Mechanism of the photocurrent generation in OTE/SnO$_2$/Mes–Acr$^+$–COOH + (C$_{60}$)$_n$ electrode [77].

The mechanism of the photocurrent generation of OTE/SnO$_2$/Mes–Acr$^+$–COOH and OTE/SnO$_2$/Mes–Acr$^+$–COOH + (C$_{60}$)$_n$ electrodes is summarized in Scheme 13.9 [77]. The photocurrent generation is initiated by photoinduced electron transfer from the Mes moiety to the singlet excited state of the Acr$^+$ moiety of Mes–Acr$^+$ to produce Mes$^{•+}$–Acr$^{•}$. The resulting acridinyl radical (Acr$^{•}$) (Acr$^+$/Acr$^{•}$ = −0.3 V vs. NHE) injects electrons into the conduction band of SnO$_2$ (0 V vs. NHE) [78], whereas the oxidized mesityl moiety (Mes/Mes$^{•+}$ = 2.0 V vs. NHE) undergoes electron-transfer reduction with the iodide (I$_3^-$/I$^-$ = 0.5 V vs. NHE) [78] in the electrolyte solution. The long lifetime of the electron-transfer state (Mes$^{•+}$–Acr$^{•}$) ensures efficient electron transfer from Acr$^{•}$ to C$_{60}$ (C$_{60}$/C$_{60}^{•-}$ = −0.2 V vs. NHE) [78] to produce C$_{60}^{•-}$, which injects an electron into the conduction band of SnO$_2$. Formation of C$_{60}^{•-}$ has been confirmed by time-resolved transient absorption spectra of 9-mesityl-10-methylacridinium ion without carboxylic acid (Mes-Acr$^+$) and fullerene in deoxygenated toluene–acetonitrile (1:1, v/v), which exhibit a broad absorption band at about 1050 nm [77]. This NIR band is diagnostic of C_{60} radical anion [79]. On the other hand, the oxidized mesityl moiety (Mes$^{+\bullet}$) undergoes electron-transfer reduction with iodide ion in the electrolyte. Enhanced IPCE values of OTE/SnO$_2$/Mes–Acr$^+$–COOH + $(C_{60})_n$ (spectrum c in Fig. 13.21) relative to the sum of the two individual IPCE values of the individual systems (spectrum d in Fig. 13.21) result from the charge-transfer interaction between the mesityl (donor) moiety of Mes–Acr$^+$–COOH and C_{60} clusters, which appears at the long wavelength region beyond 500 nm [77].

In order to improve the photoelectrochemical properties, TiO$_2$ nanoparticles have been utilized for three-dimensional control in the organization of composite nanoclusters of Mes–Acr$^+$–COOH and C_{60} [80]. TiO$_2$ nanoparticles were modified with composite nanoclusters of Mes–Acr$^+$–COOH and C_{60} in acetonitrile–toluene (3:1, v/v) and then deposited as thin films on a nanostructured SnO$_2$ electrode using an electrophoretic technique (Scheme 13.10) [80]. In the case of a monolayer system of TiO$_2$ nanocrystallites modified with Mes–Acr$^+$–COOH, no net photocurrent is observed in the photocurrent action spectrum [80]. This indicates that TiO$_2$ nanoparticles act as materials to organize composite molecules rather than to accept electrons.

The photocurrent action spectrum of the OTE/SnO$_2$/(Mes–Acr$^+$–COO–TiO$_2$ + $C_{60})_n$ electrode is compared with those of OTE/SnO$_2$/(C$_{60})_n$ and OTE/SnO$_2$/(Mes–Acr$^+$–COO–TiO$_2)_n$ electrodes in a standard three-compartment cell under a bias of 0.2 V vs. SCE (Fig. 13.22a) [80]. The photocurrent action spectrum of the OTE/SnO$_2$/(Mes–Acr$^+$–COO–TiO$_2$ + $C_{60})_n$ electrode shows a maximum IPCE value of 37% at an applied potential of 0.2 V vs. SCE (spectrum a in Fig. 13.22a). Under same experimental conditions the IPCE values are much smaller

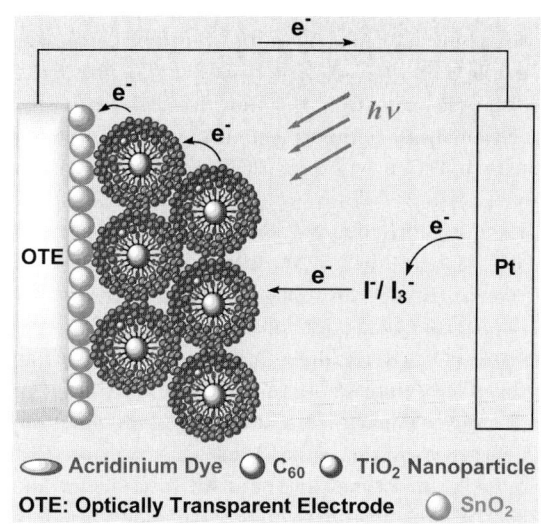

Scheme 13.10 Supramolecular photovoltaic cells based on composite molecular clusters of 9-mesityl-10-carboxymethylacridinium ion and fullerene, which is electrophoretically organized by TiO$_2$ nanoparticles [80].

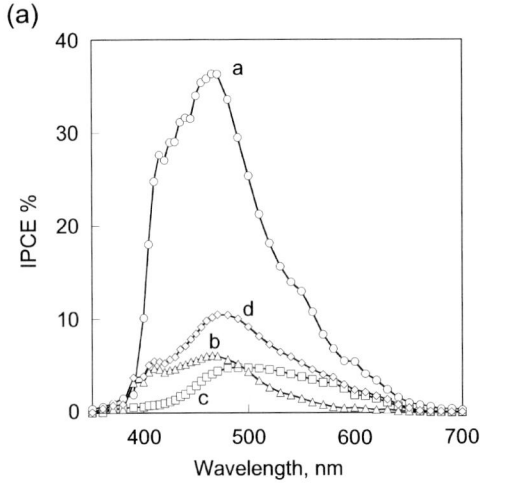

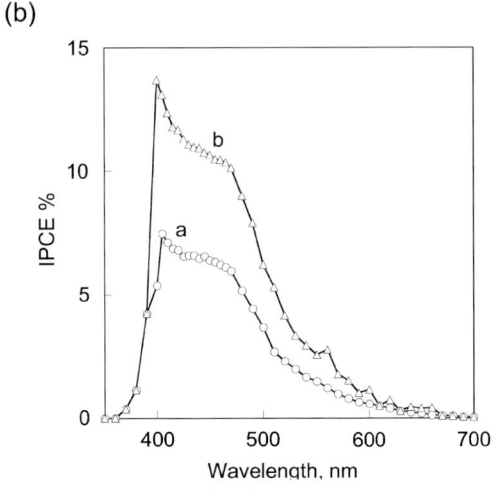

Fig. 13.22 (a) Photocurrent action spectra (IPCE vs. wavelength) of (a) OTE/SnO$_2$/(Mes–Acr$^+$–COO–TiO$_2$ + C$_{60}$)$_n$ ([Mes–Acr$^+$] = 0.025 mM, [C$_{60}$] = 0.13 mM), (b) OTE/SnO$_2$/(Mes–Acr$^+$–COO–TiO$_2$)$_n$ ([Mes–Acr$^+$] = 0.025 mM), (c) OTE/SnO$_2$/(C$_{60}$)$_n$ ([Mes–Acr$^+$] = 0.025 mM, [C$_{60}$] = 0.13 mM) and (d) the sum of the IPCE response of OTE/SnO$_2$/(Mes–Acr$^+$–COO–TiO$_2$)$_n$ (b) and OTE/SnO$_2$/(C$_{60}$)$_n$ (c) at an applied bias potential of 0.2 V vs. SCE [76]. (b) Photocurrent action spectra (IPCE vs. wavelength) of (a) OTE/SnO$_2$/(Acr$^+$–COO–TiO$_2$ + C$_{60}$)$_n$ ([Acr$^+$] = 0.025 M, [C$_{60}$] = 0.13 mM), with no applied bias potential, and (b) OTE/SnO$_2$/(Acr$^+$–COO–TiO$_2$ + C$_{60}$)$_n$ ([Acr$^+$] = 0.025 mM, [C$_{60}$] = 0.13 mM) at an applied bias potential of 0.2 V vs. SCE. Electrolyte: 0.5 M NaI and 0.01 M I$_2$ in MeCN [80].

for the single component systems, viz. OTE/SnO$_2$/(C$_{60}$)$_n$ and OTE/SnO$_2$/(Mes–Acr$^+$–COO–TiO$_2$)$_n$ (spectra b and c in Fig. 13.22a, respectively). The IPCE value obtained with the mixed cluster system (37%), viz. OTE/SnO$_2$/(Mes–Acr$^+$–COO–TiO$_2$ + C$_{60}$)$_n$ is larger than the sum of two individual IPCE values (~11%). Such enhancement in the photocurrent generation of the composite cluster systems of Mes–Acr$^+$–COO–TiO$_2$ and C$_{60}$ as compared with the single component systems results from interplay between Mes–Acr$^+$–COO–TiO$_2$ and C$_{60}$ in the supramolecular complex.

The photocurrent action spectra of the OTE/SnO$_2$/(Acr$^+$–COO–TiO$_2$ + C$_{60}$)$_n$ electrode, in which acridinium moiety (Acr$^+$) contains no donor moiety, are compared with that of the OTE/SnO$_2$/(Mes–Acr$^+$–COO–TiO$_2$ + C$_{60}$)$_n$ electrode in Fig. 13.22b. The maximum IPCE value of the OTE/SnO$_2$/(Acr$^+$–COO–TiO$_2$ + C$_{60}$)$_n$ electrode in standard two (no bias) and three (0.2 V vs. SCE) compartment cells (see above) reached 7 and 14%, respectively [80]. These IPCE values are significantly smaller than those of the OTE/SnO$_2$/(Mes–Acr$^+$–COO–TiO$_2$ + C$_{60}$)$_n$ electrodes (13 and 37%, respectively) [80]. This indicates that photoinduced electron transfer from the donor moiety (Mes) to the acceptor moiety (Acr$^+$) occurs, followed by electron transfer from the resulting acridinyl radical moiety (Acr$^\bullet$) to C$_{60}$ in the supramolecular complex, leading to enhancement of the photocurrent generation. The IPCE value of 37 % was achieved at an applied bias potential of 0.2 V vs. SCE in the Mes–Acr$^+$–COOH/C$_{60}$ composite system using TiO$_2$ nanoparticles [80].

13.8
Organic Solar Cells Composed of Multi-porphyrin/C$_{60}$ Supramolecular Assemblies

In purple photosynthetic bacteria, visible light is harvested efficiently by the antenna complexes which include a wheel-like array of chlorophylls [81]. In this context, porphyrin dendrimers are good candidates as light collectors, because the antenna function of porphyrin dendrimers resembles that of the light-harvesting antenna [82, 83]. Moreover, porphyrins and fullerenes are known to form supramolecular complexes, which contain closest contacts between one of the electron-rich 6:6 bonds of the guest fullerene and the geometric center of the host porphyrin [84–89]. The porphyrin–fullerene interaction energies are reported to be in the range from –16 to –18 kcal mol^{-1} [90]. Such a strong interaction between porphyrins and fullerenes is likely to be a good driving force for the formation of supramolecular complexes between porphyrin and C$_{60}$. Thus, a combination of porphyrin dendrimers (chromophores and electron donor) and fullerenes (electron acceptor) seems ideal for fulfilling an enhanced light-harvesting efficiency of chromophores throughout the solar spectrum and a highly efficient conversion of the harvested light into the high energy state of the charge separation by photoinduced electron transfer.

A new type of solar cells have been constructed using molecular clusters of porphyrin dendrimer (donor) and fullerene (acceptor) dye units assembled on SnO$_2$ electrodes, which have high charge-separation efficiency in addition to efficient hole and electron transport [91]. The porphyrin dendrimers (D$_n$P$_n$: n = 4, 8, 16) are shown in Fig. 13.23 together with the reference porphyrin compound (H$_2$P-ref). They are used to prepare the porphyrin and C$_{60}$ composite SnO$_2$ electrodes. Different molecular assemblies between porphyrin and C$_{60}$ make it possible to control the three-dimensional molecular structure, which is essential for efficient light energy conversion. Porphyrin dendrimers (D$_n$P$_n$) form supramolecular complexes with fullerene molecules in toluene and they are clustered in an acetonitrile–toluene mixed solvent system [91]. Then, the clusters are attached on nanostructured SnO$_2$ electrodes by an electrophoretic deposition method as in the case of the OTE/SnO$_2$/Mes–Acr$^+$–COOH + (C$_{60}$)$_n$ electrode in Scheme 13.8 to afford the modified electrodes [denoted OTE/SnO$_2$/(D$_n$P$_n$ + C$_{60}$)$_m$ (n = 4, 8 and 16)] [71].

The AFM image of OTE/SnO$_2$/(D$_4$P$_4$ + C$_{60}$)$_m$ reveals the cluster aggregation with a regular size, composed of closely packed clusters of about 100 nm size as shown in Fig. 13.24a [91]. The photoelectrochemical measurements were performed with a standard two-electrode system consisting of a working electrode and a Pt wire gauze electrode in 0.5 M NaI and 0.01 M I$_2$ in air-saturated acetonitrile [91]. Figure 13.24b shows the wavelength dependence of the incident photon-to-photocurrent efficiency (IPCE) of the OTE/SnO$_2$/(D$_n$P$_n$ + C$_{60}$)$_m$ system

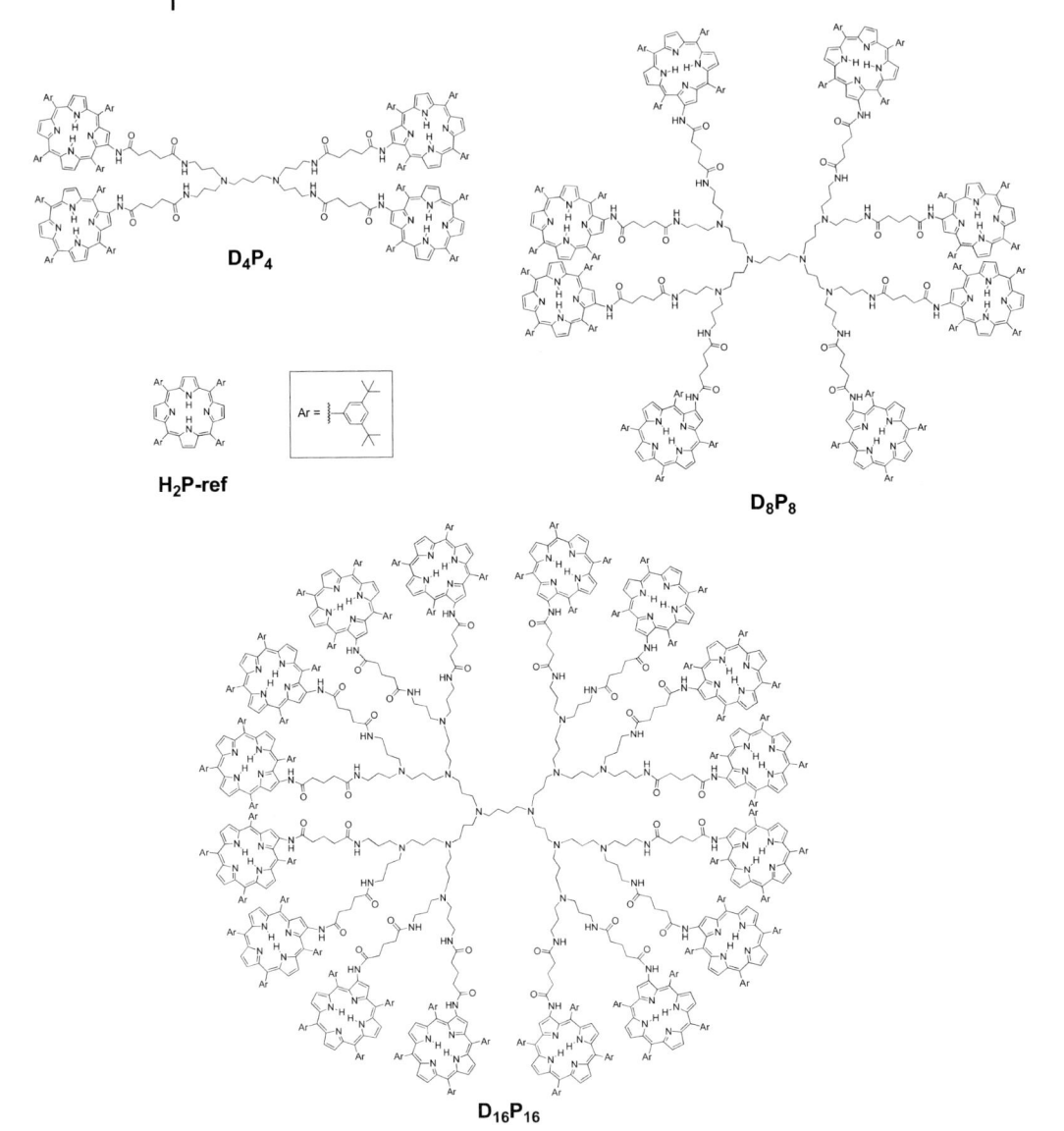

Fig. 13.23 Porphyrin dendrimers and the reference compound employed for construction of organic solar cells composed of multi-porphyrin/C_{60} supramolecular assemblies [91].

and the reference system [OTE/SnO$_2$/(H$_2$P-ref + C$_{60}$)$_m$] at a constant concentration ratio of porphyrin to C$_{60}$ [91]. The OTE/SnO$_2$/(D$_4$P$_4$ + C$_{60}$)$_m$ system has the maximum IPCE value of 15 % as well as the broad photoresponse, which extends well into the infrared (up to 1000 nm). The comparison of the photoelectrochemical

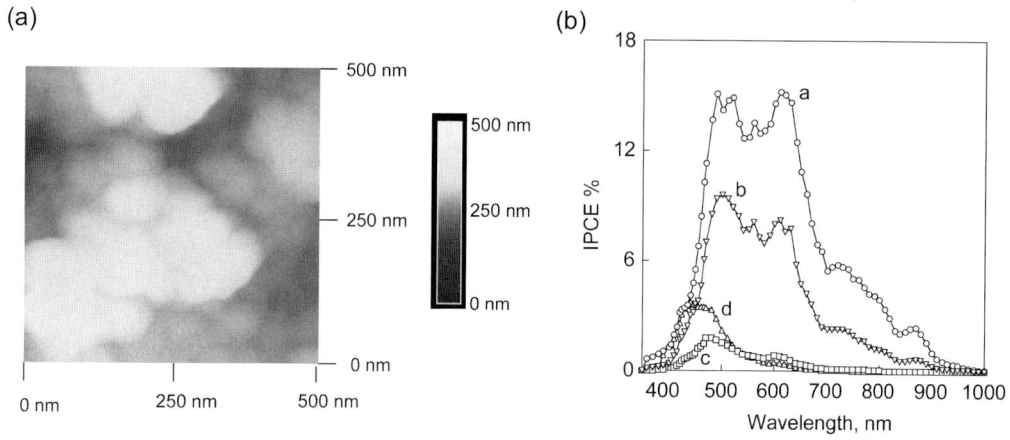

Fig. 13.24 (a) AFM images of OTE/SnO$_2$/(D$_4$P$_4$ + C$_{60}$)$_m$, [D$_4$P$_4$] = 0.048 mM; [C$_{60}$] = 0.31 mM in acetonitrile–toluene (3:1) [91]. (b) Photocurrent action spectra (IPCE vs. wavelength) of (A) OTE/SnO$_2$/(D$_n$P$_n$ + C$_{60}$)$_m$ systems [(a) n = 4, (b) n = 8 and (c) n = 16], (d) OTE/SnO$_2$/(H$_2$P-ref + C$_{60}$)$_m$ and [D$_4$P$_4$] = 0.048 mM, [D$_8$P$_8$] = 0.024 mM, [D$_{16}$P$_{16}$] = 0.012 mM and [H$_2$P-ref] = 0.19 mM; [C$_{60}$] = 0.31 mM in acetonitrile–toluene (3:1) [91]

properties between the OTE/SnO$_2$/(D$_4$P$_4$ + C$_{60}$)$_m$ system and the reference system indicates that the composite clusters of porphyrin dendrimers with fullerene exhibit remarkable enhancement in the photoelectrochemical performance due to the effective π–π interaction between porphyrins and fullerenes in the interpenetrating structure of the supramolecular clusters as compared with the reference system. Such an effective photoenergy conversion is ascribed to the dendritic structure that controls three-dimensional organization between porphyrin and C$_{60}$. However, the IPCE value decreases with increasing the number of dendrimer generation and the OTE/SnO$_2$/(D$_{16}$P$_{16}$ + C$_{60}$)$_m$ system has even smaller IPCE values compared with the reference system [OTE/SnO$_2$/(H$_2$P-ref + C$_{60}$)$_m$] (Fig. 13.24b) [91]. In the case of D$_{16}$P$_{16}$, the space between porphyrin rings may be too small for the interaction with C$_{60}$. This indicates that supramolecular assembly between electron donor and acceptor is very important in the construction of light energy conversion systems. The distance between two porphyrin rings is also an essential factor to form such a supramolecular complex.

The photocurrent generation in the present system is initiated by photoinduced charge separation from the porphyrin excited singlet state (^{1}H$_2$P*/H$_2$P$^{+•}$ = –0.7 V vs. NHE) [78] in the dendrimer to C$_{60}$ (C$_{60}$/C$_{60}^{-•}$ = –0.2 V vs. NHE) [78] in the porphyrin dendrimer–C$_{60}$ complex rather than direct electron injection to conduction band of SnO$_2$ (0 V vs. NHE) system [91]. The reduced C$_{60}$ injects electrons into the SnO$_2$ nanocrystallites, whereas the oxidized porphyrin (H$_2$P/H$_2$P$^{+•}$ = 1.2 V vs. NHE) [78] undergoes electron-transfer reduction with iodide (I$_3^-$/I$^-$ = 0.5 V vs. NHE) [78] in the electrolyte system [91].

In contrast with the porphyrin dendrimers in Fig. 13.23, porphyrin oligomers with a polypeptidic backbone in Fig. 13.25 are flexible enough to accommodate

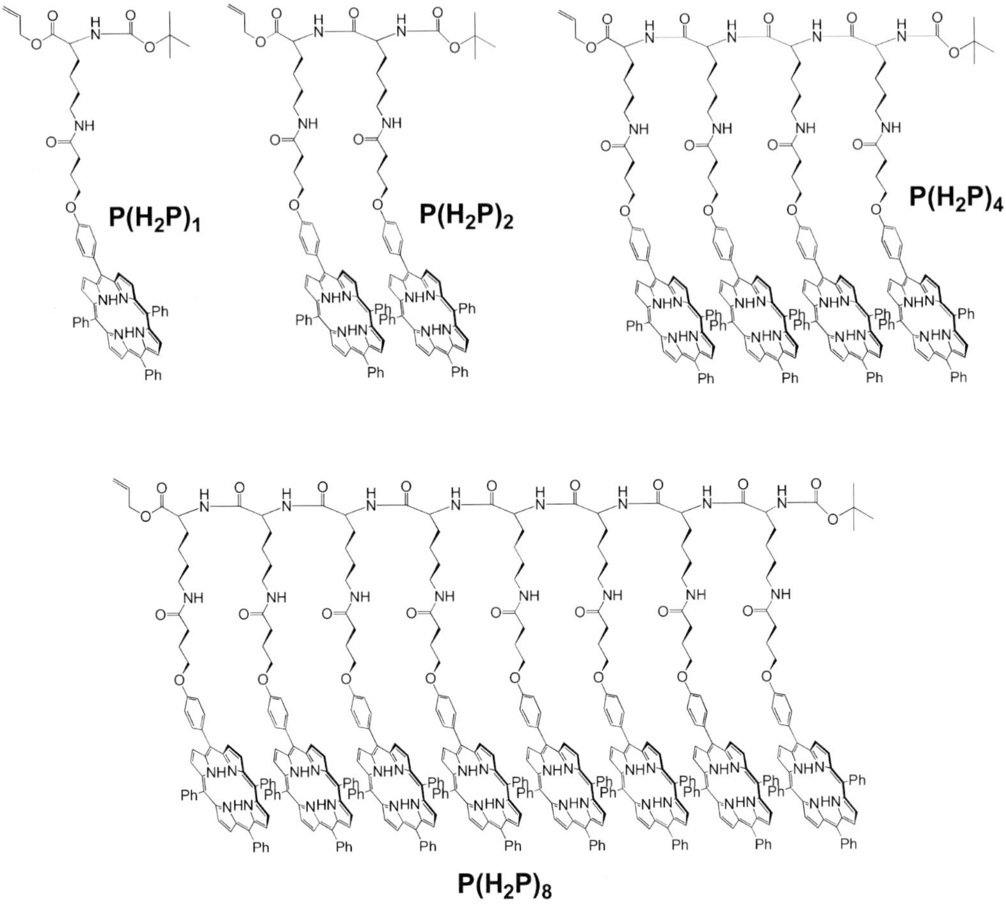

Fig. 13.25 Porphyrin–peptide oligomers [P(H$_2$P)$_n$] employed for construction of organic solar cells composed of multi-porphyrin/C$_{60}$ supramolecular assemblies [92, 93].

C$_{60}$ between the porphyrin units [92]. The organization of P(H$_2$P)$_n$ and C$_{60}$ composite clusters onto an optically transparent electrode (OTE) of a nanostructured SnO$_2$ electrode {denoted OTE/SnO$_2$/[P(H$_2$P)$_n$ + C$_{60}$]$_m$ (n = 1, 2, 4, 8)} was performed as the case of porphyrin dendrimers [93]. The IPCE value of OTE/SnO$_2$/[P(H$_2$P)$_n$ + C$_{60}$]$_m$ (n = 1, 2, 4, 8) exhibits a remarkable increase with increase in the number of porphyrins in a polypeptide unit, as shown in Fig. 13.26a [93]. The OTE/SnO$_2$/[P(H$_2$P)$_8$ + C$_{60}$]$_m$ system has the maximum IPCE value of 42% at 600 nm in addition to a broad photoresponse, extending into the infrared region (up to 1000 nm). Such an effective light energy conversion is largely ascribed to the polypeptide structure which controls the three-dimensional organization be-

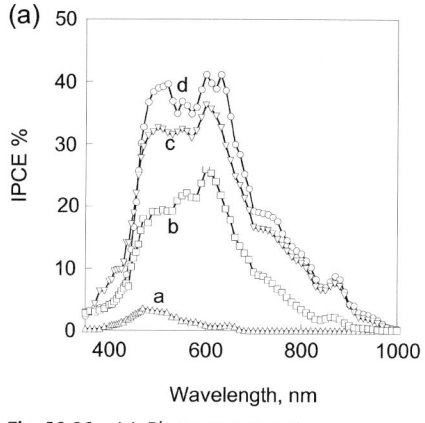

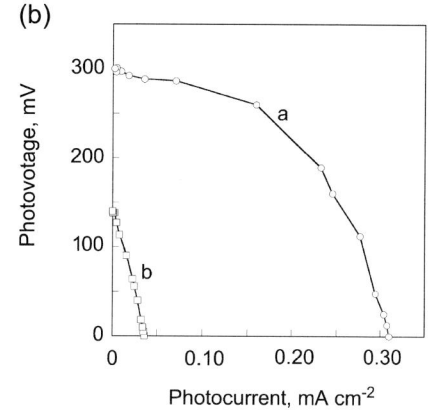

Fig. 13.26 (a) Photocurrent action spectra (IPCE vs. wavelength) of OTE/SnO$_2$/[P(H$_2$P)$_n$ + C$_{60}$]$_m$ under short-circuit conditions. (a) [P(H$_2$P)$_1$] = 0.19 mM, (b) [P(H$_2$P)$_2$] = 0.10 mM, (c) [P(H$_2$P)$_4$] = 0.048 mM and (d) [P(H$_2$P)$_8$] = 0.024 mM; [C$_{60}$] = 0.31 mM in acetonitrile–toluene (3:1) [93]. (b) Current–voltage characteristics of (a) OTE/SnO$_2$/[P(H$_2$P)$_8$ + C$_{60}$]$_m$ and (b) OTE/SnO$_2$/[P(H$_2$P)$_1$ + C$_{60}$]$_m$ electrodes. Electrolyte: 0.5 M NaI and 0.01 M I$_2$ in acetonitrile [93]. Input power: 3.4 mW cm^{-2}, $\lambda > 400$ nm. [P(H$_2$P)$_8$] = 0.024 mM; [P(H$_2$P)$_1$] = 0.19 mM; [C$_{60}$] = 0.31 mM.

tween porphyrin and C$_{60}$. This suggests that electron-transfer properties are improved with increasing the number of porphyrins in a polypeptide unit.

The current–voltage (I–V) characteristics of OTE/SnO$_2$/[P(H$_2$P)$_8$ + C$_{60}$]$_m$ and OTE/SnO$_2$/[P(H$_2$P)$_1$ + C$_{60}$]$_m$ electrodes are shown in Fig. 13.26b. The I–V characteristics of OTE/SnO$_2$/[P(H$_2$P)$_8$ + C$_{60}$]$_m$ system is remarkably enhanced (more than 30-fold) compared with the OTE/SnO$_2$/[P(H$_2$P)$_1$ + C$_{60}$]$_m$ electrode under the same experimental conditions [93]. The power conversion efficiency, η, is calculated with the equation [94]

$$\eta = FF I_{sc} V_{oc}/W_{in} \tag{7}$$

where the fill factor (FF) is defined as $FF = [IV]_{max}/I_{sc}V_{oc}$, where V_{oc} is the open-circuit photovoltage and I_{sc} is the short-circuit photocurrent. The OTE/SnO$_2$/[P(H$_2$P)$_8$ + C$_{60}$]$_m$ system has a large fill factor (FF) of 0.47, open-circuit voltage (V_{oc}) of 300 mV, short-circuit current density (I_{sc}) of 0.31 mA cm^{-2} and the overall power conversion efficiency (η) of 1.3% at an input power (W_{in}) of 3.4 mW cm^{-2} [93].

Well-organized self-assembled multi-porphyrin systems have also been obtained using porphyrin–alkanethiolate monolayer protected-gold nanoclusters with spherical shape [95–98]. Novel organic solar cells have been constructed using supramolecular complexes of porphyrin–alkanethiolate monolayer protected-gold nanoclusters with fullerenes, which are prepared using self-organization of porphyrin (donor) and fullerene (acceptor) moieties by clustering with gold nanopar-

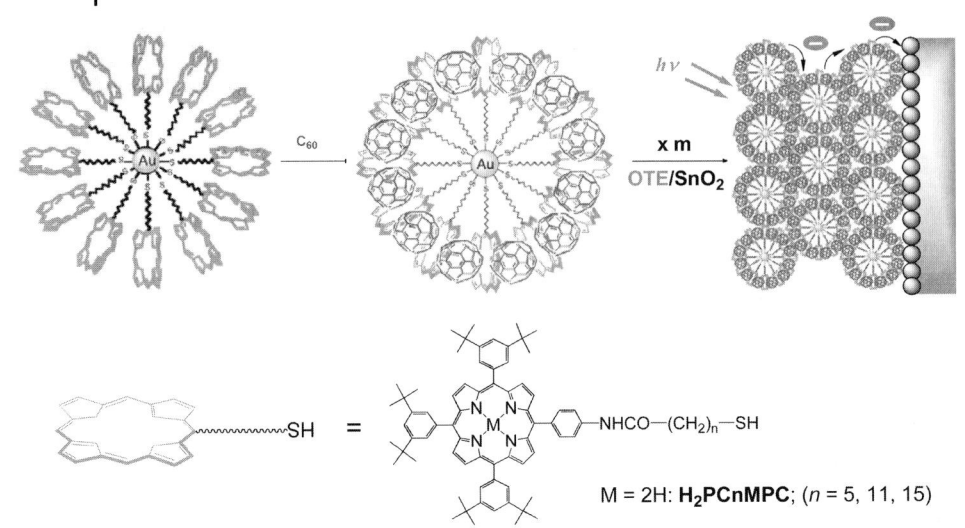

Fig. 13.27 Schematic illustration of the preparation of supramolecular complexes of porphyrin–alkanethiolate monolayer protected-gold nanoclusters with fullerenes and subsequent electrophoretic deposition on an OTE/SnO$_2$ electrode [denoted OTE/SnO$_2$/(H$_2$PCnMPC + C$_{60}$)$_m$: n = 5, 11 and 15] [99, 100].

ticles and electrophoretic deposition on SnO$_2$ electrodes, as shown in Fig. 13.27 [99].

First, porphyrin–alkanethiolate monolayer protected-gold nanoclusters with well-defined size (8–9 nm) and spherical shape [H$_2$PCnMPC (n = 5, 11, 15)] are prepared starting from porphyrin–alkanethiol. Taking the gold core as a sphere with density ϱ_{Au} (58.01 atoms nm^{-3}) covered with an outermost layer of hexagonally close-packed gold atoms (13.89 atoms nm^{-2}) with a radius of $R_{core} - R_{Au}$ (R_{Au} = 0.145 nm), the model predicts that the core of H$_2$PC11MPC contains 280 Au atoms, of which 143 lie on the Au surface [99]. Given the values for elemental analysis of H$_2$PC11MPC (H, 4.88%; C, 44.77%; N, 3.10%), there are 57 porphyrin–alkanethiolate chains on the gold surface for H$_2$PC11MPC. These nanoparticles form complexes with fullerene molecules and they are clustered in an acetonitrile–toluene mixed solvent. The structure of porphyrin–alkanethiol (n = 15) is estimated by molecular mechanics calculation [100]. Since there are 57 porphyrin–alkanethiolate chains on the gold surface, the C$_{60}$ structure with diameter 7.1 Å is used for the calculation of H$_2$PC15MPC with 21 Å diameter [100]. The closest distance between a carbon of C$_{60}$ and the center of the porphyrin ring has been reported as 2.856 Å from the X-ray crystal structure of the C$_{60}$ complex with a jaw-like bis-porphyrin [86]. The smallest center-to-center distance of two porphyrin rings which can accommodate C$_{60}$ between the rings is estimated as 12.8 Å by adding the diameter of C$_{60}$ (7.1 Å) to twice the closest distance between a carbon of C$_{60}$ and the center of the porphyrin ring (5.7 Å). Thus, the distance

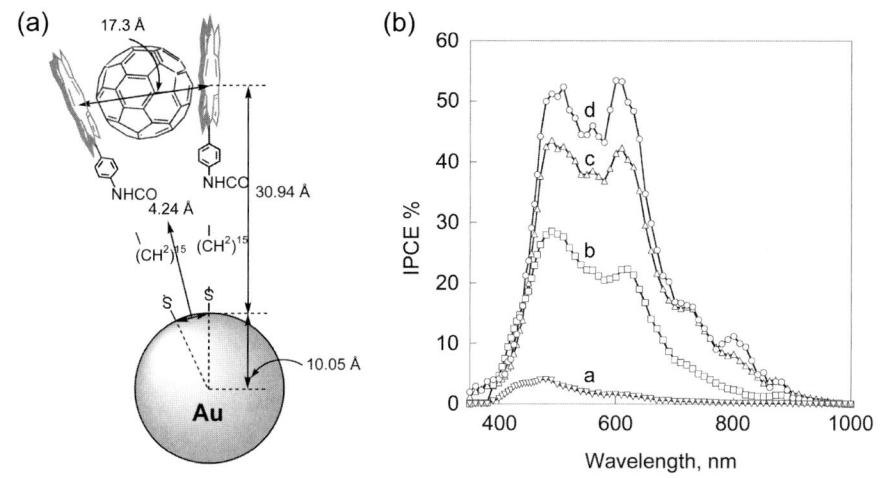

13.8 Organic Solar Cells Composed of Multi-porphyrin/C_{60} Supramolecular Assemblies

Fig. 13.28 (a) Insertion of C_{60} between the porphyrin rings of H_2PC15MPC [100]. (b) Photocurrent action spectra of OTE/SnO$_2$/(H_2PCnMPC + C_{60})$_m$ electrode. [H_2P] = 0.19 mM; (a) n = 5, [C_{60}] = 0.31 mM; (b) n = 11, [C_{60}] = 0.31 mM; (c) n = 15, [C_{60}] = 0.31 mM; (d) n = 15, [C_{60}] = 0.38 mM [100]. Electrolyte: 0.5 M NaI and 0.01 M I$_2$ in acetonitrile.

between two porphyrins in H$_2$PC15MPC is estimated as 16.6 Å, as shown in Fig. 13.28a, where the schematic structure of H$_2$PC15MPC is shown. The estimated distances between two porphyrins in H$_2$PC15MPC (16.6 Å) is long enough for the two porphyrins to accommodate C$_{60}$ between the two rings.

Figure 13.28b shows the effect of the alkanethiolate chain length on the IPCE values [100]. The action spectra indicate that a higher IPCE and a broader photoresponse are attained with the longer chain length of H$_2$PCnMPC. In particular, OTE/SnO$_2$/(H$_2$PC15MPC + C$_{60}$)$_m$+([H$_2$P] = 0.19 mM, [C$_{60}$] = 0.38 mM) exhibits a maximum IPCE value (54%) and a very broad photoresponse (up to ~1000 nm) which extends to the near-IR region. In OTE/SnO$_2$/(H$_2$PC15MPC + C$_{60}$)$_m$, a long methylene spacer of H$_2$PC15MPC allows enough space for fullerene molecules to insert them between neighboring two porphyrin rings effectively as compared with the clusters with a shorter methylene spacer, leading to more efficient photocurrent generation. The OTE/SnO$_2$/(H$_2$PC15MPC + C$_{60}$)$_m$ system has a much larger fill factor (FF) of 0.43, open-circuit voltage (V_{oc}) of 380 mV, short-circuit current density (I_{sc}) of 1.0 m cm^{-2} and an overall power conversion efficiency (η) of 1.5% at an input power (W_{in}) of 11.2 mW cm^{-2} as compared with the reference systems [OTE/SnO$_2$/(H$_2$P-ref + C$_{60}$)$_m$] [100]. An increase in the alkyl chain length (n) of porphyrin–alkanethiol from 11 to 15 in the OTE/SnO$_2$/(H$_2$P15MPC + C$_{60}$)$_m$ electrode results in a large improvement in the power conversion efficiency compared with OTE/SnO$_2$/(H$_2$PC11MPC + C$_{60}$)$_m$ (η = 0.61%) [100]. The power conversion efficiency (1.5%) of theOTE/SnO$_2$/(H$_2$PC15MPC + C$_{60}$)$_m$ system is remarkably enhanced (about 45-fold) in comparison with the OTE/SnO$_2$/(H$_2$P-ref + C$_{60}$)$_m$

system (η = 0.035%) under the same experimental conditions [100]. Such a large enhancement in the photoelectrochemical performance and the broader photoresponse in the visible and infrared region demonstrates that the present supramolecular approach for combination of porphyrin gold nanoparticles and fullerenes via π–π interaction provides a novel perspective for the development of efficient organic solar cells.

13.9
Conclusion

Functional π-compounds such as porphyrins, fullerenes and other π-chromophores exhibit excellent electron-transfer properties. The small reorganization energies of electron transfer of these functional π-compounds are essential for the design of artificial photosynthetic systems composed of functional π-compounds with fast charge separation but extremely slow charge recombination, which can be successfully applied to construct efficient light energy conversion systems such as photovoltaic devices. There are two essential factors in efficient photocurrent generation: fast charge separation between π-donors and π-acceptors and the resulting hole and electron transport in the thin film. Three-dimensional steric control in the supramolecular π-complexes between porphyrins and fullerenes particularly contributes to both the efficient formation of charge separation and the hole and electron transport in the thin film. Efficient self-exchange electron transfer between porphyrins and the corresponding radical cations [16, 17, 101, 102] and also between fullerenes and the radical anions [31, 103] has been well established. Such fast self-exchange electron transfer of porphyrins and fullerenes in the supramolecular clusters with an interpenetrating network in the thin film results in efficient hopping of holes and electrons in each network. The further development of functional π-compounds and their π–π interaction will certainly provide promising perspectives for the development of efficient light energy conversion systems.

Acknowledgments

The author gratefully acknowledges the contributions of his collaborators mentioned in the references. The author thanks the Ministry of Education, Culture, Sports, Science and Technology, Japan, for continuous support.

References

1. Blankenship, R. E., Madigan, M. T., Bauer, C. E. (Eds.), *Anoxygenic Photosynthetic Bacteria*, Kluwer, Dordrecht, **1995**.
2. Deisenhofer, J., Norris, J. R. (Eds.), *The Photosynthetic Reaction Center*, Academic Press, San Diego, **1993**.
3. Mcdermott, G., Prince, S. M., Freer, A. A., Hawthornthwaite-Lawless, A. M., Papiz, M. Z., Cogdell, R. J., Isaacs, N. W. *Nature* **1995**, *374*, 517–521.
4. Ermler, U., Fritzsch, G., Buchanan, S. K., Michel, H. *Structure* **1994**, *2*, 925.
5. Deisenhofer, J., Michel, H. *Science* **1989**, *245*, 1463.
6. Holeten, D., Kirmaier, C. *Photosynth. Res.* **1987**, *13*, 225.
7. Hoff, A. *Photochem. Photobiol.* **1986**, *43*, 727.
8. (a) Marcus, R. A. *Annu. Rev. Phys. Chem.* **1964**, *15*, 155; (b) Marcus, R. A., Sutin, N. *Biochim. Biophys. Acta* **1985**, *811*, 265; (c) Marcus, R. A. *Angew. Chem. Int. Ed. Engl.* **1993**, *32*, 1111.
9. Fukuzumi, S., in *The Porphyrin Handbook*, Kadish, K. M., Smith, K. M., Guilard, R. (Eds.), Academic Press, San Diego, **2000**, Vol. 8, pp. 115–151.
10. Kroto, H. W., Heath, J. R., O'Brien, S. C., Curl, R. F., Smalley, R. E. *Nature* **1985**, *318*, 162.
11. Echegoyen, L., Echegoyen, L. E. *Acc. Chem. Res.* **1998**, *31*, 593.
12. Guldi, D. M., Fukuzumi, S., in *Fullerenes: from Synthesis to Optoelectronic Properties*, Guldi, D. M., Martin, N. (Eds.), Kluwer, Dordrecht, **2003**, pp. 237–265.
13. Ward, R. L., Weissman, S. I. *J. Am. Chem. Soc.* **1957**, *79*, 2086–2090.
14. (a) Chang, R. *J. Chem. Educ.* **1970**, *47*, 563–568; (b) Cheng, K. S., Hirota, N., in *Investigation of Rates and Mechanisms of Reactions*, Hammes, G. G. (Ed.), Wiley-Interscience, New York, **1974**, Vol. VI, p. 565.
15. (a) Eberson, L. *Adv. Phys. Org. Chem.* **1982**, *18*, 79–185; (b) Eberson, L. *Electron Transfer Reactions in Organic Chemistry*, Springer, Berlin, **1987**.
16. Fukuzumi, S., Endo, Y., Imahori, H. *J. Am. Chem. Soc.* **2002**, *124*, 10974–10975.
17. Fukuzumi, S., Hasobe, T., Endo, Y., Ohkubo, K., Imahori, H. *J. Porphyrins Phthalocyanines* **2003**, *7*, 328–336.
18. Fukuzumi, S., Ohkubo, K., Imahori, H., Guldi, D. M. *Chem. Eur. J.* **2003**, *9*, 1585.
19. (a) Miller, J. R., Calcaterra, L. T., Closs, G. L. *J. Am. Chem. Soc.* **1984**, *106*, 3047; (b) Closs, G. L., Miller, J. R. *Science* **1988**, *240*, 440; (c) Gould, I. R., Farid, S. *Acc. Chem. Res.* **1996**, *29*, 522; (d) McLendon, G. *Acc. Chem. Res.* **1988**, *21*, 160; (e) Winkler, J. R., Gray, H. B. *Chem. Rev.* **1992**, *92*, 369; (f) McLendon, G., Hake, R. *Chem. Rev.* **1992**, *92*, 481.
20. (a) Mataga, N., Miyasaka, H., in *Electron Transfer from Isolated Molecules to Biomolecules, Part 2*, Jortner, J., Bixon, M. (Eds.), Wiley, New York, 1999, p. 431; (b) Mataga, N., Chosrowjan, H., Shibata, Y., Yoshida, N., Osuka, A., Kikuzawa, T., Okada, T. *J. Am. Chem. Soc.* **2001**, *123*, 12422; (c) Osuka, A., Noya, G., Taniguchi, S., Okada, T., Nishimura, Y., Yamazaki, I., Mataga, N. *Chem. Eur. J.* **2000**, *6*, 33
21. (a) Wasielewski, M. R., in *Photoinduced Electron Transfer, Part A*, Fox, M. A., Chanon, M. (Eds.), Elsevier, Amsterdam, **1988**, p. 161; (b) Wasielewski, M. R. *Chem. Rev.* **1992**, *92*, 435; (c) Jordan, K. D., Paddon-Row, M. N. *Chem. Rev.* **1992**, *92*, 395.
22. (a) Rehm, D., Weller, A. *Ber. Bunsenges Phys. Chem.* **1969**, *73*, 834; (b) Rehm, D., Weller, A. *Isr. J. Chem.* **1970**, *8*, 259.
23. Kavarnos, G. J. *Fundamentals of Photoinduced Electron Transfer*, Wiley-VCH, New York, **1993**.
24. (a) Bock, C. R., Meyer, T. J., Whitten, D. G. *J. Am. Chem. Soc.* **1975**, *97*, 2909; (b) Ballardini, R., Varani, G., Indelli, M. T., Scandola, F., Balzani, V. *J. Am. Chem. Soc.* **1978**, *100*, 7219; (c) Fukuzumi, S., Kuroda, S., Tanaka, T. *J. Am. Chem. Soc.* **1985**, *107*, 3020; (d) Fukuzumi, S., Koumitsu, S., Hironaka, K., Tanaka, T. *J. Am. Chem. Soc.* **1987**, *109*, 305; (e) Jayanthi, S. S., Ramamurthy, P. *J. Phys. Chem. A* **1997**, *101*, 2016.
25. (a) Weller, A., Zacharaasse, K. *Chem. Phys. Lett.* **1971**, *10*, 590; (b) Suppan, P. *Top. Curr. Chem.* **1992**, *163*, 95; (c) Efri-

ma, S., Bixon, M. *Chem. Phys. Lett.* **1974**, *25*, 34; (d) Brunschwig, B. S., Ehrenson, S., Sutin, N. *J. Am. Chem. Soc.* **1984**, *106*, 6858; (e) Barzykin, A. V., Frantsuzov, P. A., Seki, K., Tachiya, M. *Adv. Chem. Phys.* **2002**, *123*, 511.

26. Todd, W. P., Dinnocenzo, J. P., Farid, S., Goodman, J. L., Gould, I. R. *J. Am. Chem. Soc.* **1991**, *113*, 3601.

27. Fukuzumi, S., Ohkubo, K., Suenobu, T., Kato, K., Fujitsuka, M., Ito, O. *J. Am. Chem. Soc.* **2001**, *123*, 8459.

28. Steren, C. A., van Willigen, H., Biczók, L., Gupta, N., Linschitz, H. *J. Phys. Chem.* **1996**, *100*, 8920.

29. Fukuzumi, S., Suenobu, T., Patz, M., Hirasaka, T., Itoh, S., Fujitsuka, M., Ito, O. *J. Am. Chem. Soc.* **1998**, *120*, 8060.

30. Fukuzumi, S., Suenobu, T., Hirasaka, T., Sakurada, N., Arakawa. R., Fujitsuka, M. Ito, O. *J. Phys. Chem. A* **1999**, *103*, 5935.

31. Fukuzumi, S., Nakanishi, I., Suenobu, T., Kadish, K. M. *J. Am. Chem. Soc.* **1999**, *121*, 3468.

32. Guldi, D. M., Neta, P., Asmus, K.-D. *J. Phys. Chem.* **1994**, *98*, 4617.

33. Fukuzumi, S., Suenobu, T., Hirasaka, T., Arakawa, R., Kadish, K. M. *J. Am. Chem. Soc.* **1998**, *120*, 9220.

34. Boudon, C., Gisselbrecht, J.-P., Gross, M., Herrmann, A., Rüttimann, M., Crassous, J., Cardullo, F., Echegoyen, L., Diederich, F. *J. Am. Chem. Soc.* **1998**, *120*, 7860.

35. (a) Gust, D., Moore, T. A., in *The Porphyrin Handbook*, Kadish, K. M., Smith, K. M., Guilard, R. (Eds.), Academic Press, San Diego, **2000**, Vol. 8, pp. 153–190; (b) Gust, D., Moore, T. A., Moore, A. L. *Acc. Chem. Res.* **2001**, *34*, 40; (c) Paddon-Row, M. N. *Acc. Chem. Res.* **1994**, *27*, 18.

36. Fukuzumi, S., Guldi, D. M., in *Electron Transfer in Chemistry*, Balzani, V. (Ed.), Wiley-VCH, Weinheim, **2001**, Vol. 2, pp. 270–337.

37. Fukuzumi, S., Imahori, H., in *Electron Transfer in Chemistry*, Balzani, V. (Ed.), Wiley-VCH, Weinheim, **2001**, Vol. 4, pp. 927–975.

38. Guldi, D. M., Kamat, P. V., in *Fullerenes, Chemistry, Physics and Technology*, Kadish, K. M., Ruoff, R. S. (Eds.), Wiley-Interscience, New York, **2000**, pp. 225–281.

39. Fukuzumi, S., Ohkubo, K., Imahori, H., Shao, J., Ou, Z., Zheng, G., Chen, Y., Pandey, R. K., Fujitsuka, M., Ito, O., Kadish, K. M. *J. Am. Chem. Soc.* **2001**, *123*, 10676.

40. Ohkubo, K., Imahori, H., Shao, J., Ou, Z., Kadish, K. M., Chen, Y., Zheng, G., Pandey, R. K., Fujitsuka, M., Ito, O., Fukuzumi, S. *J. Phys. Chem. A* **2002**, *106*, 10991.

41. Imahori, H., Tamaki, K., Guldi, D. M., Luo, C., Fujitsuka, M., Ito, O., Sakata, Y., Fukuzumi, S. *J. Am. Chem. Soc.* **2001**, *123*, 2607.

42. Ohkubo, K., Kotani, H., Shao, J., Ou, Z., Kadish, K. M., Li, G., Pandey, R. K., Fujitsuka, M., Ito, O., Imahori, H., Fukuzumi, S. *Angew. Chem. Int. Ed.* **2004**, *43*, 853.

43. Kashiwagi, Y., Ohkubo, K., McDonald, J. A., Blake, I. M., Crossley, M. J., Araki, Y., Ito, O., Imahori, H., Fukuzumi, S. *Org. Lett.* **2003**, *5*, 2719.

44. Fukuzumi, S., Ohkubo, K., Ou, W. E. Z., Shao, J., Kadish, K. M., Hutchison, J. A., Ghiggino, K. P., Sintic, P. J., Crossley, M. J. *J. Am. Chem. Soc.* **2003**, *125*, 14984.

45. (a) Johnson, D. G., Niemczyk, M. P., Minsek, D. W., Wiederrecht, G. P., Svec, W. A., Gaines, G. L., III, Wasielewski, M. R. *J. Am. Chem. Soc.* **1993**, *115*, 5692; (b) Sakata, Y., Tsue, H., O'Neil, M. P., Wiederrecht, G. P., Wasielewski, M. R. *J. Am. Chem. Soc.* **1994**, *116*, 6904 (1994); (c) Tsue, H., Imahori, H., Kaneda, T., Tanaka, Y., Okada, T., Tamaki, K., Sakata, Y. *J. Am. Chem. Soc.* **2000**, *122*, 2279; (d) Kang, Y. K., Rubtsov, I. V., Iovine, P. M., Chen, J., Therien, M. J. *J. Am. Chem. Soc.* **2002**, *124*, 8275.

46. (a) Asahi, T., Ohkohchi, M., Matsusaka, R., Mataga, N., Zhang, R. P., Osuka, A., Maruyama, K. *J. Am. Chem. Soc.* **1993**, *113*, 5665; (b) Khundkar, L. R., Perry, J. W., Hanson, J. E., Dervan, P. B. *J. Am. Chem. Soc.* **1994**, *116*, 9700; (c) Sessler, J. L., Wang, B., Harriman, A. *J. Am. Chem. Soc.* **1993**, *115*, 10418.

47. In liquid crystals, the CR rates were attenuated into the submicrosecond timescale; see (a) Berman, A., Izraeli, E. S.,

Levanon, H., Wang, B., Sessler, J. L. *J. Am. Chem. Soc.* **1995**, *117*, 8252; (b) Berg, A., Shuali, Z., Levanon, H., Wiehe, A., Kurreck, H. *J. Phys. Chem. A* **2001**, *105*, 10060.
48. Imahori, H., Yamada, H., Guldi, D. M., Endo, Y., Shimomura, A., Kundu, S., Yamada, K., Okada, T., Sakata, Y., Fukuzumi, S. *Angew. Chem. Int. Ed.* **2002**, *41*, 2344.
49. Okamoto, K., Fukuzumi, S. *J. Phys. Chem. B* **2005**, *109*, 7713.
50. Imahori, H., Guldi, D. M., Tamaki, K., Yoshida, Y., Luo, C., Sakata, Y., Fukuzumi, S. *J. Am. Chem. Soc.* **2001**, *123*, 6617.
51. Imahori, H., Sekiguchi, Y., Kashiwagi, Y., Sato, T., Araki, Y., Ito, O., Yamada, H., Fukuzumi, S. *Chem. Eur. J.* **2004**, *10*, 3184.
52. Kikuchi, K., Sato, C., Watabe, M., Ikeda, H., Takahashi, Y., Miyashi, T. *J. Am. Chem. Soc.* **1993**, *115*, 5180.
53. Ohkubo, K., Suga, K., Morikawa, K., Fukuzumi, S. *J. Am. Chem. Soc.* **2003**, *125*, 12850.
54. Fukuzumi, S. Kotani, H., Ohkubo, K., Ogo, S., Tkachenko, N. V., Lemmetyinen, H. *J. Am. Chem. Soc.* **2004**, *126*, 1600.
55. Harriman, A. *Angew. Chem. Int. Ed.* **2004**, *43*, 4985.
56. Benniston, A. C., Harriman, A., Li, P., Rostron, J. P., Verhoeven, J. W. *Chem. Commun.* **2005**, 2701.
57. Guo, X., Gan, Z., Luo, H., Araki, Y., Zhang, D., Zhu, D., Ito, O. *J. Phys. Chem. A* **2003**, *107*, 9747.
58. Okamoto, K., Mori, Y., Yamada, H., Imahori, H., Fukuzumi, S. *Chem. Eur. J.* **2004**, *10*, 474.
59. Kotani, H., Ohkubo, K., Fukuzumi, S. *Chem. Commun.* **2005**, 4520.
60. Shida, T., *Electronic Absorption Spectra of Radical Ions*, Elsevier, Amsterdam, **1988**.
61. Kotani, H., Ohkubo, K., Fukuzumi, S. *J. Am. Chem. Soc.* **2004**, *126*, 15999.
62. Fukuzumi, S., Nakanishi, I., Tanaka, K. *J. Phys. Chem. A* **1999**, *103*, 11212.
63. Ohkubo, K., Nanjo, T., Fukuzumi, S. *Org. Lett.* **2005**, *7*, 4265.
64. (a) Schuster, G. B. *Acc. Chem. Res.* **1979**, *12*, 366; (b) Vysotski, E. S., Lee, J. *Acc. Chem. Res.* **2004**, *37*, 405.
65. Isobe, H., Takano, Y., Okumura, M., Kuramitsu, S., Yamaguchi, K. *J. Am. Chem. Soc.* **2005**, *127*, 8667.
66. Foote, C. S., Valentine, J. S., Greenberg, A., Liebman, J. F. (Eds.), *Active Oxygen in Chemistry*, Blackie, New York, **1995**, Vol. 2.
67. Adam, W., in *Four-membered Ring Peroxides: 1,2-Dioxetanes and -Peroxylactones. The Chemistry of Peroxides*, Patai, S. (Ed.), Wiley, New York, **1983**, Chapter 24, pp. 829–920.
68. Wilson, T, Schaap, A. P. *J. Am. Chem. Soc.* **1971**, *93*, 4126.
69. Frimer, A. A. *Singlet Oxygen, Volume 2: Reaction Modes and Products, Part I*, CRC Press, Boca Raton, FL, **1985**.
70. Burns, P. A., Foote, C. S. *J. Am. Chem. Soc.* **1974**, *96*, 4339.
71. Rio, G., Berthelot, J. *Bull. Chem. Soc. Fr.* **1969**, 3609.
72. Nelsen, S. F., Kapp, D. L., Gerson, F., Lopez, J. *J. Am. Chem. Soc.* **1986**, *108*, 1027.
73. (a) Hagfeldt, A., Grätzel, M. *Acc. Chem. Res.* **2000**, *33*, 269; (b) Grätzel, M. *Nature* **2001**, *414*, 338; (c) Bignozzi, C. A., Argazzi, R., Kleverlaan, C. J. *Chem. Soc. Rev.* **2000**, *29*, 87; (d) Hagfeldt, A., Grätzel, M. *Chem. Rev.* **1995**, *95*, 49.
74. (a) O'Regan, B., Grätzel, M. *Nature* **1991**, *353*, 737; (b) Bach, U., Lupo, D., Comte, P., Moser, J. E., Weissörtel, F., Salbeck, J., Spreitzer, H., Grätzel, M. *Nature* **1998**, *395*, 583.
75. Shah, A., Torres, P., Tscharner, R., Wyrsch, N., Keppner, H. *Science* **1999**, *285*, 692.
76. (a) Granström, M., Petrisch, K., Arias, A. C., Lux, A., Andersson, M. R., Friend, R. H. *Nature* **1998**, *395*, 257; (b) Halls, J. J. M., Walsh, C. A., Greenham, N. C., Marseglia, E. A., Friend, R. H., Moratti, S. C., Holmes, A. B. *Nature* **1995**, *376*, 498.
77. Hasobe, T., Hattori, S., Kotani, H., Ohkubo, K., Hosomizu, K., Imahori, H., Kamat, P. V., Fukuzumi, S. *Org. Lett.* **2004**, *6*, 3103.
78. Hasobe, T., Imahori, H., Fukuzumi, S., Kamat, P. V. *J. Phys. Chem. B* **2003**, *107*, 12105.
79. Lawson, D. R., Feldheim, D. L., Foss, C. A., Dorhout, P. K., Elliot, C. M., Martin,

C. R., Parkinson, B. *J. Electrochem. Soc.* **1992**, *139*, L68.
80. Hasobe, T., Hattori, S., Kamat, P. V., Wada, Y., Fukuzumi, S. *J. Mater. Chem.* **2005**, *15*, 372.
81. (a) McDermott, G., Prince, S. M., Freer, A. A., Hawthornthwaite-Lawless, A. M., Papiz, M. Z., Cogdell, R. J., Isaacs, N. W. *Nature* **1995**, *374*, 517; (b) Koepke, J., Hu, X., Muenke, C., Schulten, K., Michel, H. *Structure* **1996**, *4*, 581.
82. (a) Bar-Haim, A., Klafter, J., Kopelman, R. *J. Am. Chem. Soc.* **1997**, *119*, 6197; (b) Jiang, D.-L., Aida, T. *J. Am. Chem. Soc.* **1998**, *120*, 10895; (c) Choi, M.-S., Aida, T., Yamazaki, T., Yamazaki, I. *Chem. Eur. J.* **2002**, *8*, 2668.
83. (a) Gilat, S. L., Adronov, A., Fréchet, J. M. J. *Angew. Chem. Int. Ed.* **1999**, *38*, 1422; (c) Yeow, E. K. L., Ghiggino, K. P., Reek, J. N. H., Crossley, M. J., Bosman, A. W., Schenning, A. P. H. J., Meijer, E. W. *J. Phys. Chem. B* **2000**, *104*, 2596.
84. Diederich, F., Gómez-López, M. *Chem. Soc. Rev.* **1999**, *28*, 263.
85. Boyd, P. D. W., Hodgson, M. C., Rickard, C. E. F., Oliver, A. G., Chaker, L., Brothers, P. J., Bolskar, R. D., Tham, F. S., Reed, C. A. *J. Am. Chem. Soc.* **1999**, *121*, 10487.
86. (a) Sun, D., Tham, F. S., Reed, C. A., Chaker, L., Burgess, M., Boyd, P. D. W. *J. Am. Chem. Soc.* **2000**, *122*, 10704; (b) Sun, D., Tham, F. S., Reed, C. A., Chaker, L., Boyd, P. D. W. *J. Am. Chem. Soc.* **2002**, *124*, 6604; (c) Sun, D., Tham, F. S., Reed, C. A., Boyd, P. D. W. *Proc. Natl. Acad. Sci. USA* **2002**, *99*, 5088.
87. (a) Olmstead, M. M., Costa, D. A., Maitra, K., Noll, B. C., Phillips, S. L., Van Calcar, P. M., Balch, A. L. *J. Am. Chem. Soc.* **1999**, *121*, 7090; (b) Olmstead, M. M., de Bettencourt-Dias, A., Duchamp, J. C., Stevenson, S., Marciu, D., Dorn, H. C., Balch, A. L. *Angew. Chem. Int. Ed.* **2001**, *40*, 1223.
88. (a) Tashiro, K., Aida, T., Zheng, J.-Y., Kinbara, K., Saigo, K., Sakamoto, S., Yamaguchi, K. *J. Am. Chem. Soc.* **1999**, *121*, 9477; (b) Zheng, J.-Y., Tashiro, K., Hirabayashi, Y., Kinbara, K., Saigo, K., Aida, T., Sakamoto, S., Yamaguchi, K. *Angew. Chem. Int. Ed.* **2001**, *40*, 1857.

89. Guldi, D. M., Luo, C., Prato, M., Troisi, A., Zerbetto, F., Scheloske, M., Dietel, E., Bauer, W., Hirsch, A. *J. Am. Chem. Soc.* **2001**, *123*, 9166.
90. Wang, Y.-B., Lin, Z. *J. Am. Chem. Soc.* **2003**, *125*, 6072.
91. (a) Hasobe, T., Kashiwagi, Y., Absalom, M. A., Hosomizu, K., Crossley, M. J., Imahori, H., Kamat, P. V., Fukuzumi, S. *Adv. Mater.* **2004**, *16*, 975; (b) Hasobe, T., Kamat, P. V., Absalom, M. A., Kashiwagi, Y., Sly, J., Crossley, M. J., Hosomizu, K., Imahori, H., Fukuzumi, S. *J. Phys. Chem. B* **2004**, *108*, 12865.
92. Solladié, N., Hamel, A., Gross, M. *Tetrahedron Lett.* **2000**, *41*, 6075.
93. Hasobe, T., Kamat, P. V., Troiani, V., Solladié, N., Ahn, T. K., Kim, S. K., Kim, D., Kongkanand, A., Kuwabata, S., Fukuzumi, S. *J. Phys. Chem. B* **2005**, *109*, 19.
94. Kamat, P. V., Barazzouk, S., Hotchandani, S. Thomas, K. G. *Chem. Eur. J.* **2000**, *6*, 3914–3921.
95. (a) Imahori, H., Arimura, M., Hanada, T., Nishimura, Y., Yamazaki, I., Sakata, Y., Fukuzumi, S. *J. Am. Chem. Soc.* **2001**, *123*, 335–336.
96. Imahori, H., Kashiwagi, Y., Endo, Y., Hanada, T., Nishimura, Y., Yamazaki, I., Araki, Y., Ito, O., Fukuzumi, S. *Langmuir* **2004**, *20*, 73–81.
97. Imahori, H., Kashiwagi, Y., Hanada, T., Endo, Y., Nishimura, Y., Yamazaki, I., Fukuzumi, S. *J. Mater. Chem.* **2003**, *13*, 2890–2898.
98. Fukuzumi, S., Endo, Y., Kashiwagi, Y., Araki, Y., Ito, O., Imahori, H. *J. Phys. Chem. B* **2003**, *107*, 11979–11986.
99. Hasobe, T., Imahori, H., Kamat, P. V., Fukuzumi, S. *J. Am. Chem. Soc.* **2003**, *125*, 14962.
100. Hasobe, T., Imahori, H., Kamat, P. V., Ahn, T. K., Kim, D., Hanada, T., Hirakawa, T., Fukuzumi, S. *J. Am. Chem. Soc.* **2005**, *127*, 1217.
101. Crnogorac, M. M., Kostic, N. M. *Inorg. Chem.* **2000**, *39*, 5028–5035.
102. Sun, H., Smirnov, V. V., DiMagno, S. G. *Inorg. Chem.* **2003**, *42*, 6032–6040.
103. Thomas, K. G., Biju, V., Guldi, D. M., Kamat, P. V., George, M. V. *J. Phys. Chem. B* **1999**, *103*, 8864–8869.

14
Induced π-Stacking in Acenes
John E. Anthony

14.1
Introduction

Organic electronics is a rapidly developing field that is already contributing to some commercial applications [1]. Over the last decade, detailed device studies have been performed on commercially available organic materials (mostly aromatic compounds) [2], and numerous new aromatic systems are being designed to address the shortcomings of commercial molecules in these applications [3]. One of the most thoroughly studied properties of aromatic crystals and films is charge transport, a research area first investigated in detail more than 50 years ago [4], and which has undergone a resurgence in the late 1990s owing to the application of crystalline organic materials to field-effect transistors (FETs) [5]. The performance of organic FETs relies on the efficient transport of charge through thin films or single crystals of organic materials and is governed by molecular nearest-neighbor electronic coupling. The strongest form of intermolecular coupling arises from interactions between the π-electron clouds of aromatic molecules, which are maximized when molecules adopt a face-to-face orientation. The precise nature of the cofacial alignment was shown by a simple theoretical treatment in the 1990s to be a determining factor in the crystallochromy of perylene dyes [6]. More recently, higher level theory has been applied to this issue by Brédas and coworkers, showing the intimate relationship between electronic coupling and the precise nature of π-face interaction [7].

While the strong electronic coupling necessary for efficient device performance arises from π-stacking interactions between aromatic molecules, for the majority of aromatic compounds the most common interaction is between the electron-rich π-cloud and the relatively electropositive peripheral hydrogens of neighboring rings (shown for benzene in Fig. 14.1). Studies of the crystal structures of several dozen compounds by Gavezzotti and coworkers led to the rule-of-thumb that the carbon to hydrogen (C:H) ratio determines whether a material will adopt an edge-to-face or face-to-face arrangement in the solid state: the larger the carbon content, the more likely it is that a π-stacking arrangement will be observed in

Functional Organic Materials. Syntheses, Strategies, and Applications.
Edited by Thomas J.J. Müller and Uwe H.F. Bunz
Copyright © 2007 WILEY-VCH Verlag GmbH & Co. KGaA, Weinheim
ISBN: 978-3-527-31302-0

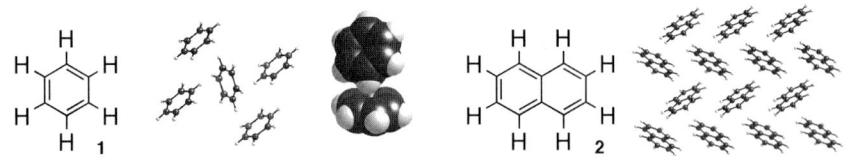

Fig. 14.1 Structures and crystalline order for benzene (**1**) and naphthalene (**2**) showing predominant edge-to-face interactions.

the crystal [8]. An alternative description of this phenomenon involves the aspect ratio of the molecule: square or circular molecules (which, owing to their shape, tend to have high C:H ratios) tend to pack in a columnar arrangement with significant π-overlap, whereas extended, rectangular-shaped molecules emphasize edge-to-face interactions in the crystal [8a].

A number of detailed studies of π-stacking interactions have been performed, but few have focused on the importance of these interactions to electronic device performance [9]. This work focuses very specifically on methods to induce π-face interactions in the solid state of linearly fused acenes. This class of aromatic compounds has the most promising electronic properties, providing excellent device performance in both FETs and organic solar cells [10], yet the packing in these molecules is dominated by edge-to-face interactions; all unfunctionalized acenes pack in a two-dimensional edge-to-face arrangement commonly referred to as a "herringbone" motif (as shown in Fig. 14.1 for naphthalene) [11]. This chapter focuses on the elucidation of functionalization strategies for the enhancement of π-face interactions in the solid state of acenes. Because the main interest in acenes lies with their electronic properties, this chapter will necessarily avoid discussion of benzannelated acenes: while these altered acenes typically do have a C:H ratio leading to improved π-face interactions, the electronic properties of the materials also change significantly upon annulation, typically undergoing marked increase in HOMO–LUMO gap, change in redox properties and increase in HOMO energy level. The functionalization approaches described in this chapter are limited those that allow the aromatic compounds maintain the unique character that qualifies them as acenes.

14.2
Anthracene

Anthracene is the smallest linearly-fused aromatic species considered to exhibit true "acene" character (Fig. 14.2) and has been the smallest acene typically studied for electronic materials applications [12]. Single-crystal anthracene does exhibit relatively high intrinsic hole mobility, with values as high as 3 cm^2 V^{-1} s^{-1} measured by the time-of-flight technique and as high as 0.02 cm^2 V^{-1} s^{-1} measured from a FET constructed on a single crystal [13]. Films of functionalized anthracene derivatives have also yielded FETs with reasonable hole mobilities (see

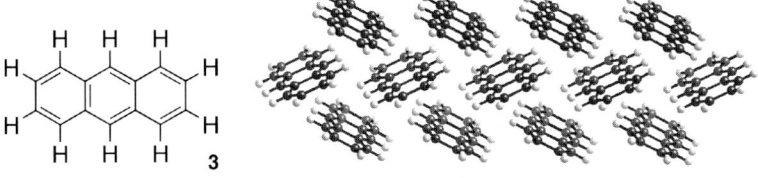

Fig. 14.2 Anthracene and its native solid-state order, showing edge-to-face interactions.

Fig. 14.3 Anthracene derivatives for use in OLEDs.

below). However, most uses of this compound in devices relate to its fluorescence properties, which are generally quenched by π-face interactions.

Anthracene and 9,10-diphenylanthracene are frequently used as reference compounds owing to their high fluorescence quantum yields {27% for anthracene (ethanol) [14], 90% for 9,10-diphenylanthracene (DPA) (cyclohexane) [15]}. Most functional groups added to anthracene typically induce a sufficient red shift in emission to yield saturated blue emission and such compounds are often exploited as blue emitters in organic light-emitting diodes (OLEDs) [16]. In these cases, elucidating functionality that disrupts π-face interactions (which further shifts and broadens the emission and often leads to a decrease in fluorescence efficiency) and yields amorphous materials (which alleviate crystallization that leads to pinholes and device defects) has been the focus of most investigations. These goals are typically simultaneously accomplished by the addition of large aromatic groups (highly substituted benzene, naphthalene, etc.) which contribute only slightly to the conjugated pathway in anthracene, shield the anthracene π-face from stacking interactions in the solid state and maintain the high fluorescence quantum yield seen in the diphenyl derivative [16, 17]. An alternative approach has been to add large diarylphosphine units to the 9,10-positions of anthracene, leading to voids in the crystal that can include solvents (Fig. 14.3). One such derivative has been postulated as a fluorescent sensor for toluene [18].

The Cambridge Crystallographic Database (CCDB, http://www.ccdc.cam.ac.uk/) holds approximately 1200 examples of structures containing the anthracene chromophore, although if charge-transfer complexes, benzannelated systems and duplicate spectra are discounted the total number of structures drops to only a few hundred. Of those, very few show anthracene in π-stacked arrangements. In gen-

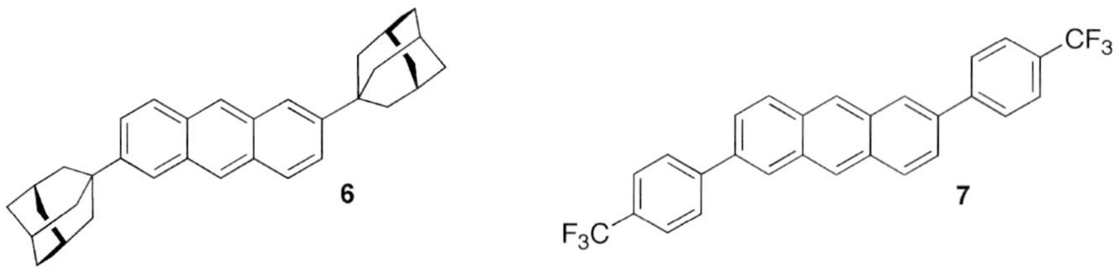

Fig. 14.4 2,6-Bis(adamantyl)anthracene (**6**) and 2,8-bis(4-trifluoromethylphenyl)anthracene (**7**).

eral, anthracenes substituted on the ends of the molecule (2,3,6- and/or 7-positions) pack in a herringbone arrangement, since substituents at these positions merely serve to exaggerate the aspect ratio of the molecule, further increasing the propensity to adopt edge-to-face arrangements (Fig. 14.4). Two examples of this effect are seen in 2,6-bis(adamantyl)anthracene (**6**) [19] and 2,6-bis(4-trifluoromethylphenyl)anthracene (**7**) [20]. The latter compound, even though it adopts a herringbone arrangement in the solid, does exhibit a reasonable electron mobility of 3.4×10^{-3} cm^2 V^{-1} s^{-1} from a vacuum-deposited film in a top-contact FET.

Because of its relatively small aromatic surface, the crystalline order of anthracene is highly sensitive to functionalization: even small substituents provide enough steric bulk to alter the aspect ratio of the molecule, leading to crystalline order with significant π-overlap. There are a number of rational approaches to the enhancement of π-stacked arrays in anthracene (which can also be applied to higher acenes), most of which consist of methods to alter the aspect ratio of the molecule or to hinder the ability of the molecule to adopt edge-to-face arrangements in the crystal. Other approaches involve altering the polarity of the anthracene to improve π-stacking and the use of hydrogen bonds to direct the self-assembly of π-stacked arrays. These general approaches are discussed in specific detail below.

14.2.1
Monosubstituted Anthracene

The addition of a single substituent to the anthracene backbone serves to interrupt edge-to-face interactions only along one edge of the acene. Hence, for most substituents, this functionalization approach does not typically lead to long-range face-to-face interactions (Fig. 14.5). An elegant study of substituent effect versus crystal packing in monosubstituted anthracenes was carried out by Sweeting et al., in order to elucidate the relationship between crystal symmetry and triboluminescence [21]. The study of a dozen anthracene-9-carboxylate esters showed only variations in the herringbone motif in their crystal packings. In the absence of nonsteric driving forces, this class of compounds most often stacks pairwise and then adopts a herringbone arrangement (a motif often referred to as a "sandwich" herringbone, Fig. 14.5). Although there is pairwise π-face inter-

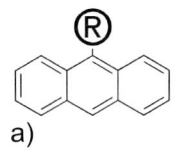

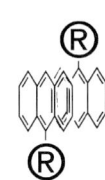

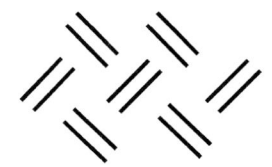

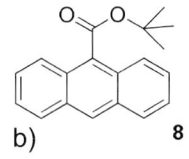

Fig. 14.5 9-substituted anthracenes typically adopt a sandwich herringbone arrangement in the crystal.

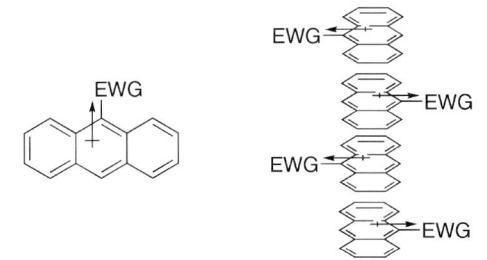

Fig. 14.6 The dipole induced by a strongly electron-withdrawing group can lead to columnar π-stacking.

action in these crystals, the absence of any long-range face-to-face or edge-to-face interaction makes such arrangements poorly suited to electronic applications.

The exception to the "single substituent" phenomenon arises when the functional group is strongly electron withdrawing. Such a substituent has two complementary effects for π-stacking in anthracene crystals. First, the substituent withdraws significant electron density from the acene π-cloud, diminishing the repulsion of the π-faces in the solid state – this effect has been observed to enhance π-stacking in a number of extended aromatic systems [22]. Second, the single electron-withdrawing substituent creates a significant dipole, causing molecules to align in order to cancel the dipole in the solid – the most efficient arrangement of molecules to accomplish this task is a π-stacked array (Fig. 14.6).

A number of anthracene derivatives have been prepared that take advantage of this approach. Two excellent examples are 9-nitroanthracene (**9**) [23] and anthryl trifluoromethyl ketone (**10**) [24]. Both of these molecules, with strongly electron-withdrawing groups, adopt a nearly columnar π-stacked arrangement in the crystal (Fig. 14.7). The separations between the π-faces of these stacked molecules (the parameter with the most dramatic impact on charge transport [7]) are 3.53 Å (**9**) and 3.56 Å (**10**).

More dramatic changes in aspect ratio leading to significant π-face interactions can be achieved by placing rigid substituents on the anthracene or substituents

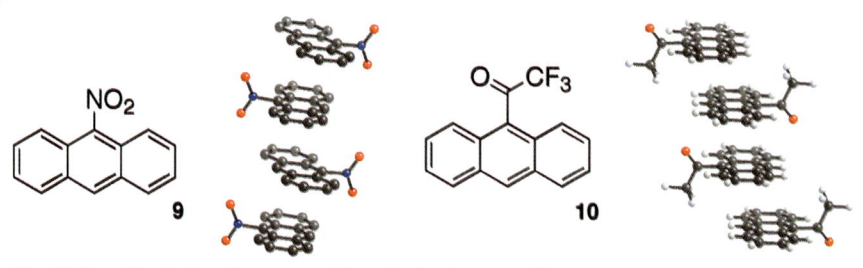

Fig. 14.7 π-Stacked anthracenes with a single electron-withdrawing group: 9-nitroanthracene (**9**) and 9-anthryl trifluoromethyl ketone (**10**).

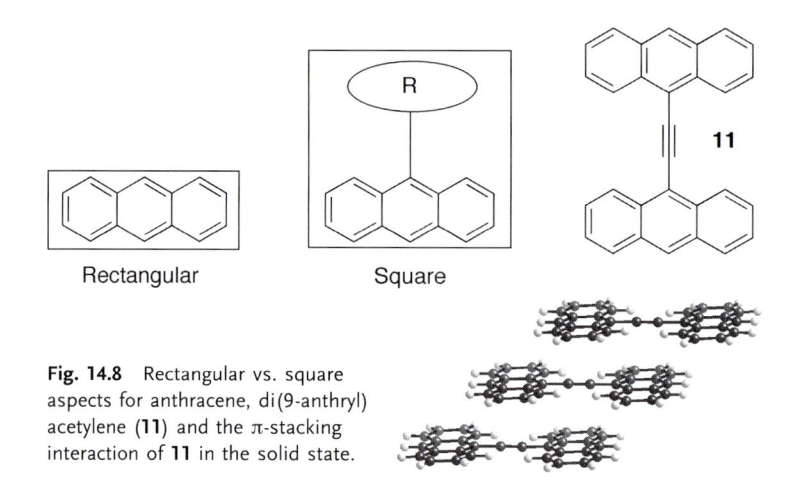

Fig. 14.8 Rectangular vs. square aspects for anthracene, di(9-anthryl)acetylene (**11**) and the π-stacking interaction of **11** in the solid state.

that cause the anthracenes to self-assemble in edge-to-edge pairs, forming a square, low aspect ratio complex. The simplest approach to the preparation of anthracene derivatives with a square perimeter, without the electronic disadvantages of benzannelation, is seen in dianthrylacetylene (**11**) (Fig. 14.8) [25]. This compound is a square-shaped molecule that adopts simple columnar arrangement in the solid state. In the crystal, the anthracene units are coplanar, with a spacing between the anthracene faces of 3.45 Å. The advantage to this functionalization approach is that the material retains many of the optical, emissive and electronic characteristics of the original anthracene chromophore.

An alternative to the use of covalent linkages to yield a square perimeter can be found in simple self-assembly processes (Fig. 14.9). Anthracene-9-carboxylic acid forms a rigid hydrogen-bonded dimer in the solid state, effectively changing the aspect ratio from rectangular (for the monomeric species) to square (for the dimer) [26]. This change leads to the formation of a columnar π-stacked array in the crystals of this molecule, similar to that observed for **11**. While the degree of π-overlap in this system is high, the spacing between the aromatic faces is also fairly large – nearly 3.6 Å. Addition of the carboxylic acid to the centermost carbon

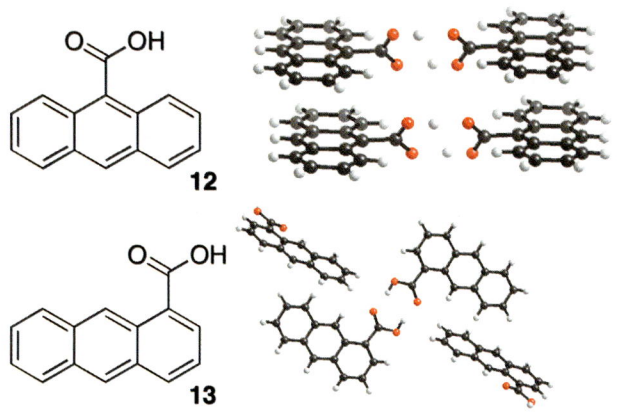

Fig. 14.9 Anthracene-9- and -1-carboxylic acids (**12** and **13**), showing the square self-assembled dimer of **12** leading to columnar π-stacking (left), whereas the "Z"-shaped dimer of **13** leads to strictly edge-to-face interactions.

on the anthracene edge is critical to the formation of this crystalline order – anthracene-1-carboxylic acid does show hydrogen-bonded dimer formation in the solid but, owing to the ability of the anthracene units to orient in opposite directions, this interaction actually increases the aspect ratio leading to herringbone arrangement in the crystal (Fig. 14.9, right) [27].

14.2.2
Disubstituted Anthracene

The addition of substituents to both the 9- and 10-positions of anthracene serves to disrupt the edge-to-face interactions of the molecule, preventing the adoption of a herringbone arrangement in the crystal. As long as the substituent groups are relatively small, the molecules will be able to adopt a π-stacked arrangement. 9,10-Dimethylanthracene (**14**) [28] and 9,10-bis(chloromethyl) anthracene (**15**) [29] are both excellent examples of the typical crystalline order of such species (Fig. 14.10).

As the size of the substituents on the central carbon atoms is increased, the longitudinal (long-axis) and lateral (short-axis) slip of the stacked molecules also increases, until a solid-state arrangement with coplanar anthracene rings that do not overlap is achieved. This arrangement is best exemplified by the well-known 9,10-diphenylanthracene (**16**) (Fig. 14.11) [30].

As the size of the substituent is increased further, the molecules typically revert to a herringbone arrangement, since the substituent is then able to fill sufficient space in the crystal to cover the π-face of the acene (e.g. **22**, see below).

Further enhancement of π-stacked order can be achieved by the use of small electron-withdrawing substituents. The 9,10-dihaloanthracenes (dichloro [31], dibromo [32] and diiodo [33]) all adopt a π-stacked arrangement in the solid, with

Fig. 14.10 9,10-Dimethyl-anthracene (**14**) and 9,10-bis(chloromethyl)anthracene (**15**), showing their solid-state π-stacking order.

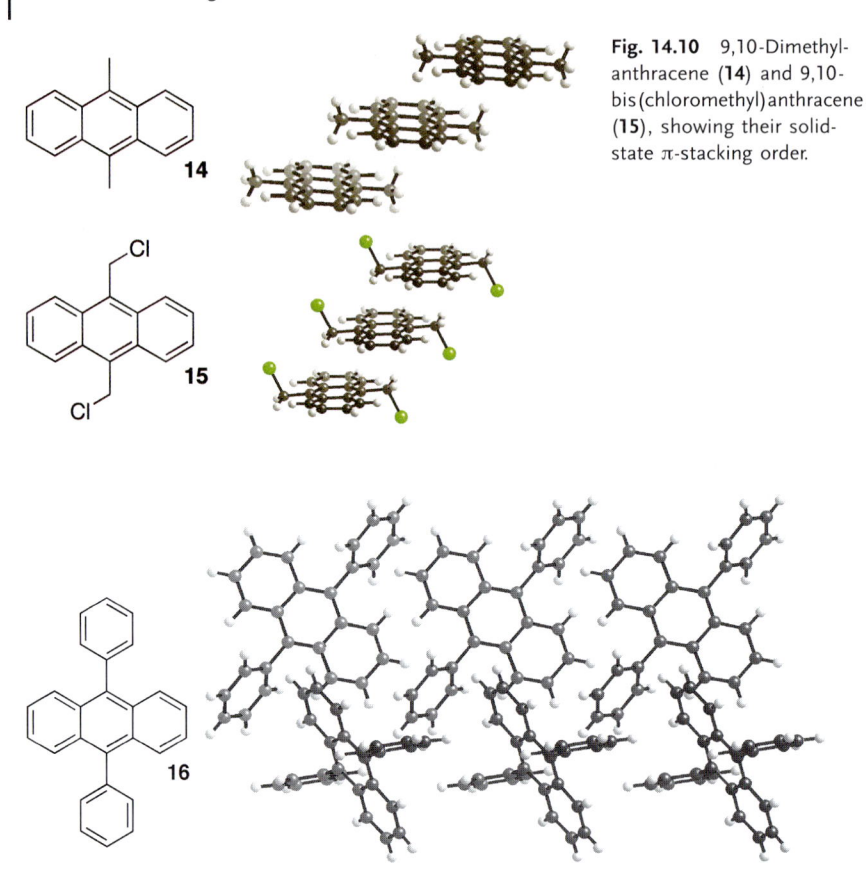

Fig. 14.11 9,10-Diphenylanthracene (**16**) and molecular order showing coplanar but non-π-interacting anthracene units.

the degree of long-axis slip directly related to the size of the halogen (Fig. 14.12). The π-face separation in these dihalo derivatives is also related to halogen diameter, ranging from 3.50 Å for the dichloro compound **17** to 3.63 Å for the diiodo compound **19**.

Dinitroanthracene also adopts a columnar π-stacking arrangement in the solid, although the spacing between π-faces in the stack is a significant 3.57 Å, due to both the size of the substituent as well as the repulsion between electron-rich oxygens in adjacent anthracene units in the crystal (Fig. 14.13) [34].

Using a rigid spacer between the anthracene chromophore and the substituent leads to an alternative and versatile approach to engineering the solid-state order of anthracene (Fig. 14.14). Diethynylanthracenes are easily synthesized from either the commercially available anthraquinone or dibromo compounds. Even with small substituents on the alkyne, the axis through the two alkynes becomes the long axis of the molecule, making interactions along this axis the driving force for crystallization. A well-studied example of a molecule in this class is 9,10-bi-

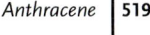

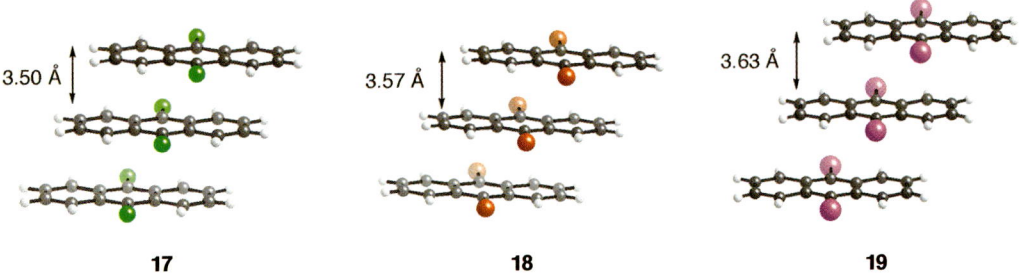

Fig. 14.12 9,10-Dichloroanthracene (**17**), 9,10-dibromoanthracene (**18**) and 9,10-diiodoanthracene (**19**) showing slipped-stacked arrangement and interplanar spacings.

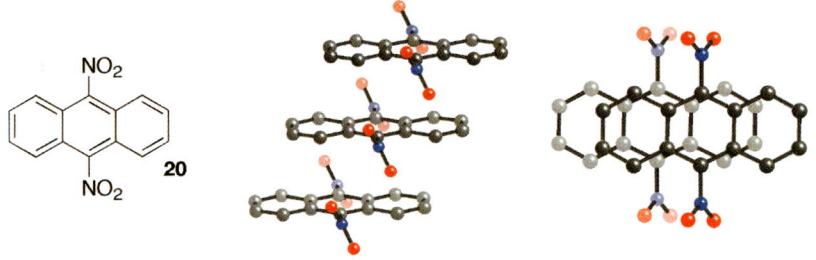

Fig. 14.13 9,10-Dinitroanthracene (**20**), showing significant intermolecular π-overlap (center, view along short axis, right, view down π-stacking axis).

s(phenylethynyl)anthracene (BPEA, **21**) [35]. This compound has a high fluorescence quantum yield and good solubility in a variety of solvents and is used as the emissive species in a variety of applications (including chemiluminescent light sticks). The crystal structure shows the solid-state order driven by strong edge-to-face interactions between the pendant phenyl groups. This interaction leads to extensive π-overlap of the anthracene units with only modest lateral offset and a separation of 3.43 Å between anthracene planes.

Groups with more three-dimensional bulk than a simple phenyl substituent lead to a change in the crystalline order from π-stacked to herringbone. The bis(trimethylsilylethynyl) derivative **22** [36], for example, shows no π-face interactions in the solid, since the size of the trimethylsilyl group is sufficient to cover a significant portion of the face of the anthracene unit, yielding an edge-to-face interaction that fills space very efficiently (Fig. 14.15). By decreasing the size of this pendant group from trimethylsilyl to *tert*-butyl, the face-to-face arrangement of anthracene is restored, but in this case with significant lateral offset of the anthracenes (although the face-to-face separation in **23** is a relatively small 3.40 Å) [37].

Switching the alkyne substituent from branched alkyl to straight-chain alkyl groups leads to crystalline order that is dominated by interactions between the alkyl chains, which arrange as an insulating layer between the anthracene chro-

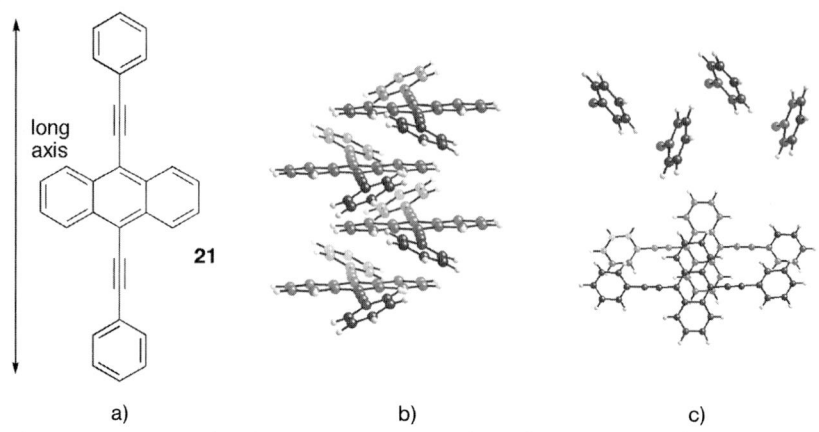

Fig. 14.14 9,10-Bis(phenylethynyl)anthracene (**21**). (a) structure, showing molecular long axis. (b) columnar arrangement of molecules in the crystal. (c, top) detail showing edge-to-face interaction of phenyl substituents on long axis. (c, bottom) detail showing degree of molecular overlap between anthracene units.

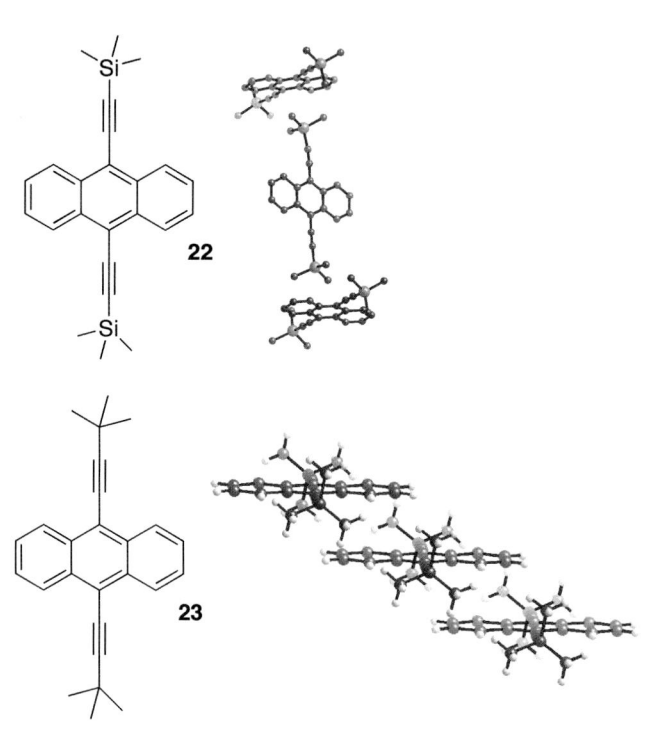

Fig. 14.15 (a) 9,10-bis(trimethylsilylethynyl)anthracene (**22**) and its solid-state order, showing predominant edge-to-face interactions. (b) 9,10-bis(*tert*-butylethynyl)anthracene (**23**) and its packing, showing face-to-face interactions in the crystal.

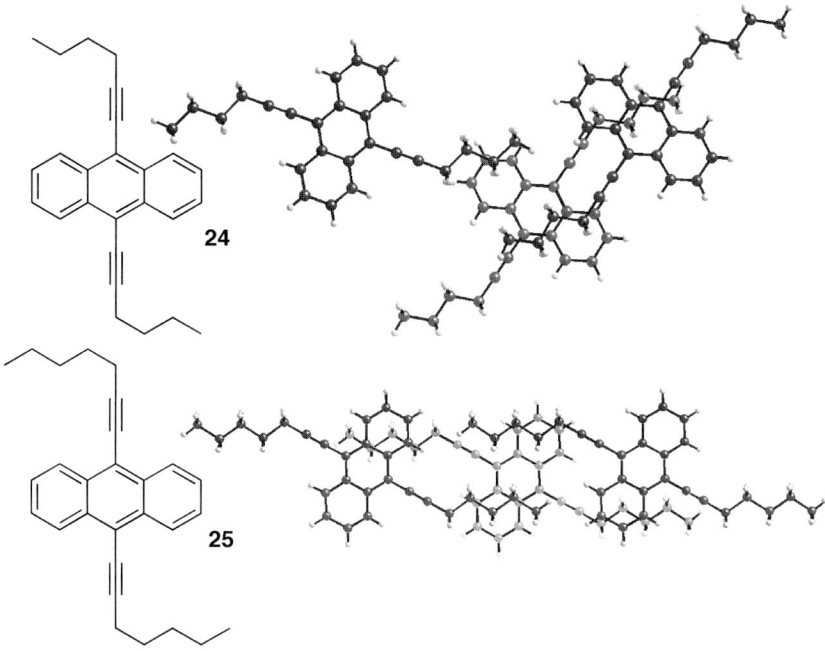

Fig. 14.16 Alkylethynylanthracenes (**24** and **25**), showing overlap of alkyl substituents with the anthracene chromophore of adjacent molecules in the crystal.

mophores (as exemplified by the hexynyl and heptynyl derivatives **24** and **25**, Fig. 14.16) [38]. Although the anthracene units are coplanar, there is no electronic interaction between the chromophores in the crystal. Although such an arrangement makes this class of anthracene derivatives unusable for use in applications requiring efficient charge transport (organic transistors or solar cells), the insulation of the fluorescent chromophore induced by this method may make these materials useful as emitters in OLEDs.

A fascinating variant of ethynylanthracenes is found in cyclophane **26** (Fig. 14.17) [39]. By effectively shrinking the size of the silyl substituent (vs. trimethylsilyl) and forcing a pair of acenes to π-stack, the assembly packs in a columnar arrangement with only slight lateral and longitudinal offset. Remarkably, the face-to-face separation between anthracene units within the cyclophane (3.48 Å) and between pairs of cyclophanes (3.49 Å) is essentially identical.

Finally, the rigid spacer between anthracene and substituent does not necessarily have to be an alkyne for this functionalization strategy to work – any rigid, conjugated system can be used to hold the substituent away from the chromophore, creating a new molecular long axis (Fig. 14.18). The benzoate ester of 9,10-anthrahydroquinone (**27**), for example, adopts a π-stacked arrangement of the anthracenes in the crystal very similar to that observed in the phenylethynyl system, with the rings of the substituent benzoyl groups showing edge-to-face

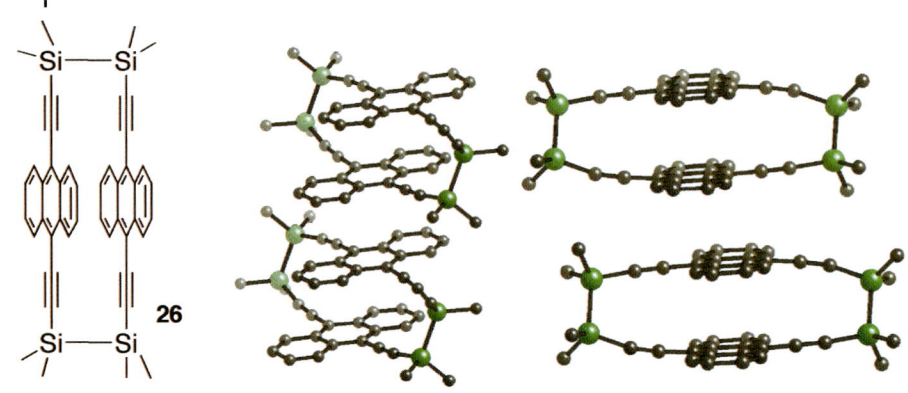

Fig. 14.17 Anthracene-based cyclophane (**26**), along with two views of crystal packing showing columnar π-overlap.

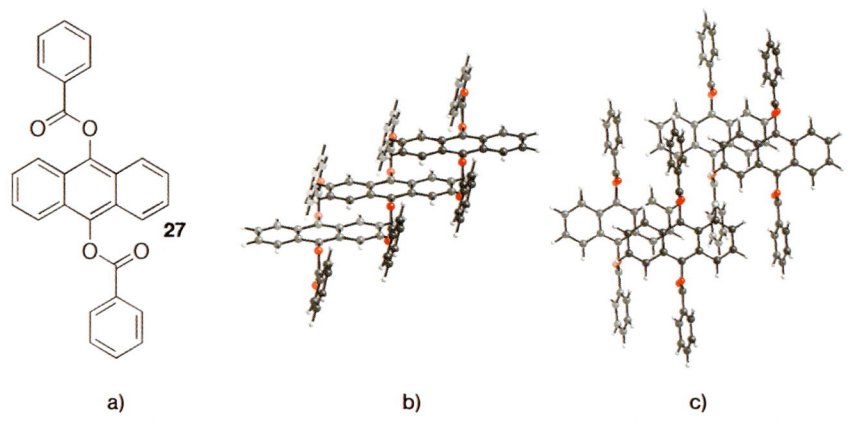

Fig. 14.18 Anthracenehydroquinone benzoate ester (**27**) (a) showing π-stacking interaction between anthracene units (b) and edge-to-face interaction between anthracene and the benzoate substituents of adjacent π-stacked column (c).

interactions with the anthracene units in adjacent stacks [40]. In this case the separation between aromatic faces is 3.45 Å, approximately the same separation observed in **21**.

14.2.3
Edge-substituted Anthracenes
(Anthracene Functionalized at the 1,8- or 1,8,9-Positions)

Functionalization of anthracene along only one edge will leave the opposite edge available for intermolecular edge-to-face interactions. Thus the majority of π-stacking acene derivatives in this class have substituents tailored to change the overall shape of the acene from rectangular to square, leading to an aspect

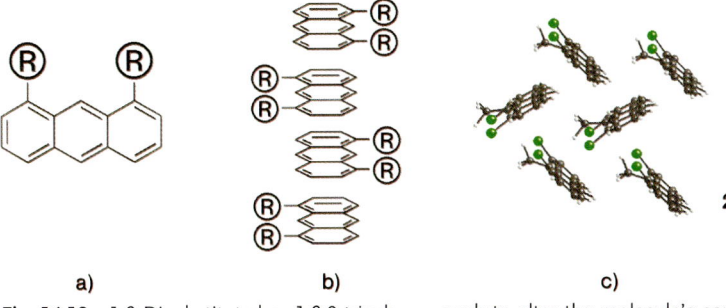

Fig. 14.19 1,8-Disubstituted or 1,8,9-trisubstituted anthracene derivatives have the potential to adopt columnar stacked arrangements (b). If the substituent is not large enough to alter the molecule's aspect ratio sufficiently, herringbone arrangements will still predominate [as in 1,8-dichloro-9-methylanthracene (**28**), (c)].

ratio more amenable to stacking. The choice of substituent in this class of acenes is critical, since the functional groups will interact strongly owing to their proximity and large groups will even perturb the shape of the acene. It is clear that molecules in this functionalization class will most commonly pack in an alternating fashion (Fig. 14.19b) and therefore will likely adopt a columnar arrangement. However, if the substituents are not chosen carefully to possess the necessary steric bulk to alter the aspect ratio of the molecule, e.g. 1,8-dichloro-9-methylanthracene (**28**) [41], herringbone and sandwich herringbone arrangements will result (Fig. 14.19c).

One derivative where the expected alternating stack structure is observed is 1,8,9-tribromoanthracene (**29**) (Fig. 14.20) [42]. The combination of the relatively small steric bulk and the electron-withdrawing nature of the bromine substituents serves to enhance the π-stacking in this crystal. It is important to note that the steric crowding along the substituted edge of the acene, along with strain in the crystal packing, lead to both a significant distortion of the acene backbone and an increase in the interplanar spacing to more than 3.7 Å.

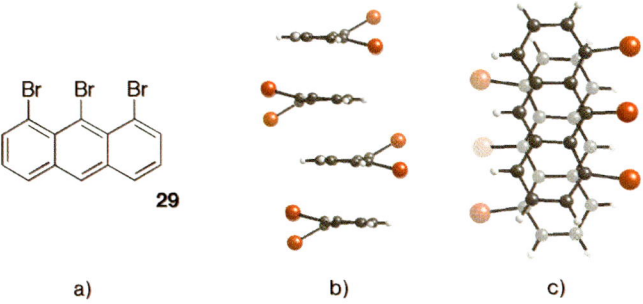

Fig. 14.20 1,8,9-Tribromoanthracene (**29**), showing columnar π-stacking (center, view down anthracene long-axis) and molecular overlap (right, view along π-stacking axis).

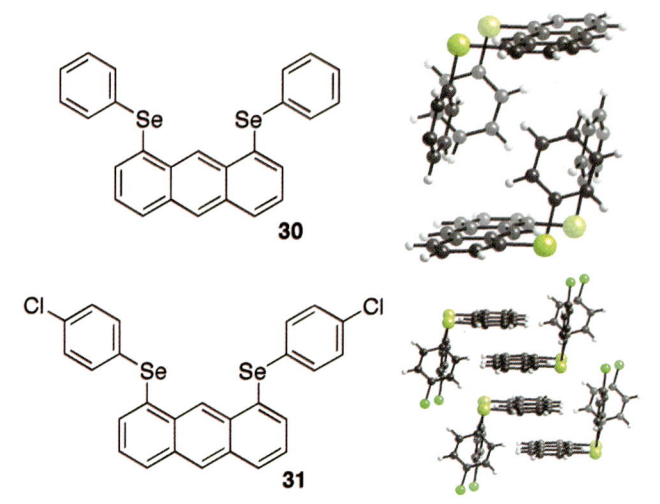

Fig. 14.21 1,8-Bis(arylselenyl)anthracenes (**30** and **31**). In the phenyl derivative **30**, there is strong edge-to-face interaction between phenyl groups on the same molecule, between phenyl groups on adjacent molecules and, most importantly, between the phenyl groups and an adjacent anthracene. In **31**, the last interaction is disrupted, leading to a columnar π-stacked arrangement of the anthracene units.

A beautiful demonstration of subtle substituent effects in edge-functionalized anthracene was reported by Nakanishi et al., who prepared a series of 1,8-bis(arylselenyl)-substituted anthracenes in an elegant study of nonbonded interactions between selenium groups in a constrained system (Fig. 14.21) [43]. The structure of 1,8-bis(selenylphenyl)anthracene (**30**) shows a pairwise arrangement, with intramolecular edge-to-face interactions between the phenyl substituents and intermolecular edge-to-face interactions between the phenyl substituents of one molecule and both the phenyl substituents and the anthracene moiety of an adjacent molecule. (Fig. 14.21, top). By disrupting the edge-to-face interaction between the phenyl substituent and the adjacent acene (by replacing the interacting *para*-hydrogen with a chlorine atom), the anthracene cores adopt a nearly columnar π-stacked arrangement. It is worth noting that in this structure the intramolecular edge-to-face interactions between the pendant aryl groups remain intact.

Phenylethynyl groups at the 1,8-positions should yield a molecule with a nearly square perimeter. However, because of the proximity of the hydrogens on the aryl substituents, these aryl groups cannot be coplanar and adopt an edge-to-face arrangement in the solid. For example, 1,8-bis[(pyrid-4-yl)ethynyl]anthracene (Fig. 14.22) shows a nearly perpendicular arrangement of the pyridyl substituents [44]. Even so, the anthracene chromophore shows strong face-to-face interactions with the pyridyl group of the adjacent molecule in the crystal (face-to-face spacing 3.46 Å), leading to some π-stacking interactions in the solid.

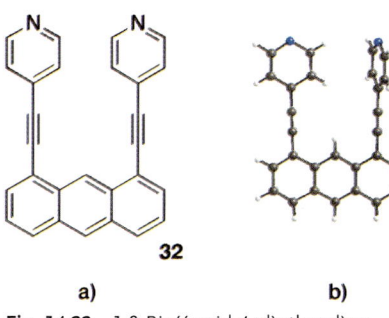

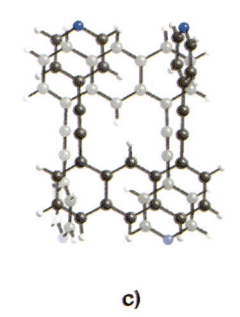

Fig. 14.22 1,8-Bis((pyrid-4-yl)ethynyl)anthracene (**32**) showing intramolecular edge-to-face interactions of the pyridyl substituents (center), leading to intermolecular edge-to-face and face-to-face interactions between the anthracene chromophore and the substituent pyridine units (right).

14.2.4
Liquid Crystalline Anthracenes

A classical approach to the enhancement of π-stacking in aromatic systems is the generation of liquid crystalline mesophases [45], and there have been a few studies of anthracene chromophores functionalized with long-chain alkyl groups (typically attached to the aromatic core via an ether linkage) designed to possess such phases. The most straightforward approach substitutes the 2,3-positions, yielding molecules with a rigid aromatic head and long, flexible alkyl tails. These materials were first investigated as non-hydrogen-bonding gelling agents, but a more systematic study of the systems did reveal liquid crystalline properties [46]. The only derivative that yielded crystals suitable for X-ray study [2,3-di(n-hexyloxy)anthracene (**33**)] showed almost exclusive edge-to-face (herringbone) interaction, with the crystallization appearing again to be driven by interactions between the alkyl chains.

An alternative arrangement of functional groups was studied by Norvez et al., who synthesized a series of 1,4,5,8-tetrasubstituted anthracene ethers (Fig. 14.24) [47]. The tetrakis(1-dodecyloxy) derivative was found to possess a smectic A phase and X-ray diffraction data from the solid implied an arrangement with interacting π-faces. However, crystallographic investigation of the tetrakis(1-butoxy) derivative **34** (the only derivative to yield crystals suitable for analysis) showed interdigitation of the alkyl groups between anthracene π-faces, similar to the alkylethynyl anthracenes **24** and **25**. It is certainly possible that this phase may be different from the liquid crystalline phase and that systems with larger alkyl groups do not exhibit this insulation of the anthracene chromophore owing to the stronger

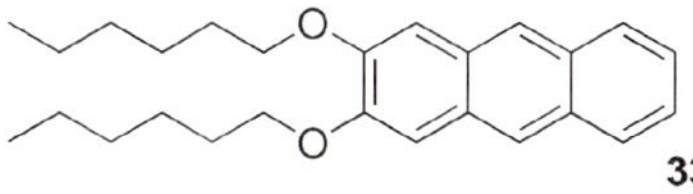

Fig. 14.23 2,3-Bis(n-hexyloxy)anthracene (**33**).

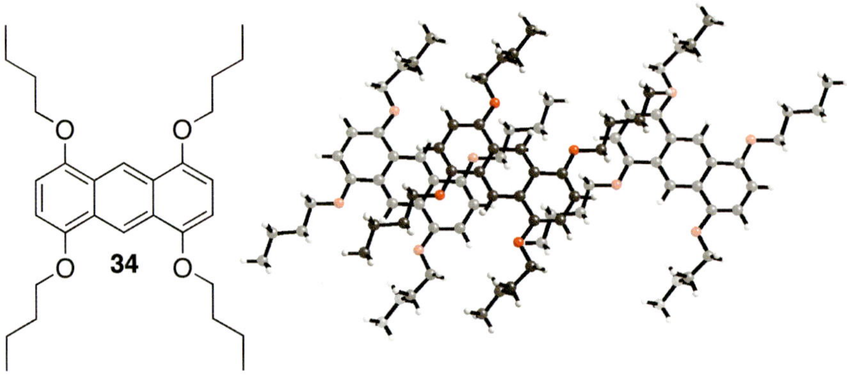

Fig. 14.24 1,4,5,8-Tetrakis(1-butoxy)anthracene (**34**) and packing showing anthracene chromophore sitting atop alkyl chains of adjacent molecules in the crystal.

interactions of the long alkyl chains, leading to a segregation of saturated and aromatic portions of the molecules.

14.2.5
Anthracene Self-assembly: Hydrogen Bonding

Hydrogen bond formation is a strong force for directing self-assembly and this interaction can be used to enforce π-stacking in a myriad of aromatic systems [48]. Directing the self-assembly of anthracenes through hydrogen bonding is a difficult endeavor, because the steric bulk of the directing group itself can retard the stacking of the aromatic cores. However, the strength and directionality of hydrogen bonds allow unusual π-stacking motifs to be created.

One such example was reported by Garcia-Garibay and coworkers, where the diethynylanthracene derivative **35** was induced to arrange in a columnar array [49]. However, in order to maximize H-bonding interactions, the anthracene units in the columns of these molecules are arranged very differently from the columns described previously (Fig. 14.25). Typically, molecules in a columnar π-stacked arrangement will overlap the outermost rings only. In the case of **35**, the anthracene chromophore rotates ±60° around an axis perpendicular to the central ring within the columnar stack, arranging so that the centermost ring of one anthracene is overlapped by the outermost ring of its neighbors. This arrangement maximizes the ability of the OH substituents to interact through hydrogen bonds (shown as dashed lines in Fig. 14.25) and leads to an interplanar spacing between adjacent anthracene faces as small as 3.29 Å.

A particularly elegant approach to the use of hydrogen bonds to induce π-stacking in crystals was reported by Lehn and coworkers (Fig. 14.26) [50]. By substituting anthracene with 2-aminopyrimidine, a base that is known to self-assemble with strong intermolecular hydrogen bonds, the arrangement of the anthracenes

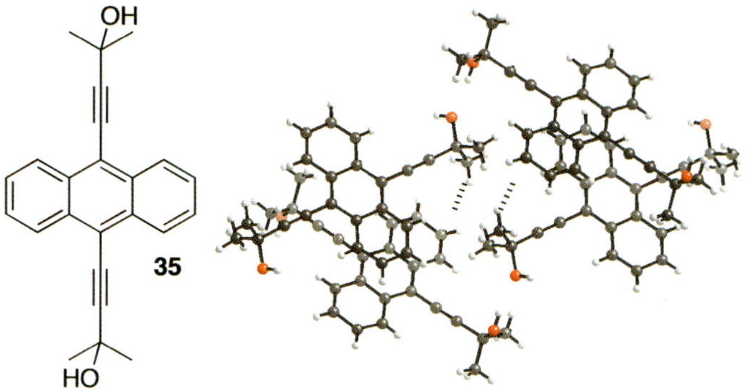

Fig. 14.25 An unusual π-stacking motif for a diethynylanthracene (**35**) created by hydrogen bonding interactions.

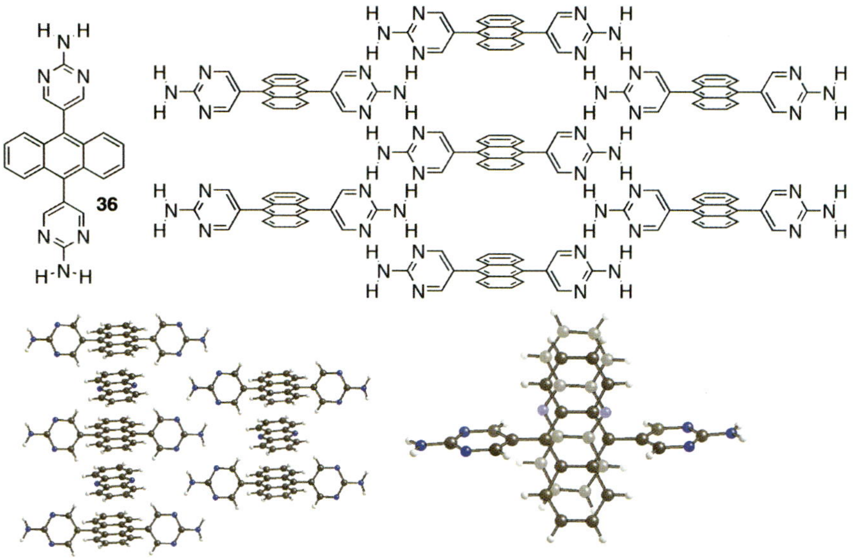

Fig. 14.26 2-Aminopyrimidine-functionalized anthracene (**36**) and the cavities formed by the hydrogen bond-directed self-assembly of this molecule (top right). These cavities incorporate a diazine molecule in the single crystal (bottom left), leading to near-perfect overlap of the aromatic moieties (bottom right).

in the solid can be influenced to leave a cavity large enough for a second acene. In that study, phenazine was used as the "guest" acene, leading to nearly columnar π-stacks of aromatic rings, formed entirely by a self-assembly process. While the interplanar spacing between aromatic faces in this clathrate was fairly large (average spacing ~3.65 Å), the alignment of the aromatic faces was good, with only 1.1 Å longitudinal and 0.16 Å lateral deviations from perfect columnar overlap, a precision of alignment difficult to attain by any other approach.

14.3
Tetracene (Naphthacene)

The general techniques described in the previous section to enhance π-stacking in the solid state for anthracene are applicable to the higher acenes, although the synthesis of these systems becomes more difficult. Thus, compared with anthracene, the solid-state engineering of tetracene (the next higher homolog in the series of acenes) has received significantly less attention. Whereas the number of structures reported for anthracene derivatives in the CCDB is more than 1000, tetracene (also called naphthacene) structures number only a few dozen. Tetracene itself arranges in the typical acene herringbone motif, with two-dimensional edge-to-face interactions (Fig. 14.27) [51]. Unlike anthracene, this material has been the subject of numerous detailed device studies regarding its application in FETs [52] and sensors [53] and as the active component in organic light-emitting FETs [54]. The ease with which tetracene makes high-quality, large single crystals by sublimation has also led to numerous studies of the transport properties in these highly ordered, essentially defect-free systems [55].

Owing to its symmetry, tetracene offers a number of strategies for functionalization, each leading to a different series of potential π-stacked arrangements. The first adds substituents symmetrically across the tetracene backbone. This arrangement can lead to either a slipped-stack arrangement (Fig. 14.28a) or a columnar arrangement (Fig. 14.28b, showing one possibility), depending on the size of the substituent. Addition of groups at the 5- and 11-positions yields a derivative that also adopts a π-stacked arrangement in the solid state (Fig. 14.28c). Finally, four of the central peri-positions can be functionalized, leading to a derivative that will pack in a manner similar to the functionalized anthracenes (Fig. 14.28d, with typically only the outermost rings participating in π-stacking interactions).

As with anthracene, the addition of a small substituent to tetracene is sufficient to prevent the typical herringbone arrangement of these molecules, owing to the inability of one edge of the molecule to participate in edge-to-face interactions, but insufficient to induce long-range π-stacking order. For example, both 5-methyltetracene [56] and 5-bromotetracene [57] adopt a sandwich-herringbone arrangement that maximizes the possible edge-to-face interactions while minimizing steric repulsion of the substituent group.

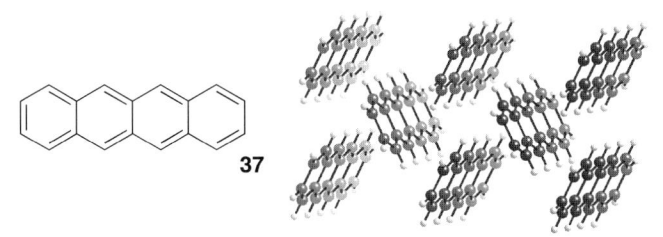

Fig. 14.27 Tetracene and its native solid-state order, showing edge-to-face interactions.

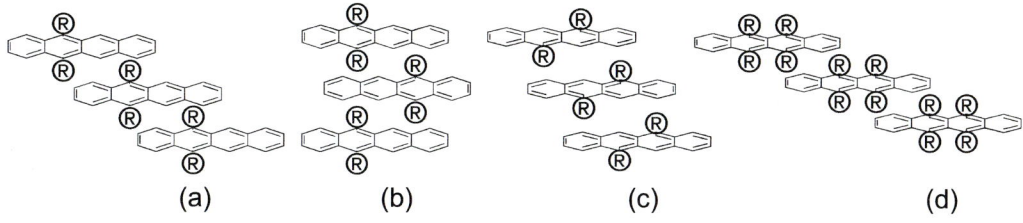

Fig. 14.28 Functionalization strategies for tetracene, leading to a variety of potential π-stacked arrangements.

A recent study of the relationship between crystalline order and device performance in organic transistors focused on halogenated tetracenes [58]. Bao and coworkers showed that both 5-chloro- and 5-bromotetracene packed in a sandwich-herringbone motif (as shown in Fig. 14.29), whereas 5,11-dichlorotetracene (**40**), with substituents blocking edge-to-face interactions along both acene edges, packed with extensive π-overlap, exhibiting only small (0.95 Å lateral, 1.3 Å longitudinal) shifts from perfect columnar order (Fig. 14.30). The separation between π-faces in **40** is ~3.5 Å (cf. the halogenated anthracenes). Experiments also showed that the π-stacked arrangement led to significantly higher carrier mobility. The hole mobility measured from devices constructed on single crystals of these halogenated tetracenes showed the hole mobility of the dichloro-substituted **40** to be 2–3 orders of magnitude higher than those measured for chlorotetracene and bromotetracene and even slightly higher than that measured for unsubstituted tetracene (**37**). The authors attributed this increased mobility to improved electronic coupling between molecules owing to their π-face interactions.

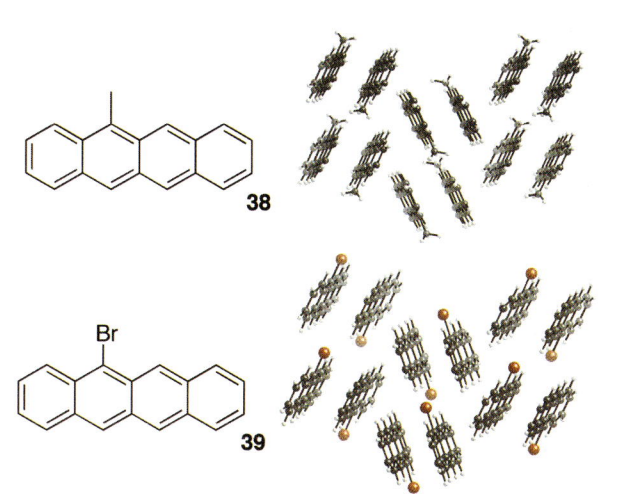

Fig. 14.29 5-Methyltetracene (**38**) and 5-bromotetracene (**39**), exhibiting sandwich herringbone packing.

Fig. 14.30 5,11-Dichlorotetracene (**40**), exhibiting a columnar π-stacked arrangement.

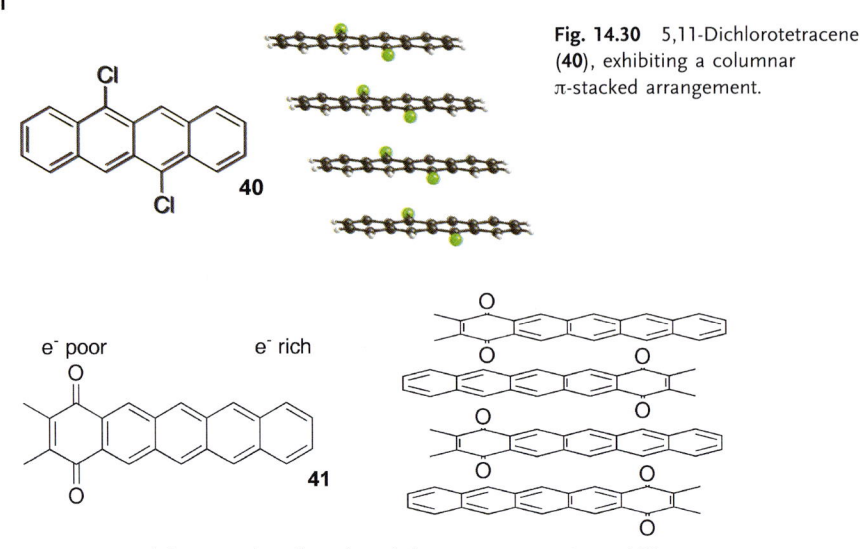

Fig. 14.31 Solid-state order of 2,3-dimethyl-1,4-pentacenequinone (**41**).

In an effort to exploit dipolar interactions to enhance the π-face interaction of tetracenes and pentacenes in the solid, Nuckolls and coworkers recently utilized an iterative synthetic method to prepare a series of 1,4-acenequinone derivatives [59]. Owing to the electron-withdrawing nature of the carbonyl groups, the molecule possesses a significant dipole, which causes the molecules to stack and align in the crystal (Fig. 14.31); although there is significant deviation from perfect overlap along the long axis of the molecule, there is little slip along the short axis and a separation between acene planes of only 3.25 Å. This principle was also applied to the corresponding hexacenequinone, yielding materials that exhibited reasonable hole mobility in a thin-film transistor formed from vapor-deposited films of the acene (unfortunately, crystallographic data were not available for this derivative).

14.3.1
Ethynyltetracenes

5,12-Diethynyltetracenes are easily prepared from commercially available tetracenequinone. Varying the size of the substituent on the alkyne leads to significant changes in the crystal packing, just as with the ethynylanthracenes. A large substituent (triisopropylsilyl derivative **42**) yields a herringbone arrangement of the molecules, with the silyl group nearly completely covering the face of adjacent tetracenes (Fig. 14.32a) [60]. Decreasing the size of the substituent slightly, to triethylsilyl (**43**), leads to an unusual arrangement where the individual molecules pair (as expected), but this pair is then surrounded by the silyl groups of adjacent pairs, preventing long-range π-face interactions (Fig. 14.32b).

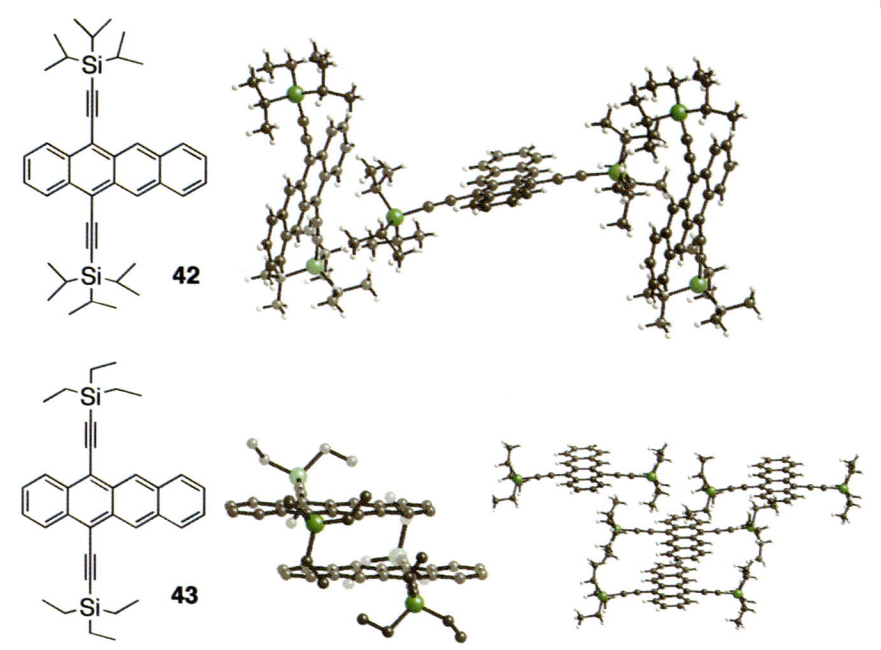

Fig. 14.32 Structure and packing for 5,12-bis(triisopropylsilylethynyl)tetracene (**42**) and 5,12-bis(triethylsilylethynyl)tetracene (**43**) (showing pairwise interaction of tetracene chromophores).

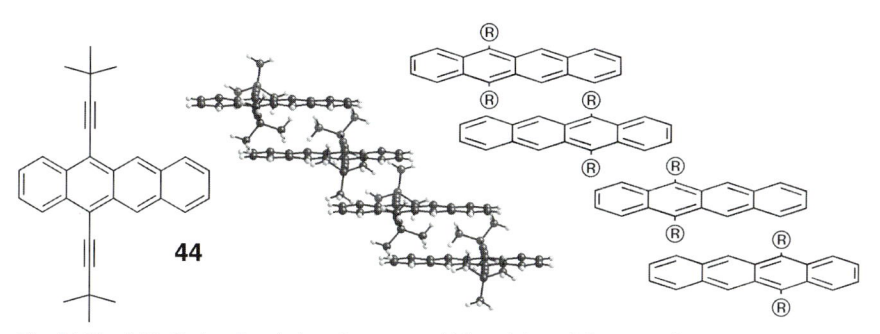

Fig. 14.33 5,12-Bis(*tert*-butylethynyl)tetracene (**44**) and its solid-state order.

In order to achieve long-range π-overlap, the substituent size must be further decreased. The *tert*-butylethynyl derivative **44** finally adopts an arrangement leading to long-range π-stacked interactions in the solid state. Compound **44** adopts an unusual arrangement similar to ethynylanthracenes such as BPEA, except that along the stacking axes the molecules alternate to overlap either three or two rings, as shown schematically in Fig. 14.33 (right). This arrangement leads to a high degree of π-overlap, with an interplanar spacing of only 3.41 Å. The measurement of transport properties in this system was hampered by the poor

14.3.2
Tetrasubstituted Tetracenes

By far the best known member of this class of tetracene derivative is commercially available rubrene [62, 63]. By clustering the phenyl groups together on the central carbons atoms (the 5-, 6-, 11- and 12-positions) of the acene periphery, the four phenyl rings essentially behave as a single, large substituent, leading to a crystalline order very similar to that adopted by 9,10-disubstituted anthracene derivatives (see above). Comparing rubrene with 9,10-bis(phenylethynyl)anthracene, we see that rubrene and anthracene **21** both exhibit an overlap comprising roughly 1.5 aromatic rings (Fig. 14.34b). Also, whereas the lateral offset for rubrene (essentially 0 Å) is significantly less than that for the anthracene derivative (~0.8 Å), the average spacing between the π-faces for anthracene **21** (3.45 Å) is smaller than the spacing observed for rubrene (~3.7 Å). Owing to the small area of overlap observed in rubrene and the comparatively large interplanar spacing, it would be expected that the charge-carrier mobility would be fairly low. In spite of these poor attributes, recent results from transistors fabricated directly on the surface of

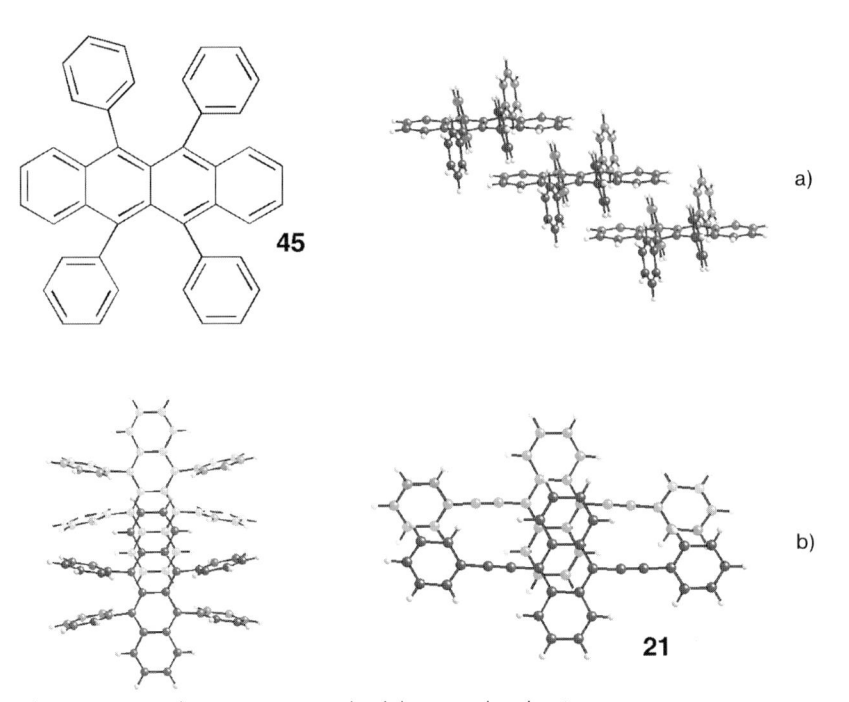

Fig. 14.34 (a) rubrene structure and solid-state order, showing slipped-stack arrangement of tetracene rings. (b) comparison of π-surface overlap between rubrene and BPEA (**21**).

high-quality rubrene crystals have shown hole mobilities as high as 20 cm^2 V^{-1} s^{-1} [64]. Theoretical investigation of this molecule by Brédas and coworkers showed that the intermolecular couplings in face-to-face stacked rubrene, although they occur over a much smaller area, are comparable in magnitude to those found in pentacene and tetracene [65]. Serendipitously, the rubrene molecules overlap in a manner that maximizes the intermolecular transfer integral. This result emphasizes the need to develop systems where the nature of the π-overlap can be fine-tuned with precision to yield maximum charge carrier mobility.

The tetrachalcogenotetracenes comprise another important class of tetrasubstituted tetracene compounds and one of the first classes of materials studied as organic conductors [66]. The myriad of structures and electronic properties of these materials have been reviewed in detail in a number of papers [67]. However, two keys to the remarkable electronic properties of this class of compounds are the ability of the chalcogen substituents to stabilize the oxidized form of the aromatic ring (making these materials powerful electron donors, to the point that they are highly susceptible to air oxidation) and the π-stacked nature of the molecules in

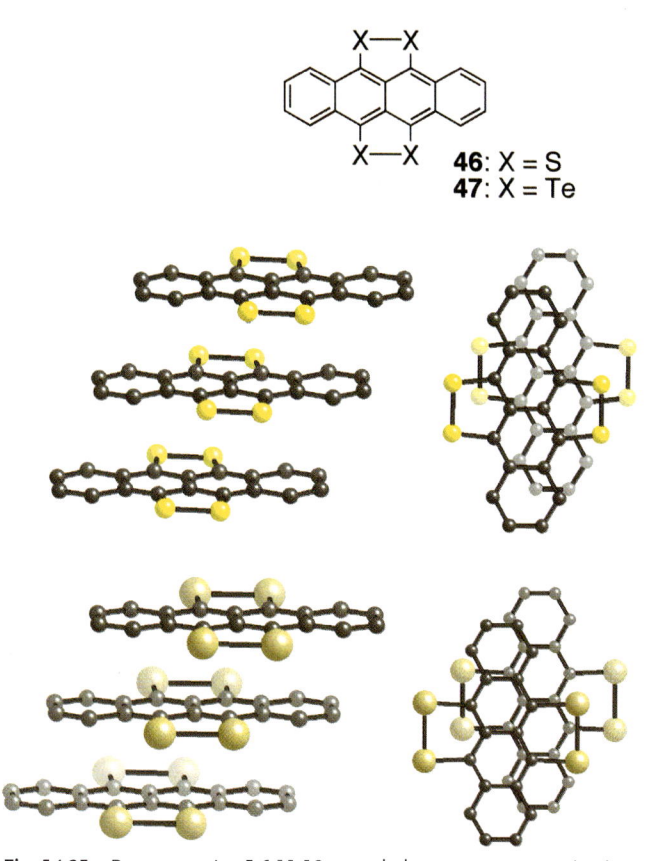

46: X = S
47: X = Te

Fig. 14.35 Representative 5,6,11,12-tetrachalcogenotetracenes (top), showing the solid-state ordering and intermolecular overlap for the sulfur derivative **46** (center) and the tellurium derivative **47** (bottom).

the crystal. Two examples are the tetrathio- and tetratellurotetracene derivatives **46** [68] and **47** [69] (Fig. 14.35). These molecules both adopt columnar π-stacked arrangements in the crystal, with small lateral and longitudinal offsets for the sulfur compound. For this derivative, the separation between π-faces is 3.49 Å. The tellurium derivative **47** undergoes further displacement from perfect columnar overlap, in order to accommodate the larger size of the tellurium functional groups (note the tellurium atom from one molecule of **47** sitting in the center of the five-membered ring formed by the tellurium atoms of the adjacent molecule, the most efficient arrangement to minimize interplanar spacing while accommodating the larger volume occupied by the tellurium atoms) and the distance between π-faces increases to 3.7 Å. It would be expected that a tetrahalotetracene derivative, such as 5,6,11,12-tetrachlorotetracene, would adopt a similar π-stacking arrangement. Although such compounds are known [69, 70], none have been studied crystallographically.

14.4
Pentacene

Pentacene (Fig. 14.36) in particular has been the subject of intense study for application in electronic devices. Recent reports of the photovoltaic [71] and transistor [72] device performance of this material have secured its place as the benchmark organic p-type semiconductor. Recurrent in all applications of this material is the importance of solid-state order. Although numerous fabrication techniques have been applied to induce improved order in these flat aromatic molecules [73, 74], little research into the effect of functionalization on either solid-state order or electronic properties of pentacenes has been reported. In fact, the CCDB lists only a handful of non-benzannelated functionalized pentacene materials. Unsubstituted pentacene crystallizes in the typical acene herringbone pattern, maximizing aromatic edge-to-face interactions [51]. This arrangement of molecules leads to imperfect π-molecular orbital overlap, with calculations suggesting that this structural motif translates to a poor dispersion of the electronic bands in the molecule, implying that pentacene's transport properties may be limited by its crystal packing [75]. Substitution at the centermost ring of pentacene leads to a number of packing arrangements, depending on the size of the substituent. 6,13-Diarylpentacenes, for example, yield microcrystalline solids that have not yet been characterized crystallographically. However, their use in efficient red-emitting OLEDs implies that no long-range π-stacking is present in the solid [76]. Alternatively, functionalization of pentacene with alkyne-based groups leads to a number of π-stacked derivatives, with the nature of the stacking depending on the size of the alkyne substituent (just as in the alkyne-substituted anthracenes and tetracenes).

The attachment of simple alkylacetylenes to pentacene in general does not yield stable materials: bis(propynyl)pentacene is too unstable to be isolated and although larger alkyl substituents (butyl, hexyl, octyl) lead to slightly more stable

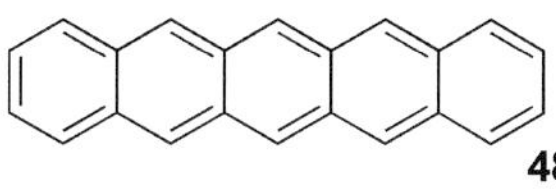

Fig. 14.36 Pentacene and its native solid-state order, showing edge-to-face interactions.

materials, in general the pentacene chromophore does not exhibit π-stacking interactions in the crystal (owing to strong interactions between the alkyl chains, cf. alkylethynylanthracenes) [77]. The one exception to this trend is the *tert*-butyl derivative **49** [78]. This material is significantly more stable than the straight-chain alkyl derivatives and the pentacene packs in a 1-D slipped-stack arrangement analogous to the related anthracene (**23**) and tetracene (**44**) derivatives.

The silylethynylpentacenes are even more stable than the alkylethynyl derivatives and the stability increases as the size of the silyl substituent increases. Both the trimethylsilyl derivative **50** and the triethylsilyl derivative **51** pack in 1-D slipped-stack arrangement similar to the *tert*-butyl derivative (Fig. 14.37). By varying the size of the alkyne substituent, the slip along both the long and short axes of pentacene in the stack can be systematically varied. In all of these molecules, the spacing between the planes changes very little from the typical value of 3.45 Å. Unfortunately, these materials did not tend to form uniform films by evaporation, so their charge-transport properties could not be determined. Further, owing to the nature of the "greasy" substituents, it is very difficult to grow single crystals of these materials by sublimation, impeding the study of charge-transport properties in the crystal.

Whereas most trialkylsilyl derivatives are roughly spherical in shape (trimethylsilyl, triisopropylsilyl, etc.), the *tert*-butyldimethylsilyl (TBDMS) group is ovoid in shape, with a short-axis diameter of 4.64 Å and a long-axis diameter of nearly 6 Å. Hence the TBDMS derivative **52** is the only compound in this series reported to crystallize in two polymorphic forms. In the major polymorph, the *tert*-butyl groups face opposite ends of the acene, yielding a C_2 symmetry axis perpendicular to the face of the centermost ring of the acene. This allows the molecule to stack in a slipped-stack arrangement similar to the previously discussed silyl derivatives, providing roughly the same degree of π-face overlap as seen in the triethylsilyl derivative. In the minor polymorph, the *tert*-butyl groups arrange pointing (approximately) towards the same end of the pentacene, placing the C_2 symmetry axis in the plane of the pentacene, equivalent to the long axis of the acene. For this molecule, a slipped-stack arrangement is not possible, owing to the poor space-filling that arises when two sets of *tert*-butyl groups have to arrange in close proximity. Instead, this polymorph adopts a columnar packing in which the bulky group alternates direction (e.g. pointing left, then pointing right; Fig. 14.38b) down the stack of molecules. Even so, there is significant distortion of both the alkynes and the pentacene backbone in this arrangement and the molecules in the column appear to be "paired" – that is, along the stacking axis there are different average distances between adjacent acenes. The short stacking distance (e.g. the middle pair of pentacenes in the stack at the bottom of Fig.

536 | *14 Induced π-Stacking in Acenes*

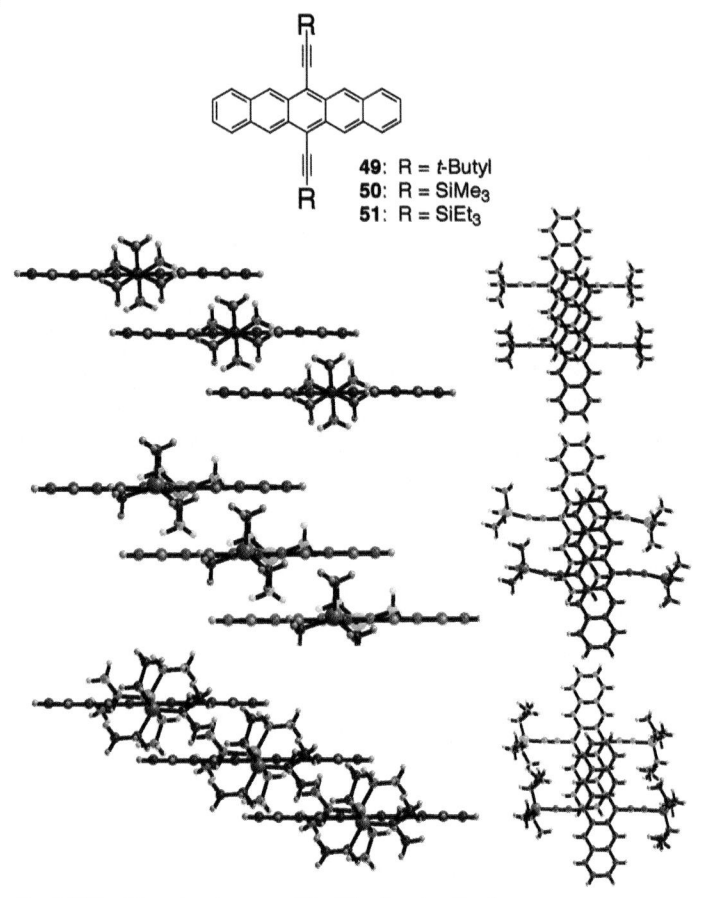

49: R = *t*-Butyl
50: R = SiMe$_3$
51: R = SiEt$_3$

Fig. 14.37 Ethynylpentacenes (**49–51**), showing the increase in longitudinal slip as the size of the alkyne substituent is increased.

14.38) is 3.38 Å, while the long stacking distance (e.g. the first and second pentacenes in the stack) is 3.45 Å [79].

A direct approach to the preparation of pentacene derivatives that will adopt a columnar arrangement similar to the minor polymorph of **52** is to move the substituents to the 5,14-positions (offset from the centermost ring), to encourage further the molecules to adopt the "flipped stack" ordering (Fig. 14.39). Again, the precise nature of the solid-state interaction is dominated by the size of the alkyne substituent. In the case of triisopropylsilyl derivative **53**, the molecules adopt the sandwich herringbone packing, since the pair of pentacenes present a well-packed cube with no space available to accommodate the silyl group of a third unit added to the stack [80]. Decreasing the size of the substituent to triethylsilyl (derivative **54**) does indeed lead to a columnar arrangement of pentacenes, in this case with no significant distortion of either alkynes or the pentacenes [78]. The

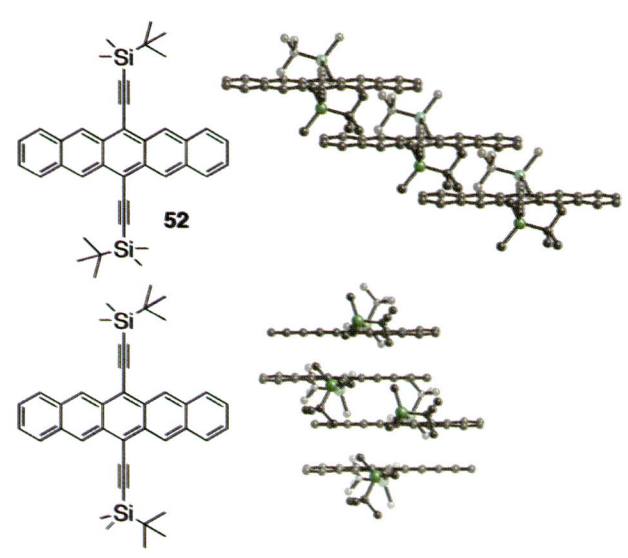

Fig. 14.38 Two polymorphs of 6,13-bis(*tert*-butyldimethylsilylethynyl)pentacene (**52**). (a) *tert*-butyl groups arrange pointing in opposite direction. (b) *tert*-butyl groups point in approximately the same direction, leading to a columnar stacking arrangement in the crystal.

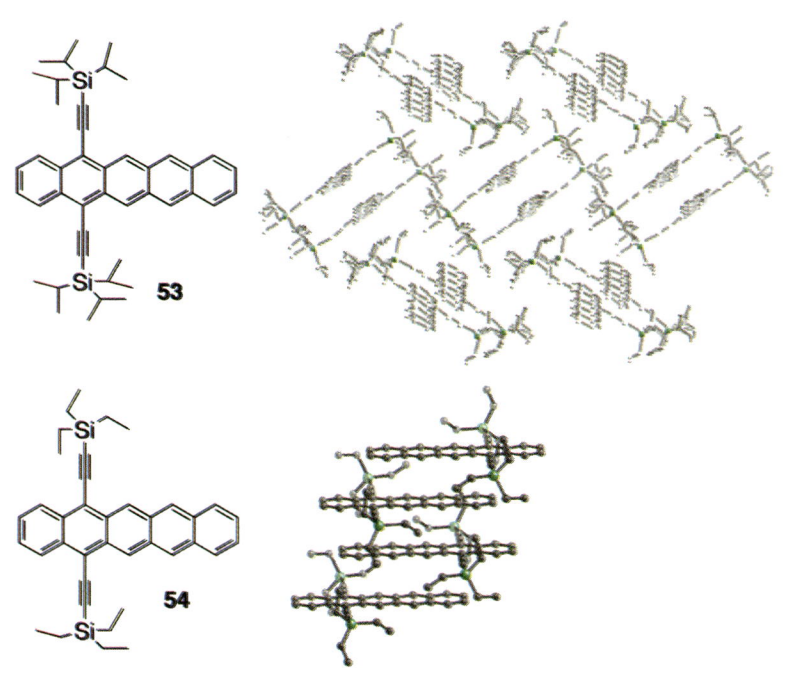

Fig. 14.39 "Offset" silylethynylated pentacenes. (a) 5,14-bis(triisopropylsilylethynyl)pentacene (**53**), showing sandwich herringbone packing. (b) 5,14-bis(triethylsilylethynyl)pentacene (**54**), showing columnar π-stacking.

interplanar spacing in this arrangement is 3.40 Å. While moving the substituents from the centermost ring leads to a number of new packing motifs, it also significantly decreases the stability of the materials: thin films of pentacenes **53** and **54** were not stable enough to measure their transport properties.

Further increasing the size of the alkyne substituent in the 6,13-disubstituted pentacene series leads to a new packing arrangement different from that of any other acene derivative. The triisopropylsilyl derivative **55** adopts a two-dimensional stacking arrangement, similar to the "running bond" pattern bricklayers use to construct walls: each pentacene unit stacks with two neighbors above and two neighbors below, rather than the one above and below seen in columnar or slipped-stack arrangements (Fig. 14.40) [80]. This packing leads to two-dimensional π-face interaction and is therefore more comparable to the two-dimensional interacting network presented by unsubstituted pentacene. This material exhibits some of the best transport properties of any pentacene derivative, yielding hole mobilities of 0.4 cm^2 V^{-1} s^{-1} from vapor-deposited films [81], and mobilities > 1.2 cm^2 V^{-1} s^{-1} from solution-cast films [82]. All-optical mobility measurements on single crystals of this compound have reported transport properties very similar to those seen in unsubstituted pentacene [83].

Additional "tuning" of this system can be accomplished by exploiting the well-known strong interaction between fluorinated aromatic systems and nonfluorinated aromatics (Fig. 14.41) [84]. Partial fluorination of the pentacene ring does not affect the nature of the general ordering of the molecules in the crystal (the fluorinated species still arrange in a two-dimensional π-stacked order nearly identical with that adopted by TIPS pentacene **55**). However, the interplanar spacing of the acenes within the π-stacked array is affected and that spacing decreases as the fluorine content increases: compared with the 3.43 Å spacing for the nonfluorinated derivative, the tetrafluoro compound **56** stacks with an average 3.36 Å spacing between aromatic planes, whereas in the octafluoro compound **57** that spacing decreases to 3.28 Å. This subtle change in packing has a significant impact on the carrier transport properties of the material. For FET devices prepared under identical (non-optimized) conditions, the nonfluorinated pentacene **55** shows a hole mobility of 0.001 cm^2 V^{-1} s^{-1}, whereas the tetrafluoro (**56**) and octafluoro (**57**) derivatives showed mobilities of 0.014 and 0.045 cm^2 V^{-1} s^{-1}, respec-

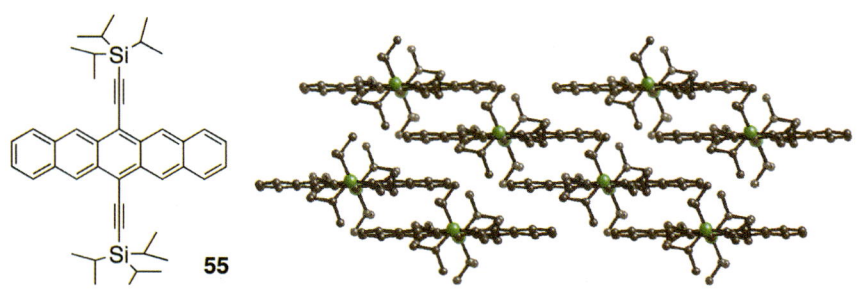

Fig. 14.40 6,13-Bis(triisopropylsilylethynyl)pentacene (**55**), showing two-dimensional π-stacked ordering in the crystal.

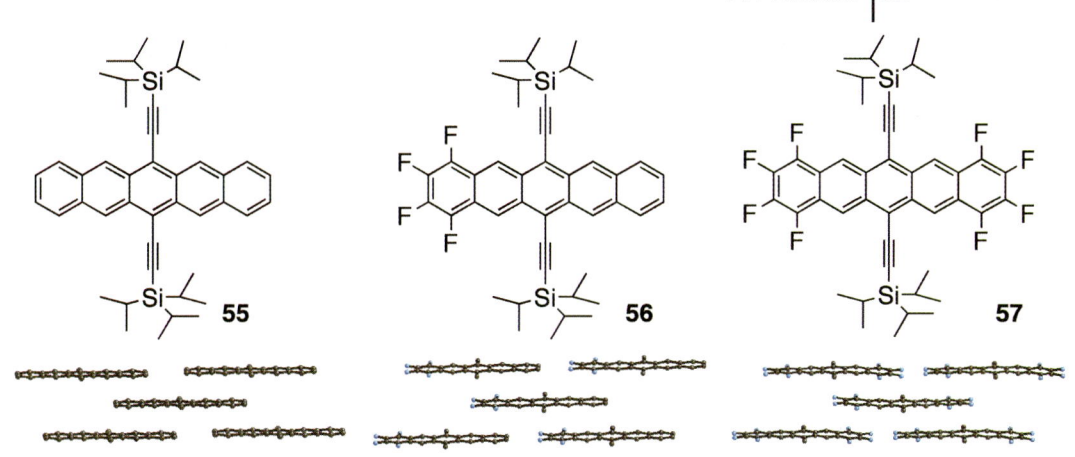

Fig. 14.41 Nonfluorinated and partially fluorinated silylethynylpentacene derivatives (**55–57**) all adopt a two-dimensional π-stacked arrangement in the crystal, with the interplanar spacing decreasing with increasing fluorine content.

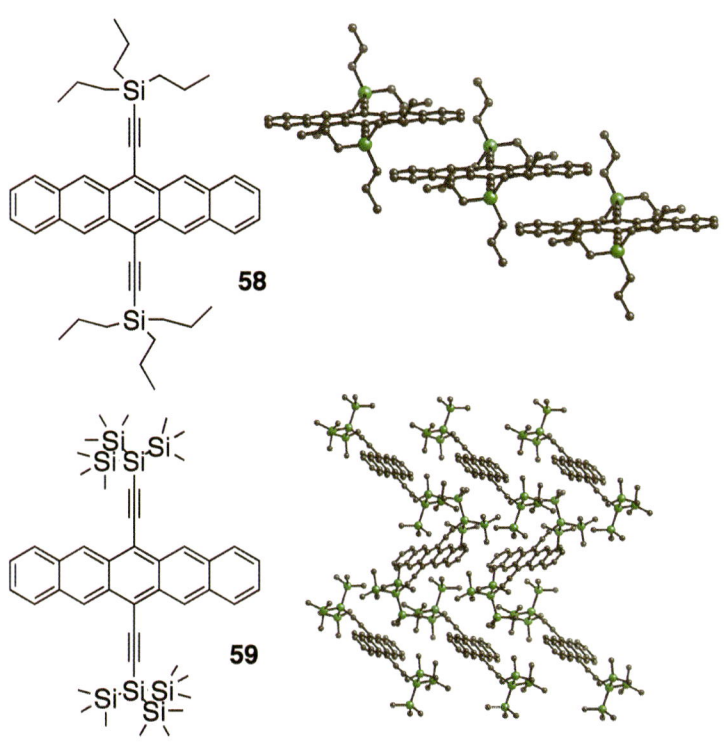

Fig. 14.42 (a) 6,13-bis(tri-n-propylsilylethynyl)pentacene (**58**) adopts a one-dimensional slipped-stack arrangement in the crystal. (b) 6,13-bis[tris-(trimethylsilylsilyl)ethynyl]pentacene (**59**) adopts a herringbone arrangement in the crystal.

tively [85]. It has been predicted that carrier mobility should be especially sensitive to interplanar separation [41b] – perfluoroaryl interactions may be a simple approach to tuning this important parameter.

Further increase in the size of the alkyne substituent in silylethynylpentacenes leads to a significant evolution in the nature of the pentacene packing. Substitution with the large tri(n-propyl)silylethynyl group causes the packing to revert back to the slipped-stack arrangement seen in the trimethylsilyl and triethylsilyl derivatives, but with significantly increased lateral and longitudinal slip (Fig. 14.42a). The interplanar spacing also increases, to 3.51 Å, which implies that this material will exhibit poor transport properties compared with the derivatives with smaller substituents. Increasing the size of the substituent still further, using the extremely bulky tris(trimethylsilyl)silyl group, causes a dramatic change in the packing, to a herringbone-like motif where each pentacene chromophore is surrounded by the silyl substituents of four neighbor molecules [cf. trimethylsilylethynylanthracene (**22**)]. The size of the silyl substituent easily occupies most of the face of the pentacene chromophore, making edge-to-face arrangements a viable space-filling strategy [86]. Clearly, in pentacene as well as in anthracene and tetracene, the ratio of the size of the alkyne substituent to the length of the acene plays a critical role in the solid-state arrangement of the molecules.

14.5
Higher Acenes

There has been little work on the functionalization of higher acenes, owing to the poor stability and very low solubility of the non-benzannelated derivatives. A poorly refined crystal structure of unsubstituted hexacene has been reported (without atomic coordinates), implying that it too arranges in a herringbone motif analogous to the lower acenes [87].

Functionalization of higher acenes with simple alkyl groups will lead to a decrease in the already low oxidation potential of these materials, making them harder to isolate in stable form. Silylethynylation has the effect of increasing oxidation potential, and so remains a viable strategy for the preparation of stable higher acenes. As was observed in the analogous pentacene series, the size of the alkyne substituent is closely related to the acene stability. As the acene chromophore increases in size, it becomes a much more potent diene and undergoes Diels–Alder addition to the alkyne of a second functionalized acene, even in the presence of the bulky triisopropylsilyl group [88]. It requires the much bulkier tri(tert-butyl)silyl substituent to retard this reaction and the only stable ethynylhexacene reported (**60**) utilizes this solubilizing group (Fig. 14.43, top) [89]. Increasing the size of the acene further requires even bulkier groups to retard the intermolecular cyclization reaction and only the tris(trimethylsilyl)silyl functional group yielded isolable functionalized heptacene [although the tri(tert-butylsilyl)ethynyl derivative of heptacene was stable for a short time in solution, it could not be isolated as a solid or even concentrated for crystallization]. The hex-

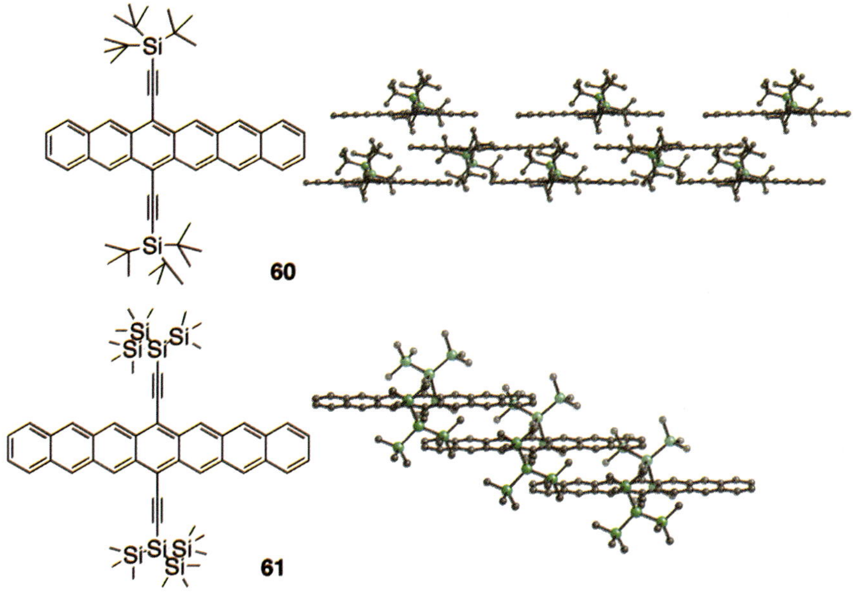

Fig. 14.43 Hexacene (**60**) and heptacene (**61**).

acene derivative **60** packs in a two-dimensional π-stacked arrangement similar to the TIPS pentacene **55**, with an interplanar spacing of 3.50 Å. The heptacene derivative **61** arranges in a slipped-stack arrangement similar to trimethylsilylpentacene derivative **50**, but with very little short-axis offset and an interplanar spacing of 3.44 Å. Neither of these materials is sufficiently stable as thin films for device fabrication and heptacene derivative **61** is stable for only a few weeks in the crystalline state. Further manipulation of the size of the alkyne substituent in these systems may lead to materials of higher stability or to stable acenes larger than heptacene.

14.6 Conclusion

Perturbing the crystalline order of acenes in order to enhance the π-face interactions desirable for electronics applications requires careful consideration of the size and placement of the added functional group. Although π-stacking interactions in the solid can lead to improved transport properties, the precise nature of the π-overlap is a critical factor that is highly sensitive to the size, polarity and hydrogen-bonding ability of the added functionality. The use of alkynes as scaffolds from which to attach crystal-modifying groups appears to be a powerful tool for the subtle control of π-face orientation in acene crystals and has the potential to fine-tune π-overlap for optimum device performance.

Acknowledgments

I am indebted to Chad Landis, Susan Odom and Marcia Payne for providing some of the X-ray crystal structures presented in this chapter, and to Dr. Sean Parkin for determining all of the structures produced by our group and assistance with searches of the CCDB.

References

1. Current commercial use of organic electronics is found mostly in the displays industry: S. Forrest, P. Burrows, M. Thompson, *IEEE Spectrum* **2000**, *37*, 8.
2. For recent reviews, see (a) Y. Sun, Y. Liu, D. Zhu, *J. Mater. Chem.* **2005**, *15*, 53, and references cited therein; (b) T. W. Kelley, P. F. Baude, C. Gerlach, D. E. Ender, D. Muyres, M. A. Haase, D. E. Vogel, S. D. Theiss, *Chem. Mater.* **2004**, *16*, 4413.
3. For example: (a) D. E. Janzen, M. W. Burand, P. C. Ewbank, T. M. Pappenfus, H. Higuchi, D. A. da Silva Fiho, V. G. Young, J.-L. Brédas, K. R. Mann, *J. Am. Chem. Soc.* **2004**, *126*, 15295; (b) H. E. Katz, *Chem. Mater.* **2004**, *16*, 4748, and references cited therein.
4. M. Pope, C. E. Swenberg, *Electronic Processes in Organic Crystals and Polymers*, Oxford University Press, New York, 2nd edn., **1999**.
5. A. Salomon, D. Cahen, S. Lindsay, J. Tomfohr, V. B. Engelkes, C. D. Frisbie, *Adv. Mater.* **2003**, *15*, 1881.
6. P. M. Kazmaier, R. Hoffmann, *J. Am. Chem. Soc.* **1994**, *116*, 9684
7. (a) J. L. Brédas, J. P. Calbert, D. A. da Silva Filho, J. Cornil, *Proc. Natl. Acad. Sci. USA* **2002**, *99*, 5804; (b) J.-L. Brédas, D. Beljonne, V. Coropceanu, J. Cornil, *Chem. Rev.* **2004**, *104*, 4971.
8. (a) G. R. Desiraju, A. Gavezzotti, *Acta Crystallogr. B* **1989**, *45*, 473; (b) J. D. Dunitz, A. Gavezzotti, *Acc. Chem. Res.* **1999**, *32*, 677; (c) A. Gavezzotti, G. R. Desiraju, *Acta Crystallogr. B* **1988**, *44*, 427.
9. There has been a recent study that proposed a model for quantizing aromatic interactions in the solid state, specifically aimed towards relating crystalline order to device performance: M. D. Curtis, J. Cao, J. W. Kampf, *J. Am. Chem. Soc.* **2004**, *126*, 4318.
10. For a recent review of advances in acene-based devices, see: M. Bendikov, F. Wudl, D. F. Perepichka, *Chem. Rev.* **2004**, *104*, 4891, and references cited therein.
11. Benzene structure: R. Fourme, D. Andre, M. Renaud, *Acta Crystallogr. B* **1971**, *27*, 1275; naphthalene structure: J. Oddershede, S. Larsen, *J. Phys. Chem. A* **2004**, *108*, 1057; for a recent review of aromatic edge-to-face interactions, see: W. B. Jennings, B. M. Farrell, J. F. Malone, *Acc. Chem. Res.* **2001**, *34*, 885.
12. J. Sworakowski, J. Ulanski, *Annu. Rep. Prog. Chem., Sect. C* **2003**, *99*, 87.
13. For an excellent review of electron and hole transport in small molecule organic solids, see: (a) N. Karl, K.-H. Kraft, J. Marktanner, M. Münch, F. Schatz, R. Stehle, H.-M. Uhde, *J. Vac. Sci. Technol. A* **1999**, *17*, 2318, and references cited therein; (b) A. N. Aleshin, J. Y. Lee, S. W. Chu, J. S. Kim, Y. W. Park, *Appl. Phys. Lett.* **2004**, *84*, 5383.
14. W. H. Melhuish, *J. Phys. Chem.* **1961**, *65*, 229.
15. S. Hamai, F. Hirayama, *J. Phys. Chem.* **1983**, *87*, 83.
16. For one of the most successful anthracene derivatives used in OLEDs, see: J. Shi, C. W. Tang, *Appl. Phys. Lett.* **2002**, *80*, 3201.
17. (a) M.-T. Lee, H.-H. Chen, C.-H. Liao, C.-H. Tsai, C. H. Chen, *Appl. Phys. Lett.* **2004**, *85*, 3301; (b) Y.-H. Kim, D.-C. Shin, S.-H. Kim, C.-H. Ko, H.-S. Yu, Y.-S. Chae, S.-K. Kwon, *Adv. Mater.* **2001**, *13*, 1690; (c) Z. L. Khang, X. Y. Jiang, W.

Q. Zhu, X. Y. Zheng, Y. Z. Wu, S. H. Xu, *Synth. Met.* **2003**, *137*, 1141; (d) X. H. Zhang, M. W. Liu, O. Y. Wong, C. S. Lee, H. L. Kwong, S. T. Lee, S. K. Wu, *Chem. Phys. Lett.* **2003**, *369*, 478; (e) T.-H. Liu, W.-J. Shen, C.-K. Yen, C.-Y. Iou, H.-H. Chen, B. Banumathy, C. H. Chen, *Synth. Met.* **2003**, *137*, 1033.
18. Z. Fei, N. Kocher, C. J. Mohrschladt, H. Ihmels, D. Stalke, *Angew. Chem. Int. Ed.* **2005**, *42*, 783.
19. A. E. Prozorovskii, V. A. Tafeenko, V. B. Ribakov, E. A. Shokova, V. V. Kovalev, *J.Struct.Chem.* **1987**, *28*, 183.
20. S. Ando, J. Nishida, E. Fujiwara, H. Tada, Y. Inoue, S. Tokito, Y. Yamashita, *Chem. Mater.* **2005**, *17*, 1261; see also: Y. Inoue, S. Tokito, K. Ito, T. Suzuki, *J. Appl. Phys.* **2004**, *95*, 5795.
21. L. M. Sweeting, A. L. Rheingold, J. M. Gingerich, A. W. Rutter, R. A. Spence, C. D. Cox, T. J. Kim, *Chem. Mater.* **1997**, *9*, 1103.
22. C. A. Hunter, J. K. M. Sanders, *J. Am. Chem. Soc.* **1990**, *112*, 5525; (b) M. Omar, C. D. Sherrill, *J. Am. Chem. Soc.* **2004**, *126*, 7690; (c) C. A. Hunter, K. R. Lawson, J. Perkins, C. J. Urch, *J. Chem. Soc., Perkin Trans. 2* **2001**, 651; (d) F. Cozzi, J. S. Siegel, *Pure Appl. Chem.* **1995**, *67*, 683.
23. J. Trotter, *Acta Crystallogr.* **1959**, *12*, 237.
24. E. J. Corey, J. O. Link, S. Sarshar, Y. Shao, *Tetrahedron Lett.* **1992**, *33*, 7103.
25. H.-D. Becker, B. W. Skelton, A. H. White, *Aust. J. Chem.* **1985**, *38*, 1567.
26. E. Heller, G. M. J. Schmidt, *Isr. J. Chem.* **1971**, *9*, 449.
27. L. J. Fitzgerald, R. E. Gerkin, *Acta Crystallogr. C* **1997**, *53*, 1080.
28. J. Iball, J. N. Low, *Acta Crystallogr. B* **1974** *30*, 2203.
29. M. Hospital, E. J. Gabe, J. P. Glusker, *Acta Crystallogr. B* **1971**, *27*, 1925.
30. (a) J. M. Adams, S. Ramdas, *Acta Crystallogr. B* **1979**, *35*, 679; (b) V. Langer, H.-D. Becker, *Z. Kristallogr.* **1992**, *199*, 313.
31. Crystal structure from Z. Burshtein, A. W. Hanson, C. F. Ingold, D. F. Williams, *J. Phys. Chem. Solids* **1978**, *39*, 1125.
32. J. Trotter, *Acta Crystallogr. C* **1986**, *42*, 862.
33. K. Peters, E.-M. Peters, K. Syassen, *Z.Kristallogr.* **1996**, *211*, 360.
34. J. Trotter, *Acta Crystallogr.* **1959**, *12*, 232.
35. For uses of BPEA: (a) J. Lukacs, R. A. Lampert, J. Metcalfe, D. Phillips, *J. Photochem. Photobiol. A* **1992**, *63*, 59; (b) P. J. Hanhela, D. B. Paul, *Aust. J. Chem.* **1984**, *37*, 553; (c) M. T. Cicerone, M, D, Ediger, *J. Chem. Phys.* **1996**, *104*, 7210; (d) D. R. Maulding, B. G. Roberts, *J. Org. Chem.* **1972**, *37*, 1458; (e) M. P. Heitz, M. Maroncelli, *J. Phys. Chem. A* **1997**, *101*, 5852; (e) D. J. Gisser, B. S. Johnson, M. D. Ediger, E. D. Von Meerwall, *Macromolecules* **1993**, *26*, 512; (f) F. R. Blackburn, C.-Y. Wang, M. D. Ediger, *J. Phys. Chem.* **1996**, *100*, 18249; for the crystal structure presented here: P. Nguyen, S. Todd, D. Van den Biggelaar, N. J. Taylor, T. B. Marder, F. Wittmann, R. H. Friend, *Synlett* **1994**, 299.
36. I. R. Butler, A. G. Callabero, G. A. Kelly, J. R. Amey, T. Kraemer, D. A. Thomas, M. E. Light, T. Gelbrich, S. J. Coles, *Tetrahedron Lett.* **2004**, *45*, 467.
37. C. A. Landis, S. R. Parkin, J. E. Anthony, *Jpn. J. Appl. Phsy.*, **2005**, *44*, 3921.
38. C. A. Landis, *PhD Dissertation*, University of Kentucky, January 2006.
39. S. A. Manhart, A. Adachi, K. Sakamaki, K. Okita, J. Ohshita, T. Ohno, T. Hamaguchi, A. Kunai, J. Kido, *J. Organomet. Chem.* **1999**, *592*, 52.
40. J. Iball, K. J. H. Mackay *Acta Crystallogr.* **1962**, *15*, 148.
41. R. J. Dellaca, B. R. Penfold, W. T. Robinson, *Acta Crystallogr. B* **1969**, *25*, 1589.
42. K. Akiba, M. Yamashita, Y. Yamamoto, S. Nagase, *J. Am. Chem. Soc.* **1999**, *121*, 10644.
43. W. Nakanishi, S. Hayashi, N. Itoh, *J. Org. Chem.* **2004**, *69*, 1676.
44. Y. K. Kryschenko, S. R. Seidel, D. C. Muddiman, A. I. Nepomuceno, P. J. Stang, *J. Am. Chem. Soc.* **2003**, *125*, 9647.
45. This approach has been exploited extensively in the area of disc-shaped aromatics; for a review, see: S. Chandrasekhar, G. S. Ranganath, *Rep. Prog. Phys.* **1990**, *53*, 57.
46. J.-L. Pozzo, J.-P. Desvergne, G. M. Clavier, H. Bouas-Laurent, P. G. Jones, J.

Perlstein, *J. Chem. Soc., Perkin Trans. 2* **2001**, 824.
47. S. Norvez, F.-G. Tournilhac, P. Bassoul, P. Herson, *Chem. Mater.* **2001**, *13*, 2552, and references cited therein.
48. See, for example: S. I. Stupp, V. LeBonheur, K. Walker, L. S. Li, K. E. Huggins, M. Keser, A. Amstutz, *Science* **1997**, *276*, 384.
49. H. Dang, M. Levitus, M. A. Garcia-Garibay, *J. Am. Chem. Soc.* **2002**, *124*, 136.
50. M. J. Krische, J.-M. Lehn, N. Kyritsakas, J. Fischer, E. K. Wegelius, K. Rissanen, *Tetrahedron* **2000**, *56*, 6701.
51. S. Sekizaki, C. Tada, H. Yamochi, G. Saito, D. E. Henn, W. G. Williams, D. J. Gibbons, D. Holmes, S. Kumaraswamy, A. J. Matzger, K. P. C. Vollhardt, *Chem. Eur. J.* **1999**, *5*, 3399.
52. D. J. Gundlach, J. A. Nichols, L. Zhou, T. N. Jackson, *Appl. Phys. Lett.* **2002**, *80*, 2925.
53. E. Botzung-Appert, V. Monnier, T. Ha Duong, R. Pansu, A. Ibanez, *Chem. Mater.* **2004**, *16*, 1609.
54. A. Hepp, H. Heil, W. Weise, M. Ahles, R. Schmechel, H. von Seggern, *Phys. Rev. Lett.* **2003**, *91*, 157406.
55. (a) R. W. I. De Boer, T. M. Klapwijk, A. F. Morpurgo, *Appl. Phys. Lett.* **2003**, *83*, 4345; (b) H. E. Katz, C. Kloc, V. Sundar, J. Zaumseil, A. L. Briseno, Z. Bao, *J. Mater. Res.* **2004**, *19*, 1995.
56. P. J. Cox, G. A. Sim, *Acta Crystallogr. B* **1979**, *35*, 404.
57. H. Moon, R. Zeis, E.-J. Borkent, C. Besnard, A. J. Lovinger, T. Siegrist, C. Kloc, Z. Bao, *J. Am. Chem. Soc.* **2004**, *126*, 15322
58. H. Moon, R. Zeis, E.-J. Borkent, C. Besnard, A. J. Lovinger, T. Siegrist, C. Kloc, Z. Bao, *J. Am. Chem. Soc.* **2004**, *126*, 15322.
59. Q. Miao, M. Lefenfeld, T.-Q. Nguyen, T. Siegrist, C. Kloc, C. Nuckolls, *Adv. Mater.* **2005**, *17*, 407.
60. S. A. Odom, S. R. Parkin, J. E. Anthony, *Org. Lett.* **2003**, *5*, 4245.
61. J. E. Anthony, *Chem. Rev.*, in press (2006).
62. J. A. Dodge, J. D. Bain, A. R. Chamberlin, *J. Org. Chem.* **1990**, *55*, 4190.
63. Rubrene crystal structure: I. Bulgarovskaya, V. Vozzhennikov, S. Aleksandrov, V. Belsky, *Latv. Zinat. Akad. Vestis, Khim. Ser.* **1983**, 53.
64. (a) V. Podzorov, E. Menard, A. Borissov, V. Kiryukhin, J. A. Rogers, M. E. Gershenson, *Phys. Rev. Lett.* **2004**, *93*, 086602, and references cited therein.
65. D. A. da Silva Filho, E. G. Kim, J. L. Brédas, *Adv. Mater.* **2005**, *17*, 1072.
66. I. F. Shchegolev, E. B. Yagubskii, in *Extended Linear Chain Compounds*, J. S. Miller (Ed.), Plenum Press, New York, **1982**, Vol. 2, p. 385.
67. (a) R. P. Shibaeva, in *Extended Linear Chain Compounds*, J. S. Miller (Ed.), Plenum Press, New York, **1982**, Vol. 2, p. 435; (b) M. Bendikov, F. Wudl, D. F. Perepichka, *Chem. Rev.* **2004**, *104*, 4891.
68. O. Dideberg, J. Toussaint, *Acta Crystallogr. B* **1974**, *30*, 2481.
69. D. J. Sandman, J. C. Stark, B. M. Foxman, *Organometallics* **1982**, *1*, 739.
70. K. A. Balodis, A. D. Livdane, R. S. Medne, O. Y. Neiland, *J. Org. Chem. USSR (Engl. Transl.)* **1979**, *15*, 343.
71. (a) A. C. Mayer, M. T. Lloyd, D. J. Herman, T. G. Kasen, G. G. Malliaras, *Appl. Phys. Lett.* **2004**, *85*, 6272; (b) S. Yoo, B. Domercq, B. Kippelen, *Appl. Phys. Lett.* **2004**, *85*, 5427.
72. For recent reviews, see: (a) C. D. Dimitrakopoulos, P. R. L. Malenfant, *Adv. Mater.* **2002**, *14*, 99; (b) G. Horowitz, *Adv. Mater.* **1998**, *10*, 365.
73. See, for example: (a) D. J. Gundlach, C. C. Kuo, S. F. Nelson, T. N. Jackson, in *Proceedings of the 57th Device Research Conference Digest*, **1999**, p. 164; (b) Z. Bao, V. Kuck, J. A. Rogers, M. A. Paczkowski, *Adv. Funct. Mater.* **2002**, *12*, 526; (c) A. Salleo, M. L. Chabinyc, M. S. Yang, R. A. Street, *Appl. Phys. Lett.* **2002**, *81*, 4383.
74. See, for example: (a) K. P. Pernistich, A. N. Rashid, S. Haas, D. Oberhoff, C. Goldmann, G. Schitter, D. J. Gundlach, B. Batlogg, *J. Appl. Phys.* **2004**, *96*, 6431; (b) I. Yagi, K. Tsukagoshi, Y. Aoyagi, *Appl. Phys. Lett.* **2005**, *86*, 103502.
75. J. Cornil, J. Ph. Calbert, J. L. Brédas, *J. Am. Chem. Soc.* **2001**, *123*, 1250.
76. (a) L. C. Picciolo, H. Murata, Z. H. Kafafi, *Appl. Phys. Lett.* **2001**, *78*, 2378; (b) M.A. Wolak, B.-B. Jang, L. C. Palilis, Z.

H. Kafafi, *J. Phys. Chem. B.* **2004**, *108*, 5492.

77. S. Subramanian, S. R. Parkin, M. Siegler, J. Chen, K. Gallup, C. Haughn, D. C. Martin, J. E. Anthony, Submitted to *Chem. Mater.* (2006).
78. J. E. Anthony, D. L. Eation, S. R. Parkin, *Org. Lett.* **2002**, *4*, 15.
79. For a detailed description of the packing and electronic properties of these molecules, see: J. E. Anthony, J. S. Brooks, D. L. Eaton, J. R. Matson, S. R. Parkin, *Proc. SPIE* **2003**, *5217*, 124.
80. J. E. Anthony, J. S. Brooks, D. L. Eaton, S. R. Parkin, *J. Am. Chem. Soc.* **2001**, *123*, 9482.
81. C. D. Sheraw, T. N. Jackson, D. L. Eaton, J. E. Anthony, *Adv. Mater.* **2003**, *15*, 2009.
82. S. K. Park, C.-C. Kuo, J. E. Anthony, T. N. Jackson, *2005 Int. Electron Dev. Mtg., Tech. Digest*, **2006**, 113.
83. F. A. Hegmann, R. R. Tykwinski, K. P. H. Lui, J. E. Bullock, J. E. Anthony, *Phys. Rev. Lett.* **2002**, *89*, 227403-(1–4); (b) O. Ostroverkhova, D. G. Cooke, S. Shcherbyna, R. F. Egerton, F. A. Hegmann, R. R. Tykwinski, J. E. Anthony, *Phys. Rev. B* **2005**, *71*, 035204.
84. (a) M. Weck, A. R. Dunn, K. Matsumoto, G. W. Coates, E. B. Lobkovsky, R. H. Grubbs, *Angew. Chem., Int. Ed.* **1999**, *38*, 2741; (b) F. Ponzini, R. Zagha, K. Hardcastle, J. S. Siegel, *Angew. Chem. Int. Ed.* **2000**, *39*, 2323; (c) R. E. Gillard, J. F. Stoddart, A. J. P. White, B. J. Williams, D. J. Williams, *J. Org. Chem.* **1996**, *61*, 4504; (d) for a case specific to acenes, see: J. C. Collings, K. P. Roscoe, R. L. Thomas, A. S. Batsanov, L. M. Stimson, J. A. K. Howard, T. B. Marder, *New J. Chem.* **2001**, *25*, 1410; (e) J. West, S. Mecozzi, D. A. Dougherty, *J. Phys. Org. Chem.* **1997**, *10*, 347.
85. C. R. Swartz, S. R. Parkin, J. E. Bullock, J. E. Anthony, A. C. Mayer, G. G. Malliaras, *Org. Lett.* **2005**, *7*, 3163.
86. M. M. Payne, S. A. Odom, S. R. Parkin, J. E. Anthony, unpublished results.
87. R. B. Campbell, J. M. Robertson, *Acta Crystallogr.* **1962**, *15*, 289.
88. M. M. Payne, S. A. Odom, S. R. Parkin, J. E. Anthony, *Org. Lett.* **2004**, *6*, 3325.
89. M. M. Payne, S. R. Parkin, J. E. Anthony, *J. Am. Chem. Soc.* **2005**, *127*, 8028.

15
Synthesis and Characterization of Novel Chiral Conjugated Materials

Andrzej Rajca and Makoto Miyasaka

15.1
Introduction

Chirality is increasingly important in the design of organic π-conjugated materials. One of the aspects of the design involves optimization of achiral properties through the introduction of chirality. In this context, the inherent three-dimensional character of chirality and the control of intermolecular interactions associated with diastereomeric recognition provide a versatile handle for the optimization of supramolecular structures, film morphology and liquid crystalline order of π-conjugated polymers and oligomers in three dimensions [1–5]. These inherently achiral properties have an impact on optoelectronic coupling, which may affect the fabrication of light-emitting diodes, field effect transistors, photodiodes, photovoltaic cells, fluorescent sensors and other devices [2, 3, 5]. Another aspect of the design involves optimization of chiral counterparts of properties in optics (circular polarization of light), electronics (chiral transport of charge carriers), etc. [6–11]. The challenge in this area is to obtain materials with inherently strong chiral properties at the macromolecular level, rather than derived from an aggregate or a supramolecular structure. Such inherently strong chiral properties, approaching or exceeding in magnitude their achiral counterparts, would facilitate exploration of chiral properties of single molecule devices and the design of novel chiral materials.

For typical π-conjugated polymers, helical conformation in the π-conjugated backbone may be induced via the chiral nonracemic pendants, complexation of a chiral nonracemic compound or polymerization in chiral nematic liquid crystal (LC) [12–14]. There are few structural motifs for nonracemic polymers with configurationally stable helical (and formally cross-conjugated) backbones such as polyisocyanides or with in-chain π-conjugated chirality [12, 15]. However, the majority, if not all, of π-conjugated polymers only display significant nonracemic helicity in an aggregated or ordered form [12].

The introduction of angular connectors (e.g. 1,3-phenylenes, 2,7-naphthylenes) in the π-conjugated backbones may provide an oligomer with a preferred helical

Functional Organic Materials. Syntheses, Strategies, and Applications.
Edited by Thomas J.J. Müller and Uwe H.F. Bunz
Copyright © 2007 WILEY-VCH Verlag GmbH & Co. KGaA, Weinheim
ISBN: 978-3-527-31302-0

conformation. The nonracemic helicity of such foldamers, which may be induced via chiral nonracemic pendants or in-chain chirality (e.g. 1,1'-binaphthyls, [4]helicenes), is also associated with an aggregated form and it is solvent dependent [16, 17].

Another class of polymers and oligomers possess highly annelated (e.g. ladder-type) chiral π-conjugated systems [18–21]. The oligomers with helical, ladder-type π-conjugated systems were found to be among molecules with the strongest chiral properties. Whereas helical polymers with ladder-type connectivity have been prepared with varied degrees of success [18, 19], a highly extended helical ladder-type π-conjugated system has not been attained.

This chapter addresses synthetic aspects of molecules with highly annelated chiral π-conjugated systems, with the focus on helicene-type oligomers [20]. In addition to synthesis, configurational stability (barriers for racemization) and selected chiroptical properties of such molecules will be discussed.

15.2
Synthetic Approaches to Highly Annelated Chiral π-Conjugated Systems

15.2.1
Helicenes

[n]Helicenes possess n angularly annelated aromatic rings, forming a helically shaped π-conjugated system, such as six ortho-annelated benzene rings in [6]helicene (1) (Fig. 15.1) [21]. For [n]helicenes with $n \geq 5$, the distortion from planarity due to steric repulsion of terminal rings is sufficient to allow for resolution of en-

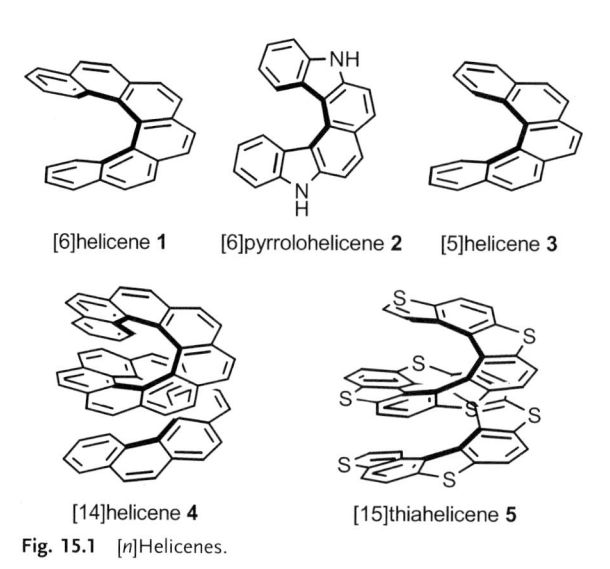

Fig. 15.1 [n]Helicenes.

antiomers. Because both high degree of annelation and significant strain have to be introduced into such π-systems, efficient syntheses of [n]helicenes are challenging.

The first [n]helicenes, [6]pyrrolohelicene (**2**) and [5]helicene (**3**), were reported in 1927 and 1933, respectively [22–24]. The synthesis of a non-racemic [n]helicene, [6]helicene (**1**), was first described by Newman and Lednicer in 1956 [25]. Photochemical syntheses developed during the 1960s and 1970s provided the longest [n]helicenes to date, that is, [n]helicenes [with up to $n = 14$ benzene rings (**4**)] and [n]thiahelicenes [with up to $n = 15$ alternating benzene and thiophene rings (**5**)] [26–30].

Recent work by Katz and coworkers led to the development of very efficient nonphotochemical, gram-scale syntheses of functionalized enantiopure [n]helicenes ($n = 5$, 6 and 7). These approaches were based on racemic syntheses and classical resolutions with a chiral auxiliary [31, 32].

The extraordinarily strong chiral properties of [n]helicenes provide an impetus for the development of synthetic approaches to nonracemic [n]helicenes for applications as organic materials. From this point of view, asymmetric syntheses of functionalized long [n]helicenes ($n > 7$), and also [n]helicene-like molecules and polymers with novel electronic structures and material properties, are important. The properties of helicenes related to materials are relatively unexplored, compared with the more synthetically accessible π-conjugated molecules and polymers. Notably, redox states of helicenes are practically unknown [33, 34]. Assembly of helicenes on surfaces, their uses as liquid crystals, chiral sensors, ligands or additives for asymmetric synthesis and helicene–biomolecule interactions are in the exploratory stages [35–43].

In general, the synthesis of [n]helicenes involves two key synthetic steps, i.e. connection and annelation. The most efficient approaches to [n]helicenes use multiple annelations, forming two or more rings in one synthetic step; however, relatively few annelation reactions have been implemented effectively.

15.2.1.1 Photochemical Syntheses

Oxidative photocyclization of stilbene to phenanthrene, which was discovered in 1960, was first applied to the synthesis of [7]helicene by Martin and coworkers in 1967 [26, 44, 45]. This approach was extended to the syntheses of long [n]helicenes ($n \leq 14$) and [n]thiahelicenes ($n \leq 15$), using both mono- and diannelations. Numerous [n]helicenes ($n \leq 13$) and [n]thiahelicenes ($n \leq 13$) were obtained in non-racemic form via the following methods: (1) seeded crystallization of conglomerate (e.g. [7]-, [8]- and [9]helicene) [46, 47], (2) resolution by chromatography (e.g. [13]thiahelicene) [48] and (3) photocyclization from a resolved precursor (e.g. [13]helicene from hexahelicene-2-carboxylic acid) [49]. The oxidative photocyclization of stilbenes is still the method of choice for the preparation of selected [n]helicenes and their heteroatom analogs [50–58].

Vollhardt's group adopted the cobalt-catalyzed photochemical cyclotrimerization of alkynes for the rapid construction of a novel class of helicenes, helical [n]phe-

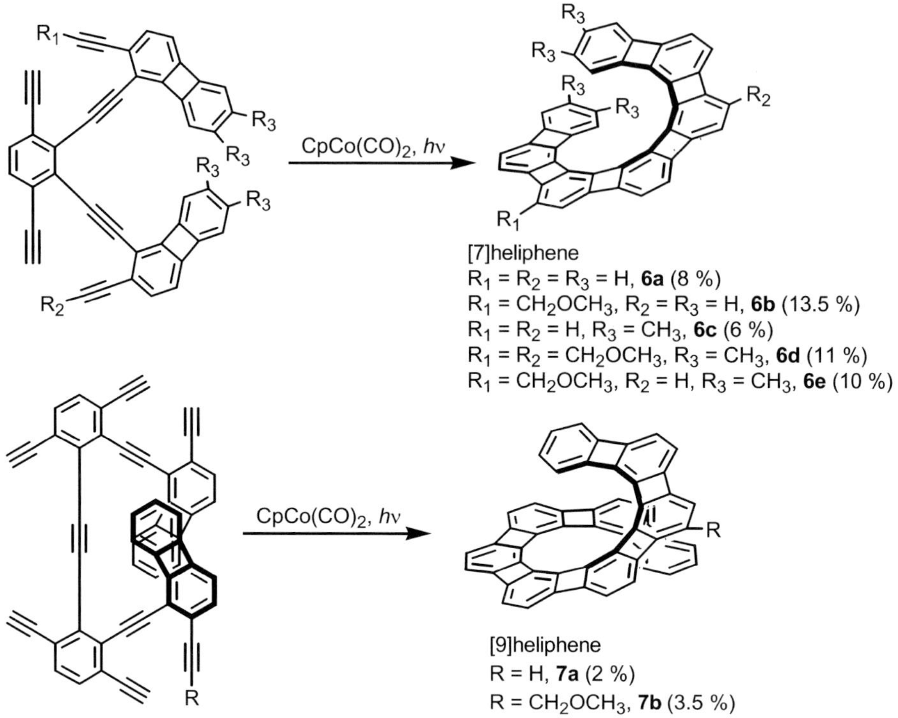

Fig. 15.2 Photochemical syntheses of [n]heliphenes via di- and triannelation.

[7]heliphene
$R_1 = R_2 = R_3 = H$, **6a** (8 %)
$R_1 = CH_2OCH_3$, $R_2 = R_3 = H$, **6b** (13.5 %)
$R_1 = R_2 = H$, $R_3 = CH_3$, **6c** (6 %)
$R_1 = R_2 = CH_2OCH_3$, $R_3 = CH_3$, **6d** (11 %)
$R_1 = CH_2OCH_3$, $R_2 = H$, $R_3 = CH_3$, **6e** (10 %)

[9]heliphene
$R = H$, **7a** (2 %)
$R = CH_2OCH_3$, **7b** (3.5 %)

nylenes or "[n]heliphenes", consisting of alternating [n]benzene units fused with [n − 1]cyclobutadiene rings [59, 60]. This class of helicenes have a significantly increased helical radius compared with benzene-based helicenes. The synthesis is based on di- and triannelations of the corresponding oligoynes, providing [7]heliphenes (**6**) and [9]heliphenes (**7**) in ~10% and 2–3.5% yields, respectively (Fig. 15.2). In the case of triannelation, nine rings were formed in one step, including six cyclobutadiene rings with an estimated strain of 300 kcal mol^{-1}! This is the only report of photochemical triannelation leading to helicene-like skeleton.

Molecular geometries from X-ray structures for [n]heliphenes with n up to 8 indicate significant bond fixation in the π-system. However, significant electron delocalization was found [59, 60]. Based on the linear plot of λ_{max} vs $1/(n + 0.5)$, the optical band gaps E_g = 2.2 eV (n = 2–9) or E_g = 2.1 eV (n = 4–9) may be estimated for the corresponding polyheliphene. Analogous plots of λ_{max} vs $1/n$ for [n]helicenes (n = 6–9) and [n]thiahelicenes (n = 5, 7, 9, 11) give somewhat higher E_g = 2.5 eV (R^2 = 0.975) and E_g = 2.4 eV (R^2 = 0.984), respectively. Heliphenes possess surprisingly low barriers for racemization and could not be resolved (see Section 15.3).

15.2.1.2 Non-photochemical Syntheses

Syntheses of [n]helicenes (and [n]thiahelicenes) via oxidative photocyclizations of stilbenes have serious constraints. They must be carried out in dilute solutions (~1 mM) and the functional groups that significantly affect relaxation of the singlet excited states of stilbenes such as bromo, iodo, keto, amino and nitro, are typically not compatible; for longer helicenes, problems with regioselectivity of photocyclization may be encountered [61]. These limitations have spurred the recent developments of nonphotochemical syntheses for efficient preparation of highly functionalized, nonracemic helicenes.

For nonphotochemical syntheses, there are three major approaches: (1) annelation of racemic intermediate leading to racemic [n]helicene, followed by resolution, (2) annelation of nonracemic intermediate giving nonracemic [n]helicene and (3) asymmetric synthesis, i.e. annelation of racemic intermediate in the presence of chiral reagent, catalyst or auxiliary.

To date, the reported nonphotochemical syntheses of racemic and enantiopure [n]helicenes are limited to $n = 9$ and 8, respectively [62, 63]. Recently, asymmetric syntheses of [n]helicenes with n up to 11 were developed in our laboratory [64].

Annelation of racemic intermediate leading to racemic [n]helicene
Larsen and Bechgaard reported the nonphotochemical synthesis of racemic [5]- and [9]thiahelicenes, relying on monoannelations of stilbene precursors [62]. Electrochemical or chemical (FeCl$_3$) oxidation was used in place of usual photooxidation, to provide thiahelicenes in 20–65 % yields. For example, racemic [9]thiahelicene **9** was obtained in ~60 % yield from stilbene **8** by oxidation with FeCl$_3$ in methylene chloride (Fig. 15.3); a similar result was obtained by classical photooxidation of stilbene **8** [51].

Annelations via Diels–Alder reaction were employed by Katz and coworkers to develop exceedingly efficient methods for multi-gram scale syntheses of [n]helicenes with n up to 7. Efficient resolutions were carried out via functionalization with the camphanate ester derivatives and then separation of diastereomers by column chromatography (> 98 % de). This synthetic approach may be illustrated by the synthesis of enantiopure [7]thiahelicene **10** [65]. Notably, the Diels–Alder diannelation and aromatization gives racemic **10** in 95 % yield. Resolution with camphanate tetraester provides enantiopure **11** (Fig. 15.4).

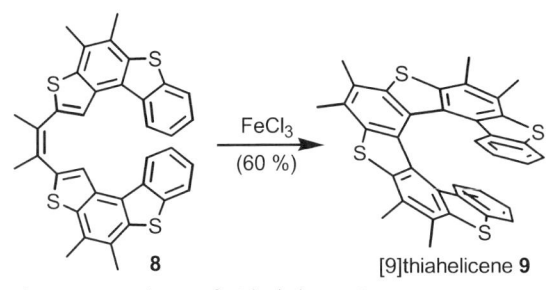

Fig. 15.3 Synthesis of [9]thiahelicene **9**.

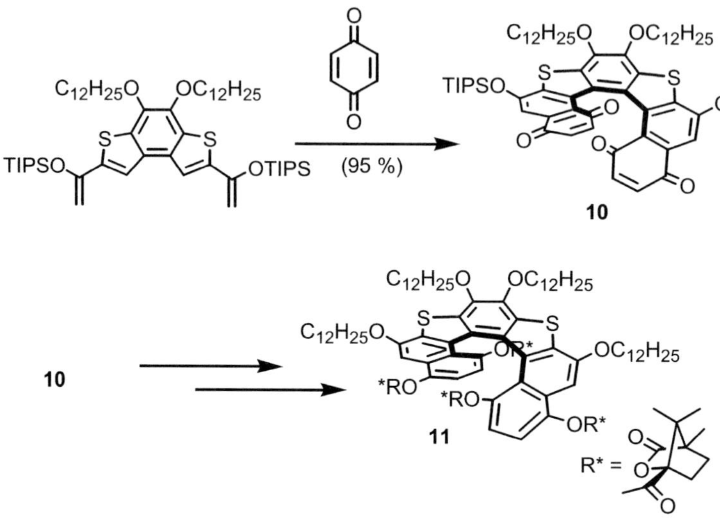

Fig. 15.4 Synthesis of [7]thiahelicene **11** via Diels–Alder diannelation.

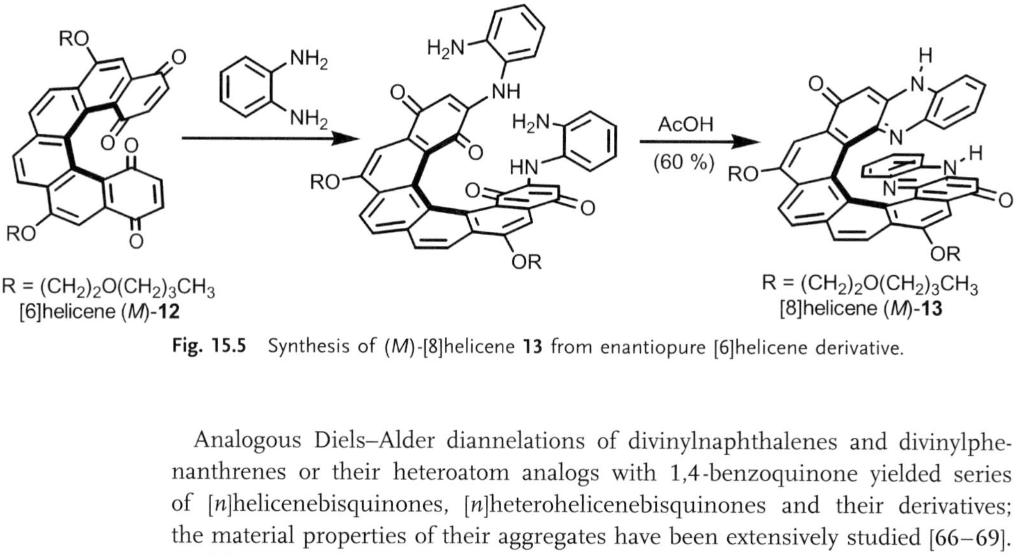

Fig. 15.5 Synthesis of (M)-[8]helicene **13** from enantiopure [6]helicene derivative.

Analogous Diels–Alder diannelations of divinylnaphthalenes and divinylphenanthrenes or their heteroatom analogs with 1,4-benzoquinone yielded series of [n]helicenebisquinones, [n]heterohelicenebisquinones and their derivatives; the material properties of their aggregates have been extensively studied [66–69].

Elaboration of enantiopure (M)-[6]helicenebisquinone **12** into (M)-[8]helicene **13** with complete transfer of ee was reported (Fig. 15.5) [63]. [8]Helicene **13** is the longest enantiopure helicene prepared via nonphotochemical synthesis.

Annelations via Friedel–Crafts acylation were applied to racemic syntheses of [n]helicenes with n up to 6. The 12-step synthesis of [6]helicene (**1**) by Newman and Ledgicer employed stepwise diannelation, followed by oxidative aromatization and then resolution with the complex of a chiral π-acceptor, α-2,4,5,7-tetranitro-9-fluorolideneaminooxypropionic acid (TAPA) and **1** [25]. The one-step diannelation was ubiquitously applied to syntheses of triarylamine [4]helicenes,

Fig. 15.6 Triarylamine [4]helicenes: synthesis and barriers for inversion.

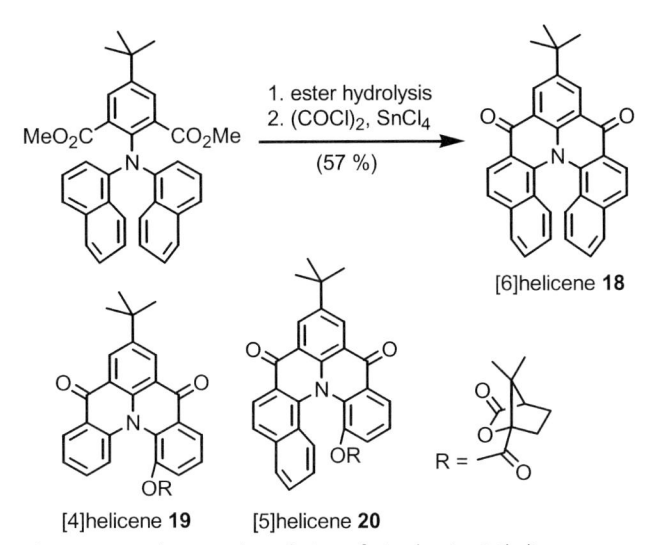

Fig. 15.7 Synthesis and resolution of triarylamine [n]helicenes.

e.g. **14–16**, by Hellwinkel and Schmidt (Fig. 15.6) [70]. Diastereotopicity of the methyl groups in **15–17** was used to determine barriers for inversion (racemization) by ^{1}H NMR spectroscopy. For 5,5,9,9-tetramethyl-5H,9H-quino[3, 2,1-*de*]acridine (**16**), Fox et al. reported an improved procedure, using 10 % P_2O_5 in methanesulfonic acid at room temperature as a mild cyclodehydrating reagent/solvent, to give a 92 % yield for one-step diannelation (Fig. 15.6) [71].

Hellwinkel and Schmidt's approach was recently extended by Venkataraman and coworkers to triarylamine [n]helicenes, with $n \leq 6$, e.g. **18** [72]. Sterically hindered [4]helicene **19** and [5]helicene **20** were resolved, using Katz's camphanate ester derivatives (Fig. 15.7) [73].

Other annelation reactions leading to racemic [n]helicenes with $n \geq 6$ include carbenoid coupling providing the parent [7]helicene [74].

Annelation of nonracemic intermediate giving nonracemic [n]helicene
Several methods for the transfer of center or axial chirality into helical chirality have been described. In many cases, practically complete transfer of enantiomeric excess was observed.

An interesting example of the transfer of center chirality to helicity is the work by Ogawa et al., based on an asymmetric aromatic oxy-Cope rearrangement to provide nonracemic [5]helicenes (Fig. 15.8) [75]. The starting material with center chirality, bicyclo[2,2,2]ketone (–)-**21** (>98 % ee), was obtained by enzymatic resolution. In the annelation step, the phenanthrene derivative was subjected to aromatic oxy-Cope rearrangement, to afford a pentacyclic product in 47 % yield. The corresponding [5]helicene **22** was obtained in 7 % overall yield (>98 % ee) after six steps.

Two approaches to nonracemic [5]helicenes, starting from axially chiral binaphthyls, provided an early stereochemical correlation between axial and helical chiralities. The correlation was based on comparison of the signs of optical rotations; the absolute configurations of binaphthyls were obtained from chemical

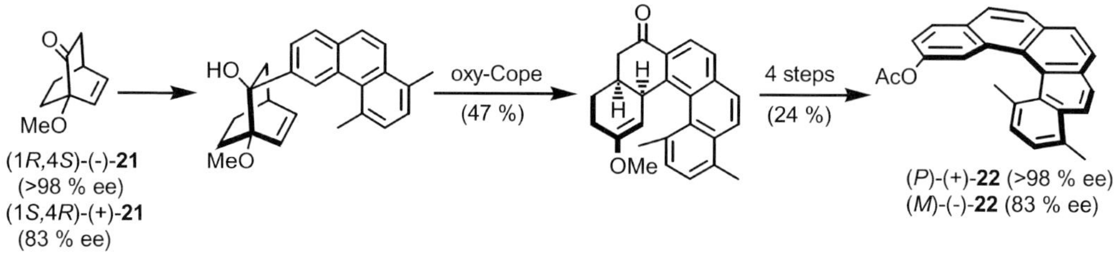

Fig. 15.8 Synthesis of nonracemic [5]helicene **22** via oxy-Cope rearrangement.

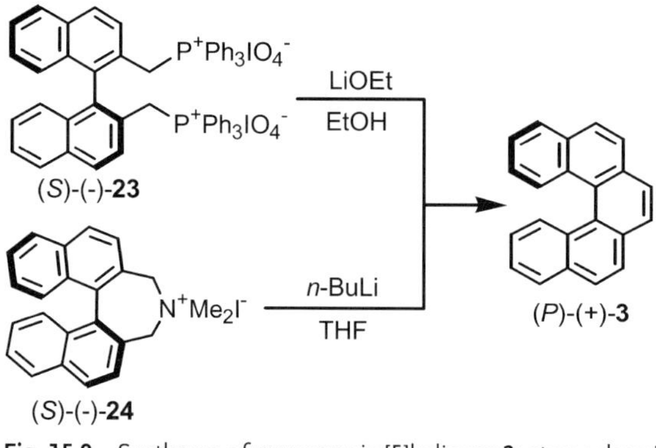

Fig. 15.9 Syntheses of nonracemic [5]helicene **3**: stereochemical correlation between axial and helical chiralities.

correlation. Starting from optically enantiopure derivatives of (S)-(–)-binaphthyl, annelation via either oxidative cyclization of the 2,2′-bis-phosphonium periodate **23** or Stevens rearrangement of quaternary ammonium salt **24** gave enantiopure (P)-(+)-[5]helicene **3** (Fig. 15.9) [76, 77].

The axial-to-helical configuration transfer was more recently applied to [7]thiahelicenes **29** and **33**, using McMurry reaction of dialdehyde or diketone in the annelation step, to form the central benzene ring [78, 79]. Both enantiomers of helicenes could be obtained.

In the synthesis of **29** [78], axially chiral precursor was obtained via biaryl coupling between benzodithiophenes with chiral oxazoline [derived from (S)-valinol] as an auxiliary, e.g. Stille cross-coupling of **25** and **26** provided bis(benzodithiophene) **27** in 68% yield with 49% *de* (Fig. 15.10). Although the diastereomers were not separable by silica gel column chromatography, crystallization of diastereomers from hexane–ethyl acetate gave crystals with an S-axial configuration. Both enantiomers were obtained according to the following sequence: ring opening and acetylation the bis(oxazolines), separation of diastereomers by silica gel chromatography, reduction with LiAlH$_4$ to diol and oxidation with PCC to dialdehyde **28**. The intramolecular McMurry reaction (TiCl$_3$–DME$_{1.5}$/Zn–Cu) of enantiomeric pure dialdehyde gave the enantiopure [7]thiahelicene **29** (>99% *ee*) (Fig. 15.10).

In the synthesis of **33** (Fig. 15.11) [79], hexathiophene **30** was converted to racemic diketone **31** and then kinetic resolution of diketone **31** was carried out using (–)-B-chlorodiisopinocampheylborane [(–)-DIP-chloride]. The unreacted (–)-diketone **31** was isolated in ~40% yield. The reduction product, mono-alcohol **32**, gave (+)-diketone **31** in ~40% yield based upon racemic diketone. The intramolecular McMurry reaction of diketone **31**, using TiCl$_3$–Zn–DME (instant method), gave [7]helicene **33** in 17–63% isolated yield. [7]Helicene diol **35** (probably the trans-isomer) was formed in complementary yields, so the overall for annelation to [7]helicene was about 70%. The yields are relatively high in view of steric hindrance of the diketone and the presence of both TMS- and Br-functionality. Removal of the TMS groups gives [7]helicene **34**; racemic **34** crystallizes as either conglomerate or racemic compound, depending on the conditions of crystallization (Fig. 15.11).

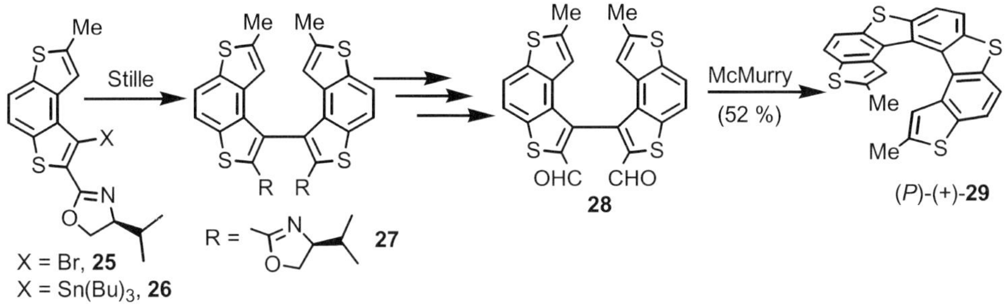

Fig. 15.10 Synthesis of enantiopure [7]thiahelicene via McMurry annelation of dialdehyde.

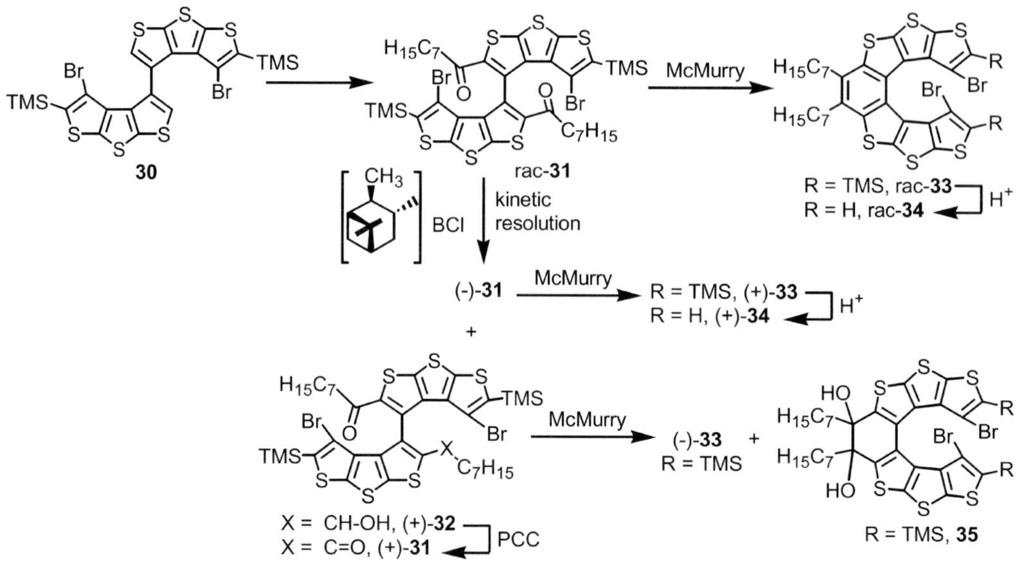

Fig. 15.11 Synthesis of highly functionalized enantiopure [7]thiahelicenes via kinetic resolution of axially chiral diketones and McMurry annelation.

Based on X-ray crystallographic determinations of the absolute configurations for (+)-enantiomers of the diketone and the [7]helicene **34**, stereochemical correlation between the *R* axial chirality of the diketone and the *M* helical chirality of the [7]helicene was unequivocally established [79]. Enantiopure [7]helicene **33** was functionalized for the design of organic chiral glasses with strong chiroptical properties (Section 15.4) and for homologation to higher [*n*]helicenes.

Asymmetric synthesis
To date, only three approaches to nonphotochemical asymmetric syntheses of [*n*]helicenes with $n \geq 6$ have been reported. The key step in such syntheses is annelation of racemic intermediate in the presence of a chiral reagent, catalyst or auxiliary.

Carreño et al. described asymmetric syntheses of [5]helicenequinone and [7]helicenebisquinone [80, 81]. This approach is illustrated for [7]helicenebisquinone **36** [80]. The key step was diannelation based on the Diels–Alder reaction between 3,6-divinyl-1,2,7,8-tetrahydrophenanthrene and enantiopure (*S*,*S*)-2-(*p*-tolylsulfinyl)-1,4-benzoquinone; subsequent elimination and aromatization steps gave **36** with high *ee* (Fig. 15.12).

Another important approach to the synthesis of helicenes is the convergent route based on energy-rich *cis,cis*-dienetriynes and their Ni(0)-catalyzed [2+2+2]-cycloisomerization, which was reported by Stará and coworkers [82, 83]. The potential of this methodology was demonstrated by the synthesis of functionalized

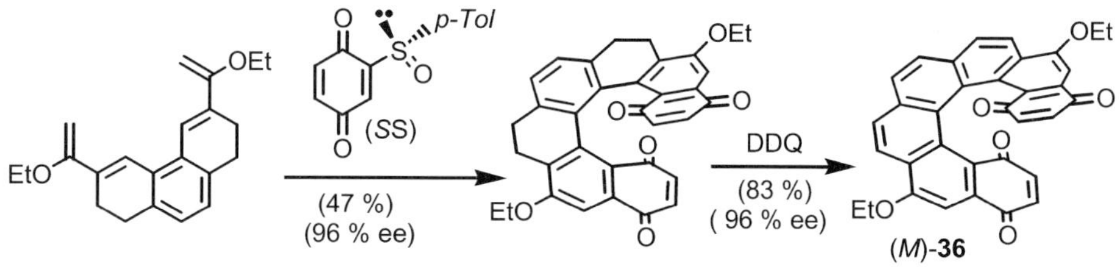

Fig. 15.12 Asymmetric synthesis of [7]helicenebisquinone **36**.

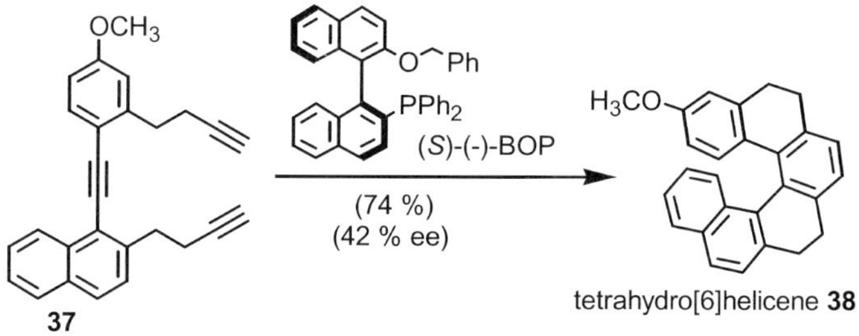

Fig. 15.13 Nickel(0)-catalyzed asymmetric synthesis of tetrahydro[6]helicene **38**.

[5]-, [6]- and [7]helicenes in 60–83% yield. However, only one example of asymmetric synthesis using this approach was reported, i.e. the Ni(0)-catalyzed annelation of triyne **37** to the corresponding tetrahydro[6]helicene **38** (74%, 42% ee) in the presence of a chiral ligand [(S)-(–)-BOP] (Fig. 15.13) [83].

A new class of β-oligothiophenes, in which n thiophenes are helically annelated to form [n]helicene, was developed recently [64, 84–87]. Such carbon–sulfur [n]helicenes are fragments of an unusual polymer such as carbon–sulfur $(C_2S)_n$ helix (Fig. 15.14). The sulfur-rich molecular periphery is expected to facilitate multiple S–S short contacts that have been associated with improved transport properties in sulfur-containing molecular solids. The carbon–carbon frameworks in $(C_2S)_n$ helices are formally cross-conjugated, in contrast to electron-delocalized π-conjugated systems of α-oligothiophenes. However, the questions of electron delocalization vs. localization and the magnitude of the band gap in the $(C_2S)_n$ helix require the synthesis of a series of sufficiently long oligomers, to account for the effect of helical distortion on the electronic structure. The synthesis of nonracemic [n]helicenes will provide an insight into the relationship between chiral properties and cross-conjugation. This is important for the development of materials with strong chirality, wide band gap and optical transparency.

Iterative racemic synthesis and asymmetric synthesis of the [7]helicene **40** and resolution of its TMS-free derivative **41** were reported [84, 85]. The racemic synthesis was based on iterative alternation of two steps: CC bond homocouplings between the β-positions of thiophenes and annelation between the α-positions of thiophenes (Fig. 15.15).

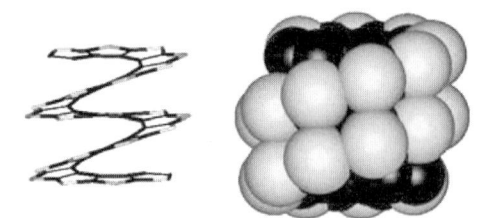

Fig. 15.14 Carbon–sulfur $(C_2S)_n$ helix illustrated as stick and space-filling plots for oligomers of 21 helically annelated thiophenes.

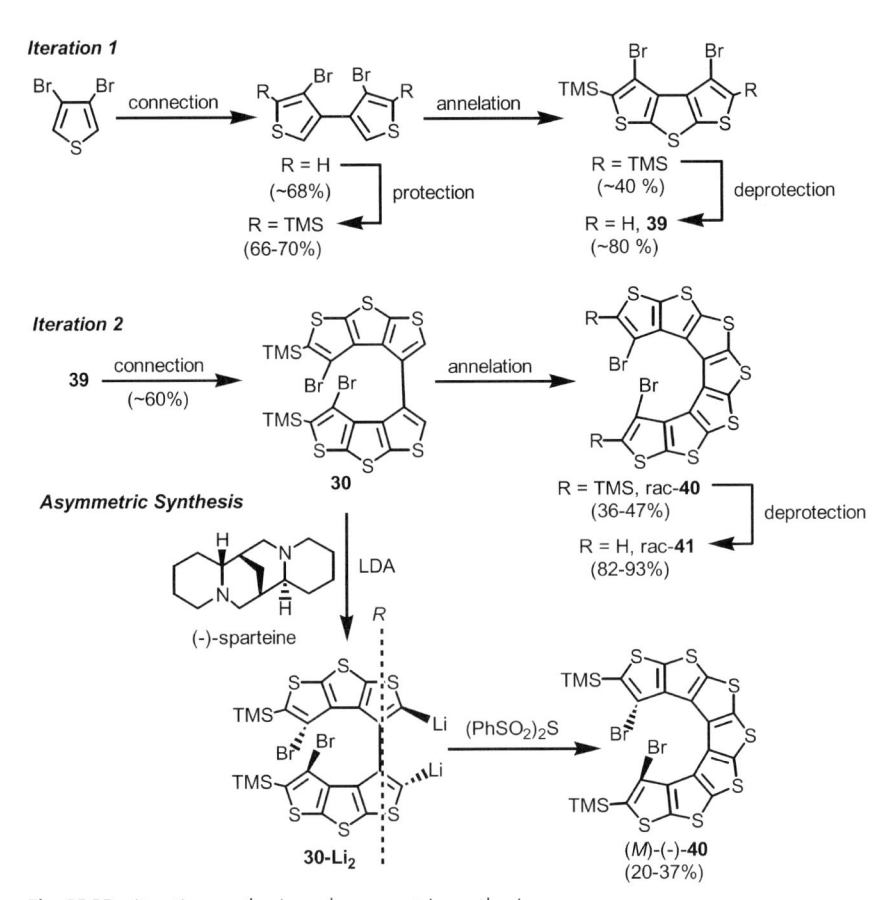

Fig. 15.15 Iterative synthesis and asymmetric synthesis of carbon–sulfur [7]helicene **40**.

Asymmetric synthesis relied upon the use of (–)-sparteine in the final annelation step leading to **40** (Fig. 15.15). Mechanistic studies revealed that (–)-sparteine-mediated stereoinduction was associated with kinetic resolution in the incomplete formation of dilithiated intermediate 30-Li$_2$ and the reaction of diastereomeric complexes of 30-Li$_2$ with bis(phenylsulfonyl) sulfide [85]. The absolute

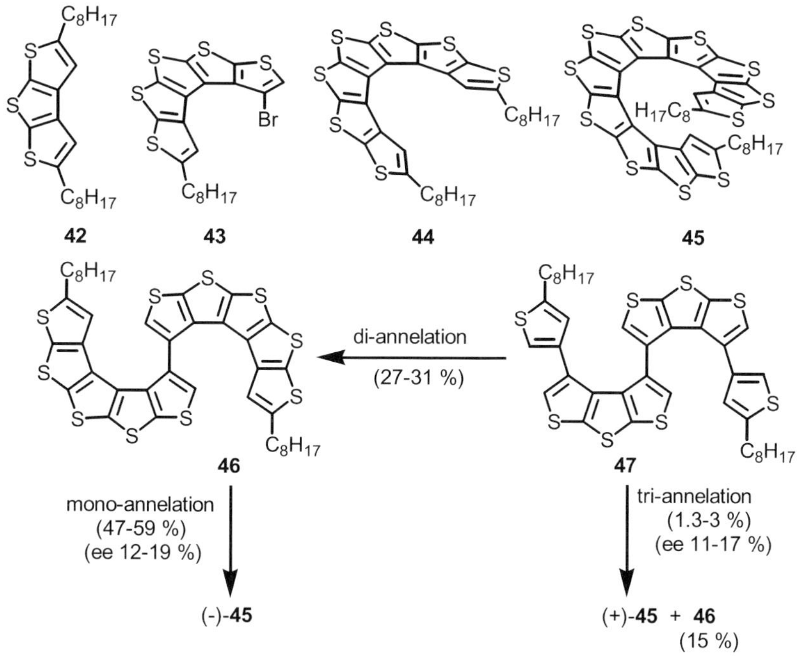

Fig. 15.16 Structures of trithiophene **42**, [5]helicene **43**, [7]helicene **44**, [11]helicene **45** and (−)-sparteine-mediated asymmetric synthesis of [11]helicene **45**. LDA (mono- and triannelation) and n-BuLi (diannelation) were used as bases.

configuration of the predominant (−)-enantiomer was established as *M* by vibrational circular dichroism (CD) studies [86]. This stereoinduction involving axial *R*-configuration was consistent with the results of asymmetric synthesis of tetraphenylenes via the Cu(II) oxidation of 2,2′-dilithiobiaryls (Section 15.2.3). Finally, resolution with menthol-based siloxanes was implemented to obtain both enantiomers of the TMS-free [7]helicene **41**.

Recently, series of carbon-sulfur [*n*]helicenes substituted with *n*-octyl groups at the α-positions of the terminal thiophene rings were prepared (Fig. 15.16) [64, 87]. The helical structures of [7]helicene **44** and [11]helicene **45** were confirmed by X-ray crystallography. Multiple short S–S contacts were found, especially for racemic [11]helicene **45**. Asymmetric synthesis of [11]helicene **45** provided enantiomeric excess of either the (−)- or the (+)-enantiomer for the monoannelation or, unprecedented, triannelation approach (Fig. 15.16). Also, selective diannelation of octathiophene **47**, followed by monoannelation of decathiophene **46**, provided an efficient synthetic route to (−)-[11]helicene **45**, avoiding protection/deprotection steps [64].

UV–Vis absorption studies revealed that the helical oligothiophenes **44** and **45** had identical absorption onset but significantly red shifted to the onset for the planar trithiophene **42**. This provided an estimate for the optical band gap,

$E_g \approx 3.5$ eV, of the $(C_2S)_n$ helix polymer and indicated that the electron localization occurs already for $n \leq 7$ [64]. These results are in contrast to electron delocalization in [n]heliphenes, [n]helicenes and [n]thiahelicenes, which possess much lower E_g values of 2.1–2.5 eV. These results are consistent with cyclic voltammetric data. [n]Helicenes **40**, **44** and **45** showed reversible cyclic voltammetric waves (1.2–1.3 V vs. SCE); the second oxidation wave for **40** was found at ~1.8 V, i.e. at a significantly more positive potential [64, 85].

15.2.2
Double Helicenes and Chiral Polycyclic Aromatic Hydrocarbons

Double and triple helicenes are π-conjugated molecules, which consist of two or three annelated helicene-like subunits. Such structures may also be considered as examples of chiral polycyclic aromatic hydrocarbons (PAHs).

The first double helicenes, consisting of head-to-tail annelated [6]helicenes, were prepared by oxidative photocyclization of stilbenes to helicenes [21]. Recently, double helicene **48**, in which two benzene rings are shared by [5]- and [7]helicene, was prepared via nonphotochemical Pd-catalyzed cyclotrimerization of 3,4-didehydrophenanthrene derived from **47** (Fig. 15.17) [88]. The ^{1}H NMR spectrum for racemic **48** showed better agreement with the calculated spectrum for the diastereomer with homochiral versus heterochiral helicenes. The characterization of double helicenes did not include X-ray crystallography [21, 88].

Recently, a conjoined double helicene, in which two hydrazine-based [5]helicenes are highly annelated in their mid-sections, was reported [89]. The conjoined double helicene **51-D_2** possesses two homochiral hydrazine-based [5]helicene-like fragments annelated in their mid-sections. The synthesis of conjoined double helicene **51-D_2** from **49** consists of two annelation steps: (1) Friedel–Crafts diannelation to provide pentacyclic diamine **50** and (2) one-step oxidative coupling forming one CC and two NN bonds [90], to give dodecacyclic structure **51-D_2** (Fig. 15.18).

The Friedel–Crafts diannelation in the first step was based on Hellwinkel and Schmidt's methodology for [4]helicenes, except for the use of alkenes instead of alcohols as the precursors to the intermediate carbocations (Section 15.2.1) [70].

In the second step, the chiral D_2-symmetric structure **51-D_2** is the kinetic product, which may be irreversibly converted to its achiral C_{2h}-symmetric diastereomer **51-C_{2h}** (Fig. 15.19). Mechanistic studies indicated that oxidation of achiral diamine

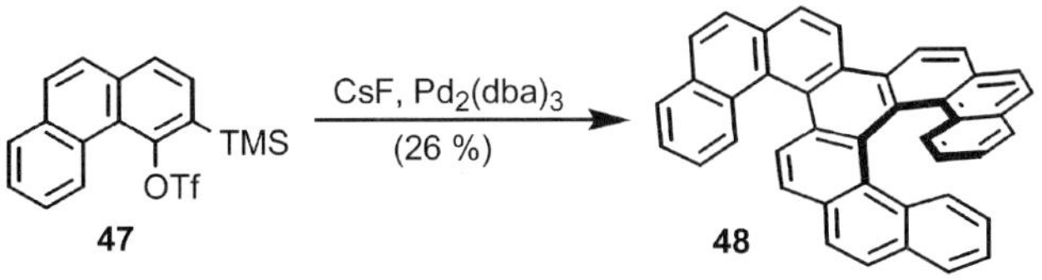

Fig. 15.17 Palladium-catalyzed cyclotrimerizations of didehydrophenanthrene leading to double helicenes.

15.2 Synthetic Approaches to Highly Annelated Chiral π-Conjugated Systems | 561

Fig. 15.18 Synthesis and X-ray structure of conjoined double helicene **51-D_2**. Each of the two homochiral [5]helicene-like fragments is shown in stick-and-ball format.

50 to chiral **51-D_2** occurs via achiral tetraamine **52**, that is, the CC bond homocoupling occurs first. Upon partial oxidation of **50**, tetraamine **52** was isolated in 59–70% yield; **52** was then further oxidized to conjoined double helicene **51-D_2** in ~75% yield. Notably, conjoined double helicene **51-D_2** could be reduced to tetraamine **52**. Thus, chiral **51-D_2** and achiral **52** are interconvertible via redox cleavage and formation of the NN bonds of the hydrazine moieties. The structures of all

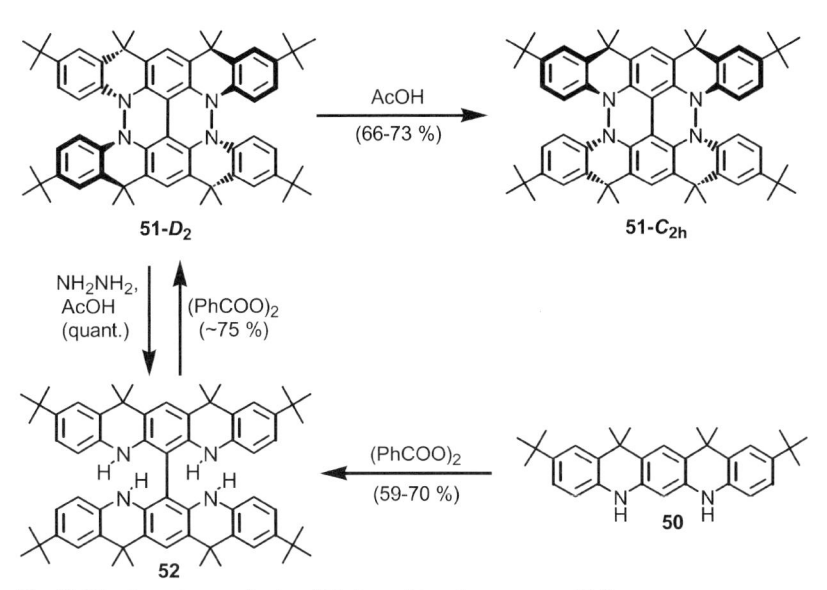

Fig. 15.19 Stepwise synthesis of **51-D_2** and its diastereomer **51-C_{2h}**.

key compounds **50**, **51-D$_2$** and **52** were established by X-ray crystallography [89, 91].

UV–Vis absorption spectra for **51-D$_2$** in *n*-heptane showed λ_{max} = 409 nm (sh 439 nm), which was significantly red shifted compared λ_{max} = 275 nm (sh 333 nm) for diamine **50**. Crystalline **51-D$_2$** showed blue–green fluorescence; in *n*-heptane, a blue fluorescence with quantum efficiency $\Phi_F \approx 15\%$ at λ^{em}_{max} = 472 nm (excitation at λ^{exc} = 289 and 409 nm) was found.

Extended C_3-symmetric PAHs, which may be viewed as graphite disks, have been prepared via efficient oxidative cyclodehydrogenations by Müllen and coworkers [92]. When the bay areas of such PAHs possess overcrowded hydrogens or when they correspond to [*n*]helicene-like units, the possibility of chiral conformations with significant barriers for racemization and/or conversion between diastereomers arises. Hexabenzotriphenylene **53** is the best studied example of chiral, overcrowded PAHs with significant steric congestion of the bay hydrogens. Hexabenzotriphenylene **53** may be viewed as a triple helicene, composed of three [5]helicene units, sharing the triphenylene core. X-ray structures for both chiral diastereomers, **53-D$_3$** and **53-C$_2$** (*D$_3$*- and *C$_2$*-symmetric, respectively) were determined [93, 94]. **53-C$_2$**, which is less thermodynamically stable than **53-D$_3$**, is the kinetic product of an efficient Pd-catalyzed cyclotrimerization of 9,10-didehydrophenanthrene (Fig. 15.20) [95]. Another recent example of a triple helicene, with three [5]helicene-like units, is an extended derivative of symmetrical triindole, a triaza analog of truxene [96].

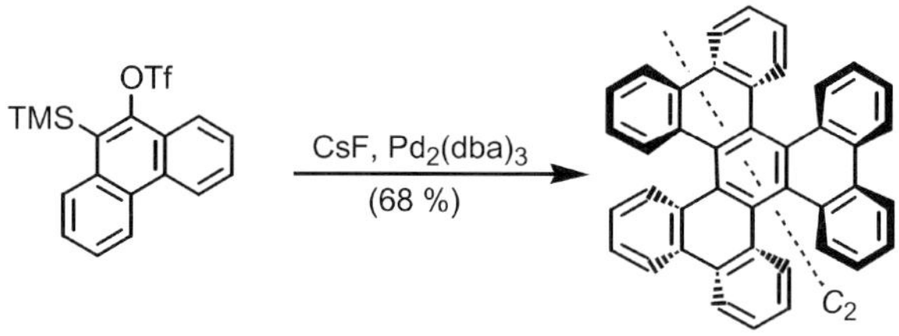

Fig. 15.20 Palladium-catalyzed cyclotrimerizations of didehydrophenanthrene leading to triple helicene.

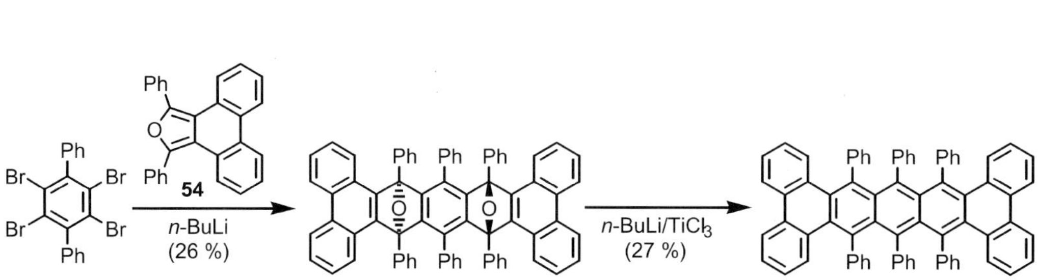

Fig. 15.21 Synthesis of PAH with a screw-type helicity.

15.2 Synthetic Approaches to Highly Annelated Chiral π-Conjugated Systems

An overcrowded PAH, 9,10,11,20,21,22-hexaphenyltetrabenzo[a,c,l,n]pentacene (55), showed an interesting screw-type helicity (Fig. 15.21) [97]. An end-to-end twist of 144° was estimated from the X-ray structure of 55. Pentacene 55 was prepared by the reaction of 1,3-diphenylphenanthro[9,10-c]furan 54 with the bisaryne equivalent generated from 1,2,4,5-tetrabromo-3,6-diphenylbenzene in the presence of n-butyllithium, followed by deoxygenation of the double adduct with low-valent titanium. Pentacene 55 could be resolved by chromatography on a chiral support, but it racemized slowly at room temperature ($t^{1/2} \approx 9$ h at 25 °C).

15.2.3
Tetraphenylenes and π-Conjugated Double Helices

Tetra-o-phenylene (tetraphenylene, Fig. 15.22) is an achiral tube-shaped molecule (D_{2d} point group). However, tetraphenylene can be readily desymmetrized, by substitution or annelation, to produce a chiral π-conjugated system with an extraordinarily high barrier for racemization (Section 15.3). In the X-ray structures of tetraphenylenes, the dihedral angles between adjacent phenylenes are in the 60–70° range, still sufficiently close to planarity to provide weak conjugation. Although tetraphenylenes were studied as inclusion compound hosts [98], relatively few were prepared in nonracemic form (Fig. 15.22) [99–103]. The absolute configuration for tetranaphthylene (R)-(+)-56 was established recently by vibrational CD spectroscopy [86].

Recent advances in the synthesis of nonracemic tetraphenylenes relied on homocoupling (or cross-coupling) of enantiopure axially chiral binaphthyls [(R)-56 and -57] [101], asymmetric synthesis [(R)-58, -59 and -60] [102] and racemic synthesis followed by resolution [(R)-, (S)-61, -62, -63 and -64] [103].

Asymmetric synthesis of tetraphenylenes 58–60 consists of two key steps, as illustrated by the synthesis of tetraphenylene 58: (1) regioselective arylation at

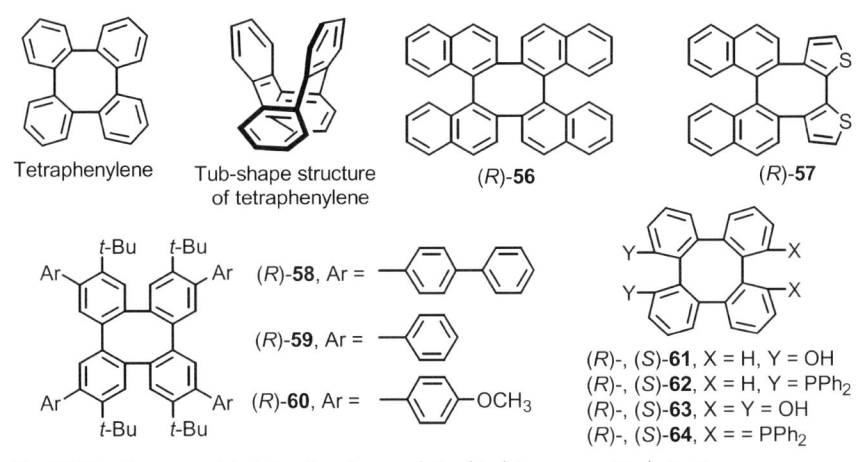

Fig. 15.22 D_{2d}-symmetric tetraphenylene and its chiral (nonracemic) derivatives.

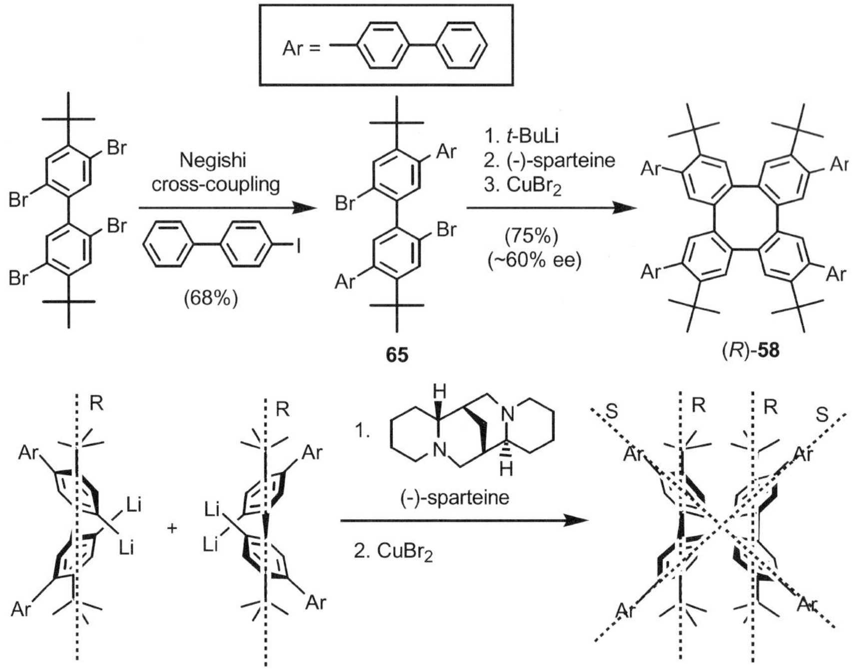

Fig. 15.23 Asymmetric synthesis of tetraphenylenes. Chiral axes for the predominant enantiomer are indicated.

the sterically hindered biaryl positions, to give **65**, and (2) annelation based on oxidative homocoupling of 2,2′-dilithiobiaryls in the presence of (–)-sparteine, to give nonracemic tetraphenylene (R)-(+)-**58** with ~60% ee (Fig. 15.23) [102]. Recrystallization provided enantiomerically pure (R)-(+)-**58**. The absolute configuration of (R)-(+)-**58** was established by CD spectroscopy [102].

Tetraphenylene may be viewed as a building block of "three-dimensional graphite", i.e. complete annelation of tetraphenylenes would give a three-dimensional carbon network (Pn3m). Such a relatively low strain structure (e.g. compared with fullerenes) was first considered by Riley and coworkers, in their studies of amorphous carbon. However, this structure remains experimentally elusive [104–106].

The achiral network of tetraphenylenes served as an inspiration for the design of a π-conjugated double helix. Double helical polymers, in which two polyphenylene helices are intertwined or tetraphenylenes are sequentially annelated, were recognized as the chiral building block of the network (Fig. 15.24) [107].

Racemic synthesis of double helical octaphenylene **67** was carried out via a convergent route. In the annelation step, the Cu(II)-mediated oxidative homocoupling of dilithiotetraphenylene, derived from the dibromotetraphenylene **66**, gave octaphenylene **67** (Fig. 15.25) [107].

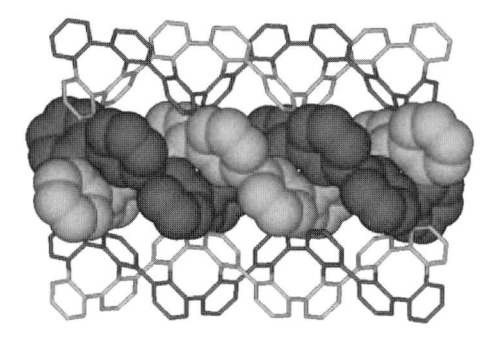

Fig. 15.24 Double helical polyphenylenes as building blocks of "three-dimensional graphite".

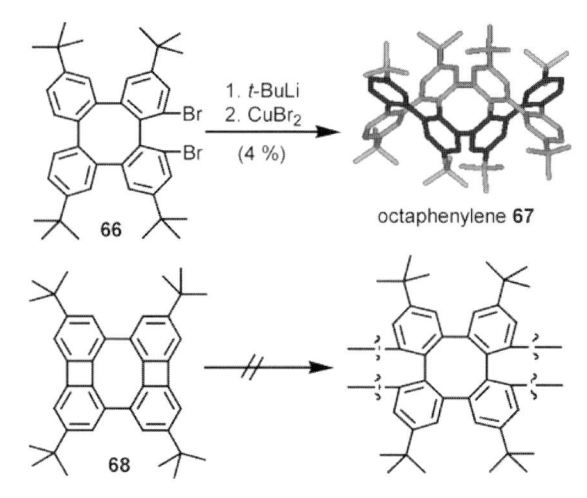

Fig. 15.25 Synthesis of double helical octaphenylene **67**.

Because starting material **66** was racemic and only homochiral tetraphenylenes may form a double helix, the yield of double helical product **67** was low and a significant amount (4%) of a single CC-homocoupling product (presumably the meso diastereomer) was isolated. Analogous results were obtained in the synthesis of other tetraphenylenes; this includes observation of duplication-like effects with nonlinear increase in enantiomeric excess and tetraphenylene yield, when the starting biaryls had higher enantiomeric excess in the synthesis of tetranaphthalene **56** [101, 108].

Biphenylene dimer **68**, a building block for another putative carbon allotrope, was prepared [109]. However, the reaction of **68** under the conditions typical for the conversion of biphenylene to tetraphenylene failed to yield the double helical polymer or corresponding oligomer (Fig. 15.25) [107].

Another racemic tetraarylene-based double helical fragment (**71**) was prepared by Marsella et al. using a divergent synthetic route [110]. The annelation step relied on Pd/Cu-mediated arylalkyne cross-coupling of tetrathienylene **69** and the dialkyne **70**, providing **71** in good yield (Fig. 15.26).

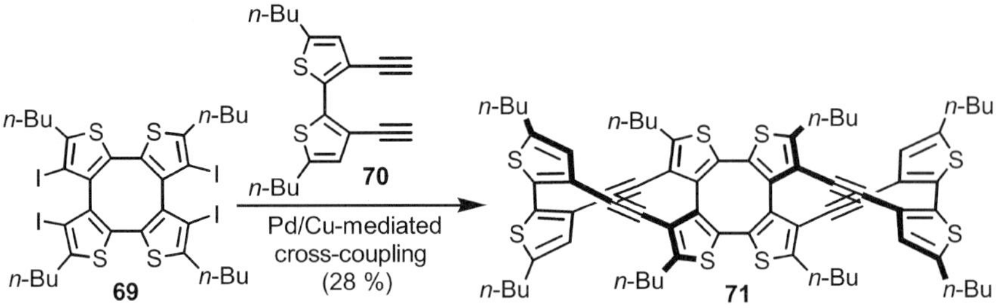

Fig. 15.26 Synthesis of double helical octathienylene **71**.

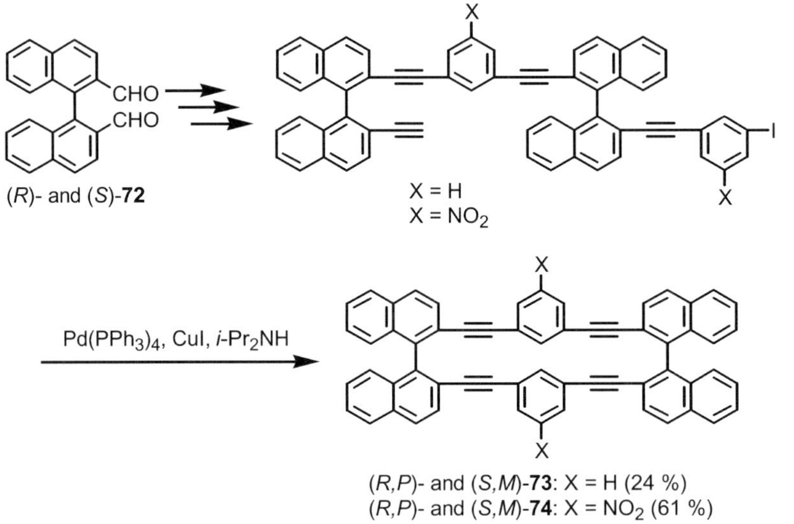

Fig. 15.27 Synthesis of double helical π-conjugated molecules starting from enantiopure binaphthyls.

Analogous Pd/Cu-mediated arylalkyne annelation was employed to prepare binaphthyl-based double helical fragments in good yields [111]. Starting from enantiopure dialdehyde **72**, both enantiomers of **73** and **74** could be obtained (Fig. 15.27). The structure of racemic **74** was confirmed by X-ray crystallography.

Recently, Wong and coworkers reported a series of rod-like D_2-symmetric, enantiopure molecules, based on Pt(II) complexes of homochiral tetraphenylenes (*R*)-**63**, (*R*)-**64** and (*R*)-BINAP. The highest homolog of such "chiral rod" had an estimated length of 4.8 nm [112].

Very few attempts at the reduction or oxidation of tetraarylenes have been made. For enantiopure tetranaphthalene (*R*)-(+)-**56**, cyclic voltammetry showed one reversible wave at −2.31 V (vs. SCE) corresponding to the reduction to the corresponding carbodianion (*R*)-**56**$^{2-}$ (Fig. 15.28) [101]. This wave was at significantly less negative potential than −2.57 V for naphthalene under identical conditions.

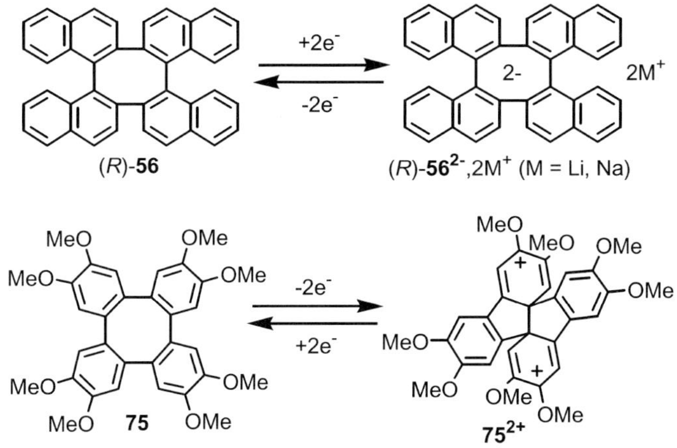

Fig. 15.28 Reversible reduction and oxidation of tetraphenylenes to the corresponding diions.

These results, and also the failure of the exciton coupling model in reproducing the CD spectra of (R)-(+)-**56**, were consistent with weak, but still significant, conjugation between naphthalene moieties. Notably, dihedral angles of ~70° were found between naphthalene moieties in the X-ray structure of enantiopure (R)-(+)-**56**. Carbodianion (R)-**56**$^{2-}$.2M$^+$ (M = Li, Na) was prepared via reduction of (R)-(+)-**56** with alkali metals gave carbodianion (R)-**56**$^{2-}$.2M$^+$ (M = Li, Na); no intermediate radical anion could be detected or isolated, in agreement with the cyclic voltammetric data [101].

Oxidation of **56** to dication gives irreversible cyclic voltammetry, which was qualitatively similar to that found by Kochi and coworkers for the achiral tetraarylene **75** [113]. The cyclic voltammetric data were consistent with the reversible CC bond formation. Dication **75**$^{2+}$.2SbCl$_6^-$ was isolated and its hexacyclic structure was unequivocally determined by X-ray crystallography (Fig. 15.28) [113].

15.3
Barriers for Racemization of Chiral π-Conjugated Systems

Configurational stability (or persistence) is one of the important properties of a chiral material. The definition of the lower limit for the free energy barrier for racemization may depend on the specific application. For optoelectronic applications, accelerated aging tests may provide very approximate guidelines [114]. For the purpose of estimating the free energy barrier for racemization, we will assume that the less than 1 % conversion of the major enantiomer to the minor enantiomer in such aging tests is tolerable, i.e. $|\Delta a|/|a| < 0.02$, where a (in units °mm^{-1}) denotes rotatory power of thin-film material. With these assumptions, the lower limit for the free energy barrier for racemization is of the order of 35 kcal mol^{-1}.

For [n]helicenes, the free energy barriers for racemization increase with n and with substitution at the inner helix sites [21, 115–117]. Thus, the free energy barriers (kcal mol^{-1} at 27 °C) for parent [n]helicenes increase in the following order: 24.1 (n = 5), 36.2 (n = 6), 41.7 (n = 7), 42.4 (n = 8), 43.5 (n = 9) [115, 117]. [n]Thiahelicenes and [n]helicenes with similar helical turns in-plane possess comparable barriers for racemization [118, 119]. For methyl-substituted [6]helicenes, the barriers (kcal mol^{-1}) are 43.8 (1-methyl), 44.0 (1,16-dimethyl), 39.5 (2,15-dimethyl). Similarly for 1-methyl[5]helicene, the barrier increases to 38.7 kcal mol^{-1}. Upon introduction of a sterically large substituent at the end of the inner helix, even [4]helicenes become configurationally stable at room temperature, e.g. [4]helicene **19** [70, 73, 120, 121]. Thus, modest steric hindrance at the ends of inner helices has a significant impact on the barrier for racemization, greater than that of extending the helix length well beyond a 360° turn angle.

Carbon sulfur [7]helicene (*M*)-(–)-**40**, which has a relatively low turning angle but possesses two bulky bromine groups at the ends of the inner helix, has a free energy barrier for racemization of 39.0 kcal mol^{-1} (half-life of 11 h at 199 °C) [85]. The relatively small helical turn in-plane (based on the X-ray structures of two racemic polymorphs) is apparently offset by two bulky bromine groups at the ends of the inner helix.

Notably, introduction of sp^3-hybridized atoms into the helicene skeleton can significantly affect the barrier, e.g. the free energy barrier for racemization of 9,10-dihydro[5]helicene is 29.9 kcal mol^{-1} [117].

Compared with [n]helicenes, [n]heliphenes possess significantly lower barriers for racemization – to the extent that the higher homologs, with the helix turn in-plane exceeding 360°, are not resolvable at room temperature. Based on ^{1}H NMR spectroscopic decoalescence temperatures of the potentially diastereotopic methylene hydrogens of the methoxymethyl substituent in [7]-, [8]- and [9]heliphene, free energy barriers for the inversion of configuration are $\Delta G^{\ddagger}_{-27\,°C}$ = 12.6, 13.4 and < 12 kcal mol^{-1}, respectively [59, 60]. These values were much lower than those for the corresponding [n]helicenes. In addition, tetramethylated [7]heliphenes **6d** and **6e** did not even show any detectable diastereotopicity for the methylene hydrogens of the methoxymethyl substituent at temperatures as low as –70 °C. In contrast to [n]helicenes, the barriers for racemization for [n]heliphenes appeared to decrease with increase in steric hindrance of the methyl substitution on the terminal rings or with the increase in ring–ring overlap. It is plausible that the greater diameter of the helix turn and/or relatively low HOMO–LUMO gaps (and E_g) in [n]heliphenes might contribute to the greater conformational flexibility and then to the low barriers for racemization.

Conjoined double helicene **51-D_2** was obtained only as a racemate. Its barrier for racemization is expected to be significantly greater than the barrier of 24.1 kcal mol^{-1} for [5]helicene. The free energy barrier of ~35 kcal mol^{-1} (half life of ~3 h at 180 °C in naphthalene solution) for isomerization of **51-D_2** to its meso diastereomer **51-C_{2h}** was determined in the absence of acid [89]. Such

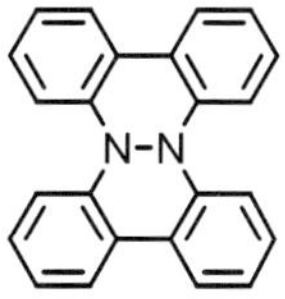

Fig. 15.29 Structures of hydrazine **76** and tetraphenylene **77**.

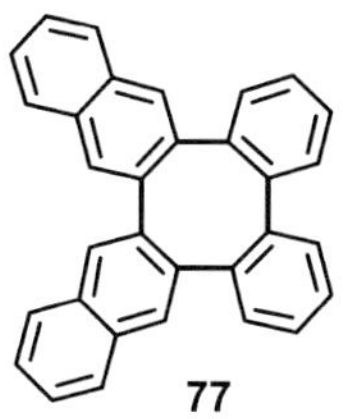

isomerization corresponds to the inversion of one [5]helicene unit, compared with the inversion of both units that is required for racemization of **51-D_2**. The relatively high free energy barrier for inversion of one of the [5]helicene units may be associated with cooperativity in conformation conversion of the pentacyclic units, corresponding to diamine **50**, from two chairs in the D_2 to two boats in the C_{2h} point group [89]. Also, double nitrogen inversion in the hydrazine moiety may contribute to the barrier [122]. Notably, double nitrogen inversion in both hydrazine moieties would be needed for racemization of **51-D_2**.

Inversion of configuration in conjoined double helicene **51-D_2** may be related to the barrier for racemization in a chiral hydrazine derivative, benzo[c]benzo[3,4]cinnolino[1,2-a]cinnoline (**76**) (Fig. 15.29) [123–125]. In hydrazine **76**, which is a diaza analog of dibenzo[g,p]chrysene, two [4]helicene fragments may be identified. Hydrazine **76** was characterized by X-ray crystallography and was readily resolvable via crystallization as a conglomerate [124]. Hydrazine **76** in decane racemized with an activation energy of 27.1 kcal mol^{-1}, a process that may correspond to double nitrogen inversion [124]. Notably, oxidation of enantiomers of **76** resulted in optically inactive radical cations [124, 126]. This is consistent with the contribution from double nitrogen inversion to the barrier for racemization in neutral **76**.

Barriers for inversion of configuration in chiral, overcrowded PAHs are relatively low. In hexabenzotriphenylene, a free energy barrier of 26.2 kcal mol^{-1} was found for isomerization of **53-C_2** to the more thermodynamically stable **53-D_3**. However, the free energy barrier for racemization in **53-C_2**, as determined from the 1H NMR coalescence temperature ($\Delta G^{\ddagger}_{rac}$ = 11.7 kcal mol^{-1}, T_c = 247 K, $\Delta \nu$ = 102 Hz at 500 MHz), was relatively low [95]. Pentacene **55** (Fig. 15.21), with a screw-type end-to-end twist of 144°, racemized slowly at room temperature ($\Delta G^{\ddagger}_{rac,25°C}$ = 23.8 kcal mol^{-1}) [97], that is, the barrier for racemization is similar to that in [5]helicene ($\Delta G^{\ddagger}_{rac}$ = 24.1 kcal mol^{-1}).

Among chiral π-conjugated systems, tetraphenylenes possess extraordinarily high barriers for racemization. This is in contrast to the relatively low barriers for PAHs, especially those based on triphenylene. Racemization of chiral tetraphenylene corresponds to ring inversion of the central cyclooctatetraene ring. The steric repulsion of the "bay hydrogen atoms" in typical tetraphenylenes is one of the reasons for the extraordinarily high barriers for racemization. For example, tetraphenylene **77** (Fig. 15.29) has a free energy barrier for racemization of 67 kcal mol^{-1}, compared with ~10 kcal mol^{-1} for cyclooctatetraene [100, 127]. Because the dianion of cyclooctatetraene itself is a planar, aromatic compound, the inversion barrier is expected to be lowered to an extent dependent upon the degree of aromatic character in the central eight-membered ring upon n-doping. This proposition was tested in tetranaphthylene **56** and its corresponding carbodianions 56^{2-}, $2M^+$ (M = Li^+, K^+) (Fig. 15.24). The lower limit of the barrier for the ring inversion of the neutral (R)-(+)-**56** is $\Delta G^\ddagger_{613\,K} > 54$ kcal mol^{-1}. This may be compared with $\Delta G^\ddagger_{363\,K} = 29$ kcal mol^{-1} for analogous process in the carbodianion (R)-56^{2-}, $2Na^+$. Hence the free energy barrier for the ring inversion in the carbodianion is lowered by at least 25 kcal mol^{-1} and probably by more than 40 kcal mol^{-1}, compared with the neutral compound [101].

15.4
Strong Chiroptical Properties in Absorption, Emission and Refraction

15.4.1
Absorption and Emission

The strength of chiroptical properties in both absorption and emission may be measured by the corresponding anisotropy factors (g-values). The factor measuring the degree of circular polarization in absorption is defined as $g_{abs} = \Delta\varepsilon/\varepsilon = (\varepsilon_L - \varepsilon_R)/0.5(\varepsilon_L + \varepsilon_R)$, where $\Delta\varepsilon$ is the difference in molar absorptivity of left- and right-handed circularly polarized light and ε is defined as average absorptivity. The maximum value of $|g_{abs}| = 2$ would be obtained when only either left- or right-circularly polarized light is absorbed. The intrinsic, molecular values of $|g_{abs}|$ are determined by the ratio of the of the magnetic transition moment (m) to the electric transition moment (μ) and the relative orientation (with cosine dependence) of these two moments. For molecules in which transitions are electric dipole and magnetic dipole allowed, $|g_{abs}| = |\Delta\varepsilon|/\varepsilon$ is of the order of 5×10^{-3}. For (+)-[6]helicene **1**, the vectors m and μ are parallel and $g_{abs} = \Delta\varepsilon/\varepsilon = +7 \times 10^{-3}$ (at 325 nm) may be estimated from the reported data [21, 128]. CD spectra of [n]helicenes may be calculated by time-dependent density functional theory (TD DFT) to provide the transition moments and the assignments of absolute configurations [129]. In cross-conjugated carbon–sulfur [7]helicene **40**, which possesses similar helical geometry but with a relatively smaller helix turn, $g_{abs} = \Delta\varepsilon/\varepsilon = -4 \times 10^{-3}$ (at 285 nm, $\Delta\varepsilon_{max} = -117$ and $\varepsilon = 3.1 \times 10^4$ L mol^{-1} cm^{-1}) is estimated based on chiroptical data for the (–)-enantiomer.

Analogous g-values may be defined for the degree of circular polarization in emission [or circularly polarized photoluminescence (CPPL)] and circularly polarized electroluminescence (CPEL), e.g. $g_{CPPL} = 2(I_L - I_R)/(I_L + I_R)$, where I_L and I_R denote the intensity of left- and right-handed circularly polarized emission, respectively. CPPL should not be confused with fluorescence-detected CD.

Materials with the highest values of $|g_{abs}|$, $|g_{CPPL}|$ and $|g_{CPEL}|$ are chirally aggregated molecules or polymers. In many cases, helical supramolecular structures (helically twisted bundles, etc.) or liquid crystalline phases were detected by microscopy techniques. Consequently, such materials may have non-negligible contributions from linear polarization (e.g. cross-terms between linear birefringence and linear dichroism), polarization-dependent scattering and the usual cholesteric LC effects (e.g. selective reflection). Hence very strong polarization properties of such LC-like materials most likely originate at the supramolecular level, analogously to the conventional low molecular weight LC materials.

Bunz and coworkers reported random poly(p-phenyleneethynylene) copolymers; maximum g-values were observed for the extensively annealed films of 1:1 copolymer [7]. Such films possessed $|g_{abs}|$ = 0.38 and $|g_{CPPL}|$ = 0.19 at λ = 432 and 443 nm, with linear dichroism <<0.01. Both wavelengths corresponded to maxima in the absorption and the emission. Both CD and CPPL spectra were monosignate in the regions of their maximum g-values, which was consistent with the absence of significant exciton coupling. Notably, the g-values were lower by a factor of more than 100 for the polymer with chiral-only pendants [7].

Among other polymers with large g-values, poly(fluorenes) with chiral pendants [e.g. 2(S)-methylbutyl, 2(R)-ethylhexyl] were perhaps most intensively studied [8, 130–133]. For poly(fluorenes) with 2(R)-ethylhexyl pendants, $|g_{CPPL}|$ = 0.28 and $|g_{CPEL}|$ = 0.25 were reported [131]. Similarly, large values of $|g_{abs}|$ were reported for other poly(fluorenes). The values of $|g_{CPEL}|$ may be compared with $|g_{CPEL}|$ = 0.0013 for the first circularly polarized LED based on chirally substituted PPV [134].

Chiroptical studies of poly(fluorenes) with 3(S),7-dimethyloctyl pendants revealed a strong dependence of $|g_{abs}|$ on the thickness of the annealed films. The sign inversion of g_{abs} at a thickness of 30 nm and the maximum of $|g_{abs}|$ ≈ 1 at a thickness exceeding 200 nm were observed. Hence the very large values of $|g_{abs}|$ in thicker films might be associated with cross-terms between linear birefringence and linear dichroism, as opposed to the intrinsic CD of the polymer chains [132].

Careful studies by Chen and coworkers on well-defined oligo(fluorenes) identified a nonamer for which acholesteric structure was observed after annealing. An order of magnitude increase in CD, disappearance of the signature of the exciton coupling and sign reversal in CPPL (and g_{CPPL} = +0.75 for the 87-nm thick film) were related to the right-handed cholesteric structure [133].

Very large values of $|g_{CPPL}|$ may be more easily obtained via doping of LCs, e.g. the near maximum value of $|g_{CPPL}|$ = 1.8 for 0.2% achiral ter(fluorene) doped in chiral nematic LC film (35-μm thick) at the wavelength range of selective reflection for the LC [135].

Katz's columnar aggregates of helicenes have relatively modest g-values, e.g. for 1 mM solutions of enantiopure derivative of [7]thiahelicene **11**, $|g_{abs}|$ ≈ $|g_{CPPL}|$ = 0.01 [65]. For enantiopure [4]helicene **19** and [5]helicene **20**, which were apparently non-aggregated, small values of $|g_{abs}|$ ≈ $|g_{CPPL}|$ = 0.0008–0.001 were reported [73].

15.4.2
Refraction

The degree of circular polarization in refraction is measured by the circular birefringence, $|n_L - n_R|$, where n_L and n_R correspond to the refractive indices for left- and right-handed polarized light, respectively. Thin films, with very large $|n_L - n_R| > 10^{-4}$, which significantly exceeds linear birefringence ($|n_o - n_e|$) in the highly transparent region, may provide new means for the control of light polarization in planar optical waveguides ("chiral waveguides") [6, 136]. Such materials should have excellent configurational stability, good processability and fast response.

For a typical chiral organic compound, such as 2-butanol, the circular birefringence, $|n_L - n_R|$, at $\lambda = 589$ nm is of the order of 10^{-7}. Even for a very high $|\alpha| = 40°$ mm^{-1} at $\lambda = 589$ nm ($|n_L - n_R| \approx 2 \times 10^{-4}$), linear birefringence, which may arise from residual ordering/aggregation of polymer chains, could easily overwhelm the circular birefringence.

Very low values of linear birefringence are generally difficult to attain in thin films of macromolecules. Although it is possible to decrease the linear birefringence by doping the polymer film with inorganic crystals with opposite sign of linear birefringence, this procedure leads to significantly lower transparency and, possibly, introduces problems with optical homogeneity [137]. Vitrification by rapid cooling is possible for many low molecular weight compounds; however, such a method is difficult to apply to macromolecules and it would be impractical for the fabrication of thin films of optical quality [138].

Organic materials with large optical rotations include cholesteric liquid crystals, molecules and polymers with chiral π-conjugated systems, especially [n]helicenes [21, 31, 139]. The most important factor contributing to their large optical rotations is anomalous optical rotatory dispersion (ORD), which is associated with the presence of absorption (or reflection) with large rotational strength (Fig. 15.30).

Rigorously, ORD and CD spectra are related through the Kronig–Kramers theorem, a well-known general relationship between refraction and absorption, i.e. $n_L - n_R$ is determined by $\varepsilon_L - \varepsilon_R$ for λ from zero to infinity [128]. (The analogous relationship between refraction and reflection applies to cholesteric liquid crystals.) Hence, in order to maximize ORD in the transparent region, Cotton effects, associated with exciton coupling (both intramolecular and intermolecular), have

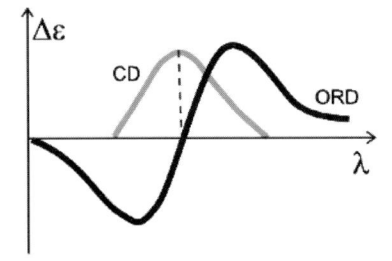

Fig. 15.30 Anomalous optical rotatory dispersion (ORD) and circular dichroism (CD) spectra for the positive Cotton effect of a single, isolated electronic transition.

to be avoided. This implies that the chromophores have to be conjugated and molecules (or macromolecules) randomly oriented.

Among non-aggregated molecules and polymers, annelated π-conjugated molecules such as [n]helicenes possess relatively large optical rotations [21]. Notably, specific rotations of helicenes and thiahelicenes increase significantly with the helix length; for (–)-[6]-, -[7]-, -[8]-, -[9]-, -[11]- and -[13]helicenes in chloroform, $[\alpha]_{589} = -3640$, $[\alpha]_{579} = -5900$, $[\alpha]_{579} = -7170$, $[\alpha]_{579} = -8150$, $[\alpha]_{579} = -9310$ and $[\alpha]_{579} = -9620$ (± 100, 10^{-1} deg cm^2 g^{-1}) were reported, respectively [26, 49]. Anomalous ORD contributes in two ways to these increases in $[\alpha]$ for longer helices: (1) the wavelength for the measurement of $[\alpha]$ (589 and 579 nm) is becoming closer to the absorption tail-off (e.g. 370 and 450 nm for [6]- and [9]-helicene, respectively) and (2) the longest wavelength Cotton effect in CD spectra is red-shifted and becomes more intense (e.g. $\Delta\varepsilon \approx 200$ L mol^{-1} cm^{-1} at $\sim$330 nm and 270 L mol^{-1} cm^{-1} at $\sim$390 nm for [6]- and [9]-helicene, respectively). The CD bandwidth increases with increase in length of the helicene further contribute to the rotational strengths for longer helicenes. In this context, cross-conjugated carbon–sulfur [n]helicenes, which become electron localized at $n \leq 7$, may provide both an excellent system for studies of intrinsic chiroptical properties (free of absorption tail-off effects) and chiroptical materials with a wide transparency range.

Substituent effects on the chiroptical properties of helicenes are relatively large but the corresponding structure–property relationships are not understood [50, 79, 85].

Enantiopure [7]thiahelicene **33** forms isotropic glassy films, with $|n_o - n_e| \approx$ 0.0003 in the range $\lambda = 630$–1550 nm. Similar results were obtained for enantiopure tetranaphthalene **56**. The isotropicity of the thin films may be associated with the tetrahedral-like shapes (Fig. 15.31), which are known to inhibit efficient crystal packing for achiral molecules. The rotatory power $|\alpha| = 11$ and 6° mm^{-1} at 670 and 850 nm, respectively, for [7]thiahelicene **33** are the highest attained to date for an isotropic material [79, 136, 140].

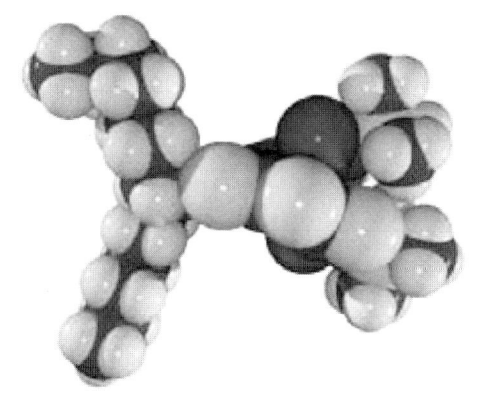

Fig. 15.31 Space-filling plot for [7]thiahelicene **33** obtained from the X-ray structure of the racemic crystal.

15.5
Conclusion

Although significant advances in the synthesis of well-defined oligomers with highly annelated chiral π-systems have been made, the synthesis of macromolecules with inherently strong chiral properties remains a challenge. This challenge is further amplified by the need to control configurational stability and achiral properties of the material.

To date, the stepwise, kinetically controlled, classical synthesis is the most effective approach to highly annelated chiral π-systems. With significant improvements in asymmetric annelation methodologies, multi-step syntheses are likely to remain the main tool in the exploration of novel chiral structures. However, the development of novel synthetic methods will be essential for the preparation of polymers with extended helical-type, ladder-type connectivity of the π-systems. Important criteria are to minimize the density of defects in the ladder connectivity and to provide conjugation pathways circumventing at least some of the defects.

Acknowledgments

We thank our collaborators who have contributed to this work and especially Dr. Suchada Rajca (University of Nebraska), Dr. Maren Pink (Indiana University) and Dr. Warren N. Herman (University of Maryland). Financial support was provided by the National Science Foundation (CHE-0414936), the Office of Naval Research (N00014-03-1-0550) and the Air Force Office of Scientific Research (FA9550-04-1-0056).

References

1. V. Percec, M. Glodde, T. K. Bera, Y. Miura, I. Shiyanovskaya, K. D. Singer, V. S. K. Balagurusamy, P. A. Heiney, I. Schnell, A. Rapp, H.-W. Spiess, S. D. Hudson, H. Duan, Self-organization of Supramolecular Helical Dendrimers into Complex Electronic Materials, *Nature* **2002**, *419*, 384–387.
2. S. Zahn, T. M. Swager, Three-dimensional Electronic Delocalization in Chiral Conjugated Polymers, *Angew. Chem. Int. Ed.* **2002**, *41*, 4225–4230.
3. (a) Ph. Leclère, M. Surin, P. Viville, R. Lazzaroni, A. F. M. Kilbinger, O. Henze, W. J. Feast, M. Cavallini, F. Biscarini, A. P. H. J. Schenning, E. W. Meijer, About Oligothiophene Self-assembly: From Aggregation in Solution to Solid-state Nanostructures, *Chem. Mater.* **2004**, *16*, 4452–4466; (b) C. R. L. P. N. Jeukens, P. Jonkheijm, F. J. P. Wijnen, J. C. Gielen, P. C. M. Christianen, A. P. H. J. Schenning, E. W. Meijer, J. C. Maan, Polarized Emission of Individual Self-assembled Oligo(*p*-phenylenevinylene)-based Nanofibers on a Solid Support, *J. Am. Chem. Soc.* **2005**, *127*, 8280–8281.
4. (a) C. Li, M. Numata, A.-H. Bae, K. Sakurai, S. Shinkai, Self-assembly of Supramolecular Chiral Insulated Molecular Wire, *J. Am. Chem. Soc.* **2005**, *127*, 4548–4549; (b) J. Bae, J.-H. Choi, Y.-S. Yoo, N.-K. Oh, B.-S. Kim, M. Lee, Helical Nanofibers from Aqueous Self-as-

sembly of an Oligo(*p*-phenylene)-based Molecular Dumbbell, *J. Am. Chem. Soc.* **2005**, *127*, 9668–9669.
5. A. Satrijo, T. M. Swager, Facile Control of Chiral Packing in Poly(*p*-phenylenevinylene) Spin-cast Films, *Macromolecules*, **2005**, *38*, 4054–4057.
6. W. N. Herman, Polarization Eccentricity of the Transverse Field for Modes in Chiral Core Planar Waveguides, *J. Opt. Soc. Am. A* **2001**, *18*, 2806–2818.
7. J. N. Wilson, W. Steffen, T. G. McKenzie, G. Lieser, M. Oda, D. Neher, U. H. F. Bunz, Chiroptical Properties of Poly(*p*-phenyleneethynylene) Copolymers in Thin Films: Large *g*-Values, *J. Am. Chem. Soc.* **2002**, *124*, 6830–6831.
8. M. Oda, H.-G. Nothofer, G. Lieser, U. Scherf, S. C. J. Meskers, D. Neher, Circularly Polarized Electroluminescence from Liquid-crystalline Chiral Polyfluorenes, *Adv. Mater.* **2000**, *12*, 362–365.
9. Y. Miyamoto, S. G. Louie, M. L. Cohen, Chiral Conductivities of Nanotubes, *Phys. Rev. Lett.* **1996**, *76*, 2121–2124.
10. G. L. J. A. Rikken, J. Fölling, P. Wyder, Electrical Magnetochiral Anisotropy, *Phys. Rev. Lett.* **2001**, *87*, 236602-1-236602-4.
11. V. Krsti, S. Roth, M. Burghard, K. Kern, G. L. J. A. Rikken, Magneto-chiral Anisotropy in Charge Transport through Single-walled Carbon Nanotubes, *J. Chem. Phys.* **2002**, *117*, 11315–11319.
12. L. Brunsveld, B. J. B. Folmer, E. W. Meijer, R. P. Sijbesma, Supramolecular Polymers, *Chem. Rev.* **2001**, *101*, 4071–4098.
13. E. Yashima, K. Maeda, T. Nishimura, Detection and Amplification of Chirality by Helical Polymers, *Chem. Eur. J.* **2004**, *10*, 42–51.
14. H. Goto, K. Akagi, Optically Active Conjugated Polymers Prepared from Achiral Monomers by Polycondensation in a Chiral Nematic Solvent, *Angew. Chem. Int. Ed.* **2005**, *44*, 4322–4328.
15. L. Pu, 1,1′-Binaphthyl Dimers, Oligomers and Polymers: Molecular Recognition, Asymmetric Catalysis and New Materials, *Chem. Rev.* **1998**, *98*, 2405–2494.
16. D. J. Hill, M. J. Mio, R. B. Prince, T. S. Hughes, J. S. Moore, A Field Guide to Foldamers, *Chem. Rev.* **2001**, *101*, 3893–4012.
17. H. Sugiura, Y. Nigorikawa, Y. Saiki, K. Nakamura, M. Yamaguchi, Marked Effect of Aromatic Solvent on Unfolding Rate of Helical Ethynylhelicene Oligomer, *J. Am. Chem. Soc.* **2004**, *126*, 14858–14864.
18. Y. Dai, T. J. Katz Synthesis of Helical Conjugated Ladder Polymers, *J. Org. Chem.* **1997**, *62*, 1274–1285.
19. T. Iwasaki, Y. Kohinata, H. Nishide, Poly(thiaheterohelicene): a Stiff Conjugated Helical Polymer Comprised of Fused Benzothiophene Rings, *Org. Lett.* **2005**, *7*, 755–758.
20. While functionalized chiral nanotubes and chiral fullerenes may provide molecularly well-defined materials in the future, their specialized chemistry is not covered in this overview.
21. P. P. Meurer, F. Vögtle, Helical Molecules in Organic Chemistry, *Top. Curr. Chem.* **1985**, *127*, 1–76.
22. W. Fuchs, F. Niszel, Über die Tautomerie der Phenole, IX.: Die Naphtho-carbazol-Bildung aus Naphtholen, *Ber. Dtsch. Chem. Ges.* **1927**, *60*, 209–212.
23. I. Pischel, S. Grimme, S. Kotila, M. Nieger, F. Vögtle, A Configurationally Stable Pyrrolohelicene: Experimental and Theoretical Structure–Chiriooptic Relationships, *Tetrahedron: Asymmetry* **1996**, *7*, 109–116.
24. J. W. Cook, Polycyclic Aromatic Hydrocarbons. Part XII. The Orientation of Derivatives of 1:2-Benzanthracene, with Notes on the Preparation of Some New Homologues and on the Isolation of 3:4:5:6-Dibenzphenanthrene, *J. Chem. Soc.* **1933**, 1952–1957.
25. M. S. Newman, D. Lednicer, The Synthesis and Resolution of Hexahelicene, *J. Am. Chem. Soc.* **1956**, *78*, 4765–4770.
26. R. H. Martin, The Helicenes, *Angew. Chem. Int. Ed. Engl.* **1974**, *13*, 649–659.
27. R. H. Martin, M. Bayes, Helicenes. Photosyntheses of [11], [12] and [14]Helicene, *Tetrahedron* **1975**, *31*, 2135–2137.

28. [11]Thiahelicene: H. Wynberg, M. B. Groen, Reaction of Optically Active Heterohelicenes. Synthesis of an Optically Active Undecaheterohelicene, *J. Am. Chem. Soc.* **1970**, *92*, 6664–6665.
29. H. Wynberg, Some Observations on the Chemical, Photochemical and Spectral Properties of Thiophenes *Acc. Chem. Res.* **1971**, *4*, 65–73.
30. K. Yamada, S. Ogashiwa, H. Tanaka, H. Nakagawa, H. Kawazura, [7], [9], [11], [13] and [15]Heterohelicenes Annelated with Alternant Thiophene and Benzene Rings. Syntheses and NMR Studies, *Chem. Lett.* **1981**, 343–346.
31. T. J. Katz, Syntheses of Functionalized and Aggregating Helical Conjugated Molecules, *Angew. Chem. Int. Ed.* **2000**, *39*, 1921–1923.
32. L. Liu, T. J. Katz, Simple Preparation of a Helical Quinone, *Tetrahedron Lett.* **1990**, *31*, 3983–3986.
33. R. Chang, S. I. Weisman, Electron Transfer between Anion and Molecule of Hexahelicene, *J. Am. Chem. Soc.* **1967**, *89*, 5968.
34. S. Bamberger, D. Hellwinkel, F. A. Neugebauer, Über verbrückte Diaryl- und Triarylamin-Radikalkationen, *Chem. Ber.* **1975**, *108*, 2416–2421.
35. K.-H. Ernst, M. Neuber, M. Grunze, U. Ellerbeck, NEXAFS Study on the Orientation of Chiral (P)-Heptahelicene on Ni(100), *J. Am. Chem. Soc.* **2001**, *123*, 493–495.
36. R. Fasel, M. Parschau, K.-H. Ernst, Chirality Transfer from Single Molecules into Self-assembled Monolayers, *Angew. Chem. Int. Ed.* **2003**, *42*, 5178–5181.
37. L. Vyklický, S. H. Eichhorn, T. L. Katz, Helical Discotic Liquid Crystals, *Chem. Mater.* **2003**, *15*, 3594–3601.
38. T. W. Bell, N. M. Hext, Supramolecular Optical Chemosensors for Organic Analytes, *Chem. Soc. Rev.* **2004**, *33*, 589–598.
39. D. J. Weix, S. D. Dreher, T. J. Katz, [5]HELOL Phosphite: a Helically Grooved Sensor of Remote Chirality, *J. Am. Chem. Soc.* **2000**, *122*, 10027–10032.
40. M. T. Reetz, S. Sostmann, 2,15-Dihydroxy-hexahelicene (HELIXOL): Synthesis and Use as an Enantioselective Fluorescent Sensor, *Tetrahedron* **2001**, *57*, 2515–2520.
41. I. Sato, R. Yamashima, K. Kadowaki, J. Yamamoto, T. Shibata, K. Soai, Asymmetric Induction by Helical Hydrocarbons: [6]- and [5]Helicenes, *Angew. Chem. Int. Ed.* **2001**, *40*, 1096–1098.
42. Y. Xu, Y. X. Zhang, H. Sugiyama, T. Umano, H. Osuga, K. Tanaka, (P)-Helicene Displays Chiral Selection in Binding to Z-DNA, *J. Am. Chem. Soc.* **2004**, *126*, 6566–6567.
43. S. Honzawa, H. Okubo, S. Anzai, M. Yamaguchi, K. Tsumoto., I. Kumagai, Chiral Recognition in the Binding of Helicenediamine to Double Strand DNA: Interactions between Low Molecular Weight Helical Compounds and a Helical Polymer, *Bioorg. Med. Chem.* **2002**, *10*, 3213–3218.
44. F. B. Mallory, C. S. Wood, J. T. Gordon, Photochemistry of Stilbenes. III. Some Aspects of the Mechanism of Photocyclization to Phenanthrenes, *J. Am. Chem. Soc.* **1964**, *86*, 3094–3102.
45. M. Flammang-Barbieux, J. Nasielski, R. H. Martin, Synthesis of Heptahelicene (1) Benzo[c]phenanthro[4,3-g]phenanthrene, *Tetrahedron Lett.* **1967**, 743–744.
46. Conglomerates of [7]-, [8]- and [9]helicene: R. H. Martin, M.-J. Marchant, Resolution and Optical Properties ($[\alpha]_{max}$, ord and cd) of Hepta-, Octa- and Nonahelicene, *Tetrahedron* **1974**, *30*, 343–345.
47. Although racemic [6]helicene 1 also may also be crystallized as a conglomerate, the effective resolution is prevented by lamellar twinning, i.e. formation of alternate layers (10–30 μm thick) of molecules with opposite configuration: B. S. Green, M. Knossow, Lamellar Twinning Explains the Nearly Racemic Composition of Chiral, Single Crystals of Hexahelicene, *Science* **1981**, *214*, 795–797.
48. H. Nakagawa, S. Ogashiwa, H. Tanaka, K. Yamada, H. Kawazura, Optical Resolution of Heterohelicenes by High

Performance Liquid Chromatography, *Bull. Chem. Soc. Jpn.* **1981**, 1903–1904.

49. R. H. Martin, V. Libert, Helicenes. The Use of Resolved Hexahelicene-2-carboxylic Acid as a Common Precursor for the Photochemical Synthesis of Optically Pure Octa-, Deca-, Undeca- and Trideca-helicenes. Thermal Racemization of Deca- and Undeca-helices, *J. Chem. Res. (S)* **1980**, 130–131.

50. C. Wachsmann, E. Weber, M. Czugler, W. Seichter, New Functional Hexahelicenes – Synthesis, Chiroptical Properties, X-ray Crystal Structures and Comparative Data Bank Analysis of Hexahelicenes, *Eur. J. Org. Chem.* **2003**, 2863–2876.

51. J. Larsen, K. Bechgaard, Thiaheterohelicenes 1. Synthesis of Unsubstituted Thia[5]-, [9]- and [13]heterohelicenes *Acta Chem. Scand.* **1996**, *50*, 71–76.

52. E. Murguly, R. McDonald, N. R. Banda, Chiral Discrimination in Hydrogen-bonded [7]Helicenes, *Org. Lett.* **2000**, *2*, 3169–3172.

53. T. Caronna, R. Sinisia, M. Catellani, S. Luzzati, L. Malpezzia, S. V. Meillea, A. Melea, C. Richter, R. Sinisi, Molecular Crystal Architecture and Optical Properties of a Thiohelicenes Series Containing 5, 7, 9 and 11 Rings Prepared via Photochemical Synthesis, *Chem. Mater.* **2001**, *13*, 3906–3914.

54. S. Maioranam, A. Papagni, E. Licandro, R. Annunziata, P. Paravidino, D. Perdicchia, C. Giannini, M. Bencini, K. Clays, A. Persoons, A Convenient Procedure for the Synthesis of Tetrathia-[7]-helicene and the Selective α-Functionalisation of Terminal Thiophene Ring, *Terahedron Lett.* **2003**, *59*, 6481–6488.

55. R. E. Abed, B. B. Hassine, J.-P. Genêt, M. Gorsane, A. Marinetti, An Alternative Procedure for the Synthesis of [5]- and [7]Carbohelicenes, *Eur. J. Org. Chem.* **2004**, 1517–1522.

56. K. Sato, S. Okazaki, T. Yamaguchi, S. Arai, The Synthesis of Azoniadithia[6]-helicene, *J. Heterocyl. Chem.* **2004**, *41*, 443–447.

57. C. Bazzini, S. Brovelli, T. Caronna, C. Gambarotti, M. Giannone, P. Macchi, F. Meinardi, A. Mele, W. Panzeri, F. Recupero, A. Sironi, R. Tubino, Synthesis and Characterization of Some Aza[5]helicenes, *Eur. J. Org. Chem.* **2005**, 1247–1257.

58. C. Baldoli, A. Bossi, C. Giannini, E. Licandro, S. Maiorana, D. Perdicchia, M. Schiavo, A Novel and Efficient Approach to (Z)-1,2-Bis(benzodithienyl)ethene Precursors of Tetrathia[7]helicenes, *Synlett* **2005**, 1137–1141.

59. S. Han, A. D. Bond, R. L. Disch, D. Holmes, J. M. Schulman, S. J. Teat, K. P. C. Vollhardt, G. D. Whitener, Total Syntheses and Structures of Angular [6]- and [7]Phenylene: the First Helical Phenylenes (Heliphenes), *Angew. Chem. Int. Ed.* **2002**, *41*, 3223–3227.

60. S. Han, D. R. Anderson, A. D. Bond, H. V. Chu, R. L. Disch, D. Holmes, J. M. Schulman, S. J. Teat, K. P. C. Vollhardt, G. D. Whitener, Total Syntheses and Structures of Angular [7]- [8]- and [9]Phenylene by Triple Cobalt-catalyzed Cycloisomerization: Remarkably Flexible Heliphenes, *Angew. Chem. Int. Ed.* **2002**, *41*, 3227–3230.

61. M. Klessinger, J. Michl, *Excited States and Photochemistry of Organic Molecules*, VCH, New York, 1995, pp. 440–442.

62. J. Larsen, K. Bechgaard, Direct Oxidative Cyclization of 1,2-Bis(benzothiophene–2-yl)ethylenes as a Replacement of Photocyclization in the Synthesis of Thiaheterohelicenes, *J. Org. Chem.* **1996**, *61*, 1151–1152.

63. J. M. Fox, T. J. Katz, Conversion of a [6]Helicene into an [8]Helicene and a Helical 1,10-Phenanthroline Ligand, *J. Org. Chem.* **1999**, *64*, 302–305.

64. M. Miyasaka, A. Rajca, M. Pink, S. Rajca, Cross-conjugated Oligothiophenes Derived from the $(C_2S)_n$ Helix: Asymmetric Synthesis and Structure of Carbon–Sulfur [11]Helicene, *J. Am. Chem. Soc.* **2005**, *127*, 13806–13807.

65. K. E. S. Phillips, T. J. Katz, S. Jockusch, A. J. Lovinger, N. J. Turro, Synthesis and Properties of an Aggregating Heterocyclic Helicene, *J. Am. Chem. Soc.* **2001**, *123*, 11899–11907.

66. C. Nuckolls, T. J. Katz, G. Katz, P. J. Collings, L. Castellanos, Synthesis and Aggregation of a Conjugated Helical

Molecule, *J. Am. Chem. Soc.* **1999**, *121*, 79–88.

67. J. M. Fox, T. J. Katz, S. V. Elshocht, T. Verbiest, M. Kauranen, A. Persoons, T. Thongpanchang, T. Krauss, L. Brus, Synthesis, Self-assembly and Nonlinear Optical Properties of Conjugated Helical Metal Phthalocyanine Derivatives, *J. Am. Chem. Soc.* **1999**, *121*, 3453–3459.

68. C. Nuckolls, T. J. Katz, T. Verbiest, S. V. Elshocht, H.-G. Kuball, S. Kiesewalter, A. J. Lovinger, A. Persoons, Circular Dichroism and UV–Visible Absorption Spectra of the Langmuir–Blodgett Films of an Aggregating Helicene, *J. Am. Chem. Soc.* **1998**, *120*, 8656–8660.

69. T. Verbiest, S. Van Elshocht, M. Kauranen, L. Hellemans, J. Snauwaert, C. Nuckolls, T. J. Katz, A. Persoons, Strong Enhancement of Nonlinear Optical Properties Through Supramolecular Chirality, *Science* **1998**, *282*, 913–915.

70. D. Hellwinkel, W. Schmidt, Zweifach *ortho*-verbrückte Triphenylamin-Derivate, *Chem. Ber.* **1980**, *113*, 358–384.

71. Improved procedure: J. L. Fox, C. H. Chen, J. F. Stenberg, An Improved Synthesis of 5,5,9,9-Tetramethyl-5H,9H-quino[3,2,1-*de*]acridine *Org. Prep. Proced. Int.* **1985**, *17*, 169–173.

72. J. E. Field, T. J. Hill, D. Venkataraman, Bridged Triarylamines: a New Class of Heterohelicenes, *J. Org. Chem.* **2003**, *68*, 6071–6078.

73. J. E. Field, G. Muller, J. P. Riehl, D. Venkataraman, Circularly Polarized Luminescence from Bridged Triarylamine Helicenes, *J. Am. Chem. Soc.* **2003**, *125*, 11808–11809.

74. M. Gingras, F. Dubois, Synthesis of Carbohelicenes and Derivatives by Carbenoid Couplings, *Tetrahedron Lett.* **1999**, *40*, 1309–1312.

75. Y. Ogawa, M. Toyama, M. Karikomi, K. Seki, K. Haga, T. Uyehara, Synthesis of Chiral [5]Helicenes Using Aromatic Oxy-Cope Rearrangement as a Key Step, *Terahedron Lett.* **2003**, *44*, 2167–2170.

76. H. J. Bestmann, W. Both, Determination of the Absolute Configuration of (+)-Pentahelicene, *Angew. Chem. Int. Ed. Engl.* **1972**, *11*, 296.

77. I. G. Stará, I. Starý, M. Tichý, J. Závada, V. Hanus, Stereochemical Dichotomy in the Stevens Rearrangement of Axially Twisted Dihydroazepinium and Dihydrothiepinium Salts. A Novel Enantioselective Synthesis of Pentahelicene, *J. Am. Chem. Soc.* **1994**, *116*, 5084–5088.

78. K. Tanaka, H. Suzuki, H. Osuga, Nonphotochemical Route to Chiral Disubstituted [7]Thiaheterohelicenes via Biaryl- and Carbonyl-coupling Reactions, *J. Org. Chem.* **1997**, *62*, 4465–4470.

79. M. Miyasaka, A. Rajca, M. Pink, S. Rajca, Chiral Molecular Glass: Synthesis and Characterization of Enantiomerically Pure Thiophene-based [7]Helicene, *Chem. Eur. J.* **2004**, *10*, 6531–6539.

80. M. C. Carreño, M González-López, A. Urbano, Efficient Asymmetric Synthesis of [7]Helicene Bisquinones, *Chem. Commun.* **2005**, 611–613.

81. A. Urbano, Recent Developments in the Synthesis of Helicene-like Molecules, *Angew. Chem. Int. Ed.* **2003**, *42*, 3986–3989.

82. F. Teplý, I. G. Stará, I. Starý, A. Kollároviè, D. Šaman, L. Rulíšek, P. Fiedler, Synthesis of [5]-, [6]- and [7]Helicene via Ni(0)- or Co(I)-Catalyzed Isomerization of Aromatic *cis,cis*-Dienetriynes, *J. Am. Chem. Soc.* **2002**, *124*, 9175–9180.

83. F. Teplý, I. G. Stará, I. Starý, A. Kollároviè, D. Šaman, Š. Vyskoèil, P. Fiedler, Synthesis of 3-Hexahelicenol and Its Transformation to 3-Hexahelicenylamines, Diphenylphosphine, Methyl Carboxylate and Dimethylthiocarbamate, *J. Org. Chem.* **2003**, *68*, 5193–5197.

84. A. Rajca, H. Wang, M. Pink, S. Rajca, Annelated Heptathiophene: a Fragment of a Carbon–Sulfur Helix, *Angew. Chem. Int. Ed.* **2000**, *39*, 4481–4483.

85. A. Rajca, M. Miyasaka, M. Pink, H. Wang, S. Rajca, Helically Annelated and Cross-conjugated Oligothiophenes: Asymmetric Synthesis, Resolution and Characterization of a Carbon–Sulfur [7]Helicene, *J. Am. Chem. Soc.* **2004**, *126*, 15211–15222.

86. T. B. Friedman, X. Cao, A. Rajca, H. Wang, L. A. Nafie, Determination of

Absolute Configuration in Two Molecules with Chiral Axes by Vibrational Circular Dichroism: A C_2-symmetric Annelated Heptathiophene and a D_2-symmetric Dimer of 1,1′-Binaphthyl, *J. Phys. Chem. A* **2003**, *107*, 7692–7696.

87. M. Miyasaka, A. Rajca, Synthesis of a Short Carbon–Sulfur Helicene: Pd-catalyzed Cross-coupling at the β-Positions of Thiophenes, *Synlett* **2004**, 177–182.

88. D. Peña, A. Cobas, D. Pérez, E. Guitián, L. Castedo, Dibenzo[a,o]phenanthro[3,4-s]pycene, a Configurationally Stable Double Helicene: Synthesis and Determination of Its Conformation by NMR and GIAO Calculations, *Org. Lett.* **2003**, *5*, 1863–1866.

89. K. Shiraishi, A. Rajca, M. Pink, S. Rajca, π-Conjugated Conjoined Double Helicene via a Sequence of Three Oxidative CC- and NN-Homocouplings, *J. Am. Chem. Soc.* **2005**, *127*, 9312–9313.

90. Two NN homocouplings in cyclodehydrogenation of tetra(benziimidazol-2-yl)benzenes: W. Wu, A. C. Grimsdale, K. Müllen, Cyclodehydrogenation of Di- and Tetra(benzimidazol-2-yl)benzenes to give Model Heteroaromatic Discotic Systems, *Chem. Commun.* **2003**, 1044–1045.

91. A. Rajca, M. Pink, unpublished data.

92. J. Wu, Z. Tomovic, V. Enkelmann, K. Müllen, From Branched Hydrocarbon Propellers to C_3-Symmetric Graphite Disks *J. Org. Chem.* **2004**, *69*, 5179–5186.

93. L. Barnett, D. M. Ho, K. K. Baldridge, R. A. Pascal Jr., The Structure of Hexabenzotriphenylene and the Problem of Overcrowded D_{3h} Polycyclic Aromatic Compounds, *J. Am. Chem. Soc.* **1999**, *121*, 727–733.

94. A. A. Bennett, M. R. Kopp, E. Wenger, A. C. Willis, Generation of Nickel(0)–Aryne and Nickel(II)–Biphenyldiyl Complexes via in situ Dehydrohalogenation of Arenas. Molecular Structures of [Ni(2,2′-$C_6H_4C_6H_4$)dcpe)] and C_2-Hexabenzotriphenylene, *J. Organomet. Chem.* **2003**, *667*, 8–15.

95. D. Peña, A. Cobas, D. Pérez, E. Guitián, L. Castedo, Kinetic Control in the Palladium-catalyzed Synthesis of C_2-Symmetric Hexabenzotriphenylene. A Conformational Study, *Org. Lett.* **2000**, *2*, 1629–1632.

96. B. Gomez-Lor, A. M. Echavarren, Synthesis of a Triaza Analogue of Crushed Fullerene by Intramolecular Palladium-catalyzed Arylation, *Org. Lett.* **2004**, *6*, 2993–2996.

97. J. Lu, D. M. Ho, N. J. Vogelaar, C. M. Kraml, R. A. Pascal, Jr., A Pentacene with a 144° Twist, *J. Am. Chem. Soc.* **2004**, *126*, 11168–11169.

98. T. C. W. Mak, H. N. C. Wong, Inclusion Properties of Tetraphenylene and Synthesis of Its Derivatives, *Top. Curr. Chem.* **1987**, *140*, 142–164.

99. Barrier for racemization of 45^+ kcal mol^{-1}: D. Gust, G. H. Senkler, Jr., K. Mislow, Resolution and Optical Stability of Tetrabenzocyclo-octatetraene Derivatives, *J. Chem. Soc., Chem. Commun.* **1972**, 1345.

100. P. Rashidi-Ranjbar, Y.-M. Man, J. Sandstrom, H. N. C. Wong, Enantiomer Resolution, Absolute Configuration and Attemted Thermal Racemization of Two Tetrabenzocyclooctatetraene (o-Tetraphenylene) Derivatives. An Exceptionally High Barrier to Ring Inversion, *J. Org. Chem.* **1989**, *54*, 4888–4892.

101. (a) A. Rajca, A. Safronov, S. Rajca, J. Wongsriratanakul, D_2-Symmetric Dimer of 1,1′-Binaphthyl and Its Chiral π-Conjugated Carbodianion, *J. Am. Chem. Soc.* **2000**, *122*, 3351–3357; (b) A. Rajca, J. Li, unpublished data.

102. A. Rajca, H. Wang, P. Bolshov, S. Rajca, Greek Cross Dodecaphenylene: Sparteine-mediated Asymmetric Synthesis of Chiral D_2-Symmetric π-Conjugated Tetra-o-phenylenes, *Tetrahedron* **2001**, *57*, 3725–3735.

103. J.-F. Wen, W. Hong, K. Yuan, T. C. W. Mak, H. C. N. Wong, Synthesis, Resolution and Applications of 1,16-Dihydroxytetraphenylene as a Novel Building Block in Molecular Recognition and Assembly, *J. Org. Chem.* **2003**, *68*, 8918–8931.

104. J. Gibson, M. Holohan, H. L. Riley, Amorphous Carbon, *J. Chem. Soc.* **1946**, 456–461.

105. H. L. Riley, Chemical and Crystallographic Factors in Carbon Combustion,

J. Chim. Phys. Phys. Chim. Biol. **1950**, 565–572.
106. F. Diederich, Y. Rubin, Synthetic Approaches Toward Molecular and Polymeric Carbon Allotropes, *Angew. Chem. Int. Ed. Engl.* **1992**, *31*, 1101–1123.
107. A. Rajca, A. Safronov, S. Rajca, R. Schoemaker, Double Helical Octaphenylene, *Angew. Chem. Int. Ed. Engl.* **1997**, *36*, 488–491.
108. Review including discussion of duplication method: H. B. Kagan, J. C. Fiaud, Kinetic Resolution, *Top. Stereochem.* **1988**, *18*, 249–330.
109. A. Rajca, A. Safronov, S. Rajca, C. R. Ross, II, J. J. Stezowski, Biphenylene Dimer. Molecular Fragment of a Two-dimensional Carbon Net and Double-stranded Polymer, *J. Am. Chem. Soc.* **1996**, *118*, 7272–7279.
110. M. J. Marsella, I. T. Kim, F. Tham, Toward Conjugated Double Helical Ladder Polymers: Cyclooctatetrathiophene as a Highly Versatile Double Helical Scaffold, *J. Am. Chem. Soc.* **2000**, *122*, 974–975.
111. D. L. An, T. Nakano, A. Orita, J. Otera, Enantiopure Double-helical Alkynyl Cyclophanes, *Angew. Chem. Int. Ed.* **2002**, *41*, 171–173.
112. H.-Y. Peng, C.-K. Lam, T. C. W. Mak, Z. Cai, W.-T. Ma, Y.-X. Li, H. N. C. Wong, Chiral Rodlike Platinum Complexes, Double Helical Chains and Potential Asymmetric Hydrogenation Ligand Based on Linear Building Blocks: 1,8,9,16-Tetrahydroxytetraphenylene and 1,8,9,16-Tetrakis(diphenylphosphino)tetraphenylene, *J. Am. Chem. Soc.* **2005**, *127*, 9603–9611.
113. R. Rathore, P. LeMagueres, S. V. Lindeman, J. K. Kochi, A Redox-controlled Molecular Switch Based on the Reversible C–C Bond Formation in Octamethoxytetraphenylene, *Angew. Chem. Int. Ed.* **2000**, *39*, 809–812.
114. Accelerated aging tests for optoelectronic devices are typically carried out at 70 or 85 °C for 2000–5000 h, e.g. based upon Telcordia (Bellcore) standards GR-468-CORE, Reliability Assurance Requirements for Optoelectronic Devices Used in Telecommunications Equipment.
115. R. H. Martin, M. J. Marchant, Thermal Racemisation of Hepta-, Octa- and Nonahelicene: Kinetic Results, Reaction Path and Experimental Proofs that the Racemisation of Hexa- and Heptahelicene Does Not Involve an Intramolecular Double Diels–Alder Reaction, *Tetrahedron* **1974**, *30*, 347–349.
116. R. H. Janke, G. Haufe, E.-U. Würthwein, J. H. Borkent, Racemization Barriers of Helicenes: a Computational Study, *J. Am. Chem. Soc.* **1996**, *118*, 6031–6035.
117. C. Goedicke, H. Stegemeyer, Resolution and Racemization of Pentahelicene, *Tetrahedron Lett.* **1970**, 937–940.
118. K. Yamada, H. Nakagawa, H. Kawazura, Thermal Racemization of Thiaheterohelicenes, *Bull. Chem. Soc. Jpn.* **1986**, *59*, 2429–2432.
119. I. Navaza, G. Tsoucaris, G. Le Bas, A. Navaza, C. de Rango, General Models for Helicenes, *Bull. Soc. Chim. Belg.* **1979**, *88*, 863–870.
120. C. Herse, D. Bas, F. C. Krebs, T. Burgi, J. Weber, T. Wesolowski, B. W. Laursen, J. Lacour, A Highly Configurationally Stable [4]Heterohelicenium Cation, *Angew. Chem. Int. Ed.* **2003**, *42*, 3162–3166.
121. B. Laleu, P. Mobian, C. Herse, B. W. Laursen, G. Hopfgartner, G. Bernardinelli, J. Lacour, Resolution of [4]Heterohelicenium Dyes with Unprecedented Pummerer-like Chemistry *Angew. Chem. Int. Ed.* **2005**, *44*, 1879–1883.
122. S. F. Nelsen, T. B. Frigo, Y. Kim, J. A. Thompson-Colon, Double Nitrogen Inversion in Sesquibicyclic Hydrazines and Their Cation Radicals, *J. Am. Chem. Soc.* **1986**, *108*, 7926–7934.
123. Benzo[c]benzo[3, 4]cinnolino[1,2-a]cinnoline: F. A. Neugebauer, S. Kuhnhäuser, A Triphenylamine Double-decker, *Angew. Chem. Int. Ed. Engl.* **1985**, *24*, 596–597.
124. H. Fischer, C. Krieger, F. A. Neugebauer, Benzo[c]benzo[3, 4]cinnolino[1,2-a]cinnoline, a Chiral Hydrazine Deriva-

tive, *Angew. Chem. Int. Ed. Engl.* **1986**, *25*, 374–375.
125. M. Dietrich, J. Heinze, C. Krieger, F. A. Neugebauer, Electrochemical Oxidation and Structural Changes of 5,6-Dihydrobenzo[c]cinnolines, *J. Am. Chem. Soc.* **1996**, *118*, 5020–5030.
126. F. A. Neugebauer, M. Bock, S. Kuhnhäuser, H. Kurreck, Darstellung, ESR- und ENDOR-Untersuchung von Radicalkationen des Tetraphenylhydrazins, des 5,6-Dihydro-5,6-diphenylbenzo[c]cinnolins und des Benzo[c]benzo[3,4]cinnolino[1,2-a]cinnolines, *Chem. Ber.* **1986**, *119*, 980–990.
127. Barrier of about 10 kcal mol−1 for the ring inversion in cyclooctatetraenes: L. A. Paquette, in *Advances in Theoretically Interesting Molecules*, Thummel, R. P. (Ed.), JAI Press, Greenwich, CT, **1992**, Vol. 2, p. 1.
128. E. L. Eliel, S. H. Wilen, *Stereochemistry of Organic Compounds*, Wiley, New York, **1994**, Chapter 13, pp. 991–1118.
129. F. Furche, R. Ahlrichs, C. Wachsmann, E. Weber, A. Sobanski, F. Vögtle, S. Grimme, Circular Dichroism of Helicenes Investigated by Time-dependent Density Functional Theory, *J. Am. Chem. Soc.* **2000**, *122*, 1717–1724.
130. G. Lieser, M. Oda, T. Miteva, A. Meisel, H.-G. Nothofer, U. Scherf, D. Neher, Ordering, Graphoepitaxial Orientation and Conformation of a Polyfluorene Derivative of the Hairy-rod Type on an Oriented Substrate of Polyimide, *Macromolecules* **2000**, *33*, 4490–4495.
131. M. Oda, H.-G. Nothofer, U. Scherf, V. Sunjic, D. Richter, W. Regenstein, D. Neher, Chiroptical Properties of Chiral Substituted Polyfluorenes, *Macromolecules* **2002**, *35*, 6792–6798.
132. M. R. Craig, P. Jonkheijm, S. C. J. Meskers, A. P. H. J. Schenning, E. W. Meijer, The Chiroptical Properties of a Thermally Annealed Film of Chiral Substituted Polyfluorene Depend on Film Thickness, *Adv. Mater.* **2003**, *15*, 1435–1438.
133. Y. Geng, A. Trajkovska, D. Katsis, J. J. Ou, S. W. Culligan, S. H. Chen, Synthesis, Characterization and Optical Properties of Monodisperse Chiral Oligofluorenes, *J. Am. Chem. Soc.* **2002**, *124*, 8337–8347.
134. E. Peeters, M. P. T. Christiaans, R. A. J. Janssen, H. F. M. Schoo, H. P. J. M. Dekkers, E. W. Meijer, Circularly Polarized Electroluminescence from a Polymer Light-emitting Diode, *J. Am. Chem. Soc.* **1997**, *119*, 9909–9910.
135. S. H. Chen, D. Katsis, A. W. Schmid, J. C. Mastrangelo, T. Tsutsui, T. N. Blanton, Circularly Polarized Light Generated by Photoexcitation of Luminophores in Glassy Liquid-crystal Films, *Nature* **1999**, *397*, 506–508.
136. Y. Kim, W. L. Cao, J. Goldhar, C. H. Lee, W. N. Herman, Optical Waveguides from Amorphous Chiral Binaphthyl Films, *Polym. Prepr.* **2002**, *43*, 594–595.
137. H. Ohkita, A. Tagaya, Y. Koike, Preparation of a Zero-birefringence Polymer Doped with a Birefringent Crystal and Analysis of Its Characteristics, *Macromolecules* **2004**, *37*, 8342–8348.
138. K. Saito, M. Massalska-Arodz, S. Ikeuchi, M. Maekawa, J. Sciesinski, E. Sciesinska, J. Mayer, T. Wasiutynski, M. Sorai, Thermodynamic Study on a Chiral Glass Former, 4-(1-Methylheptyloxy)-4′-cyanobiphenyl, *J. Phys. Chem. B* **2004**, *108*, 5785–5790.
139. McDonnell, D. G. Thermochromic Cholesteric Liquid Crystals, in *Thermotropic Liquid Crystals*, Gray, G. W. (Ed.), Wiley, Chichester, 1987, Chapter 5, pp. 120–144.
140. Y. Kim, W. N. Herman, unpublished data.

Index

a

acceptor-acceptor-donor (AAD) 267
acceptor-donor-acceptor systems 202
acenes 511 ff.
– edge-to-face arrangements 511
– face-to-face arrangements 511 ff.
acetylene dimerization 236
alkynes 210 f.
– activation 210
– fluorescence dyes 212
– propargyl activation 212
– Sonogashira coupling 210
amperometric biosensor 10
amphiphilic molecules 264
annelation 549 ff.
– Diels-Alder 552
– Friedel-Crafts acylation 552, 560
– non-racemic intermediates 554
anthracene 512 ff.
– alkyne substituents 519
– disubstituted 517 f.
– edge-substituted 522 ff.
– edge-to-face arrangements 513
– electron-withdrawing substituents 515
– π-face interactions 513
– fluorescence quantum yields 513
– herringbone arrangement 514
– higher acenes 540 ff.
– monosubstituted 514 f.
– self-assembly process 516, 526 f.
antimicrobial agents 270
arylethenes 208
– dumbbell shaped 208
– one-pot synthesis 208 f.
– star shaped 208
arylthiols 354 f.
– alligator clips 354 f.
– protecting groups 354 f.
– thiol end-capped molecules 354 f.
atom transfer radical polymerization 17

atomic force microscopy 379 f.
– conducting probe 379 f.
– substrate-molecule-particle junction 379
azo dyes 181

b

benzenoid compounds 59
– hoop-shaped 59
bilayer formation 268
bimolecular detectors 41
Bingel reaction 24, 46
biocidal nanotubes 268
biological functionalization 34
– nanomaterials 34
– nanotube-protein conjugates 34
biosensors 261
biphospholes 124
– α,α'-biphosphole cores 128
– derivatives 126
– mixed oligomers 137 f.
– properties 125 f.
– synthesis 124
bistable catenanes 297 ff.
– molecular machines 297 f.
– nanoelectromechanical systems 297
bistable rotaxanes 297 ff.
– acid-base chemistry 300 f.
– BIPY^{2+} units 301 f.
– CBPQT^{4+} 304 f.
– chemical stimuli 299
– chemically induced switching 299
– circumrotary motions 298
– condensed-phase switching 306 f.
– DB24C8 platform 301 f.
– electrochemical switching 315 f.
– microcantilever beams 309
– properties 298
– redox switching 315
– solution-phase switching 305 f.

Functional Organic Materials. Syntheses, Strategies, and Applications.
Edited by Thomas J.J. Müller and Uwe H.F. Bunz
Copyright © 2007 WILEY-VCH Verlag GmbH & Co. KGaA, Weinheim
ISBN: 978-3-527-31302-0

– switching   300 ff.
– translation movements   298
bond-order alterations   407 f.
  – dipolar chromophores   408
  – donor and acceptor strength   408
bowl-shaped arenes   59
break junctions   381 f.
BSA conjugates   14

c

C_{60} fullerenes   39, 43, 59
carbon nanotubes   3 ff., 59
  – atom transfer radical polymerization   17
  – bioconjugates   12 ff.
  – chemical derivatization   7 f.
  – covalent functionalization   4 ff.
  – cycloadditions   21 ff.
  – defect functionalization   10 ff.
  – electrochemical modification   26 f., 47
  – films   9
  – halogenation   5 ff.
  – HiPCO process   9
  – hydrogenation   19
  – multi-walled carbon nanotubes   3, 6
  – oxidative purification   8, 45
  – silylation   18 f.
  – single-walled carbon nanotubes
    – 3 f., 6, 41 f.
  – supramolecular complexes   39
  – thiolation   17 f.
caltrops   367 f.
carbodiimide technique   12
charge transport   83
chiral conjugated materials   547 ff.
  – angular connectors   547
  – annelation   548 ff.
  – applications   547
  – chiroptical properties   570 ff.
  – π-conjugated polymers   547 f.
  – free energy barriers   568
  – helicenes   548
  – inversion of configuration   569 f.
  – racemization   567 ff.
chromophores
  – charge transfer   422 f.
  – combinatorial azo coupling   181 f.
  – combinatorial condensation reactions   182 f.
  – combinatorial cross-coupling reactions   186
  – combinatorial synthesis   180 ff.
  – dendronized   416, 426

  – designs   404 ff., 420 f.
  – dipolar chromophores   418 ff.
  – donor-acceptor-donor   422
  – electrooptic applications   404
  – high-β-chromophores   414
  – multicomponent synthesis   199 f.
  – multifunctional chromophores   197
  – nonlinear optical chromophores   416
  – octupolar   410 f.
  – push-pull   210
  – quadrupolar chromophores   426
  – second-order chromophores   404 ff.
  – transition metal chromophores   409
α-chymotrypsin   34
Cl-Stetter-Paal-Knorr synthesis   213 f., 217
  – furans   213
  – pyrroles   213
co-conformation   298
  – ground-state   298
  – metastable-state   298
condensed-phase switching   306 f.
π-conjugated oligomers   83 ff.
  – binaphthyl-based oligomers   97 f.
  – bipolar tetramers   90
  – cross-like assemblage   84
  – cruciform   83
  – diaryloxadiazole   89
  – tetrahedral core unit   85 ff.
  – tetrastilbenoid oligomers   86
  – tetrasubstituted benzene core   91 ff.
  – tetrasubstituted biaryl core   95 ff.
  – tetrasubstituted biphenyl unit   102
  – thienyl-based   107 f.
σ–π conjugated species 127, 129
  – thiophene-phosphole derivatives   127
π-conjugated systems   120 ff.
  – diphosphene-containing   161 ff.
  – phosphine-containing   147 ff.
  – phosphoalkene-containing   161 ff.
  – phosphorus-derived building blocks   121 ff.
  – tailoring   120
coordination chemistry   197 ff.
  – combinatorial coordination chemistry   197
  – iridium(III) chromophores   198
  – transition metal complexes   197 f.
corannulenes   69 f.
  – absorption   70
  – emission   70
  – dibenzo-fused   69, 74
  – synthesis   69, 74
Corey-Pauling-Kultun model   304
coumarin dyes   188 ff.

- synthesis 189
crossed wires 382
cross-linked polymeric networks 282
cyanines 182
 - asymmetric cyanines 182 f.
 - rhodacyanines 185, 215
 - solid-phase synthesis 182
[n]cyclacene 59, 61 f.
 - multi-step Diels-Alder 61
[n]cyclophenacenes
 - absorption 70
 - aromaticity 67 f.
 - emission 70
 - nucleus-independent chemical shift 67
 - radialene structures 63
 - regioisomers 65
 - σ-skeleton 68
 - structure 66 ff.
 - synthesis 62, 63 ff.
cyclopropanation 24

d

DDA *see* donor-donor-acceptor
defocusing effects 393
degenerate four-wave mixing (DFWM) 403
dendrimers 10, 12
dendritic benzamides 270
dendron-modified polymers 285
diarylethene 329 ff.
 - antiferromagnetic interactions 332 f.
 - dimers 341
 - HOMO 332
 - magnetic interactions 341
 - photochromism 330 f., 333
 - photoswitches 339 f.
 - radical synthesis 333 f.
 - SOMO 332
dibenzophosphole 143
 - building block 143
difference frequency generation 393
diphosphaferrocenes 127
 - electronic properties 127
dipolar chromophores 408 ff.
 - donor-acceptor character 410
 - ferrocene-based 409
 - host polymers 411 f.
 - inorganic coordination complexes 409
 - organometallic groups 409
 - thermal stability 413
dipole moment 394 f.
dipole-dipole interactions 272
dithienophosphole 143 ff.
 - cyclotriphosphazenes 144
 - hole-transporting capability 144

 - HOMO-LUMO band gaps 144
 - Lewis basicity 144
 - photoluminescence 145
diversity-oriented synthesis 179 ff, 215
 - diarylethenes 196
DNA 441 ff.
 - base modifications 452 f.
 - base substitutes 454 f.
 - charge-transfer chemistry 443 f.
 - chromophore functionalization 448 ff.
 - π-conjugated material 442
 - electron transfer studies 456 ff.
 - electron transport 441 f.
 - excess electron transfer 446 f., 461
 - hole hopping 444
 - hole transfer 444
 - oxidative damage 444
 - phosphoramidite chemistry 446
 - reductive electron transfer 446 ff.
 - synthetic nanostructures 460 ff.
domino reactions 179
donor-acceptor dyads 482
 - acridinium ions 485 ff.
 - back electron transfer 487
 - π-donor-acceptor dyads 485
 - infinite CS lifetime 485 f.
 - long-lived CS states 482, 484
 - organic solar cells 494 f.
donor-donor-acceptor 267
DOS *see* electron density of states
drug delivery 261

e

electric-field-induced second harmonic generation (EFISH) 398
electrochemical switching 315 f.
 - neutral rotaxanes 315
electrochromic device 313
 - TTF/DNP-based 313
electroluminescence 83
electrolumiscent diodes 84
electron density of states 25
electronic devices 261
electron transfer processes 441 ff.
 - photoinduced 441 f.
electrooptic (EO) effect 393
electrooptic switching 394
electropolymerization 161
ESR spectra 330, 337
 - biradicals 347
 - weak interactions 330 f.
ethenylphosphole derivatives 138 f.
ethynylphosphole derivatives 138 f.
excitation density 344 f.

f

Fagan-Nugent method   130, 134, 139
Fermi energy   369
field-effect transistors   119, 511
finite fields   399
fluorotubes   6
foldamer formation   105
four-cylinder bundles   271
fullerenes   59, 474 ff.
– bowl-shaped conjugated systems   62
– charge delocalization   477
– electron transfer properties   474 f.
– luminescence quantum yield   71
– Rehm-Weller behaviour   475
– 60π-spherical conjugation   62
– spherical fullerenes   59
fullerene peapods   43
π-functional-systems   465 ff.
– electron transfer   465 ff.
furan   214
– four-component synthesis   214

g

glucose biosensors   25, 37
glucose oxidase   10, 25, 35

h

hairy-rod polymers   286
Heck reaction   194
– multi-component reactions   206
– one-pot reactions   206
– one-terminal wires   354 f.
– vinylsilanes   194, 206
highly-oriented pyrolytic graphite   9
Hiyama coupling   206
hoop-shaped molecules   61
hydrophobicity   264
hyper-Raleigh scattering   398

i

imino nitroxide radicals   345
indium-tin oxide anode   133
inkjet printing   83, 113
intramolecular cyclizations   247 f.
intramolecular excimers   16
intramolecular magnetic interaction   329 ff.
intramolecularization   241, 245

k

kinetic approach   227
– advantages   254
– alkyne metathesis   228
– disadvantages   254
– Stephens-Castro coupling   227
– synthesis of macrocycles   227 f.

l

Landauer equations   368
Langmuir-Blodgett monolayers   314, 353
Langmuir-Blodgett technique   411
ligand-driven light-induced spin change (LD-LISC)   333
light-induced excited spin state trapping (LIESST)   333
light-induced thermal hysteresis (LITH)   333
linear polarizability   394
linear polarization   394 f.
lipase   11
liquid crystalline network   95, 271 f.
– nematic   272
– smectic   271
liquid crystalline compounds   251
liquid crystalline materials   261
low-temperature plasma process   6
luminophore synthesis   199

m

Mach-Zehnder interferometer   414
macroscopic self-assembly   287 f.
– layer-by-layer   287
– self-assembled monolayers (SAM)   287, 354, 376
magnetic interactions   330
merocyanine dyes   199 ff., 272
– one-pot synthesis   201
– retrosynthetic analysis   200
– textile coloration   200
metal-catalyzed cross-coupling reactions   187 f.
– Pd-catalyzed   188
metallamacrocycles   241
– reductive elimination   241
metallothioneins   37
metal-to-ligand charge-transfer (MLCT)   321
molecular devices   368 f.
– classical coherent tunnelling   368
– electron transport   368
molecular electronics   261, 353
molecular elevator   301 ff.
– pH-driven   301
molecular junctions   353, 375 f.
– break junctions   382
– donor-π-acceptor molecules   377
– junction types   370 ff.
– metal-molecule-metal junction   381
– nanopore junctions   382
– nanowire junctions   385

- particle junctions 384 f.
- scanning 375
- STM junctions 380 f.
tunneling-based 375
molecular machines 295 ff.
- assembly 298
- detection of actuations 296
- interlocked molecules 297 ff.
- movements 296
- stimuli 296
molecular magnetism 333
molecular motors 295
- metal ion-based 319
- MLCT-induced switching 319 f.
- PET-induced switching 318 f.
molecular muscles 295
- chemically controlled switch 304
- interlocked molecules 296
- metal ion exchange 303 f.
- redox controlled switch 304 f.
molecular sensor 160
molecular switch 310 ff.
- benzidine/biphenol-based 310
- photochemically induced 316 f.
- photoisomerization 317 f.
- redox-driven rotaxanes 311
- rotaxanes 310 f.
- solution-phase switching 311
- TTF/DNP-based rotaxanes 311
molecular switch tunnel junctions device 314 f.
molecular wire compounds 354 f.
- Co^{2+}-bridged 361
- four-terminal wires 365 f.
- Horner-Wadsworth-Emmons reactions 355, 359
- oligophenylene-ethynylene 354, 376
- oligophenylene-vinylene 354, 376
- one-terminal wires 355 f.
- Sonogashira cross-coupling 354 f., 360
- three-terminal wires 364 f.
- two-terminal wires 359 f.
monodisperse conjugated materials 120
monophosphaferrocenes 127
MSTJs see molecular switch tunnel junctions device
multi-component reactions 179 f.
- condensations 199 f.
- cross-coupling reactions 204 f.
multifunctional ligands 274
multifunctional materials 261
multiphoton-absorption 393
- applications 393
multiple noncovalent interactions 261 ff.

- hydrogen-bonding 263
- hydrophobic interactions 263
- salt bridges 263
multi-walled carbon nanotubes (MWCNT) 3, 6

n

nanocarpet 269
nanocomposites 16
nanomechanical device 308 f.
- $CBPQT^{4+}$ 308 f.
- ground state interactions 308
- solid state 308
nanopore junctions 382
nanoscale machines 295
- nanomechanical movements 296
nanowire junctions 385
naphthacene see tetracene
negative differential resistance (NDR) 369
Newkome-type dendrons 12
nitronyl nitroxide radicals 338 f., 341 f.
- diarylethene-bridged 339
NLO-phores 159
non-covalent functionalization 32 ff.
- biomolecules 32
- charge-transfer complex 33
- endohedral functionalization 42 f.
- exohedral functionalization 32 f.
- mixed micelles 33
- nanotubes 32 ff.
- photoconversion 38
- π-stacking interactions 35 f., 48
nonlinear optical (NLO) effects 393 ff.
- measurement techniques 397 ff.
- multiphoton-absorption 393
- second-order effects 395 ff.
- third-order effects 395 ff.
nonlinear optical (NLO) materials 393 f.
- electrooptic coefficients 398
- ground-state dipole moment 399
- non-centrosymmetric assembly 397
- second-order properties 397 ff.
- third order effects 399 f.
nonlinear optics 138
nonlinear polarization 394 f.

o

OFETs see organic field effect transistors
OLED see organic light-emitting diode
oligothiophenes 189 ff.
- combinatorial synthesis 189 f.
- helical oligothiophenes 559 f.
- molecular wires 338 f.
- solid-phase synthesis 192
- structure-property relationship 191

OPE molecules 375 f.
– conductance 387
– conductance switching 377 f.
– monolayers 377 f.
– switching behaviour 377
optical rectification 396
optical waveguides 393
optically transparent electrode 494 ff.
optoelectronics 192
organic field effect transistors 84, 107, 109 f.
organic light-emitting diode 83, 119, 133, 138
– dopants 133, 135
– EL quantum yields 133
– fabrication 133, 160
– precursors 192
– single-layer 133
organic photovoltaic cells 84
organic semiconductors 83
organic solar cells 494 f.
– donor-acceptor dyads 494
– multi-porphyrin/fullerene assemblies 499 ff.
OTE see optically transparent electrode

p

2PA see two-photon absorption
PAHs see polycyclic aromatic hydrocarbons
PAM see phenylacetylene macrocycle
particle junctions 384 f.
P-branched multichromophores 160
p-conjugated oligomers 119
PEI see poly(ethylenimine)
pentacenes 534 ff.
– alkyne 536
– edge-to-face interactions 534
– partial fluorination 538 f.
– π-type semiconductor 534
– trialkylsilyl derivatives 535
peptide nucleic acids 15
peptides 263 f.
– multifunctionalization 263
– secondary structures 263 f.
PET see photo-induced electron transfer
PFO 99 f.
– films 99
– synthesis 100
phase-transfer catalyst 8
phenylacetylene macrocycle 226
– isomeric PAMs 230
– macrocyclizations 231
– non-symmetrically functionalized 251
– orthogonal reaction conditions 233

– synthesis of Moore and Zhang 229
phenylene-ethynylene macrocycles 225
p-phenylenephosphine-based oligomers 147 ff.
– cyclic voltammetry 149
– π-delocalization 149
– synthesis 147 f.
– UV/VIS spectra 147, 149
phenylene oligomers 232 ff.
– orthogonal strategy 233
– precursor synthesis 232 f.
– synthesis 232
phosphaalkyne oligomers 161 ff.
– fluorescence properties 165 f.
phosphapericyclines 152
phosphole 121 ff.
– aromaticity 122, 136
– p-conjugated systems 121 ff.
– copolymers 140
– endocyclic p-system 122
– HOMO-LUMO separations 131
– hyperconjugation 122, 136
– σ–π interaction 122
– mixed oligomers 129 ff., 138 f., 143 ff.
– nucleophilic behaviour 132
– oligophospholes 123 ff.
– spectrum 131
– thermal stability 123
– topology 136
– transition metal complexes 123
phosphole polymers 140 ff.
– arylphosphino units 155 ff.
– biphenyl-phosphole derivative 140 f.
– electropolymerization 141 f.
– higher generation dendrimers 153
p-phenylenephosphine units 147 f.
– phosphine-ethynyl units 151 ff.
– properties 142
– thienyl-capped monomers 141 f.
photocatalytic oxygenation 491 ff.
– anthracenes 491
– anthraquinones 491
– tetraphenylethylene 492
– TPE dioxetane 493
photochromic compounds 331
– properties 331
photochromic materials 147
photochromic states 342 f.
photoconversion 38
photoexcitation 38
photo-induced electron transfer 19, 319, 321, 477 f.
– chlorophyll-like donor 479

- non-adiabatic intramolecular electron transfer   477
- photocatalyst   490
- photosynthetic units   477 f.
- porphyrin-fullerene dyads   479 ff.

photonic devices   261
photophysical processes   83
photosensitizers   182
photoswitching of magnetism   333
photoswitching units   329 ff.
- bis(2-thienyl)ethane   340
- bis(3-thienyl)ethene   335 f.
- π-conjugation   340
- magnetic interaction   335
- new units   345
- photochromic molecules   341 f.
- photoswitching   329 ff.
- precursors   340
- reversed   340 f.

photosynthetic p-pigments   465 f.
photovoltaic cells   119
plastic lasers   119
PNA *see* peptide nucleic acids
polarizable π-systems   394 f.
poled-polymer systems   413 ff.
- dendrimers   416 f.
- host-guest systems   415
- nolinear chromophores   418
- photochemical stability   414
- telecommunication applications   414

poly(ethylenimine)   9
polyanilines   147
polycyclic aromatic hydrocarbons   560
polyelectrolyte   9
polyenes   412 f.
- non-ring-locked   413
- ring-locked   413

polymer-based self-assembly   275 ff.
- applications   279
- main-chain extension   275, 276 f.
- metal-trpy decomplexation   278
- orthogonal routes   280 f.
- reversibility   276
- side-chain assembly   276, 279 f.
- tuning   276
- trpy-based metal coordination   276

polymeric OLEDs   192
polymers   194
- poly(arylene-ethynylene) family   194

polyphosphacyclopolyynes   152
polyphosphazenes   121
porphyrin units   365 f., 423, 467 ff.
- electron transfer   468

- ESR spectra   469 f.
- oligomers   501
- porphyrin-alkanethiolate monolayer   504
- porphyrin dendrimers   499 f.
- porphyrin-quinone dyads   483
- self-assembled systems   503
- self-exchange electron transfer   472 f.
- Zn-porphyrins   467 ff., 473

potentiometric titration   10
pyridylphosphole ligands   135
- properties   135

pyrrole   213 f.
- four-component synthesis   214

q

quantum dots   9
quantum yield   83
quaterphospholes   124 f.
- properties   125 f.
- synthesis   124 f

quinones   466 f.

r

radialene   63, 69
recombination efficiency   83
rhodacyanines   185 f.
- structure   185
- synthesis   186

rosettes   268
helical rosette nanotubes   267 f.
rubrene   532 f.
- charge-carrier mobility   532

s

second harmonic generation   396
second-order chromophores   404 ff.
- bond-order alterations   407, 419
- design   404 f.
- dipolar chromophores   405 f.
- electrooptic measurements   405
- excitation   406
- frequence-doubling phenomena   404
- ground-state dipole moment   406, 410
- octupolar organic molecules   410
- resonance structures   407
- solvatochromism   406
- static limit   404
- sum-over-state approach   404

self-assembly   261 ff.
- bilayer-type   268 f.
- GC units   267
- hierarchy   265
- macroscopic self-assembly   287 ff.

- merocyanine dyes 272 f.
- metal coordination 273, 285
- monofunctional 279
- multifunctional ligands 274
- multi-step self-assembly 262 f.
- polymer-based 275
- receptors 280
- small molecule-based 265

self-focusing effects 393
self-sorting 283
semiconductors 4, 13 f.
sequential catalysis 204
SFG see sum frequency generation
shape-persistent macrocycles 225 ff.
- asymmetric macrocycles 246
- intermolecular cyclodimerization 234
- kinetic approach 227
- persistence length 225
- side-group orientation 226
- substituents 234
- synthetic handicaps 237 f.

SHG see second harmonic generation
siloles 122
single-molecule transistors 386 f.
singlet state 329
single-walled carbon nanotubes (SWCNT) 3 f., 6, 41 f.
small-scale electronics 353
smart materials 262
solution-phase switching 305 f.
Sonogashira reaction 194
- DNA modification 453 f.
- high-throughput screening 194
Sonogashira-Hagihara coupling 225, 246
- quadruple 246, 249
soybean peroxidase 34
spin coating 83, 113
spin coupler 329
- photochromic 331
spin crossover phenomena 333
squaraines 203, 205
square-tip junctions 383 f.
stilbazolium salts 182, 184
- preparation 185
STM measurements 380 f.
streptavidin 37
sum frequency generation 396
sum-over-states 399
- calculations 403
supramolecular building blocks 274
supramolecular chemistry 261, 322
supramolecular complexation 4
supramolecular photovoltaic cells 497
supramolecular polymers 272, 278

swivel cruciform 95, 101 ff.
- binaphthylbased 101
- extended swivel cruciform 105
- molecular shape 109
- synthesis 105 f., 108 ff.
- terphenyl dimers 103
symmetric charge transfer 422

t

template-controlled cyclizations 238 ff.
- acetylene cyclization 242
- Glaser coupling 238, 249
- phenylacetylene oligomers 242 f.
- porphyrins 239 f.
- ring precursors 245
tether length dependence 244
tetracene 528 ff.
- edge-to-face arrangements 528
- ethynyltetracenes 530 ff.
- functionalization 528 f.
- rubrene 532
- tetrachalcogenetetracenes 530 f.
- tetrasubstituted 533 f
thermodynamic approach 251 ff.
- alkyne metathesis 253
- carbazole macrocycles 253 f.
- imine formation 251 f.
- phenylene-ethynylene macrocycles 252
thiol end-capped molecules 353
- π-conjugated 354
- flat gold substrate 376
- junction types 370 ff.
- self-assembled monolayers 376
- synthetic procedures 354
thioxophosphole 133
- host material 133
third-order NLO effects 401 ff.
- nonlinear absorption 402
topological resonance method 59
trannulene systems 62
transferrin 14
tricyanofuran acceptor 414
triplet state 329
twin-wedge complex 270
twin-wedge monomers 271
two-photon absorbing chromophores 418
- essential-state models 418
- one-photon absorption 419
- static dipole moment 419
- transition diploe moment 419
two-photon absorption 394, 400, 427 f.
- acrylate polymerization 427 f.
- chromophores 418 f.
- cross section δ 401

- dyes 428
- excited-state absorption 402
- g-e excitation 421
- measurement techniques 401
- spectra 421

two-photon fluorescence 402
- fluorescence intensity 402
- quantum yields 402

u
ubiquinones 483
Ugi-type reactions 180

v
vinylsilane 194 f.
- substituted 206

z
Zerner Intermediate Neglect of Differential Overlap 399

Further Reading

Balzani, V., Credi, A., Venturi, M.

Molecular Devices and Machines

A Journey into the Nanoworld

2003
Hardcover
ISBN 3-527-30506-8

Willner, I., Katz, E. (Eds.)

Bioelectronics

From Theory to Applications

2005
Hardcover
ISBN 3-527-30690-0

Haley, M. M., Tykwinski, R. R. (Eds.)

Carbon-Rich Compounds

From Molecules to Materials

2006
Hardcover
ISBN 3-527-31224-2

Hirsch, A., Brettreich, M.

Fullerenes

Chemistry and Reactions

2005
Hardcover
ISBN 3-527-30820-2

Reich, S., Thomsen, C., Maultzsch, J.

Carbon Nanotubes

Basic Concepts and Physical Properties

2004
Hardcover
ISBN 3-527-40386-8